Springer Texts in Statistics

Advisors:
George Casella   Stephen Fienberg   Ingram Olkin

Springer
New York
Berlin
Heidelberg
Hong Kong
London
Milan
Paris
Tokyo

Springer Texts in Statistics

(continued after index)

Jun Shao

Mathematical Statistics

Second Edition

 Springer

Jun Shao
Department of Statistics
University of Wisconsin, Madison
Madison, WI 53706-1685
USA
shao@stat.wisc.edu

With 7 figures.

Library of Congress Cataloging-in-Publication Data
Shao, Jun.
    Mathematical statistics / Jun Shao.—2nd ed.
        p. cm. — (Springer texts in statistics)
    Includes bibliographical references and index.
    ISBN 0-387-95382-5 (alk. paper)
    1. Mathematical statistics. I. Title. II. Series.
    QA276.S458   2003
    519.5—dc21   2003045446

ISBN 0-387-95382-5   Printed on acid-free paper.

Printed in the United States of America.   (MVY)

9 8 7 6 5 4 3 2   SPIN 10968499

Springer-Verlag is a part of *Springer Science+Business Media*

springeronline.com

To Guang, Jason, and Annie

Preface to the First Edition

This book is intended for a course entitled *Mathematical Statistics* offered at the Department of Statistics, University of Wisconsin-Madison. This course, taught in a mathematically rigorous fashion, covers essential materials in statistical theory that a first or second year graduate student typically needs to learn as preparation for work on a Ph.D. degree in statistics. The course is designed for two 15-week semesters, with three lecture hours and two discussion hours in each week. Students in this course are assumed to have a good knowledge of advanced calculus. A course in real analysis or measure theory prior to this course is often recommended.

Chapter 1 provides a quick overview of important concepts and results in measure-theoretic probability theory that are used as tools in mathematical statistics. Chapter 2 introduces some fundamental concepts in statistics, including statistical models, the principle of sufficiency in data reduction, and two statistical approaches adopted throughout the book: statistical decision theory and statistical inference. Each of Chapters 3 through 7 provides a detailed study of an important topic in statistical decision theory and inference: Chapter 3 introduces the theory of unbiased estimation; Chapter 4 studies theory and methods in point estimation under parametric models; Chapter 5 covers point estimation in nonparametric settings; Chapter 6 focuses on hypothesis testing; and Chapter 7 discusses interval estimation and confidence sets. The classical frequentist approach is adopted in this book, although the Bayesian approach is also introduced (§2.3.2, §4.1, §6.4.4, and §7.1.3). Asymptotic (large sample) theory, a crucial part of statistical inference, is studied throughout the book, rather than in a separate chapter.

About 85% of the book covers classical results in statistical theory that are typically found in textbooks of a similar level. These materials are in the Statistics Department's Ph.D. qualifying examination syllabus. This part of the book is influenced by several standard textbooks, such as Casella and

Berger (1990), Ferguson (1967), Lehmann (1983, 1986), and Rohatgi (1976). The other 15% of the book covers some topics in modern statistical theory that have been developed in recent years, including robustness of the least squares estimators, Markov chain Monte Carlo, generalized linear models, quasi-likelihoods, empirical likelihoods, statistical functionals, generalized estimation equations, the jackknife, and the bootstrap.

In addition to the presentation of fruitful ideas and results, this book emphasizes the use of important tools in establishing theoretical results. Thus, most proofs of theorems, propositions, and lemmas are provided or left as exercises. Some proofs of theorems are omitted (especially in Chapter 1), because the proofs are lengthy or beyond the scope of the book (references are always provided). Each chapter contains a number of examples. Some of them are designed as materials covered in the discussion section of this course, which is typically taught by a teaching assistant (a senior graduate student). The exercises in each chapter form an important part of the book. They provide not only practice problems for students, but also many additional results as complementary materials to the main text.

The book is essentially based on (1) my class notes taken in 1983-84 when I was a student in this course, (2) the notes I used when I was a teaching assistant for this course in 1984-85, and (3) the lecture notes I prepared during 1997-98 as the instructor of this course. I would like to express my thanks to Dennis Cox, who taught this course when I was a student and a teaching assistant, and undoubtedly has influenced my teaching style and textbook for this course. I am also very grateful to students in my class who provided helpful comments; to Mr. Yonghee Lee, who helped me to prepare all the figures in this book; to the Springer-Verlag production and copy editors, who helped to improve the presentation; and to my family members, who provided support during the writing of this book.

Madison, Wisconsin   Jun Shao
January 1999

Preface to the Second Edition

In addition to correcting typos and errors and making a better presentation, the main effort in preparing this new edition is adding some new material to Chapter 1 (Probability Theory) and a number of new exercises to each chapter. Furthermore, two new sections are created to introduce semiparametric models and methods (§5.1.4) and to study the asymptotic accuracy of confidence sets (§7.3.4). The structure of the book remains the same.

In Chapter 1 of the new edition, moment generating and characteristic functions are treated in more detail and a proof of the uniqueness theorem is provided; some useful moment inequalities are introduced; discussions on conditional independence, Markov chains, and martingales are added, as a continuation of the discussion of conditional expectations; the concepts of weak convergence and tightness are introduced; proofs to some key results in asymptotic theory, such as the dominated convergence theorem and monotone convergence theorem, the Lévy-Cramér continuity theorem, the strong and weak laws of large numbers, and Lindeberg's central limit theorem, are included; and a new section (§1.5.6) is created to introduce Edgeworth and Cornish-Fisher expansions. As a result, Chapter 1 of the new edition is self-contained for important concepts, results, and proofs in probability theory with emphasis in statistical applications.

Since the original book was published in 1999, I have been using it as a textbook for a two-semester course in mathematical statistics. Exercise problems accumulated during my teaching are added to this new edition. Some exercises that are too trivial have been removed.

In the original book, indices on definitions, examples, theorems, propositions, corollaries, and lemmas are included in the subject index. In the new edition, they are in a separate index given in the end of the book (prior to the author index). A list of notation and a list of abbreviations, which are appendices of the original book, are given after the references.

The most significant change in notation is the notation for a vector. In the text of the new edition, a k-dimensional vector is denoted by $c = (c_1, ..., c_k)$, whether it is treated as a column or a row vector (which is not important if matrix algebra is not considered). When matrix algebra is involved, any vector c is treated as a $k \times 1$ matrix (a column vector) and its transpose c^τ is treated as a $1 \times k$ matrix (a row vector). Thus, for $c = (c_1, ..., c_k)$, $c^\tau c = c_1^2 + \cdots + c_k^2$ and cc^τ is the $k \times k$ matrix whose (i, j)th element is $c_i c_j$.

I would like to thank reviewers of this book for their constructive comments, the Springer-Verlag production and copy editors, students in my classes, and two teaching assistants, Mr. Bin Cheng and Dr. Hansheng Wang, who provided help in preparing the new edition. Any remaining errors are of course my own responsibility, and a correction of them may be found on my web page http://www.stat.wisc.edu/~shao.

Madison, Wisconsin   Jun Shao
April, 2003

Contents

Chapter 1

Probability Theory

Mathematical statistics relies on probability theory, which in turn is based on measure theory. The present chapter provides some principal concepts and notational conventions of probability theory, and some important results that are useful tools in statistics. A more complete account of probability theory can be found in a standard textbook, for example, Billingsley (1986), Chung (1974), or Loève (1977). The reader is assumed to be familiar with set operations and set functions (mappings) in advanced calculus.

1.1 Probability Spaces and Random Elements

In an elementary probability course, one defines a *random experiment* to be an experiment whose outcome cannot be predicted with certainty, and the probability of A (a collection of possible outcomes) to be the fraction of times that the outcome of the random experiment results in A in a large number of trials of the random experiment. A rigorous and logically consistent definition of probability was given by A. N. Kolmogorov in his measure-theoretic fundamental development of probability theory in 1933 (Kolmogorov, 1933).

1.1.1 σ-fields and measures

Let Ω be a set of elements of interest. For example, Ω can be a set of numbers, a subinterval of the real line, or all possible outcomes of a random experiment. In probability theory, Ω is often called the outcome space, whereas in statistical theory, Ω is called the *sample space*. This is because in probability and statistics, Ω is usually the set of all possible outcomes of a random experiment under study.

A *measure* is a natural mathematical extension of the length, area, or volume of subsets in the one-, two-, or three-dimensional Euclidean space. In a given sample space Ω, a measure is a set function defined for certain subsets of Ω. It is necessary for this collection of subsets to satisfy certain properties, which are given in the following definition.

Definition 1.1. Let $\mathcal{F}$ be a collection of subsets of a sample space Ω. $\mathcal{F}$ is called a σ-field (or σ-algebra) if and only if it has the following properties.
(i) The empty set $\emptyset \in \mathcal{F}$.
(ii) If $A \in \mathcal{F}$, then the complement $A^c \in \mathcal{F}$.
(iii) If $A_i \in \mathcal{F}$, $i = 1, 2, ...$, then their union $\cup A_i \in \mathcal{F}$.   ∎

A pair $(\Omega, \mathcal{F})$ consisting of a set Ω and a σ-field $\mathcal{F}$ of subsets of Ω is called a *measurable space*. The elements of $\mathcal{F}$ are called measurable sets in measure theory or *events* in probability and statistics.

Since $\emptyset^c = \Omega$, it follows from (i) and (ii) in Definition 1.1 that $\Omega \in \mathcal{F}$ if $\mathcal{F}$ is a σ-field on Ω. Also, it follows from (ii) and (iii) that if $A_i \in \mathcal{F}$, $i = 1, 2, ...$, and $\mathcal{F}$ is a σ-field, then the intersection $\cap A_i \in \mathcal{F}$. This can be shown using DeMorgan's law: $(\cap A_i)^c = \cup A_i^c$.

For any given Ω, there are two trivial σ-fields. The first one is the collection containing exactly two elements, $\emptyset$ and Ω. This is the smallest possible σ-field on Ω. The second one is the collection of all subsets of Ω, which is called the power set and is the largest σ-field on Ω.

Let us now consider some nontrivial σ-fields. Let A be a nonempty proper subset of Ω ($A \subset \Omega$, $A \neq \Omega$). Then (verify)

$$\{\emptyset, A, A^c, \Omega\} \tag{1.1}$$

is a σ-field. In fact, this is the smallest σ-field containing A in the sense that if $\mathcal{F}$ is any σ-field containing A, then the σ-field in (1.1) is a subcollection of $\mathcal{F}$. In general, the smallest σ-field containing $\mathcal{C}$, a collection of subsets of Ω, is denoted by $\sigma(\mathcal{C})$ and is called the σ-field generated by $\mathcal{C}$. Hence, the σ-field in (1.1) is $\sigma(\{A\})$. Note that $\sigma(\{A, A^c\})$, $\sigma(\{A, \Omega\})$, and $\sigma(\{A, \emptyset\})$ are all the same as $\sigma(\{A\})$. Of course, if $\mathcal{C}$ itself is a σ-field, then $\sigma(\mathcal{C}) = \mathcal{C}$.

On the real line $\mathcal{R}$, there is a special σ-field that will be used almost exclusively. Let $\mathcal{C}$ be the collection of all finite open intervals on $\mathcal{R}$. Then $\mathcal{B} = \sigma(\mathcal{C})$ is called the *Borel σ-field*. The elements of $\mathcal{B}$ are called *Borel sets*. The Borel σ-field $\mathcal{B}^k$ on the k-dimensional Euclidean space $\mathcal{R}^k$ can be similarly defined. It can be shown that all intervals (finite or infinite), open sets, and closed sets are Borel sets. To illustrate, we now show that, on the real line, $\mathcal{B} = \sigma(\mathcal{O})$, where $\mathcal{O}$ is the collection of all open sets. Typically, one needs to show that $\sigma(\mathcal{C}) \subset \sigma(\mathcal{O})$ and $\sigma(\mathcal{O}) \subset \sigma(\mathcal{C})$. Since an open interval is an open set, $\mathcal{C} \subset \mathcal{O}$ and, hence, $\sigma(\mathcal{C}) \subset \sigma(\mathcal{O})$ (why?). Let U be an open set. Then U can be expressed as a union of a sequence of finite open intervals (see Royden (1968, p.39)). Hence, $U \in \sigma(\mathcal{C})$ (Definition 1.1(iii)) and $\mathcal{O} \subset \sigma(\mathcal{C})$. By the definition of $\sigma(\mathcal{O})$, $\sigma(\mathcal{O}) \subset \sigma(\mathcal{C})$. This completes the proof.

Let $C \subset \mathcal{R}^k$ be a Borel set and let $\mathcal{B}_C = \{C \cap B : B \in \mathcal{B}^k\}$. Then $(C, \mathcal{B}_C)$ is a measurable space and $\mathcal{B}_C$ is called the Borel σ-field on C.

Now we can introduce the notion of a measure.

Definition 1.2. Let $(\Omega, \mathcal{F})$ be a measurable space. A set function ν defined on $\mathcal{F}$ is called a *measure* if and only if it has the following properties.
(i) $0 \leq \nu(A) \leq \infty$ for any $A \in \mathcal{F}$.
(ii) $\nu(\emptyset) = 0$.
(iii) If $A_i \in \mathcal{F}$, $i = 1, 2, ...$, and A_i's are disjoint, i.e., $A_i \cap A_j = \emptyset$ for any $i \neq j$, then

$$\nu\left(\bigcup_{i=1}^{\infty} A_i\right) = \sum_{i=1}^{\infty} \nu(A_i). \quad \blacksquare$$

The triple $(\Omega, \mathcal{F}, \nu)$ is called a *measure space*. If $\nu(\Omega) = 1$, then ν is called a *probability measure* and we usually denote it by P instead of ν, in which case $(\Omega, \mathcal{F}, P)$ is called a *probability space*.

Although measure is an extension of length, area, or volume, sometimes it can be quite abstract. For example, the following set function is a measure:

$$\nu(A) = \begin{cases} \infty & A \in \mathcal{F}, A \neq \emptyset \\ 0 & A = \emptyset. \end{cases} \tag{1.2}$$

Since a measure can take ∞ as its value, we must know how to do arithmetic with ∞. In this book, it suffices to know that (1) for any $x \in \mathcal{R}$, $\infty + x = \infty$, $x\infty = \infty$ if $x > 0$, $x\infty = -\infty$ if $x < 0$, and $0\infty = 0$; (2) $\infty + \infty = \infty$; and (3) $\infty^a = \infty$ for any $a > 0$. However, $\infty - \infty$ or ∞/∞ is not defined.

The following examples provide two very important measures in probability and statistics.

Example 1.1 (Counting measure). Let Ω be a sample space, $\mathcal{F}$ the collection of all subsets, and $\nu(A)$ the number of elements in $A \in \mathcal{F}$ ($\nu(A) = \infty$ if A contains infinitely many elements). Then ν is a measure on $\mathcal{F}$ and is called the *counting measure*. $\quad \blacksquare$

Example 1.2 (Lebesgue measure). There is a unique measure m on $(\mathcal{R}, \mathcal{B})$ that satisfies

$$m([a, b]) = b - a \tag{1.3}$$

for every finite interval $[a, b]$, $-\infty < a \leq b < \infty$. This is called the *Lebesgue measure*. If we restrict m to the measurable space $([0, 1], \mathcal{B}_{[0,1]})$, then m is a probability measure. $\quad \blacksquare$

If Ω is *countable* in the sense that there is a one-to-one correspondence between Ω and the set of all integers, then one can usually consider the trivial σ-field that contains all subsets of Ω and a measure that assigns a value to every subset of Ω. When Ω is uncountable (e.g., $\Omega = \mathcal{R}$ or $[0, 1]$), it is not possible to define a reasonable measure for every subset of Ω; for example, it is not possible to find a measure on all subsets of $\mathcal{R}$ and still satisfy property (1.3). This is why it is necessary to introduce σ-fields that are smaller than the power set.

The following result provides some basic properties of measures. Whenever we consider $\nu(A)$, it is implicitly assumed that $A \in \mathcal{F}$.

Proposition 1.1. Let $(\Omega, \mathcal{F}, \nu)$ be a measure space.
(i) (Monotonicity). If $A \subset B$, then $\nu(A) \leq \nu(B)$.
(ii) (Subadditivity). For any sequence $A_1, A_2, ...,$

$$\nu\left(\bigcup_{i=1}^{\infty} A_i\right) \leq \sum_{i=1}^{\infty} \nu(A_i).$$

(iii) (Continuity). If $A_1 \subset A_2 \subset A_3 \subset \cdots$ (or $A_1 \supset A_2 \supset A_3 \supset \cdots$ and $\nu(A_1) < \infty$), then

$$\nu\left(\lim_{n\to\infty} A_n\right) = \lim_{n\to\infty} \nu(A_n),$$

where

$$\lim_{n\to\infty} A_n = \bigcup_{i=1}^{\infty} A_i \quad \left(\text{or} = \bigcap_{i=1}^{\infty} A_i\right).$$

Proof. We prove (i) only. The proofs of (ii) and (iii) are left as exercises. Since $A \subset B$, $B = A \cup (A^c \cap B)$ and A and $A^c \cap B$ are disjoint. By Definition 1.2(iii), $\nu(B) = \nu(A) + \nu(A^c \cap B)$, which is no smaller than $\nu(A)$ since $\nu(A^c \cap B) \geq 0$ by Definition 1.2(i). ∎

There is a one-to-one correspondence between the set of all probability measures on $(\mathcal{R}, \mathcal{B})$ and a set of functions on $\mathcal{R}$. Let P be a probability measure. The *cumulative distribution function* (c.d.f.) of P is defined to be

$$F(x) = P((-\infty, x]), \quad x \in \mathcal{R}. \tag{1.4}$$

Proposition 1.2. (i) Let F be a c.d.f. on $\mathcal{R}$. Then
   (a) $F(-\infty) = \lim_{x\to-\infty} F(x) = 0$;
   (b) $F(\infty) = \lim_{x\to\infty} F(x) = 1$;
   (c) F is nondecreasing, i.e., $F(x) \leq F(y)$ if $x \leq y$;
   (d) F is right continuous, i.e., $\lim_{y\to x, y>x} F(y) = F(x)$.
(ii) Suppose that a real-valued function F on $\mathcal{R}$ satisfies (a)-(d) in part (i). Then F is the c.d.f. of a unique probability measure on $(\mathcal{R}, \mathcal{B})$. ∎

The *Cartesian product* of sets (or collections of sets) Γ_i, $i \in \mathcal{I} = \{1, ..., k\}$ (or $\{1, 2, ...\}$) is defined as the set of all $(a_1, ..., a_k)$ (or $(a_1, a_2, ...)$), $a_i \in \Gamma_i$, $i \in \mathcal{I}$, and is denoted by $\prod_{i \in \mathcal{I}} \Gamma_i = \Gamma_1 \times \cdots \times \Gamma_k$ (or $\Gamma_1 \times \Gamma_2 \times \cdots$). Let $(\Omega_i, \mathcal{F}_i)$, $i \in \mathcal{I}$, be measurable spaces. Since $\prod_{i \in \mathcal{I}} \mathcal{F}_i$ is not necessarily a σ-field, $\sigma\left(\prod_{i \in \mathcal{I}} \mathcal{F}_i\right)$ is called the *product σ-field* on the *product space* $\prod_{i \in \mathcal{I}} \Omega_i$ and $\left(\prod_{i \in \mathcal{I}} \Omega_i, \sigma\left(\prod_{i \in \mathcal{I}} \mathcal{F}_i\right)\right)$ is denoted by $\prod_{i \in \mathcal{I}}(\Omega_i, \mathcal{F}_i)$. As an example, consider $(\Omega_i, \mathcal{F}_i) = (\mathcal{R}, \mathcal{B})$, $i = 1, ..., k$. Then the product space is $\mathcal{R}^k$ and it can be shown that the product σ-field is the same as the Borel σ-field on $\mathcal{R}^k$, which is the σ-field generated by the collection of all open sets in $\mathcal{R}^k$.

In Example 1.2, the usual length of an interval $[a, b] \subset \mathcal{R}$ is the same as the Lebesgue measure of $[a, b]$. Consider a rectangle $[a_1, b_1] \times [a_2, b_2] \subset \mathcal{R}^2$. The usual area of $[a_1, b_1] \times [a_2, b_2]$ is

$$(b_1 - a_1)(b_2 - a_2) = m([a_1, b_1])m([a_2, b_2]), \tag{1.5}$$

i.e., the product of the Lebesgue measures of two intervals $[a_1, b_1]$ and $[a_2, b_2]$. Note that $[a_1, b_1] \times [a_2, b_2]$ is a measurable set by the definition of the product σ-field. Is $m([a_1, b_1])m([a_2, b_2])$ the same as the value of a measure defined on the product σ-field? The following result answers this question for any product space generated by a finite number of measurable spaces. (Its proof can be found in Billingsley (1986, pp. 235-236).) Before introducing this result, we need the following technical definition. A measure ν on $(\Omega, \mathcal{F})$ is said to be *σ-finite* if and only if there exists a sequence $\{A_1, A_2, ...\}$ such that $\cup A_i = \Omega$ and $\nu(A_i) < \infty$ for all i. Any finite measure (such as a probability measure) is clearly σ-finite. The Lebesgue measure in Example 1.2 is σ-finite, since $\mathcal{R} = \cup A_n$ with $A_n = (-n, n)$, $n = 1, 2, ...$. The counting measure in Example 1.1 is σ-finite if and only if Ω is countable. The measure defined by (1.2), however, is not σ-finite.

Proposition 1.3 (Product measure theorem). Let $(\Omega_i, \mathcal{F}_i, \nu_i)$, $i = 1, ..., k$, be measure spaces with σ-finite measures, where $k \geq 2$ is an integer. Then there exists a unique σ-finite measure on the product σ-field $\sigma(\mathcal{F}_1 \times \cdots \times \mathcal{F}_k)$, called the *product measure* and denoted by $\nu_1 \times \cdots \times \nu_k$, such that

$$\nu_1 \times \cdots \times \nu_k(A_1 \times \cdots \times A_k) = \nu_1(A_1) \cdots \nu_k(A_k)$$

for all $A_i \in \mathcal{F}_i$, $i = 1, ..., k$. ∎

In $\mathcal{R}^2$, there is a unique measure, the product measure $m \times m$, for which $m \times m([a_1, b_1] \times [a_2, b_2])$ is equal to the value given by (1.5). This measure is called the Lebesgue measure on $(\mathcal{R}^2, \mathcal{B}^2)$. The Lebesgue measure on $(\mathcal{R}^3, \mathcal{B}^3)$ is $m \times m \times m$, which equals the usual volume for a subset of the form $[a_1, b_1] \times [a_2, b_2] \times [a_3, b_3]$. The Lebesgue measure on $(\mathcal{R}^k, \mathcal{B}^k)$ for any positive integer k is similarly defined.

The concept of c.d.f. can be extended to $\mathcal{R}^k$. Let P be a probability measure on $(\mathcal{R}^k, \mathcal{B}^k)$. The c.d.f. (or *joint* c.d.f.) of P is defined by

$$F(x_1, ..., x_k) = P\left((-\infty, x_1] \times \cdots \times (-\infty, x_k]\right), \quad x_i \in \mathcal{R}. \tag{1.6}$$

Again, there is a one-to-one correspondence between probability measures and joint c.d.f.'s on $\mathcal{R}^k$. Some properties of a joint c.d.f. are given in Exercise 10 in §1.6. If $F(x_1, ..., x_k)$ is a joint c.d.f., then

$$F_i(x) = \lim_{x_j \to \infty, j=1,...,i-1,i+1,...,k} F(x_1, ..., x_{i-1}, x, x_{i+1}, ..., x_k)$$

is a c.d.f. and is called the ith *marginal* c.d.f. Apparently, marginal c.d.f.'s are determined by their joint c.d.f. But a joint c.d.f. cannot be determined by k marginal c.d.f.'s. There is one special but important case in which a joint c.d.f. F is determined by its k marginal c.d.f. F_i's through

$$F(x_1, ..., x_k) = F_1(x_1) \cdots F_k(x_k), \quad (x_1, ..., x_k) \in \mathcal{R}^k, \tag{1.7}$$

in which case the probability measure corresponding to F is the product measure $P_1 \times \cdots \times P_k$ with P_i being the probability measure corresponding to F_i.

Proposition 1.3 can be extended to cases involving infinitely many measure spaces (Billingsley, 1986). In particular, if $(\mathcal{R}^k, \mathcal{B}^k, P_i)$, $i = 1, 2, ...$, are probability spaces, then there is a product probability measure P on $\prod_{i=1}^{\infty}(\mathcal{R}^k, \mathcal{B}^k)$ such that for any positive integer l and $B_i \in \mathcal{B}^k$, $i = 1, ..., l$,

$$P(B_1 \times \cdots \times B_l \times \mathcal{R}^k \times \mathcal{R}^k \times \cdots) = P_1(B_1) \cdots P_l(B_l).$$

1.1.2 Measurable functions and distributions

Since Ω can be quite arbitrary, it is often convenient to consider a function (mapping) f from Ω to a simpler space Λ (often $\Lambda = \mathcal{R}^k$). Let $B \subset \Lambda$. Then the *inverse image* of B under f is

$$f^{-1}(B) = \{f \in B\} = \{\omega \in \Omega : f(\omega) \in B\}.$$

The inverse function f^{-1} need not exist for $f^{-1}(B)$ to be defined. The reader is asked to verify the following properties:

(a) $f^{-1}(B^c) = (f^{-1}(B))^c$ for any $B \subset \Lambda$;

(b) $f^{-1}(\cup B_i) = \cup f^{-1}(B_i)$ for any $B_i \subset \Lambda$, $i = 1, 2, ...$.

Let $\mathcal{C}$ be a collection of subsets of Λ. We define

$$f^{-1}(\mathcal{C}) = \{f^{-1}(C) : C \in \mathcal{C}\}.$$

Definition 1.3. Let $(\Omega, \mathcal{F})$ and $(\Lambda, \mathcal{G})$ be measurable spaces and f a function from Ω to Λ. The function f is called a *measurable function* from $(\Omega, \mathcal{F})$ to $(\Lambda, \mathcal{G})$ if and only if $f^{-1}(\mathcal{G}) \subset \mathcal{F}$. ∎

If $\Lambda = \mathcal{R}$ and $\mathcal{G} = \mathcal{B}$ (Borel σ-field), then f is said to be *Borel measurable* or is called a *Borel function* on $(\Omega, \mathcal{F})$ (or with respect to $\mathcal{F}$).

In probability theory, a measurable function is called a *random element* and denoted by one of X, Y, Z,.... If X is measurable from $(\Omega, \mathcal{F})$ to $(\mathcal{R}, \mathcal{B})$, then it is called a *random variable*; if X is measurable from $(\Omega, \mathcal{F})$ to $(\mathcal{R}^k, \mathcal{B}^k)$, then it is called a *random k-vector*. If $X_1, ..., X_k$ are random variables defined on a common probability space, then the vector $(X_1, ..., X_k)$ is a random k-vector. (As a notational convention, any vector $c \in \mathcal{R}^k$ is denoted by $(c_1, ..., c_k)$, where c_i is the ith component of c.)

If f is measurable from $(\Omega, \mathcal{F})$ to $(\Lambda, \mathcal{G})$, then $f^{-1}(\mathcal{G})$ is a sub-σ-field of $\mathcal{F}$ (verify). It is called the σ-field generated by f and is denoted by $\sigma(f)$.

Now we consider some examples of measurable functions. If $\mathcal{F}$ is the collection of all subsets of Ω, then any function f is measurable. Let $A \subset \Omega$. The *indicator function* for A is defined as

$$I_A(\omega) = \begin{cases} 1 & \omega \in A \\ 0 & \omega \notin A. \end{cases}$$

For any $B \subset \mathcal{R}$,

$$I_A^{-1}(B) = \begin{cases} \emptyset & 0 \notin B, 1 \notin B \\ A & 0 \notin B, 1 \in B \\ A^c & 0 \in B, 1 \notin B \\ \Omega & 0 \in B, 1 \in B. \end{cases}$$

Then $\sigma(I_A)$ is the σ-field given in (1.1). If A is a measurable set, then I_A is a Borel function.

Note that $\sigma(I_A)$ is a much smaller σ-field than the original σ-field $\mathcal{F}$. This is another reason why we introduce the concept of measurable functions and random variables, in addition to the reason that it is easy to deal with numbers. Often the σ-field $\mathcal{F}$ (such as the power set) contains too many subsets and we are only interested in some of them. One can then define a random variable X with $\sigma(X)$ containing subsets that are of interest. In general, $\sigma(X)$ is between the trivial σ-field $\{\emptyset, \Omega\}$ and $\mathcal{F}$, and contains more subsets if X is more complicated. For the simplest function I_A, we have shown that $\sigma(I_A)$ contains only four elements.

The class of *simple functions* is obtained by taking linear combinations of indicators of measurable sets, i.e.,

$$\varphi(\omega) = \sum_{i=1}^{k} a_i I_{A_i}(\omega), \tag{1.8}$$

where $A_1, ..., A_k$ are measurable sets on Ω and $a_1, ..., a_k$ are real numbers. One can show directly that such a function is a Borel function, but it follows immediately from Proposition 1.4. Let $A_1, ..., A_k$ be a partition of Ω, i.e., A_i's are disjoint and $A_1 \cup \cdots \cup A_k = \Omega$. Then the simple function φ given by (1.8) with distinct a_i's exactly characterizes this partition and $\sigma(\varphi) = \sigma(\{A_1, ..., A_k\})$.

Proposition 1.4. Let $(\Omega, \mathcal{F})$ be a measurable space.
(i) f is Borel if and only if $f^{-1}(a, \infty) \in \mathcal{F}$ for all $a \in \mathcal{R}$.
(ii) If f and g are Borel, then so are fg and $af + bg$, where a and b are real numbers; also, f/g is Borel provided $g(\omega) \neq 0$ for any $\omega \in \Omega$.
(iii) If $f_1, f_2, ...$ are Borel, then so are $\sup_n f_n$, $\inf_n f_n$, $\limsup_n f_n$, and $\liminf_n f_n$. Furthermore, the set

$$A = \left\{ \omega \in \Omega : \lim_{n \to \infty} f_n(\omega) \text{ exists} \right\}$$

is an event and the function

$$h(\omega) = \begin{cases} \lim_{n \to \infty} f_n(\omega) & \omega \in A \\ f_1(\omega) & \omega \notin A \end{cases}$$

is Borel.
(iv) Suppose that f is measurable from $(\Omega, \mathcal{F})$ to $(\Lambda, \mathcal{G})$ and g is measurable from $(\Lambda, \mathcal{G})$ to $(\Delta, \mathcal{H})$. Then the composite function $g \circ f$ is measurable from $(\Omega, \mathcal{F})$ to $(\Delta, \mathcal{H})$.
(v) Let Ω be a Borel set in $\mathcal{R}^p$. If f is a continuous function from Ω to $\mathcal{R}^q$, then f is measurable. ∎

Proposition 1.4 indicates that there are many Borel functions. In fact, it is hard to find a non-Borel function.

The following result is very useful in technical proofs. Let f be a non-negative Borel function on $(\Omega, \mathcal{F})$. Then there exists a sequence of simple functions $\{\varphi_n\}$ satisfying $0 \leq \varphi_1 \leq \varphi_2 \leq \cdots \leq f$ and $\lim_{n \to \infty} \varphi_n = f$ (Exercise 17 in §1.6).

Let $(\Omega, \mathcal{F}, \nu)$ be a measure space and f be a measurable function from $(\Omega, \mathcal{F})$ to $(\Lambda, \mathcal{G})$. The *induced measure* by f, denoted by $\nu \circ f^{-1}$, is a measure on $\mathcal{G}$ defined as

$$\nu \circ f^{-1}(B) = \nu(f \in B) = \nu\left(f^{-1}(B)\right), \quad B \in \mathcal{G}. \qquad (1.9)$$

It is usually easier to deal with $\nu \circ f^{-1}$ than to deal with ν since $(\Lambda, \mathcal{G})$ is usually simpler than $(\Omega, \mathcal{F})$. Furthermore, subsets not in $\sigma(f)$ are not involved in the definition of $\nu \circ f^{-1}$. As we discussed earlier, in some cases we are only interested in subsets in $\sigma(f)$.

If $\nu = P$ is a probability measure and X is a random variable or a random vector, then $P \circ X^{-1}$ is called the *law* or the *distribution* of X and is denoted by P_X. The c.d.f. of P_X defined by (1.4) or (1.6) is also called the c.d.f. or joint c.d.f. of X and is denoted by F_X. On the other hand, for any c.d.f. or joint c.d.f. F, there exists at least one random variable or vector (usually there are many) defined on some probability space for which $F_X = F$. The following are some examples of random variables and their c.d.f.'s. More examples can be found in §1.3.1.

Example 1.3 (Discrete c.d.f.'s). Let $a_1 < a_2 < \cdots$ be a sequence of real numbers and let p_n, $n = 1, 2, ...$, be a sequence of positive numbers such that $\sum_{n=1}^{\infty} p_n = 1$. Define

$$F(x) = \begin{cases} \sum_{i=1}^{n} p_i & a_n \leq x < a_{n+1}, \quad n = 1, 2, ... \\ 0 & -\infty < x < a_1. \end{cases} \tag{1.10}$$

Then F is a *stepwise* c.d.f. It has a jump of size p_n at each a_n and is flat between a_n and a_{n+1}, $n = 1, 2, ...$ Such a c.d.f. is called a *discrete* c.d.f. and the corresponding random variable is called a *discrete random variable*. We can easily obtain a random variable having F in (1.10) as its c.d.f. For example, let $\Omega = \{a_1, a_2, ...\}$, $\mathcal{F}$ be the collection of all subsets of Ω,

$$P(A) = \sum_{i:a_i \in A} p_i, \quad A \in \mathcal{F}, \tag{1.11}$$

and $X(\omega) = \omega$. One can show that P is a probability measure and the c.d.f. of X is F in (1.10). ∎

Example 1.4 (Continuous c.d.f.'s). Opposite to the class of discrete c.d.f.'s is the class of continuous c.d.f.'s. Without the concepts of integration and differentiation introduced in the next section, we can only provide a few examples of continuous c.d.f.'s. One such example is the *uniform* c.d.f. on the interval $[a, b]$ defined as

$$F(x) = \begin{cases} 0 & -\infty < x < a \\ \frac{x-a}{b-a} & a \leq x < b \\ 1 & b \leq x < \infty. \end{cases}$$

Another example is the *exponential* c.d.f. defined as

$$F(x) = \begin{cases} 0 & -\infty < x < 0 \\ 1 - e^{-x/\theta} & 0 \leq x < \infty, \end{cases}$$

where θ is a fixed positive constant. Note that both uniform and exponential c.d.f.'s are continuous functions. ∎

1.2 Integration and Differentiation

Differentiation and integration are two of the main components of calculus. This is also true in measure theory or probability theory, except that integration is introduced first whereas in calculus, differentiation is introduced first.

1.2.1 Integration

An important concept needed in probability and statistics is the *integration* of Borel functions with respect to (w.r.t.) a measure ν, which is a type of "average". The definition proceeds in several steps. First, we define the integral of a nonnegative simple function, i.e., a simple function φ given by (1.8) with $a_i \geq 0$, $i = 1, ..., k$.

Definition 1.4(a). The integral of a nonnegative simple function φ given by (1.8) w.r.t. ν is defined as

$$\int \varphi d\nu = \sum_{i=1}^{k} a_i \nu(A_i). \quad \blacksquare \tag{1.12}$$

The right-hand side of (1.12) is a weighted average of a_i's with $\nu(A_i)$'s as weights. Since $a\infty = \infty$ if $a > 0$ and $a\infty = 0$ if $a = 0$, the right-hand side of (1.12) is always well defined, although $\int \varphi d\nu = \infty$ is possible. Note that different a_i's and A_i's may produce the same function φ; for example, with $\Omega = \mathcal{R}$,

$$2I_{(0,1)}(x) + I_{[1,2]}(x) = I_{(0,2]}(x) + I_{(0,1)}(x).$$

However, one can show that different representations of φ in (1.8) produce the same value for $\int \varphi d\nu$ so that the integral of a nonnegative simple function is well defined.

Next, we consider a nonnegative Borel function f.

Definition 1.4(b). Let f be a nonnegative Borel function and let $\mathcal{S}_f$ be the collection of all nonnegative simple functions of the form (1.8) satisfying $\varphi(\omega) \leq f(\omega)$ for any $\omega \in \Omega$. The integral of f w.r.t. ν is defined as

$$\int f d\nu = \sup \left\{ \int \varphi d\nu : \varphi \in \mathcal{S}_f \right\}. \quad \blacksquare$$

Hence, for any Borel function $f \geq 0$, there exists a sequence of simple functions $\varphi_1, \varphi_2, ...$ such that $0 \leq \varphi_i \leq f$ for all i and $\lim_{n \to \infty} \int \varphi_n d\nu = \int f d\nu$.

Finally, for a Borel function f, we first define the positive part of f by

$$f_+(\omega) = \max\{f(\omega), 0\}$$

and the negative part of f by

$$f_-(\omega) = \max\{-f(\omega), 0\}.$$

Note that f_+ and f_- are nonnegative Borel functions, $f(\omega) = f_+(\omega) - f_-(\omega)$, and $|f(\omega)| = f_+(\omega) + f_-(\omega)$.

Definition 1.4(c). Let f be a Borel function. We say that $\int f d\nu$ exists if and only if at least one of $\int f_+ d\nu$ and $\int f_- d\nu$ is finite, in which case

$$\int f d\nu = \int f_+ d\nu - \int f_- d\nu. \tag{1.13}$$

When both $\int f_+ d\nu$ and $\int f_- d\nu$ are finite, we say that f is integrable. Let A be a measurable set and I_A be its indicator function. The integral of f over A is defined as

$$\int_A f d\nu = \int I_A f d\nu. \quad \blacksquare$$

Note that a Borel function f is integrable if and only if $|f|$ is integrable.

It is convenient to define the integral of a measurable function f from $(\Omega, \mathcal{F}, \nu)$ to $(\bar{\mathcal{R}}, \bar{\mathcal{B}})$, where $\bar{\mathcal{R}} = \mathcal{R} \cup \{-\infty, \infty\}$, $\bar{\mathcal{B}} = \sigma(\mathcal{B} \cup \{\{\infty\}, \{-\infty\}\})$. Let $A_+ = \{f = \infty\}$ and $A_- = \{f = -\infty\}$. If $\nu(A_+) = 0$, we define $\int f_+ d\nu$ to be $\int I_{A_+^c} f_+ d\nu$; otherwise $\int f_+ d\nu = \infty$. $\int f_- d\nu$ is similarly defined. If at least one of $\int f_+ d\nu$ and $\int f_- d\nu$ is finite, then $\int f d\nu$ is defined by (1.13).

The integral of f may be denoted differently whenever there is a need to indicate the variable(s) to be integrated and the integration domain; for example, $\int_\Omega f d\nu$, $\int f(\omega) d\nu$, $\int f(\omega) d\nu(\omega)$, or $\int f(\omega) \nu(d\omega)$, and so on. In probability and statistics, $\int X dP$ is usually written as EX or $E(X)$ and called the *expectation* or *expected value* of X. If F is the c.d.f. of P on $(\mathcal{R}^k, \mathcal{B}^k)$, $\int f(x) dP$ is also denoted by $\int f(x) dF(x)$ or $\int f dF$.

Example 1.5. Let Ω be a countable set, $\mathcal{F}$ be all subsets of Ω, and ν be the counting measure given in Example 1.1. For any Borel function f, it can be shown (exercise) that

$$\int f d\nu = \sum_{\omega \in \Omega} f(\omega). \quad \blacksquare \tag{1.14}$$

Example 1.6. If $\Omega = \mathcal{R}$ and ν is the Lebesgue measure, then the Lebesgue integral of f over an interval $[a, b]$ is written as $\int_{[a,b]} f(x) dx = \int_a^b f(x) dx$, which agrees with the Riemann integral in calculus when the latter is well defined. However, there are functions for which the Lebesgue integrals are defined but not the Riemann integrals. ∎

We now introduce some properties of integrals. The proof of the following result is left to the reader.

Proposition 1.5 (Linearity of integrals). Let $(\Omega, \mathcal{F}, \nu)$ be a measure space and f and g be Borel functions.
(i) If $\int f d\nu$ exists and $a \in \mathcal{R}$, then $\int (af) d\nu$ exists and is equal to $a \int f d\nu$.
(ii) If both $\int f d\nu$ and $\int g d\nu$ exist and $\int f d\nu + \int g d\nu$ is well defined, then $\int (f + g) d\nu$ exists and is equal to $\int f d\nu + \int g d\nu$. ∎

If N is an event with $\nu(N) = 0$ and a statement holds for all ω in the complement N^c, then the statement is said to hold a.e. (almost everywhere) ν (or simply a.e. if the measure ν is clear from the context). If ν is a probability measure, then a.e. may be replaced by a.s. (almost surely).

Proposition 1.6. Let $(\Omega, \mathcal{F}, \nu)$ be a measure space and f and g be Borel.
(i) If $f \leq g$ a.e., then $\int f d\nu \leq \int g d\nu$, provided that the integrals exist.
(ii) If $f \geq 0$ a.e. and $\int f d\nu = 0$, then $f = 0$ a.e.
Proof. (i) The proof for part (i) is left to the reader.
(ii) Let $A = \{f > 0\}$ and $A_n = \{f \geq n^{-1}\}$, $n = 1, 2, \ldots.$ Then $A_n \subset A$ for any n and $\lim_{n \to \infty} A_n = \cup A_n = A$ (why?). By Proposition 1.1(iii), $\lim_{n \to \infty} \nu(A_n) = \nu(A)$. Using part (i) and Proposition 1.5, we obtain that

$$n^{-1} \nu(A_n) = \int n^{-1} I_{A_n} d\nu \leq \int f I_{A_n} d\nu \leq \int f d\nu = 0$$

for any n. Hence $\nu(A) = 0$ and $f = 0$ a.e. ∎

Some direct consequences of Proposition 1.6(i) are: $|\int f d\nu| \leq \int |f| d\nu$; if $f \geq 0$ a.e., then $\int f d\nu \geq 0$; and if $f = g$ a.e., then $\int f d\nu = \int g d\nu$.

It is sometimes required to know whether the following interchange of two operations is valid:

$$\int \lim_{n \to \infty} f_n d\nu = \lim_{n \to \infty} \int f_n d\nu, \tag{1.15}$$

where $\{f_n : n = 1, 2, \ldots\}$ is a sequence of Borel functions. Note that we only require $\lim_{n \to \infty} f_n$ exists a.e. Also, $\lim_{n \to \infty} f_n$ is Borel (Proposition 1.4). The following example shows that (1.15) is not always true.

Example 1.7. Consider $(\mathcal{R}, \mathcal{B})$ and the Lebesgue measure. Define $f_n(x) = nI_{[0, n^{-1}]}(x)$, $n = 1, 2, \ldots.$ Then $\lim_{n \to \infty} f_n(x) = 0$ for all x but $x = 0$. Since the Lebesgue measure of a single point set is 0 (see Example 1.2), $\lim_{n \to \infty} f_n(x) = 0$ a.e. and $\int \lim_{n \to \infty} f_n(x) dx = 0$. On the other hand, $\int f_n(x) dx = 1$ for any n and, hence, $\lim_{n \to \infty} \int f_n(x) dx = 1$. ∎

The following result gives sufficient conditions under which (1.15) holds.

Theorem 1.1. Let $f_1, f_2, \dots$ be a sequence of Borel functions on $(\Omega, \mathcal{F}, \nu)$.
(i) (Fatou's lemma). If $f_n \geq 0$, then

$$\int \liminf_n f_n d\nu \leq \liminf_n \int f_n d\nu.$$

(ii) (Dominated convergence theorem). If $\lim_{n \to \infty} f_n = f$ a.e. and there exists an integrable function g such that $|f_n| \leq g$ a.e., then (1.15) holds.
(iii) (Monotone convergence theorem). If $0 \leq f_1 \leq f_2 \leq \cdots$ and $\lim_{n \to \infty} f_n = f$ a.e., then (1.15) holds.
Proof. The results in (i) and (iii) are equivalent (exercise). Applying Fatou's lemma to functions $g + f_n$ and $g - f_n$, we obtain that $\int (g + f) d\nu \leq \liminf_n \int (g + f_n) d\nu$ and $\int (g - f) d\nu \leq \liminf_n \int (g - f_n) d\nu$ (which is the same as $\int (f - g) d\nu \geq \limsup_n \int (f_n - g) d\nu$). Since g is integrable, these results imply that $\int f d\nu \leq \liminf_n \int f_n d\nu \leq \limsup_n \int f_n d\nu \leq \int f d\nu$.

It remains to show part (iii). Let $f, f_1, f_2, \dots$ be given in part (iii). From Proposition 1.6(i), there exists $\lim_{n \to \infty} \int f_n d\nu \leq \int f d\nu$. Let φ be a simple function with $0 \leq \varphi \leq f$ and let $A_\varphi = \{\varphi > 0\}$. Suppose that $\nu(A_\varphi) = \infty$. Let $a = \min_{w \in A_\varphi} \varphi(w)$ and $A_n = \{f_n > a\}$. Then $a > 0$, $A_1 \subset A_2 \subset \cdots$, and $A_\varphi \subset \cup A_n$ (why?). By Proposition 1.1, $\nu(A_n) \to \nu(\cup A_n) \geq \nu(A_\varphi) = \infty$ and, hence, $\int f_n d\nu \geq \int_{A_n} f_n d\nu \geq a\nu(A_n) \to \infty$. Suppose now $\nu(A_\varphi) < \infty$. By Egoroff's theorem (Exercise 20 in §1.6), for any $\epsilon > 0$, there is $B \subset A_\varphi$ with $\nu(B) < \epsilon$ such that f_n converges to f uniformly on $A_\varphi \cap B^c$. Hence, $\int f_n d\nu \geq \int_{A_\varphi \cap B^c} f_n d\nu \to \int_{A_\varphi \cap B^c} f d\nu \geq \int_{A_\varphi \cap B^c} \varphi d\nu = \int \varphi d\nu - \int_B \varphi d\nu \geq \int \varphi d\nu - \epsilon \max_{w \in A_\varphi} \varphi(w)$. Since ϵ is arbitrary, $\lim_{n \to \infty} \int f_n d\nu \geq \int \varphi d\nu$. Since φ is arbitrary, by Definition 1.4(b), $\lim_{n \to \infty} \int f_n d\nu \geq \int f d\nu$. This completes the proof. ∎

Example 1.8 (Interchange of differentiation and integration). Let $(\Omega, \mathcal{F}, \nu)$ be a measure space and, for any fixed $\theta \in \mathcal{R}$, let $f(w, \theta)$ be a Borel function on Ω. Suppose that $\partial f(w, \theta)/\partial \theta$ exists a.e. for $\theta \in (a, b) \subset \mathcal{R}$ and that $|\partial f(w, \theta)/\partial \theta| \leq g(w)$ a.e., where g is an integrable function on Ω. Then, for each $\theta \in (a, b)$, $\partial f(w, \theta)/\partial \theta$ is integrable and, by Theorem 1.1(ii),

$$\frac{d}{d\theta} \int f(w, \theta) d\nu = \int \frac{\partial f(w, \theta)}{\partial \theta} d\nu. \quad \blacksquare$$

Theorem 1.2 (Change of variables). Let f be measurable from $(\Omega, \mathcal{F}, \nu)$ to $(\Lambda, \mathcal{G})$ and g be Borel on $(\Lambda, \mathcal{G})$. Then

$$\int_\Omega g \circ f d\nu = \int_\Lambda g d(\nu \circ f^{-1}), \tag{1.16}$$

i.e., if either integral exists, then so does the other, and the two are the same. ∎

The reader is encouraged to provide a proof. A complete proof is in Billingsley (1986, p. 219). This result extends the change of variable formula for Riemann integrals, i.e., $\int g(y)dy = \int g(f(x))f'(x)dx$, $y = f(x)$.

Result (1.16) is very important in probability and statistics. Let X be a random variable on a probability space $(\Omega, \mathcal{F}, P)$. If $EX = \int_\Omega XdP$ exists, then usually it is much simpler to compute $EX = \int_\mathcal{R} xdP_X$, where $P_X = P \circ X^{-1}$ is the law of X. Let Y be a random vector from Ω to $\mathcal{R}^k$ and g be Borel from $\mathcal{R}^k$ to $\mathcal{R}$. According to (1.16), $Eg(Y)$ can be computed as $\int_{\mathcal{R}^k} g(y)dP_Y$ or $\int_\mathcal{R} xdP_{g(Y)}$, depending on which of P_Y and $P_{g(Y)}$ is easier to handle. As a more specific example, consider $k = 2$, $Y = (X_1, X_2)$, and $g(Y) = X_1 + X_2$. Using Proposition 1.5(ii), $E(X_1 + X_2) = EX_1 + EX_2$ and, hence, $E(X_1 + X_2) = \int_\mathcal{R} xdP_{X_1} + \int_\mathcal{R} xdP_{X_2}$. Then we need to handle two integrals involving P_{X_1} and P_{X_2}. On the other hand, $E(X_1 + X_2) = \int_\mathcal{R} xdP_{X_1+X_2}$, which involves one integral w.r.t. $P_{X_1+X_2}$. Unless we have some knowledge about the joint c.d.f. of (X_1, X_2), it is not easy to obtain $P_{X_1+X_2}$.

The following theorem states how to evaluate an integral w.r.t. a product measure via iterated integration. The reader is encouraged to prove this theorem. A complete proof can be found in Billingsley (1986, pp. 236-238).

Theorem 1.3 (Fubini's theorem). Let ν_i be a σ-finite measure on $(\Omega_i, \mathcal{F}_i)$, $i = 1, 2$, and let f be a Borel function on $\prod_{i=1}^2 (\Omega_i, \mathcal{F}_i)$ whose integral w.r.t. $\nu_1 \times \nu_2$ exists. Then

$$g(\omega_2) = \int_{\Omega_1} f(\omega_1, \omega_2)d\nu_1$$

exists a.e. ν_2 and defines a Borel function on Ω_2 whose integral w.r.t. ν_2 exists, and

$$\int_{\Omega_1 \times \Omega_2} f(\omega_1, \omega_2)d\nu_1 \times \nu_2 = \int_{\Omega_2} \left[\int_{\Omega_1} f(\omega_1, \omega_2)d\nu_1 \right] d\nu_2. \quad \blacksquare$$

This result can be naturally extended to the integral w.r.t. the product measure on $\prod_{i=1}^k (\Omega_i, \mathcal{F}_i)$ for any finite positive integer k.

Example 1.9. Let $\Omega_1 = \Omega_2 = \{0, 1, 2, ...\}$, and $\nu_1 = \nu_2$ be the counting measure (Example 1.1). A function f on $\Omega_1 \times \Omega_2$ defines a double sequence. If $\int fd\nu_1 \times \nu_2$ exists, then

$$\int fd\nu_1 \times \nu_2 = \sum_{i=0}^\infty \sum_{j=0}^\infty f(i, j) = \sum_{j=0}^\infty \sum_{i=0}^\infty f(i, j) \qquad (1.17)$$

(by Theorem 1.3 and Example 1.5). Thus, a double series can be summed in either order, if it is well defined. $\quad \blacksquare$

1.2.2 Radon-Nikodym derivative

Let $(\Omega, \mathcal{F}, \nu)$ be a measure space and f be a nonnegative Borel function. One can show that the set function

$$\lambda(A) = \int_A f d\nu, \quad A \in \mathcal{F}, \tag{1.18}$$

is a measure on $(\Omega, \mathcal{F})$ (verify). Note that

$$\nu(A) = 0 \quad \text{implies} \quad \lambda(A) = 0. \tag{1.19}$$

If (1.19) holds for two measures λ and ν defined on the same measurable space, then we say λ is *absolutely continuous* w.r.t. ν and write $\lambda \ll \nu$.

Formula (1.18) gives us not only a way of constructing measures, but also a method of computing measures of measurable sets. Let ν be a well-known measure (such as the Lebesgue measure or the counting measure) and λ a relatively unknown measure. If we can find a function f such that (1.18) holds, then computing $\lambda(A)$ can be done through integration. A necessary condition for (1.18) is clearly $\lambda \ll \nu$. The following result shows that $\lambda \ll \nu$ is also almost sufficient for (1.18).

Theorem 1.4 (Radon-Nikodym theorem). Let ν and λ be two measures on $(\Omega, \mathcal{F})$ and ν be σ-finite. If $\lambda \ll \nu$, then there exists a nonnegative Borel function f on Ω such that (1.18) holds. Furthermore, f is unique a.e. ν, i.e., if $\lambda(A) = \int_A g d\nu$ for any $A \in \mathcal{F}$, then $f = g$ a.e. ν. ∎

The proof of this theorem can be found in Billingsley (1986, pp. 443-444). If (1.18) holds, then the function f is called the Radon-Nikodym *derivative* or *density* of λ w.r.t. ν and is denoted by $d\lambda/d\nu$.

A useful consequence of Theorem 1.4 is that if f is Borel on $(\Omega, \mathcal{F})$ and $\int_A f d\nu = 0$ for any $A \in \mathcal{F}$, then $f = 0$ a.e.

If $\int f d\nu = 1$ for an $f \geq 0$ a.e. ν, then λ given by (1.18) is a probability measure and f is called its *probability density function* (p.d.f.) w.r.t. ν. For any probability measure P on $(\mathcal{R}^k, \mathcal{B}^k)$ corresponding to a c.d.f. F or a random vector X, if P has a p.d.f. f w.r.t. a measure ν, then f is also called the p.d.f. of F or X w.r.t. ν.

Example 1.10 (p.d.f. of a discrete c.d.f.). Consider the discrete c.d.f. F in (1.10) of Example 1.3 with its probability measure given by (1.11). Let $\Omega = \{a_1, a_2, ...\}$ and ν be the counting measure on the power set of Ω. By Example 1.5,

$$P(A) = \int_A f d\nu = \sum_{a_i \in A} f(a_i), \quad A \subset \Omega, \tag{1.20}$$

where $f(a_i) = p_i$, $i = 1, 2, ...$ That is, f is the p.d.f. of P or F w.r.t. ν. Hence, any discrete c.d.f. has a p.d.f. w.r.t. counting measure. A p.d.f. w.r.t. counting measure is called a *discrete* p.d.f. ∎

Example 1.11. Let F be a c.d.f. Assume that F is differentiable in the usual sense in calculus. Let f be the derivative of F. From calculus,

$$F(x) = \int_{-\infty}^{x} f(y)dy, \quad x \in \mathcal{R}. \tag{1.21}$$

Let P be the probability measure corresponding to F. It can be shown that $P(A) = \int_A f dm$ for any $A \in \mathcal{B}$, where m is the Lebesgue measure on $\mathcal{R}$. Hence, f is the p.d.f. of P or F w.r.t. Lebesgue measure. In this case, the Radon-Nikodym derivative is the same as the usual derivative of F in calculus.

A continuous c.d.f. may not have a p.d.f. w.r.t. Lebesgue measure. A necessary and sufficient condition for a c.d.f. F having a p.d.f. w.r.t. Lebesgue measure is that F is *absolute continuous* in the sense that for any $\epsilon > 0$, there exists a $\delta > 0$ such that for each finite collection of disjoint bounded open intervals (a_i, b_i), $\sum(b_i - a_i) < \delta$ implies $\sum[F(b_i) - F(a_i)] < \epsilon$. Absolute continuity is weaker than differentiability, but is stronger than continuity. Thus, any discontinuous c.d.f. (such as a discrete c.d.f.) is not absolute continuous. Note that every c.d.f. is differentiable a.e. Lebesgue measure (Chung, 1974, Chapter 1). Hence, if f is the p.d.f. of F w.r.t. Lebesgue measure, then f is the usual derivative of F a.e. Lebesgue measure and (1.21) holds. In such a case probabilities can be computed through integration. It can be shown that the uniform and exponential c.d.f.'s in Example 1.4 are absolute continuous and their p.d.f.'s are, respectively,

$$f(x) = \begin{cases} \frac{1}{b-a} & a \le x < b \\ 0 & \text{otherwise} \end{cases}$$

and

$$f(x) = \begin{cases} 0 & -\infty < x < 0 \\ \theta^{-1}e^{-x/\theta} & 0 \le x < \infty. \end{cases}$$ ∎

A p.d.f. w.r.t. Lebesgue measure is called a Lebesgue p.d.f.

More examples of p.d.f.'s are given in §1.3.1.

The following result provides some basic properties of Radon-Nikodym derivatives. The proof is left to the reader.

Proposition 1.7 (Calculus with Radon-Nikodym derivatives). Let ν be a σ-finite measure on a measure space $(\Omega, \mathcal{F})$. All other measures discussed in (i)-(iii) are defined on $(\Omega, \mathcal{F})$.

(i) If λ is a measure, $\lambda \ll \nu$, and $f \geq 0$, then

$$\int f d\lambda = \int f \frac{d\lambda}{d\nu} d\nu.$$

(Notice how the $d\nu$'s "cancel" on the right-hand side.)

(ii) If λ_i, $i = 1, 2$, are measures and $\lambda_i \ll \nu$, then $\lambda_1 + \lambda_2 \ll \nu$ and

$$\frac{d(\lambda_1 + \lambda_2)}{d\nu} = \frac{d\lambda_1}{d\nu} + \frac{d\lambda_2}{d\nu} \quad \text{a.e. } \nu.$$

(iii) (Chain rule). If τ is a measure, λ is a σ-finite measure, and $\tau \ll \lambda \ll \nu$, then

$$\frac{d\tau}{d\nu} = \frac{d\tau}{d\lambda} \frac{d\lambda}{d\nu} \quad \text{a.e. } \nu.$$

In particular, if $\lambda \ll \nu$ and $\nu \ll \lambda$ (in which case λ and ν are *equivalent*), then

$$\frac{d\lambda}{d\nu} = \left(\frac{d\nu}{d\lambda} \right)^{-1} \quad \text{a.e. } \nu \text{ or } \lambda.$$

(iv) Let $(\Omega_i, \mathcal{F}_i, \nu_i)$ be a measure space and ν_i be σ-finite, $i = 1, 2$. Let λ_i be a σ-finite measure on $(\Omega_i, \mathcal{F}_i)$ and $\lambda_i \ll \nu_i$, $i = 1, 2$. Then $\lambda_1 \times \lambda_2 \ll \nu_1 \times \nu_2$ and

$$\frac{d(\lambda_1 \times \lambda_2)}{d(\nu_1 \times \nu_2)}(\omega_1, \omega_2) = \frac{d\lambda_1}{d\nu_1}(\omega_1) \frac{d\lambda_2}{d\nu_2}(\omega_2) \quad \text{a.e. } \nu_1 \times \nu_2. \quad \blacksquare$$

1.3 Distributions and Their Characteristics

We now discuss some distributions useful in statistics, and their moments and generating functions.

1.3.1 Distributions and probability densities

It is often more convenient to work with p.d.f.'s than to work with c.d.f.'s. We now introduce some p.d.f.'s useful in statistics.

We first consider p.d.f.'s on $\mathcal{R}$. Most discrete p.d.f.'s are w.r.t. counting measure on the space of all nonnegative integers. Table 1.1 lists all discrete p.d.f.'s in elementary probability textbooks. For any discrete p.d.f. f, its c.d.f. $F(x)$ can be obtained using (1.20) with $A = (\infty, x]$. Values of $F(x)$ can be obtained from statistical tables or software.

Two Lebesgue p.d.f.'s are introduced in Example 1.11. Some other useful Lebesgue p.d.f.'s are listed in Table 1.2. Note that the exponential p.d.f. in Example 1.11 is a special case of that in Table 1.2 with $a = 0$. For any Lebesgue p.d.f. f, (1.21) gives its c.d.f. A few c.d.f.'s have explicit

Table 1.1. Discrete Distributions on $\mathcal{R}$

Uniform	p.d.f.	$1/m, \ x = a_1, ..., a_m$
	m.g.f.	$\sum_{j=1}^{m} e^{a_j t}/m, \ t \in \mathcal{R}$
$DU(a_1, ..., a_m)$	Expectation	$\sum_{j=1}^{m} a_j/m$
	Variance	$\sum_{j=1}^{m}(a_j - \bar{a})^2/m, \ \bar{a} = \sum_{j=1}^{m} a_j/m$
	Parameter	$a_i \in \mathcal{R}, \ m = 1, 2, ...$
Binomial	p.d.f.	$\binom{n}{x} p^x (1-p)^{n-x}, \ x = 0, 1, ..., n$
	m.g.f.	$(pe^t + 1 - p)^n, \ t \in \mathcal{R}$
$Bi(p, n)$	Expectation	np
	Variance	$np(1-p)$
	Parameter	$p \in [0, 1], \ n = 1, 2, ...$
Poisson	p.d.f.	$\theta^x e^{-\theta}/x!, \ x = 0, 1, 2, ...$
	m.g.f.	$e^{\theta(e^t - 1)}, \ t \in \mathcal{R}$
$P(\theta)$	Expectation	θ
	Variance	θ
	Parameter	$\theta > 0$
Geometric	p.d.f.	$(1-p)^{x-1}p, \ x = 1, 2, ...$
	m.g.f.	$pe^t/[1 - (1-p)e^t], \ t < -\log(1-p)$
$G(p)$	Expectation	$1/p$
	Variance	$(1-p)/p^2$
	Parameter	$p \in [0, 1]$
Hyper-geometric	p.d.f.	$\binom{n}{x}\binom{m}{r-x} / \binom{N}{r}$
		$x = 0, 1, ..., \min\{r, n\}, \ r - x \leq m$
	m.g.f.	No explicit form
$HG(r, n, m)$	Expectation	rn/N
	Variance	$rnm(N - r)/[N^2(N-1)]$
	Parameter	$r, n, m = 1, 2, ..., \ N = n + m$
Negative binomial	p.d.f.	$\binom{x-1}{r-1} p^r (1-p)^{x-r}, \ x = r, r+1, ...$
	m.g.f.	$p^r e^{rt}/[1 - (1-p)e^t]^r, \ t < -\log(1-p)$
	Expectation	r/p
$NB(p, r)$	Variance	$r(1-p)/p^2$
	Parameter	$p \in [0, 1], \ r = 1, 2, ...$
Log-distribution	p.d.f.	$-(\log p)^{-1} x^{-1}(1-p)^x, \ x = 1, 2, ...$
	m.g.f.	$\log[1 - (1-p)e^t]/\log p, \ t \in \mathcal{R}$
	Expectation	$-(1-p)/(p \log p)$
$L(p)$	Variance	$-(1-p)[1 + (1-p)/\log p]/(p^2 \log p)$
	Parameter	$p \in (0, 1)$

All p.d.f.'s are w.r.t. counting measure.

forms, whereas many others do not and they have to be evaluated numerically or computed using tables or software.

There are p.d.f.'s that are neither discrete nor Lebesgue.

Example 1.12. Let X be a random variable on $(\Omega, \mathcal{F}, P)$ whose c.d.f. F_X has a Lebesgue p.d.f. f_X and $F_X(c) < 1$, where c is a fixed constant. Let $Y = \min\{X, c\}$, i.e., Y is the smaller of X and c. Note that $Y^{-1}((-\infty, x]) = \Omega$ if $x \geq c$ and $Y^{-1}((-\infty, x]) = X^{-1}((\infty, x])$ if $x < c$. Hence Y is a random variable and the c.d.f. of Y is

$$F_Y(x) = \begin{cases} 1 & x \geq c \\ F_X(x) & x < c. \end{cases}$$

This c.d.f. is discontinuous at c, since $F_X(c) < 1$. Thus, it does not have a Lebesgue p.d.f. It is not discrete either. Does P_Y, the probability measure corresponding to F_Y, have a p.d.f. w.r.t. some measure? Define a probability measure on $(\mathcal{R}, \mathcal{B})$, called *point mass* at c, by

$$\delta_c(A) = \begin{cases} 1 & c \in A \\ 0 & c \notin A, \end{cases} \qquad A \in \mathcal{B} \qquad (1.22)$$

(which is a special case of the discrete uniform distribution in Table 1.1). Then $P_Y \ll m + \delta_c$, where m is the Lebesgue measure, and the p.d.f. of P_Y is

$$\frac{dP_Y}{d(m + \delta_c)}(x) = \begin{cases} 0 & x > c \\ 1 - F_X(c) & x = c \qquad \blacksquare \\ f_X(x) & x < c. \end{cases} \qquad (1.23)$$

A p.d.f. corresponding to a joint c.d.f. is called a joint p.d.f. The following is a joint Lebesgue p.d.f. on $\mathcal{R}^k$ that is important in statistics:

$$f(x) = (2\pi)^{-k/2}[\mathrm{Det}(\Sigma)]^{-1/2}e^{-(x-\mu)^\tau \Sigma^{-1}(x-\mu)/2}, \qquad x \in \mathcal{R}^k, \qquad (1.24)$$

where $\mu \in \mathcal{R}^k$, Σ is a positive definite $k \times k$ matrix, $\mathrm{Det}(\Sigma)$ is the determinant of Σ and, when matrix algebra is involved, any k-vector c is treated as a $k \times 1$ matrix (column vector) and c^τ denotes its transpose (row vector). The p.d.f. in (1.24) and its c.d.f. are called the k-dimensional multivariate normal p.d.f. and c.d.f., and both are denoted by $N_k(\mu, \Sigma)$. Random vectors distributed as $N_k(\mu, \Sigma)$ are also denoted by $N_k(\mu, \Sigma)$ for convenience. The normal distribution $N(\mu, \sigma^2)$ in Table 1.2 is a special case of $N_k(\mu, \Sigma)$ with $k = 1$. In particular, $N(0, 1)$ is called the *standard* normal distribution. When Σ is a nonnegative definite but singular matrix, we define X to be $N_k(\mu, \Sigma)$ if and only if $c^\tau X$ is $N(c^\tau \mu, c^\tau \Sigma c)$ for any $c \in \mathcal{R}^k$ ($N(a, 0)$ is defined to be the c.d.f. of the point mass at a), which is an important property of $N_k(\mu, \Sigma)$ with a nonsingular Σ (Exercise 81).

Another important joint p.d.f. will be introduced in Example 2.7.

Table 1.2. Distributions on $\mathcal{R}$ with Lebesgue p.d.f.'s

Uniform	p.d.f.	$(b-a)^{-1}I_{(a,b)}(x)$
	m.g.f.	$(e^{bt}-e^{at})/[(b-a)t], \ t \in \mathcal{R}$
$U(a,b)$	Expectation	$(a+b)/2$
	Variance	$(b-a)^2/12$
	Parameter	$a, \ b \in \mathcal{R}, \ a < b$
Normal	p.d.f.	$\frac{1}{\sqrt{2\pi}\sigma}e^{-(x-\mu)^2/2\sigma^2}$
	m.g.f.	$e^{\mu t+\sigma^2 t^2/2}, \ t \in \mathcal{R}$
$N(\mu,\sigma^2)$	Expectation	μ
	Variance	σ^2
	Parameter	$\mu \in \mathcal{R}, \ \sigma > 0$
Exponential	p.d.f.	$\theta^{-1}e^{-(x-a)/\theta}I_{(a,\infty)}(x)$
	m.g.f.	$e^{at}(1-\theta t)^{-1}, \ t < \theta^{-1}$
$E(a,\theta)$	Expectation	$\theta + a$
	Variance	θ^2
	Parameter	$\theta > 0, \ a \in \mathcal{R}$
Chi-square	p.d.f.	$\frac{1}{\Gamma(k/2)2^{k/2}}x^{k/2-1}e^{-x/2}I_{(0,\infty)}(x)$
	m.g.f.	$(1-2t)^{-k/2}, \ t < 1/2$
χ_k^2	Expectation	k
	Variance	$2k$
	Parameter	$k = 1, 2, \ldots$
Gamma	p.d.f.	$\frac{1}{\Gamma(\alpha)\gamma^\alpha}x^{\alpha-1}e^{-x/\gamma}I_{(0,\infty)}(x)$
	m.g.f.	$(1-\gamma t)^{-\alpha}, \ t < \gamma^{-1}$
$\Gamma(\alpha,\gamma)$	Expectation	$\alpha\gamma$
	Variance	$\alpha\gamma^2$
	Parameter	$\gamma > 0, \alpha > 0$
Beta	p.d.f.	$\frac{\Gamma(\alpha+\beta)}{\Gamma(\alpha)\Gamma(\beta)}x^{\alpha-1}(1-x)^{\beta-1}I_{(0,1)}(x)$
	m.g.f.	No explicit form
$B(\alpha,\beta)$	Expectation	$\alpha/(\alpha+\beta)$
	Variance	$\alpha\beta/[(\alpha+\beta+1)(\alpha+\beta)^2]$
	Parameter	$\alpha > 0, \ \beta > 0$
Cauchy	p.d.f.	$\frac{1}{\pi\sigma}\left[1+\left(\frac{x-\mu}{\sigma}\right)^2\right]^{-1}$
	ch.f.	$e^{\sqrt{-1}\mu t-\sigma|t|}$
$C(\mu,\sigma)$	Expectation	Does not exist
	Variance	Does not exist
	Parameter	$\mu \in \mathcal{R}, \ \sigma > 0$

Table 1.2. (continued)

t-distribution	p.d.f.	$\frac{\Gamma[(n+1)/2]}{\sqrt{n\pi}\Gamma(n/2)}\left(1+\frac{x^2}{n}\right)^{-(n+1)/2}$
	ch.f.	No explicit form
t_n	Expectation	$0,\ (n>1)$
	Variance	$n/(n-2),\ (n>2)$
	Parameter	$n=1,2,...$
F-distribution	p.d.f.	$\frac{n^{n/2}m^{m/2}\Gamma[(n+m)/2]x^{n/2-1}}{\Gamma(n/2)\Gamma(m/2)(m+nx)^{(n+m)/2}}I_{(0,\infty)}(x)$
	ch.f.	No explicit form
$F_{n,m}$	Expectation	$m/(m-2),\ (m>2)$
	Variance	$2m^2(n+m-2)/[n(m-2)^2(m-4)],$
		$(m>4)$
	Parameter	$n=1,2,...,\ m=1,2,...$
Log-normal	p.d.f.	$\frac{1}{\sqrt{2\pi}\sigma}x^{-1}e^{-(\log x-\mu)^2/2\sigma^2}I_{(0,\infty)}(x)$
	ch.f.	No explicit form
$LN(\mu,\sigma^2)$	Expectation	$e^{\mu+\sigma^2/2}$
	Variance	$e^{2\mu+\sigma^2}(e^{\sigma^2}-1)$
	Parameter	$\mu\in\mathcal{R},\ \sigma>0$
Weibull	p.d.f.	$\frac{\alpha}{\theta}x^{\alpha-1}e^{-x^\alpha/\theta}I_{(0,\infty)}(x)$
	ch.f.	No explicit form
$W(\alpha,\theta)$	Expectation	$\theta^{1/\alpha}\Gamma(\alpha^{-1}+1)$
	Variance	$\theta^{2/\alpha}\left\{\Gamma(2\alpha^{-1}+1)-\left[\Gamma(\alpha^{-1}+1)\right]^2\right\}$
	Parameter	$\theta>0,\ \alpha>0$
Double	p.d.f.	$\frac{1}{2\theta}e^{-|x-\mu|/\theta}$
Exponential	m.g.f.	$e^{\mu t}/(1-\theta^2t^2),\ t\in\mathcal{R}$
	Expectation	μ
$DE(\mu,\theta)$	Variance	$2\theta^2$
	Parameter	$\mu\in\mathcal{R},\ \theta>0$
Pareto	p.d.f.	$\theta a^\theta x^{-(\theta+1)}I_{(a,\infty)}(x)$
	ch.f.	No explicit form
$Pa(a,\theta)$	Expectation	$\theta a/(\theta-1),\ (\theta>1)$
	Variance	$\theta a^2/[(\theta-1)^2(\theta-2)],\ (\theta>2)$
	Parameter	$\theta>0,\ a>0$
Logistic	p.d.f.	$\sigma^{-1}e^{-(x-\mu)/\sigma}/[1+e^{-(x-\mu)/\sigma}]^2$
	m.g.f.	$e^{\mu t}\Gamma(1+\sigma t)\Gamma(1-\sigma t),\ |t|<\sigma$
$LG(\mu,\sigma)$	Expectation	μ
	Variance	$\sigma^2\pi^2/3$
	Parameter	$\mu\in\mathcal{R},\ \sigma>0$

If a random k-vector $(X_1, ..., X_k)$ has a joint p.d.f. f w.r.t. a product measure $\nu_1 \times \cdots \times \nu_k$ defined on $\mathcal{B}^k$, then X_i has the following marginal p.d.f. w.r.t. ν_i:

$$f_i(x) = \int_{\mathcal{R}^{k-1}} f(x_1, ..., x_{i-1}, x, x_{i+1}, ..., x_k) d\nu_1 \cdots d\nu_{i-1} d\nu_{i+1} \cdots d\nu_k.$$

Let F be the joint c.d.f. of a random k-vector $(X_1, ..., X_k)$ and F_i be the marginal c.d.f. of X_i, $i = 1, ..., k$. If (1.7) holds, then random variables $X_1, ..., X_k$ are said to be *independent*. From the discussion in the end of §1.1.1, this independence means that the probability measure corresponding to F is the product measure of the k probability measures corresponding to F_i's. The meaning of independence is further discussed in §1.4.2. If $(X_1, ..., X_k)$ has a joint p.d.f. f w.r.t. a product measure $\nu_1 \times \cdots \times \nu_k$ defined on $\mathcal{B}^k$, then $X_1, ..., X_k$ are independent if and only if

$$f(x_1, ..., x_k) = f_1(x_1) \cdots f_k(x_k), \quad (x_1, ..., x_k) \in \mathcal{R}^k, \qquad (1.25)$$

where f_i is the p.d.f. of X_i w.r.t. ν_i, $i = 1, ..., k$. For example, using (1.24), one can show (exercise) that the components of $N_k(\mu, \Sigma)$ are independent if and only if Σ is a diagonal matrix.

The following lemma is useful in considering the independence of functions of independent random variables.

Lemma 1.1. Let $X_1, ..., X_n$ be independent random variables. Then random variables $g(X_1, ..., X_k)$ and $h(X_{k+1}, ..., X_n)$ are independent, where g and h are Borel functions and k is an integer between 1 and n. ∎

Lemma 1.1 can be proved directly (exercise). But it is a simple consequence of an equivalent definition of independence introduced in §1.4.2.

Let $X_1, ..., X_k$ be random variables. If X_i and X_j are independent for every pair $i \neq j$, then $X_1, ..., X_k$ are said to be pairwise independent. If $X_1, ..., X_k$ are independent, then clearly they are pairwise independent. However, the converse is not true. The following is an example.

Example 1.13. Let X_1 and X_2 be independent random variables each assuming the values 1 and -1 with probability 0.5, and $X_3 = X_1 X_2$. Let $A_i = \{X_i = 1\}$, $i = 1, 2, 3$. Then $P(A_i) = 0.5$ for any i and $P(A_1)P(A_2)P(A_3) = 0.125$. However, $P(A_1 \cap A_2 \cap A_3) = P(A_1 \cap A_2) = P(A_1)P(A_2) = 0.25$. This implies that (1.7) does not hold and, hence, X_1, X_2, X_3 are not independent. We now show that X_1, X_2, X_3 are pairwise independent. It is enough to show that X_1 and X_3 are independent. Let $B_i = \{X_i = -1\}$, $i = 1, 2, 3$. Note that $A_1 \cap A_3 = A_1 \cap A_2$, $A_1 \cap B_3 = A_1 \cap B_2$, $B_1 \cap A_3 = B_1 \cap B_2$, and $B_1 \cap B_3 = B_1 \cap A_2$. Then the result follows from the fact that $P(A_i) = P(B_i) = 0.5$ for any i and X_1 and X_2 are independent. ∎

The random variable Y in Example 1.12 is a transformation of the random variable X. Transformations of random variables or vectors are frequently used in statistics. For a random variable or vector X, $g(X)$ is a random variable or vector as long as g is measurable (Proposition 1.4). How do we find the c.d.f. (or p.d.f.) of $g(X)$ when the c.d.f. (or p.d.f.) of X is known? In many cases, the most effective method is direct computation. Example 1.12 is one example. The following is another one.

Example 1.14. Let X be a random variable with c.d.f. F_X and Lebesgue p.d.f. f_X, and let $Y = X^2$. Since $Y^{-1}((-\infty, x])$ is empty if $x < 0$ and equals $Y^{-1}([0, x]) = X^{-1}([-\sqrt{x}, \sqrt{x}])$ if $x \geq 0$, the c.d.f. of Y is

$$
\begin{aligned}
F_Y(x) &= P \circ Y^{-1}((-\infty, x]) \\
&= P \circ X^{-1}([-\sqrt{x}, \sqrt{x}]) \\
&= F_X(\sqrt{x}) - F_X(-\sqrt{x})
\end{aligned}
$$

if $x \geq 0$ and $F_Y(x) = 0$ if $x < 0$. Clearly, the Lebesgue p.d.f. of F_Y is

$$
f_Y(x) = \frac{1}{2\sqrt{x}}[f_X(\sqrt{x}) + f_X(-\sqrt{x})]I_{(0,\infty)}(x). \tag{1.26}
$$

In particular, if

$$
f_X(x) = \frac{1}{\sqrt{2\pi}}e^{-x^2/2}, \tag{1.27}
$$

which is the Lebesgue p.d.f. of the standard normal distribution $N(0,1)$ (Table 1.2), then

$$
f_Y(x) = \frac{1}{\sqrt{2\pi x}}e^{-x/2}I_{(0,\infty)}(x),
$$

which is the Lebesgue p.d.f. for the chi-square distribution χ_1^2 (Table 1.2). This is actually an important result in statistics.   ∎

In some cases, one may apply the following general result whose proof is left to the reader.

Proposition 1.8. Let X be a random k-vector with a Lebesgue p.d.f. f_X and let $Y = g(X)$, where g is a Borel function from $(\mathcal{R}^k, \mathcal{B}^k)$ to $(\mathcal{R}^k, \mathcal{B}^k)$. Let $A_1, ..., A_m$ be disjoint sets in $\mathcal{B}^k$ such that $\mathcal{R}^k - (A_1 \cup \cdots \cup A_m)$ has Lebesgue measure 0 and g on A_j is one-to-one with a nonvanishing Jacobian, i.e., the determinant $\mathrm{Det}(\partial g(x)/\partial x) \neq 0$ on A_j, $j = 1, ..., m$. Then Y has the following Lebesgue p.d.f.:

$$
f_Y(x) = \sum_{j=1}^{m} \left| \mathrm{Det}\left(\partial h_j(x)/\partial x \right) \right| f_X\left(h_j(x) \right),
$$

where h_j is the inverse function of g on A_j, $j = 1, ..., m$.   ∎

One may apply Proposition 1.8 to obtain result (1.26) in Example 1.14, using $A_1 = (-\infty, 0)$, $A_2 = (0, \infty)$, and $g(x) = x^2$. Note that $h_1(x) = -\sqrt{x}$, $h_2(x) = \sqrt{x}$, and $|dh_j(x)/dx| = 1/(2\sqrt{x})$. Another immediate application of Proposition 1.8 is to show that $Y = AX$ is $N_k(A\mu, A\Sigma A^\tau)$ when X is $N_k(\mu, \Sigma)$, where Σ is positive definite, A is a $k \times k$ matrix of rank k, and A^τ denotes the transpose of A.

Example 1.15. Let $X = (X_1, X_2)$ be a random 2-vector having a joint Lebesgue p.d.f. f_X. Consider first the transformation $g(x) = (x_1, x_1 + x_2)$. Using Proposition 1.8, one can show that the joint p.d.f. of $g(X)$ is

$$f_{g(X)}(x_1, y) = f_X(x_1, y - x_1),$$

where $y = x_1 + x_2$ (note that the Jacobian equals 1). The marginal p.d.f. of $Y = X_1 + X_2$ is then

$$f_Y(y) = \int f_X(x_1, y - x_1) dx_1.$$

In particular, if X_1 and X_2 are independent, then

$$f_Y(y) = \int f_{X_1}(x_1) f_{X_2}(y - x_1) dx_1. \tag{1.28}$$

Next, consider the transformation $h(x_1, x_2) = (x_1/x_2, x_2)$, assuming that $X_2 \neq 0$ a.s. Using Proposition 1.8, one can show that the joint p.d.f. of $h(X)$ is

$$f_{h(X)}(z, x_2) = |x_2| f_X(zx_2, x_2),$$

where $z = x_1/x_2$. The marginal p.d.f. of $Z = X_1/X_2$ is

$$f_Z(z) = \int |x_2| f_X(zx_2, x_2) dx_2.$$

In particular, if X_1 and X_2 are independent, then

$$f_Z(z) = \int |x_2| f_{X_1}(zx_2) f_{X_2}(x_2) dx_2. \quad \blacksquare \tag{1.29}$$

A number of results can be derived from (1.28) and (1.29). For example, if X_1 and X_2 are independent and both have the standard normal p.d.f. given by (1.27), then, by (1.29), the Lebesgue p.d.f. of $Z = X_1/X_2$ is

$$\begin{aligned}
f_Z(z) &= \frac{1}{2\pi} \int |x_2| e^{-(1+z^2)x_2^2/2} dx_2 \\
&= \frac{1}{\pi} \int_0^\infty e^{-(1+z^2)x} dx \\
&= \frac{1}{\pi(1 + z^2)},
\end{aligned}$$

which is the p.d.f. of the Cauchy distribution $C(0,1)$ in Table 1.2. Another application of formula (1.29) leads to the following important result in statistics.

Example 1.16 (t-distribution and F-distribution). Let X_1 and X_2 be independent random variables having the chi-square distributions $\chi^2_{n_1}$ and $\chi^2_{n_2}$ (Table 1.2), respectively. By (1.29), the p.d.f. of $Z = X_1/X_2$ is

$$f_Z(z) = \frac{z^{n_1/2-1} I_{(0,\infty)}(z)}{2^{(n_1+n_2)/2}\Gamma(n_1/2)\Gamma(n_2/2)} \int_0^\infty x_2^{(n_1+n_2)/2-1} e^{-(1+z)x_2/2} dx_2$$

$$= \frac{\Gamma[(n_1+n_2)/2]}{\Gamma(n_1/2)\Gamma(n_2/2)} \frac{z^{n_1/2-1}}{(1+z)^{(n_1+n_2)/2}} I_{(0,\infty)}(z),$$

where the last equality follows from the fact that

$$\frac{1}{2^{(n_1+n_2)/2}\Gamma[(n_1+n_2)/2]} x_2^{(n_1+n_2)/2-1} e^{-x_2/2} I_{(0,\infty)}(x_2)$$

is the p.d.f. of the chi-square distribution $\chi^2_{n_1+n_2}$. Using Proposition 1.8, one can show that the p.d.f. of $Y = (X_1/n_1)/(X_2/n_2) = (n_2/n_1)Z$ is the p.d.f. of the F-distribution F_{n_1,n_2} given in Table 1.2.

Let U_1 be a random variable having the standard normal distribution $N(0,1)$ and U_2 a random variable having the chi-square distribution χ^2_n. Using the same argument, one can show that if U_1 and U_2 are independent, then the distribution of $T = U_1/\sqrt{U_2/n}$ is the t-distribution t_n given in Table 1.2. This result can also be derived using the result given in this example as follows. Let $X_1 = U_1^2$ and $X_2 = U_2$. Then X_1 and X_2 are independent (which can be shown directly but follows from Lemma 1.1). By Example 1.14, the distribution of X_1 is χ^2_1. Then $Y = X_1/(X_2/n)$ has the F-distribution $F_{1,n}$ and its Lebesgue p.d.f. is

$$\frac{n^{n/2}\Gamma[(n+1)/2]x^{-1/2}}{\sqrt{\pi}\Gamma(n/2)(n+x)^{(n+1)/2}} I_{(0,\infty)}(x).$$

Note that

$$T = \begin{cases} \sqrt{Y} & U_1 \geq 0 \\ -\sqrt{Y} & U_1 < 0. \end{cases}$$

The result follows from Proposition 1.8 and the fact that

$$P \circ T^{-1}\left((-\infty, -t]\right) = P \circ T^{-1}\left([t, \infty)\right), \quad t > 0. \quad \blacksquare \qquad (1.30)$$

If a random variable T satisfies (1.30), i.e., T and $-T$ have the same distribution, then T and its c.d.f. and p.d.f. (if it exists) are said to be *symmetric* about 0. If T has a Lebesgue p.d.f. f_T, then T is symmetric about 0 if and only if $f_T(x) = f_T(-x)$ for any $x > 0$. T and its c.d.f. and p.d.f. are said to be symmetric about a (or symmetric for simplicity) if and only if $T - a$ is symmetric about 0 for a fixed $a \in \mathcal{R}$. The c.d.f.'s of t-distributions are symmetric about 0 and the normal, Cauchy, and double exponential c.d.f.'s are symmetric.

The chi-square, t-, and F-distributions in the previous examples are special cases of the following *noncentral* chi-square, t-, and F-distributions, which are useful in some statistical problems.

Let $X_1, ..., X_n$ be independent random variables and $X_i = N(\mu_i, \sigma^2)$, $i = 1, ..., n$. The distribution of the random variable $Y = (X_1^2 + \cdots + X_n^2)/\sigma^2$ is called the *noncentral chi-square* distribution and denoted by $\chi_n^2(\delta)$, where $\delta = (\mu_1^2 + \cdots + \mu_n^2)/\sigma^2$ is the noncentrality parameter. The chi-square distribution χ_k^2 in Table 1.2 is a special case of the noncentral chi-square distribution $\chi_k^2(\delta)$ with $\delta = 0$ and, therefore, is called a *central* chi-square distribution. It can be shown (exercise) that Y has the following Lebesgue p.d.f.:

$$e^{-\delta/2} \sum_{j=0}^{\infty} \frac{(\delta/2)^j}{j!} f_{2j+n}(x), \qquad (1.31)$$

where $f_k(x)$ is the Lebesgue p.d.f. of the chi-square distribution χ_k^2. It follows from the definition of noncentral chi-square distributions that if $Y_1, ..., Y_k$ are independent random variables and Y_i has the noncentral chi-square distribution $\chi_{n_i}^2(\delta_i)$, $i = 1, ..., k$, then $Y = Y_1 + \cdots + Y_k$ has the noncentral chi-square distribution $\chi_{n_1 + \cdots + n_k}^2(\delta_1 + \cdots + \delta_k)$.

The result for the t-distribution in Example 1.16 can be extended to the case where U_1 has a nonzero expectation μ (U_2 still has the χ_n^2 distribution and is independent of U_1). The distribution of $T = U_1/\sqrt{U_2/n}$ is called the *noncentral* t-distribution and denoted by $t_n(\delta)$, where $\delta = \mu$ is the noncentrality parameter. Using the same argument as that in Example 1.15, one can show (exercise) that T has the following Lebesgue p.d.f.:

$$\frac{1}{2^{(n+1)/2}\Gamma(n/2)\sqrt{\pi n}} \int_0^{\infty} y^{(n-1)/2} e^{-[(x\sqrt{y/n}-\delta)^2+y]/2} dy. \qquad (1.32)$$

The t-distribution t_n in Example 1.16 is called a *central* t-distribution, since it is a special case of the noncentral t-distribution $t_n(\delta)$ with $\delta = 0$.

Similarly, the result for the F-distribution in Example 1.16 can be extended to the case where X_1 has the noncentral chi-square distribution $\chi_{n_1}^2(\delta)$, X_2 has the central chi-square distribution $\chi_{n_2}^2$, and X_1 and X_2 are independent. The distribution of $Y = (X_1/n_1)/(X_2/n_2)$ is called the *noncentral* F-distribution and denoted by $F_{n_1,n_2}(\delta)$, where δ is the noncentrality parameter. The F-distribution F_{n_1,n_2} in Example 1.16 is called a *central*

F-distribution, since it is a special case of the noncentral F-distribution $F_{n_1,n_2}(\delta)$ with $\delta = 0$. It can be shown (exercise) that the noncentral F-distribution $F_{n_1,n_2}(\delta)$ has the following Lebesgue p.d.f.:

$$e^{-\delta/2} \sum_{j=0}^{\infty} \frac{n_1(\delta/2)^j}{j!(2j+n_1)} f_{2j+n_1,n_2}\left(\frac{n_1 x}{2j+n_1}\right), \qquad (1.33)$$

where $f_{k_1,k_2}(x)$ is the Lebesgue p.d.f. of the central F-distribution F_{k_1,k_2} given in Table 1.2.

Using some results from linear algebra, we can prove the following result useful in *analysis of variance* (Scheffé, 1959; Searle, 1971).

Theorem 1.5. (Cochran's theorem). Suppose that $X = N_n(\mu, I_n)$ and

$$X^\tau X = X^\tau A_1 X + \cdots + X^\tau A_k X, \qquad (1.34)$$

where I_n is the $n \times n$ identity matrix and A_i is an $n \times n$ symmetric matrix with rank n_i, $i = 1, ..., k$. A necessary and sufficient condition that $X^\tau A_i X$ has the noncentral chi-square distribution $\chi^2_{n_i}(\delta_i)$, $i = 1, ..., k$, and $X^\tau A_i X$'s are independent is $n = n_1 + \cdots + n_k$, in which case $\delta_i = \mu^\tau A_i \mu$ and $\delta_1 + \cdots + \delta_k = \mu^\tau \mu$.

Proof. Suppose that $X^\tau A_i X$, $i = 1, ..., k$, are independent and $X^\tau A_i X$ has the $\chi^2_{n_i}(\delta_i)$ distribution. Then $X^\tau X$ has the $\chi^2_{n_1+\cdots+n_k}(\delta_1 + \cdots + \delta_k)$ distribution. By definition, $X^\tau X$ has the noncentral chi-square distribution $\chi^2_n(\mu^\tau \mu)$. By (1.34), $n = n_1 + \cdots + n_k$ and $\delta_1 + \cdots + \delta_k = \mu^\tau \mu$.

Suppose now that $n = n_1 + \cdots + n_k$. From linear algebra, for each i there exists $c_{ij} \in \mathcal{R}^n$, $j = 1, ..., n_i$, such that

$$X^\tau A_i X = \pm (c_{i1}^\tau X)^2 \pm \cdots \pm (c_{in_i}^\tau X)^2. \qquad (1.35)$$

Let C_i be the $n \times n_i$ matrix whose jth column is c_{ij}, and $C = (C_1, ..., C_k)$. By (1.34) and (1.35), $XX^\tau = X^\tau C^\tau \Delta C X$ with an $n \times n$ diagonal matrix Δ whose diagonal elements are either 1 or -1. This implies $C^\tau \Delta C = I_n$. Thus, C is of full rank and, hence, $\Delta = (C^\tau)^{-1} C^{-1}$, which is positive definite. This shows $\Delta = I_n$, which implies $C^\tau C = I_n$ and

$$X^\tau A_i X = \sum_{j=n_1+\cdots+n_{i-1}+1}^{n_1+\cdots+n_{i-1}+n_i} Y_j^2, \qquad (1.36)$$

where Y_j is the jth component of $Y = CX$. Note that $Y = N_n(C\mu, I_n)$ (Exercise 43). Hence Y_j's are independent and $Y_j = N(\lambda_j, 1)$, where λ_j is the jth component of $C\mu$. This shows that $X^\tau A_i X$ has the $\chi^2_{n_i}(\delta_i)$ distribution with $\delta_i = \lambda^2_{n_1+\cdots+n_{i-1}+1} + \cdots + \lambda^2_{n_1+\cdots+n_{i-1}+n_i}$. Letting $X = \mu$ in (1.36) and (1.34), we obtain that $\delta_i = \mu^\tau A_i \mu$ and $\delta_1 + \cdots + \delta_k = \mu^\tau C^\tau C \mu = \mu^\tau \mu$. Finally, from (1.36) and Lemma 1.1, we conclude that $X^\tau A_i X$, $i = 1, ..., k$, are independent. ∎

1.3.2 Moments and moment inequalities

We have defined the expectation of a random variable in §1.2.1. It is an important characteristic of a random variable. In this section, we introduce *moments*, which are some other important characteristics of a random variable or vector.

Let X be a random variable. If EX^k is finite, where k is a positive integer, then EX^k is called the kth *moment* of X or P_X (the distribution of X). If $E|X|^a < \infty$ for some real number a, then $E|X|^a$ is called the ath *absolute moment* of X or P_X. If $\mu = EX$ and $E(X - \mu)^k$ are finite for a positive integer k, then $E(X - \mu)^k$ is called the kth *central moment* of X or P_X. If $E|X|^a < \infty$ for an $a > 0$, then $E|X|^t < \infty$ for any positive $t < a$ and EX^k is finite for any positive integer $k \le a$ (Exercise 54).

The expectation and the second central moment (if they exist) are two important characteristics of a random variable (or its distribution) in statistics. They are listed in Tables 1.1 and 1.2 for those useful distributions. The expectation, also called the *mean* in statistics, is a measure of the central location of the distribution of a random variable. The second central moment, also called the *variance* in statistics, is a measure of dispersion or spread of a random variable. The variance of a random variable X is denoted by $\mathrm{Var}(X)$. The variance is always nonnegative. If the variance of X is 0, then X is equal to its mean a.s. (Proposition 1.6). The squared root of the variance is called the *standard deviation*, another important characteristic of a random variable in statistics.

The concept of mean and variance can be extended to random vectors. The expectation of a random matrix M with (i, j)th element M_{ij} is defined to be the matrix whose (i, j)th element is EM_{ij}. Thus, for a random k-vector $X = (X_1, ..., X_k)$, its mean is $EX = (EX_1, ..., EX_k)$. The extension of variance is the *variance-covariance matrix* of X defined as

$$\mathrm{Var}(X) = E(X - EX)(X - EX)^\tau,$$

which is a $k \times k$ symmetric matrix whose diagonal elements are variances of X_i's. The (i, j)th element of $\mathrm{Var}(X)$, $i \ne j$, is $E(X_i - EX_i)(X_j - EX_j)$, which is called the *covariance* of X_i and X_j and is denoted by $\mathrm{Cov}(X_i, X_j)$.

Let $c \in \mathcal{R}^k$ and $X = (X_1, ..., X_k)$ be a random k-vector. Then $Y = c^\tau X$ is a random variable and, by Proposition 1.5 (linearity of integrals), $EY = c^\tau EX$ if EX exists. Also, when $\mathrm{Var}(X)$ is finite (i.e., all elements of $\mathrm{Var}(X)$ are finite),

$$
\begin{aligned}
\mathrm{Var}(Y) &= E(c^\tau X - c^\tau EX)^2 \\
&= E[c^\tau (X - EX)(X - EX)^\tau c] \\
&= c^\tau [E(X - EX)(X - EX)^\tau] c \\
&= c^\tau \mathrm{Var}(X) c.
\end{aligned}
$$

Since $\text{Var}(Y) \geq 0$ for any $c \in \mathcal{R}^k$, the matrix $\text{Var}(X)$ is nonnegative definite. Consequently,

$$[\text{Cov}(X_i, X_j)]^2 \leq \text{Var}(X_i)\text{Var}(X_j), \quad i \neq j. \tag{1.37}$$

An important quantity in statistics is the *correlation coefficient* defined to be $\rho_{X_i, X_j} = \text{Cov}(X_i, X_j)/\sqrt{\text{Var}(X_i)\text{Var}(X_j)}$, which, by inequality (1.37), is always between -1 and 1. It is a measure of relationship between X_i and X_j; if ρ_{X_i, X_j} is positive (or negative), then X_i and X_j tend to be positively (or negatively) related; if $\rho_{X_i, X_j} = \pm 1$, then $P(X_i = c_1 \pm c_2 X_j) = 1$ with some constants c_1 and $c_2 > 0$; if $\rho_{X_i, X_j} = 0$ (i.e., $\text{Cov}(X_i, X_j) = 0$), then X_i and X_j are said to be *uncorrelated*. If X_i and X_j are independent, then they are uncorrelated. This follows from the following more general result. If $X_1, ..., X_n$ are independent random variables and $E|X_1 \cdots X_n| < \infty$, then, by Fubini's theorem and the fact that the joint c.d.f. of $(X_1, ..., X_n)$ corresponds to a product measure, we obtain that

$$E(X_1 \cdots X_n) = EX_1 \cdots EX_n. \tag{1.38}$$

In fact, pairwise independence of $X_1, ..., X_n$ implies that X_i's are uncorrelated, since $\text{Cov}(X_i, X_j)$ involves only a pair of random variables. However, the converse is not necessarily true: uncorrelated random variables may not be pairwise independent. Examples can be found in Exercises 60-61.

Let $R_M = \{y \in \mathcal{R}^k : y = Mx$ with some $x \in \mathcal{R}^k\}$ for any $k \times k$ symmetric matrix M. If a random k-vector X has a finite $\text{Var}(X)$, then $P(X - EX \in R_{\text{Var}(X)}) = 1$. This means that if the rank of $\text{Var}(X)$ is $r < k$, then X is in a subspace of $\mathcal{R}^k$ with dimension r. Consequently, if $P_X \ll$ Lebesgue measure on $\mathcal{R}^k$, then the rank of $\text{Var}(X)$ is k.

Example 1.17. Let X be a random k-vector having the $N_k(\mu, \Sigma)$ distribution. It can be shown (exercise) that $EX = \mu$ and $\text{Var}(X) = \Sigma$. Thus, μ and Σ in (1.24) are the mean vector and the variance-covariance matrix of X. If Σ is a diagonal matrix (i.e., all components of X are uncorrelated), then by (1.25), the components of X are independent. This shows an important property of random variables having normal distributions: they are independent if and only if they are uncorrelated. ∎

There are many useful inequalities related to moments. The inequality in (1.37) is in fact the well-known Cauchy-Schwartz inequality whose general form is

$$[E(XY)]^2 \leq EX^2 EY^2, \tag{1.39}$$

where X and Y are random variables with a well-defined $E(XY)$. Inequality (1.39) is a special case of the following Hölder's inequality:

$$E|XY| \leq (E|X|^p)^{1/p}(E|Y|^q)^{1/q}, \tag{1.40}$$

where p and q are constants satisfying $p > 1$ and $p^{-1} + q^{-1} = 1$. To show inequality (1.40), we use the following inequality (Exercise 62):

$$x^t y^{1-t} \leq tx + (1-t)y, \qquad (1.41)$$

where x and y are nonnegative real numbers and $t \in (0,1)$. If either $E|X|^p$ or $E|Y|^q$ is ∞, then (1.40) holds. Hence we can assume that both $E|X|^p$ and $E|Y|^q$ are finite. Let $a = (E|X|^p)^{1/p}$ and $b = (E|Y|^q)^{1/q}$. If either $a = 0$ or $b = 0$, then the equality in (1.40) holds because of Proposition 1.6(ii). Assume now $a \neq 0$ and $b \neq 0$. Letting $x = |X/a|^p$, $y = |Y/b|^q$, and $t = p^{-1}$ in (1.41), we obtain that

$$\left| \frac{XY}{ab} \right| \leq \frac{|X|^p}{pa^p} + \frac{|Y|^q}{qb^q}.$$

Taking expectations on both sides of this expression, we obtain that

$$\frac{E|XY|}{ab} \leq \frac{E|X|^p}{pa^p} + \frac{E|Y|^q}{qa^q} = \frac{1}{p} + \frac{1}{q} = 1,$$

which is (1.40). In fact, the equality in (1.40) holds if and only if $\alpha|X|^p = \beta|Y|^q$ a.s. for some nonzero constants α and β (Exercise 62).

Using Hölder's inequality, we can prove Liapounov's inequality

$$(E|X|^r)^{1/r} \leq (E|X|^s)^{1/s}, \qquad (1.42)$$

where r and s are constants satisfying $1 \leq r \leq s$, and Minkowski's inequality

$$(E|X+Y|^p)^{1/p} \leq (E|X|^p)^{1/p} + (E|Y|^p)^{1/p}, \qquad (1.43)$$

where X and Y are random variables and p is a constant larger than or equal to 1 (Exercise 63).

Minkowski's inequality can be extended to the case of more than two random variables (Exercise 63). The following inequality is a tightened form of Minkowski's inequality due to Esseen and von Bahr (1965). Let $X_1, ..., X_n$ be independent random variables with mean 0 and $E|X_i|^p < \infty$, $i = 1, ..., n$, where p is a constant in $[1, 2]$. Then

$$E \left| \sum_{i=1}^{n} X_i \right|^p \leq C_p \sum_{i=1}^{n} E|X_i|^p, \qquad (1.44)$$

where C_p is a constant depending only on p. When $1 < p < 2$, inequality (1.44) can be proved (Exercise 63) using inequality

$$|a+b|^p \leq |a|^p + p\,\mathrm{sgn}(a)|a|^{p-1}b + C_p|b|^p, \quad a \in \mathcal{R}, b \in \mathcal{R},$$

where $\text{sgn}(x)$ is 1 or -1 as x is positive or negative and

$$C_p = \sup_{x \in \mathcal{R}, x \neq 0} (|1 + x|^p - 1 - px)/|x|^p.$$

For $p \geq 2$, there is a similar inequality due to Marcinkiewicz and Zygmund:

$$E \left| \sum_{i=1}^{n} X_i \right|^p \leq \frac{C_p}{n^{1-p/2}} \sum_{i=1}^{n} E|X_i|^p, \tag{1.45}$$

where C_p is a constant depending only on p. A proof of inequality (1.45) can be found in Loève (1977, p. 276).

Recall from calculus that a subset A of $\mathcal{R}^k$ is *convex* if and only if $x \in A$ and $y \in A$ imply $tx + (1 - t)y \in A$ for any $t \in [0, 1]$; a function f from a convex $A \subset \mathcal{R}^k$ to $\mathcal{R}$ is *convex* if and only if

$$f(tx + (1 - t)y) \leq tf(x) + (1 - t)f(y), \quad x \in A, y \in A, t \in [0, 1]; \tag{1.46}$$

and f is *strictly convex* if and only if (1.46) holds with $\leq$ replaced by the strict inequality $<$. If f is twice differentiable on A, then a necessary and sufficient condition for f to be convex (or strictly convex) is that the $k \times k$ second-order partial derivative matrix $\partial^2 f / \partial x \partial x^\tau$, the so-called Hessian matrix, is nonnegative definite (or positive definite). For a convex function f defined on an open convex $A \subset \mathcal{R}^k$ and a random k-vector X with finite mean and $P(X \in A) = 1$, a very useful inequality in probability theory and statistics is the following Jensen's inequality:

$$f(EX) \leq Ef(X). \tag{1.47}$$

If f is strictly convex, then $\leq$ in (1.47) can be replaced by $<$ unless $P(f(X) = Ef(X)) = 1$. To prove (1.47), we use without proof the following fact for convex f on an open convex $A \subset \mathcal{R}^k$ (see, e.g., Lehmann, 1983, p. 53). For any $y \in A$, there exists a vector $a_y \in A$ such that

$$f(x) \geq f(y) + a_y(x - y)^\tau, \quad x \in A. \tag{1.48}$$

We also use the fact that $EX \in A$ (see, e.g., Ferguson, 1967, p. 74). Letting $x = X$ and $y = EX$, we obtain (1.47) by taking expectations on both sides of (1.48). If f is strictly convex, then (1.48) holds with $\geq$ replaced by $>$. By Proposition 1.6(ii), $Ef(X) > f(EX)$ unless $P(f(X) = Ef(X)) = 1$.

Example 1.18. A direct application of Jensen's inequality (1.47) is that if X is a nonconstant positive random variable with finite mean, then

$$(EX)^{-1} < E(X^{-1}) \quad \text{and} \quad E(\log X) < \log(EX),$$

since t^{-1} and $-\log t$ are convex functions on $(0, \infty)$. Another application is to prove the following inequality related to entropy. Let f and g be positive integrable functions on a measure space with a σ-finite measure ν. If $\int f d\nu \geq \int g d\nu > 0$, then one can show (exercise) that

$$\int f \log \left(\frac{f}{g} \right) d\nu \geq 0. \quad \blacksquare \tag{1.49}$$

The next inequality, Chebyshev's inequality, is almost trivial but very useful and famous. Let X be a random variable and φ a nonnegative and nondecreasing function on $[0, \infty)$ satisfying $\varphi(-t) = \varphi(t)$. Then, for each constant $t \geq 0$,

$$\varphi(t) P\left(|X| \geq t \right) \leq \int_{\{|X| \geq t\}} \varphi(X) dP \leq E\varphi(X), \tag{1.50}$$

where both inequalities in (1.50) follow from Proposition 1.6(i) and the first inequality also uses the fact that on the set $\{|X| \geq t\}$, $\varphi(X) \geq \varphi(t)$. The most familiar application of (1.50) is when $\varphi(t) = |t|^p$ for $p \in (0, \infty)$, in which case inequality (1.50) is also called Markov's inequality. Chebyshev's inequality, sometimes together with one of the moment inequalities introduced in this section, can be used to yield a desired upper bound for the "tail" probability $P(|X| \geq t)$. For example, let Y be a random variable with mean μ and variance σ^2. Then $X = (Y - \mu)/\sigma$ has mean 0 and variance 1 and, by (1.50) with $\varphi(t) = t^2$, $P(|X| \geq 2) \leq \frac{1}{4}$. This means that the probability that the random variable $|Y - \mu|$ exceeds twice its standard deviation is bounded by $\frac{1}{4}$. Similarly, we can also claim that the probability of $|Y - \mu|$ exceeding 3σ is bounded by $\frac{1}{9}$. These bounds are rough but they can be applied to *any* random variable with a finite variance. Other applications of Chebyshev's inequality can be found in §1.5.

In some cases, we need an improvement over inequality (1.50) when X is of some special form. Let $Y_1, ..., Y_n$ be independent random variables having finite variances. The following inequality is due to Hájek and Rènyi:

$$P\left(\max_{1 \leq l \leq n} c_l \left| \sum_{i=1}^{l} (Y_i - EY_i) \right| > t \right) \leq \frac{1}{t^2} \sum_{i=1}^{n} c_i^2 \mathrm{Var}(Y_i), \quad t > 0, \tag{1.51}$$

where c_i's are positive constants satisfying $c_1 \geq c_2 \geq \cdots \geq c_n$. If $c_i = 1$ for all i, then inequality (1.51) reduces to the famous Kolmogorov's inequality. A proof for (1.51) is given in Sen and Singer (1993, pp. 65-66).

1.3.3 Moment generating and characteristic functions

Moments are important characteristics of a distribution, but they do not determine a distribution in the sense that two different distributions may have the same moments of all orders. Functions that determine a distribution are introduced in the following definition.

Definition 1.5. Let X be a random k-vector.
(i) The *moment generating function* (m.g.f.) of X or P_X is defined as

$$\psi_X(t) = Ee^{t^\tau X}, \quad t \in \mathcal{R}^k.$$

(ii) The *characteristic function* (ch.f.) of X or P_X is defined as

$$\phi_X(t) = Ee^{\sqrt{-1}t^\tau X} = E[\cos(t^\tau X)] + \sqrt{-1}\, E[\sin(t^\tau X)], \quad t \in \mathcal{R}^k. \quad \blacksquare$$

Obviously $\psi_X(0) = \phi_X(0) = 1$ for any random vector X. The ch.f. is complex-valued and always well defined. In fact, any ch.f. is bounded by 1 and is a uniformly continuous function on $\mathcal{R}^k$ (exercise). The m.g.f. is nonnegative but may be ∞ everywhere except at $t = 0$ (Example 1.19). If the m.g.f. is finite in a neighborhood of $0 \in \mathcal{R}^k$, then $\phi_X(t)$ can be obtained by replacing t in $\psi_X(t)$ by $\sqrt{-1}t$. Tables 1.1 and 1.2 contain the m.g.f. (or ch.f. when the m.g.f. is ∞ everywhere except at 0) for distributions useful in statistics. For a linear transformation $Y = A^\tau X + c$, where A is a $k \times m$ matrix and $c \in \mathcal{R}^m$, it follows from Definition 1.5 that

$$\psi_Y(u) = e^{c^\tau u}\psi_X(Au) \quad \text{and} \quad \phi_Y(u) = e^{\sqrt{-1}c^\tau u}\phi_X(Au), \quad u \in \mathcal{R}^m. \tag{1.52}$$

For a random variable X, if its m.g.f. is finite at t and $-t$ for a $t \neq 0$, then X has finite moments and absolute moments of any order. To compute moments of X using its m.g.f., a condition stronger than the finiteness of the m.g.f. at some $t \neq 0$ is needed. Consider a random k-vector X. If ψ_X is finite in a neighborhood of 0, then $\mu_{r_1,\ldots,r_k} = E(X_1^{r_1} \cdots X_k^{r_k})$ is finite for any nonnegative integers $r_1, \ldots, r_k$, where X_j is the jth component of X, and ψ_X has the power series expansion

$$\psi_X(t) = \sum_{(r_1,\ldots,r_k)\in\mathcal{Z}} \frac{\mu_{r_1,\ldots,r_k} t_1^{r_1}\cdots t_k^{r_k}}{r_1!\cdots r_k!} \tag{1.53}$$

for t in the neighborhood of 0, where t_j is the jth component of t and $\mathcal{Z} \subset \mathcal{R}^k$ containing vectors whose components are nonnegative integers. Consequently, the components of X have finite moments of all orders and

$$E(X_1^{r_1} \cdots X_k^{r_k}) = \left.\frac{\partial^{r_1+\cdots+r_k}\psi_X(t)}{\partial t_1^{r_1} \cdots \partial t_k^{r_k}}\right|_{t=0},$$

which are also called moments of X. In particular,

$$\left.\frac{\partial\psi_X(t)}{\partial t}\right|_{t=0} = EX, \qquad \left.\frac{\partial^2\psi_X(t)}{\partial t\partial t^\tau}\right|_{t=0} = E(XX^\tau), \tag{1.54}$$

and, when $k = 1$ and p is a positive integer, $\psi_X^{(p)}(0) = EX^p$, where $g^{(p)}(t)$ denotes the pth order derivative of a function $g(t)$.

If $0 < \psi_X(t) < \infty$, then $\kappa_X(t) = \log \psi_X(t)$ is called the *cumulant generating function* of X or P_X. If $0 < \psi_X(t) < \infty$ for t in a neighborhood of 0, then κ_X has a power series expansion similar to that in (1.53):

$$\kappa_X(t) = \sum_{(r_1,\dots,r_k)\in\mathcal{Z}} \frac{\kappa_{r_1,\dots,r_k} t_1^{r_1} \cdots t_k^{r_k}}{r_1! \cdots r_k!}, \tag{1.55}$$

where $\kappa_{r_1,\dots,r_k}$'s are called *cumulants* of X. There is a one-to-one correspondence between the set of moments and the set of cumulants. An example for the case of $k = 1$ is given in Exercise 68.

When ψ_X is not finite, finite moments of X can be obtained by differentiating its ch.f. ϕ_X. Suppose that $E|X_1^{r_1} \cdots X_k^{r_k}| < \infty$ for some nonnegative integers $r_1, \dots, r_k$. Let $r = r_1 + \cdots + r_k$ and

$$g(t) = \frac{\partial^r e^{\sqrt{-1}t^\tau X}}{\partial t_1^{r_1} \cdots \partial t_k^{r_k}} = (-1)^{r/2} X_1^{r_1} \cdots X_k^{r_k} e^{\sqrt{-1}t^\tau X}.$$

Then $|g(t)| \leq |X_1^{r_1} \cdots X_k^{r_k}|$, which is integrable. Hence, from Example 1.8,

$$\frac{\partial^r \phi_X(t)}{\partial t_1^{r_1} \cdots \partial t_k^{r_k}} = (-1)^{r/2} E\left(X_1^{r_1} \cdots X_k^{r_k} e^{\sqrt{-1}t^\tau X} \right) \tag{1.56}$$

and

$$\frac{\partial^r \phi_X(t)}{\partial t_1^{r_1} \cdots \partial t_k^{r_k}}\bigg|_{t=0} = (-1)^{r/2} E(X_1^{r_1} \cdots X_k^{r_k}).$$

In particular,

$$\frac{\partial \phi_X(t)}{\partial t}\bigg|_{t=0} = \sqrt{-1}EX, \qquad \frac{\partial^2 \phi_X(t)}{\partial t \partial t^\tau}\bigg|_{t=0} = -E(XX^\tau),$$

and, if $k = 1$ and p is a positive integer, then $\phi_X^{(p)}(0) = (-1)^{p/2}EX^p$, provided that all moments involved are finite. In fact, when $k = 1$, if ϕ_X has a finite derivative of even order p at $t = 0$, then $EX^p < \infty$ (see, e.g., Chung, 1974, pp. 166-168).

Example 1.19. Let $X = N(\mu, \sigma^2)$. From Table 1.2, $\psi_X(t) = e^{\mu t + \sigma^2 t^2/2}$. A direct calculation shows that $EX = \psi_X'(0) = \mu$, $EX^2 = \psi_X''(0) = \sigma^2 + \mu^2$, $EX^3 = \psi_X^{(3)}(0) = 3\sigma^2\mu + \mu^3$, and $EX^4 = \psi_X^{(4)}(0) = 3\sigma^4 + 6\sigma^2\mu^2 + \mu^4$. If $\mu = 0$, then $EX^p = 0$ when p is an odd integer and $EX^p = (p-1)(p-3)\cdots 3\cdot 1\sigma^p$ when p is an even integer (exercise). The cumulant generating function of X is $\kappa_X(t) = \log \psi_X(t) = \mu t + \sigma^2 t^2/2$. Hence, $\kappa_1 = \mu$, $\kappa_2 = \sigma^2$, and $\kappa_r = 0$ for $r = 3, 4, \dots$.

We now consider a random variable X having finite moments of all order but having an m.g.f. $\psi_X(t) = \infty$ except for $t = 0$. Let P_n be the probability measure for the $N(0, n)$ distribution, $n = 1, 2, ...$ Then $P = \sum_{n=1}^{\infty} 2^{-n} P_n$ is a probability measure with Lebesgue p.d.f. $\sum_{n=1}^{\infty} 2^{-n} f_n$, where f_n is the Lebesgue p.d.f. of $N(0, n)$ (Exercise 35). Let X be a random variable having distribution P. It follows from Fubini's theorem that X has finite moments of any order but $\psi_X(t) = \infty$ if $t \neq 0$. Since the ch.f. of $N(0, n)$ is $e^{-nt^2/2}$, the ch.f. of X is $\phi_X(t) = \sum_{n=1}^{\infty} 2^{-n} e^{-nt^2/2}$ (by Fubini's theorem), which is equal to $(2e^{t^2/2} - 1)^{-1}$. Hence, the moments of X can be obtained by differentiating ϕ_X. For example, $\phi_X'(0) = 0$ and $\phi_X''(0) = -2$, which shows that $EX = 0$ and $EX^2 = 2$. ∎

A fundamental fact about ch.f.'s is that there is a one-to-one correspondence between the set of all distributions on $\mathcal{R}^k$ and the set of all ch.f.'s defined on $\mathcal{R}^k$. The same fact is true for m.g.f.'s, but we have to focus on distributions having m.g.f.'s finite in neighborhoods of 0.

Theorem 1.6. (Uniqueness). Let X and Y be random k-vectors.
(i) If $\phi_X(t) = \phi_Y(t)$ for all $t \in \mathcal{R}^k$, then $P_X = P_Y$.
(ii) If $\psi_X(t) = \psi_Y(t) < \infty$ for all t in a neighborhood of 0, then $P_X = P_Y$.
Proof. (i) The result follows from the following inversion formula whose proof can be found, for example, in Billingsley (1986, p. 395): for any $a = (a_1, ..., a_k) \in \mathcal{R}^k$, $b = (b_1, ..., b_k) \in \mathcal{R}^k$, and $(a, b] = (a_1, b_1] \times \cdots \times (a_k, b_k]$ satisfying $P_X(\text{the boundary of } (a, b]) = 0$,

$$P_X((a, b]) = \lim_{c \to \infty} \int_{-c}^{c} \cdots \int_{-c}^{c} \frac{\phi_X(t_1, ..., t_k)}{(-1)^{k/2}(2\pi)^k} \prod_{i=1}^{k} \frac{e^{-\sqrt{-1}t_i a_i} - e^{-\sqrt{-1}t_i b_i}}{t_i} dt_i.$$

(ii) First consider the case of $k = 1$. From $e^{s|x|} \leq e^{sx} + e^{-sx}$, we conclude that $|X|$ has an m.g.f. that is finite in the neighborhood $(-c, c)$ for some $c > 0$ and $|X|$ has finite moments of all order. Using the inequality $|e^{\sqrt{-1}tx}[e^{\sqrt{-1}ax} - \sum_{j=1}^{n} (\sqrt{-1}ax)^j/j!]| \leq |ax|^{n+1}/(n+1)!$, we obtain that

$$\left| \phi_X(t + a) - \sum_{j=1}^{n} \frac{a^j}{j!} E[(\sqrt{-1}X)^j e^{\sqrt{-1}tX}] \right| \leq \frac{|a|^{n+1} E|X|^{n+1}}{(n+1)!},$$

which together with (1.53) and (1.56) imply that, for any $t \in \mathcal{R}$,

$$\phi_X(t + a) = \sum_{j=1}^{\infty} \frac{\phi_X^{(j)}(t)}{j!} a^j, \qquad |a| < c. \tag{1.57}$$

Similarly, (1.57) holds with ϕ_X replaced by ϕ_Y. Under the assumption that $\psi_X = \psi_Y < \infty$ in a neighborhood of 0, X and Y have the same moments of all order. By (1.56), $\phi_X^{(j)}(0) = \phi_Y^{(j)}(0)$ for all $j = 1, 2, ...$, which and (1.57) with $t = 0$ imply that ϕ_X and ϕ_Y are the same on the interval $(-c, c)$ and hence have identical derivatives there. Considering $t = c - \epsilon$ and $-c + \epsilon$ for an arbitrarily small $\epsilon > 0$ in (1.57) shows that ϕ_X and ϕ_Y also agree on $(-2c + \epsilon, 2c - \epsilon)$ and hence on $(-2c, 2c)$. By the same argument ϕ_X and ϕ_Y are the same on $(-3c, 3c)$ and so on. Hence, $\phi_X(t) = \phi_Y(t)$ for all t and, by part (i), $P_X = P_Y$.

Consider now the general case of $k \geq 2$. If $P_X \neq P_Y$, then by part (i) there exists $t \in \mathcal{R}^k$ such that $\phi_X(t) \neq \phi_Y(t)$. Then $\phi_{t^\tau X}(1) \neq \phi_{t^\tau Y}(1)$, which implies that $P_{t^\tau X} \neq P_{t^\tau Y}$. But $\psi_X = \psi_Y < \infty$ in a neighborhood of $0 \in \mathcal{R}^k$ implies that $\psi_{t^\tau X} = \psi_{t^\tau Y} < \infty$ in a neighborhood of $0 \in \mathcal{R}$ and, by the proved result for $k = 1$, $P_{t^\tau X} = P_{t^\tau Y}$. This contradiction shows that $P_X = P_Y$. ∎

Applying result (1.38) and Lemma 1.1, we obtain that

$$\psi_{X+Y}(t) = \psi_X(t)\psi_Y(t) \quad \text{and} \quad \phi_{X+Y}(t) = \phi_X(t)\psi_Y(t), \quad t \in \mathcal{R}^k, \quad (1.58)$$

for independent random k-vectors X and Y. This result, together with Theorem 1.6, provides a useful tool to obtain distributions of sums of independent random vectors with known distributions. The following example is an illustration.

Example 1.20. Let X_i, $i = 1, ..., k$, be independent random variables and X_i have the gamma distribution $\Gamma(\alpha_i, \gamma)$ (Table 1.2), $i = 1, ..., k$. From Table 1.2, X_i has the m.g.f. $\psi_{X_i}(t) = (1 - \gamma t)^{-\alpha_i}$, $t < \gamma^{-1}$, $i = 1, ..., k$. By result (1.58), the m.g.f. of $Y = X_1 + \cdots + X_k$ is equal to $\psi_Y(t) = (1 - \gamma t)^{-(\alpha_1 + \cdots + \alpha_k)}$, $t < \gamma^{-1}$. From Table 1.2, the gamma distribution $\Gamma(\alpha_1 + \cdots + \alpha_k, \gamma)$ has the m.g.f. $\psi_Y(t)$ and, hence, is the distribution of Y (by Theorem 1.6). ∎

Similarly, result (1.52) and Theorem 1.6 can be used to determine distributions of linear transformations of random vectors with known distributions. The following is another interesting application of Theorem 1.6. Note that a random variable X is symmetric about 0 (defined according to (1.30)) if and only if X and $-X$ have the same distribution, which can then be used as the definition of a random vector X symmetric about 0. We now show that X is symmetric about 0 if and only if its ch.f. ϕ_X is real-valued. If X and $-X$ have the same distribution, then by Theorem 1.6, $\phi_X(t) = \phi_{-X}(t)$. From (1.52), $\phi_{-X}(t) = \phi_X(-t)$. Then $\phi_X(t) = \phi_X(-t)$. Since $\sin(-t^\tau X) = -\sin(t^\tau X)$ and $\cos(t^\tau X) = \cos(-t^\tau X)$, this proves $E[\sin(t^\tau X)] = 0$ and, thus, ϕ_X is real-valued. Conversely, if ϕ_X is real-valued, then $\phi_X(t) = E[\cos(t^\tau X)]$ and $\phi_{-X}(t) = \phi_X(-t) = \phi_X(t)$. By Theorem 1.6, X and $-X$ must have the same distribution.

Other applications of ch.f.'s can be found in §1.5.

1.4 Conditional Expectations

In elementary probability the conditional probability of an event B given an event A is defined as $P(B|A) = P(A \cap B)/P(A)$, provided that $P(A) > 0$. In probability and statistics, however, we sometimes need a notion of "conditional probability" even for A's with $P(A) = 0$; for example, $A = \{Y = c\}$, where $c \in \mathcal{R}$ and Y is a random variable having a continuous c.d.f. General definitions of conditional probability, expectation, and distribution are introduced in this section, and they are shown to agree with those defined in elementary probability in special cases.

1.4.1 Conditional expectations

Definition 1.6. Let X be an integrable random variable on $(\Omega, \mathcal{F}, P)$.
(i) Let $\mathcal{A}$ be a sub-σ-field of $\mathcal{F}$. The *conditional expectation* of X given $\mathcal{A}$, denoted by $E(X|\mathcal{A})$, is the a.s.-unique random variable satisfying the following two conditions:

    (a) $E(X|\mathcal{A})$ is measurable from $(\Omega, \mathcal{A})$ to $(\mathcal{R}, \mathcal{B})$;
    (b) $\int_A E(X|\mathcal{A})dP = \int_A X dP$ for any $A \in \mathcal{A}$.

(Note that the existence of $E(X|\mathcal{A})$ follows from Theorem 1.4.)
(ii) Let $B \in \mathcal{F}$. The *conditional probability* of B given $\mathcal{A}$ is defined to be $P(B|\mathcal{A}) = E(I_B|\mathcal{A})$.
(iii) Let Y be measurable from $(\Omega, \mathcal{F}, P)$ to $(\Lambda, \mathcal{G})$. The conditional expectation of X given Y is defined to be $E(X|Y) = E[X|\sigma(Y)]$. ∎

Essentially, the σ-field $\sigma(Y)$ contains "the information in Y". Hence, $E(X|Y)$ is the "expectation" of X given the information provided by $\sigma(Y)$. The following useful result shows that there is a Borel function h defined on the range of Y such that $E(X|Y) = h \circ Y$.

Lemma 1.2. Let Y be measurable from $(\Omega, \mathcal{F})$ to $(\Lambda, \mathcal{G})$ and Z a function from $(\Omega, \mathcal{F})$ to $\mathcal{R}^k$. Then Z is measurable from $(\Omega, \sigma(Y))$ to $(\mathcal{R}^k, \mathcal{B}^k)$ if and only if there is a measurable function h from $(\Lambda, \mathcal{G})$ to $(\mathcal{R}^k, \mathcal{B}^k)$ such that $Z = h \circ Y$. ∎

The function h in $E(X|Y) = h \circ Y$ is a Borel function on $(\Lambda, \mathcal{G})$. Let $y \in \Lambda$. We define

$$E(X|Y = y) = h(y)$$

to be the conditional expectation of X given $Y = y$. Note that $h(y)$ is a function on Λ, whereas $h \circ Y = E(X|Y)$ is a function on Ω.

For a random vector X, $E(X|\mathcal{A})$ is defined as the vector of conditional expectations of components of X.

Example 1.21. Let X be an integrable random variable on $(\Omega, \mathcal{F}, P)$, $A_1, A_2, ...$ be disjoint events on $(\Omega, \mathcal{F}, P)$ such that $\cup A_i = \Omega$ and $P(A_i) > 0$ for all i, and let $a_1, a_2, ...$ be distinct real numbers. Define $Y = a_1 I_{A_1} + a_2 I_{A_2} + \cdots$. We now show that

$$E(X|Y) = \sum_{i=1}^{\infty} \frac{\int_{A_i} X dP}{P(A_i)} I_{A_i}. \tag{1.59}$$

We need to verify (a) and (b) in Definition 1.6 with $\mathcal{A} = \sigma(Y)$. Since $\sigma(Y) = \sigma(\{A_1, A_2, ...\})$, it is clear that the function on the right-hand side of (1.59) is measurable on $(\Omega, \sigma(Y))$. For any $B \in \mathcal{B}$, $Y^{-1}(B) = \cup_{i:a_i \in B} A_i$. Using properties of integrals, we obtain that

$$\int_{Y^{-1}(B)} X dP = \sum_{i:a_i \in B} \int_{A_i} X dP$$

$$= \sum_{i=1}^{\infty} \frac{\int_{A_i} X dP}{P(A_i)} P\left(A_i \cap Y^{-1}(B)\right)$$

$$= \int_{Y^{-1}(B)} \left[\sum_{i=1}^{\infty} \frac{\int_{A_i} X dP}{P(A_i)} I_{A_i}\right] dP.$$

This verifies (b) and thus (1.59) holds.

Let h be a Borel function on $\mathcal{R}$ satisfying $h(a_i) = \int_{A_i} X dP / P(A_i)$. Then, by (1.59), $E(X|Y) = h \circ Y$ and $E(X|Y = y) = h(y)$.

Let $A \in \mathcal{F}$ and $X = I_A$. Then

$$P(A|Y) = E(X|Y) = \sum_{i=1}^{\infty} \frac{P(A \cap A_i)}{P(A_i)} I_{A_i},$$

which equals $P(A \cap A_i) / P(A_i) = P(A|A_i)$ if $\omega \in A_i$. Hence, the definition of conditional probability in Definition 1.6 agrees with that in elementary probability. ∎

The next result generalizes the result in Example 1.21 to conditional expectations of random variables having p.d.f.'s.

Proposition 1.9. Let X be a random n-vector and Y a random m-vector. Suppose that (X, Y) has a joint p.d.f. $f(x, y)$ w.r.t. $\nu \times \lambda$, where ν and λ are σ-finite measures on $(\mathcal{R}^n, \mathcal{B}^n)$ and $(\mathcal{R}^m, \mathcal{B}^m)$, respectively. Let $g(x, y)$ be a Borel function on $\mathcal{R}^{n+m}$ for which $E|g(X, Y)| < \infty$. Then

$$E[g(X, Y)|Y] = \frac{\int g(x, Y) f(x, Y) d\nu(x)}{\int f(x, Y) d\nu(x)} \quad \text{a.s.} \tag{1.60}$$

Proof. Denote the right-hand side of (1.60) by $h(Y)$. By Fubini's theorem, h is Borel. Then, by Lemma 1.2, $h(Y)$ is Borel on $(\Omega, \sigma(Y))$. Also, by Fubini's theorem, $f_Y(y) = \int f(x, y) d\nu(x)$ is the p.d.f. of Y w.r.t. λ. For $B \in \mathcal{B}^m$,

$$
\begin{aligned}
\int_{Y^{-1}(B)} h(Y) dP &= \int_B h(y) dP_Y \\
&= \int_B \frac{\int g(x, y) f(x, y) d\nu(x)}{\int f(x, y) d\nu(x)} f_Y(y) d\lambda(y) \\
&= \int_{\mathcal{R}^n \times B} g(x, y) f(x, y) d\nu \times \lambda \\
&= \int_{\mathcal{R}^n \times B} g(x, y) dP_{(X, Y)} \\
&= \int_{Y^{-1}(B)} g(X, Y) dP,
\end{aligned}
$$

where the first and the last equalities follow from Theorem 1.2, the second and the next to last equalities follow from the definition of h and p.d.f.'s, and the third equality follows from Theorem 1.3 (Fubini's theorem). ∎

For a random vector (X, Y) with a joint p.d.f. $f(x, y)$ w.r.t. $\nu \times \lambda$, define the *conditional* p.d.f. of X given $Y = y$ to be

$$
f_{X|Y}(x|y) = \frac{f(x, y)}{f_Y(y)}, \tag{1.61}
$$

where $f_Y(y) = \int f(x, y) d\nu(x)$ is the marginal p.d.f. of Y w.r.t. λ. One can easily check that for each fixed y with $f_Y(y) > 0$, $f_{X|Y}(x|y)$ in (1.61) is a p.d.f. w.r.t. ν. Then equation (1.60) can be rewritten as

$$
E[g(X, Y)|Y] = \int g(x, Y) f_{X|Y}(x|Y) d\nu(x).
$$

Again, this agrees with the conditional expectation defined in elementary probability (i.e., the conditional expectation of $g(X, Y)$ given Y is equal to the expectation of $g(X, Y)$ w.r.t. the conditional p.d.f. of X given Y).

Now we list some useful properties of conditional expectations. The proof is left to the reader.

Proposition 1.10. Let X, Y, $X_1, X_2, \ldots$ be integrable random variables on $(\Omega, \mathcal{F}, P)$ and $\mathcal{A}$ be a sub-σ-field of $\mathcal{F}$.
(i) If $X = c$ a.s., $c \in \mathcal{R}$, then $E(X|\mathcal{A}) = c$ a.s.
(ii) If $X \leq Y$ a.s., then $E(X|\mathcal{A}) \leq E(Y|\mathcal{A})$ a.s.
(iii) If $a \in \mathcal{R}$ and $b \in \mathcal{R}$, then $E(aX + bY|\mathcal{A}) = aE(X|\mathcal{A}) + bE(Y|\mathcal{A})$ a.s.

(iv) $E[E(X|\mathcal{A})] = EX$.

(v) $E[E(X|\mathcal{A})|\mathcal{A}_0] = E(X|\mathcal{A}_0) = E[E(X|\mathcal{A}_0)|\mathcal{A}]$ a.s., where $\mathcal{A}_0$ is a sub-σ-field of $\mathcal{A}$.

(vi) If $\sigma(Y) \subset \mathcal{A}$ and $E|XY| < \infty$, then $E(XY|\mathcal{A}) = YE(X|\mathcal{A})$ a.s.

(vii) If X and Y are independent and $E|g(X,Y)| < \infty$ for a Borel function g, then $E[g(X,Y)|Y = y] = E[g(X,y)]$ a.s. P_Y.

(viii) If $EX^2 < \infty$, then $[E(X|\mathcal{A})]^2 \leq E(X^2|\mathcal{A})$ a.s.

(ix) (Fatou's lemma). If $X_n \geq 0$ for any n, then $E\left(\liminf_n X_n | \mathcal{A}\right) \leq \liminf_n E(X_n|\mathcal{A})$ a.s.

(x) (Dominated convergence theorem). Suppose that $|X_n| \leq Y$ for any n and $X_n \to_{a.s.} X$. Then $E(X_n|\mathcal{A}) \to_{a.s.} E(X|\mathcal{A})$. ∎

Although part (vii) of Proposition 1.10 can be proved directly, it is a consequence of a more general result given in Theorem 1.7(i). Since $E(X|\mathcal{A})$ is defined only for integrable X, a version of monotone convergence theorem (i.e., $0 \leq X_1 \leq X_2 \leq \cdots$ and $X_n \to_{a.s.} X$ imply $E(X_n|\mathcal{A}) \to_{a.s.} E(X|\mathcal{A})$) becomes a special case of Proposition 1.10(x).

It can also be shown (exercise) that Hölder's inequality (1.40), Liapounov's inequality (1.42), Minkowski's inequality (1.43), and Jensen's inequality (1.47) hold a.s. with the expectation E replaced by the conditional expectation $E(\cdot|\mathcal{A})$.

As an application, we consider the following example.

Example 1.22. Let X be a random variable on $(\Omega, \mathcal{F}, P)$ with $EX^2 < \infty$ and let Y be a measurable function from $(\Omega, \mathcal{F}, P)$ to $(\Lambda, \mathcal{G})$. One may wish to predict the value of X based on an observed value of Y. Let $g(Y)$ be a predictor, i.e., $g \in \aleph = \{$all Borel functions g with $E[g(Y)]^2 < \infty\}$. Each predictor is assessed by the "mean squared prediction error" $E[X - g(Y)]^2$. We now show that $E(X|Y)$ is the best predictor of X in the sense that

$$E[X - E(X|Y)]^2 = \min_{g \in \aleph} E[X - g(Y)]^2. \tag{1.62}$$

First, Proposition 1.10(viii) implies $E(X|Y) \in \aleph$. Next, for any $g \in \aleph$,

$$\begin{aligned}
E[X - g(Y)]^2 &= E[X - E(X|Y) + E(X|Y) - g(Y)]^2 \\
&= E[X - E(X|Y)]^2 + E[E(X|Y) - g(Y)]^2 \\
&\quad + 2E\{[X - E(X|Y)][E(X|Y) - g(Y)]\} \\
&= E[X - E(X|Y)]^2 + E[E(X|Y) - g(Y)]^2 \\
&\quad + 2E\{E\{[X - E(X|Y)][E(X|Y) - g(Y)]|Y\}\} \\
&= E[X - E(X|Y)]^2 + E[E(X|Y) - g(Y)]^2 \\
&\quad + 2E\{[E(X|Y) - g(Y)]E[X - E(X|Y)|Y]\} \\
&= E[X - E(X|Y)]^2 + E[E(X|Y) - g(Y)]^2 \\
&\geq E[X - E(X|Y)]^2,
\end{aligned}$$

where the third equality follows from Proposition 1.10(iv), the fourth equality follows from Proposition 1.10(vi), and the last equality follows from Proposition 1.10(i), (iii), and (vi). ∎

1.4.2 Independence

Definition 1.7. Let $(\Omega, \mathcal{F}, P)$ be a probability space.
(i) Let $\mathcal{C}$ be a collection of subsets in $\mathcal{F}$. Events in $\mathcal{C}$ are said to be *independent* if and only if for any positive integer n and distinct events $A_1,...,A_n$ in $\mathcal{C}$,

$$P(A_1 \cap A_2 \cap \cdots \cap A_n) = P(A_1)P(A_2) \cdots P(A_n).$$

(ii) Collections $\mathcal{C}_i \subset \mathcal{F}$, $i \in \mathcal{I}$ (an index set that can be uncountable), are said to be independent if and only if events in any collection of the form $\{A_i \in \mathcal{C}_i : i \in \mathcal{I}\}$ are independent.
(iii) Random elements X_i, $i \in \mathcal{I}$, are said to be independent if and only if $\sigma(X_i)$, $i \in \mathcal{I}$, are independent. ∎

The following result is useful for checking the independence of σ-fields.

Lemma 1.3. Let $\mathcal{C}_i$, $i \in \mathcal{I}$, be independent collections of events. Suppose that each $\mathcal{C}_i$ has the property that if $A \in \mathcal{C}_i$ and $B \in \mathcal{C}_i$, then $A \cap B \in \mathcal{C}_i$. Then $\sigma(\mathcal{C}_i)$, $i \in \mathcal{I}$, are independent. ∎

An immediate application of Lemma 1.3 is to show (exercise) that random variables X_i, $i = 1,...,k$, are independent according to Definition 1.7 if and only if (1.7) holds with F being the joint c.d.f. of $(X_1,...,X_k)$ and F_i being the marginal c.d.f. of X_i. Hence, Definition 1.7(iii) agrees with the concept of independence of random variables discussed in §1.3.1.

It is easy to see from Definition 1.7 that if X and Y are independent random vectors, then so are $g(X)$ and $h(Y)$ for Borel functions g and h. Since the independence in Definition 1.7 is equivalent to the independence discussed in §1.3.1, this provides a simple proof of Lemma 1.1.

For two events A and B with $P(A) > 0$, A and B are independent if and only if $P(B|A) = P(B)$. This means that A provides no information about the probability of the occurrence of B. The following result is a useful extension.

Proposition 1.11. Let X be a random variable with $E|X| < \infty$ and let Y_i be random k_i-vectors, $i = 1, 2$. Suppose that (X, Y_1) and Y_2 are independent. Then

$$E[X|(Y_1, Y_2)] = E(X|Y_1) \quad \text{a.s.}$$

Proof. First, $E(X|Y_1)$ is Borel on $(\Omega, \sigma(Y_1, Y_2))$, since $\sigma(Y_1) \subset \sigma(Y_1, Y_2)$. Next, we need to show that for any Borel set $B \in \mathcal{B}^{k_1+k_2}$,

$$\int_{(Y_1,Y_2)^{-1}(B)} X dP = \int_{(Y_1,Y_2)^{-1}(B)} E(X|Y_1) dP. \qquad (1.63)$$

If $B = B_1 \times B_2$, where $B_i \in \mathcal{B}^{k_i}$, then $(Y_1, Y_2)^{-1}(B) = Y_1^{-1}(B_1) \cap Y_2^{-1}(B_2)$ and

$$\begin{aligned}
\int_{Y_1^{-1}(B_1) \cap Y_2^{-1}(B_2)} E(X|Y_1) dP &= \int I_{Y_1^{-1}(B_1)} I_{Y_2^{-1}(B_2)} E(X|Y_1) dP \\
&= \int I_{Y_1^{-1}(B_1)} E(X|Y_1) dP \int I_{Y_2^{-1}(B_2)} dP \\
&= \int I_{Y_1^{-1}(B_1)} X dP \int I_{Y_2^{-1}(B_2)} dP \\
&= \int I_{Y_1^{-1}(B_1)} I_{Y_2^{-1}(B_2)} X dP \\
&= \int_{Y_1^{-1}(B_1) \cap Y_2^{-1}(B_2)} X dP,
\end{aligned}$$

where the second and the next to last equalities follow from result (1.38) and the independence of (X, Y_1) and Y_2, and the third equality follows from the fact that $E(X|Y_1)$ is the conditional expectation of X given Y_1. This shows that (1.63) holds for $B = B_1 \times B_2$. We can show that the collection $\mathcal{H} = \{B \subset \mathcal{R}^{k_1+k_2} : B \text{ satisfies } (1.63)\}$ is a σ-field. Since we have already shown that $\mathcal{B}^{k_1} \times \mathcal{B}^{k_2} \subset \mathcal{H}$, $\mathcal{B}^{k_1+k_2} = \sigma(\mathcal{B}^{k_1} \times \mathcal{B}^{k_2}) \subset \mathcal{H}$ and thus the result follows. ∎

Clearly, the result in Proposition 1.11 still holds if X is replaced by $h(X)$ for any Borel h and, hence,

$$P(A|Y_1, Y_2) = P(A|Y_1) \quad \text{a.s. for any } A \in \sigma(X), \qquad (1.64)$$

if (X, Y_1) and Y_2 are independent. If Y_1 is a constant and $Y = Y_2$, (1.64) reduces to $P(A|Y) = P(A)$ a.s. for any $A \in \sigma(X)$, if X and Y are independent, i.e., $\sigma(Y)$ does not provide any additional information about the stochastic behavior of X. This actually provides another equivalent but more intuitive definition of the independence of X and Y (or two σ-fields).

With a nonconstant Y_1, we say that given Y_1, X and Y_2 are *conditionally independent* if and only if (1.64) holds. Then the result in Proposition 1.11 can be stated as: if Y_2 and (X, Y_1) are independent, then given Y_1, X and Y_2 are conditionally independent. It is important to know that the result in Proposition 1.11 may not be true if Y_2 is independent of X but not (X, Y_1) (Exercise 96).

1.4.3 Conditional distributions

The conditional p.d.f. was introduced in §1.4.1 for random variables having p.d.f.'s w.r.t. some measures. We now consider *conditional distributions* in general cases where we may not have any p.d.f.

Let X and Y be two random vectors defined on a common probability space. It is reasonable to consider $P[X^{-1}(B)|Y = y]$ as a candidate for the conditional distribution of X, given $Y = y$, where B is any Borel set. However, since conditional probability is defined almost surely, for any fixed y, $P[X^{-1}(B)|Y = y]$ may not be a probability measure. The first part of the following theorem (whose proof can be found in Billingsley (1986, pp. 460-461)) shows that there exists a version of conditional probability such that $P[X^{-1}(B)|Y = y]$ is a probability measure for any fixed y.

Theorem 1.7. (i) (Existence of conditional distributions). Let X be a random n-vector on a probability space $(\Omega, \mathcal{F}, P)$ and $\mathcal{A}$ be a sub-σ-field of $\mathcal{F}$. Then there exists a function $P(B, \omega)$ on $\mathcal{B}^n \times \Omega$ such that
    (a) $P(B, \omega) = P[X^{-1}(B)|\mathcal{A}]$ a.s. for any fixed $B \in \mathcal{B}^n$, and
    (b) $P(\cdot, \omega)$ is a probability measure on $(\mathcal{R}^n, \mathcal{B}^n)$ for any fixed $\omega \in \Omega$.
Let Y be measurable from $(\Omega, \mathcal{F}, P)$ to $(\Lambda, \mathcal{G})$. Then there exists $P_{X|Y}(B|y)$ such that
    (a) $P_{X|Y}(B|y) = P[X^{-1}(B)|Y = y]$ a.s. P_Y for any fixed $B \in \mathcal{B}^n$, and
    (b) $P_{X|Y}(\cdot|y)$ is a probability measure on $(\mathcal{R}^n, \mathcal{B}^n)$ for any fixed $y \in \Lambda$.
Furthermore, if $E|g(X, Y)| < \infty$ with a Borel function g, then

$$E[g(X,Y)|Y=y] = E[g(X,y)|Y=y] = \int_{\mathcal{R}^n} g(x,y)dP_{X|Y}(x|y) \quad \text{a.s. } P_Y.$$

(ii) Let $(\Lambda, \mathcal{G}, P_1)$ be a probability space. Suppose that P_2 is a function from $\mathcal{B}^n \times \Lambda$ to $\mathcal{R}$ and satisfies
    (a) $P_2(\cdot, y)$ is a probability measure on $(\mathcal{R}^n, \mathcal{B}^n)$ for any $y \in \Lambda$, and
    (b) $P_2(B, \cdot)$ is Borel for any $B \in \mathcal{B}^n$.
Then there is a unique probability measure P on $(\mathcal{R}^n \times \Lambda, \sigma(\mathcal{B}^n \times \mathcal{G}))$ such that, for $B \in \mathcal{B}$ and $C \in \mathcal{G}$,

$$P(B \times C) = \int_C P_2(B, y)dP_1(y). \tag{1.65}$$

Furthermore, if $(\Lambda, \mathcal{G}) = (\mathcal{R}^m, \mathcal{B}^m)$, and $X(x, y) = x$ and $Y(x, y) = y$ define the coordinate random vectors, then $P_Y = P_1$, $P_{X|Y}(\cdot|y) = P_2(\cdot, y)$, and the probability measure in (1.65) is the joint distribution of (X, Y), which has the following joint c.d.f.:

$$F(x,y) = \int_{(-\infty,y]} P_{X|Y}\big((-\infty, x]|z\big)dP_Y(z), \quad x \in \mathcal{R}^n, y \in \mathcal{R}^m, \tag{1.66}$$

where $(-\infty, a]$ denotes $(-\infty, a_1] \times \cdots \times (-\infty, a_k]$ for $a = (a_1, ..., a_k)$. ∎

For a fixed y, $P_{X|Y=y} = P_{X|Y}(\cdot|y)$ is called the conditional distribution of X given $Y = y$. Under the conditions in Theorem 1.7(i), if Y is a random m-vector and (X, Y) has a p.d.f. w.r.t. $\nu \times \lambda$ (ν and λ are σ-finite measures on $(\mathcal{R}^n, \mathcal{B}^n)$ and $(\mathcal{R}^m, \mathcal{B}^m)$, respectively), then $f_{X|Y}(x|y)$ defined in (1.61) is the p.d.f. of $P_{X|Y=y}$ w.r.t. ν for any fixed y.

The second part of Theorem 1.7 states that given a distribution on one space and a collection of conditional distributions (which are conditioned on values of the first space) on another space, we can construct a joint distribution in the product space. It is sometimes called the "two-stage experiment theorem" for the following reason. If $Y \in \mathcal{R}^m$ is selected in stage 1 of an experiment according to its marginal distribution $P_Y = P_1$, and X is chosen afterward according to a distribution $P_2(\cdot, y)$, then the combined two-stage experiment produces a jointly distributed pair (X, Y) with distribution $P_{(X,Y)}$ given by (1.65) and $P_{X|Y=y} = P_2(\cdot, y)$. This provides a way of generating dependent random variables. The following is an example.

Example 1.23. A market survey is conducted to study whether a new product is preferred over the product currently available in the market (old product). The survey is conducted by mail. Questionnaires are sent along with the sample products (both new and old) to N customers randomly selected from a population, where N is a positive integer. Each customer is asked to fill out the questionnaire and return it. Responses from customers are either 1 (new is better than old) or 0 (otherwise). Some customers, however, do not return the questionnaires. Let X be the number of ones in the returned questionnaires. What is the distribution of X?

If every customer returns the questionnaire, then (from elementary probability) X has the binomial distribution $Bi(p, N)$ in Table 1.1 (assuming that the population is large enough so that customers respond independently), where $p \in (0, 1)$ is the overall rate of customers who prefer the new product. Now, let Y be the number of customers who respond. Then Y is random. Suppose that customers respond independently with the same probability $\pi \in (0, 1)$. Then P_Y is the binomial distribution $Bi(\pi, N)$. Given $Y = y$ (an integer between 0 and N), $P_{X|Y=y}$ is the binomial distribution $Bi(p, y)$ if $y \geq 1$ and the point mass at 0 (see (1.22)) if $y = 0$. Using (1.66) and the fact that binomial distributions have p.d.f.'s w.r.t. counting measure, we obtain that the joint c.d.f. of (X, Y) is

$$F(x, y) = \sum_{k=0}^{y} P_{X|Y=k}\left((-\infty, x]\right) \binom{N}{k} \pi^k (1 - \pi)^{N-k}$$

$$= \sum_{k=0}^{y} \sum_{j=0}^{\min\{x,k\}} \binom{k}{j} p^j (1 - p)^{k-j} \binom{N}{k} \pi^k (1 - \pi)^{N-k}$$

for $x = 0, 1, ..., y$, $y = 0, 1, ..., N$. The marginal c.d.f. $F_X(x) = F(x, \infty) = F(x, N)$. The p.d.f. of X w.r.t. counting measure is

$$
\begin{aligned}
f_X(x) &= \sum_{k=x}^{N} \binom{k}{x} p^x (1-p)^{k-x} \binom{N}{k} \pi^k (1-\pi)^{N-k} \\
&= \binom{N}{x} (\pi p)^x (1 - \pi p)^{N-x} \sum_{k=x}^{N} \binom{N-x}{k-x} \left(\frac{\pi - \pi p}{1 - \pi p} \right)^{k-x} \left(\frac{1-\pi}{1-\pi p} \right)^{N-k} \\
&= \binom{N}{x} (\pi p)^x (1 - \pi p)^{N-x}
\end{aligned}
$$

for $x = 0, 1, ..., N$. It turns out that the marginal distribution of X is the binomial distribution $Bi(\pi p, N)$. ∎

1.4.4 Markov chains and martingales

As applications of conditional expectations, we introduce here two important types of dependent sequences of random variables.

Markov chains

A sequence of random vectors $\{X_n : n = 1, 2, ...\}$ is said to be a *Markov chain* or *Markov process* if and only if

$$
P(B|X_1, ..., X_n) = P(B|X_n) \text{ a.s.}, \quad B \in \sigma(X_{n+1}), \ n = 2, 3, ... \quad (1.67)
$$

Comparing (1.67) with (1.64), we conclude that (1.67) implies that X_{n+1} (tomorrow) is conditionally independent of $(X_1, ..., X_{n-1})$ (the past), given X_n (today). But $(X_1, ..., X_{n-1})$ is not necessarily independent of (X_n, X_{n+1}).

Clearly, a sequence of independent random vectors forms a Markov chain since, by Proposition 1.11, both quantities on two sides of (1.67) are equal to $P(B)$ for independent X_i's. The following example describes some Markov processes of dependent random variables.

Example 1.24 (First-order autoregressive processes). Let $\varepsilon_1, \varepsilon_2, ...$ be independent random variables defined on a probability space, $X_1 = \varepsilon_1$, and $X_{n+1} = \rho X_n + \varepsilon_{n+1}$, $n = 1, 2, ...$, where ρ is a constant in $\mathcal{R}$. Then $\{X_n\}$ is called a first-order autoregressive process. We now show that for any $B \in \mathcal{B}$ and $n = 1, 2, ...$,

$$
P(X_{n+1} \in B|X_1, ..., X_n) = P_{\varepsilon_{n+1}}(B - \rho X_n) = P(X_{n+1} \in B|X_n) \text{ a.s.},
$$

where $B - y = \{x \in \mathcal{R} : x + y \in B\}$, which implies that $\{X_n\}$ is a Markov chain. For any $y \in \mathcal{R}$,

$$P_{\varepsilon_{n+1}}(B - y) = P(\varepsilon_{n+1} + y \in B) = \int I_B(x + y)dP_{\varepsilon_{n+1}}(x)$$

and, by Fubini's theorem, $P_{\varepsilon_{n+1}}(B - y)$ is Borel. Hence, $P_{\varepsilon_{n+1}}(B - \rho X_n)$ is Borel w.r.t. $\sigma(X_n)$ and, thus, is Borel w.r.t. $\sigma(X_1, ..., X_n)$. Let $B_j \in \mathcal{B}$, $j = 1, ..., n$, and $A = \cap_{j=1}^n X_j^{-1}(B_j)$. Since $\varepsilon_{n+1} + \rho X_n = X_{n+1}$ and ε_{n+1} is independent of $(X_1, ..., X_n)$, it follows from Theorem 1.2 and Fubini's theorem that

$$\int_A P_{\varepsilon_{n+1}}(B - \rho X_n)dP = \int_{x_j \in B_j, j=1,...,n} \int_{t \in B - \rho x_n} dP_{\varepsilon_{n+1}}(t)dP_X(x)$$

$$= \int_{x_j \in B_j, j=1,...,n, x_{n+1} \in B} dP_{(X,\varepsilon_{n+1})}(x, t)$$

$$= P\left(A \cap X_{n+1}^{-1}(B)\right),$$

where X and x denote $(X_1, ..., X_n)$ and $(x_1, ..., x_n)$, respectively, and x_{n+1} denotes $\rho x_n + t$. Using this and the argument in the end of the proof for Proposition 1.11, we obtain $P(X_{n+1} \in B | X_1, ..., X_n) = P_{\varepsilon_{n+1}}(B - \rho X_n)$ a.s. The proof for $P_{\varepsilon_{n+1}}(B - \rho X_n) = P(X_{n+1} \in B | X_n)$ a.s. is similar and simpler. ∎

The following result provides some characterizations of Markov chains.

Proposition 1.12. A sequence of random vectors $\{X_n\}$ is a Markov chain if and only if one of the following three conditions holds.
(a) For any $n = 2, 3, ...$ and any integrable $h(X_{n+1})$ with a Borel function h, $E[h(X_{n+1})|X_1, ..., X_n] = E[h(X_{n+1})|X_n]$ a.s.
(b) For any $n = 1, 2, ...$ and $B \in \sigma(X_{n+1}, X_{n+2}, ...)$, $P(B|X_1, ..., X_n) = P(B|X_n)$ a.s.
(c) For any $n = 2, 3, ...$, $A \in \sigma(X_1, ..., X_n)$, and $B \in \sigma(X_{n+1}, X_{n+2}, ...)$, $P(A \cap B | X_n) = P(A|X_n)P(B|X_n)$ a.s.
Proof. (i) It is clear that (a) implies (1.67). If h is a simple function, then (1.67) and Proposition 1.10(iii) imply (a). If h is nonnegative, then by Exercise 17 there are nonnegative simple functions $h_1 \leq h_2 \leq \cdots \leq h$ such that $h_j \to h$. Then (1.67) together with Proposition 1.10(iii) and (x) imply (a). Since $h = h_+ - h_-$, we conclude that (1.67) implies (a).
(ii) It is also clear that (b) implies (1.67). We now show that (1.67) implies (b). Note that $\sigma(X_{n+1}, X_{n+2}, ...) = \sigma\left(\cup_{j=1}^\infty \sigma(X_{n+1}, ..., X_{n+j})\right)$ (Exercise 19). Hence, it suffices to show that $P(B|X_1, ..., X_n) = P(B|X_n)$ a.s. for $B \in \sigma(X_{n+1}, ..., X_{n+j})$ for any $j = 1, 2, ...$ We use induction. The result for $j = 1$ follows from (1.67). Suppose that the result holds for any $B \in$

$\sigma(X_{n+1}, ..., X_{n+j})$. To show the result for any $B \in \sigma(X_{n+1}, ..., X_{n+j+1})$, it is enough (why?) to show that for any $B_1 \in \sigma(X_{n+j+1})$ and any $B_2 \in \sigma(X_{n+1}, ..., X_{n+j})$, $P(B_1 \cap B_2 | X_1, ..., X_n) = P(B_1 \cap B_2 | X_n)$ a.s. From the proof in (i), the induction assumption implies

$$E[h(X_{n+1}, ..., X_{n+j}) | X_1, ..., X_n] = E[h(X_{n+1}, ..., X_{n+j}) | X_n] \quad (1.68)$$

for any Borel function h. The result follows from

$$
\begin{aligned}
E(I_{B_1} I_{B_2} | X_1, ..., X_n) &= E[E(I_{B_1} I_{B_2} | X_1, ..., X_{n+j}) | X_1, ..., X_n] \\
&= E[I_{B_2} E(I_{B_1} | X_1, ..., X_{n+j}) | X_1, ..., X_n] \\
&= E[I_{B_2} E(I_{B_1} | X_{n+j}) | X_1, ..., X_n] \\
&= E[I_{B_2} E(I_{B_1} | X_{n+j}) | X_n] \\
&= E[I_{B_2} E(I_{B_1} | X_n, ..., X_{n+j}) | X_n] \\
&= E[E(I_{B_1} I_{B_2} | X_n, ..., X_{n+j}) | X_n] \\
&= E(I_{B_1} I_{B_2} | X_n) \quad \text{a.s.,}
\end{aligned}
$$

where the first and last equalities follow from Proposition 1.10(v), the second and sixth equalities follow from Proposition 1.10(vi), the third and fifth equalities follow from (1.67), and the fourth equality follows from (1.68).

(iii) Let $A \in \sigma(X_1, ..., X_n)$ and $B \in \sigma(X_{n+1}, X_{n+2}, ...)$. If (b) holds, then $E(I_A I_B | X_n) = E[E(I_A I_B | X_1, ..., X_n) | X_n] = E[I_A E(I_B | X_1, ..., X_n) | X_n] = E[I_A E(I_B | X_n) | X_n] = E(I_A | X_n) E(I_B | X_n)$, which is (c).

Assume that (c) holds. Let $A_1 \in \sigma(X_n)$, $A_2 \in \sigma(X_1, ..., X_{n-1})$, and $B \in \sigma(X_{n+1}, X_{n+2}, ...)$. Then

$$
\begin{aligned}
\int_{A_1 \cap A_2} E(I_B | X_n) dP &= \int_{A_1} I_{A_2} E(I_B | X_n) dP \\
&= \int_{A_1} E[I_{A_2} E(I_B | X_n) | X_n] dP \\
&= \int_{A_1} E(I_{A_2} | X_n) E(I_B | X_n) dP \\
&= \int_{A_1} E(I_{A_2} I_B | X_n) dP \\
&= P(A_1 \cap A_2 \cap B).
\end{aligned}
$$

Since disjoint unions of events of the form $A_1 \cap A_2$ as specified above generate $\sigma(X_1, ..., X_n)$, this shows that $E(I_B | X_n) = E(I_B | X_1, ..., X_n)$ a.s., which is (b). ∎

Note that condition (b) in Proposition 1.12 can be stated as "the past and the future are conditionally independent given the present", which is a property of any Markov chain. More discussions and applications of Markov chains can be found in §4.1.4.

Martingales

Let $\{X_n\}$ be a sequence of integrable random variables on a probability space $(\Omega, \mathcal{F}, P)$ and $\mathcal{F}_1 \subset \mathcal{F}_2 \subset \cdots \subset \mathcal{F}$ be a sequence of σ-fields such that $\sigma(X_n) \subset \mathcal{F}_n$, $n = 1, 2, \dots$. The sequence $\{X_n, \mathcal{F}_n : n = 1, 2, \dots\}$ is said to be a *martingale* if and only if

$$E(X_{n+1}|\mathcal{F}_n) = X_n \quad \text{a.s.}, \quad n = 1, 2, \dots, \tag{1.69}$$

a *submartingale* if and only if (1.69) holds with $=$ replaced by $\geq$, and a *supermartingale* if and only if (1.69) holds with $=$ replaced by $\leq$. $\{X_n\}$ is said to be a martingale (submartingale or supermartingale) if and only if $\{X_n, \sigma(X_1, \dots, X_n)\}$ is a martingale (submartingale or supermartingale). From Proposition 1.10(v), if $\{X_n, \mathcal{F}_n\}$ is a martingale (submartingale or supermartingale), then so is $\{X_n\}$.

A simple property of a martingale (or a submartingale) $\{X_n, \mathcal{F}_n\}$ is that $E(X_{n+j}|\mathcal{F}_n) = X_n$ a.s. (or $E(X_{n+j}|\mathcal{F}_n) \geq X_n$ a.s.) and $EX_1 = EX_j$ (or $EX_1 \leq EX_2 \leq \cdots$) for any $j = 1, 2, \dots$ (exercise).

For any probability space $(\Omega, \mathcal{F}, P)$ and σ-fields $\mathcal{F}_1 \subset \mathcal{F}_2 \subset \cdots \subset \mathcal{F}$, we can always construct a martingale $\{E(Y|\mathcal{F}_n)\}$ by using an integrable random variable Y. Another way to construct a martingale is to use a sequence of independent integrable random variables $\{\varepsilon_n\}$ by letting $X_n = \varepsilon_1 + \cdots + \varepsilon_n$, $n = 1, 2, \dots$. Since

$$E(X_{n+1}|X_1, \dots, X_n) = E(X_n + \varepsilon_{n+1}|X_1, \dots, X_n) = X_n + E\varepsilon_{n+1} \quad \text{a.s.},$$

$\{X_n\}$ is a martingale if $E\varepsilon_n = 0$ for all n, a submartingale if $E\varepsilon_n \geq 0$ for all n, and a supermartingale if $E\varepsilon_n \leq 0$ for all n. Note that in Example 1.24 with $\rho = 1$, $\{X_n\}$ is shown to be a Markov chain.

The next example provides another example of martingales.

Example 1.25 (Likelihood ratio). Let $(\Omega, \mathcal{F}, P)$ be a probability space, Q be a probability measure on $\mathcal{F}$, and $\mathcal{F}_1 \subset \mathcal{F}_2 \subset \cdots \subset \mathcal{F}$ be a sequence of σ-fields. Let P_n and Q_n be P and Q restricted to $\mathcal{F}_n$, respectively, $n = 1, 2, \dots$. Suppose that $Q_n \ll P_n$ for each n. Then $\{X_n, \mathcal{F}_n\}$ is a martingale, where $X_n = dQ_n/dP_n$ (the Radon-Nikodym derivative of Q_n w.r.t. P_n), $n = 1, 2, \dots$ (exercise). Suppose now that $\{Y_n\}$ is a sequence of random variables on $(\Omega, \mathcal{F}, P)$, $\mathcal{F}_n = \sigma(Y_1, \dots, Y_n)$ and that there exists a σ-finite measure ν_n on $\mathcal{F}_n$ such that $P_n \ll \nu_n$ and $\nu_n \ll P_n$, $n = 1, 2, \dots$. Let $p_n(Y_1, \dots, Y_n) = dP_n/d\nu_n$ and $q_n(Y_1, \dots, Y_n) = dQ_n/d\nu_n$. By Proposition 1.7(iii), $X_n = q_n(Y_1, \dots, Y_n)/p_n(Y_1, \dots, Y_n)$, which is called a likelihood ratio in statistical terms. ∎

The following results contain some useful properties of martingales and submartingales.

Proposition 1.13. Let φ be a convex function on $\mathcal{R}$.
(i) If $\{X_n, \mathcal{F}_n\}$ is a martingale and $\varphi(X_n)$ is integrable for all n, then $\{\varphi(X_n), \mathcal{F}_n\}$ is a submartingale.
(ii) If $\{X_n, \mathcal{F}_n\}$ is a submartingale, $\varphi(X_n)$ is integrable for all n, and φ is nondecreasing, then $\{\varphi(X_n), \mathcal{F}_n\}$ is a submartingale.
Proof. (i) Note that $\varphi(X_n) = \varphi(E(X_{n+1}|\mathcal{F}_n)) \leq E[\varphi(X_{n+1}|\mathcal{F}_n)]$ a.s. by Jensen's inequality for conditional expectations (Exercise 89(c)).
(ii) Since φ is nondecreasing and $\{X_n, \mathcal{F}_n\}$ is a submartingale, $\varphi(X_n) \leq \varphi(E(X_{n+1}|\mathcal{F}_n)) \leq E[\varphi(X_{n+1}|\mathcal{F}_n)]$ a.s. ∎

An application of Proposition 1.13 shows that if $\{X_n, \mathcal{F}_n\}$ is a submartingale, then so is $\{(X_n)_+, \mathcal{F}_n\}$; if $\{X_n, \mathcal{F}_n\}$ is a martingale, then $\{|X_n|, \mathcal{F}_n\}$ is a submartingale and so are $\{|X_n|^p, \mathcal{F}_n\}$, where $p > 1$ is a constant, and $\{|X_n|(\log|X_n|)_+, \mathcal{F}_n\}$, provided that $|X_n|^p$ and $|X_n|(\log|X_n|)_+$ are integrable for all n.

Proposition 1.14 (Doob's decomposition). Let $\{X_n, \mathcal{F}_n\}$ be a submartingale. Then $X_n = Y_n + Z_n$, $n = 1, 2, ...$, where $\{Y_n, \mathcal{F}_n\}$ is a martingale, $0 = Z_1 \leq Z_2 \leq \cdots$, and $EZ_n < \infty$ for all n. Furthermore, if $\sup_n E|X_n| < \infty$, then $\sup_n E|Y_n| < \infty$ and $\sup_n EZ_n < \infty$.
Proof. Define $\eta_1 = \xi_1$, $\zeta_1 = 0$, $\eta_n = X_n - X_{n-1} - E(X_n - X_{n-1}|\mathcal{F}_{n-1})$, and $\zeta_n = E(X_n - X_{n-1}|\mathcal{F}_{n-1})$ for $n \geq 2$. Then $Y_n = \sum_{i=1}^n \eta_i$ and $Z_n = \sum_{i=1}^n \zeta_i$ satisfy $X_n = Y_n + Z_n$ and the required conditions (exercise).

Assume now that $\sup_n E|X_n| < \infty$. Since $EY_1 = EY_n$ for any n and $Z_n \leq |X_n| - Y_n$, $EZ_n \leq E|X_n| - EY_1$. Hence $\sup_n EZ_n < \infty$. Also, $|Y_n| \leq |X_n| + Z_n$. Hence $\sup_n E|Y_n| < \infty$. ∎

The following martingale convergence theorem, due to Doob, has many applications (see, e.g., Example 1.27 in §1.5.1). Its proof can be found, for example, in Billingsley (1986, pp. 490-491).

Proposition 1.15. Let $\{X_n, \mathcal{F}_n\}$ be a submartingale. If $c = \sup_n E|X_n| < \infty$, then $\lim_{n \to \infty} X_n = X$ a.s., where X is a random variable satisfying $E|X| \leq c$. ∎

1.5 Asymptotic Theory

Asymptotic theory studies limiting behavior of random variables (vectors) and their distributions. It is an important tool for statistical analysis. A more complete coverage of asymptotic theory in statistical analysis can be found in Serfling (1980), Shorack and Wellner (1986), Sen and Singer (1993), Barndorff-Nielsen and Cox (1994), and van der Vaart (1998).

1.5.1 Convergence modes and stochastic orders

There are several convergence modes for random variables/vectors. Let $r > 0$ be a constant. For any $c = (c_1, ..., c_k) \in \mathcal{R}^k$, we define $\|c\|_r = (\sum_{j=1}^{k} |c_j|^r)^{1/r}$. If $r \geq 1$, then $\|c\|_r$ is the L_r-distance between 0 and c. When $r = 2$, the subscript r is omitted and $\|c\| = \|c\|_2 = \sqrt{c^\tau c}$.

Definition 1.8. Let $X, X_1, X_2, \ldots$ be random k-vectors defined on a probability space.
(i) We say that the sequence $\{X_n\}$ converges to X almost surely (a.s.) and write $X_n \to_{a.s.} X$ if and only if $\lim_{n \to \infty} X_n = X$ a.s.
(ii) We say that $\{X_n\}$ converges to X in probability and write $X_n \to_p X$ if and only if, for every fixed $\epsilon > 0$,

$$\lim_{n \to \infty} P\left(\|X_n - X\| > \epsilon\right) = 0. \tag{1.70}$$

(iii) We say that $\{X_n\}$ converges to X in L_r (or in rth moment) and write $X_n \to_{L_r} X$ if and only if

$$\lim_{n \to \infty} E\|X_n - X\|_r^r = 0,$$

where $r > 0$ is a fixed constant.
(iv) Let $F, F_n, n = 1, 2, ...,$ be c.d.f.'s on $\mathcal{R}^k$ and $P, P_n, n = 1, ...,$ be their corresponding probability measures. We say that $\{F_n\}$ converges to F weakly (or $\{P_n\}$ converges to P weakly) and write $F_n \to_w F$ (or $P_n \to_w P$) if and only if, for each continuity point x of F,

$$\lim_{n \to \infty} F_n(x) = F(x).$$

We say that $\{X_n\}$ converges to X in distribution (or in law) and write $X_n \to_d X$ if and only if $F_{X_n} \to_w F_X$. ∎

The a.s. convergence has already been considered in previous sections. The concept of convergence in probability, convergence in L_r, or a.s. convergence represents a sense in which, for n sufficiently large, X_n and X approximate each other as functions on the original probability space. The concept of convergence in distribution in Definition 1.8(iv), however, depends only on the distributions F_{X_n} and F_X (or probability measures P_{X_n} and P_X) and does not necessitate that X_n and X are close in any sense; in fact, Definition 1.8(iv) still makes sense even if X and X_n's are not defined on the same probability space. In Definition 1.8(iv), it is *not* required that $\lim_{n \to \infty} F_n(x) = F(x)$ for every x. However, if F is a continuous function, then we have the following stronger result.

Proposition 1.16 (Pólya's theorem). If $F_n \to_w F$ and F is continuous on $\mathcal{R}^k$, then

$$\lim_{n\to\infty} \sup_{x\in\mathcal{R}^k} |F_n(x) - F(x)| = 0. \quad \blacksquare$$

A useful characterization of a.s. convergence is given in the following lemma.

Lemma 1.4. For random k-vectors $X, X_1, X_2, \ldots$ on a probability space, $X_n \to_{a.s.} X$ if and only if for every $\epsilon > 0$,

$$\lim_{n\to\infty} P\left(\bigcup_{m=n}^{\infty} \{\|X_m - X\| > \epsilon\} \right) = 0. \tag{1.71}$$

Proof. Let $A_j = \cup_{n=1}^{\infty} \cap_{m=n}^{\infty} \{\|X_m - X\| \leq j^{-1}\}$, $j = 1, 2, \ldots$. By Proposition 1.1(iii) and DeMorgan's law, (1.71) holds for every $\epsilon > 0$ if and only if $P(A_j) = 1$ for every j, which is equivalent to $P(\cap_{j=1}^{\infty} A_j) = 1$. The result follows from $\cap_{j=1}^{\infty} A_j = \{\omega : \lim_{n\to\infty} X_n(\omega) = X(\omega)\}$ (exercise). $\quad \blacksquare$

The following result describes the relationship among the four convergence modes in Definition 1.8.

Theorem 1.8. Let $X, X_1, X_2, \ldots$ be random k-vectors.
(i) If $X_n \to_{a.s.} X$, then $X_n \to_p X$.
(ii) If $X_n \to_{L_r} X$ for an $r > 0$, then $X_n \to_p X$.
(iii) If $X_n \to_p X$, then $X_n \to_d X$.
(iv) (Skorohod's theorem). If $X_n \to_d X$, then there are random vectors $Y, Y_1, Y_2, \ldots$ defined on a common probability space such that $P_Y = P_X$, $P_{Y_n} = P_{X_n}$, $n = 1, 2, \ldots$, and $Y_n \to_{a.s.} Y$.
(v) If, for every $\epsilon > 0$, $\sum_{n=1}^{\infty} P(\|X_n - X\| \geq \epsilon) < \infty$, then $X_n \to_{a.s.} X$.
(vi) If $X_n \to_p X$, then there is a subsequence $\{X_{n_j}, j = 1, 2, \ldots\}$ such that $X_{n_j} \to_{a.s.} X$ as $j \to \infty$.
(vii) If $X_n \to_d X$ and $P(X = c) = 1$, where $c \in \mathcal{R}^k$ is a constant vector, then $X_n \to_p c$.
(viii) Suppose that $X_n \to_d X$. Then, for any $r > 0$,

$$\lim_{n\to\infty} E\|X_n\|_r^r = E\|X\|_r^r < \infty \tag{1.72}$$

if and only if $\{\|X_n\|_r^r\}$ is *uniformly integrable* in the sense that

$$\lim_{t\to\infty} \sup_n E\left(\|X_n\|_r^r I_{\{\|X_n\|_r > t\}} \right) = 0. \quad \blacksquare \tag{1.73}$$

The proof of Theorem 1.8 is given after the following discussion and example.

The converse of Theorem 1.8(i), (ii), or (iii) is generally not true (see Example 1.26 and Exercise 116). Note that part (iv) of Theorem 1.8 (Skorohod's theorem) is not a converse of part (i), but it is an important result in probability theory. It is useful when we study convergence of quantities related to F_{X_n} and F_X when $X_n \to_d X$ (see, e.g., the proofs of Theorems 1.8 and 1.9). Part (v) of Theorem 1.8 indicates that the converse of part (i) is true under the additional condition that $P(\|X_n - X\| \geq \epsilon)$ tends to 0 fast enough. Part (vi) provides a partial converse of part (i) whereas part (vii) is a partial converse of part (iii). A consequence of Theorem 1.8(viii) is that if $X_n \to_p X$ and $\{\|X_n - X\|_r^r\}$ is uniformly integrable, then $X_n \to_{L_r} X$; i.e., the converse of Theorem 1.8(ii) is true under the additional condition of uniform integrability. A useful sufficient condition for uniform integrability of $\{\|X_n\|_r^r\}$ is that

$$\sup_n E\|X_n\|_r^{r+\delta} < \infty \tag{1.74}$$

for a $\delta > 0$. Some other sufficient conditions are given in Exercises 117-120.

Example 1.26. Let $\theta_n = 1 + n^{-1}$ and X_n be a random variable having the exponential distribution $E(0, \theta_n)$ (Table 1.2), $n = 1, 2, \dots$. Let X be a random variable having the exponential distribution $E(0, 1)$. For any $x > 0$,

$$F_{X_n}(x) = 1 - e^{-x/\theta_n} \to 1 - e^{-x} = F_X(x)$$

as $n \to \infty$. Since $F_{X_n}(x) \equiv 0 \equiv F_X(x)$ for $x \leq 0$, we have shown that $X_n \to_d X$.

Is it true that $X_n \to_p X$? This question cannot be answered without any further information about the random variables X and X_n. We consider two cases in which different answers can be obtained. First, suppose that $X_n \equiv \theta_n X$ (then X_n has the given c.d.f.). Note that $X_n - X = (\theta_n - 1)X = n^{-1}X$, which has the c.d.f. $(1 - e^{-nx})I_{[0,\infty)}(x)$. Hence

$$P(|X_n - X| \geq \epsilon) = e^{-n\epsilon} \to 0$$

for any $\epsilon > 0$. In fact, by Theorem 1.8(v), $X_n \to_{a.s.} X$; since $E|X_n - X|^p = n^{-p}EX^p < \infty$ for any $p > 0$, $X_n \to_{L_p} X$ for any $p > 0$. Next, suppose that X_n and X are independent random variables. Using result (1.28) and the fact that the p.d.f.'s for X_n and $-X$ are $\theta_n^{-1}e^{-x/\theta_n}I_{(0,\infty)}(x)$ and $e^x I_{(-\infty,0)}(x)$, respectively, we obtain that

$$P(|X_n - X| \leq \epsilon) = \int_{-\epsilon}^{\epsilon} \int \theta_n^{-1}e^{-x/\theta_n}e^{y-x}I_{(0,\infty)}(x)I_{(-\infty,x)}(y)dxdy,$$

which converges to (by the dominated convergence theorem)

$$\int_{-\epsilon}^{\epsilon} \int e^{-x}e^{y-x}I_{(0,\infty)}(x)I_{(-\infty,x)}(y)dxdy = 1 - e^{-\epsilon}.$$

Thus, $P(|X_n - X| \geq \epsilon) \to e^{-\epsilon} > 0$ for any $\epsilon > 0$ and, therefore, $\{X_n\}$ does not converge to X in probability. The previous discussion, however, indicates how to construct the random variables Y_n and Y in Theorem 1.8(iv) for this example. ∎

The following famous result is used in the proof of Theorem 1.8(v). Its proof is left to the reader.

Lemma 1.5. (Borel-Cantelli lemma). Let A_n be a sequence of events in a probability space and $\limsup_n A_n = \cap_{n=1}^\infty \cup_{m=n}^\infty A_m$.
(i) If $\sum_{n=1}^\infty P(A_n) < \infty$, then $P(\limsup_n A_n) = 0$.
(ii) If $A_1, A_2, ...$ are pairwise independent and $\sum_{n=1}^\infty P(A_n) = \infty$, then $P(\limsup_n A_n) = 1$. ∎

Proof of Theorem 1.8. (i) The result follows from Lemma 1.4, since (1.71) implies (1.70).
(ii) The result follows from Chebyshev's inequality with $\varphi(t) = |t|^r$.
(iii) For any $c = (c_1, ..., c_k) \in \mathcal{R}^k$, define $(-\infty, c] = (-\infty, c_1] \times \cdots \times (-\infty, c_k]$. Let x be a continuity point of F_X, $\epsilon > 0$ be given, and J_k be the k-vector of ones. Then $\{X \in (-\infty, x - \epsilon J_k], X_n \notin (-\infty, x]\} \subset \{\|X_n - X\| > \epsilon\}$ and

$$
\begin{aligned}
F_X(x - \epsilon J_k) &= P\big(X \in (-\infty, x - \epsilon J_k]\big) \\
&\leq P\big(X_n \in (-\infty, x]\big) + P\big(X \in (-\infty, x - \epsilon J_k], X_n \notin (-\infty, x]\big) \\
&\leq F_{X_n}(x) + P\left(\|X_n - X\| > \epsilon\right).
\end{aligned}
$$

Letting $n \to \infty$, we obtain that $F_X(x - \epsilon J_k) \leq \liminf_n F_{X_n}(x)$. Similarly, we can show that $F_X(x + \epsilon J_k) \geq \limsup_n F_{X_n}(x)$. Since ϵ is arbitrary and F_X is continuous at x, $F_X(x) = \lim_{n\to\infty} F_{X_n}(x)$.
(iv) The proof of this part can be found in Billingsley (1986, pp. 399-402).
(v) Let $A_n = \{\|X_n - X\| \geq \epsilon\}$. The result follows from Lemma 1.4, Lemma 1.5(i), and Proposition 1.1(iii).
(vi) From (1.70), for every $j = 1, 2, ...$, there is a positive integer n_j such that $P(\|X_{n_j} - X\| > 2^{-j}) < 2^{-j}$. For any $\epsilon > 0$, there is a k_ϵ such that for $j \geq k_\epsilon$, $P(\|X_{n_j} - X\| > \epsilon) < P(\|X_{n_j} - X\| > 2^{-j})$. Since $\sum_{j=1}^\infty 2^{-j} = 1$, it follows from the result in (v) that $X_{n_j} \to_{a.s.} X$ as $j \to \infty$.
(vii) The proof for this part is left as an exercise.
(viii) First, by part (iv), we may assume that $X_n \to_{a.s.} X$ (why?). Assume that $\{\|X_n\|_r^r\}$ is uniformly integrable. Then $\sup_n E\|X_n\|_r^r < \infty$ (why?) and by Fatou's lemma (Theorem 1.1(i)), $E\|X\|_r^r \leq \liminf_n E\|X_n\|_r^r < \infty$. Hence, (1.72) follows if we can show that

$$
\limsup_n E\|X_n\|_r^r \leq E\|X\|_r^r. \tag{1.75}
$$

For any $\epsilon > 0$ and $t > 0$, let $A_n = \{\|X_n - X\|_r \leq \epsilon\}$ and $B_n = \{\|X_n\|_r > t\}$.

Then

$$E\|X_n\|_r^r = E(\|X_n\|_r^r I_{A_n^c \cap B_n}) + E(\|X_n\|_r^r I_{A_n^c \cap B_n^c}) + E(\|X_n\|_r^r I_{A_n})$$
$$\leq E(\|X_n\|_r^r I_{B_n}) + t^r P(A_n^c) + E\|X_n I_{A_n}\|_r^r.$$

For $r \leq 1$, $\|X_n I_{A_n}\|_r^r \leq (\|X_n - X\|_r^r + \|X\|_r^r) I_{A_n}$ and

$$E\|X_n I_{A_n}\|_r^r \leq E[(\|X_n - X\|_r^r + \|X\|_r^r) I_{A_n}] \leq \epsilon^r + E\|X\|_r^r.$$

For $r > 1$, an application of Minkowski's inequality leads to

$$E\|X_n I_{A_n}\|_r^r = E\|(X_n - X)I_{A_n} + X I_{A_n}\|_r^r$$
$$\leq E\left[\|(X_n - X)I_{A_n}\|_r + \|X I_{A_n}\|_r\right]^r$$
$$\leq \left\{[E\|(X_n - X)I_{A_n}\|_r^r]^{1/r} + [E\|X I_{A_n}\|_r^r]^{1/r}\right\}^r$$
$$\leq \left\{\epsilon + [E\|X\|_r^r]^{1/r}\right\}^r.$$

In any case, since ϵ is arbitrary, $\limsup_n E\|X_n I_{A_n}\|_r^r \leq E\|X\|_r^r$. This result and the previously established inequality imply that

$$\limsup_n E\|X_n\|_r^r \leq \limsup_n E(\|X_n\|_r^r I_{B_n}) + t^r \lim_{n \to \infty} P(A_n^c)$$
$$+ \limsup_n E\|X_n I_{A_n}\|_r^r$$
$$\leq \sup_n E(\|X_n\|_r^r I_{\{\|X_n\|_r > t\}}) + E\|X\|_r^r,$$

since $P(A_n^c) \to 0$. Since $\{\|X_n\|_r^r\}$ is uniformly integrable, letting $t \to \infty$ we obtain (1.75).

Suppose now that (1.72) holds. Let $\xi_n = \|X_n\|_r^r I_{B_n^c} - \|X\|_r^r I_{B_n^c}$. Then $\xi_n \to_{a.s.} 0$ and $|\xi_n| \leq t^r + \|X\|_r^r$, which is integrable. By the dominated convergence theorem, $E\xi_n \to 0$; this and (1.72) imply that

$$E(\|X_n\|_r^r I_{B_n}) - E(\|X\|_r^r I_{B_n}) \to 0.$$

From the definition of B_n, $B_n \subset \{\|X_n - X\|_r > t/2\} \cup \{\|X\|_r > t/2\}$. Since $E\|X\|_r^r < \infty$, it follows from the dominated convergence theorem that $E(\|X\|_r^r I_{\{\|X_n - X\|_r > t/2\}}) \to 0$ as $n \to \infty$. Hence,

$$\limsup_n E(\|X_n\|_r^r I_{B_n}) \leq \limsup_n E(\|X\|_r^r I_{B_n}) \leq E(\|X\|_r^r I_{\{\|X\|_r > t/2\}}).$$

Letting $t \to \infty$, it follows from the dominated convergence theorem that

$$\lim_{t \to \infty} \limsup_n E(\|X_n\|_r^r I_{B_n}) \leq \lim_{t \to \infty} E(\|X\|_r^r I_{\{\|X\|_r > t/2\}}) = 0.$$

This proves (1.73).   ∎

Example 1.27. As an application of Theorem 1.8(viii) and Proposition 1.15, we consider again the prediction problem in Example 1.22. Suppose that we predict a random variable X by a random n-vector $Y = (Y_1, ..., Y_n)$. It is shown in Example 1.22 that $X_n = E(X|Y_1, ..., Y_n)$ is the best predictor in terms of the mean squared prediction error, when $EX^2 < \infty$. We now show that $X_n \to_{a.s.} X$ when $n \to \infty$ under the assumption that $\sigma(X) \subset \mathcal{F}_\infty = \sigma(Y_1, Y_2, ...)$ (i.e., X provides no more information than $Y_1, Y_2, ...$).

From the discussion in §1.4.4, $\{X_n\}$ is a martingale. Also, $\sup_n E|X_n| \leq \sup_n E[E(|X||Y_1, ..., Y_n)] = E|X| < \infty$. Hence, by Proposition 1.15, $X_n \to_{a.s.} Z$ for some random variable Z. We now need to show $Z = X$ a.s. Since $\sigma(X) \subset \mathcal{F}_\infty$, $X = E(X|\mathcal{F}_\infty)$ a.s. Hence, it suffices to show that $Z = E(X|\mathcal{F}_\infty)$ a.s. Since $EX_n^2 \leq EX^2 < \infty$ (why?), condition (1.74) holds for sequence $\{|X_n|\}$ and, hence, $\{|X_n|\}$ is uniformly integrable. By Theorem 1.8(viii), $E|X_n - Z| \to 0$, which implies $\int_A X_n dP \to \int_A Z dP$ for any event A. Note that if $A \in \sigma(Y_1, ..., Y_m)$, then $A \in \sigma(Y_1, ..., Y_n)$ for $n \geq m$ and $\int_A X_n dP = \int_A X dP$. This implies that for any $A \in \cup_{j=1}^\infty \sigma(Y_1, ..., Y_j)$, $\int_A X dP = \int_A Z dP$. Since $\cup_{j=1}^\infty \sigma(Y_1, ..., Y_j)$ generates $\mathcal{F}_\infty$, we conclude that $\int_A X dP = \int_A Z dP$ for any $A \in \mathcal{F}_\infty$ and thus $Z = E(X|\mathcal{F}_\infty)$ a.s.

In the proof above, the condition $EX^2 < \infty$ is used only for showing the uniform integrability of $\{|X_n|\}$. But by Exercise 120, $\{|X_n|\}$ is uniformly integrable as long as $E|X| < \infty$. Hence $X_n \to_{a.s.} X$ is still true if the condition $EX^2 < \infty$ is replaced by $E|X| < \infty$. ∎

We now introduce the notion of $O(\,\cdot\,)$, $o(\,\cdot\,)$, and stochastic $O(\,\cdot\,)$ and $o(\,\cdot\,)$. In calculus, two sequences of real numbers, $\{a_n\}$ and $\{b_n\}$, satisfy $a_n = O(b_n)$ if and only if $|a_n| \leq c|b_n|$ for all n and a constant c; and $a_n = o(b_n)$ if and only if $a_n/b_n \to 0$ as $n \to \infty$.

Definition 1.9. Let $X_1, X_2, ...$ be random vectors and $Y_1, Y_2, ...$ be random variables defined on a common probability space.
(i) $X_n = O(Y_n)$ a.s. if and only if $P(\|X_n\| = O(|Y_n|)) = 1$.
(ii) $X_n = o(Y_n)$ a.s. if and only if $X_n/Y_n \to_{a.s.} 0$.
(iii) $X_n = O_p(Y_n)$ if and only if, for any $\epsilon > 0$, there is a constant $C_\epsilon > 0$ such that $\sup_n P(\|X_n\| \geq C_\epsilon|Y_n|) < \epsilon$.
(iv) $X_n = o_p(Y_n)$ if and only if $X_n/Y_n \to_p 0$. ∎

Note that $X_n = o_p(Y_n)$ implies $X_n = O_p(Y_n)$; $X_n = O_p(Y_n)$ and $Y_n = O_p(Z_n)$ implies $X_n = O_p(Z_n)$; but $X_n = O_p(Y_n)$ does not imply $Y_n = O_p(X_n)$. The same conclusion can be obtained if $O_p(\cdot)$ and $o_p(\cdot)$ are replaced by $O(\cdot)$ a.s. and $o(\cdot)$ a.s., respectively. Some results related to O_p are given in Exercise 127. For example, if $X_n \to_d X$ for a random variable X, then $X_n = O_p(1)$. Since $a_n = O(1)$ means that $\{a_n\}$ is bounded, $\{X_n\}$ is said to be bounded in probability if $X_n = O_p(1)$.

1.5.2 Weak convergence

We now discuss more about convergence in distribution or weak convergence of probability measures. A sequence $\{P_n\}$ of probability measures on $(\mathcal{R}^k, \mathcal{B}^k)$ is *tight* if for every $\epsilon > 0$, there is a compact set $C \subset \mathcal{R}^k$ such that $\inf_n P_n(C) > 1 - \epsilon$. If $\{X_n\}$ is a sequence of random k-vectors, then the tightness of $\{P_{X_n}\}$ is the same as the boundedness of $\{\|X_n\|\}$ in probability. The proof of the following result can be found in Billingsley (1986, pp. 392-395).

Proposition 1.17. Let $\{P_n\}$ be a sequence of probability measures on $(\mathcal{R}^k, \mathcal{B}^k)$.
(i) Tightness of $\{P_n\}$ is a necessary and sufficient condition that for every subsequence $\{P_{n_i}\}$ there exists a further subsequence $\{P_{n_j}\} \subset \{P_{n_i}\}$ and a probability measure P on $(\mathcal{R}^k, \mathcal{B}^k)$ such that $P_{n_j} \to_w P$ as $j \to \infty$.
(ii) If $\{P_n\}$ is tight and if each subsequence that converges weakly at all converges to the same probability measure P, then $P_n \to_w P$. ∎

The following result gives some useful sufficient and necessary conditions for convergence in distribution.

Theorem 1.9. Let $X, X_1, X_2, \ldots$ be random k-vectors.
(i) $X_n \to_d X$ is equivalent to any one of the following conditions:
    (a) $E[h(X_n)] \to E[h(X)]$ for every bounded continuous function h;
    (b) $\limsup_n P_{X_n}(C) \leq P_X(C)$ for any closed set $C \subset \mathcal{R}^k$;
    (c) $\liminf_n P_{X_n}(O) \geq P_X(O)$ for any open set $O \subset \mathcal{R}^k$.
(ii) (Lévy-Cramér continuity theorem). Let $\phi_X, \phi_{X_1}, \phi_{X_2}, \ldots$ be the ch.f.'s of $X, X_1, X_2, \ldots$, respectively. $X_n \to_d X$ if and only if $\lim_{n \to \infty} \phi_{X_n}(t) = \phi_X(t)$ for all $t \in \mathcal{R}^k$.
(iii) (Cramér-Wold device). $X_n \to_d X$ if and only if $c^\tau X_n \to_d c^\tau X$ for every $c \in \mathcal{R}^k$.
Proof. (i) First, we show $X_n \to_d X$ implies (a). By Theorem 1.8(iv) (Skorohod's theorem), there exists a sequence of random vectors $\{Y_n\}$ and a random vector Y such that $P_{Y_n} = P_{X_n}$ for all n, $P_Y = P_X$ and $Y_n \to_{a.s.} Y$. For bounded continuous h, $h(Y_n) \to_{a.s.} h(Y)$ and, by the dominated convergence theorem, $E[h(Y_n)] \to E[h(Y)]$. Then (a) follows from $E[h(X_n)] = E[h(Y_n)]$ for all n and $E[h(X)] = E[h(Y)]$.

Next, we show (a) implies (b). Let C be a closed set and $f_C(x) = \inf\{\|x - y\| : y \in C\}$. Then f_C is continuous. For $j = 1, 2, \ldots$, define $\varphi_j(t) = I_{(-\infty, 0]} + (1 - jt)I_{(0, j^{-1}]}$. Then $h_j(x) = \varphi_j(f_C(x))$ is continuous and bounded, $h_j \geq h_{j+1}$, $j = 1, 2, \ldots$, and $h_j(x) \to I_C(x)$ as $j \to \infty$. Hence $\limsup_n P_{X_n}(C) \leq \lim_{n \to \infty} E[h_j(X_n)] = E[h_j(X)]$ for each j (by (a)). By the dominated convergence theorem, $E[h_j(X)] \to E[I_C(X)] = P_X(C)$.

This proves (b).

For any open set O, O^c is closed. Hence, (b) is equivalent to (c). Now, we show (b) and (c) imply $X_n \to_d X$. For $x = (x_1, ..., x_k) \in \mathcal{R}^k$, let $(-\infty, x] = (-\infty, x_1] \times \cdots \times (-\infty, x_k]$ and $(-\infty, x) = (-\infty, x_1) \times \cdots \times (-\infty, x_k)$. From (b) and (c), $P_X((-\infty, x)) \leq \liminf_n P_{X_n}((-\infty, x)) \leq \liminf_n F_{X_n}(x) \leq \limsup_n F_{X_n}(x) = \limsup_n P_{X_n}((-\infty, x]) \leq P_X((-\infty, x]) = F_X(x)$. If x is a continuity point of F_X, then $P_X((-\infty, x)) = F_X(x)$. This proves $X_n \to_d X$ and completes the proof of (i).

(ii) From (a) of part (i), $X_n \to_d X$ implies $\phi_{X_n}(t) \to \phi_X(t)$, since $e^{\sqrt{-1} t^\tau x} = \cos(t^\tau x) + \sqrt{-1} \sin(t^\tau x)$ and $\cos(t^\tau x)$ and $\sin(t^\tau x)$ are bounded continuous functions for any fixed t.

Suppose now that $k = 1$ and that $\phi_{X_n}(t) \to \phi_X(t)$ for every $t \in \mathcal{R}$. By Fubini's theorem,

$$
\begin{aligned}
\frac{1}{u} \int_{-u}^{u} [1 - \phi_{X_n}(t)] dt &= \int_{-\infty}^{\infty} \left[\frac{1}{u} \int_{-u}^{u} (1 - e^{\sqrt{-1} tx}) dt \right] dP_{X_n}(x) \\
&= 2 \int_{-\infty}^{\infty} \left(1 - \frac{\sin ux}{ux} \right) dP_{X_n}(x) \\
&\geq 2 \int_{\{|x| > 2u^{-1}\}} \left(1 - \frac{1}{|ux|} \right) dP_{X_n}(x) \\
&\geq P_{X_n} \left((-\infty, -2u^{-1}) \cup (2u^{-1}, \infty) \right)
\end{aligned}
$$

for any $u > 0$. Since ϕ_X is continuous at 0 and $\phi_X(0) = 1$, for any $\epsilon > 0$ there is a $u > 0$ such that $u^{-1} \int_{-u}^{u} [1 - \phi_X(t)] dt < \epsilon/2$. Since $\phi_{X_n} \to \phi_X$, by the dominated convergence theorem, $\sup_n \{ u^{-1} \int_{-u}^{u} [1 - \phi_{X_n}(t)] dt \} < \epsilon$. Hence,

$$
\inf_n P_{X_n} \left([-2u^{-1}, 2u^{-1}] \right) \geq 1 - \sup_n \left\{ \frac{1}{u} \int_{-u}^{u} [1 - \phi_{X_n}(t)] dt \right\} \geq 1 - \epsilon,
$$

i.e., $\{P_{X_n}\}$ is tight. Let $\{P_{X_{n_j}}\}$ be any subsequence that converges to a probability measure P. By the first part of the proof, $\phi_{X_{n_j}} \to \phi$, which is the ch.f. of P. By the convergence of ϕ_{X_n}, $\phi = \phi_X$. By Theorem 1.6(i), $P = P_X$. By Proposition 1.17(ii), $X_n \to_d X$.

Consider now the case where $k \geq 2$ and $\phi_{X_n} \to \phi_X$. Let Y_{nj} be the jth component of X_n and Y_j be the jth component of X. Then $\phi_{Y_{nj}} \to \phi_{Y_j}$ for each j. By the proof for the case of $k = 1$, $Y_{nj} \to_d Y_j$. By Proposition 1.17(i), $\{P_{Y_{nj}}\}$ is tight, $j = 1, ..., k$. This implies that $\{P_{X_n}\}$ is tight (why?). Then the proof for $X_n \to_d X$ is the same as that for the case of $k = 1$.

(iii) From (1.52), $\phi_{c^\tau X_n}(u) = \phi_{X_n}(uc)$ and $\phi_{c^\tau X}(u) = \phi_X(uc)$ for any $u \in \mathcal{R}$ and any $c \in \mathcal{R}^k$. Hence, convergence of ϕ_{X_n} to ϕ_X is equivalent to convergence of $\phi_{c^\tau X_n}$ to $\phi_{c^\tau X}$ for every $c \in \mathcal{R}^k$. Then the result follows from part (ii). ∎

Example 1.28. Let $X_1, ..., X_n$ be independent random variables having a common c.d.f. and $T_n = X_1 + \cdots + X_n$, $n = 1, 2, ...$ Suppose that $E|X_1| < \infty$. It follows from (1.56) and a result in calculus that the ch.f. of X_1 satisfies

$$\phi_{X_1}(t) = \phi_{X_1}(0) + \sqrt{-1}\mu t + o(|t|)$$

as $|t| \to 0$, where $\mu = EX_1$. From (1.52) and (1.58), the ch.f. of T_n/n is

$$\phi_{T_n/n}(t) = \left[\phi_{X_1}\left(\frac{t}{n}\right)\right]^n = \left[1 + \frac{\sqrt{-1}\mu t}{n} + o\left(\frac{t}{n}\right)\right]^n$$

for any $t \in \mathcal{R}$, as $n \to \infty$. Since $(1 + c_n/n)^n \to e^c$ for any complex sequence $\{c_n\}$ satisfying $c_n \to c$, we obtain that $\phi_{T_n/n}(t) \to e^{\sqrt{-1}\mu t}$, which is the ch.f. of the distribution degenerated at μ (i.e., the point mass probability measure at μ; see (1.22)). By Theorem 1.9(ii), $T_n/n \to_d \mu$. From Theorem 1.8(vii), this also shows that $T_n/n \to_p \mu$.

Similarly, $\mu = 0$ and $\sigma^2 = \text{Var}(X_1) < \infty$ imply

$$\phi_{T_n/\sqrt{n}}(t) = \left[1 - \frac{\sigma^2 t^2}{2n} + o\left(\frac{t^2}{n}\right)\right]^n$$

for any $t \in \mathcal{R}$, which implies that $\phi_{T_n/\sqrt{n}}(t) \to e^{-\sigma^2 t^2/2}$, the ch.f. of $N(0, \sigma^2)$. Hence $T_n/\sqrt{n} \to_d N(0, \sigma^2)$. (Recall that $N(\mu, \sigma^2)$ denotes a random variable having the $N(\mu, \sigma^2)$ distribution.) If $\mu \neq 0$, a transformation of $Y_i = X_i - \mu$ leads to $(T_n - n\mu)/\sqrt{n} \to_d N(0, \sigma^2)$.

Suppose now that $X_1, ..., X_n$ are random k-vectors and $\mu = EX_1$ and $\Sigma = \text{Var}(X_1)$ are finite. For any fixed $c \in \mathcal{R}^k$, it follows from the previous discussion that $(c^\tau T_n - nc^\tau \mu)/\sqrt{n} \to_d N(0, c^\tau \Sigma c)$. From Theorem 1.9(iii) and a property of the normal distribution (Exercise 81), we conclude that $(T_n - n\mu)/\sqrt{n} \to_d N_k(0, \Sigma)$. ∎

Example 1.28 shows that Theorem 1.9(ii) together with some properties of ch.f.'s can be applied to show convergence in distribution for sums of independent random variables (vectors). The following is another example.

Example 1.29. Let $X_1, ..., X_n$ be independent random variables having a common Lebesgue p.d.f. $f(x) = (1 - \cos x)/(\pi x^2)$. Then the ch.f. of X_1 is $\max\{1 - |t|, 0\}$ (Exercise 73) and the ch.f. of $T_n/n = (X_1 + \cdots + X_n)/n$ is

$$\left(\max\left\{1 - \frac{|t|}{n}, 0\right\}\right)^n \to e^{-|t|}, \qquad t \in \mathcal{R}.$$

Since $e^{-|t|}$ is the ch.f. of the Cauchy distribution $C(0, 1)$ (Table 1.2), we conclude that $T_n/n \to_d X$, where X has the Cauchy distribution $C(0, 1)$.

Does this result contradict the first result in Example 1.28? ∎

Other examples of applications of Theorem 1.9 are given in Exercises 135-140 in §1.6. The following result can be used to check whether $X_n \to_d X$ when X has a p.d.f. f and X_n has a p.d.f. f_n.

Proposition 1.18 (Scheffé's theorem). Let $\{f_n\}$ be a sequence of p.d.f.'s on $\mathcal{R}^k$ w.r.t. a measure ν. Suppose that $\lim_{n\to\infty} f_n(x) = f(x)$ a.e. ν and $f(x)$ is a p.d.f. w.r.t. ν. Then $\lim_{n\to\infty} \int |f_n(x) - f(x)| d\nu = 0$.
Proof. Let $g_n(x) = [f(x) - f_n(x)]I_{\{f \geq f_n\}}(x)$, $n = 1, 2,....$ Then

$$\int |f_n(x) - f(x)| d\nu = 2 \int g_n(x) d\nu.$$

Since $0 \leq g_n(x) \leq f(x)$ for all x and $g_n \to 0$ a.e. ν, the result follows from the dominated convergence theorem. ∎

As an example, consider the Lebesgue p.d.f. f_n of the t-distribution t_n (Table 1.2), $n = 1, 2,....$ One can show (exercise) that $f_n \to f$, where f is the standard normal p.d.f. This is an important result in statistics.

1.5.3 Convergence of transformations

Transformation is an important tool in statistics. For random vectors X_n converging to X in some sense, we often want to know whether $g(X_n)$ converges to $g(X)$ in the same sense. The following result provides an answer to this question in many problems. Its proof is left to the reader.

Theorem 1.10. Let $X, X_1, X_2, ...$ be random k-vectors defined on a probability space and g be a measurable function from $(\mathcal{R}^k, \mathcal{B}^k)$ to $(\mathcal{R}^l, \mathcal{B}^l)$. Suppose that g is continuous a.s. P_X. Then
(i) $X_n \to_{a.s.} X$ implies $g(X_n) \to_{a.s.} g(X)$;
(ii) $X_n \to_p X$ implies $g(X_n) \to_p g(X)$;
(iii) $X_n \to_d X$ implies $g(X_n) \to_d g(X)$. ∎

Example 1.30. (i) Let $X_1, X_2, ...$ be random variables. If $X_n \to_d X$, where X has the $N(0, 1)$ distribution, then $X_n^2 \to_d Y$, where Y has the chi-square distribution χ_1^2 (Example 1.14).
(ii) Let (X_n, Y_n) be random 2-vectors satisfying $(X_n, Y_n) \to_d (X, Y)$, where X and Y are independent random variables having the $N(0, 1)$ distribution, then $X_n/Y_n \to_d X/Y$, which has the Cauchy distribution $C(0, 1)$ (§1.3.1).
(iii) Under the conditions in part (ii), $\max\{X_n, Y_n\} \to_d \max\{X, Y\}$, which has the c.d.f. $[\Phi(x)]^2$ ($\Phi(x)$ is the c.d.f. of $N(0, 1)$). ∎

In Example 1.30(ii) and (iii), the condition that $(X_n, Y_n) \to_d (X, Y)$ cannot be relaxed to $X_n \to_d X$ and $Y_n \to_d Y$ (exercise); i.e., we need the convergence of the joint c.d.f. of (X_n, Y_n). This is different when $\to_d$ is replaced by $\to_p$ or $\to_{a.s.}$. The following result, which plays an important role in probability and statistics, establishes the convergence in distribution of $X_n + Y_n$ or $X_n Y_n$ when no information regarding the joint c.d.f. of (X_n, Y_n) is provided.

Theorem 1.11 (Slutsky's theorem). Let $X, X_1, X_2, ..., Y_1, Y_2, ...$ be random variables on a probability space. Suppose that $X_n \to_d X$ and $Y_n \to_p c$, where c is a fixed real number. Then
(i) $X_n + Y_n \to_d X + c$;
(ii) $Y_n X_n \to_d cX$;
(iii) $X_n / Y_n \to_d X/c$ if $c \neq 0$.
Proof. We prove (i) only. The proofs of (ii) and (iii) are left as exercises. Let $t \in \mathcal{R}$ and $\epsilon > 0$ be fixed constants. Then

$$
\begin{aligned}
F_{X_n + Y_n}(t) &= P(X_n + Y_n \leq t) \\
&\leq P(\{X_n + Y_n \leq t\} \cap \{|Y_n - c| < \epsilon\}) + P(|Y_n - c| \geq \epsilon) \\
&\leq P(X_n \leq t - c + \epsilon) + P(|Y_n - c| \geq \epsilon)
\end{aligned}
$$

and, similarly,

$$
F_{X_n + Y_n}(t) \geq P(X_n \leq t - c - \epsilon) - P(|Y_n - c| \geq \epsilon).
$$

If $t - c$, $t - c + \epsilon$, and $t - c - \epsilon$ are continuity points of F_X, then it follows from the previous two inequalities and the hypotheses of the theorem that

$$
F_X(t - c - \epsilon) \leq \liminf_n F_{X_n + Y_n}(t) \leq \limsup_n F_{X_n + Y_n}(t) \leq F_X(t - c + \epsilon).
$$

Since ϵ can be arbitrary (why?),

$$
\lim_{n \to \infty} F_{X_n + Y_n}(t) = F_X(t - c).
$$

The result follows from $F_{X+c}(t) = F_X(t - c)$. ∎

An application of Theorem 1.11 is given in the proof of the following important result.

Theorem 1.12. Let $X_1, X_2, ...$ and Y be random k-vectors satisfying

$$
a_n(X_n - c) \to_d Y, \tag{1.76}
$$

where $c \in \mathcal{R}^k$ and $\{a_n\}$ is a sequence of positive numbers with $\lim_{n \to \infty} a_n = \infty$. Let g be a function from $\mathcal{R}^k$ to $\mathcal{R}$.
(i) If g is differentiable at c, then

$$
a_n[g(X_n) - g(c)] \to_d [\nabla g(c)]^\tau Y, \tag{1.77}
$$

where $\nabla g(x)$ denotes the k-vector of partial derivatives of g at x.
(ii) Suppose that g has continuous partial derivatives of order $m > 1$ in a neighborhood of c, with all the partial derivatives of order j, $1 \leq j \leq m-1$, vanishing at c, but with the mth-order partial derivatives not all vanishing at c. Then

$$a_n^m[g(X_n) - g(c)] \to_d \frac{1}{m!} \sum_{i_1=1}^{k} \cdots \sum_{i_m=1}^{k} \frac{\partial^m g}{\partial x_{i_1} \cdots \partial x_{i_m}} \bigg|_{x=c} Y_{i_1} \cdots Y_{i_m}, \quad (1.78)$$

where Y_j is the jth component of Y.

Proof. We prove (i) only. The proof of (ii) is similar. Let

$$Z_n = a_n[g(X_n) - g(c)] - a_n[\nabla g(c)]^\tau (X_n - c).$$

If we can show that $Z_n = o_p(1)$, then by (1.76), Theorem 1.9(iii), and Theorem 1.11(i), result (1.77) holds.

The differentiability of g at c implies that for any $\epsilon > 0$, there is a $\delta_\epsilon > 0$ such that

$$|g(x) - g(c) - [\nabla g(c)]^\tau (x - c)| \leq \epsilon \|x - c\| \quad (1.79)$$

whenever $\|x - c\| < \delta_\epsilon$. Let $\eta > 0$ be fixed. By (1.79),

$$P(|Z_n| \geq \eta) \leq P(\|X_n - c\| \geq \delta_\epsilon) + P(a_n \|X_n - c\| \geq \eta/\epsilon).$$

Since $a_n \to \infty$, (1.76) and Theorem 1.11(ii) imply $X_n \to_p c$. By Theorem 1.10(iii), (1.76) implies $a_n \|X_n - c\| \to_d \|Y\|$. Without loss of generality, we can assume that η/ϵ is a continuity point of $F_{\|Y\|}$. Then

$$\limsup_n P(|Z_n| \geq \eta) \leq \lim_{n\to\infty} P(\|X_n - c\| \geq \delta_\epsilon)$$

$$+ \lim_{n\to\infty} P(a_n \|X_n - c\| \geq \eta/\epsilon)$$

$$= P(\|Y\| \geq \eta/\epsilon).$$

The proof is complete since ϵ can be arbitrary. ∎

In statistics, we often need a nondegenerated limiting distribution of $a_n[g(X_n) - g(c)]$ so that probabilities involving $a_n[g(X_n) - g(c)]$ can be approximated by the c.d.f. of $[\nabla g(c)]^\tau Y$, if (1.77) holds. Hence, result (1.77) is not useful for this purpose if $\nabla g(c) = 0$, and in such cases result (1.78) may be applied.

A useful method in statistics, called the *delta-method*, is based on the following corollary of Theorem 1.12.

Corollary 1.1. Assume the conditions of Theorem 1.12. If Y has the $N_k(0, \Sigma)$ distribution, then

$$a_n[g(X_n) - g(c)] \to_d N\left(0, [\nabla g(c)]^\tau \Sigma \nabla g(c)\right). \quad ∎$$

Example 1.31. Let $\{X_n\}$ be a sequence of random variables satisfying $\sqrt{n}(X_n - c) \to_d N(0,1)$. Consider the function $g(x) = x^2$. If $c \neq 0$, then an application of Corollary 1.1 gives that $\sqrt{n}(X_n^2 - c^2) \to_d N(0, 4c^2)$. If $c = 0$, the first-order derivative of g at 0 is 0 but the second-order derivative of $g \equiv 2$. Hence, an application of result (1.78) gives that $nX_n^2 \to_d [N(0,1)]^2$, which has the chi-square distribution χ_1^2 (Example 1.14). The last result can also be obtained by applying Theorem 1.10(iii). ∎

1.5.4 The law of large numbers

The law of large numbers concerns the limiting behavior of sums of independent random variables. The weak law of large numbers (WLLN) refers to convergence in probability, whereas the strong law of large numbers (SLLN) refers to a.s. convergence.

The following lemma is useful in establishing the SLLN. Its proof is left as an exercise.

Lemma 1.6. (Kronecker's lemma). Let $x_n \in \mathcal{R}$, $a_n \in \mathcal{R}$, $0 < a_n \leq a_{n+1}$, $n = 1, 2, ...$, and $a_n \to \infty$. If the series $\sum_{n=1}^{\infty} x_n/a_n$ converges, then $a_n^{-1} \sum_{i=1}^{n} x_i \to 0$. ∎

Our first result gives the WLLN and SLLN for a sequence of independent and identically distributed (i.i.d.) random variables.

Theorem 1.13. Let $X_1, X_2, ...$ be i.i.d. random variables.
(i) (The WLLN). A necessary and sufficient condition for the existence of a sequence of real numbers $\{a_n\}$ for which

$$\frac{1}{n} \sum_{i=1}^{n} X_i - a_n \to_p 0 \tag{1.80}$$

is that $nP(|X_1| > n) \to 0$, in which case we may take $a_n = E(X_1 I_{\{|X_1| \leq n\}})$.
(ii) (The SLLN). A necessary and sufficient condition for the existence of a constant c for which

$$\frac{1}{n} \sum_{i=1}^{n} X_i \to_{a.s.} c \tag{1.81}$$

is that $E|X_1| < \infty$, in which case $c = EX_1$ and

$$\frac{1}{n} \sum_{i=1}^{n} c_i(X_i - EX_1) \to_{a.s.} 0 \tag{1.82}$$

for any bounded sequence of real numbers $\{c_i\}$.

Proof. (i) We prove the sufficiency. The proof of necessity can be found in Petrov (1975). Consider a sequence of random variables obtained by truncating X_j's at n: $Y_{nj} = X_j I_{\{|X_j| \leq n\}}$. Let $T_n = X_1 + \cdots + X_n$ and $Z_n = Y_{n1} + \cdots + Y_{nn}$. Then

$$P(T_n \neq Z_n) \leq \sum_{j=1}^{n} P(Y_{nj} \neq X_j) = nP(|X_1| > n) \to 0. \qquad (1.83)$$

For any $\epsilon > 0$, it follows from Chebyshev's inequality that

$$P\left(\left| \frac{Z_n - EZ_n}{n} \right| > \epsilon \right) \leq \frac{\operatorname{Var}(Z_n)}{\epsilon^2 n^2} = \frac{\operatorname{Var}(Y_{n1})}{\epsilon^2 n} \leq \frac{EY_{n1}^2}{\epsilon^2 n},$$

where the last equality follows from the fact that Y_{nj}, $j = 1, ..., n$, are i.i.d. From integration by parts, we obtain that

$$\frac{EY_{n1}^2}{n} = \frac{1}{n} \int_{[0,n]} x^2 dF_{|X_1|}(x) = \frac{2}{n} \int_0^n xP(|X_1| > x)dx - nP(|X_1| > n),$$

which converges to 0 since $nP(|X_1| > n) \to 0$ (why?). This proves that $(Z_n - EZ_n)/n \to_p 0$, which together with (1.83) and the fact that $EY_{nj} = E(X_1 I_{\{|X_1| \leq n\}})$ imply the result.

(ii) For the sufficiency, let $Y_n = X_n I_{\{|X_n| \leq n\}}$, $n = 1, 2, ...$ Let $m > 0$ be an integer smaller than n. If we define $c_i = i^{-1}$ for $i \geq m$, $Z_1 = \cdots = Z_{m-1} = 0$, $Z_m = Y_1 + \cdots + Y_m$, $Z_i = Y_i$, $i = m+1, ..., n$, and apply the Hájek-Rènyi inequality (1.51) to Z_i's, then we obtain that for any $\epsilon > 0$,

$$P\left(\max_{m \leq l \leq n} |\xi_l| > \epsilon \right) \leq \frac{1}{\epsilon^2 m^2} \sum_{i=1}^{m} \operatorname{Var}(Y_i) + \frac{1}{\epsilon^2} \sum_{i=m+1}^{n} \frac{\operatorname{Var}(Y_i)}{i^2}, \qquad (1.84)$$

where $\xi_n = n^{-1} \sum_{i=1}^{n} (Z_i - EZ_i)$ $(= n^{-1} \sum_{i=1}^{n} (Y_i - EY_i)$ if $l \geq m$). Note that

$$\sum_{n=1}^{\infty} \frac{EY_n^2}{n^2} = \sum_{n=1}^{\infty} \sum_{j=1}^{n} \frac{E(X_1^2 I_{\{j-1 < |X_1| \leq j\}})}{n^2}$$

$$= \sum_{j=1}^{\infty} \sum_{n=j}^{\infty} \frac{E(X_1^2 I_{\{j-1 < |X_1| \leq j\}})}{n^2}$$

$$\leq \sum_{j=1}^{\infty} \sum_{n=j}^{\infty} \frac{jE(|X_1| I_{\{j-1 < |X_1| \leq j\}})}{n^2}$$

$$\leq \lambda \sum_{j=1}^{\infty} E(|X_1| I_{\{j-1 < |X_1| \leq j\}})$$

$$= \lambda E|X_1|,$$

where the last inequality follows from the fact that $\sum_{n=j}^{\infty} n^{-2} \leq \lambda j^{-1}$ for a constant $\lambda > 0$ and all $j = 1, 2, \ldots$. Then, letting $n \to \infty$ first and $m \to \infty$ next in (1.84), we obtain that

$$\lim_{m\to\infty} P\left(\bigcup_{l=m}^{\infty} \{|\xi_l| > \epsilon\}\right) = \lim_{m\to\infty} \lim_{n\to\infty} P\left(\max_{m\leq l\leq n} |\xi_l| > \epsilon\right)$$

$$\leq \lim_{m\to\infty} \frac{1}{\epsilon^2 m^2} \sum_{i=1}^{m} \mathrm{Var}(Y_i)$$

$$= 0,$$

where the last equality follows from Lemma 1.6. By Lemma 1.4, $\xi_n \to_{a.s.} 0$. Since $EY_n \to EX_1$, $n^{-1}\sum_{i=1}^{n} EY_i \to EX_1$ and, hence, (1.81) holds with X_i's replaced by Y_i's and $c = EX_1$. It follows from

$$\sum_{n=1}^{\infty} P(X_n \neq Y_n) = \sum_{n=1}^{\infty} P(|X_n| > n) = \sum_{n=1}^{\infty} P(|X_1| > n) < \infty$$

(Exercise 54) and Lemma 1.5(i) that $P\left(\cap_{n=1}^{\infty} \cup_{m=n}^{\infty} \{X_m \neq Y_m\}\right) = 0$, i.e., there is an event A with $P(A) = 1$ such that if $\omega \in A$, then $X_n(\omega) = Y_n(\omega)$ for sufficiently large n. This implies

$$\frac{1}{n}\sum_{i=1}^{n} X_i - \frac{1}{n}\sum_{i=1}^{n} Y_i \to_{a.s.} 0, \tag{1.85}$$

which proves the sufficiency. The proof of (1.82) is left as an exercise.

We now prove the necessity. Suppose that (1.81) holds for some $c \in \mathcal{R}$. Then

$$\frac{X_n}{n} = \frac{T_n}{n} - c - \frac{n-1}{n}\left(\frac{T_{n-1}}{n-1} - c\right) + \frac{c}{n} \to_{a.s.} 0.$$

From Exercise 114, $X_n/n \to_{a.s.} 0$ and the i.i.d. assumption on X_n's imply

$$\sum_{n=1}^{\infty} P(|X_n| \geq n) = \sum_{n=1}^{\infty} P(|X_1| \geq n) < \infty,$$

which implies $E|X_1| < \infty$ (Exercise 54). From the proved sufficiency, $c = EX_1$. ∎

If $E|X_1| < \infty$, then a_n in (1.80) converges to EX_1 and result (1.80) is actually established in Example 1.28 in a much simpler way. On the other hand, if $E|X_1| < \infty$, then the stronger result (1.81) can be obtained. Some results for the case of $E|X_1| = \infty$ can be found in Exercise 148 in §1.6 and Theorem 5.4.3 in Chung (1974).

The next result is for sequences of independent but not necessarily identically distributed random variables.

Theorem 1.14. Let $X_1, X_2, \ldots$ be independent random variables with finite expectations.
(i) (The SLLN). If there is a constant $p \in [1, 2]$ such that

$$\sum_{i=1}^{\infty} \frac{E|X_i|^p}{i^p} < \infty, \tag{1.86}$$

then

$$\frac{1}{n} \sum_{i=1}^{n} (X_i - EX_i) \to_{a.s.} 0. \tag{1.87}$$

(ii) (The WLLN). If there is a constant $p \in [1, 2]$ such that

$$\lim_{n \to \infty} \frac{1}{n^p} \sum_{i=1}^{n} E|X_i|^p = 0, \tag{1.88}$$

then

$$\frac{1}{n} \sum_{i=1}^{n} (X_i - EX_i) \to_p 0. \tag{1.89}$$

Proof. (i) Consider again the truncated X_n: $Y_n = X_n I_{\{|X_n| \leq n\}}$, $n = 1, 2, \ldots$. Since $X_n^2 I_{\{|X_n| \leq n\}} \leq n^{2-p} |X_n|^p$,

$$\sum_{n=1}^{\infty} \frac{EY_n^2}{n^2} = \sum_{n=1}^{\infty} \frac{E(X_n^2 I_{\{|X_n| \leq n\}})}{n^2} \leq \sum_{n=1}^{\infty} \frac{E|X_n|^p}{n^p} < \infty.$$

It follows from the proof of Theorem 1.13(ii) that $n^{-1} \sum_{i=1}^{n} (Y_i - EY_i) \to_{a.s.} 0$. Also,

$$\sum_{n=1}^{\infty} P(X_n \neq Y_n) = \sum_{n=1}^{\infty} P(|X_n| > n) \leq \sum_{n=1}^{\infty} \frac{E|X_n|^p}{n^p} < \infty.$$

Hence, it follows from the proof of Theorem 1.13(ii) that (1.85) holds. Finally,

$$\sum_{n=1}^{\infty} \frac{|E(X_n - Y_n)|}{n} = \sum_{n=1}^{\infty} \frac{E(|X_n| I_{\{|X_n| > n\}})}{n} \leq \sum_{n=1}^{\infty} \frac{E|X_n|^p}{n^p} < \infty,$$

which together with Lemma 1.6 imply that $n^{-1} \sum_{i=1}^{n} |E(X_i - Y_i)| \to 0$ and thus (1.87) holds.
(ii) For any $\epsilon > 0$, an application of Chebyshev's inequality and inequality (1.44) leads to

$$P\left(\frac{1}{n} \left| \sum_{i=1}^{n} (X_i - EX_i) \right| > \epsilon \right) \leq \frac{C_p}{\epsilon^p n^p} \sum_{i=1}^{n} E|X_i - EX_i|^p,$$

which converges to 0 under (1.88). This proves (1.89). ∎

Note that (1.86) implies (1.88) (Lemma 1.6). The result in Theorem 1.14(i) is called Kolmogorov's SLLN when $p = 2$ and is due to Marcinkiewicz and Zygmund when $1 \leq p < 2$. An obvious sufficient condition for (1.86) with $p \in (1, 2]$ is $\sup_n E|X_n|^p < \infty$.

For dependent random variables, a result for Markov chains introduced in §1.4.4 is discussed in §4.1.4. We now consider martingales studied in §1.4.4. First, consider the WLLN. Inequality (1.44) still holds if the independence assumption of X_i's is replaced by the martingale assumption on the sequence $\{\sum_{i=1}^{n}(X_i - EX_i)\}$ (why?). Hence, from the proof of Theorem 1.14(ii) we conclude that (1.89) still holds if the independence assumption of X_i's in Theorem 1.14 is replaced by that $\{\sum_{i=1}^{n}(X_i - EX_i)\}$ is a martingale. A result similar to the SLLN in Theorem 1.14(i) can be established if the independence assumption of X_i's is replaced by that the sequence $\{\sum_{i=1}^{n}(X_i - EX_i)\}$ is a martingale and if condition (1.86) is replaced by

$$\sum_{n=2}^{\infty} \frac{E(|X_n|^p|X_1, ..., X_{n-1})}{n^p} < \infty \quad \text{a.s.,}$$

which is the same as (1.86) if X_i's are independent. The proof of this martingale SLLN and many other versions of WLLN and SLLN can be found in standard probability textbooks, for example, Chung (1974) and Loève (1977).

The WLLN and SLLN have many applications in probability and statistics. The following is an example. Other examples can be found in later chapters.

Example 1.32. Let f and g be continuous functions on $[0, 1]$ satisfying $0 \leq f(x) \leq Cg(x)$ for all x, where $C > 0$ is a constant. We now show that

$$\lim_{n \to \infty} \int_0^1 \int_0^1 \cdots \int_0^1 \frac{\sum_{i=1}^{n} f(x_i)}{\sum_{i=1}^{n} g(x_i)} dx_1 dx_2 \cdots dx_n = \frac{\int_0^1 f(x)dx}{\int_0^1 g(x)dx} \qquad (1.90)$$

(assuming that $\int_0^1 g(x)dx \neq 0$). Let $X_1, X_2, ...$ be i.i.d. random variables having the uniform distribution on $[0, 1]$. By Theorem 1.2, $E[f(X_1)] = \int_0^1 f(x)dx < \infty$ and $E[g(X_1)] = \int_0^1 g(x)dx < \infty$. By the SLLN (Theorem 1.13(ii)),

$$\frac{1}{n}\sum_{i=1}^{n} f(X_i) \to_{a.s.} E[f(X_1)],$$

and the same result holds when f is replaced by g. By Theorem 1.10(i),

$$\frac{\sum_{i=1}^{n} f(X_i)}{\sum_{i=1}^{n} g(X_i)} \to_{a.s.} \frac{E[f(X_1)]}{E[g(X_1)]}. \qquad (1.91)$$

Since the random variable on the left-hand side of (1.91) is bounded by C, result (1.90) follows from the dominated convergence theorem and the fact that the left-hand side of (1.90) is the expectation of the random variable on the left-hand side of (1.91). ∎

Moment inequalities introduced in §1.3.2 play important roles in proving convergence theorems. They can also be used to obtain convergence rates of tail probabilities of the form $P\left(|n^{-1}\sum_{i=1}^{n}(X_i - EX_i)| > t\right)$. For example, an application of the Esseen-von Bahr, Marcinkiewicz-Zygmund, and Chebyshev inequalities produces

$$P\left(\left|\frac{1}{n}\sum_{i=1}^{n}(X_i - EX_i)\right| > t\right) \leq \begin{cases} O(n^{1-p}) & \text{if } 1 < p < 2 \\ O(n^{-p/2}) & \text{if } p \geq 2 \end{cases}$$

for independent random variables $X_1, ..., X_n$ with $\sup_n E|X_n|^p < \infty$.

1.5.5 The central limit theorem

The WLLN and SLLN may not be useful in approximating the distributions of (normalized) sums of independent random variables. We need to use the *central limit theorem* (CLT), which plays a fundamental role in statistical asymptotic theory.

Theorem 1.15 (Lindeberg's CLT). Let $\{X_{nj}, j = 1, ..., k_n\}$ be independent random variables with $0 < \sigma_n^2 = \text{Var}(\sum_{j=1}^{k_n} X_{nj}) < \infty$, $n = 1, 2, ...,$ and $k_n \to \infty$ as $n \to \infty$. If

$$\sum_{j=1}^{k_n} E\left[(X_{nj} - EX_{nj})^2 I_{\{|X_{nj}-EX_{nj}|>\epsilon\sigma_n\}}\right] = o(\sigma_n^2) \quad \text{for any } \epsilon > 0, \quad (1.92)$$

then

$$\frac{1}{\sigma_n}\sum_{j=1}^{k_n}(X_{nj} - EX_{nj}) \to_d N(0, 1). \quad (1.93)$$

Proof. Considering $(X_{nj} - EX_{nj})/\sigma_n$, without loss of generality we may assume $EX_{nj} = 0$ and $\sigma_n^2 = 1$ in this proof. Let $t \in \mathcal{R}$ be given. From the inequality $|e^{\sqrt{-1}tx} - (1 + \sqrt{-1}tx - t^2x^2/2)| \leq \min\{|tx|^2, |tx|^3\}$, the ch.f. of X_{nj} satisfies

$$\left|\phi_{X_{nj}}(t) - (1 - t^2\sigma_{nj}^2/2)\right| \leq E\left(\min\{|tX_{nj}|^2, |tX_{nj}|^3\}\right), \quad (1.94)$$

where $\sigma_{nj}^2 = \text{Var}(X_{nj})$. For any $\epsilon > 0$, the right-hand side of (1.94) is bounded by $E(|tX_{nj}|^3 I_{\{|X_{nj}|<\epsilon\}}) + E(|tX_{nj}|^2 I_{\{|X_{nj}|\geq\epsilon\}})$, which is bounded by $\epsilon|t|^3\sigma_{nj}^2 + t^2 E(X_{nj}^2 I_{\{|X_{nj}|\geq\epsilon\}})$. Summing over j and using condition (1.92), we obtain that

$$\sum_{j=1}^{k_n}\left|\phi_{X_{nj}}(t) - \left(1 - t^2\sigma_{nj}^2/2\right)\right| \to 0. \tag{1.95}$$

By condition (1.92), $\max_{j\leq k_n} \sigma_{nj}^2 \leq \epsilon^2 + \max_{j\leq k_n} E(X_{nj}^2 I_{\{|X_{nj}|>\epsilon\}}) \to \epsilon^2$ for arbitrary $\epsilon > 0$. Hence

$$\lim_{n\to\infty} \max_{j\leq k_n} \frac{\sigma_{nj}^2}{\sigma_n^2} = 0. \tag{1.96}$$

(Note that $\sigma_n^2 = 1$ is assumed for convenience.) This implies that $1 - t^2\sigma_{nj}^2$ are all between 0 and 1 for large enough n. Using the inequality

$$|a_1\cdots a_m - b_1\cdots b_m| \leq \sum_{j=1}^{m}|a_j - b_j|$$

for any complex numbers a_j's and b_j's with $|a_j| \leq 1$ and $|b_j| \leq 1$, $j = 1, ..., m$, we obtain that

$$\left|\prod_{j=1}^{k_n} e^{-t^2\sigma_{nj}^2/2} - \prod_{j=1}^{k_n}\left(1 - t^2\sigma_{nj}^2/2\right)\right| \leq \sum_{j=1}^{k_n}\left|e^{-t^2\sigma_{nj}^2/2} - \left(1 - t^2\sigma_{nj}^2/2\right)\right|,$$

which is bounded by $t^4\sum_{j=1}^{k_n}\sigma_{nj}^4 \leq t^4 \max_{j\leq k_n}\sigma_{nj}^2 \to 0$, since $|e^x - 1 - x| \leq x^2/2$ if $|x| \leq \frac{1}{2}$ and $\sum_{j=1}^{k_n}\sigma_{nj}^2 = \sigma_n^2 = 1$. Also,

$$\left|\prod_{j=1}^{k_n}\phi_{X_{nj}}(t) - \prod_{j=1}^{k_n}\left(1 - t^2\sigma_{nj}^2/2\right)\right|$$

is bounded by the quantity on the left-hand side of (1.95) and, hence, converges to 0 by (1.95). Thus,

$$\prod_{j=1}^{k_n}\phi_{X_{nj}}(t) = \prod_{j=1}^{k_n} e^{-t^2\sigma_{nj}^2/2} + o(1) = e^{-t^2/2} + o(1).$$

This shows that the ch.f. of $\sum_{j=1}^{k_n} X_{nj}$ converges to the ch.f. of $N(0,1)$ for every t. By Theorem 1.9(ii), the result follows. ∎

Condition (1.92) is called Lindeberg's condition. From the proof of Theorem 1.15, Lindeberg's condition implies (1.96), which is called Feller's condition. Feller's condition (1.96) means that all terms in the sum $\sigma_n^2 =$

$\sum_{j=1}^{k_n} \sigma_{nj}^2$ are uniformly negligible as $n \to \infty$. If Feller's condition is assumed, then Lindeberg's condition is not only sufficient but also necessary for result (1.93), which is the well-known Lindeberg-Feller CLT. A proof can be found in Billingsley (1986, pp. 373-375). Note that neither Lindeberg's condition nor Feller's condition is necessary for result (1.93) (Exercise 158).

A sufficient condition for Lindeberg's condition is the following Liapounov's condition, which is somewhat easier to verify:

$$\sum_{j=1}^{k_n} E|X_{nj} - EX_{nj}|^{2+\delta} = o(\sigma_n^{2+\delta}) \quad \text{for some } \delta > 0. \tag{1.97}$$

Example 1.33. Let $X_1, X_2, \ldots$ be independent random variables. Suppose that X_i has the binomial distribution $Bi(p_i, 1)$, $i = 1, 2, \ldots$, and that $\sigma_n^2 = \sum_{i=1}^{n} \text{Var}(X_i) = \sum_{i=1}^{n} p_i(1 - p_i) \to \infty$ as $n \to \infty$. For each i, $EX_i = p_i$ and $E|X_i - EX_i|^3 = (1 - p_i)^3 p_i + p_i^3(1 - p_i) \le 2p_i(1 - p_i)$. Hence $\sum_{i=1}^{n} E|X_i - EX_i|^3 \le 2\sigma_n^2$, i.e., Liapounov's condition (1.97) holds with $\delta = 1$. Thus, by Theorem 1.15,

$$\frac{1}{\sigma_n} \sum_{i=1}^{n} (X_i - p_i) \to_d N(0, 1). \tag{1.98}$$

It can be shown (exercise) that the condition $\sigma_n \to \infty$ is also necessary for result (1.98). ∎

The following are useful corollaries of Theorem 1.15 (and Theorem 1.9(iii)). Corollary 1.2 is in fact proved in Example 1.28. The proof of Corollary 1.3 is left as an exercise.

Corollary 1.2 (Multivariate CLT). Let $X_1, \ldots, X_n$ be i.i.d. random k-vectors with a finite $\Sigma = \text{Var}(X_1)$. Then

$$\frac{1}{\sqrt{n}} \sum_{i=1}^{n} (X_i - EX_1) \to_d N_k(0, \Sigma). \quad ∎$$

Corollary 1.3. Let $X_{ni} \in \mathcal{R}^{m_i}$, $i = 1, \ldots, k_n$, be independent random vectors with $m_i \le m$ (a fixed integer), $n = 1, 2, \ldots$, $k_n \to \infty$ as $n \to \infty$, and $\inf_{i,n} \lambda_-[\text{Var}(X_{ni})] > 0$, where $\lambda_-[A]$ is the smallest eigenvalue of A. Let $c_{ni} \in \mathcal{R}^{m_i}$ be vectors such that

$$\lim_{n \to \infty} \left(\max_{1 \le i \le k_n} \|c_{ni}\|^2 \Big/ \sum_{i=1}^{k_n} \|c_{ni}\|^2 \right) = 0.$$

(i) Suppose that $\sup_{i,n} E\|X_{ni}\|^{2+\delta} < \infty$ for some $\delta > 0$. Then

$$\sum_{i=1}^{k_n} c_{ni}^\tau (X_{ni} - EX_{ni}) \bigg/ \left[\sum_{i=1}^{k_n} \text{Var}(c_{ni}^\tau X_{ni})\right]^{1/2} \to_d N(0,1). \qquad (1.99)$$

(ii) Suppose that whenever $m_i = m_j$, $1 \le i < j \le k_n$, $n = 1, 2, ...$, X_{ni} and X_{nj} have the same distribution with $E\|X_{ni}\|^2 < \infty$. Then (1.99) holds. ∎

Applications of these corollaries can be found in later chapters.

An extension of Lindeberg's CLT is the so-called martingale CLT. In Theorem 1.15, if the independence assumption of X_{nj}, $j = 1, ..., k_n$, is replaced by that $\{Y_n\}$ is a martingale and

$$\frac{1}{\sigma_n^2} \sum_{j=1}^{k_n} E[(X_{nj} - EX_{nj})^2 | X_{n1}, ..., X_{n(j-1)}] \to_p 1,$$

where $Y_n = \sum_{j=1}^{k_n}(X_{nj} - EX_{nj})$ when $n \le k_n$, $Y_n = Y_{k_n}$ when $n > k_n$, and X_{n0} is defined to be 0, then result (1.93) still holds (see, e.g., Billingsley, 1986, p. 498 and Sen and Singer 1993, p. 120).

More results on the CLT can be found, for example, in Serfling (1980) and Shorack and Wellner (1986).

Let Y_n be a sequence of random variables, $\{\mu_n\}$ and $\{\sigma_n\}$ be sequences of real numbers such that $\sigma_n > 0$ for all n, and $(Y_n - \mu_n)/\sigma_n \to_d N(0,1)$. Then, by Proposition 1.16,

$$\lim_{n\to\infty} \sup_x |F_{(Y_n-\mu_n)/\sigma_n}(x) - \Phi(x)| = 0, \qquad (1.100)$$

where Φ is the c.d.f. of $N(0,1)$. This implies that for any sequence of real numbers $\{c_n\}$, $\lim_{n\to\infty} |P(Y_n \le c_n) - \Phi(\frac{c_n-\mu_n}{\sigma_n})| = 0$, i.e., $P(Y_n \le c_n)$ can be approximated by $\Phi(\frac{c_n-\mu_n}{\sigma_n})$, regardless of whether $\{c_n\}$ has a limit. Since $\Phi(\frac{t-\mu_n}{\sigma_n})$ is the c.d.f. of $N(\mu_n, \sigma_n^2)$, Y_n is said to be *asymptotically distributed* as $N(\mu_n, \sigma_n^2)$ or simply *asymptotically normal*. For example, $\sum_{i=1}^{k_n} c_{ni}^\tau X_{ni}$ in Corollary 1.3 is asymptotically normal. This can be extended to random vectors. For example, $\sum_{i=1}^n X_i$ in Corollary 1.2 is asymptotically distributed as $N_k(nEX_1, n\Sigma)$.

1.5.6 Edgeworth and Cornish-Fisher expansions

Let $\{Y_n\}$ be a sequence of random variables satisfying (1.100) and $W_n = (Y_n - \mu_n)/\sigma_n$. The convergence speed of (1.100) can be used to assess whether Φ provides a good approximation to the c.d.f. F_{W_n}. Also, sometimes we would like to find an approximation to F_{W_n} that is better than

Φ in terms of convergence speed. The *Edgeworth expansion* is a useful tool for these purposes.

To illustrate the idea, let $W_n = n^{-1/2} \sum_{i=1}^{n} (X_i - \mu)/\sigma$, where $X_1, X_2, ...$ are i.i.d. random variables with $EX_1 = \mu$ and $\text{Var}(X_1) = \sigma^2$. Assume that the m.g.f. of $Z = (X_1 - \mu)/\sigma$ is finite and positive in a neighborhood of 0. From (1.55), the cumulant generating function of Z has the expansion

$$\kappa(t) = \sum_{j=1}^{\infty} \frac{\kappa_j}{j!} t^j,$$

where κ_j, $j = 1, 2, ...$, are cumulants of Z (e.g., $\kappa_1 = 0$, $\kappa_2 = 1$, $\kappa_3 = EZ^3$, and $\kappa_4 = EZ^4 - 3$), and the m.g.f. of W_n is equal to

$$\psi_n(t) = \left[\exp\{\kappa(t/\sqrt{n})\} \right]^n = \exp\left\{ \frac{t^2}{2} + \sum_{j=3}^{\infty} \frac{\kappa_j t^j}{j! n^{(j-2)/2}} \right\},$$

where $\exp\{x\}$ denotes the exponential function e^x. Using the series expansion for $e^{t^2/2}$, we obtain that

$$\psi_n(t) = e^{t^2/2} + n^{-1/2} r_1(t) e^{t^2/2} + \cdots + n^{-j/2} r_j(t) e^{t^2/2} + \cdots, \quad (1.101)$$

where r_j is a polynomial of degree $3j$ depending on $\kappa_3, ..., \kappa_{j+2}$ but not on n, $j = 1, 2, ...$ For example, it can be shown (exercise) that

$$r_1(t) = \tfrac{1}{6} \kappa_3 t^3 \quad \text{and} \quad r_2(t) = \tfrac{1}{24} \kappa_4 t^4 + \tfrac{1}{72} \kappa_3^2 t^6. \quad (1.102)$$

Since $\psi_n(t) = \int e^{tx} dF_{W_n}(x)$ and $e^{t^2/2} = \int e^{tx} d\Phi(x)$, expansion (1.101) suggests the inverse expansion

$$F_{W_n}(x) = \Phi(x) + n^{-1/2} R_1(x) + \cdots + n^{-j/2} R_j(x) + \cdots,$$

where $R_j(x)$ is a function satisfying $\int e^{tx} dR_j(x) = r_j(t) e^{t^2/2}$, $j = 1, 2, ...$ Let $\nabla^j = \frac{d^j}{dx^j}$ be the differential operator and $\nabla = \nabla^1$. Then $R_j(x) = r_j(-\nabla)\Phi(x)$, $j = 1, 2, ...$, where $r_j(-\nabla)$ is interpreted as a differential operator. Thus, R_j's can be obtained once r_j's are derived. It follows from (1.102) (exercise) that

$$R_1(x) = -\tfrac{1}{6} \kappa_3 (x^2 - 1) \Phi'(x) \quad (1.103)$$

and

$$R_2(x) = -[\tfrac{1}{24} \kappa_4 x(x^2 - 3) + \tfrac{1}{72} \kappa_3^2 x(x^4 - 10x^2 + 15)] \Phi'(x). \quad (1.104)$$

A rigorous statement of the Edgeworth expansion for a more general W_n is given in the following theorem whose proof can be found in Hall (1992).

Theorem 1.16 (Edgeworth expansions). Let m be a positive integer and $X_1, X_2, \ldots$ be i.i.d. random k-vectors having finite $m+2$ moments. Consider $W_n = \sqrt{n}h(\bar{X})/\sigma_h$, where $\bar{X} = n^{-1}\sum_{i=1}^n X_i$, h is a Borel function on $\mathcal{R}^k$ that is $m + 2$ times continuously differentiable in a neighborhood of $\mu = EX_1$, $h(\mu) = 0$, and $\sigma_h^2 = [\nabla h(\mu)]^\tau \text{Var}(X_1)\nabla h(\mu) > 0$. Assume that

$$\limsup_{\|t\|\to\infty} |\phi_{X_1}(t)| < 1, \tag{1.105}$$

where ϕ_{X_1} is the ch.f. of X_1. Then, F_{W_n} admits the Edgeworth expansion

$$\sup_x \left| F_{W_n}(x) - \Phi(x) - \sum_{j=1}^m \frac{p_j(x)\Phi'(x)}{n^{j/2}} \right| = o\left(\frac{1}{n^{m/2}}\right), \tag{1.106}$$

where $p_j(x)$ is a polynomial of degree at most $3j - 1$, odd for even j and even for odd j, with coefficients depending on the first $m + 2$ moments of X_1, $j = 1, \ldots, m$. In particular,

$$p_1(x) = -c_1\sigma_h^{-1} + 6^{-1}c_2\sigma_h^{-3}(x^2 - 1) \tag{1.107}$$

with $c_1 = 2^{-1}\sum_{i=1}^k \sum_{j=1}^k a_{ij}\mu_{ij}$ and $c_2 = \sum_{i=1}^k \sum_{j=1}^k \sum_{l=1}^k a_i a_j a_l \mu_{ijl} + 3\sum_{i=1}^k \sum_{j=1}^k \sum_{l=1}^k \sum_{h=1}^k a_i a_j a_{lh}\mu_{il}\mu_{jh}$, where a_i is the ith component of $\nabla h(\mu)$, a_{ij} is the (i,j)th element of the Hessian matrix $\nabla^2 h(\mu)$, $\mu_{ij} = E(Y_iY_j)$, $\mu_{ijl} = E(Y_iY_jY_l)$, and Y_i is the ith component of $X_1 - \mu$. ∎

Condition (1.105) is Cramér's continuity condition. It is satisfied if one component of X_1 has a Lebesgue p.d.f. The polynomial p_j with $j \geq 2$ may be derived using the method in deriving (1.103) and (1.104), but the derivation is usually complicated (see Hall (1992)).

Under the conditions of Theorem 1.16, the convergence speed of (1.100) is $O(n^{-1/2})$ and, as an approximation to F_{W_n}, $\Phi + \sum_{j=1}^m n^{-j/2}p_j\Phi'$ is better than Φ, since its convergence speed is $o(n^{-m/2})$.

The results in Theorem 1.16 can be applied to many cases, as the following example indicates.

Example 1.34. Let $\bar{X} = n^{-1}\sum_{i=1}^n X_i$ with i.i.d. random variables $X_1, X_2,$ $\ldots$ satisfying condition (1.105). First, consider the normalized random variable $W_n = \sqrt{n}(\bar{X} - \mu)/\sigma$, where $\mu = EX_1$ and $\sigma^2 = \text{Var}(X_1)$. Then, Theorem 1.16 can be applied with $h(x) = x - \mu$ and $\sigma_h^2 = \sigma^2$, and the Edgeworth expansion in (1.106) holds if $E|X_1|^{m+2} < \infty$. In this case, results (1.103) and (1.104) imply that $p_j(x) = R_j(x)/\Phi'(x)$, $j = 1, 2$.

Next, consider the studentized random variable $W_n = \sqrt{n}(\bar{X} - \mu)/\hat{\sigma}$, where $\hat{\sigma}^2 = n^{-1}\sum_{i=1}^n (X_i - \bar{X})^2$. Assuming that $EX_1^{2m+4} < \infty$ and applying Theorem 1.16 to random vectors (X_i, X_i^2), $i = 1, 2, \ldots$, and $h(x, y) =$

$(x-\mu)/\sqrt{(y-x^2)}$, we obtain the Edgeworth expansion (1.106) with $\sigma_h = 1$,

$$p_1(x) = \tfrac{1}{6}\kappa_3(2x^2 + 1)$$

(exercise). Furthermore, it can be found in Hall (1992, p. 73) that

$$p_2(x) = \tfrac{1}{12}\kappa_4 x(x^2 - 3) - \tfrac{1}{18}\kappa_3^2 x(x^4 + 2x^2 - 3) - \tfrac{1}{4}x(x^2 + 3).$$

Consider now the random variable $\sqrt{n}(\hat{\sigma}^2 - \sigma^2)$. Theorem 1.16 can be applied to random vectors (X_i, X_i^2), $i = 1, 2, ...$, and $h(x, y) = (y - x^2 - \sigma^2)$. Assume that $EX_1^{2m+4} < \infty$. It can be shown (exercise) that the Edgeworth expansion in (1.106) holds with $W_n = \sqrt{n}(\hat{\sigma}^2 - \sigma^2)/\sigma_h$, $\sigma_h^2 = E(X_1 - \mu)^4 - \sigma^4$, and

$$p_1(x) = (\nu_4 - 1)^{-1/2}[1 - \tfrac{1}{6}(\nu_4 - 1)^{-1}(\nu_6 - 3\nu_4 - 6\nu_3^2 + 2)(x^2 - 1)],$$

where $\nu_j = \sigma^{-j}E(X_1 - \mu)^j$, $j = 3, ..., 6$.

Finally, consider the studentized random variable $W_n = \sqrt{n}(\hat{\sigma}^2 - \sigma^2)/\hat{\tau}$, where $\hat{\tau}^2 = n^{-1}\sum_{i=1}^{n}(X_i - \bar{X})^4 - \hat{\sigma}^4$. Theorem 1.16 can be applied to random vectors $(X_i, X_i^2, X_i^3, X_i^4)$, $i = 1, 2, ...$, and

$$h(x, y, z, w) = (y - x^2 - \sigma^2)[w - y^2 - 4xz + 8x^2y - 4x^4]^{-1/2}.$$

Assume that $EX_1^{4m+8} < \infty$. It can be shown (exercise) that the Edgeworth expansion in (1.106) holds with $\sigma_h^2 = 1$ and

$$p_1(x) = -(\nu_4-1)^{-3/2}[\tfrac{1}{2}(4\nu_3^2+\nu_4-\nu_6) + \tfrac{1}{3}(3\nu_3^2+3\nu_4-\nu_6-2)(x^2-1)]. \quad \blacksquare$$

An inverse Edgeworth expansion is referred to as a *Cornish-Fisher expansion*, which is useful in statistics (see §7.4). For $\alpha \in (0, 1)$, let $z_\alpha = \Phi^{-1}(\alpha)$. Since the c.d.f. F_{W_n} may not be strictly increasing and continuous, we define $w_{n\alpha} = \inf\{x : F_{W_n}(x) \geq \alpha\}$. The following result can be found in Hall (1992).

Theorem 1.17 (Cornish-Fisher expansions). Under the conditions of Theorem 1.16, $w_{n\alpha}$ admits the Cornish-Fisher expansion

$$\sup_{\epsilon<\alpha<1-\epsilon}\left|w_{n\alpha} - z_\alpha - \sum_{j=1}^{m}\frac{q_j(z_\alpha)}{n^{j/2}}\right| = o\left(\frac{1}{n^{m/2}}\right), \tag{1.108}$$

where ϵ is any constant in $(0, \tfrac{1}{2})$ and q_j's are polynomials depending on p_j's in (1.106). $\quad \blacksquare$

The polynomials in (1.108) can be determined using results (1.106) and (1.108). We illustrate it by deriving q_1 and q_2. Without loss of generality, assume that $F_{W_n}(w_{n\alpha}) = \alpha$ (why?). Using (1.106), (1.108), Taylor's expansions at z_α for $\Phi(w_{n\alpha})$, $p_1(w_{n\alpha})\Phi'(w_{n\alpha})$, and $p_2(w_{n\alpha})\Phi'(w_{n\alpha})$, and the fact that $\Phi''(x) = -x\Phi'(x)$, we obtain that

$$
\begin{aligned}
\alpha &= \Phi(w_{n\alpha}) + n^{-1/2}p_1(w_{n\alpha})\Phi'(w_{n\alpha}) + n^{-1}p_2(w_{n\alpha})\Phi'(w_{n\alpha}) \\
&= \Phi(z_\alpha) + \{n^{-1/2}q_1(z_\alpha) + n^{-1}q_2(z_\alpha) - \tfrac{1}{2}[n^{-1/2}q_1(z_\alpha)]^2 z_\alpha\}\Phi'(z_\alpha) \\
&\quad + n^{-1/2}\{p_1(z_\alpha) + n^{-1/2}q_1(z_\alpha)[p_1'(z_\alpha) - z_\alpha p_1(z_\alpha)]\}\Phi'(z_\alpha) \\
&\quad + n^{-1}p_2(z_\alpha)\Phi'(z_\alpha) + o(n^{-1}) \\
&= \alpha + n^{-1/2}[q_1(z_\alpha) + p_1(z_\alpha)]\Phi'(z_\alpha) + n^{-1}\{q_2(z_\alpha) - \tfrac{1}{2}z_\alpha[q_1(z_\alpha)]^2 \\
&\quad + q_1(z_\alpha)[p_1'(z_\alpha) - z_\alpha p_1(z_\alpha)] + p_2(z_\alpha)\}\Phi'(z_\alpha) + o(n^{-1}).
\end{aligned}
$$

Ignoring terms of order $o(n^{-1})$, we conclude that

$$
q_1(x) = -p_1(x)
$$

and

$$
q_2(x) = p_1(x)p_1'(x) - \tfrac{1}{2}x[p_1(x)]^2 - p_2(x).
$$

Edgeworth and Cornish-Fisher expansions for W_n in Theorem 1.16 based on non-i.i.d. X_i's or for other random variables can be found in Hall (1992), Barndorff-Nielsen and Cox (1994), and Shao and Tu (1995).

1.6 Exercises

1. Let A and B be two nonempty proper subsets of a sample space Ω, $A \neq B$ and $A \cap B \neq \emptyset$. Obtain $\sigma(\{A, B\})$, the smallest σ-field containing A and B.

2. Let $\mathcal{C}$ be a collection of subsets of Ω and let $\Gamma = \{\mathcal{F} : \mathcal{F}$ is a σ-field on Ω and $\mathcal{C} \subset \mathcal{F}\}$. Show that $\Gamma \neq \emptyset$ and $\sigma(\mathcal{C}) = \cap_{\mathcal{F} \in \Gamma}\mathcal{F}$.

3. Let $(\Omega, \mathcal{F}_j)$, $j = 1, 2, \ldots$, be measurable spaces such that $\mathcal{F}_j \subset \mathcal{F}_{j+1}$, $j = 1, 2, \ldots$. Is $\cup_j \mathcal{F}_j$ a σ-field?

4. Let $\mathcal{C}$ be the collection of intervals of the form $(a, b]$, where $-\infty < a < b < \infty$, and let $\mathcal{D}$ be the collection of closed sets on $\mathcal{R}$. Show that $\mathcal{B} = \sigma(\mathcal{C}) = \sigma(\mathcal{D})$, where $\mathcal{B}$ is the Borel σ-field on $\mathcal{R}$.

5. (π- and λ-systems). A class $\mathcal{C}$ of subsets of Ω is a π-system if and only if $A \in \mathcal{C}$ and $B \in \mathcal{C}$ imply $A \cap B \in \mathcal{C}$. A class $\mathcal{D}$ of subsets of Ω is a λ-system if and only if (i) $\Omega \in \mathcal{D}$, (ii) $A \in \mathcal{D}$ implies $A^c \in \mathcal{D}$, and (iii) $A_j \in \mathcal{D}$, $j = 1, 2, \ldots$, and A_i's are disjoint imply that $\cup_j A_j \in \mathcal{D}$. (a) Show that if $\mathcal{C}$ is a π-system and $\mathcal{D}$ is a λ-system, then $\mathcal{C} \subset \mathcal{D}$ implies $\sigma(\mathcal{C}) \subset \mathcal{D}$.

(b) Show that $\mathcal{D}$ is a λ-system if and only if the following conditions hold: (i) $\Omega \in \mathcal{D}$, (ii) $A \in \mathcal{D}$, $B \in \mathcal{D}$, and $A \subset B$ imply $A^c \cap B \in \mathcal{D}$, and (iii) $A_j \in \mathcal{D}$ and $A_j \subset A_{j+1}$, $j = 1, 2, \ldots$, imply $\cup_j A_j \in \mathcal{D}$.

6. Prove part (ii) and part (iii) of Proposition 1.1.

7. Let ν_i, $i = 1, 2, \ldots$, be measures on $(\Omega, \mathcal{F})$ and a_i, $i = 1, 2, \ldots$, be positive numbers. Show that $a_1 \nu_1 + a_2 \nu_2 + \cdots$ is a measure on $(\Omega, \mathcal{F})$.

8. Let $\{A_n\}$ be a sequence of events on a probability space $(\Omega, \mathcal{F}, P)$. Define $\limsup_n A_n = \cap_{n=1}^{\infty} \cup_{i=n}^{\infty} A_i$ and $\liminf_n A_n = \cup_{n=1}^{\infty} \cap_{i=n}^{\infty} A_i$. Show that $P(\liminf_n A_n) \leq \liminf_n P(A_n)$ and $\limsup_n P(A_n) \leq P(\limsup_n A_n)$.

9. Prove Proposition 1.2.

10. Let $F(x_1, \ldots, x_k)$ be a c.d.f. on $\mathcal{R}^k$. Show that
    (a) $F(x_1, \ldots, x_{k-1}, x_k) \leq F(x_1, \ldots, x_{k-1}, x_k')$ if $x_k \leq x_k'$.
    (b) $\lim_{x_i \to -\infty} F(x_1, \ldots, x_k) = 0$ for any $1 \leq i \leq k$.
    (c) $F(x_1, \ldots, x_{k-1}, \infty) = \lim_{x_k \to \infty} F(x_1, \ldots, x_{k-1}, x_k)$ is a c.d.f. on $\mathcal{R}^{k-1}$.

11. Let $(\Omega_i, \mathcal{F}_i) = (\mathcal{R}, \mathcal{B})$, $i = 1, \ldots, k$. Show that the product σ-field $\sigma(\mathcal{F}_1 \times \cdots \times \mathcal{F}_k)$ is the σ-field generated by all open sets in $\mathcal{R}^k$.

12. Let ν and λ be two measures on $(\Omega, \mathcal{F})$ such that $\nu(A) = \lambda(A)$ for any $A \in \mathcal{C}$, where $\mathcal{C} \subset \mathcal{F}$ and $\mathcal{C}$ is a π-system (i.e., if A and B are in $\mathcal{C}$, then so is $A \cap B$). Assume that there are $A_i \in \mathcal{C}$, $i = 1, 2, \ldots$, such that $\cup A_i = \Omega$ and $\nu(A_i) < \infty$ for all i. Show that $\nu(A) = \lambda(A)$ for any $A \in \sigma(\mathcal{C})$. This proves the uniqueness part of Proposition 1.3. (Hint: show that $\{A \in \sigma(\mathcal{C}) : \nu(A) = \lambda(A)\}$ is a σ-field.)

13. Let f be a function from Ω to Λ. Show that
    (a) $f^{-1}(B^c) = (f^{-1}(B))^c$ and $f^{-1}(\cup B_i) = \cup f^{-1}(B_i)$;
    (b) $\sigma(f^{-1}(\mathcal{C})) = f^{-1}(\sigma(\mathcal{C}))$, where $\mathcal{C}$ is a collection of subsets of Λ.

14. Prove Proposition 1.4.

15. Show that a monotone function from $\mathcal{R}$ to $\mathcal{R}$ is Borel and a c.d.f. on $\mathcal{R}^k$ is Borel.

16. Let f be a function from $(\Omega, \mathcal{F})$ to $(\Lambda, \mathcal{G})$ and $A_1, A_2, \ldots$ be disjoint events in $\mathcal{F}$ such that $\cup A_i = \Omega$. Let f_n be a function from $(A_n, \mathcal{F}_{A_n})$ to $(\Lambda, \mathcal{G})$ such that $f_n(\omega) = f(\omega)$ for any $\omega \in A_n$, $n = 1, 2, \ldots$. Show that f is measurable from $(\Omega, \mathcal{F})$ to $(\Lambda, \mathcal{G})$ if and only if f_n is measurable from $(A_n, \mathcal{F}_{A_n})$ to $(\Lambda, \mathcal{G})$ for each n.

17. Let f be a nonnegative Borel function on $(\Omega, \mathcal{F})$. Show that f is the limit of a sequence of simple functions $\{\varphi_n\}$ on $(\Omega, \mathcal{F})$ with $0 \leq \varphi_1 \leq \varphi_2 \leq \cdots \leq f$.

18. Let $\prod_{i=1}^{k}(\Omega_i, \mathcal{F}_i)$ be a product measurable space.
    (a) Let π_i be the ith *projection*, i.e., $\pi_i(\omega_1, ..., \omega_k) = \omega_i$, $\omega_i \in \Omega_i$, $i = 1, ..., k$. Show that $\pi_1, ..., \pi_k$ are measurable.
    (b) Let f be a function on $\prod_{i=1}^{k} \Omega_i$ and $g_i(\omega_i) = f(\omega_1, ..., \omega_i, ..., \omega_k)$, where ω_j is a fixed point in Ω_j, $j = 1, ..., k$ but $j \neq i$, and $i = 1, ..., k$. Show that if f is Borel on $\prod_{i=1}^{k}(\Omega_i, \mathcal{F}_i)$, then $g_1, ..., g_k$ are Borel.
    (c) In part (b), is it true that f is Borel if $g_1, ..., g_k$ are Borel?

19. Let $\{f_n\}$ be a sequence of Borel functions on a measurable space. Show that
    (a) $\sigma(f_1, f_2, ...) = \sigma\left(\cup_{j=1}^{\infty}\sigma(f_j)\right) = \sigma\left(\cup_{j=1}^{\infty}\sigma(f_1, ..., f_j)\right)$;
    (b) $\sigma(\limsup_n f_n) \subset \cap_{n=1}^{\infty}\sigma(f_n, f_{n+1}, ...)$.

20. (Egoroff's theorem). Suppose that $\{f_n\}$ is a sequence of Borel functions on a measure space $(\Omega, \mathcal{F}, \nu)$ and $f_n(\omega) \to f(\omega)$ for $\omega \in A$ with $\nu(A) < \infty$. Show that for any $\epsilon > 0$, there is a $B \subset A$ with $\nu(B) < \epsilon$ such that $f_n(\omega) \to f(\omega)$ uniformly on $A \cap B^c$.

21. Prove (1.14) in Example 1.5.

22. Prove Proposition 1.5 and Proposition 1.6(i).

23. Let ν_i, $i = 1, 2$, be measures on $(\Omega, \mathcal{F})$ and f be Borel. Show that
$$\int f d(\nu_1 + \nu_2) = \int f d\nu_1 + \int f d\nu_2,$$
i.e., if either side of the equality is well defined, then so is the other side, and the two sides are equal.

24. Let f be an integrable Borel function on $(\Omega, \mathcal{F}, \nu)$. Show that for each $\epsilon > 0$, there is a δ_ϵ such that $\nu(A) < \delta_\epsilon$ and $A \in \mathcal{F}$ imply $\int_A |f| d\nu < \epsilon$.

25. Prove that part (i) and part (iii) of Theorem 1.1 are equivalent.

26. Prove Theorem 1.2.

27. Prove Theorem 1.3. (Hint: first consider simple nonnegative f.)

28. Consider Example 1.9. Show that (1.17) does not hold for
$$f(i, j) = \begin{cases} 1 & i = j \\ -1 & i = j - 1 \\ 0 & \text{otherwise.} \end{cases}$$
Does this contradict Fubini's theorem?

29. Let f be a nonnegative Borel function on $(\Omega, \mathcal{F}, \nu)$ with a σ-finite ν, $A = \{(\omega, x) \in \Omega \times \mathcal{R} : 0 \leq x \leq f(\omega)\}$, and m be the Lebesgue measure on $(\mathcal{R}, \mathcal{B})$. Show that $A \in \sigma(\mathcal{F} \times \mathcal{B})$ and $\int_\Omega f d\nu = \nu \times m(A)$.

30. For any c.d.f. F and any $a \geq 0$, show that $\int [F(x+a) - F(x)]dx = a$.

31. (Integration by parts). Let F and G be two c.d.f.'s on $\mathcal{R}$. Show that if F and G have no common points of discontinuity in the interval $(a, b]$, then $\int_{(a,b]} G(x)dF(x) = F(b)G(b) - F(a)G(a) - \int_{(a,b]} F(x)dG(x)$.

32. Let f be a Borel function on $\mathcal{R}^2$ such that $f(x, y) = 0$ for each $x \in \mathcal{R}$ and $y \notin C_x$, where $m(C_x) = 0$ for each x and m is the Lebesgue measure. Show that $f(x, y) = 0$ for each $y \notin C$ and $x \notin B_y$, where $m(C) = 0$ and $m(B_y) = 0$ for each $y \notin C$.

33. Consider Example 1.11. Show that if (1.21) holds, then $P(A) = \int_A f(x)dx$ for any Borel set A. (Hint: $\mathcal{A} = \{A : P(A) = \int_A f(x)dx\}$ is a σ-field containing all sets of the form $(-\infty, x]$.)

34. Prove Proposition 1.7.

35. Let $\{a_n\}$ be a sequence of positive numbers satisfying $\sum_{n=1}^\infty a_n = 1$ and let $\{P_n\}$ be a sequence of probability measures on a common measurable space. Define $P = \sum_{n=1}^\infty a_n P_n$.
   (a) Show that P is a probability measure.
   (b) Show that $P_n \ll \nu$ for all n and a measure ν if and only if $P \ll \nu$ and, when $P \ll \nu$ and ν is σ-finite, $\frac{dP}{d\nu} = \sum_{n=1}^\infty a_n \frac{dP_n}{d\nu}$.
   (c) Derive the Lebesgue p.d.f. of P when P_n is the gamma distribution $\Gamma(\alpha, n^{-1})$ (Table 1.2) with $\alpha > 1$ and a_n is proportional to $n^{-\alpha}$.

36. Let F_i be a c.d.f. having a Lebesgue p.d.f. f_i, $i = 1, 2$. Assume that there is a $c \in \mathcal{R}$ such that $F_1(c) < F_2(c)$. Define

$$F(x) = \begin{cases} F_1(x) & -\infty < x < c \\ F_2(x) & c \leq x < \infty. \end{cases}$$

Show that the probability measure P corresponding to F satisfies $P \ll m + \delta_c$ and find $dP/d(m + \delta_c)$, where $m + \delta_c$ is given in (1.23).

37. Let (X, Y) be a random 2-vector with the following Lebesgue p.d.f.:

$$f(x, y) = \begin{cases} 8xy & 0 \leq x \leq y \leq 1 \\ 0 & \text{otherwise.} \end{cases}$$

Find the marginal p.d.f.'s of X and Y. Are X and Y independent?

38. Let (X, Y, Z) be a random 3-vector with the following Lebesgue p.d.f.:

$$f(x, y, z) = \begin{cases} \frac{1 - \sin x \sin y \sin z}{8\pi^3} & 0 \leq x \leq 2\pi, 0 \leq y \leq 2\pi, 0 \leq z \leq 2\pi \\ 0 & \text{otherwise.} \end{cases}$$

Show that X, Y, and Z are not independent, but are pairwise independent.

39. Prove Lemma 1.1 without using Definition 1.7 for independence.

40. Let X be a random variable having a continuous c.d.f. F. Show that $Y = F(X)$ has the uniform distribution $U(0,1)$ (Table 1.2).

41. Let U be a random variable having the uniform distribution $U(0,1)$ and let F be a c.d.f. Show that the c.d.f. of $Y = F^{-1}(U)$ is F, where $F^{-1}(t) = \inf\{x \in \mathcal{R} : F(x) \geq t\}$.

42. Prove Proposition 1.8.

43. Let $X = N_k(\mu, \Sigma)$ with a positive definite Σ.
    (a) Let $Y = AX + c$, where A is an $l \times k$ matrix of rank $l \leq k$ and $c \in \mathcal{R}^l$. Show that Y has the $N_l(A\mu + c, A\Sigma A^\tau)$ distribution.
    (b) Show that the components of X are independent if and only if Σ is a diagonal matrix.
    (c) Let Λ be positive definite and $Y = N_m(\eta, \Lambda)$ be independent of X. Show that (X, Y) has the $N_{k+m}((\mu, \eta), D)$ distribution, where D is a block diagonal matrix whose two diagonal blocks are Σ and Λ.

44. Let X be a random variable having the Lebesgue p.d.f. $\frac{2x}{\pi^2} I_{(0,\pi)}(x)$. Derive the p.d.f. of $Y = \sin X$.

45. Let X_i, $i = 1, 2, 3$, be independent random variables having the same Lebesgue p.d.f. $f(x) = e^{-x} I_{(0,\infty)}(x)$. Obtain the joint Lebesgue p.d.f. of (Y_1, Y_2, Y_3), where $Y_1 = X_1 + X_2 + X_3$, $Y_2 = X_1/(X_1 + X_2)$, and $Y_3 = (X_1 + X_2)/(X_1 + X_2 + X_3)$. Are Y_i's independent?

46. Let X_1 and X_2 be independent random variables having the standard normal distribution. Obtain the joint Lebesgue p.d.f. of (Y_1, Y_2), where $Y_1 = \sqrt{X_1^2 + X_2^2}$ and $Y_2 = X_1/X_2$. Are Y_i's independent?

47. Let X_1 and X_2 be independent random variables and $Y = X_1 + X_2$. Show that $F_Y(y) = \int F_{X_2}(y - x) dF_{X_1}(x)$.

48. Show that the Lebesgue p.d.f.'s given by (1.31) and (1.33) are the p.d.f.'s of the $\chi_n^2(\delta)$ and $F_{n_1, n_2}(\delta)$ distributions, respectively.

49. Show that the Lebesgue p.d.f. given by (1.32) is the p.d.f. of the $t_n(\delta)$ distribution.

50. Let $X = N_n(\mu, I_n)$ and A be an $n \times n$ symmetric matrix. Show that if $X^\tau AX$ has the $\chi_r^2(\delta)$ distribution, then $A^2 = A$, r is the rank of A, and $\delta = \mu^\tau A\mu$.

51. Let $X = N_n(\mu, I_n)$. Apply Cochran's theorem (Theorem 1.5) to show that if $A^2 = A$, then $X^\tau AX$ has the noncentral chi-square distribution $\chi_r^2(\delta)$, where A is an $n \times n$ symmetric matrix, r is the rank of A, and $\delta = \mu^\tau A\mu$.

52. Let $X_1, ..., X_n$ be independent and $X_i = N(0, \sigma_i^2)$, $i = 1, ..., n$. Let $\tilde{X} = \sum_{i=1}^n \sigma_i^{-2} X_i / \sum_{i=1}^n \sigma_i^{-2}$ and $\tilde{S}^2 = \sum_{i=1}^n \sigma_i^{-2}(X_i - \tilde{X})^2$. Apply Cochran's theorem to show that $\tilde{X}^2$ and $\tilde{S}^2$ are independent and that $\tilde{S}^2$ has the chi-square distribution χ_{n-1}^2.

53. Let $X = N_n(\mu, I_n)$ and A_i be an $n \times n$ symmetric matrix satisfying $A_i^2 = A_i$, $i = 1, 2$. Show that a necessary and sufficient condition that $X^\tau A_1 X$ and $X^\tau A_2 X$ are independent is $A_1 A_2 = 0$.

54. Let X be a random variable and $a > 0$. Show that $E|X|^a < \infty$ if and only if $\sum_{n=1}^\infty n^{a-1} P(|X| \geq n) < \infty$.

55. Let X be a random variable. Show that
    (a) if EX exists, then $EX = \int_0^\infty P(X > x)dx - \int_{-\infty}^0 P(X \leq x)dx$;
    (b) if X has range $\{0, 1, 2, ...\}$, then $EX = \sum_{n=1}^\infty P(X \geq n)$.

56. Let T be a random variable having the noncentral t-distribution $t_n(\delta)$. Show that
    (a) $E(T) = \delta\Gamma((n-1)/2)\sqrt{n/2}/\Gamma(n/2)$ when $n > 1$;
    (b) $\text{Var}(T) = \frac{n(1+\delta^2)}{n-2} - \frac{\delta^2 n}{2}\left[\frac{\Gamma((n-1)/2)}{\Gamma(n/2)}\right]^2$ when $n > 2$.

57. Let $\mathbf{F}$ be a random variable having the noncentral F-distribution $F_{n_1, n_2}(\delta)$. Show that
    (a) $E(\mathbf{F}) = \frac{n_2(n_1+\delta)}{n_1(n_2-2)}$ when $n_2 > 2$;
    (b) $\text{Var}(\mathbf{F}) = \frac{2n_2^2[(n_1+\delta)^2+(n_2-2)(n_1+2\delta)]}{n_1^2(n_2-2)^2(n_2-4)}$ when $n_2 > 4$.

58. Let $X = N_k(\mu, \Sigma)$ with a positive definite Σ.
    (a) Show that $EX = \mu$ and $\text{Var}(X) = \Sigma$.
    (b) Let A be an $l \times k$ matrix and B be an $m \times k$ matrix. Show that AX and BX are independent if and only if $A\Sigma B^\tau = 0$.
    (c) Suppose that $k = 2$, $X = (X_1, X_2)$, $\mu = 0$, $\text{Var}(X_1) = \text{Var}(X_2) = 1$, and $\text{Cov}(X_1, X_2) = \rho$. Show that $E(\max\{X_1, X_2\}) = \sqrt{(1-\rho)/\pi}$.

59. Let X be a random variable and g and h be nondecreasing functions on $\mathcal{R}$. Show that $\text{Cov}(g(X), h(X)) \geq 0$ when $E|g(X)h(X)| < \infty$.

60. Let X be a random variable with $EX^2 < \infty$ and let $Y = |X|$. Suppose that X has a Lebesgue p.d.f. symmetric about 0. Show that X and Y are uncorrelated, but they are not independent.

61. Let (X, Y) be a random 2-vector with the following Lebesgue p.d.f.:
$$f(x, y) = \begin{cases} \pi^{-1} & x^2 + y^2 \leq 1 \\ 0 & x^2 + y^2 > 1. \end{cases}$$
    Show that X and Y are uncorrelated, but are not independent.

62. Show that inequality (1.41) holds and that when $0 < E|X|^p < \infty$ and $0 < E|Y|^q < \infty$, the equality in (1.40) holds if and only if $\alpha|X|^p = \beta|Y|^q$ a.s. for some nonzero constants α and β.

63. Prove the following inequalities.
    (a) Liapounov's inequality (1.42).
    (b) Minkowski's inequality (1.43). (Hint: apply Hölder's inequality to random variables $|X + Y|^{p-1}$ and $|X|$.)
    (c) (C_r-inequality). $E|X+Y|^r \le C_r(E|X|^r + E|Y|^r)$, where X and Y are random variables, r is a positive constant, and $C_r = 1$ if $0 < r \le 1$ and $C_r = 2^{r-1}$ if $r > 1$.
    (d) Let X_i be a random variable with $E|X_i|^p < \infty$, $i = 1, ..., n$, where p is a constant larger than 1. Show that

    $$E\left|\frac{1}{n}\sum_{i=1}^{n} X_i\right|^p \le \min\left\{\frac{1}{n}\sum_{i=1}^{n} E|X_i|^p, \left[\frac{1}{n}\sum_{i=1}^{n}(E|X_i|^p)^{1/p}\right]^p\right\}.$$

    (e) Inequality (1.44). (Hint: prove the case of $n = 2$ first and then use induction.)
    (f) Inequality (1.49).

64. Show that the following functions of x are convex and discuss whether they are strictly convex.
    (a) $|x - a|^p$, where $p \ge 1$ and $a \in \mathcal{R}$.
    (b) x^{-p}, $x \in (0, \infty)$, where $p > 0$.
    (c) e^{cx}, where $c \in \mathcal{R}$.
    (d) $x \log x$, $x \in (0, \infty)$.
    (e) $g(\varphi(x))$, $x \in (a, b)$, where $-\infty \le a < b \le \infty$, φ is convex on (a, b), and g is convex and nondecreasing on the range of φ.
    (f) $\varphi(x) = \sum_{i=1}^{k} c_i\varphi_i(x_i)$, $x = (x_1, ..., x_k) \in \prod_{i=1}^{k} \mathcal{X}_i$, where c_i is a positive constant and φ_i is convex on $\mathcal{X}_i$, $i = 1, ..., k$.

65. Let $X = N_k(\mu, \Sigma)$ with a positive definite Σ.
    (a) Show that the m.g.f. of X is $e^{\tau^\tau \mu + t^\tau \Sigma t/2}$.
    (b) Show that $EX = \mu$ and $\text{Var}(X) = \Sigma$ by applying (1.54).
    (c) When $k = 1$ ($\Sigma = \sigma^2$), show that $EX = \psi_X'(0) = \mu$, $EX^2 = \psi_X''(0) = \sigma^2 + \mu^2$, $EX^3 = \psi_X^{(3)}(0) = 3\sigma^2\mu + \mu^3$, and $EX^4 = \psi_X^{(4)}(0) = 3\sigma^4 + 6\sigma^2\mu^2 + \mu^4$.
    (d) In part (c), show that if $\mu = 0$, then $EX^p = 0$ when p is an odd integer and $EX^p = (p-1)(p-3)\cdots 3 \cdot 1\sigma^p$ when p is an even integer.

66. Let X be a random variable having the gamma distribution $\Gamma(\alpha, \gamma)$. Find moments EX^p, $p = 1, 2, ...$, by differentiating the m.g.f. of X.

67. Let X be a random variable with finite Ee^{tX} and Ee^{-tX} for a $t \ne 0$. Show that $E|X|^a < \infty$ for any $a > 0$.

68. Let X be a random variable having $\psi_X(t) < \infty$ for t in a neighborhood of 0. Show that the moments and cumulants of X satisfy the following equations: $\mu_1 = \kappa_1$, $\mu_2 = \kappa_2 + \kappa_1^2$, $\mu_3 = \kappa_3 + 3\kappa_1\kappa_2 + \kappa_1^3$, and $\mu_4 = \kappa_4 + 3\kappa_2^2 + 4\kappa_1\kappa_3 + 6\kappa_1^2\kappa_2 + \kappa_1^4$, where μ_i and κ_i are the ith moment and cumulant of X, respectively.

69. Let X be a discrete random variable taking values 0,1,2.... The probability generating function of X is defined to be $\rho_X(t) = E(t^X)$. Show that
   (a) $\rho_X(t) = \psi_X(\log t)$, where ψ_X is the m.g.f. of X;
   (b) $\frac{d^p \rho_X(t)}{dt^p}\big|_{t=1} = E[X(X-1)\cdots(X-p+1)]$ for any positive integer p, if ρ_X is finite in a neighborhood of 1.

70. Let Y be a random variable having the noncentral chi-square distribution $\chi_k^2(\delta)$. Show that
   (a) the ch.f. of Y is $(1 - 2\sqrt{-1}t)^{-k/2} e^{\sqrt{-1}\delta t/(1-2\sqrt{-1}t)}$;
   (b) $E(Y) = k + \delta$ and $\mathrm{Var}(Y) = 2k + 4\delta$.

71. Let ϕ be a ch.f. on $\mathcal{R}^k$. Show that $|\phi| \leq 1$ and ϕ is uniformly continuous on $\mathcal{R}^k$.

72. For a complex number $z = a + \sqrt{-1}b$, where a and b are real numbers, $\bar{z}$ is defined to be $a - \sqrt{-1}b$. Show that $\sum_{i=1}^{n}\sum_{j=1}^{n}\phi(t_i - t_j)z_i\bar{z}_j \geq 0$, where ϕ is a ch.f. on $\mathcal{R}^k$, $t_1, ..., t_n$ are k-vectors, and $z_1, ..., z_n$ are complex numbers.

73. Show that the following functions of $t \in \mathcal{R}$ are ch.f.'s, where $a > 0$ and $b > 0$ are constants:
   (a) $a^2/(a^2 + t^2)$;
   (b) $(1 + ab - abe^{\sqrt{-1}t})^{-1/b}$;
   (c) $\max\{1 - |t|/a, 0\}$;
   (d) $2(1 - \cos at)/(a^2 t^2)$;
   (e) $e^{-|t|^a}$, where $0 < a \leq 2$;
   (f) $|\phi|^2$, where ϕ is a ch.f. on $\mathcal{R}$;
   (g) $\int \phi(ut)dG(u)$, where ϕ is a ch.f. on $\mathcal{R}$ and G is a c.d.f. on $\mathcal{R}$.

74. Let ϕ_n be the ch.f. of a probability measure P_n, $n = 1, 2,....$ Let $\{a_n\}$ be a sequence of nonnegative numbers with $\sum_{n=1}^{\infty} a_n = 1$. Show that $\sum_{n=1}^{\infty} a_n\phi_n$ is a ch.f. and find its corresponding probability measure.

75. Let X be a random variable whose ch.f. ϕ_X satisfies $\int |\phi_X(t)|dt < \infty$. Show that $(2\pi)^{-1} \int e^{-\sqrt{-1}xt}\phi_X(t)dt$ is the Lebesgue p.d.f. of X.

76. A random variable X or its distribution is of the *lattice type* if and only if $F_X(x) = \sum_{j=-\infty}^{\infty} p_j I_{\{a+jd\}}(x)$, $x \in \mathcal{R}$, where a, d, p_j's are

constants, $d > 0$, $p_j \geq 0$, and $\sum_{j=-\infty}^{\infty} p_j = 1$. Show that X is of the lattice type if and only if its ch.f. satisfies $|\phi_X(t)| = 1$ for some $t \neq 0$.

77. Let ϕ be a ch.f. on $\mathcal{R}$. Show that
    (a) if $|\phi(t_1)| = |\phi(t_2)| = 1$ and t_1/t_2 is an irrational number, then $\phi(t) = e^{\sqrt{-1}at}$ for some constant a;
    (b) if $t_n \to 0$, $t_n \neq 0$, and $|\phi(t_n)| = 1$, then the result in (a) holds;
    (c) $|\cos t|$ is not a ch.f., although $\cos t$ is a ch.f.

78. Let $X_1, ..., X_k$ be independent random variables and $Y = X_1 + \cdots + X_k$. Prove the following statements, using Theorem 1.6 and result (1.58).
    (a) If X_i has the binomial distribution $Bi(p, n_i)$, $i = 1, ..., k$, then Y has the binomial distribution $Bi(p, n_1 + \cdots + n_k)$.
    (b) If X_i has the Poisson distribution $P(\theta_i)$, $i = 1, ..., k$, then Y has the Poisson distribution $P(\theta_1 + \cdots + \theta_k)$.
    (c) If X_i has the negative binomial distribution $NB(p, r_i)$, $i = 1, ..., k$, then Y has the negative binomial distribution $NB(p, r_1 + \cdots + r_k)$.
    (d) If X_i has the exponential distribution $E(0, \theta)$, $i = 1, ..., k$, then Y has the gamma distribution $\Gamma(k, \theta)$.
    (e) If X_i has the Cauchy distribution $C(0, 1)$, $i = 1, ..., k$, then Y/k has the same distribution as X_1.

79. Find an example of two random variables X and Y such that X and Y are not independent but their ch.f.'s satisfy $\phi_X(t)\phi_Y(t) = \phi_{X+Y}(t)$ for all $t \in \mathcal{R}$.

80. Let $X_1, X_2, ...$ be independent random variables having the exponential distribution $E(0, \theta)$. For given $t > 0$, let Y be the maximum of n such that $T_n \leq t$, where $T_0 = 0$ and $T_n = X_1 + \cdots + X_n$, $n = 1, 2, ...$. Show that Y has the Poisson distribution $P(\theta t)$.

81. Let Σ be a $k \times k$ nonnegative definite matrix.
    (a) For a nonsingular Σ, show that X is $N_k(\mu, \Sigma)$ if and only if $c^\tau X$ is $N(c^\tau \mu, c^\tau \Sigma c)$ for any $c \in \mathcal{R}^k$.
    (b) For a singular Σ, we define X to be $N_k(\mu, \Sigma)$ if and only if $c^\tau X$ is $N(c^\tau \mu, c^\tau \Sigma c)$ for any $c \in \mathcal{R}^k$ ($N(a, 0)$ is the c.d.f. of the point mass at a). Show that the results in Exercise 43(a)-(c), Exercise 58(a)-(b), and Exercise 65(a) still hold for $X = N_k(\mu, \Sigma)$ with a singular Σ.

82. Let (X_1, X_2) be $N_k(\mu, \Sigma)$ with a $k \times k$ positive definite
$$\Sigma = \begin{pmatrix} \Sigma_{11} & \Sigma_{12} \\ \Sigma_{21} & \Sigma_{22} \end{pmatrix},$$
where X_1 is a random l-vector and Σ_{11} is an $l \times l$ matrix. Show that the conditional Lebesgue p.d.f. of X_2 given $X_1 = x_1$ is
$$N_{k-l}\left(\mu_2 + \Sigma_{21}\Sigma_{11}^{-1}(x_1 - \mu_1), \Sigma_{22} - \Sigma_{21}\Sigma_{11}^{-1}\Sigma_{12}\right),$$

where $\mu_i = EX_i$, $i = 1, 2$. (Hint: consider $X_2 - \mu_2 - \Sigma_{21}\Sigma_{11}^{-1}(X_1 - \mu_1)$ and $X_1 - \mu_1$.)

83. Let X be an integrable random variable with a Lebesgue p.d.f. f_X and let $Y = g(X)$, where g is a function with positive derivative on $(0, \infty)$ and $g(x) = g(-x)$. Find an expression for $E(X|Y)$ and verify that it is indeed the conditional expectation.

84. Prove Lemma 1.2. (Hint: first consider simple functions.)

85. Prove Proposition 1.10. (Hint for proving (ix): first show that $0 \le X_1 \le X_2 \le \cdots$ and $X_n \to_{a.s.} X$ imply $E(X_n|\mathcal{A}) \to_{a.s.} E(X|\mathcal{A})$.)

86. Let X and Y be integrable random variables on $(\Omega, \mathcal{F}, P)$ and $\mathcal{A} \subset \mathcal{F}$ be a σ-field. Show that $E[YE(X|\mathcal{A})] = E[XE(Y|\mathcal{A})]$, i.e., if either integral exists, then so does the other and the two integrals are equal.

87. Let $X, X_1, X_2, \ldots$ be a sequence of random variables on $(\Omega, \mathcal{F}, P)$ and $\mathcal{A} \subset \mathcal{F}$ be a σ-field. Suppose that $E(X_n Y) \to E(XY)$ for every integrable (or bounded) random variable Y. Show that $E[E(X_n|\mathcal{A})Y] \to E[E(X|\mathcal{A})Y]$ for every integrable (or bounded) random variable Y.

88. Let X be a nonnegative integrable random variable on $(\Omega, \mathcal{F}, P)$ and $\mathcal{A} \subset \mathcal{F}$ be a σ-field. Show that $E(X|\mathcal{A}) = \int_0^\infty P(X > t|\mathcal{A})dt$ a.s.

89. Let X and Y be random variables on $(\Omega, \mathcal{F}, P)$ and $\mathcal{A} \subset \mathcal{F}$ be a σ-field. Prove the following inequalities for conditional expectations.
    (a) If $E|X|^p < \infty$ and $E|Y|^q < \infty$ for constants p and q with $p > 1$ and $p^{-1} + q^{-1} = 1$, then $E(|XY||\mathcal{A}) \le [E(|X|^p|\mathcal{A})]^{1/p}[E(|Y|^q|\mathcal{A})]^{1/q}$ a.s.
    (b) If $E|X|^p < \infty$ and $E|Y|^p < \infty$ for a constant $p \ge 1$, then $[E(|X + Y|^p|\mathcal{A})]^{1/p} \le [E(|X|^p|\mathcal{A})]^{1/p} + [E(|Y|^p|\mathcal{A})]^{1/p}$ a.s.
    (c) If f is a convex function on $\mathcal{R}$, then $f(E(X|\mathcal{A})) \le E[f(X)|\mathcal{A}]$ a.s.

90. Let X and Y be random variables on a probability space with $Y = E(X|Y)$ a.s. and let φ be a convex function on $[0, \infty)$.
    (a) Show that if $E\varphi(|X|) < \infty$, then $E\varphi(|Y|) < \infty$.
    (b) Find an example in which $E\varphi(|Y|) < \infty$ but $E\varphi(|X|) = \infty$.
    (c) Suppose that $E\varphi(|X|) = E\varphi(|Y|) < \infty$ and φ is strictly convex and strictly increasing. Show that $X = Y$ a.s.

91. Let X, Y, and Z be random variables on a probability space. Suppose that $E|X| < \infty$ and $Y = h(Z)$ with a Borel h. Show that
    (a) if X and Z are independent and $E|Z| < \infty$, then $E(XZ|Y) = E(X)E(Z|Y)$ a.s.;
    (b) if $E[f(X)|Z] = f(Y)$ for all bounded continuous functions f on $\mathcal{R}$, then $X = Y$ a.s.;
    (c) if $E[f(X)|Z] \ge f(Y)$ for all bounded, continuous, nondecreasing functions f on $\mathcal{R}$, then $X \ge Y$ a.s.

92. Prove Lemma 1.3.

93. Show that random variables X_i, $i = 1, ..., n$, are independent according to Definition 1.7 if and only if (1.7) holds with F being the joint c.d.f. of X_i's and F_i being the marginal c.d.f. of X_i.

94. Show that a random variable X is independent of itself if and only if X is constant a.s. Can X and $f(X)$ be independent for a Borel f?

95. Let X, Y, and Z be independent random variables on a probability space and let $U = X + Z$ and $V = Y + Z$. Show that given Z, U and V are conditionally independent.

96. Show that the result in Proposition 1.11 may not be true if Y_2 is independent of X but not (X, Y_1).

97. Let X and Y be independent random variables on a probability space. Show that if $E|X|^a < \infty$ for some $a \geq 1$ and $E|Y| < \infty$, then $E|X + Y|^a \geq E|X + EY|^a$.

98. Let P_Y be a discrete distribution on $\{0, 1, 2, ...\}$ and $P_{X|Y=y}$ be the binomial distribution $Bi(p, y)$. Let (X, Y) be the random vector having the joint c.d.f. given by (1.66). Show that
    (a) if Y has the Poisson distribution $P(\theta)$, then the marginal distribution of X is the Poisson distribution $P(p\theta)$;
    (b) if $Y + r$ has the negative binomial distribution $NB(\pi, r)$, then the marginal distribution of $X + r$ is the negative binomial distribution $NB(\pi/[1 - (1 - p)(1 - \pi)], r)$.

99. Let $X_1, X_2, ...$ be i.i.d. random variables and Y be a discrete random variable taking positive integer values. Assume that Y and X_i's are independent. Let $Z = \sum_{i=1}^{Y} X_i$.
    (a) Obtain the ch.f. of Z.
    (b) Show that $EZ = EY EX_1$.
    (c) Show that $\text{Var}(Z) = EY \text{Var}(X_1) + \text{Var}(Y)(EX_1)^2$.

100. Let X, Y, and Z be random variables having a positive joint Lebesgue p.d.f. Let $f_{X|Y}(x|y)$ and $f_{X|Y,Z}(x|y, z)$ be respectively the conditional p.d.f. of X given Y and the conditional p.d.f. of X given (Y, Z), as defined by (1.61). Show that $\text{Var}(1/f_{X|Y}(X|Y)|X) \leq \text{Var}(1/f_{X|Y,Z}(X|Y, Z)|X)$ a.s., where $\text{Var}(\xi|\zeta) = E\{[\xi - E(\xi|\zeta)]^2|\zeta\}$ for any random variables ξ and ζ with $E\xi^2 < \infty$.

101. Let $\{X_n\}$ be a Markov chain. Show that if g is a one-to-one Borel function, then $\{g(X_n)\}$ is also a Markov chain. Give an example to show that $\{g(X_n)\}$ may not be a Markov chain in general.

102. A sequence of random vectors $\{X_n\}$ is said to be a Markov chain of order r for a positive integer r if $P(B|X_1, ..., X_n) = P(B|X_{n-r+1}, ..., X_n)$ a.s. for any $B \in \sigma(X_{n+1})$ and $n = r, r+1, ...$
    (a) Let $s > r$ be two positive integers. Show that if $\{X_n\}$ is a Markov chain of order r, then it is a Markov chain of order s.
    (b) Let $\{X_n\}$ be a sequence of random variables, r be a positive integer, and $Y_n = (X_n, X_{n+1}, ..., X_{n+r-1})$. Show that $\{Y_n\}$ is a Markov chain if and only if $\{X_n\}$ is a Markov chain of order r.
    (c) (Autoregressive process of order r). Let $\{\varepsilon_n\}$ be a sequence of independent random variables and r be a positive integer. Show that $\{X_n\}$ is a Markov chain of order r, where $X_n = \sum_{j=1}^{r} \rho_j X_{n-j} + \varepsilon_n$ and ρ_j's are constants.

103. Show that if $\{X_n, \mathcal{F}_n\}$ is a martingale (or a submartingale), then $E(X_{n+j}|\mathcal{F}_n) = X_n$ a.s. (or $E(X_{n+j}|\mathcal{F}_n) \geq X_n$ a.s.) and $EX_1 = EX_j$ (or $EX_1 \leq EX_2 \leq \cdots$) for any $j = 1, 2, ...$

104. Show that $\{X_n\}$ in Example 1.25 is a martingale.

105. Let $\{X_j\}$ and $\{Z_j\}$ be sequences of random variables and let f_n and g_n denote the Lebesgue p.d.f.'s of $Y_n = (X_1, ..., X_n)$ and $(Z_1, ..., Z_n)$, respectively, $n = 1, 2, ...$ Define $\lambda_n = -g_n(Y_n)/f_n(Y_n)I_{\{f_n(Y_n)>0\}}$, $n = 1, 2, ...$ Show that $\{\lambda_n\}$ is a submartingale.

106. Let $\{Y_n\}$ be a sequence of independent random variables.
    (a) Suppose that $EY_n = 0$ for all n. Let $X_1 = Y_1$ and $X_{n+1} = X_n + Y_{n+1}h_n(X_1, ..., X_n)$, $n \geq 2$, where $\{h_n\}$ is a sequence of Borel functions. Show that $\{X_n\}$ is a martingale.
    (b) Suppose that $EY_n = 0$ and $\text{Var}(Y_n) = \sigma^2$ for all n. Let $X_n = (\sum_{j=1}^{n} Y_j)^2 - n\sigma^2$. Show that $\{X_n\}$ is a martingale.
    (c) Suppose that $Y_n > 0$ and $EY_n = 1$ for all n. Let $X_n = Y_1 \cdots Y_n$. Show that $\{X_n\}$ is a martingale.

107. Prove the claims in the proof of Proposition 1.14.

108. Show that every sequence of integrable random variables is the sum of a supermartingale and a submartingale.

109. Let $\{X_n\}$ be a martingale. Show that if $\{X_n\}$ is bounded either above or below, then $\sup_n E|X_n| < \infty$.

110. Let $\{X_n\}$ be a martingale satisfying $EX_1 = 0$ and $EX_n^2 < \infty$ for all n. Show that $E(X_{n+m} - X_n)^2 = \sum_{j=1}^{m} E(X_{n+j} - X_{n+j-1})^2$ and that $\{X_n\}$ converges a.s.

111. Show that $\{X_n\}$ in Exercises 104, 105, and 106(c) converge a.s. to integrable random variables.

112. Prove Proposition 1.16.

113. In the proof of Lemma 1.4, show that $\{\omega : \lim_{n\to\infty} X_n(\omega) = X(\omega)\} = \cap_{j=1}^\infty A_j$.

114. Let $\{X_n\}$ be a sequence of independent random variables. Show that $X_n \to_{a.s.} 0$ if and only if, for any $\epsilon > 0$, $\sum_{n=1}^\infty P(|X_n| \geq \epsilon) < \infty$.

115. Let $X_1, X_2, ...$ be a sequence of identically distributed random variables with a finite $E|X_1|$ and let $Y_n = n^{-1} \max_{i\leq n} |X_i|$. Show that
    (a) $Y_n \to_{L_1} 0$;
    (b) $Y_n \to_{a.s.} 0$, assuming either $EX_1^2 < \infty$ or X_i's are independent.

116. Let $X, X_1, X_2, ...$ be random variables. Find an example for each of the following cases:
    (a) $X_n \to_p X$, but $\{X_n\}$ does not converge to X a.s.
    (b) $X_n \to_p X$, but $\{X_n\}$ does not converge to X in L_p for any $p > 0$.
    (c) $X_n \to_d X$, but $\{X_n\}$ does not converge to X in probability (do not use Example 1.26).
    (d) $X_n \to_p X$, but $\{g(X_n)\}$ does not converge to $g(X)$ in probability for some function g.

117. Let $X_1, X_2, ...$ be random variables. Show that
    (a) $\{|X_n|\}$ is uniformly integrable if and only if $\sup_n E|X_n| < \infty$ and, for any $\epsilon > 0$, there is a $\delta_\epsilon > 0$ such that $\sup_n E(|X_n|I_A) < \epsilon$ for any event A with $P(A) < \delta_\epsilon$;
    (b) $\sup_n E|X_n|^{1+\delta} < \infty$ for a $\delta > 0$ implies that $\{|X_n|\}$ is uniformly integrable.

118. Let $X, X_1, X_2, ...$ be random variables satisfying $P(|X_n| \geq c) \leq P(|X| \geq c)$ for all n and $c > 0$. Show that if $E|X| < \infty$, then $\{|X_n|\}$ is uniformly integrable.

119. Let $X_1, X_2, ...$ and $Y_1, Y_2, ...$ be random variables. Show that
    (a) if $\{|X_n|\}$ and $\{|Y_n|\}$ are uniformly integrable, then $\{|X_n + Y_n|\}$ is uniformly integrable;
    (b) if $\{|X_n|\}$ is uniformly integrable, then $\{|n^{-1}\sum_{i=1}^n X_i|\}$ is uniformly integrable.

120. Let Y be an integrable random variable and $\{\mathcal{F}_n\}$ be a sequence of σ-fields. Show that $\{|E(Y|\mathcal{F}_n)|\}$ is uniformly integrable.

121. Let $X, Y, X_1, X_2, ...$ be random variables satisfying $X_n \to_p X$ and $P(|X_n| \leq |Y|) = 1$ for all n. Show that if $E|Y|^r < \infty$ for some $r > 0$, then $X_n \to_{L_r} X$.

122. Let $X_1, X_2, ...$ be a sequence of random k-vectors. Show that $X_n \to_p 0$ if and only if $E[\|X_n\|/(1 + \|X_n\|)] \to 0$.

123. Let $X, X_1, X_2, ...$ be random variables. Show that $X_n \to_p X$ if and only if, for any subsequence $\{n_k\}$ of integers, there is a further subsequence $\{n_j\} \subset \{n_k\}$ such that $X_{n_j} \to_{a.s.} X$ as $j \to \infty$.

124. Let $X_1, X_2, ...$ be a sequence of random variables satisfying $|X_n| \le C_1$ and $\mathrm{Var}(X_n) \ge C_2$ for all n, where C_i's are positive constants. Show that $X_n \to_p 0$ does not hold.

125. Prove Lemma 1.5. (Hint for part (ii): use Chebyshev's inequality to show that $P(\sum_{n=1}^{\infty} I_{A_n} = \infty) = 1$, which can be shown to be equivalent to the result in (ii).)

126. Prove part (vii) of Theorem 1.8.

127. Let $X, X_1, X_2, ..., Y_1, Y_2, ..., Z_1, Z_2, ...$ be random variables. Prove the following statements.
    (a) If $X_n \to_d X$, then $X_n = O_p(1)$.
    (b) If $X_n = O_p(Z_n)$, then $X_n Y_n = O_p(Y_n Z_n)$.
    (c) If $X_n = O_p(Z_n)$ and $Y_n = O_p(Z_n)$, then $X_n + Y_n = O_p(Z_n)$.
    (d) If $E|X_n| = O(a_n)$, then $X_n = O_p(a_n)$, where $a_n \in (0, \infty)$.
    (e) If $X_n \to_{a.s.} X$, then $\sup_n |X_n| = O_p(1)$.

128. Let $\{X_n\}$ and $\{Y_n\}$ be two sequences of random variables such that $X_n = O_p(1)$ and $P(X_n \le t, Y_n \ge t+\epsilon) + P(X_n \ge t+\epsilon, Y_n \le t) = o(1)$ for any fixed $t \in \mathcal{R}$ and $\epsilon > 0$. Show that $X_n - Y_n = o_p(1)$.

129. Let $\{F_n\}$ be a sequence of c.d.f.'s on $\mathcal{R}$, $G_n(x) = F_n(a_n x + c_n)$, and $H_n(x) = F_n(b_n x + d_n)$, where $\{a_n\}$ and $\{b_n\}$ are sequences of positive numbers and $\{c_n\}$ and $\{d_n\}$ are sequences of real numbers. Suppose that $G_n \to_w G$ and $H_n \to_w H$, where G and H are nondegenerate c.d.f.'s. Show that $a_n/b_n \to a > 0$, $(c_n - d_n)/a_n \to b \in \mathcal{R}$, and $H(ax + b) = G(x)$ for all $x \in \mathcal{R}$.

130. Let $\{P_n\}$ be a sequence of probability measures on $(\mathcal{R}, \mathcal{B})$ and f be a nonnegative Borel function such that $\sup_n \int f dP_n < \infty$ and $f(x) \to 0$ as $|x| \to \infty$. Show that $\{P_n\}$ is tight.

131. Let $P, P_1, P_2, ...$ be probability measures on $(\mathcal{R}^k, \mathcal{B}^k)$. Show that if $P_n(O) \to P(O)$ for every open subset of $\mathcal{R}$, then $P_n(B) \to P(B)$ for every $B \in \mathcal{B}^k$.

132. Let $P, P_1, P_2, ...$ be probability measures on $(\mathcal{R}, \mathcal{B})$. Show that $P_n \to_w P$ if and only if there exists a dense subset D of $\mathcal{R}$ such that $\lim_{n \to \infty} P_n((a, b]) = P((a, b])$ for any $a < b$, $a \in D$ and $b \in D$.

133. Let F_n, $n = 0, 1, 2, ...$, be c.d.f.'s such that $F_n \to_w F_0$. Let $G_n(U) = \sup\{x : F_n(x) \le U\}$, $n = 0, 1, 2, ...$, where U is a random variable having the uniform $U(0, 1)$ distribution. Show that $G_n(U) \to_p G_0(U)$.

134. Let $P, P_1, P_2, \ldots$ be probability measures on $(\mathcal{R}, \mathcal{B})$. Suppose that $P_n \to_w P$ and $\{g_n\}$ is a sequence of bounded continuous functions on $\mathcal{R}$ converging uniformly to g. Show that $\int g_n dP_n \to \int g \, dP$.

135. Let $X, X_1, X_2, \ldots$ be random k-vectors and $Y, Y_1, Y_2, \ldots$ be random l-vectors. Suppose that $X_n \to_d X$, $Y_n \to_d Y$, and X_n and Y_n are independent for each n. Show that (X_n, Y_n) converges in distribution to a random $(k + l)$-vector.

136. Let $X_1, X_2, \ldots$ be independent random variables with $P(X_n = \pm 2^{-n}) = \frac{1}{2}$, $n = 1, 2, \ldots$. Show that $\sum_{i=1}^{n} X_i \to_d U$, where U has the uniform distribution $U(-1, 1)$.

137. Let $\{X_n\}$ and $\{Y_n\}$ be two sequences of random variables. Suppose that $X_n \to_d X$ and that $P_{Y_n | X_n = x_n} \to_w P_Y$ almost surely for every sequence of numbers $\{x_n\}$, where X and Y are independent random variables. Show that $X_n + Y_n \to_d X + Y$.

138. Let $X_1, X_2, \ldots$ be i.i.d. random variables having the ch.f. of the form $1 - c|t|^a + o(|t|^a)$ as $t \to 0$, where $0 < a \le 2$. Determine the constants b and u so that $\sum_{i=1}^{n} X_i / (bn^u)$ converges in distribution to a random variable having ch.f. $e^{-|t|^a}$.

139. Let $X, X_1, X_2, \ldots$ be random k-vectors and $A_1, A_2, \ldots$ be events. Suppose that $X_n \to_d X$. Show that $X_n I_{A_n} \to_d X$ if and only if $P(A_n) \to 1$.

140. Let X_n be a random variable having the $N(\mu_n, \sigma_n^2)$ distribution, $n = 1, 2, \ldots$, and X be a random variable having the $N(\mu, \sigma^2)$ distribution. Show that $X_n \to_d X$ if and only if $\mu_n \to \mu$ and $\sigma_n \to \sigma$.

141. Suppose that X_n is a random variable having the binomial distribution $Bi(p_n, n)$. Show that if $np_n \to \theta > 0$, then $X_n \to_d X$, where X has the Poisson distribution $P(\theta)$.

142. Let f_n be the Lebesgue p.d.f. of the t-distribution t_n, $n = 1, 2, \ldots$. Show that $f_n(x) \to f(x)$ for any $x \in \mathcal{R}$, where f is the Lebesgue p.d.f. of the standard normal distribution.

143. Prove Theorem 1.10.

144. Show by example that $X_n \to_d X$ and $Y_n \to_d Y$ does not necessarily imply that $g(X_n, Y_n) \to_d g(X, Y)$, where g is a continuous function.

145. Prove Theorem 1.11(ii)-(iii) and Theorem 1.12(ii). Extend Theorem 1.12(i) to the case where g is a function from $\mathcal{R}^p$ to $\mathcal{R}^q$ with $2 \le q \le p$.

146. Let $U_1, U_2, \ldots$ be i.i.d. random variables having the uniform distribution on $[0, 1]$ and $Y_n = (\prod_{i=1}^{n} U_i)^{-1/n}$. Show that $\sqrt{n}(Y_n - e) \to_d N(0, e^2)$.

147. Prove Lemma 1.6. (Hint: $a_n^{-1} \sum_{i=1}^n x_i = b_n - a_n^{-1} \sum_{i=0}^{n-1} b_i(a_{i+1} - a_i)$, where $b_n = \sum_{i=1}^n x_i/a_i$.)

148. In Theorem 1.13,
    (a) prove (1.82) for bounded c_i's when $E|X_1| < \infty$;
    (b) show that if $EX_1 = \infty$, then $n^{-1} \sum_{i=1}^n X_i \to_{a.s.} \infty$;
    (c) show that if $E|X_1| = \infty$, then $P(\limsup_n\{|\sum_{i=1}^n X_i| > cn\}) = P(\limsup_n\{|X_n| > cn\}) = 1$ for any fixed positive constant c, and $\limsup_n |n^{-1} \sum_{i=1}^n X_i| = \infty$ a.s.

149. Let $X_1, ..., X_n$ be i.i.d. random variables such that for $x = 3, 4, ...$, $P(X_1 = \pm x) = (2cx^2 \log x)^{-1}$, where $c = \sum_{x=3}^\infty x^{-2}/\log x$. Show that $E|X_1| - \infty$ but $n^{-1} \sum_{i=1}^n X_i \to_p 0$, using Theorem 1.13(i).

150. Let $X_1, X_2, ...$ be i.i.d. random variables satisfying $P(X_1 = 2^j) = 2^{-j}$, $j = 1, 2, ...$ Show that the WLLN does not hold for $\{X_n\}$, i.e., (1.80) does not hold for any $\{a_n\}$.

151. Let $X_1, X_2, ...$ be independent random variables. Suppose that, as $n \to \infty$, $\sum_{i=1}^n P(|X_i| > n) \to 0$ and $n^{-2} \sum_{i=1}^n E(X_i^2 I_{\{|X_i| \le n\}}) \to 0$. Show that $(T_n - b_n)/n \to_p 0$, where $T_n = \sum_{i=1}^n X_i$ and $b_n = \sum_{i=1}^n E(X_i I_{\{|X_i| \le n\}})$.

152. Let $T_n = \sum_{i=1}^n X_i$, where X_n's are independent random variables satisfying $P(X_n = \pm n^\theta) = 0.5$ and $\theta > 0$ is a constant. Show that
    (a) when $\theta < 0.5$, $T_n/n \to_{a.s.} 0$;
    (b) when $\theta \ge 1$, $T_n/n \to_p 0$ does not hold.

153. Let $X_2, X_3, ...$ be a sequence of independent random variables satisfying $P(X_n = \pm\sqrt{n/\log n}) = 0.5$. Show that (1.86) does not hold for $p \in [1, 2]$ but (1.88) is satisfied for $p = 2$ and, thus, (1.89) holds.

154. Let $X_1, ..., X_n$ be i.i.d. random variables with $\text{Var}(X_1) < \infty$. Show that $[n(n+1)]^{-1} \sum_{j=1}^n jX_j \to_p EX_1$.

155. Let $\{X_n\}$ be a sequence of random variables and let $\bar{X} = \sum_{i=1}^n X_i/n$.
    (a) Show that if $X_n \to_{a.s.} 0$, then $\bar{X} \to_{a.s.} 0$.
    (b) Show that if $X_n \to_{L_r} 0$, then $\bar{X} \to_{L_r} 0$, where $r \ge 1$ is a constant.
    (c) Show that the result in (b) may not be true for $r \in (0, 1)$.
    (d) Show that $X_n \to_p 0$ may not imply $\bar{X} \to_p 0$.

156. Let $X_1, ..., X_n$ be random variables and $\{\mu_n\}$, $\{\sigma_n\}$, $\{a_n\}$, and $\{b_n\}$ be sequences of real numbers with $\sigma_n \ge 0$ and $a_n \ge 0$. Suppose that X_n is asymptotically distributed as $N(\mu_n, \sigma_n^2)$. Show that $a_n X_n + b_n$ is asymptotically distributed as $N(\mu_n, \sigma_n^2)$ if and only if $a_n \to 1$ and $[\mu_n(a_n - 1) + b_n]/\sigma_n \to 0$.

157. Show that Liapounov's condition (1.97) implies Lindeberg's condition (1.92).

158. Let $X_1, X_2, ...$ be a sequence of independent random variables and $\sigma_n^2 = \text{Var}(\sum_{j=1}^n X_j)$.
    (a) Show that if $X_n = N(0, 2^{-n})$, $n = 1, 2, ...$, then Feller's condition (1.96) does not hold but $\sum_{j=1}^n (X_j - EX_j)/\sigma_n \to_d N(0, 1)$.
    (b) Show that the result in (a) is still true if X_1 has the uniform distribution $U(-1, 1)$ and $X_n = N(0, 2^{n-1})$, $n = 2, 3, ...$

159. In Example 1.33, show that
    (a) the condition $\sigma_n^2 \to \infty$ is also necessary for (1.98);
    (b) $n^{-1} \sum_{i=1}^n (X_i - p_i) \to_{L_r} 0$ for any constant $r > 0$;
    (c) $n^{-1} \sum_{i=1}^n (X_i - p_i) \to_{a.s.} 0$.

160. Prove Corollary 1.3.

161. Suppose that X_n is a random variable having the binomial distribution $Bi(\theta, n)$, where $0 < \theta < 1$, $n = 1, 2, ...$ Define $Y_n = \log(X_n/n)$ when $X_n \geq 1$ and $Y_n = 1$ when $X_n = 0$. Show that $Y_n \to_{a.s.} \log \theta$ and $\sqrt{n}(Y_n - \log \theta) \to_d N\left(0, \frac{1-\theta}{\theta}\right)$. Establish similar results when X_n has the Poisson distribution $P(n\theta)$.

162. Let $X_1, X_2, ...$ be independent random variables such that X_j has the uniform distribution on $[-j, j]$, $j = 1, 2, ...$ Show that Lindeberg's condition is satisfied and state the resulting CLT.

163. Let $X_1, X_2, ...$ be independent random variables such that for $j = 1, 2, ...$, $P(X_j = \pm j^a) = 6^{-1} j^{-2(a-1)}$ and $P(X_j = 0) = 1 - 3^{-1} j^{-2(a-1)}$, where $a > 1$ is a constant. Show that Lindeberg's condition is satisfied if and only if $a < 1.5$.

164. Let $X_1, X_2, ...$ be independent random variables with $P(X_j = \pm j^a) = P(X_j = 0) = 1/3$, where $a > 0$, $j = 1, 2, ...$ Can we apply Theorem 1.15 to $\{X_j\}$ by checking Liapounov's condition (1.97)?

165. Let $\{X_n\}$ be a sequence of independent random variables. Suppose that $\sum_{j=1}^n (X_j - EX_j)/\sigma_n \to_d N(0, 1)$, where $\sigma_n^2 = \text{Var}(\sum_{j=1}^n X_j)$. Show that $n^{-1} \sum_{j=1}^n (X_j - EX_j) \to_p 0$ if and only if $\sigma_n = o(n)$.

166. Consider Exercise 152. Show that $T_n/\sqrt{\text{Var}(T_n)} \to_d N(0, 1)$ and, when $0.5 \leq \theta < 1$, $T_n/n \to_p 0$ does not hold.

167. Prove (1.102)-(1.104).

168. In Example 1.34, prove $\sigma_h^2 = 1$ for $\sqrt{n}(\bar{X} - \mu)/\hat{\sigma}$ and $\sqrt{n}(\hat{\sigma}^2 - \sigma^2)/\hat{\tau}$ and derive the expressions for $p_1(x)$ in all four cases.

Chapter 2

Fundamentals of Statistics

This chapter discusses some fundamental concepts of mathematical statistics. These concepts are essential for the material in later chapters.

2.1 Populations, Samples, and Models

A typical statistical problem can be described as follows. One or a series of random experiments is performed; some data from the experiment(s) are collected; and our task is to extract information from the data, interpret the results, and draw some conclusions. In this book we do not consider the problem of planning experiments and collecting data, but concentrate on statistical analysis of the data, assuming that the data are given.

A descriptive data analysis can be performed to obtain some summary measures of the data, such as the mean, median, range, standard deviation, etc., and some graphical displays, such as the histogram and box-and-whisker diagram, etc. (see, e.g., Hogg and Tanis (1993)). Although this kind of analysis is simple and requires almost no assumptions, it may not allow us to gain enough insight into the problem. We focus on more sophisticated methods of analyzing data: *statistical inference* and *decision theory*.

2.1.1 Populations and samples

In statistical inference and decision theory, the data set is viewed as a realization or observation of a random element defined on a probability space $(\Omega, \mathcal{F}, P)$ related to the random experiment. The probability measure P is called the *population*. The data set or the random element that produces the data is called a *sample* from P. The size of the data set is called the *sample size*. A population P is *known* if and only if $P(A)$ is a known value for every event $A \in \mathcal{F}$. In a statistical problem, the population P is at least partially unknown and we would like to deduce some properties of P based on the available sample.

Example 2.1 (Measurement problems). To measure an unknown quantity θ (for example, a distance, weight, or temperature), n measurements, $x_1, ..., x_n$, are taken in an experiment of measuring θ. If θ can be measured without errors, then $x_i = \theta$ for all i; otherwise, each x_i has a possible measurement error. In descriptive data analysis, a few summary measures may be calculated, for example, the *sample mean*

$$\bar{x} = \frac{1}{n} \sum_{i=1}^{n} x_i$$

and the *sample variance*

$$s^2 = \frac{1}{n-1} \sum_{i=1}^{n} (x_i - \bar{x})^2 .$$

However, what is the relationship between $\bar{x}$ and θ? Are they close (if not equal) in some sense? The sample variance s^2 is clearly an average of squared deviations of x_i's from their mean. But, what kind of information does s^2 provide? Finally, is it enough to just look at $\bar{x}$ and s^2 for the purpose of measuring θ? These questions cannot be answered in descriptive data analysis.

In statistical inference and decision theory, the data set, $(x_1, ..., x_n)$, is viewed as an outcome of the experiment whose sample space is $\Omega = \mathcal{R}^n$. We usually assume that the n measurements are obtained in n *independent* trials of the experiment. Hence, we can define a random n-vector $X = (X_1, ..., X_n)$ on $\prod_{i=1}^{n}(\mathcal{R}, \mathcal{B}, P)$ whose realization is $(x_1, ..., x_n)$. The population in this problem is P (note that the product probability measure is determined by P) and is at least partially unknown. The random vector X is a sample and n is the sample size. Define

$$\bar{X} = \frac{1}{n} \sum_{i=1}^{n} X_i \tag{2.1}$$

and

$$S^2 = \frac{1}{n-1} \sum_{i=1}^{n} (X_i - \bar{X})^2 . \tag{2.2}$$

Then $\bar{X}$ and S^2 are random variables that produce $\bar{x}$ and s^2, respectively. Questions raised previously can be answered if some assumptions are imposed on the population P, which are discussed later.   ∎

When the sample $(X_1, ..., X_n)$ has i.i.d. components, which is often the case in applications, the population is determined by the marginal distribution of X_i.

Example 2.2 (Life-time testing problems). Let $x_1, ..., x_n$ be observed lifetimes of some electronic components. Again, in statistical inference and decision theory, $x_1, ..., x_n$ are viewed as realizations of independent random variables $X_1, ..., X_n$. Suppose that the components are of the same type so that it is reasonable to assume that $X_1, ..., X_n$ have a common marginal c.d.f. F. Then the population is F, which is often unknown. A quantity of interest in this problem is $1 - F(t)$ with a $t > 0$, which is the probability that a component does not fail at time t. It is possible that all x_i's are smaller (or larger) than t. Conclusions about $1 - F(t)$ can be drawn based on data $x_1, ..., x_n$ when certain assumptions on F are imposed. ∎

Example 2.3 (Survey problems). A survey is often conducted when one is not able to evaluate all elements in a collection $\mathcal{P} = \{y_1, ..., y_N\}$ containing N values in $\mathcal{R}^k$, where k and N are finite positive integers but N may be very large. Suppose that the quantity of interest is the *population total* $Y = \sum_{i=1}^N y_i$. In a survey, a subset s of n elements are selected from $\{1, ..., N\}$ and values y_i, $i \in s$, are obtained. Can we draw some conclusion about Y based on data y_i, $i \in s$?

How do we define some random variables that produce the survey data? First, we need to specify how s is selected. A commonly used probability sampling plan can be described as follows. Assume that every element in $\{1, ..., N\}$ can be selected at most once, i.e., we consider *sampling without replacement*. Let $\mathcal{S}$ be the collection of all subsets of n distinct elements from $\{1, ..., N\}$, $\mathcal{F}_s$ be the collection of all subsets of $\mathcal{S}$, and p be a probability measure on $(\mathcal{S}, \mathcal{F}_s)$. Any $s \in \mathcal{S}$ is selected with probability $p(s)$. Note that $p(s)$ is a known value whenever s is given. Let $X_1, ..., X_n$ be random variables such that

$$P(X_1 = y_{i_1}, ..., X_n = y_{i_n}) = \frac{p(s)}{n!}, \qquad s = \{i_1, ..., i_n\} \in \mathcal{S}. \qquad (2.3)$$

Then $(y_i, i \in s)$ can be viewed as a realization of the sample $(X_1, ..., X_n)$. If $p(s)$ is constant, then the sampling plan is called the *simple random sampling (without replacement)* and $(X_1, ..., X_n)$ is called a *simple random sample*. Although $X_1, ..., X_n$ are identically distributed, they are *not* necessarily independent. Thus, unlike in the previous two examples, the population in this problem may not be specified by the marginal distributions of X_i's. The population is determined by $\mathcal{P}$ and the known selection probability measure p. For this reason, $\mathcal{P}$ is often treated as the population. Conclusions about Y and other characteristics of $\mathcal{P}$ can be drawn based on data y_i, $i \in s$, which are discussed later. ∎

2.1.2 Parametric and nonparametric models

A *statistical model* (a set of assumptions) on the population P in a given problem is often postulated to make the analysis possible or easy. Although testing the correctness of postulated models is part of statistical inference and decision theory, postulated models are often based on knowledge of the problem under consideration.

Definition 2.1. A set of probability measures P_θ on $(\Omega, \mathcal{F})$ indexed by a *parameter* $\theta \in \Theta$ is said to be a *parametric family* if and only if $\Theta \subset \mathcal{R}^d$ for some fixed positive integer d *and* each P_θ is a *known* probability measure when θ is known. The set Θ is called the *parameter space* and d is called its *dimension*. ∎

A *parametric model* refers to the assumption that the population P is in a given parametric family. A parametric family $\{P_\theta : \theta \in \Theta\}$ is said to be *identifiable* if and only if $\theta_1 \neq \theta_2$ and $\theta_i \in \Theta$ imply $P_{\theta_1} \neq P_{\theta_2}$. In most cases an identifiable parametric family can be obtained through reparameterization. Hence, we assume in what follows that every parametric family is identifiable unless otherwise stated.

Let $\mathcal{P}$ be a family of populations and ν be a σ-finite measure on $(\Omega, \mathcal{F})$. If $P \ll \nu$ for all $P \in \mathcal{P}$, then $\mathcal{P}$ is said to be dominated by ν, in which case $\mathcal{P}$ can be identified by the family of densities $\{\frac{dP}{d\nu} : P \in \mathcal{P}\}$ (or $\{\frac{dP_\theta}{d\nu} : \theta \in \Theta\}$ for a parametric family).

Many examples of parametric families can be obtained from Tables 1.1 and 1.2 in §1.3.1. All parametric families from Tables 1.1 and 1.2 are dominated by the counting measure or the Lebesgue measure on $\mathcal{R}$.

Example 2.4 (The k-dimensional normal family). Consider the normal distribution $N_k(\mu, \Sigma)$ given by (1.24) for a fixed positive integer k. An important parametric family in statistics is the family of normal distributions

$$\mathcal{P} = \{N_k(\mu, \Sigma) : \mu \in \mathcal{R}^k, \ \Sigma \in \mathcal{M}_k\},$$

where $\mathcal{M}_k$ is a collection of $k \times k$ symmetric positive definite matrices. This family is dominated by the Lebesgue measure on $\mathcal{R}^k$.

In the measurement problem described in Example 2.1, X_i's are often i.i.d. from the $N(\mu, \sigma^2)$ distribution. Hence, we can impose a parametric model on the population, i.e., $P \in \mathcal{P} = \{N(\mu, \sigma^2) : \mu \in \mathcal{R}, \ \sigma^2 > 0\}$.

The normal parametric model is perhaps not a good model for the lifetime testing problem described in Example 2.2, since clearly $X_i \geq 0$ for all i. In practice, the normal family $\{N(\mu, \sigma^2) : \mu \in \mathcal{R}, \ \sigma^2 > 0\}$ can be used for a life-time testing problem if one puts some restrictions on μ and σ so that $P(X_i < 0)$ is negligible. Common parametric models for life-time testing problems are the exponential model (containing the exponential distributions $E(0, \theta)$ with an unknown parameter θ; see Table 1.2 in §1.3.1), the gamma model (containing the gamma distributions $\Gamma(\alpha, \gamma)$ with unknown parameters α and γ), the log-normal model (containing the log-normal distributions $LN(\mu, \sigma^2)$ with unknown parameters μ and σ), the Weibull model (containing the Weibull distributions $W(\alpha, \theta)$ with unknown parameters α and θ), and any subfamilies of these parametric families (e.g., a family containing the gamma distributions with one known parameter and one unknown parameter).

The normal family is often not a good choice for the survey problem discussed in Example 2.3. ∎

In a given problem, a parametric model is not useful if the dimension of Θ is very high. For example, the survey problem described in Example 2.3 has a natural parametric model, since the population $\mathcal{P}$ can be indexed by the parameter $\theta = (y_1, ..., y_N)$. If there is no restriction on the y-values, however, the dimension of the parameter space is kN, which is usually much larger than the sample size n. If there are some restrictions on the y-values (for example, y_i's are nonnegative integers no larger than a fixed integer m), then the dimension of the parameter space is at most $m + 1$ and the parametric model becomes useful.

A family of probability measures is said to be *nonparametric* if it is not parametric according to Definition 2.1. A *nonparametric model* refers to the assumption that the population P is in a given nonparametric family. There may be almost no assumption on a nonparametric family, for example, the family of all probability measures on $(\mathcal{R}^k, \mathcal{B}^k)$. But in many applications, we may use one or a combination of the following assumptions to form a nonparametric family on $(\mathcal{R}^k, \mathcal{B}^k)$:

(1) The joint c.d.f.'s are continuous.

(2) The joint c.d.f.'s have finite moments of order $\leq$ a fixed integer.

(3) The joint c.d.f.'s have p.d.f.'s (e.g., Lebesgue p.d.f.'s).

(4) $k = 1$ and the c.d.f.'s are symmetric.

For instance, in Example 2.1, we may assume a nonparametric model with symmetric and continuous c.d.f.'s. The symmetry assumption may not be suitable for the population in Example 2.2, but the continuity assumption seems to be reasonable.

In statistical inference and decision theory, methods designed for parametric models are called *parametric methods*, whereas methods designed for nonparametric models are called *nonparametric methods*. However, nonparametric methods are used in a parametric model when parametric methods are not effective, such as when the dimension of the parameter space is too high (Example 2.3). On the other hand, parametric methods may be applied to a *semi-parametric model*, which is a nonparametric model having a parametric component. Some examples are provided in §5.1.4.

2.1.3 Exponential and location-scale families

In this section, we discuss two types of parametric families that are of special importance in statistical inference and decision theory.

Definition 2.2 (Exponential families). A parametric family $\{P_\theta : \theta \in \Theta\}$ dominated by a σ-finite measure ν on $(\Omega, \mathcal{F})$ is called an *exponential family* if and only if

$$\frac{dP_\theta}{d\nu}(\omega) = \exp\{[\eta(\theta)]^\tau T(\omega) - \xi(\theta)\}h(\omega), \quad \omega \in \Omega, \qquad (2.4)$$

where $\exp\{x\} = e^x$, T is a random p-vector with a fixed positive integer p, η is a function from Θ to $\mathcal{R}^p$, h is a nonnegative Borel function on $(\Omega, \mathcal{F})$, and $\xi(\theta) = \log\{\int_\Omega \exp\{[\eta(\theta)]^\tau T(\omega)\}h(\omega)d\nu(\omega)\}$. ∎

In Definition 2.2, T and h are functions of ω only, whereas η and ξ are functions of θ only. Ω is usually $\mathcal{R}^k$. The representation (2.4) of an exponential family is not unique. In fact, any transformation $\tilde{\eta}(\theta) = D\eta(\theta)$ with a $p \times p$ nonsingular matrix D gives another representation (with T replaced by $\tilde{T} = (D^\tau)^{-1}T$). A change of the measure that dominates the family also changes the representation. For example, if we define $\lambda(A) = \int_A h d\nu$ for any $A \in \mathcal{F}$, then we obtain an exponential family with densities

$$\frac{dP_\theta}{d\lambda}(\omega) = \exp\{[\eta(\theta)]^\tau T(\omega) - \xi(\theta)\}. \qquad (2.5)$$

In an exponential family, consider the reparameterization $\eta = \eta(\theta)$ and

$$f_\eta(\omega) = \exp\{\eta^\tau T(\omega) - \zeta(\eta)\}h(\omega), \quad \omega \in \Omega, \qquad (2.6)$$

where $\zeta(\eta) = \log\{\int_\Omega \exp\{\eta^\tau T(\omega)\}h(\omega)d\nu(\omega)\}$. This is the *canonical form* for the family, which is not unique for the reasons discussed previously. The new parameter η is called the *natural parameter*. The new parameter space $\Xi = \{\eta(\theta) : \theta \in \Theta\}$, a subset of $\mathcal{R}^p$, is called the *natural parameter space*. An exponential family in canonical form is called a *natural exponential family*. If there is an open set contained in the natural parameter space of an exponential family, then the family is said to be of *full rank*.

Example 2.5. Let P_θ be the binomial distribution $Bi(\theta, n)$ with parameter θ, where n is a fixed positive integer. Then $\{P_\theta : \theta \in (0,1)\}$ is an exponential family, since the p.d.f. of P_θ w.r.t. the counting measure is

$$f_\theta(x) = \exp\left\{x \log \tfrac{\theta}{1-\theta} + n\log(1-\theta)\right\} \binom{n}{x} I_{\{0,1,\ldots,n\}}(x)$$

$(T(x)=x,\ \eta(\theta)=\log\tfrac{\theta}{1-\theta},\ \xi(\theta)=-n\log(1-\theta)$, and $h(x)=\binom{n}{x}I_{\{0,1,\ldots,n\}}(x))$. If we let $\eta = \log\tfrac{\theta}{1-\theta}$, then $\Xi = \mathcal{R}$ and the family with p.d.f.'s

$$f_\eta(x) = \exp\left\{x\eta - n\log(1+e^\eta)\right\} \binom{n}{x} I_{\{0,1,\ldots,n\}}(x)$$

is a natural exponential family of full rank. ∎

Example 2.6. The normal family $\{N(\mu,\sigma^2) : \mu \in \mathcal{R}, \sigma > 0\}$ is an exponential family, since the Lebesgue p.d.f. of $N(\mu,\sigma^2)$ can be written as

$$\frac{1}{\sqrt{2\pi}}\exp\left\{\frac{\mu}{\sigma^2}x - \frac{1}{2\sigma^2}x^2 - \frac{\mu^2}{2\sigma^2} - \log\sigma\right\}.$$

Hence, $T(x) = (x, -x^2)$, $\eta(\theta) = \left(\frac{\mu}{\sigma^2}, \frac{1}{2\sigma^2}\right)$, $\theta = (\mu, \sigma^2)$, $\xi(\theta) = \frac{\mu^2}{2\sigma^2} + \log\sigma$, and $h(x) = 1/\sqrt{2\pi}$. Let $\eta = (\eta_1, \eta_2) = \left(\frac{\mu}{\sigma^2}, \frac{1}{2\sigma^2}\right)$. Then $\Xi = \mathcal{R} \times (0, \infty)$ and we can obtain a natural exponential family of full rank with $\zeta(\eta) = \eta_1^2/(4\eta_2) + \log(1/\sqrt{2\eta_2})$.

A subfamily of the previous normal family, $\{N(\mu, \mu^2) : \mu \in \mathcal{R}, \mu \neq 0\}$, is also an exponential family with the natural parameter $\eta = \left(\frac{1}{\mu}, \frac{1}{2\mu^2}\right)$ and natural parameter space $\Xi = \{(x, y) : y = 2x^2,\ x \in \mathcal{R},\ y > 0\}$. This exponential family is not of full rank. ∎

For an exponential family, (2.5) implies that there is a nonzero measure λ such that

$$\frac{dP_\theta}{d\lambda}(\omega) > 0 \qquad \text{for all } \omega \text{ and } \theta. \tag{2.7}$$

We can use this fact to show that a family of distributions is not an exponential family. For example, consider the family of uniform distributions, i.e., P_θ is $U(0, \theta)$ with an unknown $\theta \in (0, \infty)$. If $\{P_\theta : \theta \in (0, \infty)\}$ is an exponential family, then from the previous discussion we have a nonzero measure λ such that (2.7) holds. For any $t > 0$, there is a $\theta < t$ such that $P_\theta([t, \infty)) = 0$, which with (2.7) implies that $\lambda([t, \infty)) = 0$. Also, for any $t \leq 0$, $P_\theta((-\infty, t]) = 0$, which with (2.7) implies that $\lambda((-\infty, t]) = 0$. Since t is arbitrary, $\lambda \equiv 0$. This contradiction implies that $\{P_\theta : \theta \in (0, \infty)\}$ cannot be an exponential family.

The reader may verify which of the parametric families from Tables 1.1 and 1.2 are exponential families. As another example, we consider an important exponential family containing multivariate discrete distributions.

Example 2.7 (The multinomial family). Consider an experiment having $k + 1$ possible outcomes with p_i as the probability for the ith outcome, $i = 0, 1, ..., k$, $\sum_{i=0}^{k} p_i = 1$. In n independent trials of this experiment, let X_i be the number of trials resulting in the ith outcome, $i = 0, 1, ..., k$. Then the joint p.d.f. (w.r.t. counting measure) of $(X_0, X_1, ..., X_k)$ is

$$f_\theta(x_0, x_1, ..., x_k) = \frac{n!}{x_0! x_1! \cdots x_k!} p_0^{x_0} p_1^{x_1} \cdots p_k^{x_k} I_B(x_0, x_1, ..., x_k),$$

where $B = \{(x_0, x_1, ..., x_k) : x_i\text{'s are integers} \geq 0, \sum_{i=0}^{k} x_i = n\}$ and $\theta = (p_0, p_1, ..., p_k)$. The distribution of $(X_0, X_1, ..., X_k)$ is called the *multinomial* distribution, which is an extension of the binomial distribution. In fact, the marginal c.d.f. of each X_i is the binomial distribution $Bi(p_i, n)$. Let $\Theta = \{\theta \in \mathcal{R}^{k+1} : 0 < p_i < 1, \sum_{i=0}^{k} p_i = 1\}$. The parametric family $\{f_\theta : \theta \in \Theta\}$ is called the multinomial family. Let $x = (x_0, x_1, ..., x_k)$, $\eta = (\log p_0, \log p_1, ..., \log p_k)$, and $h(x) = [n!/(x_0! x_1! \cdots x_k!)] I_B(x)$. Then

$$f_\theta(x_0, x_1, ..., x_k) = \exp\{\eta^\tau x\} h(x), \qquad x \in \mathcal{R}^{k+1}. \tag{2.8}$$

Hence, the multinomial family is a natural exponential family with natural parameter η. However, representation (2.8) does not provide an exponential family of full rank, since there is no open set of $\mathcal{R}^{k+1}$ contained in the natural parameter space. A reparameterization leads to an exponential family with full rank. Using the fact that $\sum_{i=0}^{k} X_i = n$ and $\sum_{i=0}^{k} p_i = 1$, we obtain that

$$f_\theta(x_0, x_1, ..., x_k) = \exp\{\eta_*^\tau x_* - \zeta(\eta_*)\} h(x), \qquad x \in \mathcal{R}^{k+1}, \tag{2.9}$$

where $x_* = (x_1, ..., x_k)$, $\eta_* = (\log(p_1/p_0), ..., \log(p_k/p_0))$, and $\zeta(\eta_*) = -n \log p_0$. The η_*-parameter space is $\mathcal{R}^k$. Hence, the family of densities given by (2.9) is a natural exponential family of full rank. ■

If $X_1, ..., X_m$ are independent random vectors with p.d.f.'s in exponential families, then the p.d.f. of $(X_1, ..., X_m)$ is again in an exponential family. The following result summarizes some other useful properties of exponential families. Its proof can be found in Lehmann (1986).

Theorem 2.1. Let $\mathcal{P}$ be a natural exponential family given by (2.6).
(i) Let $T = (Y, U)$ and $\eta = (\vartheta, \varphi)$, where Y and ϑ have the same dimension. Then, Y has the p.d.f.

$$f_\eta(y) = \exp\{\vartheta^\tau y - \zeta(\eta)\}$$

w.r.t. a σ-finite measure depending on φ. In particular, T has a p.d.f. in a natural exponential family. Furthermore, the conditional distribution of Y given $U = u$ has the p.d.f. (w.r.t. a σ-finite measure depending on u)

$$f_{\vartheta, u}(y) = \exp\{\vartheta^\tau y - \zeta_u(\vartheta)\},$$

which is in a natural exponential family indexed by ϑ.

(ii) If η_0 is an interior point of the natural parameter space, then the m.g.f. ψ_{η_0} of $P_{\eta_0} \circ T^{-1}$ is finite in a neighborhood of 0 and is given by

$$\psi_{\eta_0}(t) = \exp\{\zeta(\eta_0 + t) - \zeta(\eta_0)\}.$$

Furthermore, if f is a Borel function satisfying $\int |f| dP_{\eta_0} < \infty$, then the function

$$\int f(\omega) \exp\{\eta^\tau T(\omega)\} h(\omega) d\nu(\omega)$$

is infinitely often differentiable in a neighborhood of η_0, and the derivatives may be computed by differentiation under the integral sign. ∎

Using Theorem 2.1(ii) and the result in Example 2.5, we obtain that the m.g.f. of the binomial distribution $Bi(p, n)$ is

$$\begin{aligned} \psi_\eta(t) &= \exp\{n \log(1 + e^{\eta+t}) - n \log(1 + e^\eta)\} \\ &= \left(\frac{1 + e^\eta e^t}{1 + e^\eta}\right)^n \\ &= (1 - p + pe^t)^n, \end{aligned}$$

since $p = e^\eta/(1 + e^\eta)$.

Definition 2.3 (Location-scale families). Let P be a known probability measure on $(\mathcal{R}^k, \mathcal{B}^k)$, $\mathcal{V} \subset \mathcal{R}^k$, and $\mathcal{M}_k$ be a collection of $k \times k$ symmetric positive definite matrices. The family

$$\{P_{(\mu,\Sigma)} : \mu \in \mathcal{V}, \ \Sigma \in \mathcal{M}_k\} \tag{2.10}$$

is called a *location-scale family* (on $\mathcal{R}^k$), where

$$P_{(\mu,\Sigma)}(B) = P\left(\Sigma^{-1/2}(B - \mu)\right), \quad B \in \mathcal{B}^k,$$

$\Sigma^{-1/2}(B - \mu) = \{\Sigma^{-1/2}(x - \mu) : x \in B\} \subset \mathcal{R}^k$, and $\Sigma^{-1/2}$ is the inverse of the "square root" matrix $\Sigma^{1/2}$ satisfying $\Sigma^{1/2}\Sigma^{1/2} = \Sigma$. The parameters μ and $\Sigma^{1/2}$ are called the location and scale parameters, respectively. ∎

The following are some important examples of location-scale families. The family $\{P_{(\mu,I_k)} : \mu \in \mathcal{R}^k\}$ is called a *location family*, where I_k is the $k \times k$ identity matrix. The family $\{P_{(0,\Sigma)} : \Sigma \in \mathcal{M}_k\}$ is called a *scale family*. In some cases, we consider a location-scale family of the form $\{P_{(\mu,\sigma^2 I_k)} : \mu \in \mathcal{R}^k, \sigma > 0\}$. If $X_1, ..., X_k$ are i.i.d. with a common distribution in the location-scale family $\{P_{(\mu,\sigma^2)} : \mu \in \mathcal{R}, \sigma > 0\}$, then the joint distribution of the vector $(X_1, ..., X_k)$ is in the location-scale family $\{P_{(\mu,\sigma^2 I_k)} : \mu \in \mathcal{V}, \sigma > 0\}$ with $\mathcal{V} = \{(x, ..., x) \in \mathcal{R}^k : x \in \mathcal{R}\}$.

A location-scale family can be generated as follows. Let X be a random k-vector having a distribution P. Then the distribution of $\Sigma^{1/2}X + \mu$ is $P_{(\mu,\Sigma)}$. On the other hand, if X is a random k-vector whose distribution is in the location-scale family (2.10), then the distribution $DX + c$ is also in the same family, provided that $D\mu + c \in \mathcal{V}$ and $D\Sigma D^\tau \in \mathcal{M}_k$.

Let F be the c.d.f. of P. Then the c.d.f. of $P_{(\mu,\Sigma)}$ is $F\left(\Sigma^{-1/2}(x - \mu)\right)$, $x \in \mathcal{R}^k$. If F has a Lebesgue p.d.f. f, then the Lebesgue p.d.f. of $P_{(\mu,\Sigma)}$ is $\mathrm{Det}(\Sigma^{-1/2})f\left(\Sigma^{-1/2}(x - \mu)\right)$, $x \in \mathcal{R}^k$ (Proposition 1.8).

Many families of distributions in Table 1.2 (§1.3.1) are location, scale, or location-scale families. For example, the family of exponential distributions $E(a, \theta)$ is a location-scale family on $\mathcal{R}$ with location parameter a and scale parameter θ; the family of uniform distributions $U(0, \theta)$ is a scale family on $\mathcal{R}$ with a scale parameter θ. The k-dimensional normal family discussed in Example 2.4 is a location-scale family on $\mathcal{R}^k$.

2.2 Statistics, Sufficiency, and Completeness

Let us assume now that our data set is a realization of a sample X (a random vector) from an unknown population P on a probability space.

2.2.1 Statistics and their distributions

A measurable function of X, $T(X)$, is called a *statistic* if $T(X)$ is a known value whenever X is known, i.e., the function T is a known function. Statistical analyses are based on various statistics, for various purposes. Of course, X itself is a statistic, but it is a trivial statistic. The range of a nontrivial statistic $T(X)$ is usually simpler than that of X. For example, X may be a random n-vector and $T(X)$ may be a random p-vector with a p much smaller than n. This is desired since $T(X)$ simplifies the original data.

From a probabilistic point of view, the "information" within the statistic $T(X)$ concerning the unknown distribution of X is contained in the σ-field $\sigma(T(X))$. To see this, assume that S is any other statistic for which $\sigma(S(X)) = \sigma(T(X))$. Then, by Lemma 1.2, S is a measurable function of T, and T is a measurable function of S. Thus, once the value of S (or T) is known, so is the value of T (or S). That is, it is not the particular values of a statistic that contain the information, but the generated σ-field of the statistic. Values of a statistic may be important for other reasons.

Note that $\sigma(T(X)) \subset \sigma(X)$ and the two σ-fields are the same if and only if T is one-to-one. Usually $\sigma(T(X))$ simplifies $\sigma(X)$, i.e., a statistic provides a "reduction" of the σ-field.

Any $T(X)$ is a random element. If the distribution of X is unknown, then the distribution of T may also be unknown, although T is a known function. Finding the form of the distribution of T is one of the major problems in statistical inference and decision theory. Since T is a transformation of X, tools we learn in Chapter 1 for transformations may be useful in finding the distribution or an approximation to the distribution of $T(X)$.

Example 2.8. Let $X_1, ..., X_n$ be i.i.d. random variables having a common distribution P and $X = (X_1, ..., X_n)$. The sample mean $\bar{X}$ and sample variance S^2 defined in (2.1) and (2.2), respectively, are two commonly used statistics. Can we find the joint or the marginal distributions of $\bar{X}$ and S^2? It depends on how much we know about P.

First, let us consider the moments of $\bar{X}$ and S^2. Assume that P has a finite mean denoted by μ. Then

$$E\bar{X} = \mu.$$

If P is in a parametric family $\{P_\theta : \theta \in \Theta\}$, then $E\bar{X} = \int x dP_\theta = \mu(\theta)$ for some function $\mu(\cdot)$. Even if the form of μ is known, $\mu(\theta)$ may still be unknown when θ is unknown. Assume now that P has a finite variance denoted by σ^2. Then

$$\mathrm{Var}(\bar{X}) = \sigma^2/n,$$

which equals $\sigma^2(\theta)/n$ for some function $\sigma^2(\cdot)$ if P is in a parametric family. With a finite $\sigma^2 = \mathrm{Var}(X_1)$, we can also obtain that

$$ES^2 = \sigma^2.$$

With a finite $E|X_1|^3$, we can obtain $E(\bar{X})^3$ and $\mathrm{Cov}(\bar{X}, S^2)$, and with a finite $E|X_1|^4$, we can obtain $\mathrm{Var}(S^2)$ (exercise).

Next, consider the distribution of $\bar{X}$. If P is in a parametric family, we can often find the distribution of $\bar{X}$. See Example 1.20 and some exercises in §1.6. For example, $\bar{X}$ is $N(\mu, \sigma^2/n)$ if P is $N(\mu, \sigma^2)$; $n\bar{X}$ has the gamma distribution $\Gamma(n, \theta)$ if P is the exponential distribution $E(0, \theta)$. If P is not in a parametric family, then it is usually hard to find the exact form of the distribution of $\bar{X}$. One can, however, use the CLT (§1.5.4) to obtain an approximation to the distribution of $\bar{X}$. Applying Corollary 1.2 (for the case of $k = 1$), we obtain that

$$\sqrt{n}(\bar{X} - \mu) \to_d N(0, \sigma^2)$$

and, by (1.100), the distribution of $\bar{X}$ can be approximated by $N(\mu, \sigma^2/n)$, where μ and σ^2 are the mean and variance of P, respectively, and are assumed to be finite.

Compared to $\bar{X}$, the distribution of S^2 is harder to obtain. Assuming that P is $N(\mu, \sigma^2)$, one can show that $(n - 1)S^2/\sigma^2$ has the chi-square distribution χ^2_{n-1} (see Example 2.18). An approximate distribution for S^2 can be obtained from the approximate joint distribution of $\bar{X}$ and S^2 discussed next.

Under the assumption that P is $N(\mu, \sigma^2)$, it can be shown that $\bar{X}$ and S^2 are independent (Example 2.18). Hence, the joint distribution of $(\bar{X}, S^2)$ is the product of the marginal distributions of $\bar{X}$ and S^2 given in the previous discussion. Without the normality assumption, an approximate joint distribution can be obtained as follows. Assume again that $\mu = EX_1$, $\sigma^2 = \mathrm{Var}(X_1)$, and $E|X_1|^4$ are finite. Let $Y_i = (X_i - \mu, (X_i - \mu)^2)$, $i = 1, ..., n$. Then $Y_1, ..., Y_n$ are i.i.d. random 2-vectors with $EY_1 = (0, \sigma^2)$ and variance-covariance matrix

$$\Sigma = \begin{pmatrix} \sigma^2 & E(X_1 - \mu)^3 \\ E(X_1 - \mu)^3 & E(X_1 - \mu)^4 - \sigma^4 \end{pmatrix}.$$

Note that $\bar{Y} = n^{-1}\sum_{i=1}^n Y_i = (\bar{X} - \mu, \tilde{S}^2)$, where $\tilde{S}^2 = n^{-1}\sum_{i=1}^n (X_i - \mu)^2$. Applying the CLT (Corollary 1.2) to Y_i's, we obtain that

$$\sqrt{n}(\bar{X} - \mu, \tilde{S}^2 - \sigma^2) \to_d N_2(0, \Sigma).$$

Since

$$S^2 = \frac{n}{n-1}\left[\tilde{S}^2 - (\bar{X} - \mu)^2\right]$$

and $\bar{X} \to_{a.s.} \mu$ (the SLLN, Theorem 1.13), an application of Slutsky's theorem (Theorem 1.11) leads to

$$\sqrt{n}(\bar{X} - \mu, S^2 - \sigma^2) \to_d N_2(0, \Sigma). \quad \blacksquare$$

Example 2.9 (Order statistics). Let $X = (X_1, ..., X_n)$ with i.i.d. random components and let $X_{(i)}$ be the ith smallest value of $X_1, ..., X_n$. The statistics $X_{(1)}, ..., X_{(n)}$ are called the *order statistics*, which is a set of very useful statistics in addition to the sample mean and variance in the previous example. Suppose that X_i has a c.d.f. F having a Lebesgue p.d.f. f. Then the joint Lebesgue p.d.f. of $X_{(1)}, ..., X_{(n)}$ is

$$g(x_1, x_2, ..., x_n) = \begin{cases} n! f(x_1) f(x_2) \cdots f(x_n) & x_1 < x_2 < \cdots < x_n \\ 0 & \text{otherwise.} \end{cases}$$

The joint Lebesgue p.d.f. of $X_{(i)}$ and $X_{(j)}$, $1 \leq i < j \leq n$, is

$$g_{i,j}(x, y) = \begin{cases} \frac{n![F(x)]^{i-1}[F(y)-F(x)]^{j-i-1}[1-F(y)]^{n-j} f(x) f(y)}{(i-1)!(j-i-1)!(n-j)!} & x < y \\ 0 & \text{otherwise} \end{cases}$$

and the Lebesgue p.d.f. of $X_{(i)}$ is

$$g_i(x) = \frac{n!}{(i-1)!(n-i)!}[F(x)]^{i-1}[1 - F(x)]^{n-i} f(x). \quad \blacksquare$$

2.2.2 Sufficiency and minimal sufficiency

Having discussed the reduction of the σ-field $\sigma(X)$ by using a statistic $T(X)$, we now ask whether such a reduction results in any loss of information concerning the unknown population. If a statistic $T(X)$ is fully as informative as the original sample X, then statistical analyses can be done using $T(X)$ that is simpler than X. The next concept describes what we mean by fully informative.

Definition 2.4 (Sufficiency). Let X be a sample from an unknown population $P \in \mathcal{P}$, where $\mathcal{P}$ is a family of populations. A statistic $T(X)$ is said to be *sufficient* for $P \in \mathcal{P}$ (or for $\theta \in \Theta$ when $\mathcal{P} = \{P_\theta : \theta \in \Theta\}$ is a parametric family) if and only if the conditional distribution of X given T is *known* (does not depend on P or θ).   ∎

Definition 2.4 can be interpreted as follows. Once we observe X and compute a sufficient statistic $T(X)$, the original data X do not contain any further information concerning the unknown population P (since its conditional distribution is unrelated to P) and can be discarded. A sufficient statistic $T(X)$ contains all information about P contained in X (see Exercise 36 in §3.6 for an interpretation of this from another viewpoint) and provides a reduction of the data if T is not one-to-one. Thus, one of the questions raised in Example 2.1 can be answered as follows: it is enough to just look at $\bar{x}$ and s^2 for the problem of measuring θ if $(\bar{X}, S^2)$ is sufficient for P (or θ when θ is the only unknown parameter).

The concept of sufficiency depends on the given family $\mathcal{P}$. If T is sufficient for $P \in \mathcal{P}$, then T is also sufficient for $P \in \mathcal{P}_0 \subset \mathcal{P}$ but not necessarily sufficient for $P \in \mathcal{P}_1 \supset \mathcal{P}$.

Example 2.10. Suppose that $X = (X_1, ..., X_n)$ and $X_1, ..., X_n$ are i.i.d. from the binomial distribution with the p.d.f. (w.r.t. the counting measure)

$$f_\theta(z) = \theta^z (1 - \theta)^{1-z} I_{\{0,1\}}(z), \quad z \in \mathcal{R}, \quad \theta \in (0, 1).$$

For any realization x of X, x is a sequence of n ones and zeros. Consider the statistic $T(X) = \sum_{i=1}^n X_i$, which is the number of ones in X. Before showing that T is sufficient, we can intuitively argue that T contains all information about θ, since θ is the probability of an occurrence of a one in x. Given $T = t$ (the number of ones in x), what is left in the data set x is the redundant information about the positions of t ones. Since the random variables are discrete, it is not difficult to compute the conditional distribution of X given $T = t$. Note that

$$P(X = x | T = t) = \frac{P(X = x, T = t)}{P(T = t)}$$

and $P(T = t) = \binom{n}{t}\theta^t(1-\theta)^{n-t}I_{\{0,1,\ldots,n\}}(t)$. Let x_i be the ith component of x. If $t \neq \sum_{i=1}^n x_i$, then $P(X = x, T = t) = 0$. If $t = \sum_{i=1}^n x_i$, then

$$P(X = x, T = t) = \prod_{i=1}^n P(X_i = x_i) = \theta^t(1-\theta)^{n-t}\prod_{i=1}^n I_{\{0,1\}}(x_i).$$

Let $B_t = \{(x_1, \ldots, x_n) : x_i = 0, 1, \sum_{i=1}^n x_i = t\}$. Then

$$P(X = x | T = t) = \frac{1}{\binom{n}{t}}I_{B_t}(x)$$

is a known p.d.f. This shows that $T(X)$ is sufficient for $\theta \in (0, 1)$, according to Definition 2.4 with the family $\{f_\theta : \theta \in (0, 1)\}$. ∎

Finding a sufficient statistic by means of the definition is not convenient since it involves guessing a statistic T that might be sufficient and computing the conditional distribution of X given $T = t$. For families of populations having p.d.f.'s, a simple way of finding sufficient statistics is to use the factorization theorem. We first prove the following lemma.

Lemma 2.1. If a family $\mathcal{P}$ is dominated by a σ-finite measure, then $\mathcal{P}$ is dominated by a probability measure $Q = \sum_{i=1}^\infty c_i P_i$, where c_i's are nonnegative constants with $\sum_{i=1}^\infty c_i = 1$ and $P_i \in \mathcal{P}$.
Proof. Assume that $\mathcal{P}$ is dominated by a finite measure ν (the case of σ-finite ν is left as an exercise). Let $\mathcal{P}_0$ be the family of all measures of the form $\sum_{i=1}^\infty c_i P_i$, where $P_i \in \mathcal{P}$, $c_i \geq 0$, and $\sum_{i=1}^\infty c_i = 1$. Then, it suffices to show that there is a $Q \in \mathcal{P}_0$ such that $Q(A) = 0$ implies $P(A) = 0$ for all $P \in \mathcal{P}_0$. Let $\mathcal{C}$ be the class of events C for which there exists $P \in \mathcal{P}_0$ such that $P(C) > 0$ and $dP/d\nu > 0$ a.e. ν on C. Then there exists a sequence $\{C_i\} \subset \mathcal{C}$ such that $\nu(C_i) \to \sup_{C \in \mathcal{C}} \nu(C)$. Let C_0 be the union of all C_i's and $Q = \sum_{i=1}^\infty c_i P_i$, where P_i is the probability measure corresponding to C_i. Then $C_0 \in \mathcal{C}$ (exercise). Suppose now that $Q(A) = 0$. Let $P \in \mathcal{P}_0$ and $B = \{x : dP/d\nu > 0\}$. Since $Q(A \cap C_0) = 0$, $\nu(A \cap C_0) = 0$ and $P(A \cap C_0) = 0$. Then $P(A) = P(A \cap C_0^c \cap B)$. If $P(A \cap C_0^c \cap B) > 0$, then $\nu(C_0 \cup (A \cap C_0^c \cap B)) > \nu(C_0)$, which contradicts $\nu(C_0) = \sup_{C \in \mathcal{C}} \nu(C)$ since $A \cap C_0^c \cap B$ and therefore $C_0 \cup (A \cap C_0^c \cap B)$ is in $\mathcal{C}$. Thus, $P(A) = 0$ for all $P \in \mathcal{P}_0$. ∎

Theorem 2.2 (The factorization theorem). Suppose that X is a sample from $P \in \mathcal{P}$ and $\mathcal{P}$ is a family of probability measures on $(\mathcal{R}^n, \mathcal{B}^n)$ dominated by a σ-finite measure ν. Then $T(X)$ is sufficient for $P \in \mathcal{P}$ if and only if there are nonnegative Borel functions h (which does not depend on P) on $(\mathcal{R}^n, \mathcal{B}^n)$ and g_P (which depends on P) on the range of T such that

$$\frac{dP}{d\nu}(x) = g_P(T(x))h(x). \tag{2.11}$$

Proof. (i) Suppose that T is sufficient for $P \in \mathcal{P}$. Then, for any $A \in \mathcal{B}^n$, $P(A|T)$ does not depend on P. Let Q be the probability measure in Lemma 2.1. By Fubini's theorem and the result in Exercise 35 of §1.6,

$$
\begin{aligned}
Q(A \cap B) &= \sum_{j=1}^{\infty} c_j P_j(A \cap B) \\
&= \sum_{j=1}^{\infty} c_j \int_B P(A|T) dP_j \\
&= \int_B \sum_{j=1}^{\infty} c_j P(A|T) dP_j \\
&= \int_B P(A|T) dQ
\end{aligned}
$$

for any $B \in \sigma(T)$. Hence, $P(A|T) = E_Q(I_A|T)$ a.s. Q, where $E_Q(I_A|T)$ denotes the conditional expectation of I_A given T w.r.t. Q. Let $g_P(T)$ be the Radon-Nikodym derivative dP/dQ on the space $(\mathcal{R}^n, \sigma(T), Q)$. From Propositions 1.7 and 1.10,

$$
\begin{aligned}
P(A) &= \int P(A|T) dP \\
&= \int E_Q(I_A|T) g_P(T) dQ \\
&= \int E_Q[I_A g_P(T)|T] dQ \\
&= \int_A g_P(T) \frac{dQ}{d\nu} d\nu
\end{aligned}
$$

for any $A \in \mathcal{B}^n$. Hence, (2.11) holds with $h = dQ/d\nu$.

(ii) Suppose that (2.11) holds. Then

$$
\frac{dP}{dQ} = \frac{dP}{d\nu} \bigg/ \sum_{i=1}^{\infty} c_i \frac{dP_i}{d\nu} = g_P(T) \bigg/ \sum_{i=1}^{\infty} g_{P_i}(T) \quad \text{a.s. } Q, \tag{2.12}
$$

where the second equality follows from the result in Exercise 35 of §1.6. Let $A \in \sigma(X)$ and $P \in \mathcal{P}$. The sufficiency of T follows from

$$
P(A|T) = E_Q(I_A|T) \quad \text{a.s. } P, \tag{2.13}
$$

where $E_Q(I_A|T)$ is given in part (i) of the proof. This is because $E_Q(I_A|T)$ does not vary with $P \in \mathcal{P}$, and result (2.13) and Theorem 1.7 imply that the conditional distribution of X given T is determined by $E_Q(I_A|T)$, $A \in \sigma(X)$. By the definition of conditional probability, (2.13) follows from

$$
\int_B I_A dP = \int_B E_Q(I_A|T) dP \tag{2.14}
$$

for any $B \in \sigma(T)$. Let $B \in \sigma(T)$. By (2.12), dP/dQ is a Borel function of T. Then, by Proposition 1.7(i), Proposition 1.10(vi), and the definition of the conditional expectation, the right-hand side of (2.14) is equal to

$$\int_B E_Q(I_A|T)\frac{dP}{dQ}dQ = \int_B E_Q\left(I_A\frac{dP}{dQ}\bigg|T\right)dQ = \int_B I_A\frac{dP}{dQ}dQ,$$

which equals the left-hand side of (2.14). This proves (2.14) for any $B \in \sigma(T)$ and completes the proof. ∎

If $\mathcal{P}$ is an exponential family with p.d.f.'s given by (2.4) and $X(\omega) = \omega$, then we can apply Theorem 2.2 with $g_\theta(t) = \exp\{[\eta(\theta)]^\tau t - \xi(\theta)\}$ and conclude that T is a sufficient statistic for $\theta \in \Theta$. In Example 2.10 the joint distribution of X is in an exponential family with $T(X) = \sum_{i=1}^n X_i$. Hence, we can conclude that T is sufficient for $\theta \in (0,1)$ without computing the conditional distribution of X given T.

Example 2.11 (Truncation families). Let $\phi(x)$ be a positive Borel function on $(\mathcal{R}, \mathcal{B})$ such that $\int_a^b \phi(x)dx < \infty$ for any a and b, $-\infty < a < b < \infty$. Let $\theta = (a,b)$, $\Theta = \{(a,b) \in \mathcal{R}^2 : a < b\}$, and

$$f_\theta(x) = c(\theta)\phi(x)I_{(a,b)}(x),$$

where $c(\theta) = \left[\int_a^b \phi(x)dx\right]^{-1}$. Then $\{f_\theta : \theta \in \Theta\}$, called a truncation family, is a parametric family dominated by the Lebesgue measure on $\mathcal{R}$. Let $X_1, ..., X_n$ be i.i.d. random variables having the p.d.f. f_θ. Then the joint p.d.f. of $X = (X_1, ..., X_n)$ is

$$\prod_{i=1}^n f_\theta(x_i) = [c(\theta)]^n I_{(a,\infty)}(x_{(1)})I_{(-\infty,b)}(x_{(n)}) \prod_{i=1}^n \phi(x_i), \qquad (2.15)$$

where $x_{(i)}$ is the ith smallest value of $x_1, ..., x_n$. Let $T(X) = (X_{(1)}, X_{(n)})$, $g_\theta(t_1, t_2) = [c(\theta)]^n I_{(a,\infty)}(t_1)I_{(-\infty,b)}(t_2)$, and $h(x) = \prod_{i=1}^n \phi(x_i)$. By (2.15) and Theorem 2.2, $T(X)$ is sufficient for $\theta \in \Theta$. ∎

Example 2.12 (Order statistics). Let $X = (X_1, ..., X_n)$ and $X_1, ..., X_n$ be i.i.d. random variables having a distribution $P \in \mathcal{P}$, where $\mathcal{P}$ is the family of distributions on $\mathcal{R}$ having Lebesgue p.d.f.'s. Let $X_{(1)}, ..., X_{(n)}$ be the order statistics given in Example 2.9. Note that the joint p.d.f. of X is

$$f(x_1)\cdots f(x_n) = f(x_{(1)})\cdots f(x_{(n)}).$$

Hence, $T(X) = (X_{(1)}, ..., X_{(n)})$ is sufficient for $P \in \mathcal{P}$. The order statistics can be shown to be sufficient even when $\mathcal{P}$ is not dominated by any σ-finite measure, but Theorem 2.2 is not applicable (see Exercise 31 in §2.6). ∎

There are many sufficient statistics for a given family $\mathcal{P}$. In fact, if T is a sufficient statistic and $T = \psi(S)$, where ψ is measurable and S is another statistic, then S is sufficient. This is obvious from Theorem 2.2 if the population has a p.d.f., but it can be proved directly from Definition 2.4 (Exercise 25). For instance, in Example 2.10, $(\sum_{i=1}^{m} X_i, \sum_{i=m+1}^{n} X_i)$ is sufficient for θ, where m is any fixed integer between 1 and n. If T is sufficient and $T = \psi(S)$ with a measurable ψ that is not one-to-one, then $\sigma(T) \subset \sigma(S)$ and T is more useful than S, since T provides a further reduction of the data (or σ-field) without loss of information. Is there a sufficient statistic that provides "maximal" reduction of the data?

Before introducing the next concept, we need the following notation. If a statement holds except for outcomes in an event A satisfying $P(A) = 0$ for all $P \in \mathcal{P}$, then we say that the statement holds a.s. $\mathcal{P}$.

Definition 2.5 (Minimal sufficiency). Let T be a sufficient statistic for $P \in \mathcal{P}$. T is called a *minimal sufficient* statistic if and only if, for any other statistic S sufficient for $P \in \mathcal{P}$, there is a measurable function ψ such that $T = \psi(S)$ a.s. $\mathcal{P}$. ■

If both T and S are minimal sufficient statistics, then by definition there is a one-to-one measurable function ψ such that $T = \psi(S)$ a.s. $\mathcal{P}$. Hence, the minimal sufficient statistic is unique in the sense that two statistics that are one-to-one measurable functions of each other can be treated as one statistic.

Example 2.13. Let $X_1, ..., X_n$ be i.i.d. random variables from P_θ, the uniform distribution $U(\theta, \theta + 1)$, $\theta \in \mathcal{R}$. Suppose that $n > 1$. The joint Lebesgue p.d.f. of $(X_1, ..., X_n)$ is

$$f_\theta(x) = \prod_{i=1}^{n} I_{(\theta, \theta+1)}(x_i) = I_{(x_{(n)}-1, x_{(1)})}(\theta), \quad x = (x_1, ..., x_n) \in \mathcal{R}^n,$$

where $x_{(i)}$ denotes the ith smallest value of $x_1, ..., x_n$. By Theorem 2.2, $T = (X_{(1)}, X_{(n)})$ is sufficient for θ. Note that

$$x_{(1)} = \sup\{\theta : f_\theta(x) > 0\} \quad \text{and} \quad x_{(n)} = 1 + \inf\{\theta : f_\theta(x) > 0\}.$$

If $S(X)$ is a statistic sufficient for θ, then by Theorem 2.2, there are Borel functions h and g_θ such that $f_\theta(x) = g_\theta(S(x))h(x)$. For x with $h(x) > 0$,

$$x_{(1)} = \sup\{\theta : g_\theta(S(x)) > 0\} \quad \text{and} \quad x_{(n)} = 1 + \inf\{\theta : g_\theta(S(x)) > 0\}.$$

Hence, there is a measurable function ψ such that $T(x) = \psi(S(x))$ when $h(x) > 0$. Since $h > 0$ a.s. $\mathcal{P}$, we conclude that T is minimal sufficient. ■

Minimal sufficient statistics exist under weak assumptions, e.g., $\mathcal{P}$ contains distributions on $\mathcal{R}^k$ dominated by a σ-finite measure (Bahadur, 1957). The next theorem provides some useful tools for finding minimal sufficient statistics.

Theorem 2.3. Let $\mathcal{P}$ be a family of distributions on $\mathcal{R}^k$.
(i) Suppose that $\mathcal{P}_0 \subset \mathcal{P}$ and a.s. $\mathcal{P}_0$ implies a.s. $\mathcal{P}$. If T is sufficient for $P \in \mathcal{P}$ and minimal sufficient for $P \in \mathcal{P}_0$, then T is minimal sufficient for $P \in \mathcal{P}$.
(ii) Suppose that $\mathcal{P}$ contains p.d.f.'s $f_0, f_1, f_2, ...$, w.r.t. a σ-finite measure. Let $f_\infty(x) = \sum_{i=0}^\infty c_i f_i(x)$, where $c_i > 0$ for all i and $\sum_{i=0}^\infty c_i = 1$, and let $T_i(X) = f_i(x)/f_\infty(x)$ when $f_\infty(x) > 0$, $i = 0, 1, 2, ...$ Then $T(X) = (T_0, T_1, T_2, ...)$ is minimal sufficient for $P \in \mathcal{P}$. Furthermore, if $\{x : f_i(x) > 0\} \subset \{x : f_0(x) > 0\}$ for all i, then we may replace f_∞ by f_0, in which case $T(X) = (T_1, T_2, ...)$ is minimal sufficient for $P \in \mathcal{P}$.
(iii) Suppose that $\mathcal{P}$ contains p.d.f.'s f_P w.r.t. a σ-finite measure and that there exists a sufficient statistic $T(X)$ such that, for any possible values x and y of X, $f_P(x) = f_P(y)\phi(x, y)$ for all P implies $T(x) = T(y)$, where ϕ is a measurable function. Then $T(X)$ is minimal sufficient for $P \in \mathcal{P}$.
Proof. (i) If S is sufficient for $P \in \mathcal{P}$, then it is also sufficient for $P \in \mathcal{P}_0$ and, therefore, $T = \psi(S)$ a.s. $\mathcal{P}_0$ holds for a measurable function ψ. The result follows from the assumption that a.s. $\mathcal{P}_0$ implies a.s. $\mathcal{P}$.
(ii) Note that $f_\infty > 0$ a.s. $\mathcal{P}$. Let $g_i(T) = T_i$, $i = 0, 1, 2, ...$ Then $f_i(x) = g_i(T(x))f_\infty(x)$ a.s. $\mathcal{P}$. By Theorem 2.2, T is sufficient for $P \in \mathcal{P}$. Suppose that $S(X)$ is another sufficient statistic. By Theorem 2.2, there are Borel functions h and $\tilde{g}_i$ such that $f_i(x) = \tilde{g}_i(S(x))h(x)$, $i = 0, 1, 2, ...$ Then $T_i(x) = \tilde{g}_i(S(x))/\sum_{j=0}^\infty c_j \tilde{g}_j(S(x))$ for x's satisfying $f_\infty(x) > 0$. By Definition 2.5, T is minimal sufficient for $P \in \mathcal{P}$. The proof for the case where f_∞ is replaced by f_0 is the same.
(iii) From Bahadur (1957), there exists a minimal sufficient statistic $S(X)$. The result follows if we can show that $T(X) = \psi(S(X))$ a.s. $\mathcal{P}$ for a measurable function ψ. By Theorem 2.2, there are Borel functions g_P and h such that $f_P(x) = g_P(S(x))h(x)$ for all P. Let $A = \{x : h(x) = 0\}$. Then $P(A) = 0$ for all P. For x and y such that $S(x) = S(y)$, $x \notin A$ and $y \notin A$,

$$f_P(x) = g_P(S(x))h(x)$$
$$= g_P(S(y))h(x)h(y)/h(y)$$
$$= f_P(y)h(x)/h(y)$$

for all P. Hence $T(x) = T(y)$. This shows that there is a function ψ such that $T(x) = \psi(S(x))$ except for $x \in A$. It remains to show that ψ is measurable. Since S is minimal sufficient, $g(T(X)) = S(X)$ a.s. $\mathcal{P}$ for a measurable function g. Hence g is one-to-one and $\psi = g^{-1}$. The measurability of ψ follows from Theorem 3.9 in Parthasarathy (1967). ∎

Example 2.14. Let $\mathcal{P} = \{f_\theta : \theta \in \Theta\}$ be an exponential family with p.d.f.'s f_θ given by (2.4) and $X(\omega) = \omega$. Suppose that there exists $\Theta_0 = \{\theta_0, \theta_1, ..., \theta_p\} \subset \Theta$ such that the vectors $\eta_i = \eta(\theta_i) - \eta(\theta_0)$, $i = 1, ..., p$, are linearly independent in $\mathcal{R}^p$. (This is true if the family is of full rank.) We have shown that $T(X)$ is sufficient for $\theta \in \Theta$. We now show that T is in fact minimal sufficient for $\theta \in \Theta$. Let $\mathcal{P}_0 = \{f_\theta : \theta \in \Theta_0\}$. Note that the set $\{x : f_\theta(x) > 0\}$ does not depend on θ. It follows from Theorem 2.3(ii) with $f_\infty = f_{\theta_0}$ that

$$S(X) = \left(\exp\{\eta_1^\tau T(x) - \xi_1\}, ..., \exp\{\eta_p^\tau T(x) - \xi_p\}\right)$$

is minimal sufficient for $\theta \in \Theta_0$, where $\xi_i = \xi(\theta_i) - \xi(\theta_0)$. Since η_i's are linearly independent, there is a one-to-one measurable function ψ such that $T(X) = \psi(S(X))$ a.s. $\mathcal{P}_0$. Hence, T is minimal sufficient for $\theta \in \Theta_0$. It is easy to see that a.s. $\mathcal{P}_0$ implies a.s. $\mathcal{P}$. Thus, by Theorem 2.3(i), T is minimal sufficient for $\theta \in \Theta$. ∎

The results in Examples 2.13 and 2.14 can also be proved by using Theorem 2.3(iii) (Exercise 32).

The sufficiency (and minimal sufficiency) depends on the postulated family $\mathcal{P}$ of populations (statistical models). Hence, it may not be a useful concept if the proposed statistical model is wrong or at least one has some doubts about the correctness of the proposed model. From the examples in this section and some exercises in §2.6, one can find that for a wide variety of models, statistics such as $\bar{X}$ in (2.1), S^2 in (2.2), $(X_{(1)}, X_{(n)})$ in Example 2.11, and the order statistics in Example 2.9 are sufficient. Thus, using these statistics for data reduction and summarization does not lose any information when the true model is one of those models but we do not know exactly which model is correct.

2.2.3 Complete statistics

A statistic $V(X)$ is said to be *ancillary* if its distribution does not depend on the population P and *first-order ancillary* if $E[V(X)]$ is independent of P. A trivial ancillary statistic is the constant statistic $V(X) \equiv c \in \mathcal{R}$. If $V(X)$ is a nontrivial ancillary statistic, then $\sigma(V(X)) \subset \sigma(X)$ is a nontrivial σ-field that does not contain any information about P. Hence, if $S(X)$ is a statistic and $V(S(X))$ is a nontrivial ancillary statistic, it indicates that $\sigma(S(X))$ contains a nontrivial σ-field that does not contain any information about P and, hence, the "data" $S(X)$ may be further reduced. A sufficient statistic T appears to be most successful in reducing the data if no nonconstant function of T is ancillary or even first-order ancillary. This leads to the following concept of completeness.

Definition 2.6 (Completeness). A statistic $T(X)$ is said to be *complete* for $P \in \mathcal{P}$ if and only if, for any Borel f, $E[f(T)] = 0$ for all $P \in \mathcal{P}$ implies $f = 0$ a.s. $\mathcal{P}$. T is said to be *boundedly complete* if and only if the previous statement holds for any bounded Borel f. ∎

A complete statistic is boundedly complete. If T is complete (or boundedly complete) and $S = \psi(T)$ for a measurable ψ, then S is complete (or boundedly complete). Intuitively, a complete and sufficient statistic should be minimal sufficient, which was shown by Lehmann and Scheffé (1950) and Bahadur (1957) (see Exercise 48). However, a minimal sufficient statistic is not necessarily complete; for example, the minimal sufficient statistic $(X_{(1)}, X_{(n)})$ in Example 2.13 is not complete (Exercise 47).

Proposition 2.1. If P is in an exponential family of full rank with p.d.f.'s given by (2.6), then $T(X)$ is complete and sufficient for $\eta \in \Xi$.
Proof. We have shown that T is sufficient. Suppose that there is a function f such that $E[f(T)] = 0$ for all $\eta \in \Xi$. By Theorem 2.1(i),

$$\int f(t) \exp\{\eta^\tau t - \zeta(\eta)\} d\lambda = 0 \quad \text{for all } \eta \in \Xi,$$

where λ is a measure on $(\mathcal{R}^p, \mathcal{B}^p)$. Let η_0 be an interior point of Ξ. Then

$$\int f_+(t) e^{\eta^\tau t} d\lambda = \int f_-(t) e^{\eta^\tau t} d\lambda \quad \text{for all } \eta \in N(\eta_0), \tag{2.16}$$

where $N(\eta_0) = \{\eta \in \mathcal{R}^p : \|\eta - \eta_0\| < \epsilon\}$ for some $\epsilon > 0$. In particular,

$$\int f_+(t) e^{\eta_0^\tau t} d\lambda = \int f_-(t) e^{\eta_0^\tau t} d\lambda = c.$$

If $c = 0$, then $f = 0$ a.e. λ. If $c > 0$, then $c^{-1} f_+(t) e^{\eta_0^\tau t}$ and $c^{-1} f_-(t) e^{\eta_0^\tau t}$ are p.d.f.'s w.r.t. λ and (2.16) implies that their m.g.f.'s are the same in a neighborhood of 0. By Theorem 1.6(ii), $c^{-1} f_+(t) e^{\eta_0^\tau t} = c^{-1} f_-(t) e^{\eta_0^\tau t}$, i.e., $f = f_+ - f_- = 0$ a.e. λ. Hence T is complete. ∎

Proposition 2.1 is useful for finding a complete and sufficient statistic when the family of distributions is an exponential family of full rank.

Example 2.15. Suppose that $X_1, ..., X_n$ are i.i.d. random variables having the $N(\mu, \sigma^2)$ distribution, $\mu \in \mathcal{R}$, $\sigma > 0$. From Example 2.6, the joint p.d.f. of $X_1, ..., X_n$ is $(2\pi)^{-n/2} \exp\{\eta_1 T_1 + \eta_2 T_2 - n\zeta(\eta)\}$, where $T_1 = \sum_{i=1}^n X_i$, $T_2 = -\sum_{i=1}^n X_i^2$, and $\eta = (\eta_1, \eta_2) = \left(\frac{\mu}{\sigma^2}, \frac{1}{2\sigma^2}\right)$. Hence, the family of distributions for $X = (X_1, ..., X_n)$ is a natural exponential family of full rank ($\Xi = \mathcal{R} \times (0, \infty)$). By Proposition 2.1, $T(X) = (T_1, T_2)$ is complete and sufficient for η. Since there is a one-to-one correspondence between η

and $\theta = (\mu, \sigma^2)$, T is also complete and sufficient for θ. It can be shown that any one-to-one measurable function of a complete and sufficient statistic is also complete and sufficient (exercise). Thus, $(\bar{X}, S^2)$ is complete and sufficient for θ, where $\bar{X}$ and S^2 are the sample mean and variance given by (2.1) and (2.2), respectively. ∎

The following examples show how to find a complete statistic for a non-exponential family.

Example 2.16. Let $X_1, ..., X_n$ be i.i.d. random variables from P_θ, the uniform distribution $U(0, \theta)$, $\theta > 0$. The largest order statistic, $X_{(n)}$, is complete and sufficient for $\theta \in (0, \infty)$. The sufficiency of $X_{(n)}$ follows from the fact that the joint Lebesgue p.d.f. of $X_1, ..., X_n$ is $\theta^{-n} I_{(0,\theta)}(x_{(n)})$. From Example 2.9, $X_{(n)}$ has the Lebesgue p.d.f. $(nx^{n-1}/\theta^n) I_{(0,\theta)}(x)$ on $\mathcal{R}$. Let f be a Borel function on $[0, \infty)$ such that $E[f(X_{(n)})] = 0$ for all $\theta > 0$. Then

$$\int_0^\theta f(x) x^{n-1} dx = 0 \quad \text{for all } \theta > 0.$$

Let $G(\theta)$ be the left-hand side of the previous equation. Applying the result of differentiation of an integral (see, e.g., Royden (1968, §5.3)), we obtain that $G'(\theta) = f(\theta) \theta^{n-1}$ a.e. m_+, where m_+ is the Lebesgue measure on $([0, \infty), \mathcal{B}_{[0,\infty)})$. Since $G(\theta) = 0$ for all $\theta > 0$, $f(\theta) \theta^{n-1} = 0$ a.e. m_+ and, hence, $f(x) = 0$ a.e. m_+. Therefore, $X_{(n)}$ is complete and sufficient for $\theta \in (0, \infty)$. ∎

Example 2.17. In Example 2.12, we showed that the order statistics $T(X) = (X_{(1)}, ..., X_{(n)})$ of i.i.d. random variables $X_1, ..., X_n$ is sufficient for $P \in \mathcal{P}$, where $\mathcal{P}$ is the family of distributions on $\mathcal{R}$ having Lebesgue p.d.f.'s. We now show that $T(X)$ is also complete for $P \in \mathcal{P}$. Let $\mathcal{P}_0$ be the family of Lebesgue p.d.f.'s of the form

$$f(x) = C(\theta_1, ..., \theta_n) \exp\{-x^{2n} + \theta_1 x + \theta_2 x^2 + \cdots + \theta_n x^n\},$$

where $\theta_j \in \mathcal{R}$ and $C(\theta_1, ..., \theta_n)$ is a normalizing constant such that $\int f(x) dx = 1$. Then $\mathcal{P}_0 \subset \mathcal{P}$ and $\mathcal{P}_0$ is an exponential family of full rank. Note that the joint distribution of $X = (X_1, ..., X_n)$ is also in an exponential family of full rank. Thus, by Proposition 2.1, $U = (U_1, ..., U_n)$ is a complete statistic for $P \in \mathcal{P}_0$, where $U_j = \sum_{i=1}^n X_i^j$. Since a.s. $\mathcal{P}_0$ implies a.s. $\mathcal{P}$, $U(X)$ is also complete for $P \in \mathcal{P}$.

The result follows if we can show that there is a one-to-one correspondence between $T(X)$ and $U(X)$. Let $V_1 = \sum_{i=1}^n X_i$, $V_2 = \sum_{i<j} X_i X_j$, $V_3 = \sum_{i<j<k} X_i X_j X_k, ..., V_n = X_1 \cdots X_n$. From the identities

$$U_k - V_1 U_{k-1} + V_2 U_{k-2} - \cdots + (-1)^{k-1} V_{k-1} U_1 + (-1)^k k V_k = 0,$$

$k = 1, ..., n$, there is a one-to-one correspondence between $U(X)$ and $V(X) = (V_1, ..., V_n)$. From the identity

$$(t - X_1) \cdots (t - X_n) = t^n - V_1 t^{n-1} + V_2 t^{n-2} - \cdots + (-1)^n V_n,$$

there is a one-to-one correspondence between $V(X)$ and $T(X)$. This completes the proof and, hence, $T(X)$ is sufficient and complete for $P \in \mathcal{P}$. In fact, both $U(X)$ and $V(X)$ are sufficient and complete for $P \in \mathcal{P}$. ∎

The relationship between an ancillary statistic and a complete and sufficient statistic is characterized in the following result.

Theorem 2.4 (Basu's theorem). Let V and T be two statistics of X from a population $P \in \mathcal{P}$. If V is ancillary and T is boundedly complete and sufficient for $P \in \mathcal{P}$, then V and T are independent w.r.t. any $P \in \mathcal{P}$.

Proof. Let B be an event on the range of V. Since V is ancillary, $P(V^{-1}(B))$ is a constant. Since T is sufficient, $E[I_B(V)|T]$ is a function of T (independent of P). Since $E\{E[I_B(V)|T] - P(V^{-1}(B))\} = 0$ for all $P \in \mathcal{P}$, $P(V^{-1}(B)|T) = E[I_B(V)|T] = P(V^{-1}(B))$ a.s. $\mathcal{P}$, by the bounded completeness of T. Let A be an event on the range of T. Then, $P(T^{-1}(A) \cap V^{-1}(B)) = E\{E[I_A(T)I_B(V)|T]\} = E\{I_A(T)E[I_B(V)|T]\} = E\{I_A(T)P(V^{-1}(B))\} = P(T^{-1}(A))P(V^{-1}(B))$. Hence T and V are independent w.r.t. any $P \in \mathcal{P}$. ∎

Basu's theorem is useful in proving the independence of two statistics.

Example 2.18. Suppose that $X_1, ..., X_n$ are i.i.d. random variables having the $N(\mu, \sigma^2)$ distribution, with $\mu \in \mathcal{R}$ and a known $\sigma > 0$. It can be easily shown that the family $\{N(\mu, \sigma^2) : \mu \in \mathcal{R}\}$ is an exponential family of full rank with natural parameter $\eta = \mu/\sigma^2$. By Proposition 2.1, the sample mean $\bar{X}$ in (2.1) is complete and sufficient for η (and μ). Let S^2 be the sample variance given by (2.2). Since $S^2 = (n-1)^{-1} \sum_{i=1}^{n} (Z_i - \bar{Z})^2$, where $Z_i = X_i - \mu$ is $N(0, \sigma^2)$ and $\bar{Z} = n^{-1} \sum_{i=1}^{n} Z_i$, S^2 is an ancillary statistic (σ^2 is known). By Basu's theorem, $\bar{X}$ and S^2 are independent w.r.t. $N(\mu, \sigma^2)$ with $\mu \in \mathcal{R}$. Since σ^2 is arbitrary, $\bar{X}$ and S^2 are independent w.r.t. $N(\mu, \sigma^2)$ for any $\mu \in \mathcal{R}$ and $\sigma^2 > 0$.

Using the independence of $\bar{X}$ and S^2, we now show that $(n - 1)S^2/\sigma^2$ has the chi-square distribution χ_{n-1}^2. Note that

$$n \left(\frac{\bar{X} - \mu}{\sigma} \right)^2 + \frac{(n - 1)S^2}{\sigma^2} = \sum_{i=1}^{n} \left(\frac{X_i - \mu}{\sigma} \right)^2.$$

From the properties of the normal distributions, $n(\bar{X} - \mu)^2/\sigma^2$ has the chi-square distribution χ_1^2 with the m.g.f. $(1 - 2t)^{-1/2}$ and $\sum_{i=1}^{n} (X_i - \mu)^2/\sigma^2$

has the chi-square distribution χ_n^2 with the m.g.f. $(1 - 2t)^{-n/2}$, $t < 1/2$. By the independence of $\bar{X}$ and S^2, the m.g.f. of $(n-1)S^2/\sigma^2$ is

$$(1 - 2t)^{-n/2}/(1 - 2t)^{-1/2} = (1 - 2t)^{-(n-1)/2}$$

for $t < 1/2$. This is the m.g.f. of the chi-square distribution χ_{n-1}^2 and, therefore, the result follows. ∎

2.3 Statistical Decision Theory

In this section, we describe some basic elements in statistical decision theory. More developments are given in later chapters.

2.3.1 Decision rules, loss functions, and risks

Let X be a sample from a population $P \in \mathcal{P}$. A *statistical decision* is an *action* that we take after we observe X, for example, a conclusion about P or a characteristic of P. Throughout this section, we use $\mathbb{A}$ to denote the set of allowable actions. Let $\mathcal{F}_{\mathbb{A}}$ be a σ-field on $\mathbb{A}$. Then the measurable space $(\mathbb{A}, \mathcal{F}_{\mathbb{A}})$ is called the *action space*. Let $\mathcal{X}$ be the range of X and $\mathcal{F}_{\mathcal{X}}$ be a σ-field on $\mathcal{X}$. A *decision rule* is a measurable function (a statistic) T from $(\mathcal{X}, \mathcal{F}_{\mathcal{X}})$ to $(\mathbb{A}, \mathcal{F}_{\mathbb{A}})$. If a decision rule T is chosen, then we take the action $T(X) \in \mathbb{A}$ whence X is observed.

The construction or selection of decision rules cannot be done without any criterion about the performance of decision rules. In *statistical decision theory*, we set a criterion using a *loss function L*, which is a function from $\mathcal{P} \times \mathbb{A}$ to $[0, \infty)$ and is Borel on $(\mathbb{A}, \mathcal{F}_{\mathbb{A}})$ for each fixed $P \in \mathcal{P}$. If $X = x$ is observed and our decision rule is T, then our "loss" (in making a decision) is $L(P, T(x))$. The average loss for the decision rule T, which is called the *risk* of T, is defined to be

$$R_T(P) = E[L(P, T(X))] = \int_{\mathcal{X}} L(P, T(x)) dP_X(x). \qquad (2.17)$$

The loss and risk functions are denoted by $L(\theta, a)$ and $R_T(\theta)$ if $\mathcal{P}$ is a parametric family indexed by θ. A decision rule with small loss is preferred. But it is difficult to compare $L(P, T_1(X))$ and $L(P, T_2(X))$ for two decision rules, T_1 and T_2, since both of them are random. For this reason, the risk function (2.17) is introduced and we compare two decision rules by comparing their risks. A rule T_1 is *as good as* another rule T_2 if and only if

$$R_{T_1}(P) \le R_{T_2}(P) \quad \text{for any } P \in \mathcal{P}, \qquad (2.18)$$

and is *better* than T_2 if and only if (2.18) holds and $R_{T_1}(P) < R_{T_2}(P)$ for at least one $P \in \mathcal{P}$. Two decision rules T_1 and T_2 are *equivalent* if and only if $R_{T_1}(P) = R_{T_2}(P)$ for all $P \in \mathcal{P}$. If there is a decision rule T_* that is as good as any other rule in $\Im$, a class of allowable decision rules, then T_* is said to be $\Im$-*optimal* (or optimal if $\Im$ contains all possible rules).

Example 2.19. Consider the measurement problem in Example 2.1. Suppose that we need a decision on the value of $\theta \in \mathcal{R}$, based on the sample $X = (X_1, ..., X_n)$. If Θ is all possible values of θ, then it is reasonable to consider the action space $(\mathbb{A}, \mathcal{F}_{\mathbb{A}}) = (\Theta, \mathcal{B}_\Theta)$. An example of a decision rule is $T(X) = \bar{X}$, the sample mean defined by (2.1). A common loss function in this problem is the *squared error loss* $L(P, a) = (\theta - a)^2$, $a \in \mathbb{A}$. Then the loss for the decision rule $\bar{X}$ is the squared deviation between $\bar{X}$ and θ. Assuming that the population has mean μ and variance $\sigma^2 < \infty$, we obtain the following risk function for $\bar{X}$:

$$
\begin{aligned}
R_{\bar{X}}(P) &= E(\theta - \bar{X})^2 \\
&= (\theta - E\bar{X})^2 + E(E\bar{X} - \bar{X})^2 \\
&= (\theta - E\bar{X})^2 + \text{Var}(\bar{X}) \qquad\qquad (2.19) \\
&= (\mu - \theta)^2 + \tfrac{\sigma^2}{n}, \qquad\qquad\qquad (2.20)
\end{aligned}
$$

where result (2.20) follows from the results for the moments of $\bar{X}$ in Example 2.8. If θ is in fact the mean of the population, then the first term on the right-hand side of (2.20) is 0 and the risk is an increasing function of the population variance σ^2 and a decreasing function of the sample size n.

Consider another decision rule $T_1(X) = (X_{(1)} + X_{(n)})/2$. However, $R_{T_1}(P)$ does not have an explicit form if there is no further assumption on the population P. Suppose that $P \in \mathcal{P}$. Then, for some $\mathcal{P}$, $\bar{X}$ (or T_1) is better than T_1 (or $\bar{X}$) (exercise), whereas for some $\mathcal{P}$, neither $\bar{X}$ nor T_1 is better than the other.

A different loss function may also be considered. For example, $L(P, a) = |\theta - a|$, which is called the *absolute error loss*. However, $R_{\bar{X}}(P)$ and $R_{T_1}(P)$ do not have explicit forms unless $\mathcal{P}$ is of some specific form. ∎

The problem in Example 2.19 is a special case of a general problem called *estimation*, in which the action space is the set of all possible values of a population characteristic ϑ to be estimated. In an estimation problem, a decision rule T is called an *estimator* and result (2.19) holds with $\theta = \vartheta$ and $\bar{X}$ replaced by any estimator with a finite variance. The following example describes another type of important problem called *hypothesis testing*.

Example 2.20. Let $\mathcal{P}$ be a family of distributions, $\mathcal{P}_0 \subset \mathcal{P}$, and $\mathcal{P}_1 = \{P \in \mathcal{P} : P \notin \mathcal{P}_0\}$. A hypothesis testing problem can be formulated as that of deciding which of the following two statements is true:

$$H_0 : P \in \mathcal{P}_0 \qquad \text{versus} \qquad H_1 : P \in \mathcal{P}_1. \qquad (2.21)$$

Here, H_0 is called the *null hypothesis* and H_1 is called the *alternative hypothesis*. The action space for this problem contains only two elements, i.e., $\mathbb{A} = \{0, 1\}$, where 0 is the action of accepting H_0 and 1 is the action of rejecting H_0. A decision rule is called a *test*. Since a test $T(X)$ is a function from $\mathcal{X}$ to $\{0, 1\}$, $T(X)$ must have the form $I_C(X)$, where $C \in \mathcal{F}_{\mathcal{X}}$ is called the *rejection region* or *critical region* for testing H_0 versus H_1.

A simple loss function for this problem is the 0-1 loss: $L(P, a) = 0$ if a correct decision is made and 1 if an incorrect decision is made, i.e., $L(P, j) = 0$ for $P \in \mathcal{P}_j$ and $L(P, j) = 1$ otherwise, $j = 0, 1$. Under this loss, the risk is

$$R_T(P) = \begin{cases} P(T(X) = 1) = P(X \in C) & P \in \mathcal{P}_0 \\ P(T(X) = 0) = P(X \notin C) & P \in \mathcal{P}_1. \end{cases}$$

See Figure 2.2 on page 127 for an example of a graph of $R_T(\theta)$ for some T and P in a parametric family.

The 0-1 loss implies that the loss for two types of incorrect decisions (accepting H_0 when $P \in \mathcal{P}_1$ and rejecting H_0 when $P \in \mathcal{P}_0$) are the same. In some cases, one might assume unequal losses: $L(P, j) = 0$ for $P \in \mathcal{P}_j$, $L(P, 0) = c_0$ when $P \in \mathcal{P}_1$, and $L(P, 1) = c_1$ when $P \in \mathcal{P}_0$. ∎

In the following example the decision problem is neither an estimation nor a testing problem. Another example is given in Exercise 93 in §2.6.

Example 2.21. A hazardous toxic waste site requires clean-up when the true chemical concentration θ in the contaminated soil is higher than a given level $\theta_0 \geq 0$. Because of the limitation in resources, we would like to spend our money and efforts more in those areas that pose high risk to public health. In a particular area where soil samples are obtained, we would like to take one of these three actions: a complete clean-up (a_1), a partial clean-up (a_2), and no clean-up (a_3). Then $\mathbb{A} = \{a_1, a_2, a_3\}$. Suppose that the cost for a complete clean-up is c_1 and for a partial clean-up is $c_2 < c_1$; the risk to public health is $c_3(\theta - \theta_0)$ if $\theta > \theta_0$ and 0 if $\theta \leq \theta_0$; a complete clean-up can reduce the toxic concentration to an amount $\leq \theta_0$, whereas a partial clean-up can only reduce a fixed amount of the toxic concentration, i.e., the chemical concentration becomes $\theta - t$ after a partial clean-up, where t is a known constant. Then the loss function is given by

$L(\theta, a)$	a_1	a_2	a_3
$\theta \leq \theta_0$	c_1	c_2	0
$\theta_0 < \theta \leq \theta_0 + t$	c_1	c_2	$c_3(\theta - \theta_0)$
$\theta > \theta_0 + t$	c_1	$c_2 + c_3(\theta - \theta_0 - t)$	$c_3(\theta - \theta_0)$

The risk function can be calculated once the decision rule is specified. We discuss this example again in Chapter 4. ∎

Sometimes it is useful to consider *randomized decision rules*. Examples are given in §2.3.2, Chapters 4 and 6. A randomized decision rule is a function δ on $\mathcal{X} \times \mathcal{F}_\mathbb{A}$ such that, for every $A \in \mathcal{F}_\mathbb{A}$, $\delta(\cdot, A)$ is a Borel function and, for every $x \in \mathcal{X}$, $\delta(x, \cdot)$ is a probability measure on $(\mathbb{A}, \mathcal{F}_\mathbb{A})$. To choose an action in $\mathbb{A}$ when a randomized rule δ is used, we need to simulate a pseudorandom element of $\mathbb{A}$ according to $\delta(x, \cdot)$. Thus, an alternative way to describe a randomized rule is to specify the method of simulating the action from $\mathbb{A}$ for each $x \in \mathcal{X}$. If $\mathbb{A}$ is a subset of a Euclidean space, for example, then the result in Theorem 1.7(ii) can be applied. Also, see §7.2.3.

A nonrandomized decision rule T previously discussed can be viewed as a special randomized decision rule with $\delta(x, A) = I_A(T(x))$. Another example of a randomized rule is a discrete distribution $\delta(x, \cdot)$ assigning probability $p_j(x)$ to a nonrandomized decision rule $T_j(x)$, $j = 1, 2, \ldots$, in which case the rule δ can be equivalently defined as a rule taking value $T_j(x)$ with probability $p_j(x)$. See Exercise 64 for an example.

The loss function for a randomized rule δ is defined as

$$L(P, \delta, x) = \int_\mathbb{A} L(P, a) d\delta(x, a),$$

which reduces to the same loss function we discussed when δ is a nonrandomized rule. The risk of a randomized rule δ is then

$$R_\delta(P) = E[L(P, \delta, X)] = \int_\mathcal{X} \int_\mathbb{A} L(P, a) d\delta(x, a) dP_X(x). \qquad (2.22)$$

2.3.2 Admissibility and optimality

Consider a given decision problem with a given loss $L(P, a)$.

Definition 2.7 (Admissibility). Let $\mathfrak{S}$ be a class of decision rules (randomized or nonrandomized). A decision rule $T \in \mathfrak{S}$ is called $\mathfrak{S}$-*admissible* (or admissible when $\mathfrak{S}$ contains all possible rules) if and only if there does not exist any $S \in \mathfrak{S}$ that is better than T (in terms of the risk).   ∎

If a decision rule T is inadmissible, then there exists a rule better than T. Thus, T should not be used in principle. However, an admissible decision rule is not necessarily good. For example, in an estimation problem a silly estimator $T(X) \equiv$ a constant may be admissible (Exercise 71).

The relationship between the admissibility and the optimality defined in §2.3.1 can be described as follows. If T_* is $\mathfrak{S}$-optimal, then it is $\mathfrak{S}$-admissible; if T_* is $\mathfrak{S}$-optimal and T_0 is $\mathfrak{S}$-admissible, then T_0 is also $\mathfrak{S}$-optimal and is equivalent to T_*; if there are two $\mathfrak{S}$-admissible rules that are not equivalent, then there does not exist any $\mathfrak{S}$-optimal rule.

Suppose that we have a sufficient statistic $T(X)$ for $P \in \mathcal{P}$. Intuitively, our decision rule should be a function of T, based on the discussion in §2.2.2. This is not true in general, but the following result indicates that this is true if randomized decision rules are allowed.

Proposition 2.2. Suppose that $\mathbb{A}$ is a subset of $\mathcal{R}^k$. Let $T(X)$ be a sufficient statistic for $P \in \mathcal{P}$ and let δ_0 be a decision rule. Then

$$\delta_1(t, A) = E[\delta_0(X, A)|T = t], \qquad (2.23)$$

which is a randomized decision rule depending only on T, is equivalent to δ_0 if $R_{\delta_0}(P) < \infty$ for any $P \in \mathcal{P}$.
Proof. Note that δ_1 defined by (2.23) is a decision rule since δ_1 does not depend on the unknown P by the sufficiency of T. From (2.22),

$$
\begin{aligned}
R_{\delta_1}(P) &= E \left\{ \int_{\mathbb{A}} L(P, a) d\delta_1(X, a) \right\} \\
&= E \left\{ E \left[\int_{\mathbb{A}} L(P, a) d\delta_0(X, a) \Big| T \right] \right\} \\
&= E \left\{ \int_{\mathbb{A}} L(P, a) d\delta_0(X, a) \right\} \\
&= R_{\delta_0}(P),
\end{aligned}
$$

where the proof of the second equality is left to the reader. ∎

Note that Proposition 2.2 does not imply that δ_0 is inadmissible. Also, if δ_0 is a nonrandomized rule,

$$\delta_1(t, A) = E[I_A(\delta_0(X))|T = t] = P(\delta_0(X) \in A|T = t)$$

is still a randomized rule, unless $\delta_0(X) = h(T(X))$ a.s. P for some Borel function h (Exercise 75). Hence, Proposition 2.2 does not apply to situations where randomized rules are not allowed.

The following result tells us when nonrandomized rules are all we need and when decision rules that are not functions of sufficient statistics are inadmissible.

Theorem 2.5. Suppose that $\mathbb{A}$ is a convex subset of $\mathcal{R}^k$ and that for any $P \in \mathcal{P}$, $L(P, a)$ is a convex function of a.
(i) Let δ be a randomized rule satisfying $\int_{\mathbb{A}} \|a\| d\delta(x, a) < \infty$ for any $x \in \mathcal{X}$ and let $T_1(x) = \int_{\mathbb{A}} a d\delta(x, a)$. Then $L(P, T_1(x)) \leq L(P, \delta, x)$ (or $L(P, T_1(x)) < L(P, \delta, x)$ if L is strictly convex in a) for any $x \in \mathcal{X}$ and $P \in \mathcal{P}$.
(ii) (Rao-Blackwell theorem). Let T be a sufficient statistic for $P \in \mathcal{P}$, $T_0 \in \mathcal{R}^k$ be a nonrandomized rule satisfying $E\|T_0\| < \infty$, and $T_1 = E[T_0(X)|T]$. Then $R_{T_1}(P) \leq R_{T_0}(P)$ for any $P \in \mathcal{P}$. If L is strictly convex in a and T_0 is not a function of T, then T_0 is inadmissible. ∎

The proof of Theorem 2.5 is an application of Jensen's inequality (1.47) and is left to the reader.

The concept of admissibility helps us to eliminate some decision rules. However, usually there are still too many rules left after the elimination of some rules according to admissibility and sufficiency. Although one is typically interested in a $\Im$-optimal rule, frequently it does not exist, if $\Im$ is either too large or too small. The following examples are illustrations.

Example 2.22. Let $X_1, ..., X_n$ be i.i.d. random variables from a population $P \in \mathcal{P}$ that is the family of populations having finite mean μ and variance σ^2. Consider the estimation of μ ($\mathbb{A} = \mathcal{R}$) under the squared error loss. It can be shown that if we let $\Im$ be the class of all possible estimators, then there is no $\Im$-optimal rule (exercise). Next, let $\Im_1$ be the class of all linear functions in $X = (X_1, ..., X_n)$, i.e., $T(X) = \sum_{i=1}^{n} c_i X_i$ with known $c_i \in \mathcal{R}$, $i = 1, ..., n$. It follows from (2.19) and the discussion after Example 2.19 that

$$R_T(P) = \mu^2 \left(\sum_{i=1}^{n} c_i - 1 \right)^2 + \sigma^2 \sum_{i=1}^{n} c_i^2. \tag{2.24}$$

We now show that there does not exist $T_* = \sum_{i=1}^{n} c_i^* X_i$ such that $R_{T_*}(P) \leq R_T(P)$ for any $P \in \mathcal{P}$ and $T \in \Im_1$. If there is such a T_*, then $(c_1^*, ..., c_n^*)$ is a minimum of the function of $(c_1, ..., c_n)$ on the right-hand side of (2.24). Then $c_1^*, ..., c_n^*$ must be the same and equal to $\mu^2/(\sigma^2 + n\mu^2)$, which depends on P. Hence T_* is not a statistic. This shows that there is no $\Im_1$-optimal rule.

Consider now a subclass $\Im_2 \subset \Im_1$ with c_i's satisfying $\sum_{i=1}^{n} c_i = 1$. From (2.24), $R_T(P) = \sigma^2 \sum_{i=1}^{n} c_i^2$ if $T \in \Im_2$. Minimizing $\sigma^2 \sum_{i=1}^{n} c_i^2$ subject to $\sum_{i=1}^{n} c_i = 1$ leads to an optimal solution of $c_i = n^{-1}$ for all i. Thus, the sample mean $\bar{X}$ is $\Im_2$-optimal.

There may not be any optimal rule if we consider a small class of decision rules. For example, if $\Im_3$ contains all the rules in $\Im_2$ except $\bar{X}$, then one can show that there is no $\Im_3$-optimal rule. ∎

Example 2.23. Assume that the sample X has the binomial distribution $Bi(\theta, n)$ with an unknown $\theta \in (0, 1)$ and a fixed integer $n > 1$. Consider the hypothesis testing problem described in Example 2.20 with $H_0 : \theta \in (0, \theta_0]$ versus $H_1 : \theta \in (\theta_0, 1)$, where $\theta_0 \in (0, 1)$ is a fixed value. Suppose that we are only interested in the following class of nonrandomized decision rules: $\Im = \{T_j : j = 0, 1, ..., n-1\}$, where $T_j(X) = I_{\{j+1,...,n\}}(X)$. From Example 2.20, the risk function for T_j under the 0-1 loss is

$$R_{T_j}(\theta) = P(X > j)I_{(0,\theta_0]}(\theta) + P(X \leq j)I_{(\theta_0,1)}(\theta).$$

For any integers k and j, $0 \leq k < j \leq n - 1$,

$$R_{T_j}(\theta) - R_{T_k}(\theta) = \begin{cases} -P(k < X \leq j) < 0 & 0 < \theta \leq \theta_0 \\ P(k < X \leq j) > 0 & \theta_0 < \theta < 1. \end{cases}$$

Hence, neither T_j nor T_k is better than the other. This shows that every T_j is $\Im$-admissible and, thus, there is no $\Im$-optimal rule. ■

In view of the fact that an optimal rule often does not exist, statisticians adopt the following two approaches to choose a decision rule. The first approach is to define a class $\Im$ of decision rules that have some desirable properties (statistical and/or nonstatistical) and then try to find the best rule in $\Im$. In Example 2.22, for instance, any estimator T in $\Im_2$ has the property that T is linear in X and $E[T(X)] = \mu$. In a general estimation problem, we can use the following concept.

Definition 2.8 (Unbiasedness). In an estimation problem, the *bias* of an estimator $T(X)$ of a real-valued parameter ϑ of the unknown population is defined to be $b_T(P) = E[T(X)] - \vartheta$ (which is denoted by $b_T(\theta)$ when P is in a parametric family indexed by θ). An estimator $T(X)$ is said to be *unbiased* for ϑ if and only if $b_T(P) = 0$ for any $P \in \mathcal{P}$. ■

Thus, $\Im_2$ in Example 2.22 is the class of unbiased estimators linear in X. In Chapter 3, we discuss how to find a $\Im$-optimal estimator when $\Im$ is the class of unbiased estimators or unbiased estimators linear in X.

Another class of decision rules can be defined after we introduce the concept of *invariance*.

Definition 2.9 Let X be a sample from $P \in \mathcal{P}$.
(i) A class $\mathcal{G}$ of one-to-one transformations of X is called a *group* if and only if $g_i \in \mathcal{G}$ implies $g_1 \circ g_2 \in \mathcal{G}$ and $g_i^{-1} \in \mathcal{G}$.
(ii) We say that $\mathcal{P}$ is *invariant* under $\mathcal{G}$ if and only if $\bar{g}(P_X) = P_{g(X)}$ is a one-to-one transformation from $\mathcal{P}$ onto $\mathcal{P}$ for each $g \in \mathcal{G}$.
(iii) A decision problem is said to be *invariant* if and only if $\mathcal{P}$ is invariant under $\mathcal{G}$ and the loss $L(P, a)$ is invariant in the sense that, for every $g \in \mathcal{G}$ and every $a \in \mathbb{A}$, there exists a unique $g(a) \in \mathbb{A}$ such that $L(P_X, a) = L\left(P_{g(X)}, g(a)\right)$. (Note that $g(X)$ and $g(a)$ are different functions in general.)
(iv) A decision rule $T(x)$ is said to be *invariant* if and only if, for every $g \in \mathcal{G}$ and every $x \in \mathcal{X}$, $T(g(x)) = g(T(x))$. ■

Invariance means that our decision is not affected by one-to-one transformations of data.

In a problem where the distribution of X is in a location-scale family

$\mathcal{P}$ on $\mathcal{R}^k$, we often consider location-scale transformations of data X of the form $g(X) = AX + c$, where $c \in \mathcal{C} \subset \mathcal{R}^k$ and $A \in \mathcal{T}$, a class of invertible $k \times k$ matrices. Assume that if $A_i \in \mathcal{T}$, $i = 1, 2$, then $A_i^{-1} \in \mathcal{T}$ and $A_1 A_2 \in \mathcal{T}$, and that if $c_i \in \mathcal{C}$, $i = 1, 2$, then $-c_i \in \mathcal{C}$ and $Ac_1 + c_2 \in \mathcal{C}$ for any $A \in \mathcal{T}$. Then the collection of all transformations is a group. A special case is given in the following example.

Example 2.24. Let X have i.i.d. components from a population in a location family $\mathcal{P} = \{P_\mu : \mu \in \mathcal{R}\}$. Consider the location transformation $g_c(X) = X + cJ_k$, where $c \in \mathcal{R}$ and J_k is the k-vector whose components are all equal to 1. The group of transformation is $\mathcal{G} = \{g_c : c \in \mathcal{R}\}$, which is a location-scale transformation group with $\mathcal{T} = \{I_k\}$ and $\mathcal{C} = \{cJ_k : c \in \mathcal{R}\}$. $\mathcal{P}$ is invariant under $\mathcal{G}$ with $\bar{g}_c(P_\mu) = P_{\mu+c}$. For estimating μ under the loss $L(\mu, a) = L(\mu - a)$, where $L(\cdot)$ is a nonnegative Borel function, the decision problem is invariant with $g_c(a) = a + c$. A decision rule T is invariant if and only if $T(x + cJ_k) = T(x) + c$ for every $x \in \mathcal{R}^k$ and $c \in \mathcal{R}$. An example of an invariant decision rule is $T(x) = l^\tau x$ for some $l \in \mathcal{R}^k$ with $l^\tau J_k = 1$. Note that $T(x) = l^\tau x$ with $l^\tau J_k = 1$ is in the class $\Im_2$ in Example 2.22. ∎

In §4.2 and §6.3, we discuss the problem of finding a $\Im$-optimal rule when $\Im$ is a class of invariant decision rules.

The second approach to finding a good decision rule is to consider some characteristic R_T of $R_T(P)$, for a given decision rule T, and then minimize R_T over $T \in \Im$. The following are two popular ways to carry out this idea. The first one is to consider an average of $R_T(P)$ over $P \in \mathcal{P}$:

$$r_T(\Pi) = \int_{\mathcal{P}} R_T(P) d\Pi(P),$$

where Π is a known probability measure on $(\mathcal{P}, \mathcal{F}_{\mathcal{P}})$ with an appropriate σ-field $\mathcal{F}_{\mathcal{P}}$. $r_T(\Pi)$ is called the *Bayes risk* of T w.r.t. Π. If $T_* \in \Im$ and $r_{T_*}(\Pi) \leq r_T(\Pi)$ for any $T \in \Im$, then T_* is called a $\Im$-*Bayes rule* (or Bayes rule when $\Im$ contains all possible rules) w.r.t. Π. The second method is to consider the worst situation, i.e., $\sup_{P \in \mathcal{P}} R_T(P)$. If $T_* \in \Im$ and $\sup_{P \in \mathcal{P}} R_{T_*}(P) \leq \sup_{P \in \mathcal{P}} R_T(P)$ for any $T \in \Im$, then T_* is called a $\Im$-*minimax* rule (or minimax rule when $\Im$ contains all possible rules). Bayes and minimax rules are discussed in Chapter 4.

Example 2.25. We usually try to find a Bayes rule or a minimax rule in a parametric problem where $P = P_\theta$ for a $\theta \in \mathcal{R}^k$. Consider the special case of $k = 1$ and $L(\theta, a) = (\theta - a)^2$, the squared error loss. Note that

$$r_T(\Pi) = \int_{\mathcal{R}} E[\theta - T(X)]^2 d\Pi(\theta),$$

which is equivalent to $E[\boldsymbol{\theta} - T(X)]^2$, where $\boldsymbol{\theta}$ is a random variable having the distribution Π and, given $\boldsymbol{\theta} = \theta$, the conditional distribution of X is P_θ. Then, the problem can be viewed as a prediction problem for $\boldsymbol{\theta}$ using functions of X. Using the result in Example 1.22, the best predictor is $E(\boldsymbol{\theta}|X)$, which is the $\mathfrak{S}$-Bayes rule w.r.t. Π with $\mathfrak{S}$ being the class of rules $T(X)$ satisfying $E[T(X)]^2 < \infty$ for any θ.

As a more specific example, let $X = (X_1, ..., X_n)$ with i.i.d. components having the $N(\mu, \sigma^2)$ distribution with an unknown $\mu = \theta \in \mathcal{R}$ and a known σ^2, and let Π be the $N(\mu_0, \sigma_0^2)$ distribution with known μ_0 and σ_0^2. Then the conditional distribution of $\boldsymbol{\theta}$ given $X = x$ is $N(\mu_*(x), c^2)$ with

$$\mu_*(x) = \frac{\sigma^2}{n\sigma_0^2 + \sigma^2}\mu_0 + \frac{n\sigma_0^2}{n\sigma_0^2 + \sigma^2}\bar{x} \quad \text{and} \quad c^2 = \frac{\sigma_0^2\sigma^2}{n\sigma_0^2 + \sigma^2} \quad (2.25)$$

(exercise). The Bayes rule w.r.t. Π is $E(\boldsymbol{\theta}|X) = \mu_*(X)$.

In this special case we can show that the sample mean $\bar{X}$ is $\mathfrak{S}$-minimax with $\mathfrak{S}$ being the collection of all decision rules. For any decision rule T,

$$\sup_{\theta \in \mathcal{R}} R_T(\theta) \geq \int_{\mathcal{R}} R_T(\theta)d\Pi(\theta)$$
$$\geq \int_{\mathcal{R}} R_{\mu_*}(\theta)d\Pi(\theta)$$
$$= E\left\{[\boldsymbol{\theta} - \mu_*(X)]^2\right\}$$
$$= E\left\{E\{[\boldsymbol{\theta} - \mu_*(X)]^2|X\}\right\}$$
$$= E(c^2)$$
$$= c^2,$$

where $\mu_*(X)$ is the Bayes rule given in (2.25) and c^2 is also given in (2.25). Since this result is true for any $\sigma_0^2 > 0$ and $c^2 \to \sigma^2/n$ as $\sigma_0^2 \to \infty$,

$$\sup_{\theta \in \mathcal{R}} R_T(\theta) \geq \frac{\sigma^2}{n} = \sup_{\theta \in \mathcal{R}} R_{\bar{X}}(\theta),$$

where the equality holds because the risk of $\bar{X}$ under the squared error loss is, by (2.20), σ^2/n and independent of $\theta = \mu$. Thus, $\bar{X}$ is minimax.

A minimax rule in a general case may be difficult to obtain. It can be seen that if both μ and σ^2 are unknown in the previous discussion, then

$$\sup_{\theta \in \mathcal{R} \times (0,\infty)} R_{\bar{X}}(\theta) = \infty, \quad (2.26)$$

where $\theta = (\mu, \sigma^2)$. Hence $\bar{X}$ cannot be minimax unless (2.26) holds with $\bar{X}$ replaced by any decision rule T, in which case minimaxity becomes meaningless. ∎

2.4 Statistical Inference

The loss function plays a crucial role in statistical decision theory. Loss functions can be obtained from a utility analysis (Berger, 1985), but in many problems they have to be determined subjectively. In *statistical inference*, we make an inference about the unknown population based on the sample X and *inference procedures* without using any loss function, although any inference procedure can be cast in decision-theoretic terms as a decision rule.

There are three main types of inference procedures: *point estimators*, *hypothesis tests*, and *confidence sets*.

2.4.1 Point estimators

The problem of estimating an unknown parameter related to the unknown population is introduced in Example 2.19 and the discussion after Example 2.19 as a special statistical decision problem. In statistical inference, however, estimators of parameters are derived based on some principle (such as the unbiasedness, invariance, sufficiency, substitution principle, likelihood principle, Bayesian principle, etc.), not based on a loss or risk function. Since confidence sets are sometimes also called *interval estimators* or *set estimators*, estimators of parameters are called point estimators.

In Chapters 3 through 5, we consider how to derive a "good" point estimator based on some principle. Here we focus on how to assess performance of point estimators.

Let $\vartheta \in \tilde{\Theta} \subset \mathcal{R}$ be a parameter to be estimated, which is a function of the unknown population P or θ if P is in a parametric family. An estimator is a statistic with range $\tilde{\Theta}$. First, one has to realize that any estimator $T(X)$ of ϑ is subject to an estimation error $T(x) - \vartheta$ when we observe $X = x$. This is not just because $T(X)$ is random. In some problems $T(x)$ never equals ϑ. A trivial example is when $T(X)$ has a continuous c.d.f. so that $P(T(X) = \vartheta) = 0$. As a nontrivial example, let $X_1, ..., X_n$ be i.i.d. binary random variables (also called Bernoulli variables) with $P(X_i = 1) = p$ and $P(X_i = 0) = 1 - p$. The sample mean $\bar{X}$ is shown to be a good estimator of $\vartheta = p$ in later chapters, but $\bar{x}$ never equals ϑ if ϑ is not one of j/n, $j = 0, 1, ..., n$. Thus, we cannot assess the performance of $T(X)$ by the values of $T(x)$ with particular x's and it is also not worthwhile to do so.

The bias $b_T(P)$ and unbiasedness of a point estimator $T(X)$ is defined in Definition 2.8. Unbiasedness of $T(X)$ means that the mean of $T(X)$ is equal to ϑ. An unbiased estimator $T(X)$ can be viewed as an estimator without "systematic" error, since, on the average, it does not overestimate (i.e., $b_T(P) > 0$) or underestimate (i.e., $b_T(P) < 0$). However, an unbiased estimator $T(X)$ may have large positive and negative errors $T(x) - \vartheta$, $x \in \mathcal{X}$, although these errors cancel each other in the calculation of the bias, which is the average $\int [T(x) - \vartheta] dP_X(x)$.

Hence, for an unbiased estimator $T(X)$, it is desired that the values of $T(x)$ be highly concentrated around ϑ. The variance of $T(X)$ is commonly used as a measure of the dispersion of $T(X)$. The *mean squared error* (mse) of $T(X)$ as an estimator of ϑ is defined to be

$$\text{mse}_T(P) = E[T(X) - \vartheta]^2 = [b_T(P)]^2 + \text{Var}(T(X)), \tag{2.27}$$

which is denoted by $\text{mse}_T(\theta)$ if P is in a parametric family. $\text{mse}_T(P)$ is equal to the variance $\text{Var}(T(X))$ if and only if $T(X)$ is unbiased. Note that the mse is simply the risk of T in statistical decision theory under the squared error loss.

In addition to the variance and the mse, the following are other measures of dispersion that are often used in point estimation problems. The first one is the *mean absolute error* of an estimator $T(X)$ defined to be $E|T(X) - \vartheta|$. The second one is the probability of falling outside a stated distance of ϑ, i.e., $P(|T(X) - \vartheta| \geq \epsilon)$ with a fixed $\epsilon > 0$. Again, these two measures of dispersion are risk functions in statistical decision theory with loss functions $|\vartheta - a|$ and $I_{(\epsilon, \infty)}(|\vartheta - a|)$, respectively.

For the bias, variance, mse, and mean absolute error, we have implicitly assumed that certain moments of $T(X)$ exist. On the other hand, the dispersion measure $P(|T(X) - \vartheta| \geq \epsilon)$ depends on the choice of ϵ. It is possible that some estimators are good in terms of one measure of dispersion, but not in terms of other measures of dispersion. The mse, which is a function of bias and variance according to (2.27), is mathematically easy to handle and, hence, is used the most often in the literature. In this book, we use the mse to assess and compare point estimators unless otherwise stated.

Examples 2.19 and 2.22 provide some examples of estimators and their biases, variances, and mse's. The following are two more examples.

Example 2.26. Consider the life-time testing problem in Example 2.2. Let $X_1, ..., X_n$ be i.i.d. from an unknown c.d.f. F. Suppose that the parameter of interest is $\vartheta = 1 - F(t)$ for a fixed $t > 0$. If F is not in a parametric family, then a *nonparametric* estimator of $F(t)$ is the *empirical* c.d.f.

$$F_n(t) = \frac{1}{n} \sum_{i=1}^{n} I_{(-\infty, t]}(X_i), \qquad t \in \mathcal{R}. \tag{2.28}$$

Since $I_{(-\infty, t]}(X_1), ..., I_{(-\infty, t]}(X_n)$ are i.i.d. binary random variables with $P(I_{(-\infty, t]}(X_i) = 1) = F(t)$, the random variable $nF_n(t)$ has the binomial distribution $Bi(F(t), n)$. Consequently, $F_n(t)$ is an unbiased estimator of

$F(t)$ and $\mathrm{Var}(F_n(t)) = \mathrm{mse}_{F_n(t)}(P) = F(t)[1 - F(t)]/n$. Since any linear combination of unbiased estimators is unbiased for the same linear combination of the parameters (by the linearity of expectations), an unbiased estimator of ϑ is $U(X) = 1 - F_n(t)$, which has the same variance and mse as $F_n(t)$.

The estimator $U(X) = 1 - F_n(t)$ can be improved in terms of the mse if there is further information about F. Suppose that F is the c.d.f. of the exponential distribution $E(0, \theta)$ with an unknown $\theta > 0$. Then $\vartheta = e^{-t/\theta}$. From §2.2.2, the sample mean $\bar{X}$ is sufficient for $\theta > 0$. Since the squared error loss is strictly convex, an application of Theorem 2.5(ii) (Rao-Blackwell theorem) shows that the estimator $T(X) = E[1 - F_n(t)|\bar{X}]$, which is also unbiased, is better than $U(X)$ in terms of the mse. Figure 2.1 shows graphs of the mse's of $U(X)$ and $T(X)$, as functions of θ, in the special case of $n = 10$, $t = 2$, and $F(x) = (1 - e^{-x/\theta})I_{(0,\infty)}(x)$. ∎

Example 2.27. Consider the sample survey problem in Example 2.3 with a constant selection probability $p(s)$ and univariate y_i. Let $\vartheta = Y = \sum_{i=1}^{N} y_i$, the population total. We now show that the estimator $\hat{Y} = \frac{N}{n}\sum_{i \in s} y_i$ is an unbiased estimator of Y. Let $a_i = 1$ if $i \in s$ and $a_i = 0$ otherwise. Thus, $\hat{Y} = \frac{N}{n}\sum_{i=1}^{N} a_i y_i$. Since $p(s)$ is constant, $E(a_i) = P(a_i = 1) = n/N$ and

$$E(\hat{Y}) = E\left(\frac{N}{n}\sum_{i=1}^{N} a_i y_i\right) = \frac{N}{n}\sum_{i=1}^{N} y_i E(a_i) = \sum_{i=1}^{N} y_i = Y.$$

Note that
$$\mathrm{Var}(a_i) = E(a_i) - [E(a_i)]^2 = \frac{n}{N}\left(1 - \frac{n}{N}\right)$$

and for $i \neq j$,

$$\mathrm{Cov}(a_i, a_j) = P(a_i = 1, a_j = 1) - E(a_i)E(a_j) = \frac{n(n-1)}{N(N-1)} - \frac{n^2}{N^2}.$$

Hence, the variance or the mse of $\hat{Y}$ is

$$\mathrm{Var}(\hat{Y}) = \frac{N^2}{n^2}\mathrm{Var}\left(\sum_{i=1}^{N} a_i y_i\right)$$

$$= \frac{N^2}{n^2}\left[\sum_{i=1}^{N} y_i^2 \mathrm{Var}(a_i) + 2\sum_{1 \leq i < j \leq N} y_i y_j \mathrm{Cov}(a_i, a_j)\right]$$

$$= \frac{N}{n}\left(1 - \frac{n}{N}\right)\left(\sum_{i=1}^{N} y_i^2 - \frac{2}{N-1}\sum_{1 \leq i < j \leq N} y_i y_j\right)$$

$$= \frac{N^2}{n(N-1)}\left(1 - \frac{n}{N}\right)\sum_{i=1}^{N}\left(y_i - \frac{Y}{N}\right)^2. \quad ∎$$

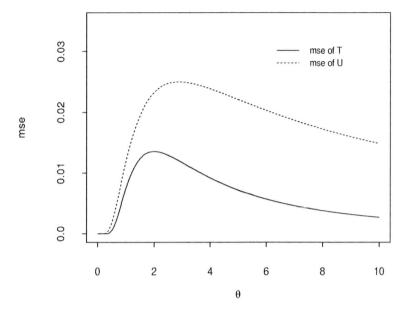

Figure 2.1: mse's of $U(X)$ and $T(X)$ in Example 2.26

2.4.2 Hypothesis tests

The basic elements of a hypothesis testing problem are described in Example 2.20. In statistical inference, tests for a hypothesis are derived based on some principles similar to those given in an estimation problem. Chapter 6 is devoted to deriving tests for various types of hypotheses. Several key ideas are discussed here.

To test the hypotheses H_0 versus H_1 given in (2.21), there are only two types of statistical errors we may commit: rejecting H_0 when H_0 is true (called the *type I error*) and accepting H_0 when H_0 is wrong (called the *type II error*). In statistical inference, a test T, which is a statistic from $\mathcal{X}$ to $\{0, 1\}$, is assessed by the probabilities of making two types of errors:

$$\alpha_T(P) = P(T(X) = 1) \qquad P \in \mathcal{P}_0 \tag{2.29}$$

and

$$1 - \alpha_T(P) = P(T(X) = 0) \qquad P \in \mathcal{P}_1, \tag{2.30}$$

which are denoted by $\alpha_T(\theta)$ and $1 - \alpha_T(\theta)$ if P is in a parametric family indexed by θ. Note that these are risks of T under the 0-1 loss in statistical decision theory. However, an optimal decision rule (test) does not exist even for a very simple problem with a very simple class of tests (Example 2.23).

That is, error probabilities in (2.29) and (2.30) cannot be minimized simultaneously. Furthermore, these two error probabilities cannot be bounded simultaneously by a fixed $\alpha \in (0,1)$ when we have a sample of a fixed size.

Therefore, a common approach to finding an "optimal" test is to assign a small bound α to one of the error probabilities, say $\alpha_T(P)$, $P \in \mathcal{P}_0$, and then to attempt to minimize the other error probability $1 - \alpha_T(P)$, $P \in \mathcal{P}_1$, subject to

$$\sup_{P \in \mathcal{P}_0} \alpha_T(P) \leq \alpha. \tag{2.31}$$

The bound α is called the *level of significance*. The left-hand side of (2.31) is called the *size* of the test T. Note that the level of significance should be positive, otherwise no test satisfies (2.31) except the silly test $T(X) \equiv 0$ a.s. $\mathcal{P}$.

Example 2.28. Let $X_1, ..., X_n$ be i.i.d. from the $N(\mu, \sigma^2)$ distribution with an unknown $\mu \in \mathcal{R}$ and a known σ^2. Consider the hypotheses

$$H_0 : \mu \leq \mu_0 \qquad \text{versus} \qquad H_1 : \mu > \mu_0,$$

where μ_0 is a fixed constant. Since the sample mean $\bar{X}$ is sufficient for $\mu \in \mathcal{R}$, it is reasonable to consider the following class of tests: $T_c(X) = I_{(c,\infty)}(\bar{X})$, i.e., H_0 is rejected (accepted) if $\bar{X} > c$ ($\bar{X} \leq c$), where $c \in \mathcal{R}$ is a fixed constant. Let Φ be the c.d.f. of $N(0,1)$. Then, by the property of the normal distributions,

$$\alpha_{T_c}(\mu) = P(T_c(X) = 1) = 1 - \Phi\left(\frac{\sqrt{n}(c - \mu)}{\sigma}\right). \tag{2.32}$$

Figure 2.2 provides an example of a graph of two types of error probabilities, with $\mu_0 = 0$. Since $\Phi(t)$ is an increasing function of t,

$$\sup_{P \in \mathcal{P}_0} \alpha_{T_c}(\mu) = 1 - \Phi\left(\frac{\sqrt{n}(c - \mu_0)}{\sigma}\right).$$

In fact, it is also true that

$$\sup_{P \in \mathcal{P}_1} [1 - \alpha_{T_c}(\mu)] = \Phi\left(\frac{\sqrt{n}(c - \mu_0)}{\sigma}\right).$$

If we would like to use an α as the level of significance, then the most effective way is to choose a c_α (a test $T_{c_\alpha}(X)$) such that

$$\alpha = \sup_{P \in \mathcal{P}_0} \alpha_{T_{c_\alpha}}(\mu),$$

in which case c_α must satisfy

$$1 - \Phi\left(\frac{\sqrt{n}(c_\alpha - \mu_0)}{\sigma}\right) = \alpha,$$

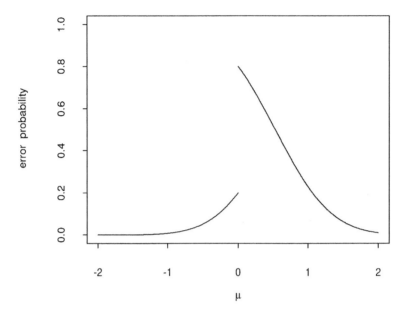

Figure 2.2: Error probabilities in Example 2.28

i.e., $c_\alpha = \sigma z_{1-\alpha}/\sqrt{n} + \mu_0$, where $z_a = \Phi^{-1}(a)$. In Chapter 6, it is shown that for any test $T(X)$ satisfying (2.31),

$$1 - \alpha_T(\mu) \geq 1 - \alpha_{T_{c_\alpha}}(\mu), \qquad \mu > \mu_0. \quad \blacksquare$$

The choice of a level of significance α is usually somewhat subjective. In most applications there is no precise limit to the size of T that can be tolerated. Standard values, such as 0.10, 0.05, or 0.01, are often used for convenience.

For most tests satisfying (2.31), a small α leads to a "small" rejection region. It is good practice to determine not only whether H_0 is rejected or accepted for a given α and a chosen test T_α, but also the smallest possible level of significance at which H_0 would be rejected for the computed $T_\alpha(x)$, i.e., $\hat{\alpha} = \inf\{\alpha \in (0,1) : T_\alpha(x) = 1\}$. Such an $\hat{\alpha}$, which depends on x and the chosen test and is a statistic, is called the *p-value* for the test T_α.

Example 2.29. Consider the problem in Example 2.28. Let us calculate the *p*-value for T_{c_α}. Note that

$$\alpha = 1 - \Phi\left(\frac{\sqrt{n}(c_\alpha - \mu_0)}{\sigma}\right) > 1 - \Phi\left(\frac{\sqrt{n}(\bar{x} - \mu_0)}{\sigma}\right)$$

if and only if $\bar{x} > c_\alpha$ (or $T_{c_\alpha}(x) = 1$). Hence

$$1 - \Phi\left(\frac{\sqrt{n}(\bar{x} - \mu_0)}{\sigma}\right) = \inf\{\alpha \in (0,1) : T_{c_\alpha}(x) = 1\} = \hat{\alpha}(x)$$

is the p-value for T_{c_α}. It turns out that $T_{c_\alpha}(x) = I_{(0,\alpha)}(\hat{\alpha}(x))$. ∎

With the additional information provided by p-values, using p-values is typically more appropriate than using fixed-level tests in a scientific problem. However, a fixed level of significance is unavoidable when acceptance or rejection of H_0 implies an imminent concrete decision. For more discussions about p-values, see Lehmann (1986) and Weerahandi (1995).

In Example 2.28, the equality in (2.31) can always be achieved by a suitable choice of c. This is, however, not true in general. In Example 2.23, for instance, it is possible to find an α such that

$$\sup_{0<\theta\leq\theta_0} P(T_j(X) = 1) \neq \alpha$$

for all T_j's. In such cases, we may consider *randomized tests*, which are introduced next.

Recall that a randomized decision rule is a probability measure $\delta(x, \cdot)$ on the action space for any fixed x. Since the action space contains only two points, 0 and 1, for a hypothesis testing problem, any randomized test $\delta(X, A)$ is equivalent to a statistic $T(X) \in [0, 1]$ with $T(x) = \delta(x, \{1\})$ and $1 - T(x) = \delta(x, \{0\})$. A nonrandomized test is obviously a special case where $T(x)$ does not take any value in $(0, 1)$.

For any randomized test $T(X)$, we define the type I error probability to be $\alpha_T(P) = E[T(X)]$, $P \in \mathcal{P}_0$, and the type II error probability to be $1 - \alpha_T(P) = E[1 - T(X)]$, $P \in \mathcal{P}_1$. For a class of randomized tests, we would like to minimize $1 - \alpha_T(P)$ subject to (2.31).

Example 2.30. Consider Example 2.23 and the following class of randomized tests:

$$T_{j,q}(X) = \begin{cases} 1 & X > j \\ q & X = j \\ 0 & X < j, \end{cases}$$

where $j = 0, 1, ..., n-1$ and $q \in [0, 1]$. Then

$$\alpha_{T_{j,q}}(\theta) = P(X > j) + qP(X = j) \qquad 0 < \theta \leq \theta_0$$

and

$$1 - \alpha_{T_{j,q}}(\theta) = P(X < j) + (1 - q)P(X = j) \qquad \theta_0 < \theta < 1.$$

It can be shown that for any $\alpha \in (0, 1)$, there exist an integer j and $q \in (0, 1)$ such that the size of $T_{j,q}$ is α (exercise). ∎

2.4.3 Confidence sets

Let ϑ be a k-vector of unknown parameters related to the unknown population $P \in \mathcal{P}$ and $C(X) \in \mathcal{B}_{\tilde{\Theta}}^k$ depending only on the sample X, where $\tilde{\Theta} \in \mathcal{B}^k$ is the range of ϑ. If

$$\inf_{P \in \mathcal{P}} P(\vartheta \in C(X)) \geq 1 - \alpha, \tag{2.33}$$

where α is a fixed constant in $(0, 1)$, then $C(X)$ is called a *confidence set* for ϑ with *level of significance* $1 - \alpha$. The left-hand side of (2.33) is called the *confidence coefficient* of $C(X)$, which is the highest possible level of significance for $C(X)$. A confidence set is a random element that covers the unknown ϑ with certain probability. If (2.33) holds, then the *coverage probability* of $C(X)$ is at least $1 - \alpha$, although $C(x)$ either covers or does not cover ϑ whence we observe $X = x$. The concepts of level of significance and confidence coefficient are very similar to the level of significance and size in hypothesis testing. In fact, it is shown in Chapter 7 that some confidence sets are closely related to hypothesis tests.

Consider a real-valued ϑ. If $C(X) = [\underline{\vartheta}(X), \overline{\vartheta}(X)]$ for a pair of real-valued statistics $\underline{\vartheta}$ and $\overline{\vartheta}$, then $C(X)$ is called a *confidence interval* for ϑ. If $C(X) = (-\infty, \overline{\vartheta}(X)]$ (or $[\underline{\vartheta}(X), \infty)$), then $\overline{\vartheta}$ (or $\underline{\vartheta}$) is called an upper (or a lower) *confidence bound* for ϑ.

A confidence set (or interval) is also called a set (or an interval) estimator of ϑ, although it is very different from a point estimator (discussed in §2.4.1).

Example 2.31. Consider Example 2.28. Suppose that a confidence interval for $\vartheta = \mu$ is needed. Again, we only need to consider $\underline{\vartheta}(\bar{X})$ and $\overline{\vartheta}(\bar{X})$, since the sample mean $\bar{X}$ is sufficient. Consider confidence intervals of the form $[\bar{X} - c, \bar{X} + c]$, where $c \in (0, \infty)$ is fixed. Note that

$$P\left(\mu \in [\bar{X} - c, \bar{X} + c]\right) = P\left(|\bar{X} - \mu| \leq c\right) = 1 - 2\Phi\left(-\sqrt{n}c/\sigma\right),$$

which is independent of μ. Hence, the confidence coefficient of $[\bar{X} - c, \bar{X} + c]$ is $1 - 2\Phi\left(-\sqrt{n}c/\sigma\right)$, which is an increasing function of c and converges to 1 as $c \to \infty$ or 0 as $c \to 0$. Thus, confidence coefficients are positive but less than 1 except for silly confidence intervals $[\bar{X}, \bar{X}]$ and $(-\infty, \infty)$. We can choose a confidence interval with an arbitrarily large confidence coefficient, but the chosen confidence interval may be so wide that it is practically useless.

If σ^2 is also unknown, then $[\bar{X} - c, \bar{X} + c]$ has confidence coefficient 0 and, therefore, is not a good inference procedure. In such a case a different confidence interval for μ with positive confidence coefficient can be derived (Exercise 97 in §2.6). ∎

This example tells us that a reasonable approach is to choose a level of significance $1 - \alpha \in (0, 1)$ (just like the level of significance in hypothesis testing) and a confidence interval or set satisfying (2.33). In Example 2.31, when σ^2 is known and c is chosen to be $\sigma z_{1-\alpha/2}/\sqrt{n}$, where $z_a = \Phi^{-1}(a)$, the confidence coefficient of the confidence interval $[\bar{X} - c, \bar{X} + c]$ is *exactly* $1 - \alpha$ for any fixed $\alpha \in (0, 1)$. This is desirable since, for all confidence intervals satisfying (2.33), the one with the shortest interval length is preferred.

For a general confidence interval $[\underline{\vartheta}(X), \overline{\vartheta}(X)]$, its length is $\overline{\vartheta}(X) - \underline{\vartheta}(X)$, which may be random. We may consider the expected (or average) length $E[\overline{\vartheta}(X) - \underline{\vartheta}(X)]$. The confidence coefficient and expected length are a pair of good measures of performance of confidence intervals. Like the two types of error probabilities of a test in hypothesis testing, however, we cannot maximize the confidence coefficient and minimize the length (or expected length) simultaneously. A common approach is to minimize the length (or expected length) subject to (2.33).

For an unbounded confidence interval, its length is ∞. Hence we have to define some other measures of performance. For an upper (or a lower) confidence bound, we may consider the distance $\overline{\vartheta}(X) - \vartheta$ (or $\vartheta - \underline{\vartheta}(X)$) or its expectation.

To conclude this section, we discuss an example of a confidence set for a two-dimensional parameter. General discussions about how to construct and assess confidence sets are given in Chapter 7.

Example 2.32. Let $X_1, ..., X_n$ be i.i.d. from the $N(\mu, \sigma^2)$ distribution with both $\mu \in \mathcal{R}$ and $\sigma^2 > 0$ unknown. Let $\theta = (\mu, \sigma^2)$ and $\alpha \in (0, 1)$ be given. Let $\bar{X}$ be the sample mean and S^2 be the sample variance. Since $(\bar{X}, S^2)$ is sufficient (Example 2.15), we focus on $C(X)$ that is a function of $(\bar{X}, S^2)$. From Example 2.18, $\bar{X}$ and S^2 are independent and $(n - 1)S^2/\sigma^2$ has the chi-square distribution χ^2_{n-1}. Since $\sqrt{n}(\bar{X} - \mu)/\sigma$ has the $N(0, 1)$ distribution (Exercise 43 in §1.6),

$$P\left(-\tilde{c}_\alpha \leq \frac{\bar{X} - \mu}{\sigma/\sqrt{n}} \leq \tilde{c}_\alpha\right) = \sqrt{1 - \alpha},$$

where $\tilde{c}_\alpha = \Phi^{-1}\left(\frac{1+\sqrt{1-\alpha}}{2}\right)$ (verify). Since the chi-square distribution χ^2_{n-1} is a known distribution, we can always find two constants $c_{1\alpha}$ and $c_{2\alpha}$ such that

$$P\left(c_{1\alpha} \leq \frac{(n - 1)S^2}{\sigma^2} \leq c_{2\alpha}\right) = \sqrt{1 - \alpha}.$$

Then

$$P\left(-\tilde{c}_\alpha \leq \frac{\bar{X} - \mu}{\sigma/\sqrt{n}} \leq \tilde{c}_\alpha, c_{1\alpha} \leq \frac{(n - 1)S^2}{\sigma^2} \leq c_{2\alpha}\right) = 1 - \alpha,$$

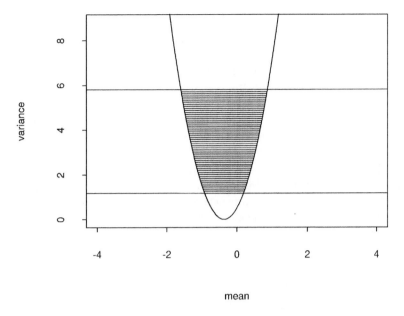

Figure 2.3: A confidence set for θ in Example 2.32

or

$$P\left(\frac{n(\bar{X} - \mu)^2}{\tilde{c}_\alpha^2} \leq \sigma^2, \frac{(n-1)S^2}{c_{2\alpha}} \leq \sigma^2 \leq \frac{(n-1)S^2}{c_{1\alpha}}\right) = 1 - \alpha. \qquad (2.34)$$

The left-hand side of (2.34) defines a set in the range of $\theta = (\mu, \sigma^2)$ bounded by two straight lines, $\sigma^2 = (n-1)S^2/c_{i\alpha}$, $i = 1, 2$, and a curve $\sigma^2 = n(\bar{X} - \mu)^2/\tilde{c}_\alpha^2$ (see the shadowed part of Figure 2.3). This set is a confidence set for θ with confidence coefficient $1 - \alpha$, since (2.34) holds for any θ. ∎

2.5 Asymptotic Criteria and Inference

We have seen that in statistical decision theory and inference, a key to the success of finding a good decision rule or inference procedure is being able to find some moments and/or distributions of various statistics. Although many examples are presented (including those in the exercises in §2.6), there are more cases in which we are not able to find exactly the moments or distributions of given statistics, especially when the problem is not parametric (see, e.g., the discussions in Example 2.8).

In practice, the sample size n is often large, which allows us to approximate the moments and distributions of statistics that are impossible to derive, using the asymptotic tools discussed in §1.5. In an asymptotic analysis, we consider a sample $X = (X_1, ..., X_n)$ not for fixed n, but as a member of a sequence corresponding to $n = n_0, n_0 + 1, ...,$ and obtain the limit of the distribution of an appropriately normalized statistic or variable $T_n(X)$ as $n \to \infty$. The limiting distribution and its moments are used as approximations to the distribution and moments of $T_n(X)$ in the situation with a large but actually finite n. This leads to some asymptotic statistical procedures and asymptotic criteria for assessing their performances, which are introduced in this section.

The asymptotic approach is not only applied to the situation where no exact method is available, but also used to provide an inference procedure simpler (e.g., in terms of computation) than that produced by the exact approach (the approach considering a fixed n). Some examples are given in later chapters.

In addition to providing more theoretical results and/or simpler inference procedures, the asymptotic approach requires less stringent mathematical assumptions than does the exact approach. The mathematical precision of the optimality results obtained in statistical decision theory, for example, tends to obscure the fact that these results are approximations in view of the approximate nature of the assumed models and loss functions. As the sample size increases, the statistical properties become less dependent on the loss functions and models. However, a major weakness of the asymptotic approach is that typically no good estimates for the precision of the approximations are available and, therefore, we cannot determine whether a particular n in a problem is large enough to safely apply the asymptotic results. To overcome this difficulty, asymptotic results are frequently used in combination with some numerical/empirical studies for selected values of n to examine the *finite sample* performance of asymptotic procedures.

2.5.1 Consistency

A reasonable point estimator is expected to perform better, at least on the average, if more information about the unknown population is available. With a fixed model assumption and sampling plan, more data (larger sample size n) provide more information about the unknown population. Thus, it is distasteful to use a point estimator T_n which, if sampling were to continue indefinitely, could possibly have a nonzero estimation error, although the estimation error of T_n for a fixed n may never equal 0 (see the discussion in §2.4.1).

Definition 2.10 (Consistency of point estimators). Let $X = (X_1, ..., X_n)$ be a sample from $P \in \mathcal{P}$ and $T_n(X)$ be a point estimator of ϑ for every n. (i) $T_n(X)$ is called *consistent* for ϑ if and only if $T_n(X) \to_p \vartheta$ w.r.t. any

$P \in \mathcal{P}$.

(ii) Let $\{a_n\}$ be a sequence of positive constants diverging to ∞. $T_n(X)$ is called a_n-*consistent* for ϑ if and only if $a_n[T_n(X) - \vartheta] = O_p(1)$ w.r.t. any $P \in \mathcal{P}$.

(iii) $T_n(X)$ is called *strongly consistent* for ϑ if and only if $T_n(X) \to_{a.s.} \vartheta$ w.r.t. any $P \in \mathcal{P}$.

(iv) $T_n(X)$ is called L_r-*consistent* for ϑ if and only if $T_n(X) \to_{L_r} \vartheta$ w.r.t. any $P \in \mathcal{P}$ for some fixed $r > 0$. ∎

Consistency is actually a concept relating to a sequence of estimators, $\{T_n, n = n_0, n_0 + 1, ...\}$, but we usually just say "consistency of T_n" for simplicity. Each of the four types of consistency in Definition 2.10 describes the convergence of $T_n(X)$ to ϑ in some sense, as $n \to \infty$. In statistics, consistency according to Definition 2.10(i), which is sometimes called *weak consistency* since it is implied by any of the other three types of consistency, is the most useful concept of convergence of T_n to ϑ. L_2-consistency is also called *consistency in mse*, which is the most useful type of L_r-consistency.

Example 2.33. Let $X_1, ..., X_n$ be i.i.d. from $P \in \mathcal{P}$. If $\vartheta = \mu$, which is the mean of P and is assumed to be finite, then by the SLLN (Theorem 1.13), the sample mean $\bar{X}$ is strongly consistent for μ and, therefore, is also consistent for μ. If we further assume that the variance of P is finite, then by (2.20), $\bar{X}$ is consistent in mse and is $\sqrt{n}$-consistent. With the finite variance assumption, the sample variance S^2 is strongly consistent for the variance of P, according to the SLLN.

Consider estimators of the form $T_n = \sum_{i=1}^n c_{ni} X_i$, where $\{c_{ni}\}$ is a double array of constants. If P has a finite variance, then by (2.24), T_n is consistent in mse if and only if $\sum_{i=1}^n c_{ni} \to 1$ and $\sum_{i=1}^n c_{ni}^2 \to 0$. If we only assume the existence of the mean of P, then T_n with $c_{ni} = c_i/n$ satisfying $n^{-1} \sum_{i=1}^n c_i \to 1$ and $\sup_i |c_i| < \infty$ is strongly consistent (Theorem 1.13(ii)). ∎

One or a combination of the law of large numbers, the CLT, Slutsky's theorem (Theorem 1.11), and the continuous mapping theorem (Theorems 1.10 and 1.12) are typically applied to establish consistency of point estimators. In particular, Theorem 1.10 implies that if T_n is (strongly) consistent for ϑ and g is a continuous function of ϑ, then $g(T_n)$ is (strongly) consistent for $g(\vartheta)$. For example, in Example 2.33 the point estimator $\bar{X}^2$ is strongly consistent for μ^2. To show that $\bar{X}^2$ is $\sqrt{n}$-consistent under the assumption that P has a finite variance σ^2, we can use the identity

$$\sqrt{n}(\bar{X}^2 - \mu^2) = \sqrt{n}(\bar{X} - \mu)(\bar{X} + \mu)$$

and the fact that $\bar{X}$ is $\sqrt{n}$-consistent for μ and $\bar{X} + \mu = O_p(1)$. (Note that

$\bar{X}^2$ may not be consistent in mse since we do not assume that P has a finite fourth moment.) Alternatively, we can use the fact that $\sqrt{n}(\bar{X}^2 - \mu^2) \to_d N(0, 4\mu^2\sigma^2)$ (by the CLT and Theorem 1.12) to show the $\sqrt{n}$-consistency of $\bar{X}^2$.

The following example shows another way to establish consistency of some point estimators.

Example 2.34. Let $X_1, ..., X_n$ be i.i.d. from an unknown P with a continuous c.d.f. F satisfying $F(\theta) = 1$ for some $\theta \in \mathcal{R}$ and $F(x) < 1$ for any $x < \theta$. Consider the largest order statistic $X_{(n)}$. For any $\epsilon > 0$, $F(\theta - \epsilon) < 1$ and

$$P(|X_{(n)} - \theta| \geq \epsilon) = P(X_{(n)} \leq \theta - \epsilon) = [F(\theta - \epsilon)]^n,$$

which imply (according to Theorem 1.8(v)) $X_{(n)} \to_{a.s.} \theta$, i.e., $X_{(n)}$ is strongly consistent for θ. If we assume that $F^{(i)}(\theta-)$, the ith-order left-hand derivative of F at θ exists and vanishes, $i = 1, ..., m$, and that $F^{(m+1)}(\theta-)$ exists and is nonzero, then

$$1 - F(X_{(n)}) = \frac{(-1)^m F^{(m+1)}(\theta-)}{(m+1)!}(\theta - X_{(n)})^{m+1} + o\left(|\theta - X_{(n)}|^{m+1}\right) \quad \text{a.s.}$$

Let

$$h_n(\theta) = \left[\frac{(-1)^m(m+1)!}{nF^{(m+1)}(\theta-)}\right]^{(m+1)^{-1}}.$$

For any $t \leq 0$, by Slutsky's theorem,

$$\lim_{n\to\infty} P\left(\frac{X_{(n)} - \theta}{h_n(\theta)} \leq t\right) = \lim_{n\to\infty} P\left(\left[\frac{\theta - X_{(n)}}{h_n(\theta)}\right]^{m+1} \geq (-t)^{m+1}\right)$$

$$= \lim_{n\to\infty} P\left(n[1 - F(X_{(n)})] \geq (-t)^{m+1}\right)$$

$$= \lim_{n\to\infty} \left[1 - (-t)^{m+1}/n\right]^n$$

$$= e^{-(-t)^{m+1}}.$$

This shows that $(X_{(n)} - \theta)/h_n(\theta) \to_d Y$, where Y is a random variable having the c.d.f. $e^{-(-t)^{m+1}} I_{(-\infty, 0)}(t)$. Thus, $X_{(n)}$ is $n^{(m+1)^{-1}}$-consistent. If $m = 0$, then $X_{(n)}$ is n-consistent, which is the most common situation. If $m = 1$, then $X_{(n)}$ is $\sqrt{n}$-consistent. ∎

It can be seen from the previous examples that there are many consistent estimators. Like the admissibility in statistical decision theory, consistency is a very essential requirement in the sense that any inconsistent estimators should not be used, but a consistent estimator is not necessarily good. Thus, consistency should be used together with one or a few more criteria.

We now discuss a situation in which finding a consistent estimator is crucial. Suppose that an estimator T_n of ϑ satisfies

$$c_n[T_n(X) - \vartheta] \to_d \sigma Y, \tag{2.35}$$

where Y is a random variable with a known distribution, $\sigma > 0$ is an unknown parameter, and $\{c_n\}$ is a sequence of constants; for example, in Example 2.33, $\sqrt{n}(\bar{X} - \mu) \to_d N(0, \sigma^2)$; in Example 2.34, (2.35) holds with $c_n = n^{(m+1)^{-1}}$ and $\sigma = [(-1)^m(m+1)!/F^{(m+1)}(\theta-)]^{(m+1)^{-1}}$. If a consistent estimator $\hat{\sigma}_n$ of σ can be found, then, by Slutsky's theorem,

$$c_n[T_n(X) - \vartheta]/\hat{\sigma}_n \to_d Y$$

and, thus, we may approximate the distribution of $c_n[T_n(X) - \vartheta]/\hat{\sigma}_n$ by the known distribution of Y.

2.5.2 Asymptotic bias, variance, and mse

Unbiasedness as a criterion for point estimators is discussed in §2.3.2 and §2.4.1. In some cases, however, there is no unbiased estimator (Exercise 84 in §2.6). Furthermore, having a "slight" bias in some cases may not be a bad idea (see Exercise 63 in §2.6). Let $T_n(X)$ be a point estimator of ϑ for every n. If ET_n exists for every n and $\lim_{n\to\infty} E(T_n - \vartheta) = 0$ for any $P \in \mathcal{P}$, then T_n is said to be *approximately unbiased*.

There are many reasonable point estimators whose expectations are not well defined. For example, consider i.i.d. $(X_1, Y_1), ..., (X_n, Y_n)$ from a bivariate normal distribution with $\mu_x = EX_1$ and $\mu_y = EY_1 \neq 0$. Let $\vartheta = \mu_x/\mu_y$ and $T_n = \bar{X}/\bar{Y}$, the ratio of two sample means. Then ET_n is not defined for any n. It is then desirable to define a concept of *asymptotic bias* for point estimators whose expectations are not well defined.

Definition 2.11. (i) Let $\xi, \xi_1, \xi_2, ...$ be random variables and $\{a_n\}$ be a sequence of positive numbers satisfying $a_n \to \infty$ or $a_n \to a > 0$. If $a_n\xi_n \to_d \xi$ and $E|\xi| < \infty$, then $E\xi/a_n$ is called an *asymptotic expectation* of ξ_n.
(ii) Let T_n be a point estimator of ϑ for every n. An asymptotic expectation of $T_n - \vartheta$, if it exists, is called an asymptotic bias of T_n and denoted by $\tilde{b}_{T_n}(P)$ (or $\tilde{b}_{T_n}(\theta)$ if P is in a parametric family). If $\lim_{n\to\infty} \tilde{b}_{T_n}(P) = 0$ for any $P \in \mathcal{P}$, then T_n is said to be *asymptotically unbiased*. ∎

Like the consistency, the asymptotic expectation (or bias) is a concept relating to sequences $\{\xi_n\}$ and $\{E\xi/a_n\}$ (or $\{T_n\}$ and $\{\tilde{b}_{T_n}(P)\}$). Note that the exact bias $b_{T_n}(P)$ is not necessarily the same as $\tilde{b}_{T_n}(P)$ when both of them exist (Exercise 115 in §2.6). The following result shows that the asymptotic expectation defined in Definition 2.11 is essentially unique.

Proposition 2.3. Let $\{\xi_n\}$ be a sequence of random variables. Suppose that both $E\xi/a_n$ and $E\eta/b_n$ are asymptotic expectations of ξ_n defined according to Definition 2.11(i). Then, one of the following three must hold: (a) $E\xi = E\eta = 0$; (b) $E\xi \neq 0$, $E\eta = 0$, and $b_n/a_n \to 0$; or $E\xi = 0$, $E\eta \neq 0$, and $a_n/b_n \to 0$; (c) $E\xi \neq 0$, $E\eta \neq 0$, and $(E\xi/a_n)/(E\eta/b_n) \to 1$.

Proof. According to Definition 2.11(i), $a_n\xi_n \to_d \xi$ and $b_n\xi_n \to_d \eta$.

(i) If both ξ and η have nondegenerate c.d.f.'s, then the result follows from Exercise 129 of §1.6.

(ii) Suppose that ξ has a nondegenerate c.d.f. but η is a constant. If $\eta \neq 0$, then by Theorem 1.11(iii), $a_n/b_n \to \xi/\eta$, which is impossible since ξ has a nondegenerate c.d.f. If $\eta = 0$, then by Theorem 1.11(ii), $b_n/a_n \to 0$.

(iii) Suppose that both ξ and η are constants. If $\xi = \eta = 0$, the result follows. If $\xi \neq 0$ and $\eta = 0$, then $b_n/a_n \to 0$. If $\xi \neq 0$ and $\eta \neq 0$, then $b_n/a_n \to \eta/\xi$. ∎

If T_n is a consistent estimator of ϑ, then $T_n = \vartheta + o_p(1)$ and, by Definition 2.11(ii), T_n is asymptotically unbiased, although T_n may not be approximately unbiased; in fact, $g(T_n)$ is asymptotically unbiased for $g(\vartheta)$ for any continuous function g. For the example of $T_n = \bar{X}/\bar{Y}$, $T_n \to_{a.s.} \mu_x/\mu_y$ by the SLLN and Theorem 1.10. Hence T_n is asymptotically unbiased, although ET_n may not be defined. In Example 2.34, $X_{(n)}$ has the asymptotic bias $\tilde{b}_{X_{(n)}}(P) = h_n(\theta)EY$, which is of order $n^{-(m+1)^{-1}}$.

When $a_n(T_n - \vartheta) \to_d Y$ with $EY = 0$ (e.g., $T_n = \bar{X}^2$ and $\vartheta = \mu^2$ in Example 2.33), a more precise order of the asymptotic bias of T_n may be obtained (for comparing different estimators in terms of their asymptotic biases). Suppose that there is a sequence of random variables $\{\eta_n\}$ such that

$$a_n\eta_n \to_d Y \quad \text{and} \quad a_n^2(T_n - \vartheta - \eta_n) \to_d W, \qquad (2.36)$$

where Y and W are random variables with finite means, $EY = 0$ and $EW \neq 0$. Then we may define a_n^{-2} to be the order of $\tilde{b}_{T_n}(P)$ or define EW/a_n^2 to be the a_n^{-2} order asymptotic bias of T_n. However, η_n in (2.36) may not be unique. Some regularity conditions have to be imposed so that the order of asymptotic bias of T_n can be uniquely defined. In the following we focus on the case where $X_1, ..., X_n$ are i.i.d. random k-vectors. Suppose that T_n has the following expansion:

$$T_n - \vartheta = \frac{1}{n}\sum_{i=1}^{n}\phi(X_i) + \frac{1}{n^2}\sum_{i=1}^{n}\sum_{j=1}^{n}\psi(X_i, X_j) + o_p\left(\frac{1}{n}\right), \qquad (2.37)$$

where ϕ and ψ are functions that may depend on P, $E\phi(X_1) = 0$, $E[\phi(X_1)]^2 < \infty$, $\psi(x, y) = \psi(y, x)$, $E\psi(x, X_1) = 0$ for all x, $E[\psi(X_i, X_j)]^2 < \infty$, $i \leq j$, and $E\psi(X_1, X_1) \neq 0$. From the result for V-statistics in §3.5.3 (Theorem

3.16 and Exercise 113 in §3.6),

$$\frac{1}{n} \sum_{i=1}^{n} \sum_{j=1}^{n} \psi(X_i, X_j) \to_d W,$$

where W is a random variable with $EW = E\psi(X_1, X_1)$. Hence (2.36) holds with $a_n = \sqrt{n}$ and $\eta_n = n^{-1} \sum_{i=1}^{n} \phi(X_i)$. Consequently, we can define $E\psi(X_1, X_1)/n$ to be the n^{-1} order asymptotic bias of T_n. Examples of estimators that have expansion (2.37) are provided in §3.5.3 and §5.2.1. In the following we consider the special case of functions of sample means.

Let $X_1, ..., X_n$ be i.i.d. random k-vectors with finite $\Sigma = \text{Var}(X_1)$, $\bar{X} = n^{-1} \sum_{i=1}^{n} X_i$, and $T_n = g(\bar{X})$, where g is a function on $\mathcal{R}^k$ that is second-order differentiable at $\mu = EX_1 \in \mathcal{R}^k$. Consider T_n as an estimator of $\vartheta = g(\mu)$. Using Taylor's expansion, we obtain expansion (2.37) with $\phi(x) = [\nabla g(\mu)]^\tau (x - \mu)$ and $\psi(x, y) = (x - \mu)^\tau \nabla^2 g(\mu)(y - \mu)/2$, where ∇g is the k-vector of partial derivatives of g and $\nabla^2 g$ is the $k \times k$ matrix of second-order partial derivatives of g. By the CLT and Theorem 1.10(iii),

$$\frac{1}{n} \sum_{i=1}^{n} \sum_{j=1}^{n} \psi(X_i, X_j) = \frac{n}{2}(\bar{X} - \mu)^\tau \nabla^2 g(\mu)(\bar{X} - \mu) \to_d \frac{Z_\Sigma^\tau \nabla^2 g(\mu) Z_\Sigma}{2},$$

where $Z_\Sigma = N_k(0, \Sigma)$. Thus,

$$\frac{E[Z_\Sigma^\tau \nabla^2 g(\mu) Z_\Sigma]}{2n} = \frac{\text{tr}\left(\nabla^2 g(\mu)\Sigma\right)}{2n} \tag{2.38}$$

is the n^{-1} order asymptotic bias of $T_n = g(\bar{X})$, where $\text{tr}(A)$ denotes the trace of the matrix A. Note that the quantity in (2.38) is the same as the leading term in the exact bias of $T_n = g(\bar{X})$ obtained under a much more stringent condition on the derivatives of g (Lehmann, 1983, Theorem 2.5.1).

Example 2.35. Let $X_1, ..., X_n$ be i.i.d. binary random variables with $P(X_i = 1) = p$, where $p \in (0, 1)$ is unknown. Consider first the estimation of $\vartheta = p(1-p)$. Since $\text{Var}(\bar{X}) = p(1-p)/n$, the n^{-1} order asymptotic bias of $T_n = \bar{X}(1 - \bar{X})$ according to (2.38) with $g(x) = x(1-x)$ is $-p(1-p)/n$. On the other hand, a direct computation shows $E[\bar{X}(1 - \bar{X})] = E\bar{X} - E\bar{X}^2 = p - (E\bar{X})^2 - \text{Var}(\bar{X}) = p(1-p) - p(1-p)/n$. Hence, the exact bias of T_n is the same as the n^{-1} order asymptotic bias.

Consider next the estimation of $\vartheta = p^{-1}$. In this case, there is no unbiased estimator of p^{-1} (Exercise 84 in §2.6). Let $T_n = \bar{X}^{-1}$. Then, an n^{-1} order asymptotic bias of T_n according to (2.38) with $g(x) = x^{-1}$ is $(1 - p)/(p^2 n)$. On the other hand, $ET_n = \infty$ for every n. ∎

Like the bias, the mse of an estimator T_n of ϑ, $\text{mse}_{T_n}(P) = E(T_n - \vartheta)^2$, is not well defined if the second moment of T_n does not exist. We now define a version of *asymptotic mean squared error* (amse) and a measure of assessing different point estimators of a common parameter.

Definition 2.12. Let T_n be an estimator of ϑ for every n and $\{a_n\}$ be a sequence of positive numbers satisfying $a_n \to \infty$ or $a_n \to a > 0$. Assume that $a_n(T_n - \vartheta) \to_d Y$ with $0 < EY^2 < \infty$.
(i) The asymptotic mean squared error of T_n, denoted by $\text{amse}_{T_n}(P)$ or $\text{amse}_{T_n}(\theta)$ if P is in a parametric family indexed by θ, is defined to be the asymptotic expectation of $(T_n - \vartheta)^2$, i.e., $\text{amse}_{T_n}(P) = EY^2/a_n^2$. The asymptotic variance of T_n is defined to be $\sigma_{T_n}^2(P) = \text{Var}(Y)/a_n^2$.
(ii) Let T_n' be another estimator of ϑ. The *asymptotic relative efficiency* of T_n' w.r.t. T_n is defined to be $e_{T_n', T_n}(P) = \text{amse}_{T_n}(P)/\text{amse}_{T_n'}(P)$.
(iii) T_n is said to be *asymptotically more efficient* than T_n' if and only if $\limsup_n e_{T_n', T_n}(P) \leq 1$ for any P and < 1 for some P. ∎

The amse and asymptotic variance are the same if and only if $EY = 0$. By Proposition 2.3, the amse or the asymptotic variance of T_n is essentially unique and, therefore, the concept of asymptotic relative efficiency in Definition 2.12(ii)-(iii) is well defined.

In Example 2.33, $\text{amse}_{\bar{X}^2}(P) = \sigma_{\bar{X}^2}^2(P) = 4\mu^2\sigma^2/n$. In Example 2.34, $\sigma_{X_{(n)}}^2(P) = [h_n(\theta)]^2\text{Var}(Y)$ and $\text{amse}_{X_{(n)}}(P) = [h_n(\theta)]^2EY^2$.

When both $\text{mse}_{T_n}(P)$ and $\text{mse}_{T_n'}(P)$ exist, one may compare T_n and T_n' by evaluating the relative efficiency $\text{mse}_{T_n}(P)/\text{mse}_{T_n'}(P)$. However, this comparison may be different from the one using the asymptotic relative efficiency in Definition 2.12(ii), since the mse and amse of an estimator may be different (Exercise 115 in §2.6). The following result shows that when the exact mse of T_n exists, it is no smaller than the amse of T_n. It also provides a condition under which the exact mse and the amse are the same.

Proposition 2.4. Let T_n be an estimator of ϑ for every n and $\{a_n\}$ be a sequence of positive numbers satisfying $a_n \to \infty$ or $a_n \to a > 0$. Suppose that $a_n(T_n - \vartheta) \to_d Y$ with $0 < EY^2 < \infty$. Then
(i) $EY^2 \leq \liminf_n E[a_n^2(T_n - \vartheta)^2]$ and
(ii) $EY^2 = \lim_{n\to\infty} E[a_n^2(T_n - \vartheta)^2]$ if and only if $\{a_n^2(T_n - \vartheta)^2\}$ is uniformly integrable.
Proof. (i) By Theorem 1.10(iii),

$$\min\{a_n^2(T_n - \vartheta)^2, t\} \to_d \min\{Y^2, t\}$$

for any $t > 0$. Since $\min\{a_n^2(T_n - \vartheta)^2, t\}$ is bounded by t,

$$\lim_{n\to\infty} E(\min\{a_n^2(T_n - \vartheta)^2, t\}) = E(\min\{Y^2, t\})$$

(Theorem 1.8(viii)). Then

$$
\begin{aligned}
EY^2 &= \lim_{t \to \infty} E(\min\{Y^2, t\}) \\
&= \lim_{t \to \infty} \lim_{n \to \infty} E(\min\{a_n^2 (T_n - \vartheta)^2, t\}) \\
&= \liminf_{t,n} E(\min\{a_n^2 (T_n - \vartheta)^2, t\}) \\
&\leq \liminf_n E[a_n^2 (T_n - \vartheta)^2],
\end{aligned}
$$

where the third equality follows from the fact that $E(\min\{a_n^2 (T_n - \vartheta)^2, t\})$ is nondecreasing in t for any fixed n.
(ii) The result follows from Theorem 1.8(viii). ∎

Example 2.36. Let $X_1, ..., X_n$ be i.i.d. from the Poisson distribution $P(\theta)$ with an unknown $\theta > 0$. Consider the estimation of $\vartheta = P(X_i = 0) = e^{-\theta}$. Let $T_{1n} = F_n(0)$, where F_n is the empirical c.d.f. defined in (2.28). Then T_{1n} is unbiased and has $\mathrm{mse}_{T_{1n}}(\theta) = e^{-\theta}(1 - e^{-\theta})/n$. Also, $\sqrt{n}(T_{1n} - \vartheta) \to_d N(0, e^{-\theta}(1 - e^{-\theta}))$ by the CLT. Thus, in this case $\mathrm{amse}_{T_{1n}}(\theta) = \mathrm{mse}_{T_{1n}}(\theta)$.

Next, consider $T_{2n} = e^{-\bar{X}}$. Note that $ET_{2n} = e^{n\theta(e^{-1/n} - 1)}$. Hence $nb_{T_{2n}}(\theta) \to \theta e^{-\theta}/2$. Using Theorem 1.12 and the CLT, we can show that $\sqrt{n}(T_{2n} - \vartheta) \to_d N(0, e^{-2\theta}\theta)$. By Definition 2.12(i), $\mathrm{amse}_{T_{2n}}(\theta) = e^{-2\theta}\theta/n$. Thus, the asymptotic relative efficiency of T_{1n} w.r.t. T_{2n} is

$$
e_{T_{1n}, T_{2n}}(\theta) = \theta/(e^\theta - 1),
$$

which is always less than 1. This shows that T_{2n} is asymptotically more efficient than T_{1n}. ∎

The result for T_{2n} in Example 2.36 is a special case (with $U_n = \bar{X}$) of the following general result.

Theorem 2.6. Let g be a function on $\mathcal{R}^k$ that is differentiable at $\theta \in \mathcal{R}^k$ and let U_n be a k-vector of statistics satisfying $a_n(U_n - \theta) \to_d Y$ for a random k-vector Y with $0 < E\|Y\|^2 < \infty$ and a sequence of positive numbers $\{a_n\}$ satisfying $a_n \to \infty$. Let $T_n = g(U_n)$ be an estimator of $\vartheta = g(\theta)$. Then, the amse and asymptotic variance of T_n are, respectively, $E\{[\nabla g(\theta)]^\tau Y\}^2/a_n^2$ and $[\nabla g(\theta)]^\tau \mathrm{Var}(Y)\nabla g(\theta)/a_n^2$. ∎

2.5.3 Asymptotic inference

Statistical inference based on asymptotic criteria and approximations is called *asymptotic statistical inference* or simply *asymptotic inference*. We have previously considered asymptotic estimation. We now focus on asymptotic hypothesis tests and confidence sets.

Definition 2.13. Let $X = (X_1, ..., X_n)$ be a sample from $P \in \mathcal{P}$ and $T_n(X)$ be a test for $H_0 : P \in \mathcal{P}_0$ versus $H_1 : P \in \mathcal{P}_1$.

(i) If $\limsup_n \alpha_{T_n}(P) \leq \alpha$ for any $P \in \mathcal{P}_0$, then α is an *asymptotic significance level* of T_n.

(ii) If $\lim_{n \to \infty} \sup_{P \in \mathcal{P}_0} \alpha_{T_n}(P)$ exists, then it is called the *limiting size* of T_n.

(iii) T_n is called *consistent* if and only if the type II error probability converges to 0, i.e., $\lim_{n \to \infty}[1 - \alpha_{T_n}(P)] = 0$, for any $P \in \mathcal{P}_1$.

(iv) T_n is called *Chernoff-consistent* if and only if T_n is consistent *and* the type I error probability converges to 0, i.e., $\lim_{n \to \infty} \alpha_{T_n}(P) = 0$, for any $P \in \mathcal{P}_0$. T_n is called *strongly Chernoff-consistent* if and only if T_n is consistent and the limiting size of T_n is 0. ∎

Obviously if T_n has size (or significance level) α for all n, then its limiting size (or asymptotic significance level) is α. If the limiting size of T_n is $\alpha \in (0,1)$, then for any $\epsilon > 0$, T_n has size $\alpha + \epsilon$ for all $n \geq n_0$, where n_0 is independent of P. Hence T_n has level of significance $\alpha + \epsilon$ for any $n \geq n_0$. However, if $\mathcal{P}_0$ is not a parametric family, it is likely that the limiting size of T_n is 1 (see, e.g., Example 2.37). This is the reason why we consider the weaker requirement in Definition 2.13(i). If T_n has asymptotic significance level α, then for any $\epsilon > 0$, $\alpha_{T_n}(P) < \alpha + \epsilon$ for all $n \geq n_0(P)$ but $n_0(P)$ depends on $P \in \mathcal{P}_0$; and there is no guarantee that T_n has significance level $\alpha + \epsilon$ for any n.

The consistency in Definition 2.13(iii) only requires that the type II error probability converge to 0. We may define uniform consistency to be $\lim_{n \to \infty} \sup_{P \in \mathcal{P}_1}[1 - \alpha_{T_n}(P)] = 0$, but it is not satisfied in most problems. If $\alpha \in (0,1)$ is a pre-assigned level of significance for the problem, then a consistent test T_n having asymptotic significance level α is called *asymptotically correct*, and a consistent test having limiting size α is called *strongly asymptotically correct*.

The Chernoff-consistency (or strong Chernoff-consistency) in Definition 2.13(iv) requires that both types of error probabilities converge to 0. Mathematically, Chernoff-consistency (or strong Chernoff-consistency) is better than asymptotic correctness (or strongly asymptotic correctness). After all, both types of error probabilities should decrease to 0 if sampling can be continued indefinitely. However, if α is chosen to be small enough so that error probabilities smaller than α can be practically treated as 0, then the asymptotic correctness (or strongly asymptotic correctness) is enough, and is probably preferred, since requiring an unnecessarily small type I error probability usually results in an unnecessary increase in the type II error probability, as the following example illustrates.

Example 2.37. Consider the testing problem $H_0 : \mu \leq \mu_0$ versus $H_1 :$

$\mu > \mu_0$ based on i.i.d. $X_1, ..., X_n$ with $EX_1 = \mu \in \mathcal{R}$. If each X_i has the $N(\mu, \sigma^2)$ distribution with a known σ^2, then the test T_{c_α} given in Example 2.28 with $c_\alpha = \sigma z_{1-\alpha}/\sqrt{n} + \mu_0$ and $\alpha \in (0, 1)$ has size α (and, therefore, limiting size α). It also follows from (2.32) that for any $\mu > \mu_0$,

$$1 - \alpha_{T_{c_\alpha}}(\mu) = \Phi\left(z_{1-\alpha} + \frac{\sqrt{n}(\mu_0 - \mu)}{\sigma}\right) \to 0 \qquad (2.39)$$

as $n \to \infty$. This shows that T_{c_α} is consistent and, hence, is strongly asymptotically correct. Note that the convergence in (2.39) is not uniform in $\mu > \mu_0$, but is uniform in $\mu > \mu_1$ for any fixed $\mu_1 > \mu_0$.

Since the size of T_{c_α} is α for all n, T_{c_α} is not Chernoff-consistent. A strongly Chernoff-consistent test can be obtained as follows. Let

$$\alpha_n = 1 - \Phi(\sqrt{n}a_n), \qquad (2.40)$$

where a_n's are positive numbers satisfying $a_n \to 0$ and $\sqrt{n}a_n \to \infty$. Let T_n be T_{c_α} with $\alpha = \alpha_n$ for each n. Then, T_n has size α_n. Since $\alpha_n \to 0$, The limiting size of T_n is 0. On the other hand, (2.39) still holds with α replaced by α_n. This follows from the fact that

$$z_{1-\alpha_n} + \frac{\sqrt{n}(\mu_0 - \mu)}{\sigma} = \sqrt{n}\left(a_n + \frac{\mu_0 - \mu}{\sigma}\right) \to -\infty$$

for any $\mu > \mu_0$. Hence T_n is strongly Chernoff-consistent. However, if $\alpha_n < \alpha$, then, from the left-hand side of (2.39), $1 - \alpha_{T_{c_\alpha}}(\mu) < 1 - \alpha_{T_n}(\mu)$ for any $\mu > \mu_0$.

We now consider the case where the population P is not in a parametric family. We still assume that $\sigma^2 = \text{Var}(X_i)$ is known. Using the CLT, we can show that for $\mu > \mu_0$,

$$\lim_{n \to \infty} [1 - \alpha_{T_{c_\alpha}}(\mu)] = \lim_{n \to \infty} \Phi\left(z_{1-\alpha} + \frac{\sqrt{n}(\mu_0 - \mu)}{\sigma}\right) = 0,$$

i.e., T_{c_α} is still consistent. For $\mu \leq \mu_0$,

$$\lim_{n \to \infty} \alpha_{T_{c_\alpha}}(\mu) = 1 - \lim_{n \to \infty} \Phi\left(z_{1-\alpha} + \frac{\sqrt{n}(\mu_0 - \mu)}{\sigma}\right),$$

which equals α if $\mu = \mu_0$ and 0 if $\mu < \mu_0$. Thus, the asymptotic significance level of T_{c_α} is α. Combining these two results, we know that T_{c_α} is asymptotically correct. However, if P contains all possible populations on $\mathcal{R}$ with finite second moments, then one can show that the limiting size of T_{c_α} is 1 (exercise). For α_n defined by (2.40), we can show that $T_n = T_{c_\alpha}$ with $\alpha = \alpha_n$ is Chernoff-consistent (exercise). But T_n is not strongly Chernoff-consistent if P contains all possible populations on $\mathcal{R}$ with finite second moments. ∎

Definition 2.14. Let $X = (X_1, ..., X_n)$ be a sample from $P \in \mathcal{P}$, ϑ be a k-vector of parameters related to P, and $C(X)$ be a confidence set for ϑ.
(i) If $\liminf_n P(\vartheta \in C(X)) \geq 1 - \alpha$ for any $P \in \mathcal{P}$, then $1 - \alpha$ is an *asymptotic significance level* of $C(X)$.
(ii) If $\lim_{n \to \infty} \inf_{P \in \mathcal{P}} P(\vartheta \in C(X))$ exists, then it is called the *limiting confidence coefficient* of $C(X)$. ∎

Note that the asymptotic significance level and limiting confidence coefficient of a confidence set are very similar to the asymptotic significance level and limiting size of a test, respectively. Some conclusions are also similar. For example, in a parametric problem one can often find a confidence set having limiting confidence coefficient $1 - \alpha \in (0, 1)$, which implies that for any $\epsilon > 0$, the confidence coefficient of $C(X)$ is $1 - \alpha - \epsilon$ for all $n \geq n_0$, where n_0 is independent of P; in a nonparametric problem the limiting confidence coefficient of $C(X)$ might be 0, whereas $C(X)$ may have asymptotic significance level $1 - \alpha \in (0, 1)$, but for any fixed n, the confidence coefficient of $C(X)$ might be 0.

The confidence interval in Example 2.31 with $c = \sigma z_{1-\alpha/2}/\sqrt{n}$ and the confidence set in Example 2.32 have confidence coefficient $1 - \alpha$ for any n and, therefore, have limiting confidence coefficient $1 - \alpha$. If we drop the normality assumption and assume $EX_i^4 < \infty$, then these confidence sets have asymptotic significance level $1-\alpha$; their limiting confidence coefficients may be 0 (exercise).

2.6 Exercises

1. Consider Example 2.3. Suppose that $p(s)$ is constant. Show that X_i and X_j, $i \neq j$, are not uncorrelated and, hence, $X_1, ..., X_n$ are not independent. Furthermore, when y_i's are either 0 or 1, show that $Z = \sum_{i=1}^n X_i$ has a hypergeometric distribution and compute the mean of Z.

2. Consider Example 2.3. Suppose that we do not require that the elements in s be distinct, i.e., we consider sampling with replacement. Define a probability measure p and a sample $(X_1, ..., X_n)$ such that (2.3) holds. If $p(s)$ is constant, are $X_1, ..., X_n$ independent? If $p(s)$ is constant and y_i's are either 0 or 1, what are the distribution and mean of $Z = \sum_{i=1}^n X_i$?

3. Show that $\{P_\theta : \theta \in \Theta\}$ is an exponential family and find its canonical form and natural parameter space, when
   (a) P_θ is the Poisson distribution $P(\theta)$, $\theta \in \Theta = (0, \infty)$;
   (b) P_θ is the negative binomial distribution $NB(\theta, r)$ with a fixed r,

$\theta \in \Theta = (0, 1)$;

(c) P_θ is the exponential distribution $E(a, \theta)$ with a fixed a, $\theta \in \Theta = (0, \infty)$;

(d) P_θ is the gamma distribution $\Gamma(\alpha, \gamma)$, $\theta = (\alpha, \gamma) \in \Theta = (0, \infty) \times (0, \infty)$;

(e) P_θ is the beta distribution $B(\alpha, \beta)$, $\theta = (\alpha, \beta) \in \Theta = (0, 1) \times (0, 1)$;

(f) P_θ is the Weibull distribution $W(\alpha, \theta)$ with a fixed $\alpha > 0$, $\theta \in \Theta = (0, \infty)$.

4. Show that the family of exponential distributions $E(a, \theta)$ with two unknown parameters a and θ is not an exponential family.

5. Show that the family of negative binomial distributions $NB(p, r)$ with two unknown parameters p and r is not an exponential family.

6. Show that the family of Cauchy distributions $C(\mu, \sigma)$ with two unknown parameters μ and σ is not an exponential family.

7. Show that the family of Weibull distributions $W(\alpha, \theta)$ with two unknown parameters α and θ is not an exponential family.

8. Is the family of log-normal distributions $LN(\mu, \sigma^2)$ with two unknown parameters μ and σ^2 an exponential family?

9. Show that the family of double exponential distributions $DE(\mu, \theta)$ with two unknown parameters μ and θ is not an exponential family, but the family of double exponential distributions $DE(\mu, \theta)$ with a fixed μ and an unknown parameter θ is an exponential family.

10. Show that the k-dimensional normal family discussed in Example 2.4 is an exponential family. Identify the functions T, η, ξ, and h.

11. Obtain the variance-covariance matrix for $(X_1, ..., X_k)$ in Example 2.7, using (a) Theorem 2.1(ii) and (b) direct computation.

12. Show that the m.g.f. of the gamma distribution $\Gamma(\alpha, \gamma)$ is $(1 - \gamma t)^{-\alpha}$, $t < \gamma^{-1}$, using Theorem 2.1(ii).

13. A discrete random variable X with

$$P(X = x) = \gamma(x)\theta^x / c(\theta), \quad x = 0, 1, 2, ...,$$

where $\gamma(x) \geq 0$, $\theta > 0$, and $c(\theta) = \sum_{x=0}^{\infty} \gamma(x)\theta^x$, is called a random variable with a *power series* distribution.

(a) Show that $\{\gamma(x)\theta^x / c(\theta) : \theta > 0\}$ is an exponential family.

(b) Suppose that $X_1, ..., X_n$ are i.i.d. with a power series distribution $\gamma(x)\theta^x / c(\theta)$. Show that $\sum_{i=1}^{n} X_i$ has the power series distribution $\gamma_n(x)\theta^x / [c(\theta)]^n$, where $\gamma_n(x)$ is the coefficient of θ^x in the power series expansion of $[c(\theta)]^n$.

14. Let X be a random variable with a p.d.f. f_θ in an exponential family $\{P_\theta : \theta \in \Theta\}$ and let A be a Borel set. Show that the distribution of X truncated on A (i.e., the conditional distribution of X given $X \in A$) has a p.d.f. $f_\theta I_A / P_\theta(A)$ that is in an exponential family.

15. Let $\{P_{(\mu,\Sigma)} : \mu \in \mathcal{R}^k, \Sigma \in \mathcal{M}_k\}$ be a location-scale family on $\mathcal{R}^k$. Suppose that $P_{(0,I_k)}$ has a Lebesgue p.d.f. that is always positive and that the mean and variance-covariance matrix of $P_{(0,I_k)}$ are 0 and I_k, respectively. Show that the mean and variance-covariance matrix of $P_{(\mu,\Sigma)}$ are μ and Σ, respectively.

16. Show that if the distribution of a positive random variable X is in a scale family, then the distribution of $\log X$ is in a location family.

17. Let X be a random variable having the gamma distribution $\Gamma(\alpha, \gamma)$ with a known α and an unknown $\gamma > 0$ and let $Y = \sigma \log X$.
    (a) Show that if $\sigma > 0$ is unknown, then the distribution of Y is in a location-scale family.
    (b) Show that if $\sigma > 0$ is known, then the distribution of Y is in an exponential family.

18. Let $X_1, ..., X_n$ be i.i.d. random variables having a finite $E|X_1|^4$ and let $\bar{X}$ and S^2 be the sample mean and variance defined by (2.1) and (2.2). Express $E(\bar{X}^3)$, $\text{Cov}(\bar{X}, S^2)$, and $\text{Var}(S^2)$ in terms of $\mu_k = EX_1^k$, $k = 1, 2, 3, 4$. Find a condition under which $\bar{X}$ and S^2 are uncorrelated.

19. Let $X_1, ..., X_n$ be i.i.d. random variables having the gamma distribution $\Gamma(\alpha, \gamma_x)$ and $Y_1, ..., Y_n$ be i.i.d. random variables having the gamma distribution $\Gamma(\alpha, \gamma_y)$, where $\alpha > 0$, $\gamma_x > 0$, and $\gamma_y > 0$. Derive the distribution of the statistic $\bar{X}/\bar{Y}$, where $\bar{X}$ and $\bar{Y}$ are sample means based on X_i's and Y_i's, respectively.

20. Let $X_1, ..., X_n$ be i.i.d. random variables having the exponential distribution $E(a, \theta)$, $a \in \mathcal{R}$, and $\theta > 0$. Show that the smallest order statistic, $X_{(1)}$, has the exponential distribution $E(a, \theta/n)$ and that $2\sum_{i=1}^n (X_i - X_{(1)})/\theta$ has the chi-square distribution χ^2_{2n-2}.

21. Let $(X_1, Y_1), ..., (X_n, Y_n)$ be i.i.d. random 2-vectors. Suppose that X_1 has the Cauchy distribution $C(0, 1)$ and given $X_1 = x$, Y_1 has the Cauchy distribution $C(\beta x, 1)$, where $\beta \in \mathcal{R}$. Let $\bar{X}$ and $\bar{Y}$ be the sample means based on X_i's and Y_i's, respectively. Obtain the marginal distributions of $\bar{Y}$, $\bar{Y} - \beta\bar{X}$, and $\bar{Y}/\bar{X}$.

22. Let $X_i = (Y_i, Z_i)$, $i = 1, ..., n$, be i.i.d. random 2-vectors. The sample correlation coefficient is defined to be

$$T(X) = \frac{1}{(n-1)\sqrt{S_Y^2 S_Z^2}} \sum_{i=1}^{n} (Y_i - \bar{Y})(Z_i - \bar{Z}),$$

where $\bar{Y} = n^{-1} \sum_{i=1}^{n} Y_i$, $\bar{Z} = n^{-1} \sum_{i=1}^{n} Z_i$, $S_Y^2 = (n-1)^{-1} \sum_{i=1}^{n} (Y_i - \bar{Y})^2$, and $S_Z^2 = (n-1)^{-1} \sum_{i=1}^{n} (Z_i - \bar{Z})^2$.
(a) Assume that $E|Y_i|^4 < \infty$ and $E|Z_i|^4 < \infty$. Show that

$$\sqrt{n}[T(X) - \rho] \to_d N(0, c^2),$$

where ρ is the correlation coefficient between Y_1 and Z_1 and c is a constant depending on some unknown parameters.
(b) Assume that Y_i and Z_i are independently distributed as $N(\mu_1, \sigma_1^2)$ and $N(\mu_2, \sigma_2^2)$, respectively. Show that T has the Lebesgue p.d.f.

$$f(t) = \frac{\Gamma\left(\frac{n-1}{2}\right)}{\sqrt{\pi}\Gamma\left(\frac{n-2}{2}\right)} (1 - t^2)^{(n-4)/2} I_{(-1,1)}(t).$$

(c) Assume the conditions in (b). Obtain the result in (a) using Scheffé's theorem (Proposition 1.18).

23. Let $X_1, ..., X_n$ be i.i.d. random variables with $EX_1^4 < \infty$, $T = (Y, Z)$, and $T_1 = Y/\sqrt{Z}$, where $Y = n^{-1} \sum_{i=1}^{n} |X_i|$ and $Z = n^{-1} \sum_{i=1}^{n} X_i^2$.
(a) Show that $\sqrt{n}(T - \theta) \to_d N_2(0, \Sigma)$ and $\sqrt{n}(T_1 - \vartheta) \to_d N(0, c^2)$. Identify θ, Σ, ϑ, and c^2 in terms of moments of X_1.
(b) Repeat (a) when X_1 has the normal distribution $N(0, \sigma^2)$.
(c) Repeat (a) when X_1 has the double exponential distribution $D(0, \sigma)$.

24. Prove the claims in Example 2.9 for the distributions related to order statistics.

25. Show that if T is a sufficient statistic and $T = \psi(S)$, where ψ is measurable and S is another statistic, then S is sufficient.

26. In the proof of Lemma 2.1, show that $C_0 \in \mathcal{C}$. Also, prove Lemma 2.1 when $\mathcal{P}$ is dominated by a σ-finite measure.

27. Let $X_1, ..., X_n$ be i.i.d. random variables from $P_\theta \in \{P_\theta : \theta \in \Theta\}$. In the following cases, find a sufficient statistic for $\theta \in \Theta$ that has the same dimension as θ.
(a) P_θ is the Poisson distribution $P(\theta)$, $\theta \in (0, \infty)$.
(b) P_θ is the negative binomial distribution $NB(\theta, r)$ with a known r, $\theta \in (0, 1)$.

(c) P_θ is the exponential distribution $E(0, \theta)$, $\theta \in (0, \infty)$.

(d) P_θ is the gamma distribution $\Gamma(\alpha, \gamma)$, $\theta = (\alpha, \gamma) \in (0, \infty) \times (0, \infty)$.

(e) P_θ is the beta distribution $B(\alpha, \beta)$, $\theta = (\alpha, \beta) \in (0, 1) \times (0, 1)$.

(f) P_θ is the log-normal distribution $LN(\mu, \sigma^2)$, $\theta = (\mu, \sigma^2) \in \mathcal{R} \times (0, \infty)$.

(g) P_θ is the Weibull distribution $W(\alpha, \theta)$ with a known $\alpha > 0$, $\theta \in (0, \infty)$.

28. Let $X_1, ..., X_n$ be i.i.d. random variables from $P_{(a,\theta)}$, where $(a, \theta) \in \mathcal{R}^2$ is a parameter. Find a two-dimensional sufficient statistic for (a, θ) in the following cases.

(a) $P_{(a,\theta)}$ is the exponential distribution $E(a, \theta)$, $a \in \mathcal{R}$, $\theta \in (0, \infty)$.

(b) $P_{(a,\theta)}$ is the Pareto distribution $Pa(a, \theta)$, $a \in (0, \infty)$, $\theta \in (0, \infty)$.

29. In Example 2.11, show that $X_{(1)}$ (or $X_{(n)}$) is sufficient for a (or b) if we consider a subfamily $\{f_{(a,b)} : a < b\}$ with a fixed b (or a).

30. Let X and Y be two random variables such that Y has the binomial distribution $Bi(\pi, N)$ and, given $Y = y$, X has the binomial distribution $Bi(p, y)$.

(a) Suppose that $p \in (0, 1)$ and $\pi \in (0, 1)$ are unknown and N is known. Show that (X, Y) is minimal sufficient for (p, π).

(b) Suppose that π and N are known and $p \in (0, 1)$ is unknown. Show whether X is sufficient for p and whether Y is sufficient for p.

31. Let $X_1, ..., X_n$ be i.i.d. random variables having a distribution $P \in \mathcal{P}$, where $\mathcal{P}$ is the family of distributions on $\mathcal{R}$ having continuous c.d.f.'s. Let $T = (X_{(1)}, ..., X_{(n)})$ be the vector of order statistics. Show that, given T, the conditional distribution of $X = (X_1, ..., X_n)$ is a discrete distribution putting probability $1/n!$ on each of the $n!$ points $(X_{i_1}, ..., X_{i_n}) \in \mathcal{R}^n$, where $\{i_1, ..., i_n\}$ is a permutation of $\{1, ..., n\}$; hence, T is sufficient for $P \in \mathcal{P}$.

32. In Example 2.13 and Example 2.14, show that T is minimal sufficient for θ by using Theorem 2.3(iii).

33. A coin has probability p of coming up heads and $1 - p$ of coming up tails, where $p \in (0, 1)$. The first stage of an experiment consists of tossing this coin a known total of M times and recording X, the number of heads. In the second stage, the coin is tossed until a total of $X + 1$ tails have come up. The number Y of heads observed in the second stage along the way to getting the $X + 1$ tails is then recorded. This experiment is repeated independently a total of n times and the two counts (X_i, Y_i) for the ith experiment are recorded, $i = 1, ..., n$. Obtain a statistic that is minimal sufficient for p and derive its distribution.

34. Let $X_1, ..., X_n$ be i.i.d. random variables having the Lebesgue p.d.f.

$$f_\theta(x) = \exp\left\{ - \left(\tfrac{x-\mu}{\sigma}\right)^4 - \xi(\theta) \right\},$$

where $\theta = (\mu, \sigma) \in \Theta = \mathcal{R} \times (0, \infty)$. Show that $\mathcal{P} = \{P_\theta : \theta \in \Theta\}$ is an exponential family, where P_θ is the joint distribution of $X_1, ..., X_n$, and that the statistic $T = \left(\sum_{i=1}^n X_i, \sum_{i=1}^n X_i^2, \sum_{i=1}^n X_i^3, \sum_{i=1}^n X_i^4\right)$ is minimal sufficient for $\theta \in \Theta$.

35. Let $X_1, ..., X_n$ be i.i.d. random variables having the Lebesgue p.d.f.

$$f_\theta(x) = (2\theta)^{-1} \left[I_{(0,\theta)}(x) + I_{(2\theta, 3\theta)}(x) \right].$$

Find a minimal sufficient statistic for $\theta \in (0, \infty)$.

36. Let $X_1, ..., X_n$ be i.i.d. random variables having the Cauchy distribution $C(\mu, \sigma)$ with unknown $\mu \in \mathcal{R}$ and $\sigma > 0$. Show that the vector of order statistics is minimal sufficient for (μ, σ).

37. Let $X_1, ..., X_n$ be i.i.d. random variables having the double exponential distribution $DE(\mu, \theta)$ with unknown $\mu \in \mathcal{R}$ and $\theta > 0$. Show that the vector of order statistics is minimal sufficient for (μ, θ).

38. Let $X_1, ..., X_n$ be i.i.d. random variables having the Weibull distribution $W(\alpha, \theta)$ with unknown $\alpha > 0$ and $\theta > 0$. Show that the vector of order statistics is minimal sufficient for (α, θ).

39. Let $X_1, ..., X_n$ be i.i.d. random variables having the beta distribution $B(\beta, \beta)$ with an unknown $\beta > 0$. Find a minimal sufficient statistic for β.

40. Let $X_1, ..., X_n$ be i.i.d. random variables having a population P in a parametric family indexed by (θ, j), where $\theta > 0$, $j = 1, 2$, and $n \geq 2$. When $j = 1$, P is the $N(0, \theta^2)$ distribution. When $j = 2$, P is the double exponential distribution $DE(0, \theta)$. Show that $T = \left(\sum_{i=1}^n X_i^2, \sum_{i=1}^n |X_i|\right)$ is minimal sufficient for (θ, j).

41. Let $X_1, ..., X_n$ be i.i.d. random variables having a population P in a parametric family indexed by (θ, j), where $\theta \in (0, 1)$, $j = 1, 2$, and $n \geq 2$. When $j = 1$, P is the Poisson distribution $P(\theta)$. When $j = 2$, P is the binomial distribution $Bi(\theta, 1)$.
(a) Show that $T = \sum_{i=1}^n X_i$ is not sufficient for (θ, j).
(b) Find a two-dimensional minimal sufficient statistic for (θ, j).

42. Let X be a sample from $P \in \mathcal{P} = \{f_{\theta,j} : \theta \in \Theta, j = 1, ..., k\}$, where $f_{\theta,j}$'s are p.d.f.'s w.r.t. a common σ-finite measure and Θ is a set of parameters. Assume that $\{x : f_{\theta,j}(x) > 0\} \subset \{x : f_{\theta,k}(x) > 0\}$ for all

θ and $j = 1, ..., k - 1$. Suppose that for each fixed j, $T = T(X)$ is a statistic sufficient for θ.

(a) Obtain a k-dimensional statistic that is sufficient for (θ, j).

(b) Derive a sufficient condition under which T is minimal sufficient for (θ, j).

43. A box has an unknown odd number of balls labeled consecutively as $-\theta, -(\theta - 1), ..., -2, -1, 0, 1, 2, ..., (\theta - 1), \theta$, where θ is an unknown nonnegative integer. A simple random sample $X_1, ..., X_n$ is taken without replacement, where X_i is the label on the ith ball selected and $n < 2\theta + 1$.

(a) Find a statistic that is minimal sufficient for θ and derive its distribution.

(b) Show that the minimal sufficient statistic in (a) is also complete.

44. Let $X_1, ..., X_n$ be i.i.d. random variables having the Lebesgue p.d.f. $\theta^{-1}e^{-(x-\theta)/\theta}I_{(\theta,\infty)}(x)$, where $\theta > 0$ is an unknown parameter.

(a) Find a statistic that is minimal sufficient for θ.

(b) Show whether the minimal sufficient statistic in (a) is complete.

45. Let $X_1, ..., X_n$ ($n \geq 2$) be i.i.d. random variables having the normal distribution $N(\theta, 2)$ when $\theta = 0$ and the normal distribution $N(\theta, 1)$ when $\theta \in \mathcal{R}$ and $\theta \neq 0$. Show that the sample mean $\bar{X}$ is a complete statistic for θ but it is not a sufficient statistic for θ.

46. Let X be a random variable with a distribution P_θ in $\{P_\theta : \theta \in \Theta\}$, f_θ be the p.d.f. of P_θ w.r.t. a measure ν, A be an event, and $\mathcal{P}_A = \{f_\theta I_A / P_\theta(A) : \theta \in \Theta\}$.

(a) Show that if $T(X)$ is sufficient for $P_\theta \in \mathcal{P}$, then it is sufficient for $P_\theta \in \mathcal{P}_A$.

(b) Show that if T is sufficient and complete for $P_\theta \in \mathcal{P}$, then it is complete for $P_\theta \in \mathcal{P}_A$.

47. Show that $(X_{(1)}, X_{(n)})$ in Example 2.13 is not complete.

48. Let T be a complete (or boundedly complete) and sufficient statistic. Suppose that there is a minimal sufficient statistic S. Show that T is minimal sufficient and S is complete (or boundedly complete).

49. Let T and S be two statistics such that $S = \psi(T)$ for a measurable ψ. Show that

(a) if T is complete, then S is complete;

(b) if T is complete and sufficient and ψ is one-to-one, then S is complete and sufficient;

(c) the results in (a) and (b) still hold if the completeness is replaced by the bounded completeness.

50. Find complete and sufficient statistics for the families in Exercises 27 and 28.

51. Show that $(X_{(1)}, X_{(n)})$ in Example 2.11 is complete.

52. Let $(X_1, Y_1), ..., (X_n, Y_n)$ be i.i.d. random 2-vectors having the following Lebesgue p.d.f.

$$f_\theta(x, y) = (2\pi\gamma^2)^{-1} I_{(0,\gamma)}\left(\sqrt{(x-a)^2 + (y-b)^2}\right), \quad (x, y) \in \mathcal{R}^2,$$

where $\theta = (a, b, \gamma) \in \mathcal{R}^2 \times (0, \infty)$.
(a) If $a = 0$ and $b = 0$, find a complete and sufficient statistic for γ.
(b) If all parameters are unknown, show that the convex hull of the sample points is a sufficient statistic for θ.

53. Let X be a discrete random variable with p.d.f.

$$f_\theta(x) = \begin{cases} \theta & x = 0 \\ (1-\theta)^2 \theta^{x-1} & x = 1, 2, ... \\ 0 & \text{otherwise}, \end{cases}$$

where $\theta \in (0, 1)$. Show that X is boundedly complete, but not complete.

54. Show that the sufficient statistic T in Example 2.10 is also complete without using Proposition 2.1.

55. Let $Y_1, ..., Y_n$ be i.i.d. random variables having the Lebesgue p.d.f. $\lambda x^{\lambda-1} I_{(0,1)}(x)$ with an unknown $\lambda > 0$ and let $Z_1, ..., Z_n$ be i.i.d. discrete random variables having the power series distribution given in Exercise 13 with an unknown $\theta > 0$. Assume that Y_i's and Z_j's are independent. Let $X_i = Y_i + Z_i$, $i = 1, ..., n$. Find a complete and sufficient statistic for the unknown parameter (θ, λ) based on the sample $X = (X_1, ..., X_n)$.

56. Suppose that $(X_1, Y_1), ..., (X_n, Y_n)$ are i.i.d. random 2-vectors and X_i and Y_i are independently distributed as $N(\mu, \sigma_X^2)$ and $N(\mu, \sigma_Y^2)$, respectively, with $\theta = (\mu, \sigma_X^2, \sigma_Y^2) \in \mathcal{R} \times (0, \infty) \times (0, \infty)$. Let $\bar{X}$ and S_X^2 be the sample mean and variance given by (2.1) and (2.2) for X_i's and $\bar{Y}$ and S_Y^2 be the sample mean and variance for Y_i's. Show that $T = (\bar{X}, \bar{Y}, S_X^2, S_Y^2)$ is minimal sufficient for θ but T is not boundedly complete.

57. Let $X_1, ..., X_n$ be i.i.d. from the $N(\theta, \theta^2)$ distribution, where $\theta > 0$ is a parameter. Find a minimal sufficient statistic for θ and show whether it is complete.

58. Suppose that $(X_1, Y_1), ..., (X_n, Y_n)$ are i.i.d. random 2-vectors having the normal distribution with $EX_1 = EY_1 = 0$, $\mathrm{Var}(X_1) = \mathrm{Var}(Y_1) = 1$, and $\mathrm{Cov}(X_1, Y_1) = \theta \in (-1, 1)$.
    (a) Find a minimal sufficient statistic for θ.
    (b) Show whether the minimal sufficient statistic in (a) is complete or not.
    (c) Prove that $T_1 = \sum_{i=1}^{n} X_i^2$ and $T_2 = \sum_{i=1}^{n} Y_i^2$ are both ancillary but (T_1, T_2) is not ancillary.

59. Let $X_1, ..., X_n$ be i.i.d. random variables having the exponential distribution $E(a, \theta)$.
    (a) Show that $\sum_{i=1}^{n}(X_i - X_{(1)})$ and $X_{(1)}$ are independent for any (a, θ).
    (b) Show that $Z_i = (X_{(n)} - X_{(i)})/(X_{(n)} - X_{(n-1)})$, $i = 1, ..., n - 2$, are independent of $(X_{(1)}, \sum_{i=1}^{n}(X_i - X_{(1)}))$.

60. Let $X_1, ..., X_n$ be i.i.d. random variables having the gamma distribution $\Gamma(\alpha, \gamma)$. Show that $\sum_{i=1}^{n} X_i$ and $\sum_{i=1}^{n}[\log X_i - \log X_{(1)}]$ are independent for any (α, γ).

61. Let $X_1, ..., X_n$ be i.i.d. random variables having the uniform distribution on the interval (a, b), where $-\infty < a < b < \infty$. Show that $(X_{(i)} - X_{(1)})/(X_{(n)} - X_{(1)})$, $i = 2, ..., n - 1$, are independent of $(X_{(1)}, X_{(n)})$ for any a and b.

62. Consider Example 2.19. Assume that $n > 2$.
    (a) Show that $\bar{X}$ is better than T_1 if $P = N(\theta, \sigma^2)$, $\theta \in \mathcal{R}$, $\sigma > 0$.
    (b) Show that T_1 is better than $\bar{X}$ if P is the uniform distribution on the interval $(\theta - \frac{1}{2}, \theta + \frac{1}{2})$, $\theta \in \mathcal{R}$.
    (c) Find a family $\mathcal{P}$ for which neither $\bar{X}$ nor T_1 is better than the other.

63. Let $X_1, ..., X_n$ be i.i.d. from the $N(\mu, \sigma^2)$ distribution, where $\mu \in \mathcal{R}$ and $\sigma > 0$. Consider the estimation of σ^2 with the squared error loss. Show that $\frac{n-1}{n} S^2$ is better than S^2, the sample variance. Can you find an estimator of the form cS^2 with a nonrandom c such that it is better than $\frac{n-1}{n} S^2$?

64. Let $X_1, ..., X_n$ be i.i.d. binary random variables with $P(X_i = 1) = \theta \in (0, 1)$. Consider estimating θ with the squared error loss. Calculate the risks of the following estimators:
    (a) the nonrandomized estimators $\bar{X}$ (the sample mean) and

$$T_0(X) = \begin{cases} 0 & \text{if more than half of } X_i\text{'s are } 0 \\ 1 & \text{if more than half of } X_i\text{'s are } 1 \\ \frac{1}{2} & \text{if exactly half of } X_i\text{'s are } 0; \end{cases}$$

(b) the randomized estimators

$$T_1(X) = \begin{cases} \bar{X} & \text{with probability } \frac{1}{2} \\ T_0 & \text{with probability } \frac{1}{2} \end{cases}$$

and

$$T_2(X) = \begin{cases} \bar{X} & \text{with probability } \bar{X} \\ \frac{1}{2} & \text{with probability } 1 - \bar{X}. \end{cases}$$

65. Let $X_1, ..., X_n$ be i.i.d. random variables having the exponential distribution $E(0, \theta)$, $\theta \in (0, \infty)$. Consider estimating θ with the squared error loss. Calculate the risks of the sample mean $\bar{X}$ and $cX_{(1)}$, where c is a positive constant. Is $\bar{X}$ better than $cX_{(1)}$ for some c?

66. Consider the estimation of an unknown parameter $\theta \geq 0$ under the squared error loss. Show that if T and U are two estimators such that $T \leq U$ and $R_T(P) < R_U(P)$, then $R_{T_+}(P) < R_{U_+}(P)$, where $R_T(P)$ is the risk of an estimator T and T_+ denotes the positive part of T.

67. Let $X_1, ..., X_n$ be i.i.d. random variables having the exponential distribution $E(0, \theta)$, $\theta \in (0, \infty)$. Consider the hypotheses

$$H_0 : \theta \leq \theta_0 \quad \text{versus} \quad H_1 : \theta > \theta_0,$$

where $\theta_0 > 0$ is a fixed constant. Obtain the risk function (in terms of θ) of the test rule $T_c(X) = I_{(c,\infty)}(\bar{X})$, under the 0-1 loss.

68. Let $X_1, ..., X_n$ be i.i.d. random variables having the Cauchy distribution $C(\mu, \sigma)$ with unknown $\mu \in \mathcal{R}$ and $\sigma > 0$. Consider the hypotheses

$$H_0 : \mu \leq \mu_0 \quad \text{versus} \quad H_1 : \mu > \mu_0,$$

where μ_0 is a fixed constant. Obtain the risk function of the test rule $T_c(X) = I_{(c,\infty)}(\bar{X})$, under the 0-1 loss.

69. Let $X_1, ..., X_n$ be i.i.d. binary random variables with $P(X_i = 1) = \theta$, where $\theta \in (0, 1)$ is unknown and n is an even integer. Consider the problem of testing $H_0 : \theta \leq 0.5$ versus $H_1 : \theta > 0.5$ with action space $\{0, 1\}$ (0 means H_0 is accepted and 1 means H_1 is accepted). Let the loss function be $L(\theta, a) = 0$ if H_j is true and $a = j$, $j = 0, 1$; $L(\theta, 0) = C_0$ when $\theta > 0.5$; and $L(\theta, 1) = C_1$ when $\theta \leq 0.5$, where $C_0 > C_1 > 0$ are some constants. Calculate the risk function of the following randomized test (decision rule):

$$T = \begin{cases} 0 & \text{if more than half of } X_i\text{'s are } 0 \\ 1 & \text{if more than half of } X_i\text{'s are } 1 \\ \frac{1}{2} & \text{if exactly half of } X_i\text{'s are } 0. \end{cases}$$

70. Consider Example 2.21. Suppose that our decision rule, based on a sample $X = (X_1, ..., X_n)$ with i.i.d. components from the $N(\theta, 1)$ distribution with an unknown $\theta > 0$, is

$$
T(X) = \begin{cases} a_1 & b_1 < \bar{X} \\ a_2 & b_0 < \bar{X} \leq b_1 \\ a_3 & \bar{X} \leq b_0. \end{cases}
$$

Express the risk of T in terms of θ.

71. Consider an estimation problem with $\mathcal{P} = \{P_\theta : \theta \in \Theta\}$ (a parametric family), $\mathbb{A} = \Theta$, and the squared error loss. If $\theta_0 \in \Theta$ satisfies that $P_\theta \ll P_{\theta_0}$ for any $\theta \in \Theta$, show that the estimator $T \equiv \theta_0$ is admissible.

72. Let $\mathfrak{F}$ be a class of decision rules. A subclass $\mathfrak{F}_0 \subset \mathfrak{F}$ is called $\mathfrak{F}$-complete if and only if, for any $T \in \mathfrak{F}$ and $T \notin \mathfrak{F}_0$, there is a $T_0 \in \mathfrak{F}_0$ that is better than T, and $\mathfrak{F}_0$ is called $\mathfrak{F}$-minimal complete if and only if $\mathfrak{F}_0$ is $\mathfrak{F}$-complete and no proper subclass of $\mathfrak{F}_0$ is $\mathfrak{F}$-complete. Show that if a $\mathfrak{F}$-minimal complete class exists, then it is exactly the class of $\mathfrak{F}$-admissible rules.

73. Let $X_1, ..., X_n$ be i.i.d. random variables having a distribution $P \in \mathcal{P}$. Assume that $EX_1^2 < \infty$. Consider estimating $\mu = EX_1$ under the squared error loss.
    (a) Show that any estimator of the form $a\bar{X} + b$ is inadmissible, where $\bar{X}$ is the sample mean, a and b are constants, and $a > 1$.
    (b) Show that any estimator of the form $\bar{X} + b$ is inadmissible, where $b \neq 0$ is a constant.

74. Consider an estimation problem with $\vartheta \in [c, d] \subset \mathcal{R}$, where c and d are known. Suppose that the action space is $\mathbb{A} \supset [c, d]$ and the loss function is $L(|\vartheta - a|)$, where $L(\cdot)$ is an increasing function on $[0, \infty)$. Show that any decision rule T with $P(T(X) \notin [c, d]) > 0$ for some $P \in \mathcal{P}$ is inadmissible.

75. Suppose that the action space is $(\Omega, \mathcal{B}_\Omega^k)$, where $\Omega \in \mathcal{B}^k$. Let X be a sample from $P \in \mathcal{P}$, $\delta_0(X)$ be a nonrandomized rule, and T be a sufficient statistic for $P \in \mathcal{P}$. Show that if $E[I_A(\delta_0(X))|T]$ is a nonrandomized rule, i.e., $E[I_A(\delta_0(X))|T] = I_A(h(T))$ for any $A \in \mathcal{B}_\Omega^k$, where h is a Borel function, then $\delta_0(X) = h(T(X))$ a.s. P.

76. Let T, δ_0, and δ_1 be as given in the statement of Proposition 2.2. Show that

$$
\int_{\mathbb{A}} L(P, a) d\delta_1(X, a) = E\left[\int_{\mathbb{A}} L(P, a) d\delta_0(X, a) \middle| T\right] \quad \text{a.s. } P.
$$

77. Prove Theorem 2.5.

78. In Exercise 64, use Theorem 2.5 to find decision rules that are better than T_j, $j = 0, 1, 2$.

79. In Exercise 65, use Theorem 2.5 to find a decision rule better than $cX_{(1)}$.

80. Consider Example 2.22.
    (a) Show that there is no optimal rule if $\Im$ contains all possible estimators. (Hint: consider constant estimators.)
    (b) Find a $\Im_2$-optimal rule if $X_1, ..., X_n$ are independent random variables having a common mean μ and $\text{Var}(X_i) = \sigma^2/a_i$ with known a_i, $i = 1, ..., n$.
    (c) Find a $\Im_2$-optimal rule if $X_1, ..., X_n$ are identically distributed but are correlated with a common correlation coefficient ρ.

81. Let $X_{ij} = \mu + a_i + \epsilon_{ij}$, $i = 1, ..., m$, $j = 1, ..., n$, where a_i's and ϵ_{ij}'s are independent random variables, a_i is $N(0, \sigma_a^2)$, ϵ_{ij} is $N(0, \sigma_e^2)$, and μ, σ_a^2, and σ_e^2 are unknown parameters. Define $\bar{X}_i = n^{-1} \sum_{j=1}^{n} X_{ij}$, $\bar{X} = m^{-1} \sum_{i=1}^{m} \bar{X}_i$, MSA $= n(m-1)^{-1} \sum_{i=1}^{m} (\bar{X}_i - \bar{X})^2$, and MSE $= m^{-1}(n-1)^{-1} \sum_{i=1}^{m} \sum_{j=1}^{n} (X_{ij} - \bar{X}_i)^2$. Assume that $m(n-1) > 4$. Consider the following class of estimators of $\theta = \sigma_a^2/\sigma_e^2$:

$$\left\{ \hat{\theta}(\delta) = \frac{1}{n} \left[(1 - \delta) \frac{\text{MSA}}{\text{MSE}} - 1 \right] : \delta \in \mathcal{R} \right\}.$$

    (a) Show that MSA and MSE are independent.
    (b) Obtain a $\delta \in \mathcal{R}$ such that $\hat{\theta}(\delta)$ is unbiased for θ.
    (c) Show that the risk of $\hat{\theta}(\delta)$ under the squared error loss is a function of (δ, θ).
    (d) Show that there is a constant δ^* such that for any fixed θ, the risk of $\hat{\theta}(\delta)$ is strictly decreasing in δ for $\delta < \delta^*$ and strictly increasing for $\delta > \delta^*$.
    (e) Show that the unbiased estimator of θ derived in (b) is inadmissible.

82. Let $T_0(X)$ be an unbiased estimator of ϑ in an estimation problem. Show that any unbiased estimator of ϑ is of the form $T(X) = T_0(X) - U(X)$, where $U(X)$ is an "unbiased estimator" of 0.

83. Let X be a discrete random variable with

$$P(X = -1) = p, \quad P(X = k) = (1 - p)^2 p^k, \quad k = 0, 1, 2, ...,$$

    where $p \in (0, 1)$ is unknown.
    (a) Show that $U(X)$ is an unbiased estimator of 0 if and only if $U(k) =$

ak for all $k = -1, 0, 1, 2, \ldots$ and some a.

(b) Show that $T_0(X) = I_{\{0\}}(X)$ is unbiased for $\vartheta = (1-p)^2$ and that, under the squared error loss, T_0 is a $\Im$-optimal rule, where $\Im$ is the class of all unbiased estimators of ϑ.

(c) Show that $T_0(X) = I_{\{-1\}}(X)$ is unbiased for $\vartheta = p$ and that, under the squared error loss, there is no $\Im$-optimal rule, where $\Im$ is the class of all unbiased estimators of ϑ.

84. (Nonexistence of an unbiased estimator). Let X be a random variable having the binomial distribution $Bi(p, n)$ with an unknown $p \in (0, 1)$ and a known n. Consider the problem of estimating $\vartheta = p^{-1}$. Show that there is no unbiased estimator of ϑ.

85. Let $X_1, \ldots, X_n$ be i.i.d. random variables having the normal distribution $N(\theta, 1)$, where $\theta = 0$ or 1. Consider the estimation of θ.

(a) Let $\Im$ be the class of nonrandomized rules (estimators), i.e., estimators that take values 0 and 1 only. Show that there does not exist any unbiased estimator of θ in $\Im$.

(b) Find a randomized rule (estimator) that is unbiased for θ.

86. Let $X_1, \ldots, X_n$ be i.i.d. from the Poisson distribution $P(\theta)$ with an unknown $\theta > 0$. Find the bias and mse of $T_n = (1 - a/n)^{n\bar{X}}$ as an estimator of $\vartheta = e^{-a\theta}$, where $a \neq 0$ is a known constant.

87. Let $X_1, \ldots, X_n$ be i.i.d. $(n \geq 3)$ from $N(\mu, \sigma^2)$, where $\mu > 0$ and $\sigma > 0$ are unknown parameters. Let $T_1 = \bar{X}/S$ be an estimator of μ/σ and $T_2 = \bar{X}^2$ be an estimator of μ^2, where $\bar{X}$ and S^2 are the sample mean and variance, respectively. Calculate the mse's of T_1 and T_2.

88. Consider a location family $\{P_\mu : \mu \in \mathcal{R}^k\}$ on $\mathcal{R}^k$, where $P_\mu = P_{(\mu, I_k)}$ is given in (2.10). Let $l_0 \in \mathcal{R}^k$ be a fixed vector and $L(P, a) = L(\|\mu - a\|)$, where $a \in \mathbb{A} = \mathcal{R}^k$ and $L(\cdot)$ is a nonnegative Borel function on $[0, \infty)$. Show that the family is invariant and the decision problem is invariant under the transformation $g(X) = X + cl_0$, $c \in \mathcal{R}$. Find an invariant decision rule.

89. Let $X_1, \ldots, X_n$ be i.i.d. from the $N(\mu, \sigma^2)$ distribution with unknown $\mu \in \mathcal{R}$ and $\sigma^2 > 0$. Consider the scale transformation aX, $a \in (0, \infty)$.

(a) For estimating σ^2 under the loss function $L(P, a) = (1 - a/\sigma^2)^2$, show that the problem is invariant and that the sample variance S^2 is invariant.

(b) For testing $H_0 : \mu \leq 0$ versus $H_1 : \mu > 0$ under the loss

$$L(P, 0) = \frac{\mu}{\sigma} I_{(0, \infty)}(\mu) \quad \text{and} \quad L(P, 1) = \frac{|\mu|}{\sigma} I_{(-\infty, 0]}(\mu),$$

show that the problem is invariant and any test that is a function of $\bar{X}/\sqrt{S^2/n}$ is invariant.

90. Let $X_1, ..., X_n$ be i.i.d. random variables having the c.d.f. $F(x - \theta)$, where F is symmetric about 0 and $\theta \in \mathcal{R}$ is unknown.
   (a) Show that the c.d.f. of $\sum_{i=1}^{n} w_i X_{(i)} - \theta$ is symmetric about 0, where $X_{(i)}$ is the ith order statistic and w_i's are constants satisfying $w_i = w_{n-i+1}$ and $\sum_{i=1}^{n} w_i = 1$.
   (b) Show that $\sum_{i=1}^{n} w_i X_{(i)}$ in (a) is unbiased for θ if the mean of F exists.
   (c) Show that $\sum_{i=1}^{n} w_i X_{(i)}$ is location invariant when $\sum_{i=1}^{n} w_i = 1$.

91. In Example 2.25, show that the conditional distribution of θ given $X = x$ is $N(\mu_*(x), c^2)$ with $\mu_*(x)$ and c^2 given by (2.25).

92. A *median* of a random variable Y (or its distribution) is any value m such that $P(Y \leq m) \geq \frac{1}{2}$ and $P(Y \geq m) \geq \frac{1}{2}$.
   (a) Show that the set of medians is a closed interval $[m_0, m_1]$.
   (b) Suppose that $E|Y| < \infty$. If c is not a median of Y, show that $E|Y - c| \geq E|Y - m|$ for any median m of Y.
   (c) Let X be a sample from P_θ, where $\theta \in \Theta \subset \mathcal{R}$. Consider the estimation of θ under the absolute error loss function $|a - \theta|$. Let Π be a given distribution on Θ with finite mean. Find the $\mathfrak{S}$-Bayes rule w.r.t. Π, where $\mathfrak{S}$ is the class of all rules.

93. (Classification). Let X be a sample having a p.d.f. $f_j(x)$ w.r.t. a σ-finite measure ν, where j is unknown and $j \in \{1, ..., J\}$ with a known integer $J \geq 2$. Consider a decision problem in which the action space $\mathbb{A} = \{1, ..., J\}$ and the loss function is

$$L(j, a) = \begin{cases} 0 & \text{if } a = j \\ 1 & \text{if } a \neq j. \end{cases}$$

   (a) Let $\mathfrak{S}$ be the class of all nonrandomized decision rules. Obtain the risk of a $\delta \in \mathfrak{S}$.
   (b) Let Π be a probability measure on $\{1, ..., J\}$ with $\Pi(\{j\}) = \pi_j$, $j = 1, ..., J$. Obtain the Bayes risk of $\delta \in \mathfrak{S}$ w.r.t. Π.
   (c) Obtain a $\mathfrak{S}$-Bayes rule w.r.t. Π in (b).
   (d) Assume that $J = 2$, $\pi_1 = \pi_2 = 0.5$, and $f_j(x) = \phi(x - \mu_j)$, where $\phi(x)$ is the p.d.f. of the standard normal distribution and μ_j, $j = 1, 2$, are known constants. Obtain the Bayes rule in (c) and compute the Bayes risk.
   (e) Obtain the risk and the Bayes risk (w.r.t. Π in (b)) of a randomized decision rule.
   (f) Obtain a Bayes rule w.r.t. Π.
   (g) Obtain a minimax rule.

94. Let $\hat{\theta}$ be an unbiased estimator of an unknown $\theta \in \mathcal{R}$.
   (a) Under the squared error loss, show that the estimator $\hat{\theta} + c$ is not

minimax, where $c \neq 0$ is a known constant.

(b) Under the squared error loss, show that the estimator $c\hat{\theta}$ is not minimax unless $\sup_\theta R_T(\theta) = \infty$ for any estimator T, where $c \in (0,1)$ is a known constant.

(c) Consider the loss function $L(\theta, a) = (a - \theta)^2/\theta^2$ (assuming $\theta \neq 0$). Show that $\hat{\theta}$ is not minimax unless $\sup_\theta R_T(\theta) = \infty$ for any T.

95. Let X be a binary observation with $P(X = 1) = \theta_1$ or θ_2, where $0 < \theta_1 < \theta_2 < 1$ are known values. Consider the estimation of θ with action space $\{a_1, a_2\}$ and loss function $L(\theta_i, a_j) = l_{ij}$, where $l_{21} \geq l_{12} > l_{11} = l_{22} = 0$. For a decision rule $\delta(X)$, the vector $(R_\delta(\theta_1), R_\delta(\theta_2))$ is defined to be its risk point.

(a) Show that the set of risk points of all decision rules is the convex hull of the set of risk points of all nonrandomized rules.

(b) Find a minimax rule.

(c) Let Π be a distribution on $\{\theta_1, \theta_2\}$. Obtain the class of all Bayes rules w.r.t. Π. Discuss when there is a unique Bayes rule.

96. Consider the decision problem in Example 2.23.

(a) Let Π be the uniform distribution on $(0, 1)$. Show that a $\Im$-Bayes rule w.r.t. Π is $T_{j^*}(X)$, where j^* is the smallest integer in $\{0, 1, ..., n-1\}$ such that $B_{j+1,n-j+1}(\theta_0) \geq \frac{1}{2}$ and $B_{a,b}(\cdot)$ denotes the c.d.f. of the beta distribution $B(a, b)$.

(b) Derive a $\Im$-minimax rule.

97. Let $X_1, ..., X_n$ be i.i.d. from the $N(\mu, \sigma^2)$ distribution with unknown $\mu \in \mathcal{R}$ and $\sigma^2 > 0$. To test the hypotheses

$$H_0 : \mu \leq \mu_0 \qquad \text{versus} \qquad H_1 : \mu > \mu_0,$$

where μ_0 is a fixed constant, consider a test of the form $T_c(X) = I_{(c,\infty)}(T_{\mu_0})$, where $T_{\mu_0} = (\bar{X} - \mu_0)/\sqrt{S^2/n}$ and c is a fixed constant.

(a) Find the size of T_c. (Hint: T_{μ_0} has the t-distribution t_{n-1}.)

(b) If α is a given level of significance, find a c_α such that T_{c_α} has size α.

(c) Compute the p-value for T_{c_α} derived in (b).

(d) Find a c_α such that $[\bar{X} - c_\alpha\sqrt{S^2/n}, \bar{X} + c_\alpha\sqrt{S^2/n}]$ is a confidence interval for μ with confidence coefficient $1 - \alpha$. What is the expected interval length?

98. In Exercise 67, calculate the size of $T_c(X)$; find a c_α such that T_{c_α} has size α, a given level of significance; and find the p-value for T_{c_α}.

99. In Exercise 68, assume that σ is known. Calculate the size of $T_c(X)$; find a c_α such that T_{c_α} has size α, a given level of significance; and find the p-value for T_{c_α}.

100. Let $\alpha \in (0, 1)$ be given and $T_{j,q}(X)$ be the test given in Example 2.30. Show that there exist integer j and $q \in (0, 1)$ such that the size of $T_{j,q}$ is α.

101. Let $X_1, ..., X_n$ be i.i.d. from the exponential distribution $E(a, \theta)$ with unknown $a \in \mathcal{R}$ and $\theta > 0$. Let $\alpha \in (0, 1)$ be given.
(a) Using $T_1(X) = \sum_{i=1}^{n}(X_i - X_{(1)})$, construct a confidence interval for θ with confidence coefficient $1 - \alpha$ and find the expected interval length.
(b) Using $T_1(X)$ and $T_2(X) = X_{(1)}$, construct a confidence interval for a with confidence coefficient $1 - \alpha$ and find the expected interval length.
(c) Using the method in Example 2.32, construct a confidence set for the two-dimensional parameter (a, θ) with confidence coefficient $1 - \alpha$.

102. Suppose that X is a sample and a statistic $T(X)$ has a distribution in a location family $\{P_\mu : \mu \in \mathcal{R}\}$. Using $T(X)$, derive a confidence interval for μ with level of significance $1 - \alpha$ and obtain the expected interval length. Show that if the c.d.f. of $T(X)$ is continuous, then we can always find a confidence interval for μ with confidence coefficient $1 - \alpha$ for any $\alpha \in (0, 1)$.

103. Let $X = (X_1, ..., X_n)$ be a sample from P_θ, where $\theta \in \{\theta_1, ..., \theta_k\}$ with a fixed integer k. Let $T_n(X)$ be an estimator of θ with range $\{\theta_1, ..., \theta_k\}$.
(a) Show that $T_n(X)$ is consistent if and only if $P_\theta(T_n(X) = \theta) \to 1$.
(b) Show that if $T_n(X)$ is consistent, then it is a_n-consistent for any $\{a_n\}$.

104. Let $X_1, ..., X_n$ be i.i.d. from the uniform distribution on $(\theta - \frac{1}{2}, \theta + \frac{1}{2})$, where $\theta \in \mathcal{R}$ is unknown. Show that $(X_{(1)} + X_{(n)})/2$ is strongly consistent for θ and also consistent in mse.

105. Let $X_1, ..., X_n$ be i.i.d. from a population with the Lebesgue p.d.f. $f_\theta(x) = 2^{-1}(1 + \theta x)I_{(-1,1)}(x)$, where $\theta \in (-1, 1)$ is an unknown parameter. Find a consistent estimator of θ. Is your estimator $\sqrt{n}$-consistent?

106. Let $X_1, ..., X_n$ be i.i.d. observations. Suppose that T_n is an unbiased estimator of ϑ based on $X_1, ..., X_n$ such that for any n, $\text{Var}(T_n) < \infty$ and $\text{Var}(T_n) \le \text{Var}(U_n)$ for any other unbiased estimator U_n of ϑ based on $X_1, ..., X_n$. Show that T_n is consistent in mse.

107. Consider the Bayes rule $\mu_*(X)$ in Example 2.25. Show that $\mu_*(X)$ is a strongly consistent, $\sqrt{n}$-consistent, and L_2-consistent estimator of μ. What is the order of the bias of $\mu_*(X)$ as an estimator of μ?

108. In Exercise 21, show that
    (a) $\bar{Y}/\bar{X}$ is an inconsistent estimator of β;
    (b) $\hat{\beta} = Z_{(m)}$ is a consistent estimator of β, where $m = n/2$ when n is even, $m = (n+1)/2$ when n is odd, and $Z_{(i)}$ is the ith smallest value of Y_i/X_i, $i = 1, ..., n$.

109. Show that the estimator T_0 of θ in Exercise 64 is inconsistent.

110. Let $g_1, g_2,...$ be continuous functions on $(a, b) \subset \mathcal{R}$ such that $g_n(x) \to g(x)$ uniformly for x in any closed subinterval of (a, b). Let T_n be a consistent estimator of $\theta \in (a, b)$. Show that $g_n(T_n)$ is consistent for $\vartheta = g(\theta)$.

111. Let $X_1, ..., X_n$ be i.i.d. from P with unknown mean $\mu \in \mathcal{R}$ and variance $\sigma^2 > 0$, and let $g(\mu) = 0$ if $\mu \neq 0$ and $g(0) = 1$. Find a consistent estimator of $\vartheta = g(\mu)$.

112. Establish results for the smallest order statistic $X_{(1)}$ (based on i.i.d. random variables $X_1, ..., X_n$) similar to those in Example 2.34.

113. (Consistency for finite population). In Example 2.27, show that $\hat{Y} \to_p Y$ as $n \to N$ for any fixed N and population. Is $\hat{Y}$ still consistent if sampling is with replacement?

114. Assume that $X_i = \theta t_i + e_i$, $i = 1, ..., n$, where $\theta \in \Theta$ is an unknown parameter, Θ is a closed subset of $\mathcal{R}$, e_i's are i.i.d. on the interval $[-\tau, \tau]$ with some unknown $\tau > 0$ and $Ee_i = 0$, and t_i's are fixed constants. Let
$$T_n = S_n(\tilde{\theta}_n) = \min_{\gamma \in \Theta} S_n(\gamma),$$
where
$$S_n(\gamma) = 2 \max_{i \leq n} |X_i - \gamma t_i| / \sqrt{1 + \gamma^2}.$$

    (a) Assume that $\sup_i |t_i| < \infty$ and $\sup_i t_i - \inf_i t_i > 2\tau$. Show that the sequence $\{\tilde{\theta}_n, n = 1, 2, ...\}$ is bounded a.s.
    (b) Let $\theta_n \in \Theta$, $n = 1, 2, ...$ If $\theta_n \to \theta$, show that
$$S_n(\theta_n) - S_n(\theta) = O(|\theta_n - \theta|) \quad \text{a.s.}$$

    (c) Under the conditions in (a), show that T_n is a strongly consistent estimator of $\vartheta = \min_{\gamma \in \Theta} S(\gamma)$, where $S(\gamma) = \lim_{n \to \infty} S_n(\gamma)$ a.s.

115. Let $X_1, ..., X_n$ be i.i.d. random variables with $EX_1^2 < \infty$ and $\bar{X}$ be the sample mean. Consider the estimation of $\mu = EX_1$.
    (a) Let $T_n = \bar{X} + \xi_n/\sqrt{n}$, where ξ_n is a random variable satisfying $\xi_n = 0$ with probability $1 - n^{-1}$ and $\xi_n = n^{3/2}$ with probability n^{-1}.

Show that $b_{T_n}(P) \neq \tilde{b}_{T_n}(P)$ for any P.

(b) Let $T_n = \bar{X} + \eta_n/\sqrt{n}$, where η_n is a random variable that is independent of $X_1, ..., X_n$ and equals 0 with probability $1 - 2n^{-1}$ and $\pm\sqrt{n}$ with probability n^{-1}. Show that $\mathrm{amse}_{T_n}(P) = \mathrm{amse}_{\bar{X}}(P) = \mathrm{mse}_{\bar{X}}(P)$ and $\mathrm{mse}_{T_n}(P) > \mathrm{amse}_{T_n}(P)$ for any P.

116. Let $X_1, ..., X_n$ be i.i.d. random variables with finite $\theta = EX_1$ and $\mathrm{Var}(X_1) = \theta$, where $\theta > 0$ is unknown. Consider the estimation of $\vartheta = \sqrt{\theta}$. Let $T_{1n} = \sqrt{\bar{X}}$ and $T_{2n} = \bar{X}/S$, where $\bar{X}$ and S^2 are the sample mean and sample variance.
(a) Obtain the n^{-1} order asymptotic biases of T_{1n} and T_{2n} according to (2.38).
(b) Obtain the asymptotic relative efficiency of T_{1n} w.r.t. T_{2n}.

117. Let $X_1, ..., X_n$ be i.i.d. according to $N(\mu, 1)$ with an unknown $\mu \in \mathcal{R}$. Let $\vartheta = P(X_1 \leq c)$ for a fixed constant c. Consider the following estimators of ϑ: $T_{1n} = F_n(c)$, where F_n is the empirical c.d.f. defined in (2.28), and $T_{2n} = \Phi(c - \bar{X})$, where Φ is the c.d.f. of $N(0, 1)$.
(a) Find the n^{-1} order asymptotic bias of T_{2n} according to (2.38).
(b) Find the asymptotic relative efficiency of T_{1n} w.r.t. T_{2n}.

118. Let $X_1, ..., X_n$ be i.i.d. from the $N(0, \sigma^2)$ distribution with an unknown $\sigma > 0$. Consider the estimation of $\vartheta = \sigma$. Find the asymptotic relative efficiency of $\sqrt{\pi/2} \sum_{i=1}^n |X_i|/n$ w.r.t. $(\sum_{i=1}^n X_i^2/n)^{1/2}$.

119. Let $X_1, ..., X_n$ be i.i.d. from P with $EX_1^4 < \infty$ and unknown mean $\mu \in \mathcal{R}$ and variance $\sigma^2 > 0$. Consider the estimation of $\vartheta = \mu^2$ and the following three estimators: $T_{1n} = \bar{X}^2$, $T_{2n} = \bar{X}^2 - S^2/n$, $T_{3n} = \max\{0, T_{2n}\}$, where $\bar{X}$ and S^2 are the sample mean and variance. Show that the amse's of T_{jn}, $j = 1, 2, 3$, are the same when $\mu \neq 0$ but may be different when $\mu = 0$. Which estimator is the best in terms of the asymptotic relative efficiency when $\mu = 0$?

120. Prove Theorem 2.6.

121. Let $X_1, ..., X_n$ be i.i.d. with $EX_i = \mu$, $\mathrm{Var}(X_i) = 1$, and $EX_i^4 < \infty$. Let $T_{1n} = n^{-1} \sum_{i=1}^n X_i^2 - 1$ and $T_{2n} = \bar{X}^2 - n^{-1}$ be estimators of $\vartheta = \mu^2$.
(a) Find the asymptotic relative efficiency of T_{1n} w.r.t. T_{2n}.
(b) Show that $e_{T_{1n}, T_{2n}}(P) \leq 1$ if the c.d.f. of $X_i - \mu$ is symmetric about 0 and $\mu \neq 0$.
(c) Find a distribution P for which $e_{T_{1n}, T_{2n}}(P) > 1$.

122. Let $X_1, ..., X_n$ be i.i.d. binary random variables with unknown $p = P(X_i = 1) \in (0, 1)$. Consider the estimation of p. Let a and b be two positive constants. Find the asymptotic relative efficiency of the estimator $(a + n\bar{X})/(a + b + n)$ w.r.t. $\bar{X}$.

123. Let $X_1, ..., X_n$ be i.i.d. from $N(\mu, \sigma^2)$ with an unknown $\mu \in \mathcal{R}$ and a known σ^2. Let $T_1 = \bar{X}$ be the sample mean and $T_2 = \mu_*(X)$ be the Bayes estimator given in (2.25). Assume that $EX_1^4 < \infty$.
(a) Calculate the exact mse of both estimators. Can you conclude that one estimator is better than the other in terms of the mse?
(b) Find the asymptotic relative efficiency of T_1 w.r.t. T_2.

124. In Example 2.37, show that
(a) the limiting size of T_{c_α} is 1 if $\mathcal{P}$ contains all possible populations on $\mathcal{R}$ with finite second moments;
(b) $T_n = T_{c_\alpha}$ with $\alpha = \alpha_n$ (given by (2.40)) is Chernoff-consistent;
(c) T_n in (b) is not strongly Chernoff-consistent if $\mathcal{P}$ contains all possible populations on $\mathcal{R}$ with finite second moments.

125. Let $X_1, ..., X_n$ be i.i.d. with unknown mean $\mu \in \mathcal{R}$ and variance $\sigma^2 > 0$. For testing $H_0 : \mu \leq \mu_0$ versus $H_1 : \mu > \mu_0$, consider the test T_{c_α} obtained in Exercise 97(b).
(a) Show that T_{c_α} has asymptotic significance level α and is consistent.
(b) Find a test that is Chernoff-consistent.

126. Consider the test T_j in Example 2.23. For each n, find a $j = j_n$ such that T_{j_n} has asymptotic significance level $\alpha \in (0, 1)$.

127. Show that the test T_{c_α} in Exercise 98 is consistent, but T_{c_α} in Exercise 99 is not consistent.

128. In Example 2.31, suppose that we drop the normality assumption but assume that $\mu = EX_i$ and $\sigma^2 = \text{Var}(X_i)$ are finite.
(a) Show that when σ^2 is known, the asymptotic significance level of the confidence interval $[\bar{X} - c_\alpha, \bar{X} + c_\alpha]$ is $1 - \alpha$, where $c_\alpha = \sigma z_{1-\alpha/2}/\sqrt{n}$ and $z_a = \Phi^{-1}(a)$.
(b) Show that when σ^2 is known, the limiting confidence coefficient of the interval in (a) might be 0 if $\mathcal{P}$ contains all possible populations on $\mathcal{R}$.
(c) Show that the confidence interval in Exercise 97(d) has asymptotic significance level $1 - \alpha$.

129. Let $X_1, ..., X_n$ be i.i.d. with unknown mean $\mu \in \mathcal{R}$ and variance $\sigma^2 > 0$. Assume that $EX_1^4 < \infty$. Using the sample variance S^2, construct a confidence interval for σ^2 that has asymptotic significance level $1 - \alpha$.

130. Consider the sample correlation coefficient T defined in Exercise 22. Construct a confidence interval for ρ that has asymptotic significance level $1 - \alpha$, assuming that (Y_i, Z_i) is normally distributed. (Hint: show that the asymptotic variance of T is $(1 - \rho^2)^2$.)

Chapter 3

Unbiased Estimation

Unbiased or asymptotically unbiased estimation plays an important role in point estimation theory. Unbiasedness of point estimators is defined in §2.3.2. In this chapter, we discuss in detail how to derive unbiased estimators and, more importantly, how to find the best unbiased estimators in various situations. Although an unbiased estimator (even the best unbiased estimator if it exists) is not necessarily better than a slightly biased estimator in terms of their mse's (see Exercise 63 in §2.6), unbiased estimators can be used as "building blocks" for the construction of better estimators. Furthermore, one may give up the exact unbiasedness, but cannot give up asymptotic unbiasedness since it is necessary for consistency (see §2.5.2). Properties and the construction of asymptotically unbiased estimators are studied in the last part of this chapter.

3.1 The UMVUE

Let X be a sample from an unknown population $P \in \mathcal{P}$ and ϑ be a real-valued parameter related to P. Recall that an estimator $T(X)$ of ϑ is unbiased if and only if $E[T(X)] = \vartheta$ for any $P \in \mathcal{P}$. If there exists an unbiased estimator of ϑ, then ϑ is called an *estimable* parameter.

Definition 3.1. An unbiased estimator $T(X)$ of ϑ is called the *uniformly minimum variance unbiased estimator* (UMVUE) if and only if $\text{Var}(T(X)) \leq \text{Var}(U(X))$ for any $P \in \mathcal{P}$ and any other unbiased estimator $U(X)$ of ϑ. ∎

Since the mse of any unbiased estimator is its variance, a UMVUE is $\Im$-optimal in mse with $\Im$ being the class of all unbiased estimators. One can similarly define the uniformly minimum risk unbiased estimator in statistical decision theory when we use an arbitrary loss instead of the squared error loss that corresponds to the mse.

3.1.1 Sufficient and complete statistics

The derivation of a UMVUE is relatively simple if there exists a sufficient and complete statistic for $P \in \mathcal{P}$.

Theorem 3.1 (Lehmann-Scheffé theorem). Suppose that there exists a sufficient and complete statistic $T(X)$ for $P \in \mathcal{P}$. If ϑ is estimable, then there is a unique unbiased estimator of ϑ that is of the form $h(T)$ with a Borel function h. (Two estimators that are equal a.s. $\mathcal{P}$ are treated as one estimator.) Furthermore, $h(T)$ is the unique UMVUE of ϑ.   ∎

This theorem is a consequence of Theorem 2.5(ii) (Rao-Blackwell theorem). One can easily extend this theorem to the case of the uniformly minimum risk unbiased estimator under any loss function $L(P, a)$ that is strictly convex in a. The uniqueness of the UMVUE follows from the completeness of $T(X)$.

There are two typical ways to derive a UMVUE when a sufficient and complete statistic T is available. The first one is solving for h when the distribution of T is available. The following are two typical examples.

Example 3.1. Let $X_1, ..., X_n$ be i.i.d. from the uniform distribution on $(0, \theta)$, $\theta > 0$. Let $\vartheta = g(\theta)$, where g is a differentiable function on $(0, \infty)$. Since the sufficient and complete statistic $X_{(n)}$ has the Lebesgue p.d.f. $n\theta^{-n}x^{n-1}I_{(0,\theta)}(x)$, an unbiased estimator $h(X_{(n)})$ of ϑ must satisfy

$$\theta^n g(\theta) = n \int_0^\theta h(x)x^{n-1}dx \qquad \text{for all } \theta > 0.$$

Differentiating both sizes of the previous equation and applying the result of differentiation of an integral (Royden (1968, §5.3)) lead to

$$n\theta^{n-1}g(\theta) + \theta^n g'(\theta) = nh(\theta)\theta^{n-1}.$$

Hence, the UMVUE of ϑ is $h(X_{(n)}) = g(X_{(n)}) + n^{-1}X_{(n)}g'(X_{(n)})$. In particular, if $\vartheta = \theta$, then the UMVUE of θ is $(1 + n^{-1})X_{(n)}$.   ∎

Example 3.2. Let $X_1, ..., X_n$ be i.i.d. from the Poisson distribution $P(\theta)$ with an unknown $\theta > 0$. Then $T(X) = \sum_{i=1}^n X_i$ is sufficient and complete for $\theta > 0$ and has the Poisson distribution $P(n\theta)$. Suppose that $\vartheta = g(\theta)$, where g is a smooth function such that $g(x) = \sum_{j=0}^\infty a_j x^j$, $x > 0$. An unbiased estimator $h(T)$ of ϑ must satisfy

$$\sum_{t=0}^{\infty} \frac{h(t)n^t}{t!}\theta^t = e^{n\theta}g(\theta)$$

$$= \sum_{k=0}^{\infty} \frac{n^k}{k!}\theta^k \sum_{j=0}^{\infty} a_j\theta^j$$

$$= \sum_{t=0}^{\infty}\left(\sum_{j,k:j+k=t} \frac{n^k a_j}{k!}\right)\theta^t$$

for any $\theta > 0$. Thus, a comparison of coefficients in front of θ^t leads to

$$h(t) = \frac{t!}{n^t}\sum_{j,k:j+k=t} \frac{n^k a_j}{k!},$$

i.e., $h(T)$ is the UMVUE of ϑ. In particular, if $\vartheta = \theta^r$ for some fixed integer $r \geq 1$, then $a_r = 1$ and $a_k = 0$ if $k \neq r$ and

$$h(t) = \begin{cases} 0 & t < r \\ \frac{t!}{n^r(t-r)!} & t \geq r. \end{cases} \quad \blacksquare$$

The second method of deriving a UMVUE when there is a sufficient and complete statistic $T(X)$ is conditioning on T, i.e., if $U(X)$ is any unbiased estimator of ϑ, then $E[U(X)|T]$ is the UMVUE of ϑ. To apply this method, we do not need the distribution of T, but need to work out the conditional expectation $E[U(X)|T]$. From the uniqueness of the UMVUE, it does not matter which $U(X)$ is used and, thus, we should choose $U(X)$ so as to make the calculation of $E[U(X)|T]$ as easy as possible.

Example 3.3. Consider the estimation problem in Example 2.26, where $\vartheta = 1 - F_\theta(t)$ and $F_\theta(x) = (1 - e^{-x/\theta})I_{(0,\infty)}(x)$. Since $\bar{X}$ is sufficient and complete for $\theta > 0$ and $I_{(t,\infty)}(X_1)$ is unbiased for ϑ,

$$T(X) = E[I_{(t,\infty)}(X_1)|\bar{X}] = P(X_1 > t|\bar{X})$$

is the UMVUE of ϑ. If the conditional distribution of X_1 given $\bar{X}$ is available, then we can calculate $P(X_1 > t|\bar{X})$ directly. But the following technique can be applied to avoid the derivation of conditional distributions. By Basu's theorem (Theorem 2.4), $X_1/\bar{X}$ and $\bar{X}$ are independent. By Proposition 1.10(vii),

$$P(X_1 > t|\bar{X} = \bar{x}) = P(X_1/\bar{X} > t/\bar{X}|\bar{X} = \bar{x}) = P(X_1/\bar{X} > t/\bar{x}).$$

To compute this unconditional probability, we need the distribution of

$$X_1 \bigg/ \sum_{i=1}^{n} X_i = X_1 \bigg/ \left(X_1 + \sum_{i=2}^{n} X_i \right).$$

Using the transformation technique discussed in §1.3.1 and the fact that $\sum_{i=2}^{n} X_i$ is independent of X_1 and has a gamma distribution, we obtain that $X_1 / \sum_{i=1}^{n} X_i$ has the Lebesgue p.d.f. $(n-1)(1-x)^{n-2}I_{(0,1)}(x)$. Hence

$$P(X_1 > t | \bar{X} = \bar{x}) = (n-1) \int_{t/(n\bar{x})}^{1} (1-x)^{n-2} dx = \left(1 - \frac{t}{n\bar{x}} \right)^{n-1}$$

and the UMVUE of ϑ is

$$T(X) = \left(1 - \frac{t}{n\bar{X}} \right)^{n-1}. \quad \blacksquare$$

We now show more examples of applying these two methods to find UMVUE's.

Example 3.4. Let $X_1, ..., X_n$ be i.i.d. from $N(\mu, \sigma^2)$ with unknown $\mu \in \mathcal{R}$ and $\sigma^2 > 0$. From Example 2.18, $T = (\bar{X}, S^2)$ is sufficient and complete for $\theta = (\mu, \sigma^2)$ and $\bar{X}$ and $(n-1)S^2/\sigma^2$ are independent and have the $N(\mu, \sigma^2/n)$ and chi-square distribution χ_{n-1}^2, respectively. Using the method of solving for h directly, we find that the UMVUE for μ is $\bar{X}$; the UMVUE of μ^2 is $\bar{X}^2 - S^2/n$; the UMVUE for σ^r with $r > 1-n$ is $k_{n-1,r}S^r$, where

$$k_{n,r} = \frac{n^{r/2}\Gamma(n/2)}{2^{r/2}\Gamma\left(\frac{n+r}{2}\right)}$$

(exercise); and the UMVUE of μ/σ is $k_{n-1,-1}\bar{X}/S$, if $n > 2$.

Suppose that ϑ satisfies $P(X_1 \leq \vartheta) = p$ with a fixed $p \in (0,1)$. Let Φ be the c.d.f. of the standard normal distribution. Then $\vartheta = \mu + \sigma\Phi^{-1}(p)$ and its UMVUE is $\bar{X} + k_{n-1,1}S\Phi^{-1}(p)$.

Let c be a fixed constant and $\vartheta = P(X_1 \leq c) = \Phi\left(\frac{c-\mu}{\sigma}\right)$. We can find the UMVUE of ϑ using the method of conditioning and the technique used in Example 3.3. Since $I_{(-\infty,c)}(X_1)$ is an unbiased estimator of ϑ, the UMVUE of ϑ is $E[I_{(-\infty,c)}(X_1)|T] = P(X_1 \leq c|T)$. By Basu's theorem, the ancillary statistic $Z(X) = (X_1 - \bar{X})/S$ is independent of $T = (\bar{X}, S^2)$. Then, by Proposition 1.10(vii),

$$P\left(X_1 \leq c | T = (\bar{x}, s^2)\right) = P\left(Z \leq \frac{c - \bar{X}}{S} \bigg| T = (\bar{x}, s^2) \right)$$

$$= P\left(Z \leq \frac{c - \bar{x}}{s} \right).$$

It can be shown that Z has the Lebesgue p.d.f.

$$f(z) = \frac{\sqrt{n}\Gamma\left(\frac{n-1}{2}\right)}{\sqrt{\pi}(n-1)\Gamma\left(\frac{n-2}{2}\right)} \left[1 - \frac{nz^2}{(n-1)^2}\right]^{(n/2)-2} I_{(0,(n-1)/\sqrt{n})}(|z|) \quad (3.1)$$

(exercise). Hence the UMVUE of ϑ is

$$P(X_1 \leq c|T) = \int_{-(n-1)/\sqrt{n}}^{(c-\bar{X})/S} f(z)dz \quad (3.2)$$

with f given by (3.1).

Suppose that we would like to estimate $\vartheta = \frac{1}{\sigma}\Phi'\left(\frac{c-\mu}{\sigma}\right)$, the Lebesgue p.d.f. of X_1 evaluated at a fixed c, where Φ' is the first-order derivative of Φ. By (3.2), the conditional p.d.f. of X_1 given $\bar{X} = \bar{x}$ and $S^2 = s^2$ is $s^{-1}f\left(\frac{x-\bar{x}}{s}\right)$. Let f_T be the joint p.d.f. of $T = (\bar{X}, S^2)$. Then

$$\vartheta = \int\int \frac{1}{s}f\left(\frac{c-\bar{x}}{s}\right) f_T(t)dt = E\left[\frac{1}{S}f\left(\frac{c-\bar{X}}{S}\right)\right].$$

Hence the UMVUE of ϑ is

$$\frac{1}{S}f\left(\frac{c-\bar{X}}{S}\right). \quad \blacksquare$$

Example 3.5. Let $X_1, ..., X_n$ be i.i.d. from a power series distribution (see Exercise 13 in §2.6), i.e.,

$$P(X_i = x) = \gamma(x)\theta^x/c(\theta), \qquad x = 0, 1, 2, ...,$$

with a known function $\gamma(x) \geq 0$ and an unknown parameter $\theta > 0$. It turns out that the joint distribution of $X = (X_1, ..., X_n)$ is in an exponential family with a sufficient and complete statistic $T(X) = \sum_{i=1}^{n} X_i$. Furthermore, the distribution of T is also in a power series family, i.e.,

$$P(T = t) = \gamma_n(t)\theta^t/[c(\theta)]^n, \qquad t = 0, 1, 2, ...,$$

where $\gamma_n(t)$ is the coefficient of θ^t in the power series expansion of $[c(\theta)]^n$ (Exercise 13 in §2.6). This result can help us to find the UMVUE of $\vartheta = g(\theta)$. For example, by comparing both sides of

$$\sum_{t=0}^{\infty} h(t)\gamma_n(t)\theta^t = [c(\theta)]^{n-p}\theta^r,$$

we conclude that the UMVUE of $\theta^r/[c(\theta)]^p$ is

$$h(T) = \begin{cases} 0 & T < r \\ \frac{\gamma_{n-p}(T-r)}{\gamma_n(T)} & T \geq r, \end{cases}$$

where r and p are nonnegative integers. In particular, the case of $p = 1$ produces the UMVUE $\gamma(r)h(T)$ of the probability $P(X_1 = r) = \gamma(r)\theta^r/c(\theta)$ for any nonnegative integer r. ∎

Example 3.6. Let $X_1, ..., X_n$ be i.i.d. from an unknown population P in a nonparametric family $\mathcal{P}$. We have discussed in §2.2 that in many cases the vector of order statistics, $T = (X_{(1)}, ..., X_{(n)})$, is sufficient and complete for $P \in \mathcal{P}$. Note that an estimator $\varphi(X_1, ..., X_n)$ is a function of T if and only if the function φ is symmetric in its n arguments. Hence, if T is sufficient and complete, then a symmetric unbiased estimator of any estimable ϑ is the UMVUE. For example, $\bar{X}$ is the UMVUE of $\vartheta = EX_1$; S^2 is the UMVUE of $\text{Var}(X_1)$; $n^{-1}\sum_{i=1}^n X_i^2 - S^2$ is the UMVUE of $(EX_1)^2$; and $F_n(t)$ is the UMVUE of $P(X_1 \leq t)$ for any fixed t.

Note that these conclusions are not true if T is *not* sufficient and complete for $P \in \mathcal{P}$. For example, if $\mathcal{P}$ contains all symmetric distributions having Lebesgue p.d.f.'s and finite means, then there is no UMVUE for $\vartheta = EX_1$ (exercise). ∎

More discussions of UMVUE's in nonparametric problems are provided in §3.2.

3.1.2 A necessary and sufficient condition

When a complete and sufficient statistic is not available, it is usually very difficult to derive a UMVUE. In some cases, the following result can be applied, if we have enough knowledge about unbiased estimators of 0.

Theorem 3.2. Let $\mathcal{U}$ be the set of all unbiased estimators of 0 with finite variances and T be an unbiased estimator of ϑ with $E(T^2) < \infty$.
(i) A necessary and sufficient condition for $T(X)$ to be a UMVUE of ϑ is that $E[T(X)U(X)] = 0$ for any $U \in \mathcal{U}$ and any $P \in \mathcal{P}$.
(ii) Suppose that $T = h(\tilde{T})$, where $\tilde{T}$ is a sufficient statistic for $P \in \mathcal{P}$ and h is a Borel function. Let $\mathcal{U}_{\tilde{T}}$ be the subset of $\mathcal{U}$ consisting of Borel functions of $\tilde{T}$. Then a necessary and sufficient condition for T to be a UMVUE of ϑ is that $E[T(X)U(X)] = 0$ for any $U \in \mathcal{U}_{\tilde{T}}$ and any $P \in \mathcal{P}$.
Proof. (i) Suppose that T is a UMVUE of ϑ. Then $T_c = T + cU$, where $U \in \mathcal{U}$ and c is a fixed constant, is also unbiased for ϑ and, thus,

$$\text{Var}(T_c) \geq \text{Var}(T) \qquad c \in \mathcal{R}, \ P \in \mathcal{P},$$

which is the same as

$$c^2\text{Var}(U) + 2c\text{Cov}(T, U) \geq 0 \qquad c \in \mathcal{R}, \ P \in \mathcal{P}.$$

This is impossible unless $\text{Cov}(T, U) = E(TU) = 0$ for any $P \in \mathcal{P}$.

Suppose now $E(TU) = 0$ for any $U \in \mathcal{U}$ and $P \in \mathcal{P}$. Let T_0 be another unbiased estimator of ϑ with $\text{Var}(T_0) < \infty$. Then $T - T_0 \in \mathcal{U}$ and, hence,

$$E[T(T - T_0)] = 0 \qquad P \in \mathcal{P},$$

which with the fact that $ET = ET_0$ implies that

$$\text{Var}(T) = \text{Cov}(T, T_0) \qquad P \in \mathcal{P}.$$

By inequality (1.37), $[\text{Cov}(T, T_0)]^2 \leq \text{Var}(T)\text{Var}(T_0)$. Hence $\text{Var}(T) \leq \text{Var}(T_0)$ for any $P \in \mathcal{P}$.

(ii) It suffices to show that $E(TU) = 0$ for any $U \in \mathcal{U}_{\tilde{T}}$ and $P \in \mathcal{P}$ implies that $E(TU) = 0$ for any $U \in \mathcal{U}$ and $P \in \mathcal{P}$. Let $U \in \mathcal{U}$. Then $E(U|\tilde{T}) \in \mathcal{U}_{\tilde{T}}$ and the result follows from the fact that $T = h(\tilde{T})$ and

$$E(TU) = E[E(TU|\tilde{T})] = E[E(h(\tilde{T})U|\tilde{T})] = E[h(\tilde{T})E(U|\tilde{T})]. \quad \blacksquare$$

Theorem 3.2 can be used to find a UMVUE, to check whether a particular estimator is a UMVUE, and to show the nonexistence of any UMVUE. If there is a sufficient statistic, then by Rao-Blackwell's theorem, we only need to focus on functions of the sufficient statistic and, hence, Theorem 3.2(ii) is more convenient to use.

Example 3.7. Let $X_1, ..., X_n$ be i.i.d. from the uniform distribution on the interval $(0, \theta)$. In Example 3.1, $(1 + n^{-1})X_{(n)}$ is shown to be the UMVUE for θ when the parameter space is $\Theta = (0, \infty)$. Suppose now that $\Theta = [1, \infty)$. Then $X_{(n)}$ is not complete, although it is still sufficient for θ. Thus, Theorem 3.1 does not apply. We now illustrate how to use Theorem 3.2(ii) to find a UMVUE of θ. Let $U(X_{(n)})$ be an unbiased estimator of 0. Since $X_{(n)}$ has the Lebesgue p.d.f. $n\theta^{-n}x^{n-1}I_{(0,\theta)}(x)$,

$$0 = \int_0^1 U(x)x^{n-1}dx + \int_1^\theta U(x)x^{n-1}dx$$

for all $\theta \geq 1$. This implies that $U(x) = 0$ a.e. Lebesgue measure on $[1, \infty)$ and

$$\int_0^1 U(x)x^{n-1}dx = 0.$$

Consider $T = h(X_{(n)})$. To have $E(TU) = 0$, we must have

$$\int_0^1 h(x)U(x)x^{n-1}dx = 0.$$

Thus, we may consider the following function:

$$h(x) = \begin{cases} c & 0 \leq x \leq 1 \\ bx & x > 1, \end{cases}$$

where c and b are some constants. From the previous discussion,

$$E[h(X_{(n)})U(X_{(n)})] = 0, \qquad \theta \geq 1.$$

Since $E[h(X_{(n)})] = \theta$, we obtain that

$$\theta = cP(X_{(n)} \leq 1) + bE[X_{(n)}I_{(1,\infty)}(X_{(n)})]$$
$$= c\theta^{-n} + [bn/(n+1)](\theta - \theta^{-n}).$$

Thus, $c = 1$ and $b = (n+1)/n$. The UMVUE of θ is then

$$T = \begin{cases} 1 & 0 \leq X_{(n)} \leq 1 \\ (1+n^{-1})X_{(n)} & X_{(n)} > 1. \end{cases}$$

This estimator is better than $(1 + n^{-1})X_{(n)}$, which is the UMVUE when $\Theta = (0, \infty)$ and does not make use of the information about $\theta \geq 1$. ∎

Example 3.8. Let X be a sample (of size 1) from the uniform distribution $U(\theta - \frac{1}{2}, \theta + \frac{1}{2})$, $\theta \in \mathcal{R}$. We now apply Theorem 3.2 to show that there is no UMVUE of $\vartheta = g(\theta)$ for any nonconstant function g. Note that an unbiased estimator $U(X)$ of 0 must satisfy

$$\int_{\theta - \frac{1}{2}}^{\theta + \frac{1}{2}} U(x)dx = 0 \qquad \text{for all } \theta \in \mathcal{R}.$$

Differentiating both sizes of the previous equation and applying the result of differentiation of an integral lead to $U(x) = U(x+1)$ a.e. m, where m is the Lebesgue measure on $\mathcal{R}$. If T is a UMVUE of $g(\theta)$, then $T(X)U(X)$ is unbiased for 0 and, hence, $T(x)U(x) = T(x+1)U(x+1)$ a.e. m, where $U(X)$ is any unbiased estimator of 0. Since this is true for all U, $T(x) = T(x+1)$ a.e. m. Since T is unbiased for $g(\theta)$,

$$g(\theta) = \int_{\theta - \frac{1}{2}}^{\theta + \frac{1}{2}} T(x)dx \qquad \text{for all } \theta \in \mathcal{R}.$$

Differentiating both sizes of the previous equation and applying the result of differentiation of an integral, we obtain that

$$g'(\theta) = T\left(\theta + \frac{1}{2}\right) - T\left(\theta - \frac{1}{2}\right) = 0 \quad \text{a.e. } m. \quad ∎$$

As a consequence of Theorem 3.2, we have the following useful result.

Corollary 3.1. (i) Let T_j be a UMVUE of ϑ_j, $j = 1, ..., k$, where k is a fixed positive integer. Then $\sum_{j=1}^{k} c_j T_j$ is a UMVUE of $\vartheta = \sum_{j=1}^{k} c_j \vartheta_j$ for any constants $c_1, ..., c_k$.
(ii) Let T_1 and T_2 be two UMVUE's of ϑ. Then $T_1 = T_2$ a.s. P for any $P \in \mathcal{P}$. ∎

3.1.3 Information inequality

Suppose that we have a lower bound for the variances of all unbiased estimators of ϑ and that there is an unbiased estimator T of ϑ whose variance is always the same as the lower bound. Then T is a UMVUE of ϑ. Although this is not an effective way to find UMVUE's (compared with the methods introduced in §3.1.1 and §3.1.2), it provides a way of assessing the performance of UMVUE's. The following result provides such a lower bound in some cases.

Theorem 3.3 (Cramér-Rao lower bound). Let $X = (X_1, ..., X_n)$ be a sample from $P \in \mathcal{P} = \{P_\theta : \theta \in \Theta\}$, where Θ is an open set in $\mathcal{R}^k$. Suppose that $T(X)$ is an estimator with $E[T(X)] = g(\theta)$ being a differentiable function of θ; P_θ has a p.d.f. f_θ w.r.t. a measure ν for all $\theta \in \Theta$; and f_θ is differentiable as a function of θ and satisfies

$$\frac{\partial}{\partial \theta} \int h(x) f_\theta(x) d\nu = \int h(x) \frac{\partial}{\partial \theta} f_\theta(x) d\nu, \qquad \theta \in \Theta, \qquad (3.3)$$

for $h(x) \equiv 1$ and $h(x) = T(x)$. Then

$$\text{Var}(T(X)) \geq \left[\frac{\partial}{\partial \theta} g(\theta)\right]^\tau [I(\theta)]^{-1} \frac{\partial}{\partial \theta} g(\theta), \qquad (3.4)$$

where

$$I(\theta) = E\left\{ \frac{\partial}{\partial \theta} \log f_\theta(X) \left[\frac{\partial}{\partial \theta} \log f_\theta(X) \right]^\tau \right\} \qquad (3.5)$$

is assumed to be positive definite for any $\theta \in \Theta$.
Proof. We prove the univariate case ($k = 1$) only. The proof for the multivariate case ($k > 1$) is left to the reader. When $k = 1$, (3.4) reduces to

$$\text{Var}(T(X)) \geq \frac{[g'(\theta)]^2}{E\left[\frac{\partial}{\partial \theta} \log f_\theta(X) \right]^2}. \qquad (3.6)$$

From inequality (1.37), we only need to show that

$$E\left[\frac{\partial}{\partial \theta} \log f_\theta(X) \right]^2 = \text{Var}\left(\frac{\partial}{\partial \theta} \log f_\theta(X) \right)$$

and

$$g'(\theta) = \text{Cov}\left(T(X), \frac{\partial}{\partial \theta} \log f_\theta(X) \right).$$

These two results are consequences of condition (3.3). ∎

The $k \times k$ matrix $I(\theta)$ in (3.5) is called the *Fisher information matrix*. The greater $I(\theta)$ is, the easier it is to distinguish θ from neighboring values and, therefore, the more accurately θ can be estimated. In fact, if the equality in (3.6) holds for an unbiased estimator $T(X)$ of $g(\theta)$ (which is then a UMVUE), then the greater $I(\theta)$ is, the smaller $\text{Var}(T(X))$ is. Thus, $I(\theta)$ is a measure of the information that X contains about the unknown θ. The inequalities in (3.4) and (3.6) are called *information inequalities*.

The following result is helpful in finding the Fisher information matrix.

Proposition 3.1. (i) Let X and Y be independent with the Fisher information matrices $I_X(\theta)$ and $I_Y(\theta)$, respectively. Then, the Fisher information about θ contained in (X, Y) is $I_X(\theta) + I_Y(\theta)$. In particular, if $X_1, ..., X_n$ are i.i.d. and $I_1(\theta)$ is the Fisher information about θ contained in a single X_i, then the Fisher information about θ contained in $X_1, ..., X_n$ is $nI_1(\theta)$. (ii) Suppose that X has the p.d.f. f_θ that is twice differentiable in θ and that (3.3) holds with $h(x) \equiv 1$ and f_θ replaced by $\partial f_\theta / \partial \theta$. Then

$$I(\theta) = -E\left[\frac{\partial^2}{\partial\theta\partial\theta^\tau}\log f_\theta(X)\right]. \tag{3.7}$$

Proof. Result (i) follows from the independence of X and Y and the definition of the Fisher information. Result (ii) follows from the equality

$$\frac{\partial^2}{\partial\theta\partial\theta^\tau}\log f_\theta(X) = \frac{\frac{\partial^2}{\partial\theta\partial\theta^\tau}f_\theta(X)}{f_\theta(X)} - \frac{\partial}{\partial\theta}\log f_\theta(X)\left[\frac{\partial}{\partial\theta}\log f_\theta(X)\right]^\tau. \quad\blacksquare$$

The following example provides a formula for the Fisher information matrix for many parametric families with a two-dimensional parameter θ.

Example 3.9. Let $X_1, ..., X_n$ be i.i.d. with the Lebesgue p.d.f. $\frac{1}{\sigma}f\left(\frac{x-\mu}{\sigma}\right)$, where $f(x) > 0$ and $f'(x)$ exists for all $x \in \mathcal{R}$, $\mu \in \mathcal{R}$, and $\sigma > 0$ (a location-scale family). Let $\theta = (\mu, \sigma)$. Then, the Fisher information about θ contained in $X_1, ..., X_n$ is (exercise)

$$I(\theta) = \frac{n}{\sigma^2}\begin{pmatrix} \int \frac{[f'(x)]^2}{f(x)}dx & \int x\frac{[f'(x)]^2}{f(x)}dx \\ \int x\frac{[f'(x)]^2}{f(x)}dx & \int \frac{[xf'(x)+f(x)]^2}{f(x)}dx \end{pmatrix}. \quad\blacksquare$$

Note that $I(\theta)$ depends on the particular parameterization. If $\theta = \psi(\eta)$ and ψ is differentiable, then the Fisher information that X contains about η is

$$\frac{\partial}{\partial\eta}\psi(\eta)I(\psi(\eta))\left[\frac{\partial}{\partial\eta}\psi(\eta)\right]^\tau.$$

However, it is easy to see that the Cramér-Rao lower bound in (3.4) or (3.6) is not affected by any one-to-one reparameterization.

If we use inequality (3.4) or (3.6) to find a UMVUE $T(X)$, then we obtain a formula for $\text{Var}(T(X))$ at the same time. On the other hand, the Cramér-Rao lower bound in (3.4) or (3.6) is typically not sharp. Under some regularity conditions, the Cramér-Rao lower bound is attained if and only if f_θ is in an exponential family; see Propositions 3.2 and 3.3 and the discussion in Lehmann (1983, p. 123). Some improved information inequalities are available (see, e.g., Lehmann (1983, Sections 2.6 and 2.7)).

Proposition 3.2. Suppose that the distribution of X is from an exponential family $\{f_\theta : \theta \in \Theta\}$, i.e., the p.d.f. of X w.r.t. a σ-finite measure is

$$f_\theta(x) = \exp\{[\eta(\theta)]^\tau T(x) - \xi(\theta)\}c(x) \tag{3.8}$$

(see §2.1.3), where Θ is an open subset of $\mathcal{R}^k$.
(i) The regularity condition (3.3) is satisfied for any h with $E|h(X)| < \infty$ and (3.7) holds.
(ii) If $\underline{I}(\eta)$ is the Fisher information matrix for the natural parameter η, then the variance-covariance matrix $\text{Var}(T) = \underline{I}(\eta)$.
(iii) If $\overline{I}(\vartheta)$ is the Fisher information matrix for the parameter $\vartheta = E[T(X)]$, then $\text{Var}(T) = [\overline{I}(\vartheta)]^{-1}$.
Proof. (i) This is a direct consequence of Theorem 2.1.
(ii) From (2.6), the p.d.f. under the natural parameter η is

$$f_\eta(x) = \exp\left\{\eta^\tau T(x) - \zeta(\eta)\right\} c(x).$$

From Theorem 2.1 and result (1.54) in §1.3.3, $E[T(X)] = \frac{\partial}{\partial \eta}\zeta(\eta)$. The result follows from

$$\frac{\partial}{\partial \eta} \log f_\eta(x) = T(x) - \frac{\partial}{\partial \eta}\zeta(\eta).$$

(iii) Since $\vartheta = E[T(X)] = \frac{\partial}{\partial \eta}\zeta(\eta)$,

$$\underline{I}(\eta) = \frac{\partial \vartheta}{\partial \eta}\overline{I}(\vartheta)\left(\frac{\partial \vartheta}{\partial \eta}\right)^\tau = \frac{\partial^2}{\partial \eta \partial \eta^\tau}\zeta(\eta)\overline{I}(\vartheta)\left[\frac{\partial^2}{\partial \eta \partial \eta^\tau}\zeta(\eta)\right]^\tau.$$

By Theorem 2.1, result (1.54), and the result in (ii), $\frac{\partial^2}{\partial \eta \partial \eta^\tau}\zeta(\eta) = \text{Var}(T) = \underline{I}(\eta)$. Hence

$$\overline{I}(\vartheta) = [\underline{I}(\eta)]^{-1}\underline{I}(\eta)[\underline{I}(\eta)]^{-1} = [\underline{I}(\eta)]^{-1} = [\text{Var}(T)]^{-1}. \quad \blacksquare$$

A direct consequence of Proposition 3.2(ii) is that the variance of any linear function of T in (3.8) attains the Cramér-Rao lower bound. The following result gives a necessary condition for $\text{Var}(U(X))$ of an estimator $U(X)$ to attain the Cramér-Rao lower bound.

Proposition 3.3. Assume that the conditions in Theorem 3.3 hold with $T(X)$ replaced by $U(X)$ and that $\Theta \subset \mathcal{R}$.
(i) If $\text{Var}(U(X))$ attains the Cramér-Rao lower bound in (3.6), then

$$a(\theta)[U(X) - g(\theta)] = g'(\theta)\frac{\partial}{\partial \theta} \log f_\theta(X) \quad \text{a.s. } P_\theta$$

for some function $a(\theta)$, $\theta \in \Theta$.
(ii) Let f_θ and T be given by (3.8). If $\text{Var}(U(X))$ attains the Cramér-Rao lower bound, then $U(X)$ is a linear function of $T(X)$ a.s. P_θ, $\theta \in \Theta$. ∎

Example 3.10. Let $X_1, ..., X_n$ be i.i.d. from the $N(\mu, \sigma^2)$ distribution with an unknown $\mu \in \mathcal{R}$ and a known σ^2. Let f_μ be the joint distribution of $X = (X_1, ..., X_n)$. Then

$$\frac{\partial}{\partial \mu} \log f_\mu(X) = \sum_{i=1}^{n} (X_i - \mu)/\sigma^2.$$

Thus, $I(\mu) = n/\sigma^2$. It is obvious that $\text{Var}(\bar{X})$ attains the Cramér-Rao lower bound in (3.6). Consider now the estimation of $\vartheta = \mu^2$. Since $E\bar{X}^2 = \mu^2 + \sigma^2/n$, the UMVUE of ϑ is $h(\bar{X}) = \bar{X}^2 - \sigma^2/n$. A straightforward calculation shows that

$$\text{Var}(h(\bar{X})) = \frac{4\mu^2\sigma^2}{n} + \frac{2\sigma^4}{n^2}.$$

On the other hand, the Cramér-Rao lower bound in this case is $4\mu^2\sigma^2/n$. Hence $\text{Var}(h(\bar{X}))$ does not attain the Cramér-Rao lower bound. The difference is $2\sigma^4/n^2$. ∎

Condition (3.3) is a key regularity condition for the results in Theorem 3.3 and Proposition 3.3. If f_θ is not in an exponential family, then (3.3) has to be checked. Typically, it does not hold if the set $\{x : f_\theta(x) > 0\}$ depends on θ (Exercise 37). More discussions can be found in Pitman (1979).

3.1.4 Asymptotic properties of UMVUE's

UMVUE's are typically consistent (see Exercise 106 in §2.6). If there is an unbiased estimator of ϑ whose mse is of the order a_n^{-2}, where $\{a_n\}$ is a sequence of positive numbers diverging to ∞, then the UMVUE of ϑ (if it exists) has an mse of order a_n^{-2} and is a_n-consistent. For instance, in Example 3.3, the mse of $U(X) = 1 - F_n(t)$ is $F_\theta(t)[1 - F_\theta(t)]/n$; hence the UMVUE $T(X)$ is $\sqrt{n}$-consistent and its mse is of the order n^{-1}.

UMVUE's are exactly unbiased so that there is no need to discuss their asymptotic biases. Their variances (or mse's) are finite, but amse's can be used to assess their performance if the exact forms of mse's are difficult to obtain. In many cases, although the variance of a UMVUE T_n does not attain the Cramér-Rao lower bound, the limit of the ratio of the amse (or mse) of T_n over the Cramér-Rao lower bound (if it is not 0) is 1. For instance, in Example 3.10,

$$\frac{\text{Var}(\bar{X}^2 - \sigma^2/n)}{\text{the Cramér-Rao lower bound}} = 1 + \frac{\sigma^2}{2\mu^2 n} \to 1$$

if $\mu \neq 0$. In general, under the conditions in Theorem 3.3, if $T_n(X)$ is unbiased for $g(\theta)$ and if, for any $\theta \in \Theta$,

$$T_n(X) - g(\theta) = \left[\frac{\partial}{\partial \theta} g(\theta)\right]^\tau [I(\theta)]^{-1} \frac{\partial}{\partial \theta} \log f_\theta(X) [1 + o_p(1)] \quad \text{a.s. } P_\theta, \quad (3.9)$$

then

$$\text{amse}_{T_n}(\theta) = \text{the Cramér-Rao lower bound} \quad (3.10)$$

whenever the Cramér-Rao lower bound is not 0. Note that the case of zero Cramér-Rao lower bound is not of interest since a zero lower bound does not provide any information on the performance of estimators.

Consider the UMVUE $T_n = \left(1 - \frac{t}{n\bar{X}}\right)^{n-1}$ of $e^{-t/\theta}$ in Example 3.3. Using the fact that

$$\log(1 - x) = -\sum_{j=1}^{\infty} \frac{x^j}{j}, \qquad |x| \leq 1,$$

we obtain that

$$T_n - e^{-t/\bar{X}} = O_p\left(n^{-1}\right).$$

Using Taylor's expansion, we obtain that

$$e^{-t/\bar{X}} - e^{-t/\theta} = g'(\theta)(\bar{X} - \theta)[1 + o_p(1)],$$

where $g(\theta) = e^{-t/\theta}$. On the other hand,

$$[I(\theta)]^{-1} \frac{\partial}{\partial \theta} \log f_\theta(X) = \bar{X} - \theta.$$

Hence (3.9) and (3.10) hold. Note that the exact variance of T_n is not easy to obtain. In this example, it can be shown that $\{n[T_n - g(\theta)]^2\}$ is uniformly integrable and, therefore,

$$\lim_{n\to\infty} n\text{Var}(T_n) = \lim_{n\to\infty} n[\text{amse}_{T_n}(\theta)]$$
$$= \lim_{n\to\infty} n[g'(\theta)]^2 [I(\theta)]^{-1}$$
$$= \frac{t^2 e^{-2t/\theta}}{\theta^2}.$$

It is shown in Chapter 4 that if (3.10) holds, then T_n is asymptotically optimal in some sense. Hence UMVUE's satisfying (3.9), which is often true, are asymptotically optimal, although they may be improved in terms of the exact mse's.

3.2 U-Statistics

Let $X_1, ..., X_n$ be i.i.d. from an unknown population P in a nonparametric family $\mathcal{P}$. In Example 3.6 we argued that if the vector of order statistic is sufficient and complete for $P \in \mathcal{P}$, then a symmetric unbiased estimator of any estimable ϑ is the UMVUE of ϑ. In a large class of problems, parameters to be estimated are of the form

$$\vartheta = E[h(X_1, ..., X_m)]$$

with a positive integer m and a Borel function h that is symmetric and satisfies $E|h(X_1, ..., X_m)| < \infty$ for any $P \in \mathcal{P}$. It is easy to see that a symmetric unbiased estimator of ϑ is

$$U_n = \binom{n}{m}^{-1} \sum_c h(X_{i_1}, ..., X_{i_m}), \tag{3.11}$$

where $\sum_c$ denotes the summation over the $\binom{n}{m}$ combinations of m distinct elements $\{i_1, ..., i_m\}$ from $\{1, ..., n\}$.

Definition 3.2. The statistic U_n in (3.11) is called a *U-statistic* with kernel h of order m. ∎

3.2.1 Some examples

The use of U-statistics is an effective way of obtaining unbiased estimators. In nonparametric problems, U-statistics are often UMVUE's, whereas in parametric problems, U-statistics can be used as initial estimators to derive more efficient estimators.

If $m = 1$, U_n in (3.11) is simply a type of sample mean. Examples include the empirical c.d.f. (2.28) evaluated at a particular t and the *sample moments* $n^{-1} \sum_{i=1}^n X_i^k$ for a positive integer k. We now consider some examples with $m > 1$.

Consider the estimation of $\vartheta = \mu^m$, where $\mu = EX_1$ and m is a positive integer. Using $h(x_1, ..., x_m) = x_1 \cdots x_m$, we obtain the following U-statistic unbiased for $\vartheta = \mu^m$:

$$U_n = \binom{n}{m}^{-1} \sum_c X_{i_1} \cdots X_{i_m}. \tag{3.12}$$

Consider next the estimation of $\vartheta = \sigma^2 = \text{Var}(X_1)$. Since

$$\sigma^2 = [\text{Var}(X_1) + \text{Var}(X_2)]/2 = E[(X_1 - X_2)^2/2],$$

we obtain the following U-statistic with kernel $h(x_1, x_2) = (x_1 - x_2)^2/2$:

$$U_n = \frac{2}{n(n-1)} \sum_{1 \leq i < j \leq n} \frac{(X_i - X_j)^2}{2} = \frac{1}{n-1} \left(\sum_{i=1}^{n} X_i^2 - n\bar{X}^2 \right) = S^2,$$

which is the sample variance in (2.2).

In some cases, we would like to estimate $\vartheta = E|X_1 - X_2|$, a measure of concentration. Using kernel $h(x_1, x_2) = |x_1 - x_2|$, we obtain the following U-statistic unbiased for $\vartheta = E|X_1 - X_2|$:

$$U_n = \frac{2}{n(n-1)} \sum_{1 \leq i < j \leq n} |X_i - X_j|,$$

which is known as *Gini's mean difference*.

Let $\vartheta = P(X_1 + X_2 \leq 0)$. Using kernel $h(x_1, x_2) = I_{(-\infty,0]}(x_1 + x_2)$, we obtain the following U-statistic unbiased for ϑ:

$$U_n = \frac{2}{n(n-1)} \sum_{1 \leq i < j \leq n} I_{(-\infty,0]}(X_i + X_j),$$

which is known as the *one-sample Wilcoxon statistic*.

Let $T_n = T_n(X_1, ..., X_n)$ be a given statistic and let r and d be two positive integers such that $r + d = n$. For any $s = \{i_1, ..., i_r\} \subset \{1, ..., n\}$, define

$$T_{r,s} = T_r(X_{i_1}, ..., X_{i_r}),$$

which is the statistic T_n computed after X_i, $i \notin s$, are deleted from the original sample. Let

$$U_n = \binom{n}{r}^{-1} \sum_c \frac{r}{d}(T_{r,s} - T_n)^2. \tag{3.13}$$

Then U_n is a U-statistic with kernel

$$h_n(x_1, ..., x_r) = \frac{r}{d}[T_r(x_1, ..., x_r) - T_n(x_1, ..., x_n)]^2.$$

Unlike the kernels in the previous examples, the kernel in this example depends on n. The order of the kernel, r, may also depend on n. The statistic U_n in (3.13) is known as the *delete-d jackknife variance estimator* for T_n (see, e.g., Shao and Tu (1995)), since it is often true that

$$E[h_n(X_1, ..., X_r)] \approx \text{Var}(T_n).$$

It can be shown that if $T_n = \bar{X}$, then nU_n in (3.13) is exactly the same as the sample variance S^2 (exercise).

3.2.2 Variances of U-statistics

If $E[h(X_1, ..., X_m)]^2 < \infty$, then the variance of U_n in (3.11) with kernel
h has an explicit form. To derive $\text{Var}(U_n)$, we need some notation. For
$k = 1, ..., m$, let

$$h_k(x_1, ..., x_k) = E[h(X_1, ..., X_m)|X_1 = x_1, ..., X_k = x_k]$$
$$= E[h(x_1, ..., x_k, X_{k+1}, ..., X_m)].$$

Note that $h_m = h$. It can be shown that

$$h_k(x_1, ..., x_k) = E[h_{k+1}(x_1, ..., x_k, X_{k+1})]. \qquad (3.14)$$

Define

$$\tilde{h}_k = h_k - E[h(X_1, ..., X_m)], \qquad (3.15)$$

$k = 1, ..., m$, and $\tilde{h} = \tilde{h}_m$. Then, for any U_n defined by (3.11),

$$U_n - E(U_n) = \binom{n}{m}^{-1} \sum_c \tilde{h}(X_{i_1}, ..., X_{i_m}). \qquad (3.16)$$

Theorem 3.4 (Hoeffding's theorem). For a U-statistic U_n given by (3.11)
with $E[h(X_1, ..., X_m)]^2 < \infty$,

$$\text{Var}(U_n) = \binom{n}{m}^{-1} \sum_{k=1}^{m} \binom{m}{k}\binom{n-m}{m-k} \zeta_k,$$

where

$$\zeta_k = \text{Var}(h_k(X_1, ..., X_k)).$$

Proof. Consider two sets $\{i_1, ..., i_m\}$ and $\{j_1, ..., j_m\}$ of m distinct integers
from $\{1, ..., n\}$ with exactly k integers in common. The number of distinct
choices of two such sets is $\binom{n}{m}\binom{m}{k}\binom{n-m}{m-k}$. By the symmetry of $\tilde{h}_m$ and
independence of $X_1, ..., X_n$,

$$E[\tilde{h}(X_{i_1}, ..., X_{i_m})\tilde{h}(X_{j_1}, ..., X_{j_m})] = \zeta_k \qquad (3.17)$$

for $k = 1, ..., m$ (exercise). Then, by (3.16),

$$\text{Var}(U_n) = \binom{n}{m}^{-2} \sum_c \sum_c E[\tilde{h}(X_{i_1}, ..., X_{i_m})\tilde{h}(X_{j_1}, ..., X_{j_m})]$$
$$= \binom{n}{m}^{-2} \sum_{k=1}^{m} \binom{n}{m}\binom{m}{k}\binom{n-m}{m-k} \zeta_k.$$

This proves the result. ∎

Corollary 3.2. Under the condition of Theorem 3.4,
(i) $\frac{m^2}{n}\zeta_1 \le \text{Var}(U_n) \le \frac{m}{n}\zeta_m$;
(ii) $(n+1)\text{Var}(U_{n+1}) \le n\text{Var}(U_n)$ for any $n > m$;
(iii) For any fixed m and $k = 1, ..., m$, if $\zeta_j = 0$ for $j < k$ and $\zeta_k > 0$, then

$$\text{Var}(U_n) = \frac{k!\binom{m}{k}^2\zeta_k}{n^k} + O\left(\frac{1}{n^{k+1}}\right). \quad \blacksquare$$

It follows from Corollary 3.2 that a U-statistic U_n as an estimator of its mean is consistent in mse (under the finite second moment assumption on h). In fact, for any fixed m, if $\zeta_j = 0$ for $j < k$ and $\zeta_k > 0$, then the mse of U_n is of the order n^{-k} and, therefore, U_n is $n^{k/2}$-consistent.

Example 3.11. Consider first $h(x_1, x_2) = x_1 x_2$, which leads to a U-statistic unbiased for μ^2, $\mu = EX_1$. Note that $h_1(x_1) = \mu x_1$, $\tilde{h}_1(x_1) = \mu(x_1 - \mu)$, $\zeta_1 = E[\tilde{h}_1(X_1)]^2 = \mu^2\text{Var}(X_1) = \mu^2\sigma^2$, $h(x_1, x_2) = x_1 x_2 - \mu^2$, and $\zeta_2 = \text{Var}(X_1 X_2) = E(X_1 X_2)^2 - \mu^4 = (\mu^2 + \sigma^2)^2 - \mu^4$. By Theorem 3.4, for $U_n = \binom{n}{2}^{-1}\sum_{1 \le i < j \le n} X_i X_j$,

$$\text{Var}(U_n) = \binom{n}{2}^{-1}\left[\binom{2}{1}\binom{n-2}{1}\zeta_1 + \binom{2}{2}\binom{n-2}{0}\zeta_2\right]$$

$$= \frac{2}{n(n-1)}\left[2(n-2)\mu^2\sigma^2 + (\mu^2+\sigma^2)^2 - \mu^4\right]$$

$$= \frac{4\mu^2\sigma^2}{n} + \frac{2\sigma^4}{n(n-1)}.$$

Comparing U_n with $\bar{X}^2 - \sigma^2/n$ in Example 3.10, which is the UMVUE under the normality and known σ^2 assumption, we find that

$$\text{Var}(U_n) - \text{Var}(\bar{X}^2 - \sigma^2/n) = \frac{2\sigma^4}{n^2(n-1)}.$$

Next, consider $h(x_1, x_2) = I_{(-\infty,0]}(x_1 + x_2)$, which leads to the one-sample Wilcoxon statistic. Note that $h_1(x_1) = P(x_1 + X_2 \le 0) = F(-x_1)$, where F is the c.d.f. of P. Then $\zeta_1 = \text{Var}(F(-X_1))$. Let $\vartheta = E[h(X_1, X_2)]$. Then $\zeta_2 = \text{Var}(h(X_1, X_2)) = \vartheta(1 - \vartheta)$. Hence, for U_n being the one-sample Wilcoxon statistic,

$$\text{Var}(U_n) = \frac{2}{n(n-1)}\left[2(n-2)\zeta_1 + \vartheta(1-\vartheta)\right].$$

If F is continuous and symmetric about 0, then ζ_1 can be simplified as

$$\zeta_1 = \text{Var}(F(-X_1)) = \text{Var}(1 - F(X_1)) = \text{Var}(F(X_1)) = \tfrac{1}{12},$$

since $F(X_1)$ has the uniform distribution on $[0, 1]$.

Finally, consider $h(x_1, x_2) = |x_1 - x_2|$, which leads to Gini's mean difference. Note that

$$h_1(x_1) = E|x_1 - X_2| = \int |x_1 - y| dP(y),$$

and

$$\zeta_1 = \text{Var}(h_1(X_1)) = \int \left[\int |x - y| dP(y) \right]^2 dP(x) - \vartheta^2,$$

where $\vartheta = E|X_1 - X_2|$. ∎

3.2.3 The projection method

Since $\mathcal{P}$ is nonparametric, the exact distribution of any U-statistic is hard to derive. In this section, we study asymptotic distributions of U-statistics by using the method of *projection*.

Definition 3.3. Let T_n be a given statistic based on $X_1, ..., X_n$. The projection of T_n on k_n random elements $Y_1, ..., Y_{k_n}$ is defined to be

$$\check{T}_n = E(T_n) + \sum_{i=1}^{k_n} [E(T_n|Y_i) - E(T_n)].$$ ∎

Let $\psi_n(X_i) = E(T_n|X_i)$. If T_n is symmetric (as a function of $X_1, ..., X_n$), then $\psi_n(X_1), ..., \psi_n(X_n)$ are i.i.d. with mean $E[\psi_n(X_i)] = E[E(T_n|X_i)] = E(T_n)$. If $E(T_n^2) < \infty$ and $\text{Var}(\psi_n(X_i)) > 0$, then

$$\frac{1}{\sqrt{n \text{Var}(\psi_n(X_1))}} \sum_{i=1}^{n} [\psi_n(X_i) - E(T_n)] \to_d N(0, 1) \qquad (3.18)$$

by the CLT. Let $\check{T}_n$ be the projection of T_n on $X_1, ..., X_n$. Then

$$T_n - \check{T}_n = T_n - E(T_n) - \sum_{i=1}^{n} [\psi_n(X_i) - E(T_n)]. \qquad (3.19)$$

If we can show that $T_n - \check{T}_n$ has a negligible order of magnitude, then we can derive the asymptotic distribution of T_n by using (3.18)-(3.19) and Slutsky's theorem. The order of magnitude of $T_n - \check{T}_n$ can be obtained with the help of the following lemma.

Lemma 3.1. Let T_n be a symmetric statistic with $\text{Var}(T_n) < \infty$ for every n and $\check{T}_n$ be the projection of T_n on $X_1, ..., X_n$. Then $E(T_n) = E(\check{T}_n)$ and

$$E(T_n - \check{T}_n)^2 = \text{Var}(T_n) - \text{Var}(\check{T}_n).$$

Proof. Since $E(T_n) = E(\check{T}_n)$,

$$E(T_n - \check{T}_n)^2 = \text{Var}(T_n) + \text{Var}(\check{T}_n) - 2\text{Cov}(T_n, \check{T}_n).$$

From Definition 3.3 with $Y_i = X_i$ and $k_n = n$,

$$\text{Var}(\check{T}_n) = n\text{Var}(E(T_n|X_i)).$$

The result follows from

$$\begin{aligned}
\text{Cov}(T_n, \check{T}_n) &= E(T_n\check{T}_n) - [E(T_n)]^2 \\
&= nE[T_nE(T_n|X_i)] - n[E(T_n)]^2 \\
&= nE\{E[T_nE(T_n|X_i)|X_i]\} - n[E(T_n)]^2 \\
&= nE\{[E(T_n|X_i)]^2\} - n[E(T_n)]^2 \\
&= n\text{Var}(E(T_n|X_i)) \\
&= \text{Var}(\check{T}_n). \quad \blacksquare
\end{aligned}$$

This method of deriving the asymptotic distribution of T_n is known as the method of projection and is particularly effective for U-statistics. For a U-statistic U_n given by (3.11), one can show (exercise) that

$$\check{U}_n = E(U_n) + \frac{m}{n}\sum_{i=1}^{n}\tilde{h}_1(X_i), \tag{3.20}$$

where $\check{U}_n$ is the projection of U_n on $X_1, ..., X_n$ and $\tilde{h}_1$ is defined by (3.15). Hence

$$\text{Var}(\check{U}_n) = m^2\zeta_1/n$$

and, by Corollary 3.2 and Lemma 3.1,

$$E(U_n - \check{U}_n)^2 = O(n^{-2}).$$

If $\zeta_1 > 0$, then (3.18) holds with $\psi_n(X_i) = mh_1(X_i)$, which leads to the result in Theorem 3.5(i) stated later.

If $\zeta_1 = 0$, then $\tilde{h}_1 \equiv 0$ and we have to use another projection of U_n. Suppose that $\zeta_1 = \cdots = \zeta_{k-1} = 0$ and $\zeta_k > 0$ for an integer $k > 1$. Consider the projection $\check{U}_{kn}$ of U_n on $\binom{n}{k}$ random vectors $\{X_{i_1}, ..., X_{i_k}\}$, $1 \le i_1 < \cdots < i_k \le n$. We can establish a result similar to that in Lemma 3.1 (exercise) and show that

$$E(U_n - \check{U}_n)^2 = O(n^{-(k+1)}).$$

Also, see Serfling (1980, §5.3.4).

With these results, we obtain the following theorem.

Theorem 3.5. Let U_n be given by (3.11) with $E[h(X_1, ..., X_m)]^2 < \infty$.
(i) If $\zeta_1 > 0$, then

$$\sqrt{n}[U_n - E(U_n)] \to_d N(0, m^2\zeta_1).$$

(ii) If $\zeta_1 = 0$ but $\zeta_2 > 0$, then

$$n[U_n - E(U_n)] \to_d \frac{m(m-1)}{2} \sum_{j=1}^{\infty} \lambda_j(\chi_{1j}^2 - 1), \qquad (3.21)$$

where χ_{1j}^2's are i.i.d. random variables having the chi-square distribution χ_1^2 and λ_j's are some constants (which may depend on P) satisfying $\sum_{j=1}^{\infty} \lambda_j^2 = \zeta_2$. ∎

We have actually proved Theorem 3.5(i). A proof for Theorem 3.5(ii) is given in Serfling (1980, §5.5.2). One may derive results for the cases where $\zeta_2 = 0$, but the case of either $\zeta_1 > 0$ or $\zeta_2 > 0$ is the most interesting case in applications.

If $\zeta_1 > 0$, it follows from Theorem 3.5(i) and Corollary 3.2(iii) that $\text{amse}_{U_n}(P) = m^2\zeta_1/n = \text{Var}(U_n) + O(n^{-2})$. By Proposition 2.4(ii), $\{n[U_n - E(U_n)]^2\}$ is uniformly integrable.

If $\zeta_1 = 0$ but $\zeta_2 > 0$, it follows from Theorem 3.5(ii) that $\text{amse}_{U_n}(P) = EY^2/n^2$, where Y denotes the random variable on the right-hand side of (3.21). The following result provides the value of EY^2.

Lemma 3.2. Let Y be the random variable on the right-hand side of (3.21). Then $EY^2 = \frac{m^2(m-1)^2}{2}\zeta_2$.
Proof. Define

$$Y_k = \frac{m(m-1)}{2} \sum_{j=1}^{k} \lambda_j(\chi_{1j}^2 - 1), \quad k = 1, 2, ...$$

It can be shown (exercise) that $\{Y_k^2\}$ is uniformly integrable. Since $Y_k \to_d Y$ as $k \to \infty$, $\lim_{k\to\infty} EY_k^2 = EY^2$ (Theorem 1.8(viii)). Since χ_{1j}^2's are independent chi-square random variables with $E\chi_{1j}^2 = 1$ and $\text{Var}(\chi_{1j}^2) = 2$, $EY_k = 0$ for any k and

$$EY_k^2 = \frac{m^2(m-1)^2}{4} \sum_{j=1}^{k} \lambda_j^2 \text{Var}(\chi_{1j}^2)$$

$$= \frac{m^2(m-1)^2}{4} \left(2 \sum_{j=1}^{k} \lambda_j^2\right)$$

$$\to \frac{m^2(m-1)^2}{2}\zeta_2. \quad ∎$$

It follows from Corollary 3.2(iii) and Lemma 3.2 that $\text{amse}_{U_n}(P) = \frac{m^2(m-1)^2}{2}\zeta_2/n^2 = \text{Var}(U_n) + O(n^{-3})$ if $\zeta_1 = 0$. Again, by Proposition 2.4(ii), the sequence $\{n^2[U_n - E(U_n)]^2\}$ is uniformly integrable.

We now apply Theorem 3.5 to the U-statistics in Example 3.11. For $U_n = \frac{2}{n(n-1)}\sum_{1\le i<j\le n} X_i X_j$, $\zeta_1 = \mu^2\sigma^2$. Thus, if $\mu \ne 0$, the result in Theorem 3.5(i) holds with $\zeta_1 = \mu^2\sigma^2$. If $\mu = 0$, then $\zeta_1 = 0$, $\zeta_2 = \sigma^4 > 0$, and Theorem 3.5(ii) applies. However, it is not convenient to use Theorem 3.5(ii) to find the limiting distribution of U_n. We may derive this limiting distribution using the following technique, which is further discussed in §3.5. By the CLT and Theorem 1.10,

$$n\bar{X}^2/\sigma^2 \to_d \chi_1^2$$

when $\mu = 0$, where χ_1^2 is a random variable having the chi-square distribution χ_1^2. Note that

$$\frac{n\bar{X}^2}{\sigma^2} = \frac{1}{\sigma^2 n}\sum_{i=1}^n X_i^2 + \frac{(n-1)U_n}{\sigma^2}.$$

By the SLLN, $\frac{1}{\sigma^2 n}\sum_{i=1}^n X_i^2 \to_{a.s.} 1$. An application of Slutsky's theorem leads to

$$nU_n/\sigma^2 \to_d \chi_1^2 - 1.$$

Since $\mu = 0$, this implies that the right-hand side of (3.21) is $\sigma^2(\chi_1^2 - 1)$, i.e., $\lambda_1 = \sigma^2$ and $\lambda_j = 0$ when $j > 1$.

For the one-sample Wilcoxon statistic, $\zeta_1 = \text{Var}(F(-X_1)) > 0$ unless F is degenerate. Similarly, for Gini's mean difference, $\zeta_1 > 0$ unless F is degenerate. Hence Theorem 3.5(i) applies to these two cases.

Theorem 3.5 does not apply to U_n defined by (3.13) if r, the order of the kernel, depends on n and diverges to ∞ as $n \to \infty$. We consider the simple case where

$$T_n = \frac{1}{n}\sum_{i=1}^n \psi(X_i) + R_n \tag{3.22}$$

for some R_n satisfying $E(R_n^2) = o(n^{-1})$. Note that (3.22) is satisfied for T_n being a U-statistic (exercise). Assume that r/d is bounded. Let $S_\psi^2 = (n-1)^{-1}\sum_{i=1}^n[\psi(X_i) - n^{-1}\sum_{i=1}^n \psi(X_i)]^2$. Then

$$nU_n = S_\psi^2 + o_p(1) \tag{3.23}$$

(exercise). Under (3.22), if $0 < E[\psi(X_i)]^2 < \infty$, then $\text{amse}_{T_n}(P) = E[\psi(X_i)]^2/n$. Hence, the jackknife estimator U_n in (3.13) provides a consistent estimator of $\text{amse}_{T_n}(P)$, i.e., $U_n/\text{amse}_{T_n}(P) \to_p 1$.

3.3 The LSE in Linear Models

One of the most useful statistical models for non-i.i.d. data in applications
is the general linear model

$$X_i = \beta^\tau Z_i + \varepsilon_i, \qquad i = 1, ..., n, \tag{3.24}$$

where X_i is the ith observation and is often called the ith response; β
is a p-vector of unknown parameters, $p < n$; Z_i is the ith value of a p-
vector of explanatory variables (or covariates); and $\varepsilon_1, ..., \varepsilon_n$ are random
errors. Our data in this case are $(X_1, Z_1), ..., (X_n, Z_n)$ (ε_i's are not observed). Throughout this book Z_i's are considered to be nonrandom or
given values of a random p-vector, in which case our analysis is conditioned
on $Z_1, ..., Z_n$. Each ε_i can be viewed as a random measurement error in
measuring the unknown mean of X_i when the covariate vector is equal to
Z_i. The main parameter of interest is β. More specific examples of model
(3.24) are provided in this section. Other examples and examples of data
from model (3.24) can be found in many standard books for linear models,
for example, Draper and Smith (1981) and Searle (1971).

3.3.1 The LSE and estimability

Let $X = (X_1, ..., X_n)$, $\varepsilon = (\varepsilon_1, ..., \varepsilon_n)$, and Z be the $n \times p$ matrix whose ith
row is the vector Z_i, $i = 1, ..., n$. Then, a matrix form of model (3.24) is

$$X = Z\beta + \varepsilon. \tag{3.25}$$

Definition 3.4. Suppose that the range of β in model (3.25) is $B \subset \mathcal{R}^p$.
A *least squares estimator* (LSE) of β is defined to be any $\hat{\beta} \in B$ such that

$$\|X - Z\hat{\beta}\|^2 = \min_{b \in B} \|X - Zb\|^2. \tag{3.26}$$

For any $l \in \mathcal{R}^p$, $l^\tau \hat{\beta}$ is called an LSE of $l^\tau \beta$. ∎

Throughout this book, we consider $B = \mathcal{R}^p$ unless otherwise stated.
Differentiating $\|X - Zb\|^2$ w.r.t. b, we obtain that any solution of

$$Z^\tau Zb = Z^\tau X \tag{3.27}$$

is an LSE of β. If the rank of the matrix Z is p, in which case $(Z^\tau Z)^{-1}$
exists and Z is said to be of full rank, then there is a unique LSE, which is

$$\hat{\beta} = (Z^\tau Z)^{-1} Z^\tau X. \tag{3.28}$$

If Z is not of full rank, then there are infinitely many LSE's of β. It can be shown (exercise) that any LSE of β is of the form

$$\hat{\beta} = (Z^\tau Z)^- Z^\tau X, \tag{3.29}$$

where $(Z^\tau Z)^-$ is called a *generalized inverse* of $Z^\tau Z$ and satisfies

$$Z^\tau Z (Z^\tau Z)^- Z^\tau Z = Z^\tau Z.$$

Generalized inverse matrices are not unique unless Z is of full rank, in which case $(Z^\tau Z)^- = (Z^\tau Z)^{-1}$ and (3.29) reduces to (3.28).

To study properties of LSE's of β, we need some assumptions on the distribution of X. Since Z_i's are nonrandom, assumptions on the distribution of X can be expressed in terms of assumptions on the distribution of ε. Several commonly adopted assumptions are stated as follows.

Assumption A1: ε is distributed as $N_n(0, \sigma^2 I_n)$ with an unknown $\sigma^2 > 0$.

Assumption A2: $E(\varepsilon) = 0$ and $\text{Var}(\varepsilon) = \sigma^2 I_n$ with an unknown $\sigma^2 > 0$.

Assumption A3: $E(\varepsilon) = 0$ and $\text{Var}(\varepsilon)$ is an unknown matrix.

Assumption A1 is the strongest and implies a parametric model. We may assume a slightly more general assumption that ε has the $N_n(0, \sigma^2 D)$ distribution with unknown σ^2 but a known positive definite matrix D. Let $D^{-1/2}$ be the inverse of the square root matrix of D. Then model (3.25) with assumption A1 holds if we replace X, Z, and ε by the transformed variables $\tilde{X} = D^{-1/2}X$, $\tilde{Z} = D^{-1/2}Z$, and $\tilde{\varepsilon} = D^{-1/2}\varepsilon$, respectively. A similar conclusion can be made for assumption A2.

Under assumption A1, the distribution of X is $N_n(Z\beta, \sigma^2 I_n)$, which is in an exponential family $\mathcal{P}$ with parameter $\theta = (\beta, \sigma^2) \in \mathcal{R}^p \times (0, \infty)$. However, if the matrix Z is not of full rank, then $\mathcal{P}$ is not identifiable (see §2.1.2), since $Z\beta_1 = Z\beta_2$ does not imply $\beta_1 = \beta_2$.

Suppose that the rank of Z is $r \le p$. Then there is an $n \times r$ submatrix Z_* of Z such that

$$Z = Z_* Q \tag{3.30}$$

and Z_* is of rank r, where Q is a fixed $r \times p$ matrix. Then

$$Z\beta = Z_* Q\beta$$

and $\mathcal{P}$ is identifiable if we consider the reparameterization $\tilde{\beta} = Q\beta$. Note that the new parameter $\tilde{\beta}$ is in a subspace of $\mathcal{R}^p$ with dimension r.

In many applications, we are interested in estimating some linear functions of β, i.e., $\vartheta = l^\tau \beta$ for some $l \in \mathcal{R}^p$. From the previous discussion, however, estimation of $l^\tau \beta$ is meaningless unless $l = Q^\tau c$ for some $c \in \mathcal{R}^r$ so that

$$l^\tau \beta = c^\tau Q\beta = c^\tau \tilde{\beta}.$$

The following result shows that $l^\tau\beta$ is estimable if $l = Q^\tau c$, which is also necessary for $l^\tau\beta$ to be estimable under assumption A1.

Theorem 3.6. Assume model (3.25) with assumption A3.
(i) A necessary and sufficient condition for $l \in \mathcal{R}^p$ being $Q^\tau c$ for some $c \in \mathcal{R}^r$ is $l \in \mathcal{R}(Z) = \mathcal{R}(Z^\tau Z)$, where Q is given by (3.30) and $\mathcal{R}(A)$ is the smallest linear subspace containing all rows of A.
(ii) If $l \in \mathcal{R}(Z)$, then the LSE $l^\tau\hat\beta$ is unique and unbiased for $l^\tau\beta$.
(iii) If $l \notin \mathcal{R}(Z)$ and assumption A1 holds, then $l^\tau\beta$ is not estimable.
Proof. (i) Note that $a \in \mathcal{R}(A)$ if and only if $a = A^\tau b$ for some vector b. If $l = Q^\tau c$, then

$$l = Q^\tau c = Q^\tau Z_*^\tau Z_*(Z_*^\tau Z_*)^{-1}c = Z^\tau[Z_*(Z_*^\tau Z_*)^{-1}c].$$

Hence $l \in \mathcal{R}(Z)$. If $l \in \mathcal{R}(Z)$, then $l = Z^\tau\zeta$ for some ζ and

$$l = (Z_*Q)^\tau\zeta = Q^\tau c$$

with $c = Z_*^\tau\zeta$.
(ii) If $l \in \mathcal{R}(Z) = \mathcal{R}(Z^\tau Z)$, then $l = Z^\tau Z\zeta$ for some ζ and by (3.29),

$$\begin{aligned}
E(l^\tau\hat\beta) &= E[l^\tau(Z^\tau Z)^- Z^\tau X] \\
&= \zeta^\tau Z^\tau Z(Z^\tau Z)^- Z^\tau Z\beta \\
&= \zeta^\tau Z^\tau Z\beta \\
&= l^\tau\beta.
\end{aligned}$$

If $\bar\beta$ is any other LSE of β, then, by (3.27),

$$l^\tau\hat\beta - l^\tau\bar\beta = \zeta^\tau(Z^\tau Z)(\hat\beta - \bar\beta) = \zeta^\tau(Z^\tau X - Z^\tau X) = 0.$$

(iii) Under assumption A1, if there is an estimator $h(X, Z)$ unbiased for $l^\tau\beta$, then

$$l^\tau\beta = \int_{\mathcal{R}^n} h(x, Z)(2\pi)^{-n/2}\sigma^{-n} \exp\left\{-\tfrac{1}{2\sigma^2}\|x - Z\beta\|^2\right\} dx.$$

Differentiating w.r.t. β and applying Theorem 2.1 lead to

$$l^\tau = Z^\tau \int_{\mathcal{R}^n} h(x, Z)(2\pi)^{-n/2}\sigma^{-n-2}(x - Z\beta) \exp\left\{-\tfrac{1}{2\sigma^2}\|x - Z\beta\|^2\right\} dx,$$

which implies $l \in \mathcal{R}(Z)$. ∎

Theorem 3.6 shows that LSE's are unbiased for estimable parameters $l^\tau\beta$. If Z is of full rank, then $\mathcal{R}(Z) = \mathcal{R}^p$ and, therefore, $l^\tau\beta$ is estimable for any $l \in \mathcal{R}^p$.

Example 3.12 (Simple linear regression). Let $\beta = (\beta_0, \beta_1) \in \mathcal{R}^2$ and $Z_i = (1, t_i)$, $t_i \in \mathcal{R}$, $i = 1, ..., n$. Then model (3.24) or (3.25) is called a *simple linear regression* model. It turns out that

$$Z^\tau Z = \begin{pmatrix} n & \sum_{i=1}^n t_i \\ \sum_{i=1}^n t_i & \sum_{i=1}^n t_i^2 \end{pmatrix}.$$

This matrix is invertible if and only if some t_i's are different. Thus, if some t_i's are different, then the unique unbiased LSE of $l^\tau \beta$ for any $l \in \mathcal{R}^2$ is $l^\tau (Z^\tau Z)^{-1} Z^\tau X$, which has the normal distribution if assumption A1 holds.

The result can be easily extended to the case of *polynomial regression* of order p in which $\beta = (\beta_0, \beta_1, ..., \beta_{p-1})$ and $Z_i = (1, t_i, ..., t_i^{p-1})$. ∎

Example 3.13 (One-way ANOVA). Suppose that $n = \sum_{j=1}^m n_j$ with m positive integers $n_1, ..., n_m$ and that

$$X_i = \mu_j + \varepsilon_i, \qquad i = k_{j-1} + 1, ..., k_j, \ j = 1, ..., m,$$

where $k_0 = 0$, $k_j = \sum_{l=1}^j n_l$, $j = 1, ..., m$, and $(\mu_1, ..., \mu_m) = \beta$. Let J_m be the m-vector of ones. Then the matrix Z in this case is a block diagonal matrix with J_{n_j} as the jth diagonal column. Consequently, $Z^\tau Z$ is an $m \times m$ diagonal matrix whose jth diagonal element is n_j. Thus, $Z^\tau Z$ is invertible and the unique LSE of β is the m-vector whose jth component is $n_j^{-1} \sum_{i=k_{j-1}+1}^{k_j} X_i$, $j = 1, ..., m$.

Sometimes it is more convenient to use the following notation:

$$X_{ij} = X_{k_{i-1}+j}, \ \varepsilon_{ij} = \varepsilon_{k_{i-1}+j}, \qquad j = 1, ..., n_i, i = 1, ..., m,$$

and

$$\mu_i = \mu + \alpha_i, \qquad i = 1, ..., m.$$

Then our model becomes

$$X_{ij} = \mu + \alpha_i + \varepsilon_{ij}, \qquad j = 1, ..., n_i, i = 1, ..., m, \tag{3.31}$$

which is called a *one-way analysis of variance* (ANOVA) model. Under model (3.31), $\beta = (\mu, \alpha_1, ..., \alpha_m) \in \mathcal{R}^{m+1}$. The matrix Z under model (3.31) is not of full rank (exercise). An LSE of β under model (3.31) is

$$\hat\beta = (\bar{X}, \bar{X}_{1\cdot} - \bar{X}, ..., \bar{X}_{m\cdot} - \bar{X}),$$

where $\bar{X}$ is still the sample mean of X_{ij}'s and $\bar{X}_{i\cdot}$ is the sample mean of the ith group $\{X_{ij}, j = 1, ..., n_i\}$. The problem of finding the form of $l \in \mathcal{R}(Z)$ under model (3.31) is left as an exercise. ∎

The notation used in model (3.31) allows us to generalize the one-way ANOVA model to any s-way ANOVA model with a positive integer s under the so-called factorial experiments. The following example is for the two-way ANOVA model.

Example 3.14 (Two-way balanced ANOVA). Suppose that

$$X_{ijk} = \mu + \alpha_i + \beta_j + \gamma_{ij} + \varepsilon_{ijk}, \quad i = 1, ..., a, j = 1, ..., b, k = 1, ..., c, \quad (3.32)$$

where a, b, and c are some positive integers. Model (3.32) is called a two-way balanced ANOVA model. If we view model (3.32) as a special case of model (3.25), then the parameter vector β is

$$\beta = (\mu, \alpha_1, ..., \alpha_a, \beta_1, ..., \beta_b, \gamma_{11}, ..., \gamma_{1b}, ..., \gamma_{a1}, ..., \gamma_{ab}). \quad (3.33)$$

One can obtain the matrix Z and show that it is $n \times p$, where $n = abc$ and $p = 1 + a + b + ab$, and is of rank $ab < p$ (exercise). It can also be shown (exercise) that an LSE of β is given by the right-hand side of (3.33) with μ, α_i, β_j, and γ_{ij} replaced by $\hat{\mu}$, $\hat{\alpha}_i$, $\hat{\beta}_j$, and $\hat{\gamma}_{ij}$, respectively, where $\hat{\mu} = \bar{X}_{...}$, $\hat{\alpha}_i = \bar{X}_{i..} - \bar{X}_{...}$, $\hat{\beta}_j = \bar{X}_{.j.} - \bar{X}_{...}$, $\hat{\gamma}_{ij} = \bar{X}_{ij.} - \bar{X}_{i..} - \bar{X}_{.j.} + \bar{X}_{...}$, and a dot is used to denote averaging over the indicated subscript, e.g.,

$$\bar{X}_{.j.} = \frac{1}{ac} \sum_{i=1}^{a} \sum_{k=1}^{c} X_{ijk}$$

with a fixed j. ∎

3.3.2 The UMVUE and BLUE

We now study UMVUE's in model (3.25) with assumption A1.

Theorem 3.7. Consider model (3.25) with assumption A1.
(i) The LSE $l^\tau \hat{\beta}$ is the UMVUE of $l^\tau \beta$ for any estimable $l^\tau \beta$.
(ii) The UMVUE of σ^2 is $\hat{\sigma}^2 = (n - r)^{-1} \|X - Z\hat{\beta}\|^2$, where r is the rank of Z.
Proof. (i) Let $\hat{\beta}$ be an LSE of β. By (3.27),

$$(X - Z\hat{\beta})^\tau Z(\hat{\beta} - \beta) = (X^\tau Z - X^\tau Z)(\hat{\beta} - \beta) = 0$$

and, hence,

$$\begin{aligned} \|X - Z\beta\|^2 &= \|X - Z\hat{\beta} + Z\hat{\beta} - Z\beta\|^2 \\ &= \|X - Z\hat{\beta}\|^2 + \|Z\hat{\beta} - Z\beta\|^2 \\ &= \|X - Z\hat{\beta}\|^2 - 2\beta^\tau Z^\tau X + \|Z\beta\|^2 + \|Z\hat{\beta}\|^2. \end{aligned}$$

Using this result and assumption A1, we obtain the following joint Lebesgue p.d.f. of X:

$$(2\pi\sigma^2)^{-n/2} \exp\left\{ \frac{\beta^\tau Z^\tau x}{\sigma^2} - \frac{\|x - Z\hat{\beta}\|^2 + \|Z\hat{\beta}\|^2}{2\sigma^2} - \frac{\|Z\beta\|^2}{2\sigma^2} \right\}.$$

By Proposition 2.1 and the fact that $Z\hat{\beta} = Z(Z^{\tau}Z)^{-}Z^{\tau}X$ is a function of $Z^{\tau}X$, $(Z^{\tau}X, \|X - Z\hat{\beta}\|^2)$ is complete and sufficient for $\theta = (\beta, \sigma^2)$. Note that $\hat{\beta}$ is a function of $Z^{\tau}X$ and, hence, a function of the complete sufficient statistic. If $l^{\tau}\beta$ is estimable, then $l^{\tau}\hat{\beta}$ is unbiased for $l^{\tau}\beta$ (Theorem 3.6) and, hence, $l^{\tau}\hat{\beta}$ is the UMVUE of $l^{\tau}\beta$.

(ii) From $\|X - Z\beta\|^2 = \|X - Z\hat{\beta}\|^2 + \|Z\hat{\beta} - Z\beta\|^2$ and $E(Z\hat{\beta}) = Z\beta$ (Theorem 3.6),

$$
\begin{aligned}
E\|X - Z\hat{\beta}\|^2 &= E(X - Z\beta)^{\tau}(X - Z\beta) - E(\beta - \hat{\beta})^{\tau}Z^{\tau}Z(\beta - \hat{\beta}) \\
&= \operatorname{tr}\left(\operatorname{Var}(X) - \operatorname{Var}(Z\hat{\beta})\right) \\
&= \sigma^2[n - \operatorname{tr}\left(Z(Z^{\tau}Z)^{-}Z^{\tau}Z(Z^{\tau}Z)^{-}Z^{\tau})\right] \\
&= \sigma^2[n - \operatorname{tr}\left((Z^{\tau}Z)^{-}Z^{\tau}Z\right)].
\end{aligned}
$$

Since each row of $Z \in \mathcal{R}(Z)$, $Z\hat{\beta}$ does not depend on the choice of $(Z^{\tau}Z)^{-}$ in $\hat{\beta} = (Z^{\tau}Z)^{-}Z^{\tau}X$ (Theorem 3.6). Hence, we can evaluate $\operatorname{tr}((Z^{\tau}Z)^{-}Z^{\tau}Z)$ using a particular $(Z^{\tau}Z)^{-}$. From the theory of linear algebra, there exists a $p \times p$ matrix C such that $CC^{\tau} = I_p$ and

$$
C^{\tau}(Z^{\tau}Z)C = \begin{pmatrix} \Lambda & 0 \\ 0 & 0 \end{pmatrix},
$$

where Λ is an $r \times r$ diagonal matrix whose diagonal elements are positive. Then, a particular choice of $(Z^{\tau}Z)^{-}$ is

$$
(Z^{\tau}Z)^{-} = C \begin{pmatrix} \Lambda^{-1} & 0 \\ 0 & 0 \end{pmatrix} C^{\tau} \tag{3.34}
$$

and

$$
(Z^{\tau}Z)^{-}Z^{\tau}Z = C \begin{pmatrix} I_r & 0 \\ 0 & 0 \end{pmatrix} C^{\tau}
$$

whose trace is r. Hence $\hat{\sigma}^2$ is the UMVUE of σ^2, since it is a function of the complete sufficient statistic and

$$
E\hat{\sigma}^2 = (n - r)^{-1}E\|X - Z\hat{\beta}\|^2 = \sigma^2. \quad \blacksquare
$$

In general,

$$
\operatorname{Var}(l^{\tau}\hat{\beta}) = l^{\tau}(Z^{\tau}Z)^{-}Z^{\tau}\operatorname{Var}(\varepsilon)Z(Z^{\tau}Z)^{-}l. \tag{3.35}
$$

If $l \in \mathcal{R}(Z)$ and $\operatorname{Var}(\varepsilon) = \sigma^2 I_n$ (assumption A2), then the use of the generalized inverse matrix in (3.34) leads to $\operatorname{Var}(l^{\tau}\hat{\beta}) = \sigma^2 l^{\tau}(Z^{\tau}Z)^{-}l$, which attains the Cramér-Rao lower bound under assumption A1 (Proposition 3.2).

The vector $X - Z\hat{\beta}$ is called the *residual vector* and $\|X - Z\hat{\beta}\|^2$ is called the *sum of squared residuals* and is denoted by SSR. The estimator $\hat{\sigma}^2$ is then equal to $SSR/(n - r)$.

Since $X - Z\hat{\beta} = [I_n - Z(Z^\tau Z)^- Z^\tau]X$ and $l^\tau \hat{\beta} = l^\tau (Z^\tau Z)^- Z^\tau X$ are linear in X, they are normally distributed under assumption A1. Also, using the generalized inverse matrix in (3.34), we obtain that

$$[I_n - Z(Z^\tau Z)^- Z^\tau]Z(Z^\tau Z)^- = Z(Z^\tau Z)^- - Z(Z^\tau Z)^- Z^\tau Z(Z^\tau Z)^- = 0,$$

which implies that $\hat{\sigma}^2$ and $l^\tau \hat{\beta}$ are independent (Exercise 58 in §1.6) for any estimable $l^\tau \beta$. Furthermore,

$$[Z(Z^\tau Z)^- Z^\tau]^2 = Z(Z^\tau Z)^- Z^\tau$$

(i.e., $Z(Z^\tau Z)^- Z^\tau$ is a projection matrix) and

$$SSR = X^\tau [I_n - Z(Z^\tau Z)^- Z^\tau]X.$$

The rank of $Z(Z^\tau Z)^- Z^\tau$ is $\text{tr}(Z(Z^\tau Z)^- Z^\tau) = r$. Similarly, the rank of the projection matrix $I_n - Z(Z^\tau Z)^- Z^\tau$ is $n - r$. From

$$X^\tau X = X^\tau [Z(Z^\tau Z)^- Z^\tau]X + X^\tau [I_n - Z(Z^\tau Z)^- Z^\tau]X$$

and Theorem 1.5 (Cochran's theorem), SSR/σ^2 has the chi-square distribution $\chi^2_{n-r}(\delta)$ with

$$\delta = \sigma^{-2} \beta^\tau Z^\tau [I_n - Z(Z^\tau Z)^- Z^\tau]Z\beta = 0.$$

Thus, we have proved the following result.

Theorem 3.8. Consider model (3.25) with assumption A1. For any estimable parameter $l^\tau \beta$, the UMVUE's $l^\tau \hat{\beta}$ and $\hat{\sigma}^2$ are independent; the distribution of $l^\tau \hat{\beta}$ is $N(l^\tau \beta, \sigma^2 l^\tau (Z^\tau Z)^- l)$; and $(n - r)\hat{\sigma}^2/\sigma^2$ has the chi-square distribution χ^2_{n-r}. ∎

Example 3.15. In Examples 3.12-3.14, UMVUE's of estimable $l^\tau \beta$ are the LSE's $l^\tau \hat{\beta}$, under assumption A1. In Example 3.13,

$$SSR = \sum_{i=1}^m \sum_{j=1}^{n_i} (X_{ij} - \bar{X}_{i\cdot})^2;$$

in Example 3.14, if $c > 1$,

$$SSR = \sum_{i=1}^a \sum_{j=1}^b \sum_{k=1}^c (X_{ijk} - \bar{X}_{ij\cdot})^2. \quad ∎$$

We now study properties of $l^\tau\hat\beta$ and $\hat\sigma^2$ under assumption A2, i.e., without the normality assumption on ε. From Theorem 3.6 and the proof of Theorem 3.7(ii), $l^\tau\hat\beta$ (with an $l \in \mathcal{R}(Z)$) and $\hat\sigma^2$ are still unbiased without the normality assumption. In what sense are $l^\tau\hat\beta$ and $\hat\sigma^2$ optimal beyond being unbiased? We have the following result for the LSE $l^\tau\hat\beta$. Some discussion about $\hat\sigma^2$ can be found, for example, in Rao (1973, p. 228).

Theorem 3.9. Consider model (3.25) with assumption A2.
(i) A necessary and sufficient condition for the existence of a linear unbiased estimator of $l^\tau\beta$ (i.e., an unbiased estimator that is linear in X) is $l \in \mathcal{R}(Z)$.
(ii) (Gauss-Markov theorem). If $l \in \mathcal{R}(Z)$, then the LSE $l^\tau\hat\beta$ is the *best linear unbiased estimator* (BLUE) of $l^\tau\beta$ in the sense that it has the minimum variance in the class of linear unbiased estimators of $l^\tau\beta$.
Proof. (i) The sufficiency has been established in Theorem 3.6. Suppose now a linear function of X, $c^\tau X$ with $c \in \mathcal{R}^n$, is unbiased for $l^\tau\beta$. Then

$$l^\tau\beta = E(c^\tau X) = c^\tau EX = c^\tau Z\beta.$$

Since this equality holds for all β, $l = Z^\tau c$, i.e., $l \in \mathcal{R}(Z)$.
(ii) Let $l \in \mathcal{R}(Z) = \mathcal{R}(Z^\tau Z)$. Then $l = (Z^\tau Z)\zeta$ for some ζ and $l^\tau\hat\beta = \zeta^\tau(Z^\tau Z)\hat\beta = \zeta^\tau Z^\tau X$ by (3.27). Let $c^\tau X$ be any linear unbiased estimator of $l^\tau\beta$. From the proof of (i), $Z^\tau c = l$. Then

$$
\begin{aligned}
\mathrm{Cov}(\zeta^\tau Z^\tau X, c^\tau X - \zeta^\tau Z^\tau X) &= E(X^\tau Z\zeta c^\tau X) - E(X^\tau Z\zeta\zeta^\tau Z^\tau X) \\
&= \sigma^2 \mathrm{tr}(Z\zeta c^\tau) + \beta^\tau Z^\tau Z\zeta c^\tau Z\beta \\
&\quad - \sigma^2 \mathrm{tr}(Z\zeta\zeta^\tau Z^\tau) - \beta^\tau Z^\tau Z\zeta\zeta^\tau Z^\tau Z\beta \\
&= \sigma^2\zeta^\tau l + (l^\tau\beta)^2 - \sigma^2\zeta^\tau l - (l^\tau\beta)^2 \\
&= 0.
\end{aligned}
$$

Hence

$$
\begin{aligned}
\mathrm{Var}(c^\tau X) &= \mathrm{Var}(c^\tau X - \zeta^\tau Z^\tau X + \zeta^\tau Z^\tau X) \\
&= \mathrm{Var}(c^\tau X - \zeta^\tau Z^\tau X) + \mathrm{Var}(\zeta^\tau Z^\tau X) \\
&\quad + 2\mathrm{Cov}(\zeta^\tau Z^\tau X, c^\tau X - \zeta^\tau Z^\tau X) \\
&= \mathrm{Var}(c^\tau X - \zeta^\tau Z^\tau X) + \mathrm{Var}(l^\tau\hat\beta) \\
&\geq \mathrm{Var}(l^\tau\hat\beta). \quad \blacksquare
\end{aligned}
$$

3.3.3 Robustness of LSE's

Consider now model (3.25) under assumption A3. An interesting question is under what conditions on $\mathrm{Var}(\varepsilon)$ is the LSE of $l^\tau\beta$ with $l \in \mathcal{R}(Z)$ still the BLUE. If $l^\tau\hat\beta$ is still the BLUE, then we say that $l^\tau\hat\beta$, considered as a BLUE, is *robust* against violation of assumption A2. In general, a statistical procedure having certain properties under an assumption is said to be robust against violation of the assumption if and only if the statistical procedure still has the same properties when the assumption is (slightly) violated. For example, the LSE of $l^\tau\beta$ with $l \in \mathcal{R}(Z)$, as an unbiased estimator, is robust against violation of assumption A1 or A2, since the LSE is unbiased as long as $E(\varepsilon) = 0$, which can be always assumed without loss of generality. On the other hand, the LSE as a UMVUE may not be robust against violation of assumption A1 (see §3.5).

Theorem 3.10. Consider model (3.25) with assumption A3. The following are equivalent.
(a) $l^\tau\hat{\beta}$ is the BLUE of $l^\tau\beta$ for any $l \in \mathcal{R}(Z)$.
(b) $E(l^\tau\hat{\beta}\eta^\tau X) = 0$ for any $l \in \mathcal{R}(Z)$ and any η such that $E(\eta^\tau X) = 0$.
(c) $Z^\tau\text{Var}(\varepsilon)U = 0$, where U is a matrix such that $Z^\tau U = 0$ and $\mathcal{R}(U^\tau) + \mathcal{R}(Z^\tau) = \mathcal{R}^n$.
(d) $\text{Var}(\varepsilon) = Z\Lambda_1 Z^\tau + U\Lambda_2 U^\tau$ for some Λ_1 and Λ_2.
(e) The matrix $Z(Z^\tau Z)^- Z^\tau\text{Var}(\varepsilon)$ is symmetric.
Proof. We first show that (a) and (b) are equivalent, which is an analogue of Theorem 3.2(i). Suppose that (b) holds. Let $l \in \mathcal{R}(Z)$. If $c^\tau X$ is unbiased for $l^\tau\beta$, then $E(\eta^\tau X) = 0$ with $\eta = c - Z(Z^\tau Z)^- l$. Hence

$$\begin{aligned}
\text{Var}(c^\tau X) &= \text{Var}(c^\tau X - l^\tau\hat{\beta} + l^\tau\hat{\beta}) \\
&= \text{Var}(c^\tau X - l^\tau(Z^\tau Z)^- Z^\tau X + l^\tau\hat{\beta}) \\
&= \text{Var}(\eta^\tau X + l^\tau\hat{\beta}) \\
&= \text{Var}(\eta^\tau X) + \text{Var}(l^\tau\hat{\beta}) + 2\text{Cov}(\eta^\tau X, l^\tau\hat{\beta}) \\
&= \text{Var}(\eta^\tau X) + \text{Var}(l^\tau\hat{\beta}) + 2E(l^\tau\hat{\beta}\eta^\tau X) \\
&= \text{Var}(\eta^\tau X) + \text{Var}(l^\tau\hat{\beta}) \\
&\geq \text{Var}(l^\tau\hat{\beta}).
\end{aligned}$$

Suppose now that there are $l \in \mathcal{R}(Z)$ and η such that $E(\eta^\tau X) = 0$ but $\delta = E(l^\tau\hat{\beta}\eta^\tau X) \neq 0$. Let $c_t = t\eta + Z(Z^\tau Z)^- l$. From the previous proof,

$$\text{Var}(c_t^\tau X) = t^2\text{Var}(\eta^\tau X) + \text{Var}(l^\tau\hat{\beta}) + 2\delta t.$$

As long as $\delta \neq 0$, there exists a t such that $\text{Var}(c_t^\tau X) < \text{Var}(l^\tau\hat{\beta})$. This shows that $l^\tau\hat{\beta}$ cannot be a BLUE and, therefore, (a) implies (b).

We next show that (b) implies (c). Suppose that (b) holds. Since $l \in \mathcal{R}(Z)$, $l = Z^\tau\gamma$ for some γ. Let $\eta \in \mathcal{R}(U^\tau)$. Then $E(\eta^\tau X) = \eta^\tau Z\beta = 0$ and, hence,

$$0 = E(l^\tau\hat{\beta}\eta^\tau X) = E[\gamma^\tau Z(Z^\tau Z)^- Z^\tau X X^\tau \eta] = \gamma^\tau Z(Z^\tau Z)^- Z^\tau\text{Var}(\varepsilon)\eta.$$

Since this equality holds for all $l \in \mathcal{R}(Z)$, it holds for all γ. Thus,

$$Z(Z^\tau Z)^- Z^\tau\text{Var}(\varepsilon)U = 0,$$

which implies

$$Z^\tau Z(Z^\tau Z)^- Z^\tau \text{Var}(\varepsilon) U = Z^\tau \text{Var}(\varepsilon) U = 0,$$

since $Z^\tau Z(Z^\tau Z)^- Z^\tau = Z^\tau$. Thus, (c) holds.

To show that (c) implies (d), we need to use the following facts from the theory of linear algebra: there exists a nonsingular matrix C such that $\text{Var}(\varepsilon) = CC^\tau$ and $C = ZC_1 + UC_2$ for some matrices C_j (since $\mathcal{R}(U^\tau) + \mathcal{R}(Z^\tau) = \mathcal{R}^n$). Let $\Lambda_1 = C_1 C_1^\tau$, $\Lambda_2 = C_2 C_2^\tau$, and $\Lambda_3 = C_1 C_2^\tau$. Then

$$\text{Var}(\varepsilon) = Z\Lambda_1 Z^\tau + U\Lambda_2 U^\tau + Z\Lambda_3 U^\tau + U\Lambda_3^\tau Z^\tau \qquad (3.36)$$

and $Z^\tau \text{Var}(\varepsilon) U = Z^\tau Z\Lambda_3 U^\tau U$, which is 0 if (c) holds. Hence, (c) implies

$$0 = Z(Z^\tau Z)^- Z^\tau Z\Lambda_3 U^\tau U(U^\tau U)^- U^\tau = Z\Lambda_3 U^\tau,$$

which with (3.36) implies (d).

If (d) holds, then $Z(Z^\tau Z)^- Z^\tau \text{Var}(\varepsilon) = Z\Lambda_1 Z^\tau$, which is symmetric. Hence (d) implies (e). To complete the proof, we need to show that (e) implies (b), which is left as an exercise. ∎

As a corollary of this theorem, the following result shows when the UMVUE's in model (3.25) with assumption A1 are robust against the violation of $\text{Var}(\varepsilon) = \sigma^2 I_n$.

Corollary 3.3. Consider model (3.25) with a full rank Z, $\varepsilon = N_n(0, \Sigma)$, and an unknown positive definite matrix Σ. Then $l^\tau \hat{\beta}$ is a UMVUE of $l^\tau \beta$ for any $l \in \mathcal{R}^p$ if and only if one of (b)-(e) in Theorem 3.10 holds. ∎

Example 3.16. Consider model (3.25) with β replaced by a random vector $\boldsymbol{\beta}$ that is independent of ε. Such a model is called a linear model with random coefficients. Suppose that $\text{Var}(\varepsilon) = \sigma^2 I_n$ and $E(\boldsymbol{\beta}) = \beta$. Then

$$X = Z\beta + Z(\boldsymbol{\beta} - \beta) + \varepsilon = Z\beta + e, \qquad (3.37)$$

where $e = Z(\boldsymbol{\beta} - \beta) + \varepsilon$ satisfies $E(e) = 0$ and

$$\text{Var}(e) = Z\text{Var}(\boldsymbol{\beta})Z^\tau + \sigma^2 I_n.$$

Since

$$Z(Z^\tau Z)^- Z^\tau \text{Var}(e) = Z\text{Var}(\boldsymbol{\beta})Z^\tau + \sigma^2 Z(Z^\tau Z)^- Z^\tau$$

is symmetric, by Theorem 3.10, the LSE $l^\tau \hat{\beta}$ under model (3.37) is the BLUE for any $l^\tau \beta$, $l \in \mathcal{R}(Z)$. If Z is of full rank and ε is normal, then, by Corollary 3.3, $l^\tau \hat{\beta}$ is the UMVUE of $l^\tau \beta$ for any $l \in \mathcal{R}^p$. ∎

Example 3.17 (Random effects models). Suppose that

$$X_{ij} = \mu + A_i + e_{ij}, \quad j = 1, ..., n_i, i = 1, ..., m, \qquad (3.38)$$

where $\mu \in \mathcal{R}$ is an unknown parameter, A_i's are i.i.d. random variables having mean 0 and variance σ_a^2, e_{ij}'s are i.i.d. random errors with mean 0 and variance σ^2, and A_i's and e_{ij}'s are independent. Model (3.38) is called a one-way *random effects* model and A_i's are unobserved random effects. Let $\varepsilon_{ij} = A_i + e_{ij}$. Then (3.38) is a special case of the general model (3.25) with

$$\text{Var}(\varepsilon) = \sigma_a^2 \Sigma + \sigma^2 I_n,$$

where Σ is a block diagonal matrix whose ith block is $J_{n_i} J_{n_i}^\tau$ and J_k is the k-vector of ones. Under this model, $Z = J_n$, $n = \sum_{i=1}^m n_i$, and $Z(Z^\tau Z)^- Z^\tau = n^{-1} J_n J_n^\tau$. Note that

$$J_n J_n^\tau \Sigma = \begin{pmatrix} n_1 J_{n_1} J_{n_1}^\tau & n_2 J_{n_1} J_{n_2}^\tau & \cdots & n_m J_{n_1} J_{n_m}^\tau \\ n_1 J_{n_2} J_{n_1}^\tau & n_2 J_{n_2} J_{n_2}^\tau & \cdots & n_m J_{n_2} J_{n_m}^\tau \\ \cdots\cdots & \cdots\cdots & \cdots & \cdots\cdots \\ n_1 J_{n_m} J_{n_1}^\tau & n_2 J_{n_m} J_{n_2}^\tau & \cdots & n_m J_{n_m} J_{n_m}^\tau \end{pmatrix},$$

which is symmetric if and only if $n_1 = n_2 = \cdots = n_m$. Since $J_n J_n^\tau \text{Var}(\varepsilon)$ is symmetric if and only if $J_n J_n^\tau \Sigma$ is symmetric, a necessary and sufficient condition for the LSE of μ to be the BLUE is that all n_i's are the same. This condition is also necessary and sufficient for the LSE of μ to be the UMVUE when ε_{ij}'s are normal. ∎

In some cases, we are interested in some (not all) linear functions of β. For example, consider $l^\tau \beta$ with $l \in \mathcal{R}(H)$, where H is an $n \times p$ matrix such that $\mathcal{R}(H) \subset \mathcal{R}(Z)$. We have the following result.

Proposition 3.4. Consider model (3.25) with assumption A3. Suppose that H is a matrix such that $\mathcal{R}(H) \subset \mathcal{R}(Z)$. A necessary and sufficient condition for the LSE $l^\tau \hat{\beta}$ to be the BLUE of $l^\tau \beta$ for any $l \in \mathcal{R}(H)$ is $H(Z^\tau Z)^- Z^\tau \text{Var}(\varepsilon)U = 0$, where U is the same as that in (c) of Theorem 3.10. ∎

Example 3.18. Consider model (3.25) with assumption A3 and $Z = (H_1 \ H_2)$, where $H_1^\tau H_2 = 0$. Suppose that under the reduced model

$$X = H_1 \beta_1 + \varepsilon,$$

$l^\tau \hat{\beta}_1$ is the BLUE for any $l^\tau \beta_1$, $l \in \mathcal{R}(H_1)$, and that under the reduced model

$$X = H_2 \beta_2 + \varepsilon,$$

$l^\tau \hat{\beta}_2$ is not a BLUE for some $l^\tau \beta_2$, $l \in \mathcal{R}(H_2)$, where $\beta = (\beta_1, \beta_2)$ and $\hat{\beta}_j$'s are LSE's under the reduced models. Let $H = (H_1\ 0)$ be $n \times p$. Note that

$$H(Z^\tau Z)^- Z^\tau \text{Var}(\varepsilon)U = H_1(H_1^\tau H_1)^- H_1^\tau \text{Var}(\varepsilon)U,$$

which is 0 by Theorem 3.10 for the U given in (c) of Theorem 3.10, and

$$Z(Z^\tau Z)^- Z^\tau \text{Var}(\varepsilon)U = H_2(H_2^\tau H_2)^- H_2^\tau \text{Var}(\varepsilon)U,$$

which is not 0 by Theorem 3.10. This implies that some LSE $l^\tau \hat{\beta}$ is not a BLUE of $l^\tau \beta$ but $l^\tau \hat{\beta}$ is the BLUE of $l^\tau \beta$ if $l \in \mathcal{R}(H)$. ∎

Finally, we consider model (3.25) with $\text{Var}(\varepsilon)$ being a diagonal matrix whose ith diagonal element is σ_i^2, i.e., ε_i's are uncorrelated but have unequal variances. A straightforward calculation shows that condition (e) in Theorem 3.10 holds if and only if, for all $i \neq j$, $\sigma_i^2 \neq \sigma_j^2$ only when $h_{ij} = 0$, where h_{ij} is the (i, j)th element of the projection matrix $Z(Z^\tau Z)^- Z^\tau$. Thus, an LSE is not a BLUE in general, although it is still unbiased for estimable $l^\tau \beta$.

Suppose that the unequal variances of ε_i's are caused by some small perturbations, i.e., $\varepsilon_i = e_i + u_i$, where $\text{Var}(e_i) = \sigma^2$, $\text{Var}(u_i) = \delta_i$, and e_i and u_i are independent so that $\sigma_i^2 = \sigma^2 + \delta_i$. From (3.35),

$$\text{Var}(l^\tau \hat{\beta}) = l^\tau (Z^\tau Z)^- \sum_{i=1}^n \sigma_i^2 Z_i Z_i^\tau (Z^\tau Z)^- l.$$

If $\delta_i = 0$ for all i (no perturbations), then assumption A2 holds and $l^\tau \hat{\beta}$ is the BLUE of any estimable $l^\tau \beta$ with $\text{Var}(l^\tau \hat{\beta}) = \sigma^2 l^\tau (Z^\tau Z)^- l$. Suppose that $0 < \delta_i \leq \sigma^2 \delta$. Then

$$\text{Var}(l^\tau \hat{\beta}) \leq (1 + \delta)\sigma^2 l^\tau (Z^\tau Z)^- l.$$

This indicates that the LSE is robust in the sense that its variance increases slightly when there is a slight violation of the equal variance assumption (small δ).

3.3.4 Asymptotic properties of LSE's

We consider first the consistency of the LSE $l^\tau \hat{\beta}$ with $l \in \mathcal{R}(Z)$ for every n.

Theorem 3.11. Consider model (3.25) with assumption A3. Suppose that $\sup_n \lambda_+[\text{Var}(\varepsilon)] < \infty$, where $\lambda_+[A]$ is the largest eigenvalue of the matrix A, and that $\lim_{n \to \infty} \lambda_+[(Z^\tau Z)^-] = 0$. Then $l^\tau \hat{\beta}$ is consistent in mse for any $l \in \mathcal{R}(Z)$.

Proof. The result follows from the fact that $l^\tau \hat{\beta}$ is unbiased and

$$\text{Var}(l^\tau \hat{\beta}) = l^\tau (Z^\tau Z)^- Z^\tau \text{Var}(\varepsilon) Z (Z^\tau Z)^- l$$
$$\leq \lambda_+ [\text{Var}(\varepsilon)] l^\tau (Z^\tau Z)^- l. \quad \blacksquare$$

Without the normality assumption on ε, the exact distribution of $l^\tau \hat{\beta}$ is very hard to obtain. The asymptotic distribution of $l^\tau \hat{\beta}$ is derived in the following result.

Theorem 3.12. Consider model (3.25) with assumption A3. Suppose that $0 < \inf_n \lambda_- [\text{Var}(\varepsilon)]$, where $\lambda_- [A]$ is the smallest eigenvalue of the matrix A, and that

$$\lim_{n \to \infty} \max_{1 \leq i \leq n} Z_i^\tau (Z^\tau Z)^- Z_i = 0. \tag{3.39}$$

Suppose further that $n = \sum_{j=1}^k m_j$ for some integers k, m_j, $j = 1, ..., k$, with m_j's bounded by a fixed integer m, $\varepsilon = (\xi_1, ..., \xi_k)$, $\xi_j \in \mathcal{R}^{m_j}$, and ξ_j's are independent.

(i) If $\sup_i E|\varepsilon_i|^{2+\delta} < \infty$, then for any $l \in \mathcal{R}(Z)$,

$$l^\tau (\hat{\beta} - \beta) \Big/ \sqrt{\text{Var}(l^\tau \hat{\beta})} \to_d N(0, 1). \tag{3.40}$$

(ii) Suppose that when $m_i = m_j$, $1 \leq i < j \leq k$, ξ_i and ξ_j have the same distribution. Then result (3.40) holds for any $l \in \mathcal{R}(Z)$.

Proof. Let $l \in \mathcal{R}(Z)$. Then

$$l^\tau (Z^\tau Z)^- Z^\tau Z \beta - l^\tau \beta = 0$$

and

$$l^\tau (\hat{\beta} - \beta) = l^\tau (Z^\tau Z)^- Z^\tau \varepsilon = \sum_{j=1}^k c_{nj}^\tau \xi_j,$$

where c_{nj} is the m_j-vector whose components are $l^\tau (Z^\tau Z)^- Z_i$, $i = k_{j-1} + 1, ..., k_j$, $k_0 = 0$, and $k_j = \sum_{t=1}^j m_t$, $j = 1, ..., k$. Note that

$$\sum_{j=1}^k \|c_{nj}\|^2 = l^\tau (Z^\tau Z)^- Z^\tau Z (Z^\tau Z)^- l = l^\tau (Z^\tau Z)^- l. \tag{3.41}$$

Also,

$$\max_{1 \leq j \leq k} \|c_{nj}\|^2 \leq m \max_{1 \leq i \leq n} [l^\tau (Z^\tau Z)^- Z_i]^2$$
$$\leq m l^\tau (Z^\tau Z)^- l \max_{1 \leq i \leq n} Z_i^\tau (Z^\tau Z)^- Z_i,$$

which, together with (3.41) and condition (3.39), implies that

$$\lim_{n\to\infty} \left(\max_{1\le j\le k} \|c_{nj}\|^2 \Big/ \sum_{j=1}^k \|c_{nj}\|^2 \right) = 0.$$

The results then follow from Corollary 1.3. ∎

Under the conditions of Theorem 3.12, $\text{Var}(\varepsilon)$ is a diagonal block matrix with $\text{Var}(\xi_j)$ as the jth diagonal block, which includes the case of independent ε_i's as a special case.

The following lemma tells us how to check condition (3.39).

Lemma 3.3. The following are sufficient conditions for (3.39).
(a) $\lambda_+[(Z^\tau Z)^-] \to 0$ and $Z_n^\tau (Z^\tau Z)^- Z_n \to 0$, as $n \to \infty$.
(b) There is an increasing sequence $\{a_n\}$ such that $a_n \to \infty$ and $Z^\tau Z/a_n$ converges to a positive definite matrix. ∎

If $n^{-1} \sum_{i=1}^n t_i^2 \to c$ in the simple linear regression model (Example 3.12), where c is a positive constant, then condition (b) in Lemma 3.3 is satisfied with $a_n = n$ and, therefore, Theorem 3.12 applies. In the one-way ANOVA model (Example 3.13),

$$\max_{1\le i\le n} Z_i^\tau (Z^\tau Z)^- Z_i = \lambda_+[(Z^\tau Z)^-] = \max_{1\le j\le m} n_j^{-1}.$$

Hence conditions related to Z in Theorem 3.12 are satisfied if and only if $\min_j n_j \to \infty$. Some similar conclusions can be drawn in the two-way ANOVA model (Example 3.14).

3.4 Unbiased Estimators in Survey Problems

In this section, we consider unbiased estimation for another type of non-i.i.d. data often encountered in applications: survey data from finite populations. A description of the problem is given in Example 2.3 of §2.1.1. Examples and a fuller account of theoretical aspects of survey sampling can be found, for example, in Cochran (1977) and Särndal, Swensson, and Wretman (1992).

3.4.1 UMVUE's of population totals

We use the same notation as in Example 2.3. Let $X = (X_1, ..., X_n)$ be a sample from a finite population $\mathcal{P} = \{y_1, ..., y_N\}$ with

$$P(X_1 = y_{i_1}, ..., X_n = y_{i_n}) = p(s)/n!,$$

where $s = \{i_1, ..., i_n\}$ is a subset of distinct elements of $\{1, ..., N\}$ and p is a selection probability measure. We consider univariate y_i, although most of our conclusions are valid for the case of multivariate y_i. In many survey problems the parameter to be estimated is $Y = \sum_{i=1}^{N} y_i$, the population total.

In Example 2.27, it is shown that $\hat{Y} = N\bar{X} = \frac{N}{n} \sum_{i \in s} y_i$ is unbiased for Y if $p(s)$ is constant (simple random sampling); a formula of $\text{Var}(\hat{Y})$ is also given. We now show that $\hat{Y}$ is in fact the UMVUE of Y under simple random sampling. Let $\mathcal{Y}$ be the range of y_i, $\theta = (y_1, ..., y_N)$ and $\Theta = \prod_{i=1}^{N} \mathcal{Y}$. Under simple random sampling, the population under consideration is a parametric family indexed by $\theta \in \Theta$.

Theorem 3.13 (Watson-Royall theorem). (i) If $p(s) > 0$ for all s, then the vector of order statistics $X_{(1)} \leq \cdots \leq X_{(n)}$ is complete for $\theta \in \Theta$.
(ii) Under simple random sampling, the vector of order statistics is sufficient for $\theta \in \Theta$.
(iii) Under simple random sampling, for any estimable function of θ, its unique UMVUE is the unbiased estimator $g(X_1, ..., X_n)$, where g is symmetric in its n arguments.
Proof. (i) Let $h(X)$ be a function of the order statistics. Then h is symmetric in its n arguments. We need to show that if

$$E[h(X)] = \sum_{s = \{i_1, ..., i_n\} \subset \{1, ..., N\}} p(s) h(y_{i_1}, ..., y_{i_n}) / n! = 0 \qquad (3.42)$$

for all $\theta \in \Theta$, then $h(y_{i_1}, ..., y_{i_n}) = 0$ for all $y_{i_1}, ..., y_{i_n}$. First, suppose that all N elements of θ are equal to $a \in \mathcal{Y}$. Then (3.42) implies $h(a, ..., a) = 0$. Next, suppose that $N - 1$ elements in θ are equal to a and one is $b > a$. Then (3.42) reduces to

$$q_1 h(a, ..., a) + q_2 h(a, ..., a, b),$$

where q_1 and q_2 are some known numbers in $(0, 1)$. Since $h(a, ..., a) = 0$ and $q_2 \neq 0$, $h(a, ..., a, b) = 0$. Using the same argument, we can show that $h(a, ..., a, b, ..., b) = 0$ for any k a's and $n - k$ b's. Suppose next that elements of θ are equal to a, b, or c, $a < b < c$. Then we can show that $h(a, ..., a, b, ..., b, c, ..., c) = 0$ for any k a's, l b's, and $n-k-l$ c's. Continuing inductively, we see that $h(y_1, ..., y_n) = 0$ for all possible $y_1, ..., y_n$. This completes the proof of (i).
(ii) The result follows from the factorization theorem (Theorem 2.2), the fact that $p(s)$ is constant under simple random sampling, and

$$P(X_1 = y_{i_1}, ..., X_n = y_{i_n}) = P(X_{(1)} = y_{(i_1)}, ..., X_{(n)} = y_{(i_n)})/n!,$$

where $y_{(i_1)} \leq \cdots \leq y_{(i_n)}$ are the ordered values of $y_{i_1}, ..., y_{i_n}$.
(iii) The result follows directly from (i) and (ii).   ∎

It is interesting to note the following two issues. (1) Although we have a parametric problem under simple random sampling, the sufficient and complete statistic is the same as that in a nonparametric problem (Example 2.17). (2) For the completeness of the order statistics, we do not need the assumption of simple random sampling.

Example 3.19. From Example 2.27, $\hat{Y} = N\bar{X}$ is unbiased for Y. Since $\hat{Y}$ is symmetric in its arguments, it is the UMVUE of Y. We now derive the UMVUE for $\text{Var}(\hat{Y})$. From Example 2.27,

$$\text{Var}(\hat{Y}) = \frac{N^2}{n}\left(1 - \frac{n}{N}\right)\sigma^2, \tag{3.43}$$

where

$$\sigma^2 = \frac{1}{N-1}\sum_{i=1}^{N}\left(y_i - \frac{Y}{N}\right)^2.$$

It can be shown (exercise) that $E(S^2) = \sigma^2$, where S^2 is the usual sample variance

$$S^2 = \frac{1}{n-1}\sum_{i=1}^{n}(X_i - \bar{X})^2 = \frac{1}{n-1}\sum_{i\in s}\left(y_i - \frac{\hat{Y}}{N}\right)^2.$$

Since S^2 is symmetric in its arguments, $\frac{N^2}{n}\left(1 - \frac{n}{N}\right)S^2$ is the UMVUE of $\text{Var}(\hat{Y})$. ∎

Simple random sampling is simple and easy to use, but it is inefficient unless the population is fairly homogeneous w.r.t. the y_i's. A sampling plan often used in practice is the *stratified sampling* plan, which can be described as follows. The population $\mathcal{P}$ is divided into nonoverlapping sub-populations $\mathcal{P}_1, ..., \mathcal{P}_H$ called strata; a sample is drawn from each stratum $\mathcal{P}_h$, independently across the strata. There are many reasons for stratification: (1) it may produce a gain in precision in parameter estimation when a heterogeneous population is divided into strata, each of which is internally homogeneous; (2) sampling problems may differ markedly in different parts of the population; and (3) administrative considerations may also lead to stratification. More discussions can be found, for example, in Cochran (1977).

In stratified sampling, if a simple random sample (without replacement), $X_h = (X_{h1}, ..., X_{hn_h})$, is drawn from each stratum, where n_h is the sample size in stratum h, then the joint distribution of $X = (X_1, ..., X_H)$ is in a parametric family indexed by $\theta = (\theta_1, ..., \theta_H)$, where $\theta_h = (y_i, i \in \mathcal{P}_h)$, $h = 1, ..., H$. Let $\mathcal{Y}_h$ be the range of y_i's in stratum h and $\Theta_h = \prod_{i=1}^{N_h}\mathcal{Y}_h$, where N_h is the size of $\mathcal{P}_h$. We assume that the parameter space is $\Theta = \prod_{i=1}^{H}\Theta_h$. The following result is similar to Theorem 3.13.

Theorem 3.14. Let X be a sample obtained using the stratified simple random sampling plan described previously.
(i) For each h, let Z_h be the vector of the ordered values of the sample in stratum h. Then $(Z_1, ..., Z_H)$ is sufficient and complete for $\theta \in \Theta$.
(ii) For any estimable function of θ, its unique UMVUE is the unbiased estimator $g(X)$ that is symmetric in its first n_1 arguments, symmetric in its second n_2 arguments,..., and symmetric in its last n_H arguments. ∎

Example 3.20. Consider the estimation of the population total Y based on a sample $X = (X_{hi}, i = 1, ..., n_h, h = 1, ..., H)$ obtained by stratified simple random sampling. Let Y_h be the population total of the hth stratum and let $\hat{Y}_h = N_h \bar{X}_{h\cdot}$, where $\bar{X}_{h\cdot}$ is the sample mean of the sample from stratum h, $h = 1, ..., H$. From Example 2.27, each $\hat{Y}_h$ is an unbiased estimator of Y_h. Let

$$\hat{Y}_{st} = \sum_{h=1}^{H} \hat{Y}_h = \sum_{h=1}^{H} \sum_{i=1}^{n_h} \frac{N_h}{n_h} X_{hi}.$$

Then, by Theorem 3.14, $\hat{Y}_{st}$ is the UMVUE of Y. Since $\hat{Y}_1, ..., \hat{Y}_H$ are independent, it follows from (3.43) that

$$\text{Var}(\hat{Y}_{st}) = \sum_{h=1}^{H} \frac{N_h^2}{n_h} \left(1 - \frac{n_h}{N_h} \right) \sigma_h^2, \tag{3.44}$$

where $\sigma_h^2 = (N_h - 1)^{-1} \sum_{i \in P_h} (y_i - Y_h/N_h)^2$. An argument similar to that in Example 3.19 shows that the UMVUE of $\text{Var}(\hat{Y}_{st})$ is

$$S_{st}^2 = \sum_{h=1}^{H} \frac{N_h^2}{n_h} \left(1 - \frac{n_h}{N_h} \right) S_h^2, \tag{3.45}$$

where S_h^2 is the usual sample variance based on $X_{h1}, ..., X_{hn_h}$.

It is interesting to compare the mse of the UMVUE $\hat{Y}_{st}$ with the mse of the UMVUE $\hat{Y}$ under simple random sampling (Example 3.19). Let σ^2 be given in (3.43). Then

$$(N - 1)\sigma^2 = \sum_{h=1}^{H} (N_h - 1)\sigma_h^2 + \sum_{h=1}^{H} N_h (\mu_h - \mu)^2,$$

where $\mu_h = Y_h/N_h$ is the population mean of the hth stratum and $\mu = Y/N$ is the overall population mean. By (3.43), (3.44), and (3.45), $\text{Var}(\hat{Y}) \geq \text{Var}(\hat{Y}_{st})$ if and only if

$$\sum_{h=1}^{H} \frac{N^2 N_h}{n(N-1)} \left(1 - \frac{n}{N} \right) (\mu_h - \mu)^2 \geq \sum_{h=1}^{H} \left[\frac{N_h^2}{n_h} \left(1 - \frac{n_h}{N_h} \right) - \frac{N^2 (N_h - 1)}{n(N-1)} \left(1 - \frac{n}{N} \right) \right] \sigma_h^2.$$

This means that stratified simple random sampling is better than simple random sampling if the deviations $\mu_h - \mu$ are sufficiently large. If $\frac{n_h}{N_h} \equiv \frac{n}{N}$ (proportional allocation), then this condition simplifies to

$$\sum_{h=1}^{H} N_h(\mu_h - \mu)^2 \geq \sum_{h=1}^{H} \left(1 - \frac{N_h}{N}\right) \sigma_h^2, \qquad (3.46)$$

which is usually true when μ_h's are different and some N_h's are large.

Note that the variances $\text{Var}(\hat{Y})$ and $\text{Var}(\hat{Y}_{st})$ are w.r.t. different sampling plans under which $\hat{Y}$ and $\hat{Y}_{st}$ are obtained. ∎

3.4.2 Horvitz-Thompson estimators

If some elements of the finite population $\mathcal{P}$ are groups (called clusters) of subunits, then sampling from $\mathcal{P}$ is *cluster sampling*. Cluster sampling is used often because of administrative convenience or economic considerations. Although sometimes the first intention may be to use the subunits as sampling units, it is found that no reliable list of the subunits in the population is available. For example, in many countries there are no complete lists of the people or houses in a region. From the maps of the region, however, it can be divided into units such as cities or blocks in the cities.

In cluster sampling, one may greatly increase the precision of estimation by using sampling with probability proportional to cluster size. Thus, unequal probability sampling is often used.

Suppose that a sample of clusters is obtained. If subunits within a selected cluster give similar results, then it may be uneconomical to measure them all. A sample of the subunits in any chosen cluster may be selected. This is called two-stage sampling. One can continue this process to have a multistage sampling (e.g., cities → blocks → houses → people). Of course, at each stage one may use stratified sampling and/or unequal probability sampling.

When the sampling plan is complex, so is the structure of the observations. We now introduce a general method of deriving unbiased estimators of population totals, which are called *Horvitz-Thompson estimators*.

Theorem 3.15. Let $X = \{y_i, i \in s\}$ denote a sample from $\mathcal{P} = \{y_1, ..., y_N\}$ that is selected, without replacement, by some method. Define

$$\pi_i = \text{probability that } i \in s, \quad i = 1, ..., N.$$

(i) (Horvitz-Thompson). If $\pi_i > 0$ for $i = 1, ..., N$ and π_i is known when $i \in s$, then $\hat{Y}_{ht} = \sum_{i \in s} y_i / \pi_i$ is an unbiased estimator of the population total Y.

(ii) Define

$$\pi_{ij} = \text{probability that } i \in s \text{ and } j \in s, \quad i = 1, ..., N, \; j = 1, ..., N.$$

Then

$$\text{Var}(\hat{Y}_{ht}) = \sum_{i=1}^{N} \frac{1 - \pi_i}{\pi_i} y_i^2 + 2 \sum_{i=1}^{N} \sum_{j=i+1}^{N} \frac{\pi_{ij} - \pi_i \pi_j}{\pi_i \pi_j} y_i y_j \qquad (3.47)$$

$$= \sum_{i=1}^{N} \sum_{j=i+1}^{N} (\pi_i \pi_j - \pi_{ij}) \left(\frac{y_i}{\pi_i} - \frac{y_j}{\pi_j} \right)^2. \qquad (3.48)$$

Proof. (i) Let $a_i = 1$ if $i \in s$ and $a_i = 0$ if $i \notin s$, $i = 1, ..., N$. Then $E(a_i) = \pi_i$ and

$$E(\hat{Y}_{ht}) = E \left(\sum_{i=1}^{N} \frac{a_i y_i}{\pi_i} \right) = \sum_{i=1}^{N} y_i = Y.$$

(ii) Since $a_i^2 = a_i$,

$$\text{Var}(a_i) = E(a_i) - [E(a_i)]^2 = \pi_i(1 - \pi_i).$$

For $i \neq j$,

$$\text{Cov}(a_i, a_j) = E(a_i a_j) - E(a_i)E(a_j) = \pi_{ij} - \pi_i \pi_j.$$

Then

$$\text{Var}(\hat{Y}_{ht}) = \text{Var} \left(\sum_{i=1}^{N} \frac{a_i y_i}{\pi_i} \right)$$

$$= \sum_{i=1}^{N} \frac{y_i^2}{\pi_i^2} \text{Var}(a_i) + 2 \sum_{i=1}^{N} \sum_{j=i+1}^{N} \frac{y_i y_j}{\pi_i \pi_j} \text{Cov}(a_i, a_j)$$

$$= \sum_{i=1}^{N} \frac{1 - \pi_i}{\pi_i} y_i^2 + 2 \sum_{i=1}^{N} \sum_{j=i+1}^{N} \frac{\pi_{ij} - \pi_i \pi_j}{\pi_i \pi_j} y_i y_j.$$

Hence (3.47) follows. To show (3.48), note that

$$\sum_{i=1}^{N} \pi_i = n \qquad \text{and} \qquad \sum_{j=1,...,N, j \neq i} \pi_{ij} = (n - 1)\pi_i,$$

which implies

$$\sum_{j=1,...,N, j \neq i} (\pi_{ij} - \pi_i \pi_j) = (n - 1)\pi_i - \pi_i(n - \pi_i) = -\pi_i(1 - \pi_i).$$

Hence

$$\sum_{i=1}^{N} \frac{1-\pi_i}{\pi_i} y_i^2 = \sum_{i=1}^{N} \sum_{j=1,\dots,N, j\neq i} (\pi_i \pi_j - \pi_{ij}) \frac{y_i^2}{\pi_i^2}$$

$$= \sum_{i=1}^{N} \sum_{j=i+1}^{N} (\pi_i \pi_j - \pi_{ij}) \left(\frac{y_i^2}{\pi_i^2} + \frac{y_j^2}{\pi_j^2} \right)$$

and, by (3.47),

$$\text{Var}(\hat{Y}_{ht}) = \sum_{i=1}^{N} \sum_{j=i+1}^{N} (\pi_{ij} - \pi_i \pi_j) \left(\frac{y_i^2}{\pi_i^2} + \frac{y_j^2}{\pi_j^2} - \frac{2 y_i y_j}{\pi_i \pi_j} \right)$$

$$= \sum_{i=1}^{N} \sum_{j=i+1}^{N} (\pi_i \pi_j - \pi_{ij}) \left(\frac{y_i}{\pi_i} - \frac{y_j}{\pi_j} \right)^2. \quad \blacksquare$$

Using the same idea, we can obtain unbiased estimators of $\text{Var}(\hat{Y}_{ht})$. Suppose that $\pi_{ij} > 0$ for all i and j and π_{ij} is known when $i \in s$ and $j \in s$. By (3.47), an unbiased estimator of $\text{Var}(\hat{Y}_{ht})$ is

$$v_1 = \sum_{i \in s} \frac{1 - \pi_i}{\pi_i^2} y_i^2 + 2 \sum_{i \in s} \sum_{j \in s, j > i} \frac{\pi_{ij} - \pi_i \pi_j}{\pi_i \pi_j \pi_{ij}} y_i y_j. \qquad (3.49)$$

By (3.48), an unbiased estimator of $\text{Var}(\hat{Y}_{ht})$ is

$$v_2 = \sum_{i \in s} \sum_{j \in s, j > i} \frac{\pi_i \pi_j - \pi_{ij}}{\pi_{ij}} \left(\frac{y_i}{\pi_i} - \frac{y_j}{\pi_j} \right)^2. \qquad (3.50)$$

Variance estimators v_1 and v_2 may not be the same in general, but they are the same in some special cases (Exercise 92). A more serious problem is that they may take negative values. Some discussions about deriving better estimators of $\text{Var}(\hat{Y}_{ht})$ are provided in Cochran (1977, Chapter 9A).

Some special cases of Theorem 3.15 are considered as follows.

Under simple random sampling, $\pi_i = n/N$. Thus, $\hat{Y}$ in Example 3.19 is the Horvitz-Thompson estimator.

Under stratified simple random sampling, $\pi_i = n_h/N_h$ if unit i is in stratum h. Hence, the estimator $\hat{Y}_{st}$ in Example 3.20 is the Horvitz-Thompson estimator.

Suppose now each $y_i \in \mathcal{P}$ is a cluster, i.e., $y_i = (y_{i1}, \dots, y_{iM_i})$, where M_i is the size of the ith cluster, $i = 1, \dots, N$. The total number of units in $\mathcal{P}$ is then $M = \sum_{i=1}^{N} M_i$. Consider a single-stage sampling plan, i.e., if y_i is selected, then every y_{ij} is observed. If simple random sampling is used, then $\pi_i = k/N$, where k is the first-stage sample size (the total sample size is $n = \sum_{i=1}^{k} M_i$), and the Horvitz-Thompson estimator is

$$\hat{Y}_s = \frac{N}{k} \sum_{i \in \mathbf{s}_1} \sum_{j=1}^{M_i} y_{ij} = \frac{N}{k} \sum_{i \in \mathbf{s}_1} Y_i,$$

where $\mathbf{s}_1$ is the index set of first-stage sampled clusters and Y_i is the total of the ith cluster. In this case,

$$\text{Var}(\hat{Y}_s) = \frac{N^2}{k(N-1)} \left(1 - \frac{k}{N}\right) \sum_{i=1}^{N} \left(Y_i - \frac{Y}{N}\right)^2.$$

If the selection probability is proportional to the cluster size, then $\pi_i = kM_i/M$ and the Horvitz-Thompson estimator is

$$\hat{Y}_{pps} = \frac{M}{k} \sum_{i \in \mathbf{s}_1} \frac{1}{M_i} \sum_{j=1}^{M_i} y_{ij} = \frac{M}{k} \sum_{i \in \mathbf{s}_1} \frac{Y_i}{M_i}$$

whose variance is given by (3.47) or (3.48). Usually $\text{Var}(\hat{Y}_{pps})$ is smaller than $\text{Var}(\hat{Y}_s)$; see the discussions in Cochran (1977, Chapter 9A).

Consider next a two-stage sampling in which k first-stage clusters are selected and a simple random sample of size m_i is selected from each sampled cluster y_i, where sampling is independent across clusters. If the first-stage sampling plan is simple random sampling, then $\pi_i = km_i/(NM_i)$ and the Horvitz-Thompson estimator is

$$\hat{Y}_s = \frac{N}{k} \sum_{i \in \mathbf{s}_1} \frac{M_i}{m_i} \sum_{j \in \mathbf{s}_{2i}} y_{ij},$$

where $\mathbf{s}_{2i}$ denotes the second-stage sample from cluster i. If the first-stage selection probability is proportional to the cluster size, then $\pi_i = km_i/M$ and the Horvitz-Thompson estimator is

$$\hat{Y}_{pps} = \frac{M}{k} \sum_{i \in \mathbf{s}_1} \frac{1}{m_i} \sum_{j \in \mathbf{s}_{2i}} y_{ij}.$$

Finally, let us consider another popular sampling method called *systematic sampling*. Suppose that $\mathcal{P} = \{y_1, ..., y_N\}$ and the population size $N = nk$ for two integers n and k. To select a sample of size n, we first draw a j randomly from $\{1, ..., k\}$. Our sample is then

$$\{y_j, y_{j+k}, y_{j+2k}, ..., y_{j+(n-1)k}\}.$$

Systematic sampling is used mainly because it is easier to draw a systematic sample and often easier to execute without mistakes. It is also likely that systematic sampling provides more efficient point estimators than simple random sampling or even stratified sampling, since the sample units are spread more evenly over the population. Under systematic sampling, $\pi_i = k^{-1}$ for every i and the Horvitz-Thompson estimator of the population total is

$$\hat{Y}_{sy} = k \sum_{t=1}^{n} y_{j+(t-1)k}.$$

The unbiasedness of this estimator is a direct consequence of Theorem 3.15, but it can be easily shown as follows. Since j takes value $i \in \{1, ..., k\}$ with probability k^{-1},

$$E(\hat{Y}_{sy}) = k \left(\frac{1}{k} \sum_{i=1}^{k} \sum_{t=1}^{n} y_{i+(t-1)k} \right) = \sum_{i=1}^{N} y_i = Y.$$

The variance of $\hat{Y}_{sy}$ is simply

$$\mathrm{Var}(\hat{Y}_{sy}) = \frac{N^2}{k} \sum_{i=1}^{k} (\mu_i - \mu)^2,$$

where $\mu_i = n^{-1} \sum_{t=1}^{n} y_{i+(t-1)k}$ and $\mu = k^{-1} \sum_{i=1}^{k} \mu_i = Y/N$. Let σ^2 be given in (3.43) and

$$\sigma_{sy}^2 = \frac{1}{k(n-1)} \sum_{i=1}^{k} \sum_{t=1}^{n} (y_{i+(t-1)k} - \mu_i)^2.$$

Then

$$(N-1)\sigma^2 = n \sum_{i=1}^{k} (\mu_i - \mu)^2 + \sum_{i=1}^{k} \sum_{t=1}^{n} (y_{i+(t-1)k} - \mu_i)^2.$$

Thus,

$$(N-1)\sigma^2 = N^{-1} \mathrm{Var}(\hat{Y}_{sy}) + k(n-1)\sigma_{sy}^2$$

and

$$\mathrm{Var}(\hat{Y}_{sy}) = N(N-1)\sigma^2 - N(N-k)\sigma_{sy}^2.$$

Since the variance of the Horvitz-Thompson estimator of the population total under simple random sampling is, by (3.43),

$$\frac{N^2}{n} \left(1 - \frac{n}{N} \right) \sigma^2 = N(k-1)\sigma^2,$$

the Horvitz-Thompson estimator under systematic sampling has a smaller variance if and only if $\sigma_{sy}^2 > \sigma^2$.

3.5 Asymptotically Unbiased Estimators

As we discussed in §2.5, we often need to consider biased but asymptotically unbiased estimators. A large and useful class of such estimators are smooth functions of some exactly unbiased estimators such as UMVUE's, U-statistics, LSE's, and Horvitz-Thompson estimators. Some other methods of constructing asymptotically unbiased estimators are also introduced in this section.

3.5.1 Functions of unbiased estimators

If the parameter to be estimated is $\vartheta = g(\theta)$ with a vector-valued parameter θ and U_n is a vector of unbiased estimators of components of θ (i.e., $EU_n = \theta$), then $T_n = g(U_n)$ is often asymptotically unbiased for ϑ. Assume that g is differentiable and $c_n(U_n - \theta) \to_d Y$. Then

$$\text{amse}_{T_n}(P) = E\{[\nabla g(\theta)]^\tau Y\}^2 / c_n^2$$

(Theorem 2.6). Hence, T_n has a good performance in terms of amse if U_n is optimal in terms of mse (such as the UMVUE).

The following are some examples.

Example 3.21 (Ratio estimators). Let $(X_1, Y_1), ..., (X_n, Y_n)$ be i.i.d. random 2-vectors with $EX_1 = \mu_x$ and $EY_1 = \mu_y$. Consider the estimation of the ratio of two population means: $\vartheta = \mu_y / \mu_x$ ($\mu_x \neq 0$). Note that $(\bar{Y}, \bar{X})$, the vector of sample means, is unbiased for (μ_y, μ_x). The sample means are UMVUE's under some statistical models (§3.1 and §3.2) and are BLUE's in general (Example 2.22). The *ratio* estimator is $T_n = \bar{Y}/\bar{X}$. Assume that $\sigma_x^2 = \text{Var}(X_1)$, $\sigma_y^2 = \text{Var}(Y_1)$, and $\sigma_{xy} = \text{Cov}(X_1, Y_1)$ exist. A direct calculation shows that the n^{-1} order asymptotic bias of T_n according to (2.38) is

$$\tilde{b}_{T_n}(P) = \frac{\vartheta \sigma_x^2 - \sigma_{xy}}{\mu_x^2 n}$$

(verify). Using the CLT and the delta-method (Corollary 1.1), we obtain that

$$\sqrt{n}(T_n - \vartheta) \to_d N\left(0, \frac{\sigma_y^2 - 2\vartheta\sigma_{xy} + \vartheta^2 \sigma_x^2}{\mu_x^2}\right)$$

(verify), which implies

$$\text{amse}_{T_n}(P) = \frac{\sigma_y^2 - 2\vartheta\sigma_{xy} + \vartheta^2 \sigma_x^2}{\mu_x^2 n}.$$

In some problems, we are not interested in the ratio, but the use of a ratio estimator to improve an estimator of a marginal mean. For example,

suppose that μ_x is known and we are interested in estimating μ_y. Consider the following estimator:
$$\hat{\mu}_y = (\bar{Y}/\bar{X})\mu_x.$$

Note that $\hat{\mu}_y$ is not unbiased; its n^{-1} order asymptotic bias according to (2.38) is
$$\tilde{b}_{\hat{\mu}_y}(P) = \frac{\vartheta\sigma_x^2 - \sigma_{xy}}{\mu_x n};$$

and
$$\mathrm{amse}_{\hat{\mu}_y}(P) = \frac{\sigma_y^2 - 2\vartheta\sigma_{xy} + \vartheta^2\sigma_x^2}{n}.$$

Comparing $\hat{\mu}_y$ with the unbiased estimator $\bar{Y}$, we find that $\hat{\mu}_y$ is asymptotically more efficient if and only if
$$2\vartheta\sigma_{xy} > \vartheta^2\sigma_x^2,$$

which means that $\hat{\mu}_y$ is a better estimator if and only if the correlation between X_1 and Y_1 is large enough to pay off the extra variability caused by using $\mu_x/\bar{X}$. ∎

Another example related to a bivariate sample is the sample correlation coefficient defined in Exercise 22 in §2.6.

Example 3.22. Consider a polynomial regression of order p:
$$X_i = \beta^\tau Z_i + \varepsilon_i, \qquad i = 1, ..., n,$$

where $\beta = (\beta_0, \beta_1, ..., \beta_{p-1})$, $Z_i = (1, t_i, ..., t_i^{p-1})$, and ε_i's are i.i.d. with mean 0 and variance $\sigma^2 > 0$. Suppose that the parameter to be estimated is $t_\beta \in \mathcal{T} \subset \mathcal{R}$ such that
$$\sum_{j=0}^{p-1} \beta_j t_\beta^j = \max_{t \in \mathcal{T}} \sum_{j=0}^{p-1} \beta_j t^j.$$

Note that $t_\beta = g(\beta)$ for some function g. Let $\hat{\beta}$ be the LSE of β. Then the estimator $\hat{t}_\beta = g(\hat{\beta})$ is asymptotically unbiased and its amse can be derived under some conditions (Exercise 98). ∎

Example 3.23. In the study of the reliability of a system component, we assume that
$$X_{ij} = \theta_i^\tau z(t_j) + \varepsilon_{ij}, \quad i = 1, ..., k, \ j = 1, ..., m.$$

Here X_{ij} is the measurement of the ith sample component at time t_j; $z(t)$ is a q-vector whose components are known functions of the time t; θ_i's are unobservable random q-vectors that are i.i.d. from $N_q(\theta, \Sigma)$, where θ and Σ are unknown; ε_{ij}'s are i.i.d. measurement errors with mean zero and variance σ^2; and θ_i's and ε_{ij}'s are independent. As a function of t, $\theta^\tau z(t)$ is the degradation curve for a particular component and $\theta^\tau z(t)$ is the mean degradation curve. Suppose that a component will fail to work if $\theta^\tau z(t) < \eta$, a given critical value. Assume that $\theta^\tau z(t)$ is always a decreasing function of t. Then the reliability function of a component is

$$R(t) = P(\theta^\tau z(t) > \eta) = \Phi\left(\frac{\theta^\tau z(t) - \eta}{s(t)}\right),$$

where $s(t) = \sqrt{[z(t)]^\tau \Sigma z(t)}$ and Φ is the standard normal distribution function. For a fixed t, estimators of $R(t)$ can be obtained by estimating θ and Σ, since Φ is a known function. It can be shown (exercise) that the BLUE of θ is the LSE

$$\hat{\theta} = (Z^\tau Z)^{-1} Z^\tau \bar{X},$$

where Z is the $m \times q$ matrix whose jth row is the vector $z(t_j)$, $X_i = (X_{i1}, ..., X_{im})$, and $\bar{X}$ is the sample mean of X_i's. The estimation of Σ is more difficult. It can be shown (exercise) that a consistent (as $k \to \infty$) estimator of Σ is

$$\hat{\Sigma} = \frac{1}{k} \sum_{i=1}^{k} (Z^\tau Z)^{-1} Z^\tau (X_i - \bar{X})(X_i - \bar{X})^\tau Z(Z^\tau Z)^{-1} - \hat{\sigma}^2 (Z^\tau Z)^{-1},$$

where

$$\hat{\sigma}^2 = \frac{1}{k(m-q)} \sum_{i=1}^{k} [X_i^\tau X_i - X_i^\tau Z(Z^\tau Z)^{-1} Z^\tau X_i].$$

Hence an estimator of $R(t)$ is

$$\hat{R}(t) = \Phi\left(\frac{\hat{\theta}^\tau z(t) - \eta}{\hat{s}(t)}\right),$$

where

$$\hat{s}(t) = \sqrt{[z(t)]^\tau \hat{\Sigma} z(t)}.$$

If we define $Y_{i1} = X_i^\tau Z(Z^\tau Z)^{-1} z(t)$, $Y_{i2} = [X_i^\tau Z(Z^\tau Z)^{-1} z(t)]^2$, $Y_{i3} = [X_i^\tau X_i - X_i^\tau Z(Z^\tau Z)^{-1} Z^\tau X_i]/(m-q)$, and $Y_i = (Y_{i1}, Y_{i2}, Y_{i3})$, then it is apparent that $\hat{R}(t)$ can be written as $g(\bar{Y})$ for a function

$$g(y_1, y_2, y_3) = \Phi\left(\frac{y_1 - \eta}{\sqrt{y_2 - y_1^2 - y_3[z(t)]^\tau (Z^\tau Z)^{-1} z(t)}}\right).$$

Suppose that ε_{ij} has a finite fourth moment, which implies the existence of $\mathrm{Var}(Y_i)$. The amse of $\hat{R}(t)$ can be derived (exercise). ∎

3.5.2 The method of moments

The method of moments is the oldest method of deriving point estimators. It almost always produces some asymptotically unbiased estimators, although they may not be the best estimators.

Consider a parametric problem where $X_1, ..., X_n$ are i.i.d. random variables from P_θ, $\theta \in \Theta \subset \mathcal{R}^k$, and $E|X_1|^k < \infty$. Let $\mu_j = EX_1^j$ be the jth moment of P and let

$$\hat{\mu}_j = \frac{1}{n} \sum_{i=1}^n X_i^j$$

be the jth *sample moment*, which is an unbiased estimator of μ_j, $j = 1, ..., k$. Typically,

$$\mu_j = h_j(\theta), \qquad j = 1, ..., k, \tag{3.51}$$

for some functions h_j on $\mathcal{R}^k$. By substituting μ_j's on the left-hand side of (3.51) by the sample moments $\hat{\mu}_j$, we obtain a *moment estimator* $\hat{\theta}$, i.e., $\hat{\theta}$ satisfies

$$\hat{\mu}_j = h_j(\hat{\theta}), \qquad j = 1, ..., k,$$

which is a sample analogue of (3.51). This method of deriving estimators is called the *method of moments*. Note that an important statistical principle, the *substitution principle*, is applied in this method.

Let $\hat{\mu} = (\hat{\mu}_1, ..., \hat{\mu}_k)$ and $h = (h_1, ..., h_k)$. Then $\hat{\mu} = h(\hat{\theta})$. If the inverse function h^{-1} exists, then the unique moment estimator of θ is $\hat{\theta} = h^{-1}(\hat{\mu})$. When h^{-1} does not exist (i.e., h is not one-to-one), any solution of $\hat{\mu} = h(\hat{\theta})$ is a moment estimator of θ; if possible, we always choose a solution $\hat{\theta}$ in the parameter space Θ. In some cases, however, a moment estimator does not exist (see Exercise 111).

Assume that $\hat{\theta} = g(\hat{\mu})$ for a function g. If h^{-1} exists, then $g = h^{-1}$. If g is continuous at $\mu = (\mu_1, ..., \mu_k)$, then $\hat{\theta}$ is strongly consistent for θ, since $\hat{\mu}_j \to_{a.s.} \mu_j$ by the SLLN. If g is differentiable at μ and $E|X_1|^{2k} < \infty$, then $\hat{\theta}$ is asymptotically normal, by the CLT and Theorem 1.12, and

$$\text{amse}_{\hat{\theta}}(\theta) = n^{-1} [\nabla g(\mu)]^\tau V_\mu \nabla g(\mu),$$

where V_μ is a $k \times k$ matrix whose (i, j)th element is $\mu_{i+j} - \mu_i \mu_j$. Furthermore, it follows from (2.38) that the n^{-1} order asymptotic bias of $\hat{\theta}$ is

$$(2n)^{-1} \text{tr} \left(\nabla^2 g(\mu) V_\mu \right).$$

Example 3.24. Let $X_1, ..., X_n$ be i.i.d. from a population P_θ indexed by the parameter $\theta = (\mu, \sigma^2)$, where $\mu = EX_1 \in \mathcal{R}$ and $\sigma^2 = \text{Var}(X_1) \in (0, \infty)$. This includes cases such as the family of normal distributions, double exponential distributions, or logistic distributions (Table 1.2, page 20). Since $EX_1 = \mu$ and $EX_1^2 = \text{Var}(X_1) + (EX_1)^2 = \sigma^2 + \mu^2$, setting $\hat\mu_1 = \mu$ and $\hat\mu_2 = \sigma^2 + \mu^2$ we obtain the moment estimator

$$\hat\theta = \left(\bar X,\ \frac{1}{n}\sum_{i=1}^n (X_i - \bar X)^2\right) = \left(\bar X,\ \frac{n-1}{n}S^2\right).$$

Note that $\bar X$ is unbiased, but $\frac{n-1}{n}S^2$ is not. If X_i is normal, then $\hat\theta$ is suffi-cient and is nearly the same as an optimal estimator such as the UMVUE. On the other hand, if X_i is from a double exponential or logistic distribu-tion, then $\hat\theta$ is not sufficient and can often be improved.

Consider now the estimation of σ^2 when we know that $\mu = 0$. Obviously we cannot use the equation $\hat\mu_1 = \mu$ to solve the problem. Using $\hat\mu_2 = \mu_2 = \sigma^2$, we obtain the moment estimator $\hat\sigma^2 = \hat\mu_2 = n^{-1}\sum_{i=1}^n X_i^2$. This is still a good estimator when X_i is normal, but is not a function of sufficient statistic when X_i is from a double exponential distribution. For the double exponential case one can argue that we should first make a transformation $Y_i = |X_i|$ and then obtain the moment estimator based on the transformed data. The moment estimator of σ^2 based on the transformed data is $\bar Y^2 = (n^{-1}\sum_{i=1}^n |X_i|)^2$, which is sufficient for σ^2. Note that this estimator can also be obtained based on absolute moment equations. ∎

Example 3.25. Let $X_1, ..., X_n$ be i.i.d. from the uniform distribution on (θ_1, θ_2), $-\infty < \theta_1 < \theta_2 < \infty$. Note that

$$EX_1 = (\theta_1 + \theta_2)/2$$

and

$$EX_1^2 = (\theta_1^2 + \theta_2^2 + \theta_1\theta_2)/3.$$

Setting $\hat\mu_1 = EX_1$ and $\hat\mu_2 = EX_1^2$ and substituting θ_1 in the second equa-tion by $2\hat\mu_1 - \theta_2$ (the first equation), we obtain that

$$(2\hat\mu_1 - \theta_2)^2 + \theta_2^2 + (2\hat\mu_1 - \theta_2)\theta_2 = 3\hat\mu_2,$$

which is the same as

$$(\theta_2 - \hat\mu_1)^2 = 3(\hat\mu_2 - \hat\mu_1^2).$$

Since $\theta_2 > EX_1$, we obtain that

$$\hat\theta_2 = \hat\mu_1 + \sqrt{3(\hat\mu_2 - \hat\mu_1^2)} = \bar X + \sqrt{\tfrac{3(n-1)}{n}S^2}$$

and

$$\hat\theta_1 = \hat\mu_1 - \sqrt{3(\hat\mu_2 - \hat\mu_1^2)} = \bar X - \sqrt{\tfrac{3(n-1)}{n}S^2}.$$

These estimators are not functions of the sufficient and complete statistic $(X_{(1)}, X_{(n)})$. ∎

Example 3.26. Let $X_1, ..., X_n$ be i.i.d. from the binomial distribution $Bi(p, k)$ with unknown parameters $k \in \{1, 2, ...\}$ and $p \in (0, 1)$. Since

$$EX_1 = kp$$

and

$$EX_1^2 = kp(1 - p) + k^2 p^2,$$

we obtain the moment estimators

$$\hat{p} = (\hat{\mu}_1 + \hat{\mu}_1^2 - \hat{\mu}_2)/\hat{\mu}_1 = 1 - \tfrac{n-1}{n} S^2/\bar{X}$$

and

$$\hat{k} = \hat{\mu}_1^2/(\hat{\mu}_1 + \hat{\mu}_1^2 - \hat{\mu}_2) = \bar{X}/(1 - \tfrac{n-1}{n} S^2/\bar{X}).$$

The estimator $\hat{p}$ is in the range of $(0, 1)$. But $\hat{k}$ may not be an integer. It can be improved by an estimator that is $\hat{k}$ rounded to the nearest positive integer. ∎

Example 3.27. Suppose that $X_1, ..., X_n$ are i.i.d. from the Pareto distribution $Pa(a, \theta)$ with unknown $a > 0$ and $\theta > 2$ (Table 1.2, page 20). Note that

$$EX_1 = \theta a/(\theta - 1)$$

and

$$EX_1^2 = \theta a^2/(\theta - 2).$$

From the moment equation,

$$\frac{(\theta-1)^2}{\theta(\theta-2)} = \hat{\mu}_2/\hat{\mu}_1^2.$$

Note that $\frac{(\theta-1)^2}{\theta(\theta-2)} - 1 = \frac{1}{\theta(\theta-2)}$. Hence

$$\theta(\theta - 2) = \hat{\mu}_1^2/(\hat{\mu}_2 - \hat{\mu}_1^2).$$

Since $\theta > 2$, there is a unique solution in the parameter space:

$$\hat{\theta} = 1 + \sqrt{\hat{\mu}_2/(\hat{\mu}_2 - \hat{\mu}_1^2)} = 1 + \sqrt{1 + \tfrac{n}{n-1} \bar{X}^2/S^2}$$

and

$$\hat{a} = \frac{\hat{\mu}_1(\hat{\theta} - 1)}{\hat{\theta}}$$
$$= \bar{X}\sqrt{1 + \tfrac{n}{n-1}\bar{X}^2/S^2} \Big/ \left(1 + \sqrt{1 + \tfrac{n}{n-1}\bar{X}^2/S^2}\right). ∎$$

The method of moments can also be applied to nonparametric problems. Consider, for example, the estimation of the central moments

$$c_j = E(X_1 - \mu_1)^j, \qquad j = 2, ..., k.$$

Since

$$c_j = \sum_{t=0}^{j} \binom{j}{t} (-\mu_1)^t \mu_{j-t},$$

the moment estimator of c_j is

$$\hat{c}_j = \sum_{t=0}^{j} \binom{j}{t} (-\bar{X})^t \hat{\mu}_{j-t},$$

where $\hat{\mu}_0 = 1$. It can be shown (exercise) that

$$\hat{c}_j = \frac{1}{n} \sum_{i=1}^{n} (X_i - \bar{X})^j, \qquad j = 2, ..., k, \tag{3.52}$$

which are sample central moments. From the SLLN, $\hat{c}_j$'s are strongly consistent. If $E|X_1|^{2k} < \infty$, then

$$\sqrt{n} \, (\hat{c}_2 - c_2, ..., \hat{c}_k - c_k) \rightarrow_d N_{k-1}(0, D) \tag{3.53}$$

(exercise), where the (i, j)th element of the $(k - 1) \times (k - 1)$ matrix D is

$$c_{i+j+2} - c_{i+1} c_{j+1} - (i + 1) c_i c_{j+2} - (j + 1) c_{i+2} c_j + (i + 1)(j + 1) c_i c_j c_2.$$

3.5.3 V-statistics

Let $X_1, ..., X_n$ be i.i.d. from P. For every U-statistic U_n defined in (3.11) as an estimator of $\vartheta = E[h(X_1, ..., X_m)]$, there is a closely related *V-statistic* defined by

$$V_n = \frac{1}{n^m} \sum_{i_1=1}^{n} \cdots \sum_{i_m=1}^{n} h(X_{i_1}, ..., X_{i_m}). \tag{3.54}$$

As an estimator of ϑ, V_n is biased; but the bias is small asymptotically as the following results show. For a fixed sample size n, V_n may be better than U_n in terms of their mse's. Consider, for example, the kernel $h(x_1, x_2) = (x_1 - x_2)^2/2$ in §3.2.1, which leads to $\vartheta = \sigma^2 = \text{Var}(X_1)$ and $U_n = S^2$, the sample variance. The corresponding V-statistic is

$$\frac{1}{n^2} \sum_{i=1}^{n} \sum_{j=1}^{n} \frac{(X_i - X_j)^2}{2} = \frac{1}{n^2} \sum_{1 \leq i < j \leq n} (X_i - X_j)^2 = \frac{n-1}{n} S^2,$$

which is the moment estimator of σ^2 discussed in Example 3.24. In Exercise 63 in §2.6, $\frac{n-1}{n}S^2$ is shown to have a smaller mse than S^2 when X_i is normally distributed. Of course, there are situations where U-statistics are better than their corresponding V-statistics.

The following result provides orders of magnitude of the bias and variance of a V-statistic as an estimator of ϑ.

Proposition 3.5. Let V_n be defined by (3.54).
(i) Assume that $E|h(X_{i_1}, ..., X_{i_m})| < \infty$ for all $1 \le i_1 \le \cdots \le i_m \le m$. Then the bias of V_n satisfies

$$b_{V_n}(P) = O(n^{-1}).$$

(ii) Assume that $E[h(X_{i_1}, ..., X_{i_m})]^2 < \infty$ for all $1 \le i_1 \le \cdots \le i_m \le m$. Then the variance of V_n satisfies

$$\text{Var}(V_n) = \text{Var}(U_n) + O(n^{-2}),$$

where U_n is given by (3.11).
Proof. (i) Note that

$$U_n - V_n = \left[1 - \frac{n!}{n^m(n-m)!}\right](U_n - W_n), \tag{3.55}$$

where W_n is the average of all terms $h(X_{i_1}, ..., X_{i_m})$ with at least one equality $i_m = i_l$, $m \ne l$. The result follows from $E(U_n - W_n) = O(1)$.
(ii) The result follows from $E(U_n - W_n)^2 = O(1)$, $E[W_n(U_n - \vartheta)] = O(n^{-1})$ (exercise), and (3.55).   ∎

To study the asymptotic behavior of a V-statistic, we consider the following representation of V_n in (3.54):

$$V_n = \sum_{j=1}^{m} \binom{m}{j} V_{nj},$$

where

$$V_{nj} = \vartheta + \frac{1}{n^j} \sum_{i_1=1}^{n} \cdots \sum_{i_j=1}^{n} g_j(X_{i_1}, ..., X_{i_j})$$

is a "V-statistic" with

$$g_j(x_1, ..., x_j) = h_j(x_1, ..., x_j) - \sum_{i=1}^{j} \int h_j(x_1, ..., x_j) dP(x_i)$$

$$+ \sum_{1 \le i_1 < i_2 \le j} \int \int h_j(x_1, ..., x_j) dP(x_{i_1}) dP(x_{i_2}) - \cdots$$

$$+ (-1)^j \int \cdots \int h_j(x_1, ..., x_j) dP(x_1) \cdots dP(x_j)$$

and $h_j(x_1, ..., x_j) = E[h(x_1, ..., x_j, X_{j+1}, ..., X_m)]$. Using an argument similar to the proof of Theorem 3.4, we can show (exercise) that

$$EV_{nj}^2 = O(n^{-j}), \qquad j = 1, ..., m, \tag{3.56}$$

provided that $E[h(X_{i_1}, ..., X_{i_m})]^2 < \infty$ for all $1 \le i_1 \le \cdots \le i_m \le m$. Thus,

$$V_n - \vartheta = mV_{n1} + \frac{m(m-1)}{2}V_{n2} + o_p(n^{-1}), \tag{3.57}$$

which leads to the following result similar to Theorem 3.5.

Theorem 3.16. Let V_n be given by (3.54) with $E[h(X_{i_1}, ..., X_{i_m})]^2 < \infty$ for all $1 \le i_1 \le \cdots \le i_m \le m$.
(i) If $\zeta_1 = \text{Var}(h_1(X_1)) > 0$, then

$$\sqrt{n}(V_n - \vartheta) \to_d N(0, m^2\zeta_1).$$

(ii) If $\zeta_1 = 0$ but $\zeta_2 = \text{Var}(h_2(X_1, X_2)) > 0$, then

$$n(V_n - \vartheta) \to_d \frac{m(m-1)}{2} \sum_{j=1}^{\infty} \lambda_j \chi_{1j}^2,$$

where χ_{1j}^2's and λ_j's are the same as those in (3.21).   ∎

Result (3.57) and Theorem 3.16 imply that V_n has expansion (2.37) and, therefore, the n^{-1} order asymptotic bias of V_n is $E[g_2(X_1, X_1)]/n = nEV_{n2} = m(m-1)\sum_{j=1}^{\infty} \lambda_j/(2n)$ (exercise).

Theorem 3.16 shows that if $\zeta_1 > 0$, then the amse's of U_n and V_n are the same. If $\zeta_1 = 0$ but $\zeta_2 > 0$, then an argument similar to that in the proof of Lemma 3.2 leads to

$$\text{amse}_{V_n}(P) = \frac{m^2(m-1)^2\zeta_2}{2n^2} + \frac{m^2(m-1)^2}{4n^2}\left(\sum_{j=1}^{\infty} \lambda_j\right)^2$$

$$= \text{amse}_{U_n}(P) + \frac{m^2(m-1)^2}{4n^2}\left(\sum_{j=1}^{\infty} \lambda_j\right)^2$$

(see Lemma 3.2). Hence U_n is asymptotically more efficient than V_n, unless $\sum_{j=1}^{\infty} \lambda_j = 0$. Technically, the proof of the asymptotic results for V_n also requires moment conditions stronger than those for U_n.

Example 3.28. Consider the estimation of μ^2, where $\mu = EX_1$. From the results in §3.2, the U-statistic $U_n = \frac{1}{n(n-1)}\sum_{1 \le i < j \le n} X_i X_j$ is unbiased for

μ^2. The corresponding V-statistic is simply $V_n = \bar{X}^2$. If $\mu \neq 0$, then $\zeta_1 \neq 0$ and the asymptotic relative efficiency of V_n w.r.t. U_n is 1. If $\mu = 0$, then

$$nV_n \to_d \sigma^2 \chi_1^2 \qquad \text{and} \qquad nU_n \to_d \sigma^2(\chi_1^2 - 1),$$

where χ_1^2 is a random variable having the chi-square distribution χ_1^2. Hence the asymptotic relative efficiency of V_n w.r.t. U_n is

$$E(\chi_1^2 - 1)^2/E(\chi_1^2)^2 = 2/3. \quad \blacksquare$$

3.5.4 The weighted LSE

In linear model (3.25), the unbiased LSE of $l^\tau\beta$ may be improved by a slightly biased estimator when $\mathrm{Var}(\varepsilon)$ is not $\sigma^2 I_n$ and the LSE is not BLUE.

Assume that Z in (3.25) is of full rank so that every $l^\tau\beta$ is estimable. For simplicity, let us denote $\mathrm{Var}(\varepsilon)$ by V. If V is known, then the BLUE of $l^\tau\beta$ is $l^\tau\check{\beta}$, where

$$\check{\beta} = (Z^\tau V^{-1} Z)^{-1} Z^\tau V^{-1} X \tag{3.58}$$

(see the discussion after the statement of assumption A3 in §3.3.1). If V is unknown and $\hat{V}$ is an estimator of V, then an application of the substitution principle leads to a *weighted least squares estimator*

$$\hat{\beta}_w = (Z^\tau \hat{V}^{-1} Z)^{-1} Z^\tau \hat{V}^{-1} X. \tag{3.59}$$

The weighted LSE is not linear in X and not necessarily unbiased for β. If the distribution of ε is symmetric about 0 and $\hat{V}$ remains unchanged when ε changes to $-\varepsilon$ (Examples 3.29 and 3.30), then the distribution of $\hat{\beta}_w - \beta$ is symmetric about 0 and, if $E\hat{\beta}_w$ is well defined, $\hat{\beta}_w$ is unbiased for β. In such a case the LSE $l^\tau\hat{\beta}$ may not be a UMVUE (when ε is normal), since $\mathrm{Var}(l^\tau\hat{\beta}_w)$ may be smaller than $\mathrm{Var}(l^\tau\hat{\beta})$.

Asymptotic properties of the weighted LSE depend on the asymptotic behavior of $\hat{V}$. We say that $\hat{V}$ is consistent for V if and only if

$$\|\hat{V}^{-1}V - I_n\|_{\max} \to_p 0, \tag{3.60}$$

where $\|A\|_{\max} = \max_{i,j}|a_{ij}|$ for a matrix A whose (i,j)th element is a_{ij}.

Theorem 3.17. Consider model (3.25) with a full rank Z. Let $\check{\beta}$ and $\hat{\beta}_w$ be defined by (3.58) and (3.59), respectively, with a $\hat{V}$ consistent in the sense of (3.60). Assume the conditions in Theorem 3.12. Then

$$l^\tau(\hat{\beta}_w - \beta)/a_n \to_d N(0, 1),$$

where $l \in \mathcal{R}^p$, $l \neq 0$, and

$$a_n^2 = \text{Var}(l^\tau \breve{\beta}) = l^\tau (Z^\tau V^{-1} Z)^{-1} l.$$

Proof. Using the same argument as in the proof of Theorem 3.12, we obtain that

$$l^\tau (\breve{\beta} - \beta)/a_n \to_d N(0,1).$$

By Slutsky's theorem, the result follows from

$$l^\tau \hat{\beta}_w - l^\tau \breve{\beta} = o_p(a_n). \tag{3.61}$$

Define

$$\xi_n = l^\tau (Z^\tau \hat{V}^{-1} Z)^{-1} Z^\tau (\hat{V}^{-1} - V^{-1})\varepsilon$$

and

$$\zeta_n = l^\tau [(Z^\tau \hat{V}^{-1} Z)^{-1} - (Z^\tau V^{-1} Z)^{-1}] Z^\tau V^{-1} \varepsilon.$$

Then

$$l^\tau \hat{\beta}_w - l^\tau \breve{\beta} = \xi_n + \zeta_n.$$

Let $B_n = (Z^\tau \hat{V}^{-1} Z)^{-1} Z^\tau V^{-1} Z - I_p$ and $C_n = \hat{V}^{1/2} V^{-1} \hat{V}^{1/2} - I_n$. By (3.60), $\|C_n\|_{\max} = o_p(1)$. For any matrix A, denote $\sqrt{\text{tr}(A^\tau A)}$ by $\|A\|$. Then

$$\begin{aligned}
\|B_n\|^2 &= \|(Z^\tau \hat{V}^{-1} Z)^{-1} Z^\tau \hat{V}^{-1/2} C_n \hat{V}^{-1/2} Z\|^2 \\
&= \text{tr}\left((Z^\tau \hat{V}^{-1} Z)^{-1} (Z^\tau \hat{V}^{-1/2} C_n \hat{V}^{-1/2} Z)^2 (Z^\tau \hat{V}^{-1} Z)^{-1} \right) \\
&\leq \|C_n\|_{\max}^2 \text{tr}\left((Z^\tau \hat{V}^{-1} Z)^{-1} (Z^\tau \hat{V}^{-1} Z)^2 (Z^\tau \hat{V}^{-1} Z)^{-1} \right) \\
&= o_p(1)\text{tr}(I_p).
\end{aligned}$$

This proves that $\|B_n\|_{\max} = o_p(1)$. Let $A_n = V^{1/2} \hat{V}^{-1} V^{1/2} - I_n$. Using inequality (1.37) and the previous results, we obtain that

$$\begin{aligned}
\xi_n^2 &= [l^\tau (Z^\tau \hat{V}^{-1} Z)^{-1} Z^\tau V^{-1/2} A_n V^{-1/2} \varepsilon]^2 \\
&\leq \varepsilon V^{-1} \varepsilon^\tau l^\tau (Z^\tau \hat{V}^{-1} Z)^{-1} Z^\tau V^{-1/2} A_n^2 V^{-1/2} Z (Z^\tau \hat{V}^{-1} Z)^{-1} l \\
&\leq O_p(1) \|A_n\|_{\max}^2 l^\tau (Z^\tau \hat{V}^{-1} Z)^{-1} Z^\tau V^{-1} Z (Z^\tau \hat{V}^{-1} Z)^{-1} l \\
&= o_p(1) l^\tau (B_n + I_p)^2 (Z^\tau V^{-1} Z)^{-1} l \\
&= o_p(a_n).
\end{aligned}$$

Since $E\|(Z^\tau V^{-1} Z)^{-1/2} Z^\tau V^{-1} \varepsilon\|^2 = p$, $\|(Z^\tau V^{-1} Z)^{-1/2} Z^\tau V^{-1} \varepsilon\| = O_p(1)$. Define $B_{1n} = (Z^\tau \hat{V}^{-1} Z)^{1/2} B_n (Z^\tau \hat{V}^{-1} Z)^{-1/2}$. Then

$$\begin{aligned}
B_{1n} &= (Z^\tau \hat{V}^{-1} Z)^{-1/2} Z^\tau \hat{V}^{-1/2} C_n \hat{V}^{-1/2} Z (Z^\tau \hat{V}^{-1} Z)^{-1/2} \\
&\leq \|C_n\|_{\max} (Z^\tau \hat{V}^{-1} Z)^{-1/2} Z^\tau \hat{V}^{-1} Z (Z^\tau \hat{V}^{-1} Z)^{-1/2} \\
&= o_p(1) I_p.
\end{aligned}$$

Let $B_{2n} = (Z^\tau V^{-1}Z)^{1/2}(Z^\tau \hat{V}^{-1}Z)^{-1/2}$. Since

$$
\begin{aligned}
\|B_{2n}\|^2 &= \mathrm{tr}\left((Z^\tau V^{-1}Z)^{1/2}(Z^\tau \hat{V}^{-1}Z)^{-1}(Z^\tau V^{-1}Z)^{1/2}\right) \\
&= \mathrm{tr}\left((Z^\tau \hat{V}^{-1}Z)^{-1}Z^\tau V^{-1}Z\right) \\
&= \mathrm{tr}(B_n + I_p) \\
&= p + o_p(1),
\end{aligned}
$$

we obtain that

$$
\|B_{2n}B_{1n}B_{2n}^\tau\| = o_p(1).
$$

Then

$$
\begin{aligned}
\zeta_n^2 &= [l^\tau B_n(Z^\tau V^{-1}Z)^{-1}Z^\tau V^{-1}\varepsilon]^2 \\
&= [l^\tau (Z^\tau V^{-1}Z)^{-1/2}B_{2n}B_{1n}B_{2n}^\tau(Z^\tau V^{-1}Z)^{-1/2}Z^\tau V^{-1}\varepsilon]^2 \\
&\le l^\tau (Z^\tau V^{-1}Z)^{-1}l\|B_{2n}B_{1n}B_{2n}^\tau\|^2\|(Z^\tau V^{-1}Z)^{-1/2}Z^\tau V^{-1}\varepsilon\|^2 \\
&= o_p(a_n^2).
\end{aligned}
$$

This proves (3.61) and thus completes the proof. ∎

Theorem 3.17 shows that as long as $\hat{V}$ is consistent in the sense of (3.60), the weighted LSE $\hat{\beta}_w$ is asymptotically as efficient as $\breve{\beta}$, which is the BLUE if V is known. If V is known and ε is normal, then $\mathrm{Var}(l^\tau\breve{\beta})$ attains the Cramér-Rao lower bound (Proposition 3.2) and, thus, (3.10) holds with $T_n = l^\tau\hat{\beta}_w$.

By Theorems 3.12 and 3.17, the asymptotic relative efficiency of the LSE $l^\tau\hat{\beta}$ w.r.t. the weighted LSE $l^\tau\hat{\beta}_w$ is

$$
\frac{l^\tau(Z^\tau V^{-1}Z)^{-1}l}{l^\tau(Z^\tau Z)^{-1}Z^\tau V Z(Z^\tau Z)^{-1}l},
$$

which is always less than 1 and equals 1 if $l^\tau\hat{\beta}$ is a BLUE (in which case $\hat{\beta} = \breve{\beta}$).

Finding a consistent $\hat{V}$ is possible when V has a certain type of structure. We consider three examples.

Example 3.29. Consider model (3.25). Suppose that $V = \mathrm{Var}(\varepsilon)$ is a block diagonal matrix with the ith diagonal block

$$
\sigma^2 I_{m_i} + U_i \Sigma U_i^\tau, \qquad i = 1, ..., k, \tag{3.62}
$$

where m_i's are integers bounded by a fixed integer m, $\sigma^2 > 0$ is an unknown parameter, Σ is a $q \times q$ unknown nonnegative definite matrix, U_i is an $m_i \times q$

full rank matrix whose columns are in $\mathcal{R}(W_i)$, $q < \inf_i m_i$, and W_i is the $p \times m_i$ matrix such that $Z^\tau = (\ W_1\ W_2\ ...\ W_k\ )$. Under (3.62), a consistent $\hat{V}$ can be obtained if we can obtain consistent estimators of σ^2 and Σ.

Let $X = (Y_1, ..., Y_k)$, where Y_i is an m_i-vector, and let R_i be the matrix whose columns are linearly independent rows of W_i. Then

$$\hat{\sigma}^2 = \frac{1}{n - kq} \sum_{i=1}^{k} Y_i^\tau [I_{m_i} - R_i(R_i^\tau R_i)^{-1} R_i^\tau] Y_i \qquad (3.63)$$

is an unbiased estimator of σ^2. Assume that Y_i's are independent and that $\sup_i E|\varepsilon_i|^{2+\delta} < \infty$ for some $\delta > 0$. Then $\hat{\sigma}^2$ is consistent for σ^2 (exercise). Let $r_i = Y_i - W_i^\tau \hat{\beta}$ and

$$\hat{\Sigma} = \frac{1}{k} \sum_{i=1}^{k} \left[(U_i^\tau U_i)^{-1} U_i^\tau r_i r_i^\tau U_i (U_i^\tau U_i)^{-1} - \hat{\sigma}^2 (U_i^\tau U_i)^{-1} \right]. \qquad (3.64)$$

It can be shown (exercise) that $\hat{\Sigma}$ is consistent for Σ in the sense that $\|\hat{\Sigma} - \Sigma\|_{\max} \to_p 0$ or, equivalently, $\|\hat{\Sigma} - \Sigma\| \to_p 0$ (see Exercise 116).   ∎

Example 3.30. Suppose that V is a block diagonal matrix with the ith diagonal block matrix V_{m_i}, $i = 1, ..., k$, where V_t is an unknown $t \times t$ matrix and $m_i \in \{1, ..., m\}$ with a fixed positive integer m. Thus, we need to obtain consistent estimators of at most m different matrices $V_1, ..., V_m$. It can be shown (exercise) that the following estimator is consistent for V_t when $k_t \to \infty$ as $k \to \infty$:

$$\hat{V}_t = \frac{1}{k_t} \sum_{i \in B_t} r_i r_i^\tau, \quad t = 1, ..., m,$$

where r_i is the same as that in Example 3.29, B_t is the set of i's such that $m_i = t$, and k_t is the number of i's in B_t.   ∎

Example 3.31. Suppose that V is diagonal with the ith diagonal element $\sigma_i^2 = \psi(Z_i)$, where ψ is an unknown function. The simplest case is $\psi(t) = \theta_0 + \theta_1 v(Z_i)$ for a known function v and some unknown θ_0 and θ_1. One can then obtain a consistent estimator $\hat{V}$ by using the LSE of θ_0 and θ_1 under the "model"

$$Er_i^2 = \theta_0 + \theta_1 v(Z_i), \qquad i = 1, ..., n, \qquad (3.65)$$

where $r_i = X_i - Z_i^\tau \hat{\beta}$ (exercise). If ψ is nonlinear or nonparametric, some results are given in Carroll (1982) and Müller and Stadtmüller (1987).   ∎

Finally, if $\hat{V}$ is not consistent (i.e., (3.60) does not hold), then the weighted LSE $l^\tau \hat{\beta}_w$ can still be consistent and asymptotically normal, but its asymptotic variance is not $l^\tau(Z^\tau V^{-1}Z)^{-1}l$; in fact, $l^\tau\hat{\beta}_w$ may not be asymptotically as efficient as the LSE $l^\tau\hat{\beta}$ (Carroll and Cline, 1988; Chen and Shao, 1993). For example, if

$$\|\hat{V}^{-1}U - I_n\|_{\max} \to_p 0,$$

where U is positive definite, $0 < \inf_n \lambda_-[U] \le \sup_n \lambda_+[U] < \infty$, and $U \ne V$ (i.e., $\hat{V}$ is inconsistent for V), then, using the same argument as that in the proof of Theorem 3.17, we can show (exercise) that

$$l^\tau(\hat{\beta}_w - \beta)/b_n \to_d N(0,1) \tag{3.66}$$

for any $l \ne 0$, where $b_n^2 = l^\tau(Z^\tau U^{-1}Z)^{-1}Z^\tau U^{-1}VU^{-1}Z(Z^\tau U^{-1}Z)^{-1}l$. Hence, the asymptotic relative efficiency of the LSE $l^\tau\hat{\beta}$ w.r.t. $l^\tau\hat{\beta}_w$ can be less than 1 or larger than 1.

3.6 Exercises

1. Let $X_1, ..., X_n$ be i.i.d. binary random variables with $P(X_i = 1) = p \in (0,1)$.
   (a) Find the UMVUE of p^m, $m \le n$.
   (b) Find the UMVUE of $P(X_1 + \cdots + X_m = k)$, where m and k are positive integers $\le n$.
   (c) Find the UMVUE of $P(X_1 + \cdots + X_{n-1} > X_n)$.

2. Let $X_1, ..., X_n$ be i.i.d. having the $N(\mu, \sigma^2)$ distribution with an unknown $\mu \in \mathcal{R}$ and a known $\sigma^2 > 0$.
   (a) Find the UMVUE's of μ^3 and μ^4.
   (b) Find the UMVUE's of $P(X_1 \le t)$ and $\frac{d}{dt}P(X_1 \le t)$ with a fixed $t \in \mathcal{R}$.

3. In Example 3.4,
   (a) show that the UMVUE of σ^r is $k_{n-1,r}S^r$, where $r > 1 - n$;
   (b) prove that $(X_1 - \bar{X})/S$ has the p.d.f. given by (3.1);
   (c) show that $(X_1 - \bar{X})/S \to_d N(0,1)$ by using (i) the SLLN and (ii) Scheffé's theorem (Proposition 1.18).

4. Let $X_1, ..., X_m$ be i.i.d. having the $N(\mu_x, \sigma_x^2)$ distribution and let $Y_1, ..., Y_n$ be i.i.d. having the $N(\mu_y, \sigma_y^2)$ distribution. Assume that X_i's and Y_j's are independent.
   (a) Assume that $\mu_x \in \mathcal{R}$, $\mu_y \in \mathcal{R}$, $\sigma_x^2 > 0$, and $\sigma_y^2 > 0$. Find the UMVUE's of $\mu_x - \mu_y$ and $(\sigma_x/\sigma_y)^r$, $r > 0$.
   (b) Assume that $\mu_x \in \mathcal{R}$, $\mu_y \in \mathcal{R}$, and $\sigma_x^2 = \sigma_y^2 > 0$. Find the UMVUE's of σ_x^2 and $(\mu_x - \mu_y)/\sigma_x$.

(c) Assume that $\mu_x = \mu_y \in \mathcal{R}$, $\sigma_x^2 > 0$, $\sigma_y^2 > 0$, and $\sigma_x^2/\sigma_y^2 = \gamma$ is known. Find the UMVUE of μ_x.

(d) Assume that $\mu_x = \mu_y \in \mathcal{R}$, $\sigma_x^2 > 0$, and $\sigma_y^2 > 0$. Show that a UMVUE of μ_x does not exist.

(e) Assume that $\mu_x \in \mathcal{R}$, $\mu_y \in \mathcal{R}$, $\sigma_x^2 > 0$, and $\sigma_y^2 > 0$. Find the UMVUE of $P(X_1 \leq Y_1)$.

(f) Repeat (e) under the assumption that $\sigma_x = \sigma_y$.

5. Let $X_1, ..., X_n$ be i.i.d. having the uniform distribution on the interval $(\theta_1 - \theta_2, \theta_1 + \theta_2)$, where $\theta_j \in \mathcal{R}$, $j = 1, 2$. Find the UMVUE's of θ_j, $j = 1, 2$, and θ_1/θ_2.

6. Let $X_1, ..., X_n$ be i.i.d. having the exponential distribution $E(a, \theta)$ with parameters $\theta > 0$ and $a \in \mathcal{R}$.
   (a) Find the UMVUE of a when θ is known.
   (b) Find the UMVUE of θ when a is known.
   (c) Find the UMVUE's of θ and a.
   (d) Assume that θ is known. Find the UMVUE of $P(X_1 \geq t)$ and $\frac{d}{dt} P(X_1 \geq t)$ for a fixed $t > 0$.
   (e) Find the UMVUE of $P(X_1 \geq t)$ for a fixed $t > 0$.

7. Let $X_1, ..., X_n$ be i.i.d. having the Pareto distribution $Pa(a, \theta)$ with $\theta > 0$ and $a > 0$.
   (a) Find the UMVUE of θ when a is known.
   (b) Find the UMVUE of a when θ is known.
   (c) Find the UMVUE's of a and θ.

8. Consider Exercise 52(a) of §2.6. Find the UMVUE of γ.

9. Let $X_1, ..., X_m$ be i.i.d. having the exponential distribution $E(a_x, \theta_x)$ with $\theta_x > 0$ and $a_x \in \mathcal{R}$ and $Y_1, ..., Y_n$ be i.i.d. having the exponential distribution $E(a_y, \theta_y)$ with $\theta_y > 0$ and $a_y \in \mathcal{R}$. Assume that X_i's and Y_j's are independent.
   (a) Find the UMVUE's of $a_x - a_y$ and θ_x/θ_y.
   (b) Suppose that $\theta_x = \theta_y$ but it is unknown. Find the UMVUE's of θ_x and $(a_x - a_y)/\theta_x$.
   (c) Suppose that $a_x = a_y$ but it is unknown. Show that a UMVUE of a_x does not exist.
   (d) Suppose that $n = m$ and $a_x = a_y = 0$ and that our sample is $(Z_1, \Delta_1), ..., (Z_n, \Delta_n)$, where $Z_i = \min\{X_i, Y_i\}$ and $\Delta_i = 1$ if $X_i \geq Y_i$ and 0 otherwise, $i = 1, ..., n$. Find the UMVUE of $\theta_x - \theta_y$.

10. Let $X_1, ..., X_m$ be i.i.d. having the uniform distribution $U(0, \theta_x)$ and $Y_1, ..., Y_n$ be i.i.d. having the uniform distribution $U(0, \theta_y)$. Suppose that X_i's and Y_j's are independent and that $\theta_x > 0$ and $\theta_y > 0$. Find the UMVUE of θ_x/θ_y when $n > 1$.

11. Let X be a random variable having the negative binomial distribution $NB(p, r)$ with an unknown $p \in (0, 1)$ and a known r.
    (a) Find the UMVUE of p^t, $t < r$.
    (b) Find the UMVUE of $\text{Var}(X)$.
    (c) Find the UMVUE of $\log p$.

12. Let $X_1, ..., X_n$ be i.i.d. random variables having the Poisson distribution $P(\theta)$ truncated at 0, i.e., $P(X_i = x) = (e^\theta - 1)^{-1}\theta^x/x!$, $x = 1, 2, ..., \theta > 0$. Find the UMVUE of θ when $n = 1, 2$.

13. Let X be a random variable having the negative binomial distribution $NB(p, r)$ truncated at r, where r is known and $p \in (0, 1)$ is unknown. Let k be a fixed positive integer $> r$. For $r = 1, 2, 3$, find the UMVUE of p^k.

14. Let $X_1, ..., X_n$ be i.i.d. having the log-distribution $L(p)$ with an unknown $p \in (0, 1)$. Let k be a fixed positive integer.
    (a) For $n = 1, 2, 3$, find the UMVUE of p^k.
    (b) For $n = 1, 2, 3$, find the UMVUE of $P(X = k)$.

15. Consider Exercise 43 of §2.6.
    (a) Show that the estimator $U = 2(|X_1| - \frac{1}{4})I_{\{X_1 \neq 0\}}$ is unbiased for θ.
    (b) Derive the UMVUE of θ.

16. Derive the UMVUE of p in Exercise 33 of §2.6.

17. Derive the UMVUE's of θ and λ in Exercise 55 of §2.6, based on data $X_1, ..., X_n$.

18. Suppose that $(X_0, X_1, ..., X_k)$ has the multinomial distribution in Example 2.7 with $p_i \in (0, 1)$, $\sum_{j=0}^{k} p_j = 1$. Find the UMVUE of $p_0^{r_0} \cdots p_k^{r_k}$, where r_j's are nonnegative integers with $r_0 + \cdots + r_k \leq n$.

19. Let $Y_1, ..., Y_n$ be i.i.d. from the uniform distribution $U(0, \theta)$ with an unknown $\theta \in (1, \infty)$.
    (a) Suppose that we only observe

    $$X_i = \begin{cases} Y_i & \text{if } Y_i \geq 1 \\ 1 & \text{if } Y_i < 1, \end{cases} \quad i = 1, ..., n.$$

    Derive a UMVUE of θ.
    (b) Suppose that we only observe

    $$X_i = \begin{cases} Y_i & \text{if } Y_i \leq 1 \\ 1 & \text{if } Y_i > 1, \end{cases} \quad i = 1, ..., n.$$

    Derive a UMVUE of the probability $P(Y_1 > 1)$.

20. Let $(X_1, Y_1), ..., (X_n, Y_n)$ be i.i.d. random 2-vectors distributed as bi-variate normal with $EX_i = EY_i = \beta z_i$, $\text{Var}(X_i) = \text{Var}(Y_i) = \sigma^2$, and $\text{Cov}(X_i, Y_i) = \rho\sigma^2$, $i = 1, ..., n$, where $\beta \in \mathcal{R}$, $\sigma > 0$, and $\rho \in (-1, 1)$ are unknown parameters, and z_i's are known constants.
    (a) Obtain a UMVUE of β and calculate its variance.
    (b) Obtain a UMVUE of σ^2 and calculate its variance.

21. Let $(X_1, Y_1), ..., (X_n, Y_n)$ be i.i.d. random 2-vectors from a population $P \in \mathcal{P}$ that is the family of all bivariate populations with Lebesgue p.d.f.'s.
    (a) Show that the set of n pairs (X_i, Y_i) ordered according to the value of their first coordinate constitutes a sufficient and complete statistic for $P \in \mathcal{P}$.
    (b) A statistic T is a function of the complete and sufficient statistic if and only if T is invariant under permutation of the n pairs.
    (c) Show that $(n - 1)^{-1} \sum_{i=1}^{n} (X_i - \bar{X})(Y_i - \bar{Y})$ is the UMVUE of $\text{Cov}(X_1, Y_1)$.
    (d) Find the UMVUE's of $P(X_i \le Y_i)$ and $P(X_i \le X_j \text{ and } Y_i \le Y_j)$, $i \ne j$.

22. Let $X_1, ..., X_n$ be i.i.d. from $P \in \mathcal{P}$ containing all symmetric c.d.f.'s with finite means and with Lebesgue p.d.f.'s on $\mathcal{R}$. Show that there is no UMVUE of $\mu = EX_1$.

23. Prove Corollary 3.1.

24. Suppose that T is a UMVUE of an unknown parameter ϑ. Show that T^k is a UMVUE of $E(T^k)$, where k is any positive integer for which $E(T^{2k}) < \infty$.

25. Consider the problem in Exercise 83 of §2.6. Use Theorem 3.2 to show that $I_{\{0\}}(X)$ is a UMVUE of $(1 - p)^2$ and that there is no UMVUE of p.

26. Let $X_1, ..., X_n$ be i.i.d. from a discrete distribution with

$$P(X_i = \theta - 1) = P(X_i = \theta) = P(X_i = \theta + 1) = \tfrac{1}{3},$$

where θ is an unknown integer. Show that no nonconstant function of θ has a UMVUE.

27. Let X be a random variable having the Lebesgue p.d.f.

$$[(1 - \theta) + \theta/(2\sqrt{x})]I_{(0,1)}(x),$$

where $\theta \in [0, 1]$. Show that there is no UMVUE of θ.

28. Let X be a discrete random variable with $P(X = -1) = 2p(1 - p)$
    and $P(X = k) = p^k(1 - p)^{3-k}$, $k = 0, 1, 2, 3$, where $p \in (0, 1)$.
    (a) Determine whether there is a UMVUE of p.
    (b) Determine whether there is a UMVUE of $p(1 - p)$.

29. Let $X_1, ..., X_n$ be i.i.d. observations. Obtain a UMVUE of a in the
    following cases.
    (a) X_i has the exponential distribution $E(a, \theta)$ with a known θ and
    an unknown $a \leq 0$.
    (b) X_i has the Pareto distribution $Pa(a, \theta)$ with a known $\theta > 1$ and
    an unknown $a \in (0, 1]$.

30. In Exercise 41 of §2.6, find a UMVUE of θ and show that it is unique
    a.s.

31. Prove Theorem 3.3 for the multivariate case ($k > 1$).

32. Let X be a single sample from P_θ. Find the Fisher information $I(\theta)$
    in the following cases.
    (a) P_θ is the $N(\mu, \sigma^2)$ distribution with $\theta = \mu \in \mathcal{R}$.
    (b) P_θ is the $N(\mu, \sigma^2)$ distribution with $\theta = \sigma^2 > 0$.
    (c) P_θ is the $N(\mu, \sigma^2)$ distribution with $\theta = \sigma > 0$.
    (d) P_θ is the $N(\sigma, \sigma^2)$ distribution with $\theta = \sigma > 0$.
    (e) P_θ is the $N(\mu, \sigma^2)$ distribution with $\theta = (\mu, \sigma^2) \in \mathcal{R} \times (0, \infty)$.
    (f) P_θ is the negative binomial distribution $NB(\theta, r)$ with $\theta \in (0, 1)$.
    (g) P_θ is the gamma distribution $\Gamma(\alpha, \gamma)$ with $\theta = (\alpha, \gamma) \in (0, \infty) \times (0, \infty)$.
    (h) P_θ is the beta distribution $B(\alpha, \beta)$ with $\theta = (\alpha, \beta) \in (0, 1) \times (0, 1)$.

33. Find a function of θ for which the amount of information is indepen-
    dent of θ, when P_θ is
    (a) the Poisson distribution $P(\theta)$ with $\theta > 0$;
    (b) the binomial distribution $Bi(\theta, r)$ with $\theta \in (0, 1)$;
    (c) the gamma distribution $\Gamma(\alpha, \theta)$ with $\theta > 0$.

34. Prove the result in Example 3.9.

35. Obtain the Fisher information matrix for a random variable with
    (a) the Cauchy distribution $C(\mu, \sigma)$, $\mu \in \mathcal{R}$, $\sigma > 0$;
    (b) the double exponential distribution $DE(\mu, \theta)$, $\mu \in \mathcal{R}$, $\theta > 0$;
    (c) the logistic distribution $LG(\mu, \sigma)$, $\mu \in \mathcal{R}$, $\sigma > 0$;
    (d) the c.d.f. $F_r\left(\frac{x-\mu}{\sigma}\right)$, where F_r is the c.d.f. of the t-distribution t_r
    with a known r, $\mu \in \mathcal{R}$, $\sigma > 0$;
    (e) the Lebesgue p.d.f. $f_\theta(x) = (1 - \epsilon)\phi(x - \mu) + \frac{\epsilon}{\sigma}\phi\left(\frac{x-\mu}{\sigma}\right)$, $\theta = (\mu, \sigma, \epsilon) \in \mathcal{R} \times (0, \infty) \times (0, 1)$, where ϕ is the standard normal p.d.f.

36. Let X be a sample having a p.d.f. satisfying the conditions in Theorem 3.3, where θ is a k-vector of unknown parameters, and let $T(X)$ be a statistic. If T has a p.d.f. g_θ satisfying the conditions in Theorem 3.3, then we define $I_T(\theta) = E\{\frac{\partial}{\partial\theta} \log g_\theta(T)[\frac{\partial}{\partial\theta} \log g_\theta(T)]^\tau\}$ to be the Fisher information about θ contained in T.
    (a) Show that $I_X(\theta) - I_T(\theta)$ is nonnegative definite, where $I_X(\theta)$ is the Fisher information about θ contained in X.
    (b) Show that $I_X(\theta) = I_T(\theta)$ if T is sufficient for θ.

37. Let $X_1, ..., X_n$ be i.i.d. from the uniform distribution $U(0, \theta)$ with $\theta > 0$.
    (a) Show that condition (3.3) does not hold for $h(X) = X_{(n)}$.
    (b) Show that the inequality in (3.6) does not hold for the UMVUE of θ.

38. Prove Proposition 3.3.

39. Let X be a single sample from the double exponential distribution $DE(\mu, \theta)$ with $\mu = 0$ and $\theta > 0$. Find the UMVUE's of the following parameters and, in each case, determine whether the variance of the UMVUE attains the Cramér-Rao lower bound.
    (a) $\vartheta = \theta$;
    (b) $\vartheta = \theta^r$, where $r > 1$;
    (c) $\vartheta = (1 + \theta)^{-1}$.

40. Let $X_1, ..., X_n$ be i.i.d. binary random variables with $P(X_i = 1) = p \in (0, 1)$.
    (a) Show that the UMVUE of $p(1 - p)$ is $T_n = n\bar{X}(1 - \bar{X})/(n - 1)$.
    (b) Show that $\text{Var}(T_n)$ does not attain the Cramér-Rao lower bound.
    (c) Show that (3.10) holds.

41. Let $X_1, ..., X_n$ be i.i.d. having the Poisson distribution $P(\theta)$ with $\theta > 0$. Find the amse of the UMVUE of $e^{-t\theta}$ with a fixed $t > 0$ and show that (3.10) holds.

42. Let $X_1, ..., X_n$ be i.i.d. having the $N(\mu, \sigma^2)$ distribution with an unknown $\mu \in \mathcal{R}$ and a known $\sigma^2 > 0$.
    (a) Find the UMVUE of $\vartheta = e^{t\mu}$ with a fixed $t \neq 0$.
    (b) Determine whether the variance of the UMVUE in (a) attains the Cramér-Rao lower bound.
    (c) Show that (3.10) holds.

43. Show that if $X_1, ..., X_n$ are i.i.d. binary random variables, U_n in (3.12) equals $T(T - 1) \cdots (T - m + 1)/[n(n - 1) \cdots (n - m + 1)]$, where $T = \sum_{i=1}^n X_i$.

44. Show that if $T_n = \bar{X}$, then U_n in (3.13) is the same as the sample variance S^2 in (2.2). Show that (3.23) holds for T_n given by (3.22) with $E(R_n^2) = o(n^{-1})$.

45. Prove (3.14), (3.16), and (3.17).

46. Let ζ_k be given in Theorem 3.4. Show that $\zeta_1 \leq \zeta_2 \leq \cdots \leq \zeta_m$.

47. Prove Corollary 3.2.

48. Prove (3.20) and show that $U_n - \check{U}_n$ is also a U-statistic.

49. Let T_n be a symmetric statistic with $\mathrm{Var}(T_n) < \infty$ for every n and $\check{T}_n$ be the projection of T_n on $\binom{n}{k}$ random vectors $\{X_{i_1}, ..., X_{i_k}\}$, $1 \leq i_1 < \cdots < i_k \leq n$. Show that $E(T_n) = E(\check{T}_n)$ and calculate $E(T_n - \check{T}_n)^2$.

50. Let Y_k be defined in Lemma 3.2. Show that $\{Y_k^2\}$ is uniformly integrable.

51. Show that (3.22) with $E(R_n^2) = o(n^{-1})$ is satisfied for T_n being a U-statistic with $E[h(X_1, ..., X_m)]^2 < \infty$.

52. Let S^2 be the sample variance given by (2.2), which is also a U-statistic (§3.2.1). Find the corresponding h_1, h_2, ζ_1, and ζ_2. Discuss how to apply Theorem 3.5 to this case.

53. Let $h(x_1, x_2, x_3) = I_{(-\infty,0)}(x_1 + x_2 + x_3)$. Define the U-statistic with this kernel and find h_k and ζ_k, $k = 1, 2, 3$.

54. Let $X_1, ..., X_n$ be i.i.d. random variables having finite $\mu = EX_1$ and $\bar{\mu} = EX_1^{-1}$. Find a U-statistic that is an unbiased estimator of $\mu\bar{\mu}$ and derive its variance and asymptotic distribution.

55. Show that $\hat{\beta}$ is an LSE of β if and only if it is given by (3.29).

56. Obtain explicit forms for the LSE's of β_j, $j = 0, 1$, and SSR, under the simple linear regression model in Example 3.11, assuming that some t_i's are different.

57. Consider the polynomial model
$$X_i = \beta_0 + \beta_1 t_i + \beta_2 t_i^2 + \varepsilon_i, \quad i = 1, ..., n.$$
Find explicit forms for the LSE's of β_j, $j = 0, 1, 2$, and SSR, assuming that some t_i's are different.

58. Suppose that
$$X_{ij} = \alpha_i + \beta t_{ij} + \varepsilon_{ij}, \quad i = 1, ..., a, \ j = 1, ..., b.$$
Find explicit forms for the LSE's of β, α_i, $i = 1, ..., a$, and SSR.

59. Consider the polynomial model

$$X_i = \beta_0 + \beta_1 t_i + \beta_2 t_i^2 + \beta_3 t_i^3 + \varepsilon_i, \quad i = 1, ..., n,$$

where ε's are i.i.d. from $N(0, \sigma^2)$. Suppose that $n = 12$, $t_i = -1$, $i = 1, ..., 4$, $t_i = 0$, $i = 5, ..., 8$, and $t_i = 1$, $i = 9, ..., 12$.
(a) Obtain the matrix $Z^\tau Z$ when this polynomial model is considered as a special case of model (3.24).
(b) Show whether the following parameters are estimable: $\beta_0 + \beta_2$, β_1, $\beta_0 - \beta_1$, $\beta_1 + \beta_3$, and $\beta_0 + \beta_1 + \beta_2 + \beta_3$.

60. Find the matrix Z, $Z^\tau Z$, and the form of $l \in \mathcal{R}(Z)$ under the one-way ANOVA model (3.31).

61. Obtain the matrix Z under the two-way balanced ANOVA model (3.32). Show that the rank of Z is ab. Verify the form of the LSE of β given in Example 3.14. Find the form of $l \in \mathcal{R}(Z)$.

62. Consider the following model as a special case of model (3.25):

$$X_{ijk} = \mu + \alpha_i + \beta_j + \varepsilon_{ijk}, \quad i = 1, ..., a, j = 1, ..., b, k = 1, ..., c.$$

Obtain the matrix Z, the parameter vector β, and the form of LSE's of β. Discuss conditions under which $l \in \mathcal{R}(Z)$.

63. Under model (3.25) and assumption A1, find the UMVUE's of $(l^\tau \beta)^2$, $l^\tau \beta / \sigma$, and $(l^\tau \beta / \sigma)^2$ for an estimable $l^\tau \beta$.

64. Verify the formulas for SSR's in Example 3.15.

65. Consider the one-way random effects model in Example 3.17. Assume that $n_i = n$ for all i and that A_i's and e_{ij}'s are normally distributed. Show that the family of populations is an exponential family with sufficient and complete statistics $\bar{X}_{..}$, $S_A = n \sum_{i=1}^m (\bar{X}_{i.} - \bar{X}_{..})^2$, and $S_E = \sum_{i=1}^m \sum_{j=1}^n (X_{ij} - \bar{X}_{i.})^2$. Find the UMVUE's of μ, σ_a^2, and σ^2.

66. Consider model (3.25). Suppose that ε_i's are i.i.d. with $E\varepsilon_i = 0$ and a Lebesgue p.d.f. $\sigma^{-1} f(x/\sigma)$, where f is a known Lebesgue p.d.f. and $\sigma > 0$ is unknown.
(a) Show that X is from a location-scale family given by (2.10).
(b) Find the Fisher information about (β, σ) contained in X_i.
(c) Find the Fisher information about (β, σ) contained in X.

67. Consider model (3.25) with assumption A2. Let $c \in \mathcal{R}^p$. Show that if the equation $c = Z^\tau y$ has a solution, then there is a unique solution $y_0 \in \mathcal{R}(Z)$ such that $\text{Var}(y_0^\tau X) \leq \text{Var}(y^\tau X)$ for any other solution of $c = Z^\tau y$.

68. Consider model (3.25). Show that the number of independent linear functions of X with mean 0 is $n - r$, where r is the rank of Z.

69. Consider model (3.25) with assumption A2. Let $\hat{X}_i = Z_i^\tau \hat{\beta}$, which is called the least squares *prediction* of X_i. Let h_{ij} be the (i,j)th element of $Z(Z^\tau Z)^- Z^\tau$ and $h_i = h_{ii}$. Show that
    (a) $\text{Var}(\hat{X}_i) = \sigma^2 h_i$;
    (b) $\text{Var}(X_i - \hat{X}_i) = \sigma^2(1 - h_i)$;
    (c) $\text{Cov}(\hat{X}_i, \hat{X}_j) = \sigma^2 h_{ij}$;
    (d) $\text{Cov}(X_i - \hat{X}_i, X_j - \hat{X}_j) = -\sigma^2 h_{ij}$, $i \neq j$;
    (e) $\text{Cov}(\hat{X}_i, X_j - \hat{X}_j) = 0$.

70. Consider model (3.25) with assumption A2. Let $Z = (Z_1, Z_2)$ and $\beta = (\beta_1, \beta_2)$, where Z_j is $n \times p_j$ and β_j is a p_j-vector, $j = 1, 2$. Assume that $(Z_1^\tau Z_1)^{-1}$ and $[Z_2^\tau Z_2 - Z_2^\tau Z_1(Z_1^\tau Z_1)^{-1}Z_1^\tau Z_2]^{-1}$ exist.
    (a) Derive the LSE of β in terms of Z_1, Z_2, and X.
    (b) Let $\hat{\beta} = (\hat{\beta}_1, \hat{\beta}_2)$ be the LSE in (a). Calculate the covariance between $\hat{\beta}_1$ and $\hat{\beta}_2$.
    (c) Suppose that it is known that $\beta_2 = 0$. Let $\tilde{\beta}_1$ be the LSE of β_1 under the reduced model $X = Z_1\beta_1 + \varepsilon$. Show that, for any $l \in \mathcal{R}^{p_1}$, $l^\tau \tilde{\beta}_1$ is better than $l^\tau \hat{\beta}_1$ in terms of their mse's.

71. Prove that (e) implies (b) in Theorem 3.10.

72. Show that (a) in Theorem 3.10 is equivalent to either
    (f) $\text{Var}(\varepsilon)Z = ZB$ for some matrix B, or
    (g) $\mathcal{R}(Z)$ is generated by r eigenvectors of $\text{Var}(\varepsilon)$, where r is the rank of Z.

73. Prove Corollary 3.3.

74. Suppose that
$$X = \mu J_n + H\xi + e,$$
where $\mu \in \mathcal{R}$ is an unknown parameter, J_n is the n-vector of 1's, H is an $n \times p$ known matrix of full rank, ξ is a random p-vector with $E(\xi) = 0$ and $\text{Var}(\xi) = \sigma_\xi^2 I_p$, e is a random n-vector with $E(e) = 0$ and $\text{Var}(e) = \sigma^2 I_n$, and ξ and e are independent. Show that the LSE of μ is the BLUE if and only if the row totals of HH^τ are the same.

75. Consider a special case of model (3.25):
$$X_{ij} = \mu + \alpha_i + \beta_j + \varepsilon_{ij}, \quad i = 1, ..., a, j = 1, ..., b,$$
where μ, α_i's, and β_j's are unknown parameters, $E(\varepsilon_{ij}) = 0$, $\text{Var}(\varepsilon_{ij}) = \sigma^2$, $\text{Cov}(\varepsilon_{ij}, \varepsilon_{i'j'}) = 0$ if $i \neq i'$, and $\text{Cov}(\varepsilon_{ij}, \varepsilon_{ij'}) = \sigma^2\rho$ if $j \neq j'$. Show that the LSE of $l^\tau \beta$ is the BLUE for any $l \in \mathcal{R}(Z)$.

76. Consider model (3.25) under assumption A3 with $\text{Var}(\varepsilon) = $ a block diagonal matrix whose ith block diagonal V_i is $n_i \times n_i$ and has a single eigenvalue λ_i with eigenvector J_{n_i} (the n_i-vector of 1's) and a repeated eigenvalue ρ_i with multiplicity $n_i - 1$, $i = 1, ..., k$, $\sum_{i=1}^{k} n_i = n$. Let U be the $n \times k$ matrix whose ith column is U_i, where $U_1 = (J_{n_1}^\tau, 0, ..., 0)$, $U_2 = (0, J_{n_2}^\tau, ..., 0)$,..., $U_k = (0, 0, ..., J_{n_k}^\tau)$.
    (a) If $\mathcal{R}(Z^\tau) \subset \mathcal{R}(U^\tau)$ and $\lambda_i \equiv \lambda$, show that $l^\tau \hat{\beta}$ is the BLUE of $l^\tau \beta$ for any $l \in \mathcal{R}(Z)$.
    (b) If $Z^\tau U_i = 0$ for all i and $\rho_i \equiv \rho$, show that $l^\tau \hat{\beta}$ is the BLUE of $l^\tau \beta$ for any $l \in \mathcal{R}(Z)$.

77. Prove Proposition 3.4.

78. Show that the condition $\sup_n \lambda_+[\text{Var}(\varepsilon)] < \infty$ is equivalent to the condition $\sup_i \text{Var}(\varepsilon_i) < \infty$.

79. Find a condition under which the mse of $l^\tau \hat{\beta}$ is of the order n^{-1}. Apply it to problems in Exercises 56, 58, and 60-62.

80. Consider model (3.25) with i.i.d. $\varepsilon_1, ..., \varepsilon_n$ having $E(\varepsilon_i) = 0$ and $\text{Var}(\varepsilon_i) = \sigma^2$. Let $\hat{X}_i = Z_i^\tau \hat{\beta}$ and $h_i = Z_i^\tau (Z^\tau Z)^- Z_i$.
    (a) Show that for any $\epsilon > 0$,
    $$P(|\hat{X}_i - E\hat{X}_i| \geq \epsilon) \geq \min\{P(\varepsilon_i \geq \epsilon/h_i), P(\varepsilon_i \leq -\epsilon/h_i)\}.$$
    (Hint: for independent random variables X and Y, $P(|X+Y| \geq \epsilon) \geq P(X \geq \epsilon)P(Y \geq 0) + P(X \leq -\epsilon)P(Y < 0)$.)
    (b) Show that $\hat{X}_i - E\hat{X}_i \to_p 0$ if and only if $h_i \to 0$.

81. Prove Lemma 3.3 and show that condition (a) is implied by $\{\|Z_i\|\}$ being bounded and $\lambda_+(Z^\tau Z)^- \to 0$.

82. Consider the problem in Exercise 58. Suppose that $\{t_{ij}\}$ is bounded. Find a condition under which (3.39) holds.

83. Under the two-way ANOVA models in Example 3.14 and Exercise 62, find sufficient conditions for (3.39).

84. Consider the one-way random effects model in Example 3.17. Assume that $\{n_i\}$ is bounded and $E|e_{ij}|^{2+\delta} < \infty$ for some $\delta > 0$. Show that the LSE $\hat{\mu}$ of μ is asymptotically normal and derive an explicit form of $\text{Var}(\hat{\mu})$.

85. Suppose that
    $$X_i = \rho t_i + \varepsilon_i, \quad i = 1, ..., n,$$
    where $\rho \in \mathcal{R}$ is an unknown parameter, t_i's are known and in (a, b), a and b are known positive constants, and ε_i's are independent random

variables satisfying $E(\varepsilon_i) = 0$, $E|\varepsilon_i|^{2+\delta} < \infty$ for some $\delta > 0$, and $\mathrm{Var}(\varepsilon_i) = \sigma^2 t_i$ with an unknown $\sigma^2 > 0$.

(a) Obtain the LSE of ρ.

(b) Obtain the BLUE of ρ.

(c) Show that both the LSE and BLUE are asymptotically normal and obtain the asymptotic relative efficiency of the BLUE w.r.t. the LSE.

86. In Example 3.19, show that $E(S^2) = \sigma^2$ given in (3.43).

87. Suppose that $X = (X_1, ..., X_n)$ is a simple random sample (without replacement) from a finite population $\mathcal{P} = \{y_1, ..., y_N\}$ with univariate y_i.

(a) Show that a necessary condition for $h(\theta)$ to be estimable is that h is symmetric in its N arguments.

(b) Find the UMVUE of Y^m, where m is a fixed positive integer $< n$ and Y is the population total.

(c) Find the UMVUE of $P(X_i \leq X_j)$, $i \neq j$.

(d) Find the UMVUE of $\mathrm{Cov}(X_i, X_j)$, $i \neq j$.

88. Prove Theorem 3.14.

89. Under stratified simple random sampling described in §3.4.1, show that the vector of ordered values of all X_{hi}'s is neither sufficient nor complete for $\theta \in \Theta$.

90. Let $\mathcal{P} = \{y_1, ..., y_N\}$ be a population with univariate y_i. Define the population c.d.f. by $F(t) = N^{-1} \sum_{i=1}^{N} I_{(-\infty, t]}(y_i)$. Find the UMVUE of $F(t)$ under (a) simple random sampling and (b) stratified simple random sampling.

91. Consider the estimation of $F(t)$ in the previous exercise. Suppose that a sample of size n is selected with $\pi_i > 0$. Find the Horvitz-Thompson estimator of $F(t)$. Is it a c.d.f.?

92. Show that v_1 in (3.49) and v_2 in (3.50) are unbiased estimators of $\mathrm{Var}(\hat{Y}_{ht})$. Prove that $v_1 = v_2$ under (a) simple random sampling and (b) stratified simple random sampling.

93. Consider the following two-stage stratified sampling plan. In the first stage, the population is stratified into H strata and k_h clusters are selected from stratum h with probability proportional to cluster size, where sampling is independent across strata. In the second stage, a sample of m_{hi} units is selected from sampled cluster i in stratum h, and sampling is independent across clusters. Find π_i and the Horvitz-Thompson estimator $\hat{Y}_{ht}$ of the population total.

94. In the previous exercise, prove the unbiasedness of $\hat{Y}_{ht}$ directly (without using Theorem 3.15).

95. Under systematic sampling, show that $\mathrm{Var}(\hat{Y}_{sy})$ is equal to

$$\left(1 - \frac{1}{N}\right)\frac{\sigma^2}{n} + \frac{2}{nN}\sum_{i=1}^{k}\sum_{1 \le t < u \le n}\left(y_{i+(t-1)k} - \frac{Y}{N}\right)\left(y_{i+(u-1)k} - \frac{Y}{N}\right).$$

96. In Exercise 91, discuss how to obtain a consistent (as $n \to N$) estimator $\hat{F}(t)$ of $F(t)$ such that $\hat{F}$ is a c.d.f.

97. Derive the n^{-1} order asymptotic bias of the sample correlation coefficient defined in Exercise 22 in §2.6.

98. Derive the n^{-1} order asymptotic bias and amse of $\hat{t}_\beta$ in Example 3.22, assuming that $\sum_{j=0}^{p-1}\beta_j t^j$ is convex in $t \in \mathcal{T}$.

99. Consider Example 3.23.
    (a) Show that $\hat{\theta}$ is the BLUE of θ.
    (b) Show that $\hat{\sigma}^2$ is unbiased for σ^2.
    (c) Show that $\hat{\Sigma}$ is consistent for Σ as $k \to \infty$.
    (d) Derive the amse of $\hat{R}(t)$ as $k \to \infty$.

100. Let $X_1, ..., X_n$ be i.i.d. from $N(\mu, \sigma^2)$, where $\mu \in \mathcal{R}$ and $\sigma^2 > 0$. Consider the estimation of $\vartheta = E\Phi(a + bX_1)$, where Φ is the standard normal c.d.f. and a and b are known constants. Obtain an explicit form of a function $g(\mu, \sigma^2) = \vartheta$ and the amse of $\hat{\vartheta} = g(\bar{X}, S^2)$.

101. Let $X_1, ..., X_n$ be i.i.d. with mean μ, variance σ^2, and finite $\mu_j = EX_1^j$, $j = 2, 3, 4$. The sample coefficient of variation is defined to be $S/\bar{X}$, where S is the squared root of the sample variance S^2.
    (a) If $\mu \ne 0$, show that $\sqrt{n}(S/\bar{X} - \sigma/\mu) \to_d N(0, \tau)$ and obtain an explicit formula of τ in terms of μ, σ^2, and μ_j.
    (b) If $\mu = 0$, show that $n^{-1/2}S/\bar{X} \to_d [N(0, 1)]^{-1}$.

102. Prove (3.52) and (3.53).

103. Let $X_1, ..., X_n$ be i.i.d. from P in a parametric family. Obtain moment estimators of parameters in the following cases.
    (a) P is the gamma distribution $\Gamma(\alpha, \gamma)$, $\alpha > 0$, $\gamma > 0$.
    (b) P is the exponential distribution $E(a, \theta)$, $a \in \mathcal{R}$, $\theta > 0$.
    (c) P is the beta distribution $B(\alpha, \beta)$, $\alpha > 0$, $\beta > 0$.
    (d) P is the log-normal distribution $LN(\mu, \sigma^2)$, $\mu \in \mathcal{R}$, $\sigma > 0$.
    (e) P is the uniform distribution $U(\theta - \frac{1}{2}, \theta + \frac{1}{2})$, $\theta \in \mathcal{R}$.
    (f) P is the negative binomial distribution $NB(p, r)$, $p \in (0, 1)$, $r = 1, 2, ...$.

(g) P is the log-distribution $L(p)$, $p \in (0, 1)$.

(h) P is the log-normal distribution $LN(\mu, \sigma^2)$, $\mu \in \mathcal{R}$, $\sigma = 1$.

(i) P is the chi-square distribution χ_k^2 with an unknown $k = 1, 2, ...$

104. Obtain moment estimators of λ and p in Exercise 55 of §2.6, based on data $X_1, ..., X_n$.

105. Obtain the asymptotic distributions of the moment estimators in Exercise 103(a), (c), (e), and (g), and the asymptotic relative efficiencies of moment estimators w.r.t. UMVUE's in Exercise 103(b) and (h).

106. In Exercise 19(a), find a moment estimator of θ and derive its asymptotic distribution. In Exercise 19(b), obtain a moment estimator of θ^{-1} and its asymptotic relative efficiency w.r.t. the UMVUE of θ^{-1}.

107. Let $X_1, ..., X_n$ be i.i.d. random variables having the Lebesgue p.d.f. $f_{\alpha, \beta}(x) = \alpha(x/\theta)^{\alpha-1} I_{(0, \beta)}(x)$, where $\alpha > 0$ and $\beta > 0$ are unknown.

(a) Obtain moment estimators of α and β.

(b) Obtain the asymptotic distribution of the moment estimators of α and β derived in (a).

108. Let $X_1, ..., X_n$ be i.i.d. from the following discrete distribution:

$$P(X_1 = 1) = \frac{2(1 - \theta)}{2 - \theta}, \quad P(X_1 = 2) = \frac{\theta}{2 - \theta},$$

where $\theta \in (0, 1)$ is unknown.

(a) Obtain an estimator of θ using the method of moments.

(b) Obtain the amse of the moment estimator in (a).

109. Let $X_1, ..., X_n$ ($n > 1$) be i.i.d. from a population having the Lebesgue p.d.f.

$$f_\theta(x) = (1 - \epsilon)\phi(x - \mu) + \frac{\epsilon}{\sigma}\phi\left(\frac{x - \mu}{\sigma}\right),$$

where ϕ is the standard normal p.d.f., $\theta = (\mu, \sigma) \in \mathcal{R} \times (0, \infty)$ is unknown, and $\epsilon \in (0, 1)$ is a known constant.

(a) Obtain an estimator of θ using the method of moments.

(b) Obtain the asymptotic distribution of the moment estimator in part (a).

110. Let $X_1, ..., X_n$ be i.i.d. random variables having the Lebesgue p.d.f.

$$f_{\theta_1, \theta_2}(x) = \begin{cases} (\theta_1 + \theta_2)^{-1} e^{-x/\theta_1} & x > 0 \\ (\theta_1 + \theta_2)^{-1} e^{x/\theta_2} & x \leq 0, \end{cases}$$

where $\theta_1 > 0$ and $\theta_2 > 0$ are unknown.

(a) Obtain an estimator of (θ_1, θ_2) using the method of moments.

(b) Obtain the asymptotic distribution of the moment estimator in part (a).

111. (Nonexistence of a moment estimator). Consider $X_1, ..., X_n$ and the parametric family indexed by $(\theta, j) \in (0, 1) \times \{1, 2\}$ in Exercise 41 of §2.6. Let $h_i(\theta, j) = EX_1^i$, $i = 1, 2$. Show that

$$P(\hat{\mu}_i = h_i(\theta, j) \text{ has a solution}) \to 0$$

as $n \to \infty$, when X_i's are from the Poisson distribution $P(\theta)$.

112. In the proof of Proposition 3.5, show that $E[W_n(U_n - \vartheta)] = O(n^{-1})$.

113. Assume the conditions of Theorem 3.16.
    (a) Prove (3.56).
    (b) Show that $E[g_2(X_1, X_1)]/n = nEV_{n2} = m(m-1)\sum_{j=1}^{\infty}\lambda_j/(2n)$.

114. Let $X_1, ..., X_n$ be i.i.d. with a c.d.f. F and U_n and V_n be the U- and V-statistics with kernel $\int[I_{(-\infty, y]}(x_1) - F_0(y)][I_{(-\infty, y]}(x_2) - F_0(y)]dF_0$, where F_0 is a known c.d.f.
    (a) Obtain the asymptotic distributions of U_n and V_n when $F \neq F_0$.
    (b) Obtain the asymptotic relative efficiency of U_n w.r.t. V_n when $F = F_0$.

115. Let $X_1, ..., X_n$ be i.i.d. with a c.d.f. F having a finite sixth moment. Consider the estimation of μ^3, where $\mu = EX_1$. When $\mu = 0$, find $\text{amse}_{\bar{X}^3}(P)/\text{amse}_{U_n}(P)$, where $U_n = \binom{n}{3}^{-1}\sum_{1 \leq i < j < k \leq n} X_i X_j X_k$.

116. Let A_n, $n = 1, 2, ...$, be a sequence of $k \times k$ matrices, where k is a fixed integer.
    (a) Show that $\|A_n\|_{\max} \to 0$ if and only if $\|A_n\| \to 0$, where $\|A_n\|_{\max}$ is defined in (3.60) and $\|A_n\|^2 = \text{tr}(A_n^\tau A_n)$.
    (b) Show that if A_n's are nonnegative definite, then $\|A_n\| \to 0$ if and only if $\lambda_+[A_n] \to 0$, where $\lambda_+[A_n]$ is the largest eigenvalue of A_n.
    (c) Show that the result in (a) is not always true if k varies with n.

117. Prove that $\hat{\sigma}^2$ in (3.63) is unbiased and consistent for σ^2 under model (3.25) with (3.62) and $\sup_i E|\varepsilon_i|^{2+\delta} < \infty$ for some $\delta > 0$. Under the same conditions, show that $\hat{\Sigma}$ in (3.64) is consistent for Σ in the sense that $\|\hat{\Sigma} - \Sigma\|_{\max} \to_p 0$.

118. In Example 3.30, show that $\hat{V}_t$ is consistent for V_t when $k_t \to \infty$ as $k \to \infty$.

119. Show how to use equation (3.65) to obtain consistent estimators of θ_0 and θ_1.

120. Prove (3.66) under the assumed conditions in §3.5.4.

Chapter 4

Estimation in Parametric Models

In this chapter, we consider point estimation methods in parametric models. One such method, the moment method, has been introduced in §3.5.2. It is assumed in this chapter that the sample X is from a population in a parametric family $\mathcal{P} = \{P_\theta : \theta \in \Theta\}$, where $\Theta \subset \mathcal{R}^k$ for a fixed integer $k \geq 1$.

4.1 Bayes Decisions and Estimators

Bayes rules are introduced in §2.3.2 as decision rules minimizing the average risk w.r.t. a given probability measure Π on Θ. Bayes rules, however, are optimal rules in the *Bayesian approach*, which is fundamentally different from the classical frequentist approach that we have been adopting.

4.1.1 Bayes actions

In the Bayesian approach, θ is viewed as a realization of a random vector $\boldsymbol{\theta}$ whose *prior* distribution is Π. The prior distribution is based on past experience, past data, or a statistician's belief and thus may be very subjective. A sample X is drawn from $P_\theta = P_{x|\theta}$, which is viewed as the conditional distribution of X given $\boldsymbol{\theta} = \theta$. The sample $X = x$ is then used to obtain an updated prior distribution, which is called the *posterior* distribution and can be derived as follows. By Theorem 1.7, the joint distribution of X and $\boldsymbol{\theta}$ is a probability measure on $\mathcal{X} \times \Theta$ determined by

$$P(A \times B) = \int_B P_{x|\theta}(A) d\Pi(\theta), \qquad A \in \mathcal{B}_{\mathcal{X}}, \ B \in \mathcal{B}_\Theta,$$

where $\mathcal{X}$ is the range of X. The posterior distribution of $\boldsymbol{\theta}$, given $X = x$, is the conditional distribution $P_{\theta|x}$ whose existence is guaranteed by Theorem 1.7 for almost all $x \in \mathcal{X}$. When $P_{x|\theta}$ has a p.d.f., the following result provides a formula for the p.d.f. of the posterior distribution $P_{\theta|x}$.

Theorem 4.1 (Bayes formula). Assume that $\mathcal{P} = \{P_{x|\theta} : \theta \in \Theta\}$ is dominated by a σ-finite measure ν and $f_\theta(x) = \frac{dP_{x|\theta}}{d\nu}(x)$ is a Borel function on $(\mathcal{X} \times \Theta, \sigma(\mathcal{B}_{\mathcal{X}} \times \mathcal{B}_\Theta))$. Let Π be a prior distribution on Θ. Suppose that $m(x) = \int_\Theta f_\theta(x) d\Pi > 0$.
(i) The posterior distribution $P_{\theta|x} \ll \Pi$ and
$$\frac{dP_{\theta|x}}{d\Pi} = \frac{f_\theta(x)}{m(x)}.$$
(ii) If $\Pi \ll \lambda$ and $\frac{d\Pi}{d\lambda} = \pi(\theta)$ for a σ-finite measure λ, then
$$\frac{dP_{\theta|x}}{d\lambda} = \frac{f_\theta(x)\pi(\theta)}{m(x)}. \tag{4.1}$$

Proof. Result (ii) follows from result (i) and Proposition 1.7(iii). To show (i), we first show that $m(x) < \infty$ a.e. ν. Note that
$$\int_\mathcal{X} m(x) d\nu = \int_\mathcal{X} \int_\Theta f_\theta(x) d\Pi d\nu = \int_\Theta \int_\mathcal{X} f_\theta(x) d\nu d\Pi = 1, \tag{4.2}$$
where the second equality follows from Fubini's theorem. Thus, $m(x)$ is integrable w.r.t. ν and $m(x) < \infty$ a.e. ν.

For $x \in \mathcal{X}$ with $m(x) < \infty$, define
$$P(B, x) = \frac{1}{m(x)} \int_B f_\theta(x) d\Pi, \qquad B \in \mathcal{B}_\Theta.$$

Then $P(\cdot, x)$ is a probability measure on Θ a.e. ν. By Theorem 1.7, it remains to show that
$$P(B, x) = P(\boldsymbol{\theta} \in B | X = x).$$

By Fubini's theorem, $P(B, \cdot)$ is a measurable function of x. Let $P_{x,\theta}$ denote the "joint" distribution of $(X, \boldsymbol{\theta})$. For any $A \in \sigma(X)$,
$$\int_{A \times \Theta} I_B(\theta) dP_{x,\theta} = \int_A \int_B f_\theta(x) d\Pi d\nu$$
$$= \int_A \left[\int_B \frac{f_\theta(x)}{m(x)} d\Pi \right] \left[\int_\Theta f_\theta(x) d\Pi \right] d\nu$$
$$= \int_\Theta \int_A \left[\int_B \frac{f_\theta(x)}{m(x)} d\Pi \right] f_\theta(x) d\nu d\Pi$$
$$= \int_{A \times \Theta} P(B, x) dP_{x,\theta},$$

where the third equality follows from Fubini's theorem. This completes the proof. ∎

Because of (4.2), $m(x)$ is called the marginal p.d.f. of X w.r.t. ν. If $m(x) = 0$ for an $x \in \mathcal{X}$, then $f_\theta(x) = 0$ a.s. Π. Thus, either x should be eliminated from $\mathcal{X}$ or the prior Π is incorrect and a new prior should be specified. Therefore, without loss of generality we may assume that the assumption of $m(x) > 0$ in Theorem 4.1 is always satisfied.

If both X and θ are discrete and ν and λ are the counting measures, then (4.1) becomes

$$P(\theta = \theta | X = x) = \frac{P(X = x | \theta = \theta) P(\theta = \theta)}{\sum_{\theta \in \Theta} P(X = x | \theta = \theta) P(\theta = \theta)},$$

which is the Bayes formula that appears in elementary probability.

In the Bayesian approach, the posterior distribution $P_{\theta|x}$ contains all the information we have about θ and, therefore, statistical decisions and inference should be made based on $P_{\theta|x}$, conditional on the observed $X = x$. In the problem of estimating θ, $P_{\theta|x}$ can be viewed as a randomized decision rule under the approach discussed in §2.3.

Definition 4.1. Let $\mathbb{A}$ be an action space in a decision problem and $L(\theta, a) \geq 0$ be a loss function. For any $x \in \mathcal{X}$, a *Bayes action* w.r.t. Π is any $\delta(x) \in \mathbb{A}$ such that

$$E[L(\theta, \delta(x)) | X = x] = \min_{a \in \mathbb{A}} E[L(\theta, a) | X = x], \qquad (4.3)$$

where the expectation is w.r.t. the posterior distribution $P_{\theta|x}$. ∎

The existence and uniqueness of Bayes actions can be discussed under some conditions on the loss function and the action space.

Proposition 4.1. Assume that the conditions in Theorem 4.1 hold; $L(\theta, a)$ is convex in a for each fixed θ; and for each $x \in \mathcal{X}$, $E[L(\theta, a) | X = x] < \infty$ for some a.
(i) If $\mathbb{A}$ is a compact subset of $\mathcal{R}^p$ for some integer $p \geq 1$, then a Bayes action $\delta(x)$ exists for each $x \in \mathcal{X}$.
(ii) If $\mathbb{A} = \mathcal{R}^p$ and $L(\theta, a)$ tends to ∞ as $\|a\| \to \infty$ uniformly in $\theta \in \Theta_0 \subset \Theta$ with $\Pi(\Theta_0) > 0$, then a Bayes action $\delta(x)$ exists for each $x \in \mathcal{X}$.
(iii) In (i) or (ii), if $L(\theta, a)$ is strictly convex in a for each fixed θ, then the Bayes action is unique.
Proof. The convexity of the loss function implies the convexity and continuity of $E[L(\theta, a) | X = x]$ as a function of a with any fixed x. Then, the result in (i) follows from the fact that any continuous function on a compact

set attains its minimum. The result in (ii) follows from the fact that

$$\lim_{\|a\| \to \infty} E[L(\boldsymbol{\theta}, a)|X = x] \geq \lim_{\|a\| \to \infty} \int_{\Theta_0} L(\theta, a) dP_{\theta|x} = \infty$$

under the assumed condition in (ii). Finally, the result in (iii) follows from the fact that $E[L(\boldsymbol{\theta}, a)|X = x]$ is strictly convex in a for any fixed x under the assumed conditions. ∎

Other conditions on L under which a Bayes action exists can be found, for example, in Lehmann (1983, §1.6 and §4.1).

Example 4.1. Consider the estimation of $\vartheta = g(\theta)$ for some real-valued function g such that $\int_{\Theta}[g(\theta)]^2 d\Pi < \infty$. Suppose that $\mathbb{A} =$ the range of $g(\theta)$ and $L(\theta, a) = [g(\theta) - a]^2$ (squared error loss). Using the same argument as in Example 1.22, we obtain the Bayes action

$$\delta(x) = \frac{\int_{\Theta} g(\theta) f_\theta(x) d\Pi}{m(x)} = \frac{\int_{\Theta} g(\theta) f_\theta(x) d\Pi}{\int_{\Theta} f_\theta(x) d\Pi}, \tag{4.4}$$

which is the posterior expectation of $g(\boldsymbol{\theta})$, given $X = x$.

More specifically, let us consider the case where $g(\theta) = \theta^j$ for some integer $j \geq 1$, $f_\theta(x) = e^{-\theta} \theta^x I_{\{0,1,2,\dots\}}(x)/x!$ (the Poisson distribution) with $\theta > 0$, and Π has a Lebesgue p.d.f. $\pi(\theta) = \theta^{\alpha-1} e^{-\theta/\gamma} I_{(0,\infty)}(\theta)/[\Gamma(\alpha)\gamma^\alpha]$ (the gamma distribution $\Gamma(\alpha, \gamma)$ with known $\alpha > 0$ and $\gamma > 0$). Then, for $x = 0, 1, 2, \dots$,

$$\frac{f_\theta(x)\pi(\theta)}{m(x)} = c(x)\theta^{x+\alpha-1} e^{-\theta(\gamma+1)/\gamma} I_{(0,\infty)}(\theta), \tag{4.5}$$

where $c(x)$ is some function of x. By using Theorem 4.1 and matching the right-hand side of (4.5) with that of the p.d.f. of the gamma distribution, we know that the posterior is the gamma distribution $\Gamma(x + \alpha, \gamma/(\gamma + 1))$. Hence, without actually working out the integral $m(x)$, we know that $c(x) = (1 + \gamma^{-1})^{x+\alpha}/\Gamma(x + \alpha)$. Then

$$\delta(x) = c(x) \int_0^\infty \theta^{j+x+\alpha-1} e^{-\theta(\gamma+1)/\gamma} d\theta.$$

Note that the integrand is proportional to the p.d.f. of the gamma distribution $\Gamma(j + x + \alpha, \gamma/(\gamma + 1))$. Hence

$$\delta(x) = c(x)\Gamma(j + x + \alpha)/(1 + \gamma^{-1})^{j+x+\alpha}$$
$$= (j + x + \alpha - 1) \cdots (x + \alpha)/(1 + \gamma^{-1})^j.$$

In particular, $\delta(x) = (x + \alpha)\gamma/(\gamma + 1)$ when $j = 1$. ∎

An interesting phenomenon in Example 4.1 is that the prior and the posterior are in the same parametric family of distributions. Such a prior is called a *conjugate* prior. Under a conjugate prior, Bayes actions often have explicit forms (in x) when the loss function is simple. Whether a prior is conjugate involves a pair of families; one is the family $\mathcal{P} = \{f_\theta : \theta \in \Theta\}$ and the other is the family from which Π is chosen. Example 4.1 shows that the Poisson family and the gamma family produce conjugate priors. It can be shown (exercise) that many pairs of families in Table 1.1 (page 18) and Table 1.2 (pages 20-21) produce conjugate priors.

In general, numerical methods have to be used in evaluating the integrals in (4.4) or Bayes actions under general loss functions. Even under a conjugate prior, the integral in (4.4) involving a general g may not have an explicit form. More discussions on the computation of Bayes actions are given in §4.1.4.

As an example of deriving a Bayes action in a general decision problem, we consider Example 2.21.

Example 4.2. Consider the decision problem in Example 2.21. Let $P_{\theta|x}$ be the posterior distribution of θ, given $X = x$. In this problem, $\mathbb{A} = \{a_1, a_2, a_3\}$, which is compact in $\mathcal{R}$. By Proposition 4.1, we know that there is a Bayes action if the mean of $P_{\theta|x}$ is finite. Let $E_{\theta|x}$ be the expectation w.r.t. $P_{\theta|x}$. Since $\mathbb{A}$ contains only three elements, a Bayes action can be obtained by comparing

$$E_{\theta|x}[L(\theta, a_j)] = \begin{cases} c_1 & j = 1 \\ c_2 + c_3 E_{\theta|x}[\psi(\theta, t)] & j = 2 \\ c_3 E_{\theta|x}[\psi(\theta, 0)] & j = 3, \end{cases}$$

where $\psi(\theta, t) = (\theta - \theta_0 - t)I_{(\theta_0+t,\infty)}(\theta)$. ∎

The minimization problem (4.3) is the same as the minimization problem

$$\int_\Theta L(\theta, \delta(x)) f_\theta(x) d\Pi = \min_{a \in \mathbb{A}} \int_\Theta L(\theta, a) f_\theta(x) d\Pi. \qquad (4.6)$$

The minimization problem (4.6) is still defined even if Π is not a probability measure but a σ-finite measure on Θ, in which case $m(x)$ may not be finite. If $\Pi(\Theta) \neq 1$, Π is called an *improper prior*. A prior with $\Pi(\Theta) = 1$ is then called a *proper prior*. An action $\delta(x)$ that satisfies (4.6) with an improper prior is called a *generalized Bayes action*.

The following is a reason why we need to discuss improper priors and generalized Bayes actions. In many cases, one has no past information and has to choose a prior subjectively. In such cases, one would like to select a *noninformative* prior that tries to treat all parameter values in Θ

equitably. A noninformative prior is often improper. We only provide one example here. For more detailed discussions of the use of improper priors, see Jeffreys (1939, 1948, 1961), Box and Tiao (1973), and Berger (1985).

Example 4.3. Suppose that $X = (X_1, ..., X_n)$ and X_i's are i.i.d. from $N(\mu, \sigma^2)$, where $\mu \in \Theta \subset \mathcal{R}$ is unknown and σ^2 is known. Consider the estimation of $\vartheta = \mu$ under the squared error loss. If $\Theta = [a, b]$ with $-\infty < a < b < \infty$, then a noninformative prior that treats all parameter values equitably is the uniform distribution on $[a, b]$. If $\Theta = \mathcal{R}$, however, the corresponding "uniform distribution" is the Lebesgue measure on $\mathcal{R}$, which is an improper prior. If Π is the Lebesgue measure on $\mathcal{R}$, then

$$(2\pi\sigma^2)^{-n/2} \int_{-\infty}^{\infty} \mu^2 \exp\left\{ -\sum_{i=1}^{n} \frac{(x_i - \mu)^2}{2\sigma^2} \right\} d\mu < \infty.$$

By differentiating a in

$$(2\pi\sigma^2)^{-n/2} \int_{-\infty}^{\infty} (\mu - a)^2 \exp\left\{ -\sum_{i=1}^{n} \frac{(x_i - \mu)^2}{2\sigma^2} \right\} d\mu$$

and using the fact that $\sum_{i=1}^{n}(x_i - \mu)^2 = \sum_{i=1}^{n}(x_i - \bar{x})^2 + n(\bar{x} - \mu)^2$, where $\bar{x}$ is the sample mean of the observations $x_1, ..., x_n$, we obtain that

$$\delta(x) = \frac{\int_{-\infty}^{\infty} \mu \exp\left\{ -n(\bar{x} - \mu)^2/(2\sigma^2) \right\} d\mu}{\int_{-\infty}^{\infty} \exp\left\{ -n(\bar{x} - \mu)^2/(2\sigma^2) \right\} d\mu} = \bar{x}.$$

Thus, the sample mean is a generalized Bayes action under the squared error loss. From Example 2.25 and Exercise 91 in §2.6, if Π is $N(\mu_0, \sigma_0^2)$, then the Bayes action is $\mu_*(x)$ in (2.25). Note that in this case $\bar{x}$ is a limit of $\mu_*(x)$ as $\sigma_0^2 \to \infty$. ∎

4.1.2 Empirical and hierarchical Bayes methods

A Bayes action depends on the chosen prior that may depend on some parameters called *hyperparameters*. In §4.1.1, hyperparameters are assumed to be known. If hyperparameters are unknown, one way to solve the problem is to estimate them using data $x_1, ..., x_n$; the resulting Bayes action is called an *empirical Bayes* action.

The simplest empirical Bayes method is to estimate prior parameters by viewing $x = (x_1, ..., x_n)$ as a "sample" from the marginal distribution

$$P_{x|\xi}(A) = \int_{\Theta} P_{x|\theta}(A) d\Pi_{\theta|\xi}, \qquad A \in \mathcal{B}_x,$$

where $\Pi_{\theta|\xi}$ is a prior depending on an unknown vector ξ of hyperparameters, or from the marginal p.d.f. $m(x)$ in (4.2), if $P_{x|\theta}$ has a p.d.f. f_θ. The method of moments introduced in §3.5.3, for example, can be applied to estimate ξ. We consider an example.

Example 4.4. Let $X = (X_1, ..., X_n)$ and X_i's be i.i.d. from $N(\mu, \sigma^2)$ with an unknown $\mu \in \mathcal{R}$ and a known σ^2. Consider the prior $\Pi_{\mu|\xi} = N(\mu_0, \sigma_0^2)$ with $\xi = (\mu_0, \sigma_0^2)$. To obtain a moment estimate of ξ, we need to calculate

$$\int_{\mathcal{R}^n} x_1 m(x) dx \qquad \text{and} \qquad \int_{\mathcal{R}^n} x_1^2 m(x) dx,$$

where $x = (x_1, ..., x_n)$. These two integrals can be obtained without calculating $m(x)$. Note that

$$\int_{\mathcal{R}^n} x_1 m(x) dx = \int_{\mathcal{R}^n} \int_{\Theta} x_1 f_\mu(x) dx d\Pi_{\mu|\xi} = \int_{\mathcal{R}} \mu d\Pi_{\mu|\xi} = \mu_0$$

and

$$\int_{\mathcal{R}^n} x_1^2 m(x) dx = \int_{\mathcal{R}^n} \int_{\Theta} x_1^2 f_\mu(x) dx d\Pi_{\mu|\xi} = \sigma^2 + \int_{\mathcal{R}} \mu^2 d\Pi_{\mu|\xi} = \sigma^2 + \mu_0^2 + \sigma_0^2.$$

Thus, by viewing $x_1, ..., x_n$ as a sample from $m(x)$, we obtain the moment estimates

$$\hat{\mu}_0 = \bar{x} \qquad \text{and} \qquad \hat{\sigma}_0^2 = \frac{1}{n} \sum_{i=1}^n (x_i - \bar{x})^2 - \sigma^2,$$

where $\bar{x}$ is the sample mean of x_i's. Replacing μ_0 and σ_0^2 in formula (2.25) (Example 2.25) by $\hat{\mu}_0$ and $\hat{\sigma}_0^2$, respectively, we find that the empirical Bayes action under the squared error loss is simply the sample mean $\bar{x}$ (which is a generalized Bayes action; see Example 4.3). ∎

Note that $\hat{\sigma}_0^2$ in Example 4.4 can be negative. Better empirical Bayes methods can be found, for example, in Berger (1985, §4.5). The following method, called the *hierarchical Bayes* method, is generally better than empirical Bayes methods.

Instead of estimating hyperparameters, in the hierarchical Bayes approach we put a prior on hyperparameters. Let $\Pi_{\theta|\xi}$ be a (first-stage) prior with a hyperparameter vector ξ and let Λ be a prior on Ξ, the range of ξ. Then the "marginal" prior for θ is defined by

$$\Pi(B) = \int_{\Xi} \Pi_{\theta|\xi}(B) d\Lambda(\xi), \qquad B \in \mathcal{B}_{\Theta}. \tag{4.7}$$

If the second-stage prior Λ also depends on some unknown hyperparameters, then one can go on to consider a third-stage prior. In most applications, however, two-stage priors are sufficient, since misspecifying a second-stage prior is much less serious than misspecifying a first-stage prior (Berger, 1985, §4.6). In addition, the second-stage prior can be chosen to be noninformative (improper).

Bayes actions can be obtained in the same way as before using the prior in (4.7). Thus, the hierarchical Bayes method is simply a Bayes method with a hierarchical prior. Empirical Bayes methods, however, deviate from the Bayes method since $x_1, ..., x_n$ are used to estimate hyperparameters.

Suppose that X has a p.d.f. $f_\theta(x)$ w.r.t. a σ-finite measure ν and $P_{\theta|\xi}$ has a p.d.f. $\pi_{\theta|\xi}(\theta)$ w.r.t. a σ-finite measure κ. Then the prior Π in (4.7) has a p.d.f.

$$\pi(\theta) = \int_\Xi \pi_{\theta|\xi}(\theta) d\Lambda(\xi)$$

w.r.t. κ and

$$m(x) = \int_\Theta \int_\Xi f_\theta(x) \pi_{\theta|\xi}(\theta) d\Lambda d\kappa.$$

Let $P_{\theta|x,\xi}$ be the posterior distribution of $\boldsymbol{\theta}$ given x and ξ (or ξ is assumed known) and

$$m_{x|\xi}(x) = \int_\Theta f_\theta(x) \pi_{\theta|\xi}(\theta) d\kappa,$$

which is the marginal of X given θ and ξ (or ξ is assumed known). Then the posterior distribution $P_{\theta|x}$ has a p.d.f.

$$\frac{dP_{\theta|x}}{d\kappa} = \frac{f_\theta(x)\pi(\theta)}{m(x)}$$

$$= \int_\Xi \frac{f_\theta(x)\pi_{\theta|\xi}(\theta)}{m(x)} d\Lambda(\xi)$$

$$= \int_\Xi \frac{f_\theta(x)\pi_{\theta|\xi}(\theta)}{m_{x|\xi}(x)} \frac{m_{x|\xi}(x)}{m(x)} d\Lambda(\xi)$$

$$= \int_\Xi \frac{dP_{\theta|x,\xi}}{d\kappa} dP_{\xi|x},$$

where $P_{\xi|x}$ is the posterior distribution of ξ given x. Thus, under the estimation problem considered in Example 4.1, the (hierarchical) Bayes action is

$$\delta(x) = \int_\Xi \delta(x,\xi) dP_{\xi|x}, \tag{4.8}$$

where $\delta(x,\xi)$ is the Bayes action when ξ is known. A result similar to (4.8) is given in Lemma 4.1.

Example 4.5. Consider Example 4.4 again. Suppose that one of the parameters in the first-stage prior $N(\mu_0, \sigma_0^2)$, μ_0, is unknown and σ_0^2 is known. Let the second-stage prior for $\xi = \mu_0$ be the Lebesgue measure on $\mathcal{R}$ (improper prior). From Example 2.25,

$$\delta(x, \xi) = \frac{\sigma^2}{n\sigma_0^2 + \sigma^2}\xi + \frac{n\sigma_0^2}{n\sigma_0^2 + \sigma^2}\bar{x}.$$

To obtain the Bayes action $\delta(x)$, it suffices to calculate $E_{\xi|x}(\xi)$, where the expectation is w.r.t. $P_{\xi|x}$. Note that the p.d.f. of $P_{\xi|x}$ is proportional to

$$\psi(\xi) = \int_{-\infty}^{\infty} \exp\left\{-\frac{n(\bar{x}-\mu)^2}{2\sigma^2} - \frac{(\mu-\xi)^2}{2\sigma_0^2}\right\}d\mu.$$

Using the properties of normal distributions, one can show that

$$\psi(\xi) = C_1\exp\left\{\left(\frac{n}{2\sigma^2} + \frac{1}{2\sigma_0^2}\right)^{-1}\left(\frac{n\bar{x}}{2\sigma^2} + \frac{\xi}{2\sigma_0^2}\right)^2 - \frac{\xi^2}{2\sigma_0^2}\right\}$$

$$= C_2\exp\left\{-\frac{n\xi^2}{2(n\sigma_0^2+\sigma^2)} + \frac{n\bar{x}\xi}{n\sigma_0^2+\sigma^2}\right\}$$

$$= C_3\exp\left\{-\frac{n(\xi-\bar{x})^2}{2(n\sigma_0^2+\sigma^2)}\right\},$$

where C_1, C_2, and C_3 are quantities not depending on ξ. Hence $E_{\xi|x}(\xi) = \bar{x}$. The (hierarchical) generalized Bayes action is then

$$\delta(x) = \frac{\sigma^2}{n\sigma_0^2 + \sigma^2}E_{\xi|x}(\xi) + \frac{n\sigma_0^2}{n\sigma_0^2 + \sigma^2}\bar{x} = \bar{x}. \quad \blacksquare$$

4.1.3 Bayes rules and estimators

The discussion in §4.1.1 and §4.1.2 is more general than point estimation and adopts an approach that is different from the frequentist approach used in the rest of this book. In the frequentist approach, if a Bayes action $\delta(x)$ is a measurable function of x, then $\delta(X)$ is a nonrandomized decision rule. It can be shown (exercise) that $\delta(X)$ defined in Definition 4.1 (if it exists for $X = x \in A$ with $\int_\Theta P_\theta(A)d\Pi = 1$) also minimizes the Bayes risk

$$r_T(\Pi) = \int_\Theta R_T(\theta)d\Pi$$

over all decision rules T (randomized or nonrandomized), where $R_T(\theta)$ is the risk function of T defined in (2.22). Thus, $\delta(X)$ is a Bayes rule (§2.3.2). In an estimation problem, a Bayes rule is called a *Bayes estimator*.

Generalized Bayes risks, generalized Bayes rules (or estimators), and empirical Bayes rules (or estimators) can be defined similarly.

In view of the discussion in §2.3.2, even if we do not adopt the Bayesian approach, the method described in §4.1.1 can be used as a way of generating decision rules. In this section, we study a Bayes rule or estimator in terms of its risk (and bias and consistency for a Bayes estimator).

Bayes rules are typically admissible since, if there is a rule better than a Bayes rule, then that rule has the same Bayes risk as the Bayes rule and, therefore, is itself a Bayes rule. This actually proves part (i) of the following result. The proof of the other parts of the following result is left as an exercise.

Theorem 4.2. In a decision problem, let $\delta(X)$ be a Bayes rule w.r.t. a prior Π.
(i) If $\delta(X)$ is a unique Bayes rule, then $\delta(X)$ is admissible.
(ii) If Θ is a countable set, the Bayes risk $r_\delta(\Pi) < \infty$, and Π gives positive probability to each $\theta \in \Theta$, then $\delta(X)$ is admissible.
(iii) Let $\Im$ be the class of decision rules having continuous risk functions. If $\delta(X) \in \Im$, $r_\delta(\Pi) < \infty$, and Π gives positive probability to any open subset of Θ, then $\delta(X)$ is $\Im$-admissible. ∎

Generalized Bayes rules or estimators are not necessarily admissible. Many generalized Bayes rules are limits of Bayes rules (see Examples 4.3 and 4.7). Limits of Bayes rules are often admissible (Farrell, 1968a,b). The following result shows a technique of proving admissibility using limits of generalized Bayes risks.

Theorem 4.3. Suppose that Θ is an open set of $\mathcal{R}^k$. In a decision problem, let $\Im$ be the class of decision rules having continuous risk functions. A decision rule $T \in \Im$ is $\Im$-admissible if there exists a sequence $\{\Pi_j\}$ of (possibly improper) priors such that (a) the generalized Bayes risks $r_T(\Pi_j)$ are finite for all j; (b) for any $\theta_0 \in \Theta$ and $\eta > 0$,

$$\lim_{j \to \infty} \frac{r_T(\Pi_j) - r_j^*(\Pi_j)}{\Pi_j(O_{\theta_0, \eta})} = 0,$$

where $r_j^*(\Pi_j) = \inf_{T \in \Im} r_T(\Pi_j)$ and $O_{\theta_0, \eta} = \{\theta \in \Theta : \|\theta - \theta_0\| < \eta\}$ with $\Pi_j(O_{\theta_0, \eta}) < \infty$ for all j.
Proof. Suppose that T is not $\Im$-admissible. Then there exists $T_0 \in \Im$ such that $R_{T_0}(\theta) \leq R_T(\theta)$ for all θ and $R_{T_0}(\theta_0) < R_T(\theta_0)$ for a $\theta_0 \in \Theta$. From the continuity of the risk functions, we conclude that $R_{T_0}(\theta) < R_T(\theta) - \epsilon$ for all $\theta \in O_{\theta_0, \eta}$ and some constants $\epsilon > 0$ and $\eta > 0$. Then, for any j,

$$r_T(\Pi_j) - r_j^*(\Pi_j) \geq r_T(\Pi_j) - r_{T_0}(\Pi_j)$$

$$\geq \int_{O_{\theta_0, \eta}} [R_T(\theta) - R_{T_0}(\theta)] d\Pi_j(\theta)$$

$$\geq \epsilon \Pi_j(O_{\theta_0, \eta}),$$

which contradicts condition (b). Hence, T is $\Im$-admissible. ∎

Example 4.6. Consider Example 4.3 and the estimation of μ under the squared error loss. From Theorem 2.1, the risk function of any decision rule is continuous in μ if the risk is finite. We now apply Theorem 4.3 to show that the sample mean $\bar{X}$ is admissible. Let $\Pi_j = N(0, j)$. Since $R_{\bar{X}}(\mu) = \sigma^2/n$, $r_{\bar{X}}^*(\Pi_j) = \sigma^2/n$ for any j. Hence, condition (a) in Theorem 4.3 is satisfied. From Example 2.25, the Bayes estimator w.r.t. Π_j is $\delta_j(X) = \frac{nj}{nj+\sigma^2}\bar{X}$ (see formula (2.25)). Thus,

$$R_{\delta_j}(\mu) = \frac{\sigma^2 n j^2 + \sigma^4 \mu^2}{(nj + \sigma^2)^2}$$

and

$$r_j^*(\Pi_j) = \int R_{\delta_j}(\mu) d\Pi_j = \frac{\sigma^2 j}{nj + \sigma^2}.$$

For any $O_{\mu_0, \eta} = \{\mu : |\mu - \mu_0| < \eta\}$,

$$\Pi_j(O_{\mu_0, \eta}) = \Phi\left(\frac{\mu_0 + \eta}{\sqrt{j}}\right) - \Phi\left(\frac{\mu_0 - \eta}{\sqrt{j}}\right) = \frac{2\eta \Phi'(\xi_j)}{\sqrt{j}}$$

for some ξ_j satisfying $(\mu_0 - \eta)/\sqrt{j} \leq \xi_j \leq (\mu_0 + \eta)/\sqrt{j}$, where Φ is the standard normal c.d.f. and Φ' is its derivative. Since $\Phi'(\xi_j) \to \Phi'(0) = (2\pi)^{-1/2}$,

$$\frac{r_{\bar{X}}(\Pi_j) - r_j^*(\Pi_j)}{\Pi_j(O_{\mu_0, \eta})} = \frac{\sigma^4 \sqrt{j}}{2\eta \Phi'(\xi_j) n (nj + \sigma^2)} \to 0$$

as $j \to \infty$. Thus, condition (b) in Theorem 4.3 is satisfied and, hence, the sample mean $\bar{X}$ is admissible. ∎

More results in admissibility can be found in §4.2 and §4.3.

The following result concerns the bias of a Bayes estimator.

Proposition 4.2. Let $\delta(X)$ be a Bayes estimator of $\vartheta = g(\theta)$ under the squared error loss. Then $\delta(X)$ is not unbiased unless the Bayes risk $r_\delta(\Pi) = 0$.

Proof. Suppose that $\delta(X)$ is unbiased, i.e., $E[\delta(X)|\theta] = g(\theta)$. Conditioning on θ and using Proposition 1.10, we obtain that

$$E[g(\theta)\delta(X)] = E\{g(\theta)E[\delta(X)|\theta]\} = E[g(\theta)]^2.$$

Since $\delta(X) = E[g(\theta)|X]$, conditioning on X and using Proposition 1.10, we obtain that

$$E[g(\theta)\delta(X)] = E\{\delta(X)E[g(\theta)|X]\} = E[\delta(X)]^2.$$

Then

$$r_\delta(\Pi) = E[\delta(X) - g(\theta)]^2 = E[\delta(X)]^2 + E[g(\theta)]^2 - 2E[g(\theta)\delta(X)] = 0. ∎$$

Since $r_\delta(\Pi) = 0$ occurs usually in some trivial cases, a Bayes estimator is typically not unbiased. Hence, Proposition 4.2 can be used to check whether an estimator can be a Bayes estimator w.r.t. some prior under the squared error loss. However, a generalized Bayes estimator may be unbiased; see, for instance, Examples 4.3 and 4.7.

Bayes estimators are usually consistent and approximately unbiased. In a particular problem, it is usually easy to check directly whether Bayes estimators are consistent and approximately unbiased (Examples 4.7-4.9), especially when Bayes estimators have explicit forms. Bayes estimators also have some other good asymptotic properties, which are studied in §4.5.3.

Let us consider some examples.

Example 4.7. Let $X = (X_1, ..., X_n)$ and X_i's be i.i.d. from the exponential distribution $E(0, \theta)$ with an unknown $\theta > 0$. Let the prior be such that θ^{-1} has the gamma distribution $\Gamma(\alpha, \gamma)$ with known $\alpha > 0$ and $\gamma > 0$. Then the posterior of $\omega = \theta^{-1}$ is the gamma distribution $\Gamma(n + \alpha, (n\bar{X} + \gamma^{-1})^{-1})$ (verify), where $\bar{X}$ is the sample mean.

Consider first the estimation of $\theta = \omega^{-1}$. The Bayes estimator of θ under the squared error loss is

$$\delta(X) = \frac{(n\bar{X} + \gamma^{-1})^{n+\alpha}}{\Gamma(n+\alpha)} \int_0^\infty \omega^{n+\alpha-2} e^{-(n\bar{X}+\gamma^{-1})\omega} d\omega = \frac{n\bar{X} + \gamma^{-1}}{n + \alpha - 1}.$$

The bias of $\delta(X)$ is

$$\frac{n\theta + \gamma^{-1}}{n + \alpha - 1} - \theta = \frac{\gamma^{-1} - (\alpha - 1)\theta}{n + \alpha - 1} = O\left(\frac{1}{n}\right).$$

It is also easy to see that $\delta(X)$ is consistent. The UMVUE of θ is $\bar{X}$. Since $\text{Var}(\bar{X}) = \theta^2/n$, $r_{\bar{x}}(\Pi) > 0$ for any Π and, hence, $\bar{X}$ is not a Bayes estimator. In this case, $\bar{X}$ is the generalized Bayes estimator w.r.t. the improper prior $\frac{d\Pi}{d\omega} = I_{(0,\infty)}(\omega)$ and is a limit of Bayes estimators $\delta(X)$ as $\alpha \to 1$ and $\gamma \to \infty$ (exercise). The admissibility of $\delta(X)$ is considered in Exercises 32 and 80.

Consider next the estimation of $e^{-t/\theta} = e^{-t\omega}$ (see Examples 2.26 and 3.3). The Bayes estimator under the squared error loss is

$$\delta_t(X) = \frac{(n\bar{X} + \gamma^{-1})^{n+\alpha}}{\Gamma(n+\alpha)} \int_0^\infty \omega^{n+\alpha-1} e^{-(n\bar{X}+\gamma^{-1}+t)\omega} d\omega$$

$$= \left(1 + \frac{t}{n\bar{X} + \gamma^{-1}}\right)^{-(n+\alpha)}.$$

Again, this estimator is biased and it is easy to show that $\delta_t(X)$ is consistent as $n \to \infty$. In this case, the UMVUE given in Example 3.3 is neither a Bayes estimator nor a limit of $\delta_t(X)$. ∎

Example 4.8. Let $X = (X_1, ..., X_n)$ and X_i's be i.i.d. from $N(\mu, \sigma^2)$ with unknown $\mu \in \mathcal{R}$ and $\sigma^2 > 0$. Let the prior for $\omega = (2\sigma^2)^{-1}$ be the gamma distribution $\Gamma(\alpha, \gamma)$ with known α and γ and let the prior for μ be $N(\mu_0, \sigma_0^2/\omega)$ (conditional on ω). Then the posterior p.d.f. of (μ, ω) is proportional to

$$\omega^{(n+1)/2+\alpha-1} \exp\left\{ -\left[\gamma^{-1} + Y + n(\bar{X} - \mu)^2 + \frac{(\mu-\mu_0)^2}{2\sigma_0^2} \right] \omega \right\},$$

where $Y = \sum_{i=1}^n (X_i - \bar{X})^2$ and $\bar{X}$ is the sample mean. Note that

$$n(\bar{X} - \mu)^2 + \frac{(\mu-\mu_0)^2}{2\sigma_0^2} = \left(n + \frac{1}{2\sigma_0^2} \right) \mu^2 - 2 \left(n\bar{X} + \frac{\mu_0}{2\sigma_0^2} \right) \mu + n\bar{X}^2 + \frac{\mu_0^2}{2\sigma_0^2}.$$

Hence, the posterior p.d.f. of (μ, ω) is proportional to

$$\omega^{(n+1)/2+\alpha-1} \exp\left\{ -\left[\gamma^{-1} + W + \left(n + \frac{1}{2\sigma_0^2} \right) (\mu - \zeta(X))^2 \right] \omega \right\},$$

where

$$\zeta(X) = \frac{n\bar{X} + \frac{\mu_0}{2\sigma_0^2}}{n + \frac{1}{2\sigma_0^2}} \quad \text{and} \quad W = Y + n\bar{X}^2 + \frac{\mu_0^2}{2\sigma_0^2} - \left(n + \frac{1}{2\sigma_0^2} \right) [\zeta(X)]^2.$$

Thus, the posterior of ω is the gamma distribution $\Gamma(n/2+\alpha, (\gamma^{-1}+W)^{-1})$ and the posterior of μ (given ω and X) is $N(\zeta(X), [(2n+\sigma_0^{-2})\omega]^{-1})$. Under the squared error loss, the Bayes estimator of μ is $\zeta(X)$ and the Bayes estimator of $\sigma^2 = (2\omega)^{-1}$ is $(\gamma^{-1}+W)/(n+2\alpha-2)$, provided that $n+2\alpha > 2$. Apparently, these Bayes estimators are biased but the biases are of the order n^{-1}; and they are consistent as $n \to \infty$. ∎

To consider the last example, we need the following useful lemma whose proof is similar to the proof of result (4.8).

Lemma 4.1. Suppose that X has a p.d.f. $f_\theta(x)$ w.r.t. a σ-finite measure ν. Suppose that $\theta = (\theta_1, \theta_2)$, $\theta_j \in \Theta_j$, and that the prior has a p.d.f.

$$\pi(\theta) = \pi_{\theta_1|\theta_2}(\theta_1)\pi_{\theta_2}(\theta_2),$$

where $\pi_{\theta_2}(\theta_2)$ is a p.d.f. w.r.t. a σ-finite measure ν_2 on Θ_2 and for any given θ_2, $\pi_{\theta_1|\theta_2}(\theta_1)$ is a p.d.f. w.r.t. a σ-finite measure ν_1 on Θ_1. Suppose further that if θ_2 is given, the Bayes estimator of $h(\theta_1) = g(\theta_1, \theta_2)$ under the squared error loss is $\delta(X, \theta_2)$. Then the Bayes estimator of $g(\theta_1, \theta_2)$ under the squared error loss is $\delta(X)$ with

$$\delta(x) = \int_{\Theta_2} \delta(x, \theta_2)p_{\theta_2|x}(\theta_2)d\nu_2,$$

where $p_{\theta_2|x}(\theta_2)$ is the posterior p.d.f. of θ_2 given $X = x$. ∎

Example 4.9. Consider a linear model

$$X_{ij} = \beta^\tau Z_i + \varepsilon_{ij}, \qquad j = 1, ..., n_i, \ i = 1, ..., k,$$

where $\beta \in \mathcal{R}^p$ is unknown, Z_i's are known vectors, ε_{ij}'s are independent, and ε_{ij} is $N(0, \sigma_i^2)$, $j = 1, ..., n_i$, $i = 1, ..., k$. Let X be the sample vector containing all X_{ij}'s. The parameter vector is then $\theta = (\beta, \omega)$, where $\omega = (\omega_1, ..., \omega_k)$ and $\omega_i = (2\sigma_i^2)^{-1}$. Assume that the prior for θ has the Lebesgue p.d.f.

$$c\,\pi(\beta) \prod_{i=1}^{k} \omega_i^\alpha e^{-\omega_i/\gamma}, \tag{4.9}$$

where $\alpha > 0$, $\gamma > 0$, and $c > 0$ are known constants and $\pi(\beta)$ is a known Lebesgue p.d.f. on $\mathcal{R}^p$. The posterior p.d.f. of θ is then proportional to

$$h(X, \theta) = \pi(\beta) \prod_{i=1}^{k} \omega_i^{n_i/2 + \alpha} e^{-[\gamma^{-1} + v_i(\beta)]\omega_i},$$

where $v_i(\beta) = \sum_{j=1}^{n_i} (X_{ij} - \beta^\tau Z_i)^2$. If β is known, the Bayes estimator of σ_i^2 under the squared error loss is

$$\int \frac{1}{2\omega_i} \frac{h(X, \theta)}{\int h(X, \theta)d\omega} d\omega = \frac{\gamma^{-1} + v_i(\beta)}{2\alpha + n_i}.$$

By Lemma 4.1, the Bayes estimator of σ_i^2 is

$$\hat{\sigma}_i^2 = \int \frac{\gamma^{-1} + v_i(\beta)}{2\alpha + n_i} f_{\beta|X}(\beta)d\beta, \tag{4.10}$$

where

$$f_{\beta|X}(\beta) \propto \int h(X, \theta)d\omega$$

$$\propto \pi(\beta) \prod_{i=1}^{k} \int \omega_i^{\alpha + n_i/2} e^{-[\gamma^{-1} + v_i(\beta)]\omega_i} d\omega_i$$

$$\propto \pi(\beta) \prod_{i=1}^{k} \left[\gamma^{-1} + v_i(\beta)\right]^{-(\alpha + 1 + n_i/2)} \tag{4.11}$$

is the posterior p.d.f. of β. The Bayes estimator of $l^\tau \beta$ for any $l \in \mathcal{R}^p$ is then the posterior mean of $l^\tau \beta$ w.r.t. the p.d.f. $f_{\beta|X}(\beta)$.

In this problem, Bayes estimators do not have explicit forms. A numerical method (such as one of those in §4.1.4) has to be used to evaluate Bayes estimators (see Example 4.10).

Let $\bar{X}_{i\cdot}$ and S_i^2 be the sample mean and variance of X_{ij}, $j = 1, ..., n_i$ (S_i^2 is defined to be 0 if $n_i = 1$), and let $\sigma_0^2 = (2\alpha\gamma)^{-1}$ (the prior mean of σ_i^2). Then the Bayes estimator $\hat{\sigma}_i^2$ in (4.10) can be written as

$$\frac{2\alpha}{2\alpha + n_i}\sigma_0^2 + \frac{n_i - 1}{2\alpha + n_i}S_i^2 + \frac{n_i}{2\alpha + n_i}\int(\bar{X}_{i\cdot} - \beta^\tau Z_i)^2 f_{\beta|X}(\beta)d\beta. \qquad (4.12)$$

The Bayes estimator in (4.12) is a weighted average of prior information, "within group" variation, and averaged squared "residuals".

If $n_i \to \infty$, then the first term in (4.12) converges to 0 and the second term in (4.12) is consistent and approximately unbiased for σ_i^2. Hence, the Bayes estimator $\hat{\sigma}_i^2$ is consistent and approximately unbiased for σ_i^2 if the mean of the last term in (4.12) tends to 0, which is true under some conditions (see, e.g., Exercise 36). It is easy to see that $\hat{\sigma}_i^2$ is consistent and approximately unbiased for σ_i^2 w.r.t. the joint distribution of $(X, \boldsymbol{\theta})$, since the mean of the last term in (4.12) w.r.t. the joint distribution of $(X, \boldsymbol{\theta})$ is bounded by σ_0^2/n_i. ∎

4.1.4 Markov chain Monte Carlo

As we discussed previously, Bayes actions or estimators have to be computed numerically in many applications. Typically we need to compute an integral of the form

$$E_p(g) = \int_\Theta g(\theta)p(\theta)d\nu$$

with some function g, where $p(\theta)$ is a p.d.f. w.r.t. a σ-finite measure ν on $(\Theta, \mathcal{B}_\Theta)$ and $\Theta \subset \mathcal{R}^k$. For example, if g is an indicator function of $A \in \mathcal{B}_\Theta$ and $p(\theta)$ is the posterior p.d.f. of θ given $X = x$, then $E_p(g)$ is the posterior probability of A; under the squared error loss, $E_p(g)$ is the Bayes action (4.4) if $p(\theta)$ is the posterior p.d.f.

There are many numerical methods for computing integrals $E_p(g)$; see, for example, §4.5.3 and Berger (1985, §4.9). In this section, we discuss the *Markov chain Monte Carlo* (MCMC) methods, which are powerful numerical methods not only for Bayesian computations, but also for general statistical computing (see, e.g., §4.4.1).

We start with the simple Monte Carlo method, which can be viewed as a special case of the MCMC. Suppose that we can generate i.i.d. $\theta^{(1)}, ..., \theta^{(m)}$ from a p.d.f. $h(\theta) > 0$ w.r.t. ν. By the SLLN (Theorem 1.13(ii)), as $m \to \infty$,

$$\hat{E}_p(g) = \frac{1}{m}\sum_{j=1}^m \frac{g(\theta^{(j)})p(\theta^{(j)})}{h(\theta^{(j)})} \to_{a.s.} \int_\Theta \frac{g(\theta)p(\theta)}{h(\theta)}h(\theta)d\nu = E_p(g).$$

Hence $\hat{E}_p(g)$ can be used as a numerical approximation to $E_p(g)$. The process of generating $\theta^{(j)}$ according to h is called *importance sampling* and

$h(\theta)$ is called the *importance function*. More discussions on importance sampling can be found, for example, in Berger (1985), Geweke (1989), Shao (1989), and Tanner (1996). When $p(\theta)$ is intractable or complex, it is often difficult to choose a function h that is simple enough for importance sampling and results in a fast convergence of $\hat{E}_p(g)$ as well.

The simple Monte Carlo method, however, may not work well when k, the dimension of Θ, is large. This is because, when k is large, the convergence of $\hat{E}_p(g)$ requires a very large m; generating a random vector from a k-dimensional distribution is usually expensive, if not impossible. More sophisticated MCMC methods are different from the simple Monte Carlo in two aspects: generating random vectors can be done using distributions whose dimensions are much lower than k; and $\theta^{(1)}, ..., \theta^{(m)}$ are not independent, but form a Markov chain.

Let $\{Y^{(t)} : t = 0, 1, ...\}$ be a Markov chain (§1.4.4) taking values in $\mathcal{Y} \subset \mathcal{R}^k$. $\{Y^{(t)}\}$ is homogeneous if and only if

$$P(Y^{(t+1)} \in A | Y^{(t)}) = P(Y^{(1)} \in A | Y^{(0)})$$

for any t. For a homogeneous Markov chain $\{Y^{(t)}\}$, define

$$P(y, A) = P(Y^{(1)} \in A | Y^{(0)} = y), \qquad y \in \mathcal{Y}, \ A \in \mathcal{B}_{\mathcal{Y}},$$

which is called the transition kernel of the Markov chain. Note that $P(y, \cdot)$ is a probability measure for every $y \in \mathcal{Y}$; $P(\cdot, A)$ is a Borel function for every $A \in \mathcal{B}_{\mathcal{Y}}$; and the distribution of a homogeneous Markov chain is determined by $P(y, A)$ and the distribution of $Y^{(0)}$ (initial distribution). MCMC approximates an integral of the form $\int_{\mathcal{Y}} g(y)p(y)d\nu$ by $m^{-1} \sum_{t=1}^{m} g(Y^{(t)})$ with a Markov chain $\{Y^{(t)} : t = 0, 1, ...\}$. The basic justification of the MCMC approximation is given in the following result.

Theorem 4.4. Let $p(y)$ be a p.d.f. on $\mathcal{Y}$ w.r.t. a σ-finite measure ν and g be a Borel function on $\mathcal{Y}$ with $\int_{\mathcal{Y}} |g(y)|p(y)d\nu < \infty$. Let $\{Y^{(t)} : t = 0, 1, ...\}$ be a homogeneous Markov chain taking values on $\mathcal{Y} \subset \mathcal{R}^k$ with the transition kernel $P(y, A)$. Then

$$\frac{1}{m} \sum_{t=1}^{m} g(Y^{(t)}) \to_{a.s.} \int_{\mathcal{Y}} g(y)p(y)d\nu \tag{4.13}$$

and, as $t \to \infty$,

$$P^t(y, A) = P(Y^{(t)} \in A | Y^{(0)} = y) \to_{a.s.} \int_A p(y)d\nu, \tag{4.14}$$

provided that
(a) the Markov chain is *aperiodic* in the sense that there does not exist $d \geq 2$

nonempty disjoint events $A_0, ..., A_{d-1}$ in $\mathcal{B}_y$ such that for all $i = 0, ..., d-1$ and all $y \in A_i$, $P(y, A_j) = 1$ for $j = i+1 \pmod d$;
(b) the Markov chain is p-*invariant* in the sense that $\int P(y, A)p(y)d\nu = \int_A p(y)d\nu$ for all $A \in \mathcal{B}_y$;
(c) the Markov chain is p-*irreducible* in the sense that for any $y \in \mathcal{Y}$ and any A with $\int_A p(y)d\nu > 0$, there exists a positive integer t such that $P^t(y, A)$ in (4.14) is positive; and
(d) the Markov chain is *Harris recurrent* in the sense that for any A with $\int_A p(y)d\nu > 0$, $P\left(\sum_{t=1}^{\infty} I_A(Y^{(t)}) = \infty | Y^{(0)} = y\right) = 1$ for all y.   ∎

The proof of these results is beyond the scope of this book and, hence, is omitted. It can be found, for example, in Nummelin (1984), Chan (1993), and Tierney (1994). A homogeneous Markov chain satisfying conditions (a)-(d) in Theorem 4.4 is called *ergodic* with *equilibrium distribution p*. Result (4.13) means that the MCMC approximation is consistent and result (4.14) indicates that p is the limiting p.d.f. of the Markov chain.

One of the key issues in MCMC is the choice of the kernel $P(y, A)$. The first requirement on $P(y, A)$ is that conditions (a)-(d) in Theorem 4.4 be satisfied. Condition (a) is usually easy to check for any given $P(y, A)$. In the following, we consider two popular MCMC methods satisfying conditions (a)-(d).

Gibbs sampler

One way to construct a p-invariant homogeneous Markov chain is to use conditioning. Suppose that Y has the p.d.f. $p(y)$. Let Y_i (or y_i) be the ith component of Y (or y) and let Y_{-i} (or y_{-i}) be the $(k-1)$-vector containing all components of Y (or y) except Y_i (or y_i). Then

$$P_i(y_{-i}, A) = P(Y \in A | Y_{-i} = y_{-i})$$

is a transition kernel for any i. The MCMC method using this kernel is called the *single-site Gibbs sampler*. Note that

$$\int P_i(y_{-i}, A)p(y)d\nu = E[P(Y \in A|Y_{-i})] = P(Y \in A) = \int_A p(y)d\nu$$

and, therefore, the chain with kernel $P_i(y_{-i}, A)$ is p-invariant. However, this chain is not p-irreducible since $P_i(y_{-i}, \cdot)$ puts all its mass on the set $\psi_i^{-1}(y_{-i})$, where $\psi_i(y) = y_{-i}$. Gelfand and Smith (1990) considered a *systematic scan Gibbs sampler* whose kernel $P(y, A)$ is a composite of k kernels $P_i(y_{-i}, A)$, $i = 1, ..., k$. More precisely, the chain is defined as follows. Given $Y^{(t-1)} = y^{(t-1)}$, we generate $y_1^{(t)}$ from $P_1(y_2^{(t-1)}, ..., y_k^{(t-1)}, \cdot), ..., y_j^{(t)}$ from $P_j(y_1^{(t)}, ..., y_{j-1}^{(t)}, y_{j+1}^{(t-1)}, ..., y_k^{(t-1)}, \cdot), ..., y_k^{(t)}$ from $P_k(y_1^{(t)}, ..., y_{k-1}^{(t)}, \cdot)$. It can be shown that this Markov chain is still p-invariant. We illustrate this with the case of $k = 2$. Note that $Y_1^{(1)}$ is generated from $P_2(y_2^{(0)}, \cdot)$, the conditional distribution of Y given $Y_2 = y_2^{(0)}$. Hence $(Y_1^{(1)}, Y_2^{(0)})$ has p.d.f. p. Similarly, we can show that $Y^{(1)} = (Y_1^{(1)}, Y_2^{(1)})$ has p.d.f. p. Thus,

$$
\int P(y, A)p(y)d\nu = \int P(Y^{(1)} \in A | Y^{(0)} = y)p(y)d\nu
$$
$$
= E[P(Y^{(1)} \in A | Y^{(0)})]
$$
$$
= P(Y^{(1)} \in A)
$$
$$
= \int_A p(y)d\nu.
$$

This Markov chain is also p-irreducible and aperiodic if $p(y) > 0$ for all $y \in \mathcal{Y}$; see, for example, Chan (1993). Finally, if $p(y) > 0$ for all $y \in \mathcal{Y}$, then $P(y, A) \ll$ the distribution with p.d.f. p for all y and, by Corollary 1 of Tierney (1994), the Markov chain is Harris recurrent. Thus, Theorem 4.4 applies and (4.13) and (4.14) hold.

The previous Gibbs sampler can obviously be extended to the case where y_i's are subvectors (of possibly different dimensions) of y.

Let us now return to Bayesian computation and consider the following example.

Example 4.10. Consider Example 4.9. Under the given prior for $\theta = (\beta, \omega)$, it is difficult to generate random vectors directly from the posterior p.d.f., given $X = x$ (which does not have a familiar form). To apply a Gibbs sampler with $y = \theta$, $y_1 = \beta$, and $y_2 = \omega$, we need to generate random vectors from the posterior of β, given x and ω, and the posterior of ω, given x and β. From (4.9) and (4.11), the posterior of $\omega = (\omega_1, ..., \omega_k)$, given x and β, is a product of marginals of ω_i's that are the gamma distributions $\Gamma(\alpha + 1 + n_i/2, [\gamma^{-1} + v_i(\beta)]^{-1})$, $i = 1, ..., k$. Assume now that $\pi(\beta) \equiv 1$ (noninformative prior for β). It follows from (4.9) that the posterior p.d.f. of β, given x and ω, is proportional to

$$
\prod_{i=1}^{k} e^{-\omega_i v_i(\beta)} \propto e^{-\|W^{1/2}Z\beta - W^{1/2}X\|^2},
$$

where W is the diagonal block matrix whose ith block on the diagonal is $\omega_i I_{n_i}$. Let $n = \sum_{i=1}^{k} n_i$. Then, the posterior of $W^{1/2}Z\beta$, given X and ω, is $N_n(W^{1/2}X, 2^{-1}I_n)$ and the posterior of β, given X and ω, is $N_p((Z^\tau W Z)^{-1}Z^\tau W X, 2^{-1}(Z^\tau W Z)^{-1})$ ($Z^\tau W Z$ is assumed of full rank for simplicity), since $\beta = [(Z^\tau W Z)^{-1}Z^\tau W^{1/2}]W^{1/2}Z\beta$. Note that random generation using these two posterior distributions is fairly easy. ∎

The Metropolis algorithm

A large class of MCMC methods are obtained using the *Metropolis algorithm* (Metropolis et al., 1953). We introduce Hastings' version of the algorithm. Let $Q(y, A)$ be a transition kernel of a homogeneous Markov chain satisfying

$$Q(y, A) = \int_A q(y, z) d\nu(z)$$

for a measurable function $q(y, z) \geq 0$ on $\mathcal{Y} \times \mathcal{Y}$ and a σ-finite measure ν on $(\mathcal{Y}, \mathcal{B}_\mathcal{Y})$. Without loss of generality, assume that $\int_\mathcal{Y} p(y) d\nu = 1$ and that p is not concentrated on a single point. Define

$$\alpha(y, z) = \begin{cases} \min\left\{\frac{p(z)q(z,y)}{p(y)q(y,z)}, 1\right\} & p(y)q(y, z) > 0 \\ 1 & p(y)q(y, z) = 0 \end{cases}$$

and

$$p(y, z) = \begin{cases} q(y, z)\alpha(y, z) & y \neq z \\ 0 & y = z. \end{cases}$$

The Metropolis kernel $P(y, A)$ is defined by

$$P(y, A) = \int_A p(y, z) d\nu(z) + r(y)\delta_y(A), \tag{4.15}$$

where $r(y) = 1 - \int p(y, z) d\nu(z)$ and δ_y is the point mass at y defined in (1.22). The corresponding Markov chain can be described as follows. If the chain is currently at a point $Y^{(t)} = y$, then it generates a candidate value z for the next location $Y^{(t+1)}$ from $Q(y, \cdot)$. With probability $\alpha(y, z)$, the chain moves to $Y^{(t+1)} = z$. Otherwise, the chain remains at $Y^{(t+1)} = y$.

Note that this algorithm only depends on $p(y)$ through $p(y)/p(z)$. Thus, it can be used when $p(y)$ is known up to a normalizing constant, which often occurs in Bayesian analysis.

We now show that a Markov chain with a Metropolis kernel $P(y, A)$ is p-invariant. First, by the definition of $p(y, z)$ and $\alpha(y, z)$,

$$p(y)p(y, z) = p(z)p(z, y)$$

for any y and z. Then, for any $A \in \mathcal{B}_\mathcal{Y}$,

$$\int P(y, A)p(y) d\nu = \int \left[\int_A p(y, z) d\nu(z)\right] p(y) d\nu(y) + \int r(y)\delta_y(A)p(y) d\nu(y)$$

$$= \int_A \left[\int p(y, z)p(y) d\nu(y)\right] d\nu(z) + \int_A r(y)p(y) d\nu(y)$$

$$= \int_A \left[\int p(z, y)p(z) d\nu(y)\right] d\nu(z) + \int_A r(y)p(y) d\nu(y)$$

$$= \int_A [1 - r(z)]p(z)d\nu(z) + \int_A r(z)p(z)d\nu(z)$$

$$= \int_A p(z)d\nu(z).$$

If a Markov chain with a Metropolis kernel defined by (4.15) is p-irreducible and $\int_{r(y)>0} p(y)d\nu > 0$, then, by the results of Nummelin (1984, §2.4), the chain is aperiodic; by Corollary 2 of Tierney (1994), the chain is Harris recurrent. Hence, to apply Theorem 4.4 to a Markov chain with a Metropolis kernel, it suffices to show that the chain is p-irreducible.

Lemma 4.2. Suppose that $Q(y, A)$ is the transition kernel of a p-irreducible Markov chain and that either $q(y, z) > 0$ for all y and z or $q(y, z) = q(z, y)$ for all y and z. Then the chain with the Metropolis kernel $p(y, A)$ in (4.15) is p-irreducible.

Proof. It can be shown (exercise) that if Q is any transition kernel of a homogeneous Markov chain, then

$$Q^t(y, A) = \int_A \int \cdots \int \prod_{j=1}^t q(z_{n-j+1}, z_{n-j})d\nu(z_{n-j}), \qquad (4.16)$$

where $z_n = y$, $y \in \mathcal{Y}$, and $A \in \mathcal{B}_y$. Let $y \in \mathcal{Y}$, $A \in \mathcal{B}_y$ with $\int_A p(z)d\nu > 0$, and $B_y = \{z : \alpha(y, z) = 1\}$. If $\int_{A \cap B_y^c} p(z)d\nu > 0$, then

$$P(y, A) \geq \int_{A \cap B_y^c} q(y, z)\alpha(y, z)d\nu(z) = \int_{A \cap B_y^c} \frac{q(z, y)p(z)}{p(y)}d\nu(z) > 0,$$

which follows from either $q(z, y) > 0$ or $q(z, y) = q(y, z) > 0$ on B_y^c. If $\int_{A \cap B_y^c} p(z)d\nu = 0$, then $\int_{A \cap B_y} p(z)d\nu > 0$. From the irreducibility of $Q(y, A)$, there exists a $t \geq 1$ such that $Q^t(y, A \cap B_y) > 0$. Then, by (4.15) and (4.16),

$$P^t(y, A) \geq P^t(y, A \cap B_y) \geq Q^t(y, A \cap B_y) > 0. \quad \blacksquare$$

Two examples of $q(y, z)$ given by Tierney (1994) are $q(y, z) = f(z - y)$ with a Lebesgue p.d.f. f on $\mathcal{R}^k$, which corresponds to a *random walk chain*, and $q(y, z) = f(z)$ with a p.d.f. f, which corresponds to an *independence chain* and is closely related to the importance sampling discussed earlier.

Although the MCMC methods have been used over the last 50 years, the research on the theory of MCMC is still very active. Important topics include the choice of the transition kernel for MCMC; the rate of the convergence in (4.13); the choice of the Monte Carlo size m; and the estimation of the errors due to Monte Carlo. See more results and discussions in Tierney (1994), Basag et al. (1995), Tanner (1996), and the references therein.

4.2 Invariance

The concept of invariance is introduced in §2.3.2 (Definition 2.9). In this section, we study the best invariant estimators and their properties in one-parameter location families (§4.2.1), in one-parameter scale families (§4.2.2), and in general location-scale families (§4.2.3). Note that invariant estimators are also called equivariant estimators.

4.2.1 One-parameter location families

Assume that the sample $X = (X_1, ..., X_n)$ has a joint distribution P_μ with a Lebesgue p.d.f.

$$f(x_1 - \mu, ..., x_n - \mu), \tag{4.17}$$

where f is known and $\mu \in \mathcal{R}$ is an unknown location parameter. The family $\mathcal{P} = \{P_\mu : \mu \in \mathcal{R}\}$ is called a one-parameter location family, a special case of the general location-scale family described in Definition 2.3. It is invariant under the location transformations $g_c(X) = (X_1 + c, ..., X_n + c)$, $c \in \mathcal{R}$.

We consider the estimation of μ as a statistical decision problem with action space $\mathbb{A} = \mathcal{R}$ and loss function $L(\mu, a)$. It is natural to consider the same transformation in the action space, i.e., if X_i is transformed to $X_i + c$, then our action a is transformed to $a + c$. Consequently, the decision problem is invariant under location transformation if and only if

$$L(\mu, a) = L(\mu + c, a + c) \qquad \text{for all } c \in \mathcal{R},$$

which is equivalent to

$$L(\mu, a) = L(a - \mu) \tag{4.18}$$

for a Borel function $L(\cdot)$ on $\mathcal{R}$.

According to Definition 2.9 (see also Example 2.24), an estimator T (decision rule) of μ is *location invariant* if and only if

$$T(X_1 + c, ..., X_n + c) = T(X_1, ..., X_n) + c. \tag{4.19}$$

Many estimators of μ, such as the sample mean and weighted average of the order statistics, are location invariant. The following result provides a characterization of location invariant estimators.

Proposition 4.3. Let T_0 be a location invariant estimator of μ. Let $d_i = x_i - x_n$, $i = 1, ..., n - 1$, and $d = (d_1, ..., d_{n-1})$. A necessary and sufficient condition for an estimator T to be location invariant is that there exists a Borel function u on $\mathcal{R}^{n-1}$ ($u \equiv$ a constant if $n = 1$) such that

$$T(x) = T_0(x) - u(d) \qquad \text{for all } x \in \mathcal{R}^n. \tag{4.20}$$

Proof. It is easy to see that T given by (4.20) satisfies (4.19) and, therefore, is location invariant. Suppose that T is location invariant. Let $\tilde{u}(x) = T(x) - T_0(x)$ for any $x \in \mathcal{R}^n$. Then

$$\tilde{u}(x_1 + c, ..., x_n + c) = T(x_1 + c, ..., x_n + c) - T_0(x_1 + c, ..., x_n + c)$$
$$= T(x_1, ..., x_n) - T_0(x_1, ..., x_n)$$

for all $c \in \mathcal{R}$ and $x_i \in \mathcal{R}$. Putting $c = -x_n$ leads to

$$\tilde{u}(x_1 - x_n, ..., x_{n-1} - x_n, 0) = T(x) - T_0(x), \quad x \in \mathcal{R}^n.$$

The result follows with $u(d_1, ..., d_{n-1}) = \tilde{u}(x_1 - x_n, ..., x_{n-1} - x_n, 0)$. ∎

Therefore, once we have a location invariant estimator T_0 of μ, any other location invariant estimator of μ can be constructed by taking the difference between T_0 and a Borel function of the ancillary statistic $D = (X_1 - X_n, ..., X_{n-1} - X_n)$.

The next result states an important property of location invariant estimators.

Proposition 4.4. Let X be distributed with the p.d.f. given by (4.17) and let T be a location invariant estimator of μ under the loss function given by (4.18). If the bias, variance, and risk of T are well defined, then they are all constant (do not depend on μ).
Proof. The result for the bias follows from

$$b_T(\mu) = \int T(x) f(x_1 - \mu, ..., x_n - \mu) dx - \mu$$
$$= \int T(x_1 + \mu, ..., x_n + \mu) f(x) dx - \mu$$
$$= \int [T(x) + \mu] f(x) dx - \mu$$
$$= \int T(x) f(x) dx.$$

The proof of the result for variance or risk is left as an exercise. ∎

An important consequence of this result is that the problem of finding the best location invariant estimator reduces to comparing constants instead of risk functions. The following definition can be used not only for location invariant estimators, but also for general invariant estimators.

Definition 4.2. Consider an invariant estimation problem in which all invariant estimators have constant risks. An invariant estimator T is called the *minimum risk invariant estimator* (MRIE) if and only if T has the smallest risk among all invariant estimators. ∎

Theorem 4.5. Let X be distributed with the p.d.f. given by (4.17) and consider the estimation of μ under the loss function given by (4.18). Suppose that there is a location invariant estimator T_0 of μ with finite risk. Let $D = (X_1 - X_n, ..., X_{n-1} - X_n)$.
(i) Assume that for each d there exists a $u_*(d)$ that minimizes

$$h(d) = E_0[L(T_0(X) - u(d))|D = d]$$

over all functions u, where the expectation E_0 is calculated under the assumption that X has p.d.f. $f(x_1, ..., x_n)$. Then an MRIE exists and is given by

$$T_*(X) = T_0(X) - u_*(D).$$

(ii) The function u_* in (i) exists if $L(t)$ is convex and not monotone; it is unique if L is strictly convex.
(iii) If T_0 and D are independent, then u_* is a constant that minimizes $E_0[L(T_0(X) - u)]$. If, in addition, the distribution of T_0 is symmetric about μ and L is convex and even, then $u_* = 0$.
Proof. By Theorem 1.7 and Propositions 4.3 and 4.4,

$$R_T(\mu) = E_0[h(D)],$$

where $T(X) = T_0(X) - u(D)$. This proves part (i). If L is (strictly) convex and not monotone, then $E_0[L(T_0(x) - a)|D = d]$ is (strictly) convex and not monotone in a (exercise). Hence $\lim_{|a| \to \infty} E_0[L(T_0(x) - a)|D = d] = \infty$. This proves part (ii). The proof of part (iii) is left as an exercise. ∎

Theorem 4.6. Assume the conditions of Theorem 4.5 and that the loss is the squared error loss.
(i) The unique MRIE of μ is

$$T_*(X) = \frac{\int_{-\infty}^{\infty} tf(X_1 - t, ..., X_n - t)dt}{\int_{-\infty}^{\infty} f(X_1 - t, ..., X_n - t)dt},$$

which is known as the *Pitman estimator* of μ.
(ii) The MRIE of μ is unbiased.
Proof. (i) Under the squared error loss,

$$u_*(d) = E_0[T_0(X)|D = d] \tag{4.21}$$

(exercise). Let $T_0(X) = X_n$ (the nth observation). Then X_n is location invariant. If there exists a location invariant estimator of μ with finite risk, then $E_0(X_n|D = d)$ is finite a.s. $\mathcal{P}$ (exercise). By Proposition 1.8, when $\mu = 0$, the joint Lebesgue p.d.f. of (D, X_n) is $f(d_1 + x_n, ..., d_{n-1} + x_n, x_n)$, $d = (d_1, ..., d_{n-1})$. The conditional p.d.f. of X_n given $D = d$ is then

$$\frac{f(d_1 + x_n, ..., d_{n-1} + x_n, x_n)}{\int_{-\infty}^{\infty} f(d_1 + t, ..., d_{n-1} + t, t)dt}$$

(see (1.61)). By Proposition 1.9,

$$
\begin{aligned}
E_0(X_n|D=d) &= \frac{\int_{-\infty}^{\infty} t f(d_1+t,...,d_{n-1}+t,t)dt}{\int_{-\infty}^{\infty} f(d_1+t,...,d_{n-1}+t,t)dt} \\
&= \frac{\int_{-\infty}^{\infty} t f(x_1-x_n+t,...,x_{n-1}-x_n+t,t)dt}{\int_{-\infty}^{\infty} f(x_1-x_n+t,...,x_{n-1}-x_n+t,t)dt} \\
&= x_n - \frac{\int_{-\infty}^{\infty} u f(x_1-u,...,x_n-u)du}{\int_{-\infty}^{\infty} f(x_1-u,...,x_n-u)du}
\end{aligned}
$$

by letting $u = x_n - t$. The result in (i) follows from $T_*(X) = X_n - E(X_n|D)$ (Theorem 4.5).
(ii) Let b be the constant bias of T_* (Proposition 4.4). Then $T_1(X) = T_*(X) - b$ is a location invariant estimator of μ and

$$
R_{T_1} = E[T_*(X) - b - \mu]^2 = \text{Var}(T_*) \le \text{Var}(T_*) + b^2 = R_{T_*}.
$$

Since T_* is the MRIE, $b = 0$, i.e., T_* is unbiased. ∎

Theorem 4.6(ii) indicates that we only need to consider unbiased location invariant estimators in order to find the MRIE, if the loss is the squared error loss. In particular, a location invariant UMVUE is an MRIE.

Example 4.11. Let $X_1,...,X_n$ be i.i.d. from $N(\mu,\sigma^2)$ with an unknown $\mu \in \mathcal{R}$ and a known σ^2. Note that $\bar{X}$ is location invariant. Since $\bar{X}$ is the UMVUE of μ (§2.1), it is the MRIE under the squared error loss. Since the distribution of $\bar{X}$ is symmetric about μ and $\bar{X}$ is independent of D (Basu's theorem), it follows from Theorem 4.5(iii) that $\bar{X}$ is an MRIE if L is convex and even. ∎

Example 4.12. Let $X_1,...,X_n$ be i.i.d. from the exponential distribution $E(\mu,\theta)$, where θ is known and $\mu \in \mathcal{R}$ is unknown. Since $X_{(1)} - \theta/n$ is location invariant and is the UMVUE of μ, it is the MRIE under the squared error loss. Note that $X_{(1)}$ is independent of D (Basu's theorem). By Theorem 4.5(iii), an MRIE is of the form $X_{(1)} - u_*$ with a constant u_*. For the absolute error loss, $X_{(1)} - \theta \log 2/n$ is an MRIE (exercise). ∎

Example 4.13. Let $X_1,...,X_n$ be i.i.d. from the uniform distribution on $(\mu - \frac{1}{2}, \mu + \frac{1}{2})$ with an unknown $\mu \in \mathcal{R}$. Consider the squared error loss. Note that

$$
f(x_1-\mu,...,x_n-\mu) = \begin{cases} 1 & \mu - \frac{1}{2} \le x_{(1)} \le x_{(n)} \le \mu + \frac{1}{2} \\ 0 & \text{otherwise.} \end{cases}
$$

By Theorem 4.6(i), the MRIE of μ is

$$T_*(X) = \int_{X_{(n)}-\frac{1}{2}}^{X_{(1)}+\frac{1}{2}} t\,dt \bigg/ \int_{X_{(n)}-\frac{1}{2}}^{X_{(1)}+\frac{1}{2}} dt = \frac{X_{(1)} + X_{(n)}}{2}. \quad \blacksquare$$

We end this section with a brief discussion of the admissibility of MRIE's in a one-parameter location problem. Under the squared error loss, the MRIE (Pitman's estimator) is admissible if there exists a location invariant estimator T_0 with $E|T_0(X)|^3 < \infty$ (Stein, 1959). Under a general loss function, an MRIE is admissible when it is a unique MRIE (under some other minor conditions). See Farrell (1964), Brown (1966), and Brown and Fox (1974) for further discussions.

4.2.2 One-parameter scale families

Assume that the sample $X = (X_1, ..., X_n)$ has a joint distribution P_σ with a Lebesgue p.d.f.

$$\frac{1}{\sigma^n} f\left(\frac{x_1}{\sigma}, ..., \frac{x_n}{\sigma}\right), \tag{4.22}$$

where f is known and $\sigma > 0$ is an unknown scale parameter. The family $\mathcal{P} = \{P_\sigma : \sigma > 0\}$ is called a one-parameter scale family and is a special case of the general location-scale family in Definition 2.3. This family is invariant under the scale transformations $g_r(X) = rX$, $r > 0$.

We consider the estimation of σ^h with $\mathbb{A} = [0, \infty)$, where h is a nonzero constant. The transformation g_r induces the transformation $g_r(\sigma^h) = r^h \sigma^h$. Hence, a loss function L is scale invariant if and only if

$$L(r\sigma, r^h a) = L(\sigma, a) \qquad \text{for all } r > 0,$$

which is equivalent to

$$L(\sigma, a) = L\left(\frac{a}{\sigma^h}\right) \tag{4.23}$$

for a Borel function $L(\cdot)$ on $[0, \infty)$. An example of a loss function satisfying (4.23) is

$$L(\sigma, a) = \left|\frac{a}{\sigma^h} - 1\right|^p = \frac{|a - \sigma^h|^p}{\sigma^{ph}}, \tag{4.24}$$

where $p \geq 1$ is a constant. However, the squared error loss does not satisfy (4.23).

An estimator T of σ^h is *scale invariant* if and only if

$$T(rX_1, ..., rX_n) = r^h T(X_1, ..., X_n).$$

Examples of scale invariant estimators are the sample variance S^2 (for $h = 2$), the sample standard deviation $S = \sqrt{S^2}$ (for $h = 1$), the sample range

$X_{(n)} - X_{(1)}$ (for $h = 1$), and the sample mean deviation $n^{-1} \sum_{i=1}^{n} |X_i - \bar{X}|$ (for $h = 1$).

The following result is an analogue of Proposition 4.3. Its proof is left as an exercise.

Proposition 4.5. Let T_0 be a scale invariant estimator of σ^h. A necessary and sufficient condition for an estimator T to be scale invariant is that there exists a positive Borel function u on $\mathcal{R}^n$ such that

$$T(x) = T_0(x)/u(z) \qquad \text{for all } x \in \mathcal{R}^n,$$

where $z = (z_1, ..., z_n)$, $z_i = x_i/x_n$, $i = 1, ..., n - 1$, and $z_n = x_n/|x_n|$. ∎

The next result is similar to Proposition 4.4. It applies to any invariant problem defined in Definition 2.9. We use the notation in Definition 2.9.

Theorem 4.7. Let $\mathcal{P}$ be a family invariant under $\mathcal{G}$ (a group of transformations). Suppose that the loss function is invariant and T is an invariant decision rule. Then the risk function of T is a constant. ∎

The proof is left as an exercise. Note that a special case of Theorem 4.7 is that any scale invariant estimator of σ^h has a constant risk and, therefore, an MRIE (Definition 4.2) of σ^h usually exists. However, Proposition 4.4 is not a special case of Theorem 4.7, since the bias of a scale invariant estimator may not be a constant in general. For example, the bias of the sample standard deviation is a function of σ.

The next result and its proof are analogues of those of Theorem 4.5.

Theorem 4.8. Let X be distributed with the p.d.f. given by (4.22) and consider the estimation of σ^h under the loss function given by (4.23). Suppose that there is a scale invariant estimator T_0 of σ^h with finite risk. Let $Z = (Z_1, ..., Z_n)$ with $Z_i = X_i/X_n$, $i = 1, ..., n - 1$, and $Z_n = X_n/|X_n|$.
(i) Assume that for each z there exists a $u_*(z)$ that minimizes

$$E_1[L(T_0(X)/u(z))|Z = z]$$

over all positive Borel functions u, where the conditional expectation E_1 is calculated under the assumption that X has p.d.f. $f(x_1, ..., x_n)$. Then, an MRIE exists and is given by

$$T_*(X) = T_0(X)/u_*(Z).$$

(ii) The function u_* in (i) exists if $\gamma(t) = L(e^t)$ is convex and not monotone; it is unique if $\gamma(t)$ is strictly convex. ∎

The loss function given by (4.24) satisfies the condition in Theorem 4.8(ii). A loss function corresponding to the squared error loss in this problem is the loss function (4.24) with $p = 2$. We have the following result similar to Theorem 4.6 (its proof is left as an exercise).

Corollary 4.1. Under the conditions of Theorem 4.8 and the loss function (4.24) with $p = 2$, the unique MRIE of σ^h is

$$T_*(X) = \frac{T_0(X)E_1[T_0(X)|Z]}{E_1\{[T_0(X)]^2|Z\}} = \frac{\int_0^\infty t^{n+h-1}f(tX_1,...,tX_n)dt}{\int_0^\infty t^{n+2h-1}f(tX_1,...,tX_n)dt},$$

which is known as the Pitman estimator of σ^h. ∎

Example 4.14. Let $X_1,...,X_n$ be i.i.d. from $N(0,\sigma^2)$ and consider the estimation of σ^2. Then $T_0 = \sum_{i=1}^n X_i^2$ is scale invariant. By Basu's theorem, T_0 is independent of Z. Hence u_* in Theorem 4.8 is a constant minimizing $E_1[L(T_0/u)]$ over $u > 0$. When the loss is given by (4.24) with $p = 2$, by Corollary 4.1, the MRIE (Pitman's estimator) is

$$T_*(X) = \frac{T_0(X)E_1[T_0(X)]}{E_1[T_0(X)]^2} = \frac{1}{n+2}\sum_{i=1}^n X_i^2,$$

since T_0 has the chi-square distribution χ_n^2 when $\sigma = 1$. Note that the UMVUE of σ^2 is T_0/n, which is different from the MRIE. ∎

Example 4.15. Let $X_1,...,X_n$ be i.i.d. from the uniform distribution on $(0,\sigma)$ and consider the estimation of σ. By Basu's theorem, the scale invariant estimator $X_{(n)}$ is independent of Z. Hence u_* in Theorem 4.8 is a constant minimizing $E_1[L(X_{(n)}/u)]$ over $u > 0$. When the loss is given by (4.24) with $p = 2$, by Corollary 4.1, the MRIE (Pitman's estimator) is

$$T_*(X) = \frac{X_{(n)}E_1X_{(n)}}{E_1X_{(n)}^2} = \frac{(n+2)X_{(n)}}{n+1}. \blacksquare$$

4.2.3 General location-scale families

Assume that $X = (X_1,...,X_n)$ has a joint distribution P_θ with a Lebesgue p.d.f.

$$\frac{1}{\sigma^n}f\left(\frac{x_1-\mu}{\sigma},...,\frac{x_n-\mu}{\sigma}\right), \tag{4.25}$$

where f is known, $\theta = (\mu,\sigma) \in \Theta$, and $\Theta = \mathcal{R} \times (0,\infty)$. The family $\mathcal{P} = \{P_\theta : \theta \in \Theta\}$ is a location-scale family defined by Definition 2.3 and is invariant under the location-scale transformations of the form $g_{c,r}(X) = (rX_1 + c,...,rX_n + c)$, $c \in \mathcal{R}$, $r > 0$, which induce similar transformations on Θ: $g_{c,r}(\theta) = (r\mu + c, r\sigma)$, $c \in \mathcal{R}$, $r > 0$.

Consider the estimation of σ^h with a fixed $h \neq 0$ under the loss function (4.23), which is invariant under the location-scale transformations $g_{c,r}$. An estimator T of σ^h is *location-scale* invariant if and only if

$$T(rX_1 + c, ..., rX_n + c) = r^h T(X_1, ..., X_n). \qquad (4.26)$$

By Theorem 4.7, any location-scale invariant T has a constant risk. Letting $r = 1$ in (4.26), we obtain that

$$T(X_1 + c, ..., X_n + c) = T(X_1, ..., X_n)$$

for all $c \in \mathcal{R}$. Therefore, T is a function of $D = (D_1, ..., D_{n-1})$, $D_i = X_i - X_n$, $i = 1, ..., n - 1$. From (4.25), the joint Lebesgue p.d.f. of D is

$$\frac{1}{\sigma^{n-1}} \int_{-\infty}^{\infty} f\left(\frac{d_1}{\sigma} + t, ..., \frac{d_{n-1}}{\sigma} + t, t\right) dt, \qquad (4.27)$$

which is of the form (4.22) with n replaced by $n-1$ and x_i's replaced by d_i's. It follows from Theorem 4.8 that if $T_0(D)$ is any finite risk scale invariant estimator of σ^h based on D, then an MRIE of σ^h is

$$T_*(D) = T_0(D)/u_*(W), \qquad (4.28)$$

where $W = (W_1, ..., W_{n-1})$, $W_i = D_i/D_{n-1}$, $i = 1, ..., n - 2$, $W_{n-1} = D_{n-1}/|D_{n-1}|$, $u_*(w)$ is any number minimizing $\tilde{E}_1[L(T_0(D)/u(w))|W = w]$ over all positive Borel functions u, and $\tilde{E}_1$ is the conditional expectation calculated under the assumption that D has p.d.f. (4.27) with $\sigma = 1$.

Consider next the estimation of μ. Under the location-scale transformation $g_{c,r}$, it can be shown (exercise) that a loss function is invariant if and only if it is of the form

$$L\left(\frac{a-\mu}{\sigma}\right). \qquad (4.29)$$

An estimator T of μ is location-scale invariant if and only if

$$T(rX_1 + c, ..., rX_n + c) = rT(X_1, ..., X_n) + c.$$

Again, by Theorem 4.7, the risk of an invariant T is a constant.

The following result is an analogue of Proposition 4.3 or 4.5.

Proposition 4.6. Let T_0 be any estimator of μ invariant under location-scale transformation and let T_1 be any estimator of σ satisfying (4.26) with $h = 1$ and $T_1 > 0$. Then an estimator T of μ is location-scale invariant if and only if there is a Borel function u on $\mathcal{R}^{n-1}$ such that

$$T(X) = T_0(X) - u(W)T_1(X),$$

where W is given in (4.28). ∎

The proofs of Proposition 4.6 and the next result, an analogue of Theorem 4.5 or 4.8, are left as exercises.

Theorem 4.9. Let X be distributed with p.d.f. given by (4.25) and consider the estimation of μ under the loss function given by (4.29). Suppose that there is a location-scale invariant estimator T_0 of μ with finite risk. Let T_1 be given in Proposition 4.6. Then an MRIE of μ is

$$T_*(X) = T_0(X) - u_*(W)T_1(X),$$

where W is given in (4.28), $u_*(w)$ is any number minimizing

$$E_{0,1}[L(T_0(X) - u(w)T_1(X))|W = w]$$

over all Borel functions u, and $E_{0,1}$ is computed under the assumption that X has the p.d.f. (4.25) with $\mu = 0$ and $\sigma = 1$. ∎

Corollary 4.2. Under the conditions of Theorem 4.9 and the loss function $(a - \mu)^2/\sigma^2$, $u_*(w)$ in Theorem 4.9 is equal to

$$u_*(w) = \frac{E_{0,1}[T_0(X)T_1(X)|W = w]}{E_{0,1}\{[T_1(X)]^2|W = w\}}.$$ ∎

Example 4.16. Let $X_1, ..., X_n$ be i.i.d. from $N(\mu, \sigma^2)$, where $\mu \in \mathcal{R}$ and $\sigma^2 > 0$ are unknown. Consider first the estimation of σ^2 under loss function (4.23). The sample variance S^2 is location-scale invariant and is independent of W in (4.28) (Basu's theorem). Thus, by (4.28), S^2/u_* is an MRIE, where u_* is a constant minimizing $\tilde{E}_1[L(S^2/u)]$ over all $u > 0$. If the loss function is given by (4.24) with $p = 2$, then by Corollary 4.1, the MRIE of σ^2 is

$$T_*(X) = \frac{S^2\tilde{E}_1(S^2)}{\tilde{E}_1(S^2)^2} = \frac{S^2}{(n^2-1)/(n-1)^2} = \frac{1}{n+1}\sum_{i=1}^n (X_i - \bar{X})^2,$$

since $(n-1)S^2$ has a chi-square distribution χ^2_{n-1} when $\sigma = 1$.

Next, consider the estimation of μ under the loss function (4.29). Since $\bar{X}$ is a location-scale invariant estimator of μ and is independent of W in (4.28) (Basu's theorem), by Theorem 4.9, an MRIE of μ is

$$T_*(X) = \bar{X} - u_*S^2,$$

where u_* is a constant. If L in (4.29) is convex and even, then $u_* = 0$ (see Theorem 4.5(iii)) and, hence, $\bar{X}$ is an MRIE of μ. ∎

Example 4.17. Let $X_1, ..., X_n$ be i.i.d. from the uniform distribution on $(\mu - \frac{1}{2}\sigma, \mu + \frac{1}{2}\sigma)$, where $\mu \in \mathcal{R}$ and $\sigma > 0$ are unknown. Consider first the estimation of σ under the loss function (4.24) with $p = 2$. The sample range $X_{(n)} - X_{(1)}$ is a location-scale invariant estimator of σ and is independent of W in (4.28) (Basu's theorem). By (4.28) and Corollary 4.1, the MRIE of σ is

$$T_*(X) = \frac{(X_{(n)} - X_{(1)})\tilde{E}_1(X_{(n)} - X_{(1)})}{\tilde{E}_1(X_{(n)} - X_{(1)})^2} = \frac{(n+2)(X_{(n)} - X_{(1)})}{n}.$$

Consider now the estimation of μ under the loss function (4.29). Since $(X_{(1)} + X_{(n)})/2$ is a location-scale invariant estimator of μ and is independent of W in (4.28) (Basu's theorem), by Theorem 4.9, an MRIE of μ is

$$T_*(X) = \frac{X_{(1)} + X_{(n)}}{2} - u_*(X_{(n)} - X_{(1)}),$$

where u_* is a constant. If L in (4.29) is convex and even, then $u_* = 0$ (see Theorem 4.5(iii)) and, hence, $(X_{(1)} + X_{(n)})/2$ is an MRIE of μ.   ∎

Finding MRIE's in various location-scale families under transformations $AX + c$, where $A \in T$ and $c \in C$ with given T and C, can be done in a similar way. We only provide some brief discussions for two important cases. The first case is the two-sample location-scale problem in which two samples, $X = (X_1, ..., X_m)$ and $Y = (Y_1, ..., Y_n)$, are taken from a distribution with Lebesgue p.d.f.

$$\frac{1}{\sigma_x^m \sigma_y^n} f\left(\frac{x_1 - \mu_x}{\sigma_x}, ..., \frac{x_m - \mu_x}{\sigma_x}, \frac{y_1 - \mu_y}{\sigma_y}, ..., \frac{y_n - \mu_y}{\sigma_y}\right), \qquad (4.30)$$

where f is known, $\mu_x \in \mathcal{R}$ and $\mu_y \in \mathcal{R}$ are unknown location parameters, and $\sigma_x > 0$ and $\sigma_y > 0$ are unknown scale parameters. The family of distributions is invariant under the transformations

$$g(X, Y) = (rX_1 + c, ..., rX_m + c, r'Y_1 + c', ..., r'Y_n + c'), \qquad (4.31)$$

where $r > 0$, $r' > 0$, $c \in \mathcal{R}$, and $c' \in \mathcal{R}$. The parameters to be estimated in this problem are usually $\Delta = \mu_y - \mu_x$ and $\eta = (\sigma_y/\sigma_x)^h$ with a fixed $h \neq 0$. If X and Y are from two populations, Δ and η are measures of the difference between the two populations. For estimating η, results similar to those in this section can be established. For estimating Δ, MRIE's can be obtained under some conditions. See Exercises 63-65.

The second case is the general linear model (3.25) under the assumption that ε_i's are i.i.d. with the p.d.f. $\sigma^{-1}f(x/\sigma)$, where f is a known Lebesgue p.d.f. The family of populations is invariant under the transformations

$$g(X) = rX + Zc, \qquad r \in (0, \infty), \ c \in \mathcal{R}^p \qquad (4.32)$$

(exercise). The estimation of $l^\tau \beta$ with $l \in \mathcal{R}(Z)$ is invariant under the loss function $L\left(\frac{a-l^\tau \beta}{\sigma}\right)$ and the LSE $l^\tau \hat{\beta}$ is an invariant estimator of $l^\tau \beta$ (exercise). When f is normal, the following result can be established using an argument similar to that in Example 4.16.

Theorem 4.10. Consider model (3.25) with assumption A1.
(i) Under transformations (4.32) and the loss function $L\left(\frac{a-l^\tau \beta}{\sigma}\right)$, where L is convex and even, the LSE $l^\tau \hat{\beta}$ is an MRIE of $l^\tau \beta$ for any $l \in \mathcal{R}(Z)$.
(ii) Under transformations (4.32) and the loss function $(a - \sigma^2)^2/\sigma^4$, the MRIE of σ^2 is $SSR/(n - r + 2)$, where SSR is given by (3.35) and r is the rank of Z. ∎

MRIE's in a parametric family with a multi-dimensional θ are often inadmissible. See Lehmann (1983, p. 285) for more discussions.

4.3 Minimaxity and Admissibility

Consider the estimation of a real-valued $\vartheta = g(\theta)$ based on a sample X from P_θ, $\theta \in \Theta$, under a given loss function. A *minimax estimator* minimizes the maximum risk $\sup_{\theta \in \Theta} R_T(\theta)$ over all estimators T (see §2.3.2).

A unique minimax estimator is admissible, since any estimator better than a minimax estimator is also minimax. This indicates that we should consider minimaxity and admissibility together. The situation is different for a UMVUE (or an MRIE), since if a UMVUE (or an MRIE) is inadmissible, it is dominated by an estimator that is not unbiased (or invariant).

4.3.1 Estimators with constant risks

By minimizing the maximum risk, a minimax estimator tries to do as well as possible in the worst case. Such an estimator can be very unsatisfactory. However, if a minimax estimator has some other good properties (e.g., it is a Bayes estimator), then it is often a reasonable estimator. Here we study when estimators having constant risks (e.g., MRIE's) are minimax.

Theorem 4.11. Let Π be a proper prior on Θ and δ be a Bayes estimator of ϑ w.r.t. Π. Let $\Theta_\Pi = \{\theta : R_\delta(\theta) = \sup_{\theta \in \Theta} R_\delta(\theta)\}$. If $\Pi(\Theta_\Pi) = 1$, then δ is minimax. If, in addition, δ is the unique Bayes estimator w.r.t. Π, then it is the unique minimax estimator.
Proof. Let T be any other estimator of ϑ. Then

$$\sup_{\theta \in \Theta} R_T(\theta) \geq \int_{\Theta_\Pi} R_T(\theta) d\Pi \geq \int_{\Theta_\Pi} R_\delta(\theta) d\Pi = \sup_{\theta \in \Theta} R_\delta(\theta).$$

If δ is the unique Bayes estimator, then the second inequality in the previous expression should be replaced by $>$ and, therefore, δ is the unique minimax estimator. ∎

The condition of Theorem 4.11 essentially means that δ has a constant risk. Thus, a Bayes estimator having constant risk is minimax.

Example 4.18. Let $X_1, ..., X_n$ be i.i.d. binary random variables with $P(X_1 = 1) = p \in (0, 1)$. Consider the estimation of p under the squared error loss. The UMVUE $\bar{X}$ has risk $p(1-p)/n$ which is not constant. In fact, $\bar{X}$ is not minimax (Exercise 67). To find a minimax estimator by applying Theorem 4.11, we consider the Bayes estimator w.r.t. the beta distribution $B(\alpha, \beta)$ with known α and β (Exercise 1):

$$\delta(X) = (\alpha + n\bar{X})/(\alpha + \beta + n).$$

A straightforward calculation shows that

$$R_\delta(p) = [np(1-p) + (\alpha - \alpha p - \beta p)^2]/(\alpha + \beta + n)^2.$$

To apply Theorem 4.11, we need to find values of $\alpha > 0$ and $\beta > 0$ such that $R_\delta(p)$ is constant. It can be shown that $R_\delta(p)$ is constant if and only if $\alpha = \beta = \sqrt{n}/2$, which leads to the unique minimax estimator

$$T(X) = (n\bar{X} + \sqrt{n}/2)/(n + \sqrt{n}).$$

The risk of T is $R_T = 1/[4(1 + \sqrt{n})^2]$.

Note that T is a Bayes estimator and has some good properties. Comparing the risk of T with that of $\bar{X}$, we find that T has smaller risk if and only if

$$p \in \left(\frac{1}{2} - \frac{1}{2}\sqrt{1 - \frac{n}{(1+\sqrt{n})^2}}, \; \frac{1}{2} + \frac{1}{2}\sqrt{1 - \frac{n}{(1+\sqrt{n})^2}} \right). \quad (4.33)$$

Thus, for a small n, T is better (and can be much better) than $\bar{X}$ for most of the range of p (Figure 4.1). When $n \to \infty$, the interval in (4.33) shrinks toward $\frac{1}{2}$. Hence, for a large (and even moderate) n, $\bar{X}$ is better than T for most of the range of p (Figure 4.1). The limit of the asymptotic relative efficiency of T w.r.t. $\bar{X}$ is $4p(1-p)$, which is always smaller than 1 when $p \neq \frac{1}{2}$ and equals 1 when $p = \frac{1}{2}$.

The minimax estimator depends strongly on the loss function. To see this, let us consider the loss function $L(p, a) = (a-p)^2/[p(1-p)]$. Under this loss function, $\bar{X}$ has constant risk and is the unique Bayes estimator w.r.t. the uniform prior on $(0, 1)$. By Theorem 4.11, $\bar{X}$ is the unique minimax estimator. On the other hand, the risk of T is equal to $1/[4(1+\sqrt{n})^2 p(1-p)]$, which is unbounded. ∎

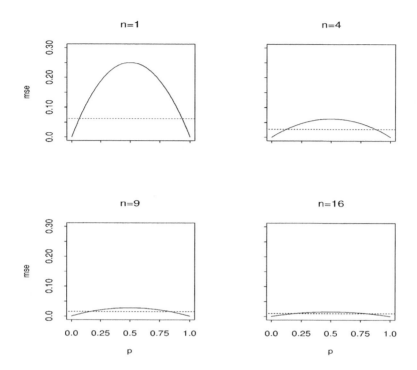

Figure 4.1: mse's of $\bar{X}$ (curve) and $T(X)$ (straight line) in Example 4.18

In many cases a constant risk estimator is not a Bayes estimator (e.g., an unbiased estimator under the squared error loss), but a limit of Bayes estimators w.r.t. a sequence of priors. Then the following result may be used to find a minimax estimator.

Theorem 4.12. Let Π_j, $j = 1, 2, ...$, be a sequence of priors and r_j be the Bayes risk of a Bayes estimator of ϑ w.r.t. Π_j. Let T be a constant risk estimator of ϑ. If $\liminf_j r_j \geq R_T$, then T is minimax. ∎

The proof of this theorem is similar to that of Theorem 4.11. Although Theorem 4.12 is more general than Theorem 4.11 in finding minimax estimators, it does not provide uniqueness of the minimax estimator even when there is a unique Bayes estimator w.r.t. each Π_j.

In Example 2.25, we actually applied the result in Theorem 4.12 to show the minimaxity of $\bar{X}$ as an estimator of $\mu = EX_1$ when $X_1, ..., X_n$ are i.i.d. from a normal distribution with a known $\sigma^2 = \text{Var}(X_1)$, under the squared error loss. To discuss the minimaxity of $\bar{X}$ in the case where σ^2 is unknown, we need the following lemma.

Lemma 4.3. Let Θ_0 be a subset of Θ and T be a minimax estimator of ϑ when Θ_0 is the parameter space. Then T is a minimax estimator if

$$\sup_{\theta \in \Theta} R_T(\theta) = \sup_{\theta \in \Theta_0} R_T(\theta).$$

Proof. If there is an estimator T_0 with $\sup_{\theta \in \Theta} R_{T_0}(\theta) < \sup_{\theta \in \Theta} R_T(\theta)$, then

$$\sup_{\theta \in \Theta_0} R_{T_0}(\theta) \leq \sup_{\theta \in \Theta} R_{T_0}(\theta) < \sup_{\theta \in \Theta} R_T(\theta) = \sup_{\theta \in \Theta_0} R_T(\theta),$$

which contradicts the minimaxity of T when Θ_0 is the parameter space. Hence, T is minimax when Θ is the parameter space. ∎

Example 4.19. Let $X_1, ..., X_n$ be i.i.d. from $N(\mu, \sigma^2)$ with unknown $\theta = (\mu, \sigma^2)$. Consider the estimation of μ under the squared error loss. Suppose first that $\Theta = \mathcal{R} \times (0, c]$ with a constant $c > 0$. Let $\Theta_0 = \mathcal{R} \times \{c\}$. From Example 2.25, $\bar{X}$ is a minimax estimator of μ when the parameter space is Θ_0. An application of Lemma 4.3 shows that $\bar{X}$ is also minimax when the parameter space is Θ. Although σ^2 is assumed to be bounded by c, the minimax estimator $\bar{X}$ does not depend on c.

Consider next the case where $\Theta = \mathcal{R} \times (0, \infty)$, i.e., σ^2 is unbounded. Let T be any estimator of μ. For any fixed σ^2,

$$\frac{\sigma^2}{n} \leq \sup_{\mu \in \mathcal{R}} R_T(\theta),$$

since σ^2/n is the risk of $\bar{X}$ that is minimax when σ^2 is known (Example 2.25). Letting $\sigma^2 \to \infty$, we obtain that $\sup_\theta R_T(\theta) = \infty$ for any estimator T. Thus, minimaxity is meaningless (any estimator is minimax). ∎

Theorem 4.13. Suppose that T as an estimator of ϑ has constant risk and is admissible. Then T is minimax. If the loss function is strictly convex, then T is the unique minimax estimator.

Proof. By the admissibility of T, if there is another estimator T_0 with $\sup_\theta R_{T_0}(\theta) \leq R_T$, then $R_{T_0}(\theta) = R_T$ for all θ. This proves that T is minimax. If the loss function is strictly convex and T_0 is another minimax estimator, then

$$R_{(T+T_0)/2}(\theta) < (R_{T_0} + R_T)/2 = R_T$$

for all θ and, therefore, T is inadmissible. This shows that T is unique if the loss is strictly convex. ∎

Combined with Theorem 4.7, Theorem 4.13 tells us that if an MRIE is admissible, then it is minimax. From the discussion at the end of §4.2.1, MRIE's in one-parameter location families (such as Pitman's estimators) are usually minimax.

4.3.2 Results in one-parameter exponential families

The following result provides a sufficient condition for the admissibility of a class of estimators when the population P_θ is in a one-parameter exponential family. Using this result and Theorem 4.13, we can obtain a class of minimax estimators. The proof of this result is an application of the information inequality introduced in §3.1.3.

Theorem 4.14. Suppose that X has the p.d.f. $c(\theta)e^{\theta T(x)}$ w.r.t. a σ-finite measure ν, where $T(x)$ is real-valued and $\theta \in (\theta_-, \theta_+) \subset \mathcal{R}$. Consider the estimation of $\vartheta = E[T(X)]$ under the squared error loss. Let $\lambda \geq 0$ and γ be known constants and let $T_{\lambda,\gamma}(X) = (T + \gamma\lambda)/(1 + \lambda)$. Then a sufficient condition for the admissibility of $T_{\lambda,\gamma}$ is that

$$\int_{\theta_0}^{\theta_+} \frac{e^{-\gamma\lambda\theta}}{[c(\theta)]^\lambda}d\theta = \int_{\theta_-}^{\theta_0} \frac{e^{-\gamma\lambda\theta}}{[c(\theta)]^\lambda}d\theta = \infty, \tag{4.34}$$

where $\theta_0 \in (\theta_-, \theta_+)$.

Proof. From Theorem 2.1, $\vartheta = E[T(X)] = -c'(\theta)/c(\theta)$ and $\frac{d\vartheta}{d\theta} = \text{Var}(T) = I(\theta)$, the Fisher information defined in (3.5). Suppose that there is an estimator δ of ϑ such that for all θ,

$$R_\delta(\theta) \leq R_{T_{\lambda,\gamma}}(\theta) = [I(\theta) + \lambda^2(\vartheta - \gamma)^2]/(1 + \lambda)^2.$$

Let $b_\delta(\theta)$ be the bias of δ. From the information inequality (3.6),

$$R_\delta(\theta) \geq [b_\delta(\theta)]^2 + [I(\theta) + b'_\delta(\theta)]^2/I(\theta).$$

Let $h(\theta) = b_\delta(\theta) - \lambda(\gamma - \vartheta)/(1 + \lambda)$. Then

$$[h(\theta)]^2 - \frac{2\lambda h(\theta)(\vartheta - \gamma) + 2h'(\theta)}{1 + \lambda} + \frac{[h'(\theta)]^2}{I(\theta)} \leq 0,$$

which implies

$$[h(\theta)]^2 - \frac{2\lambda h(\theta)(\vartheta - \gamma) + 2h'(\theta)}{1 + \lambda} \leq 0. \tag{4.35}$$

Let $a(\theta) = h(\theta)[c(\theta)]^\lambda e^{\gamma\lambda\theta}$. Differentiation of $a(\theta)$ reduces (4.35) to

$$\frac{[a(\theta)]^2 e^{-\gamma\lambda\theta}}{[c(\theta)]^\lambda} + \frac{2a'(\theta)}{1 + \lambda} \leq 0. \tag{4.36}$$

Suppose that $a(\theta_0) < 0$ for some $\theta_0 \in (\theta_-, \theta_+)$. From (4.36), $a'(\theta) \leq 0$ for all θ. Hence $a(\theta) < 0$ for all $\theta \geq \theta_0$ and, for $\theta > \theta_0$, (4.36) can be written as

$$\frac{d}{d\theta}\left[\frac{1}{a(\theta)}\right] \geq \frac{(1 + \lambda)e^{-\gamma\lambda\theta}}{2[c(\theta)]^\lambda}.$$

Integrating both sides from θ_0 to θ gives

$$\frac{1+\lambda}{2} \int_{\theta_0}^{\theta} \frac{e^{-\gamma\lambda\theta}}{[c(\theta)]^\lambda} d\theta \le \frac{1}{a(\theta)} - \frac{1}{a(\theta_0)} \le -\frac{1}{a(\theta_0)}.$$

Letting $\theta \to \theta_+$, the left-hand side of the previous expression diverges to ∞ by condition (4.34), which is impossible. This shows that $a(\theta) \ge 0$ for all θ. Similarly, we can show that $a(\theta) \le 0$ for all θ. Thus, $a(\theta) = 0$ for all θ. This means that $h(\theta) = 0$ for all θ and $b'_\delta(\theta) = -\lambda\vartheta'/(1+\lambda) = -\lambda I(\theta)/(1+\lambda)$, which implies $R_\delta(\theta) \equiv R_{T_{\lambda,\gamma}}(\theta)$. This proves the admissibility of $T_{\lambda,\gamma}$. ∎

The reason why $T_{\lambda,\gamma}$ is considered is that it is often a Bayes estimator w.r.t. some prior; see, for example, Examples 2.25, 4.1, 4.7, and 4.8. To find minimax estimators, we may use the following result.

Corollary 4.3. Assume that X has the p.d.f. as described in Theorem 4.14 with $\theta_- = -\infty$ and $\theta_+ = \infty$.
(i) As an estimator of $\vartheta = E(T)$, $T(X)$ is admissible under the squared error loss and the loss $(a - \vartheta)^2/\text{Var}(T)$.
(ii) T is the unique minimax estimator of ϑ under the loss $(a - \vartheta)^2/\text{Var}(T)$.
Proof. (i) With $\lambda = 0$, condition (4.34) is clearly satisfied. Hence, Theorem 4.14 applies under the squared error loss. The admissibility of T under the loss $(a - \vartheta)^2/\text{Var}(T)$ follows from the fact that T is admissible under the squared error loss and $\text{Var}(T) \ne 0$.
(ii) This is a consequence of part (i) and Theorem 4.13. ∎

Example 4.20. Let $X_1, ..., X_n$ be i.i.d. from $N(0, \sigma^2)$ with an unknown $\sigma^2 > 0$. Let $Y = \sum_{i=1}^n X_i^2$. From Example 4.14, $Y/(n+2)$ is the MRIE of σ^2 and has constant risk under the loss $(a - \sigma^2)^2/\sigma^4$. We now apply Theorem 4.14 to show that $Y/(n+2)$ is admissible. Note that the joint p.d.f. of X_i's is of the form $c(\theta)e^{\theta T(x)}$ with $\theta = -n/(4\sigma^2)$, $c(\theta) = (-2\theta/n)^{n/2}$, $T(X) = 2Y/n$, $\theta_- = -\infty$, and $\theta_+ = 0$. By Theorem 4.14, $T_{\lambda,\gamma} = (T + \gamma\lambda)/(1 + \lambda)$ is admissible under the squared error loss if

$$\int_{-\infty}^{-c} e^{-\gamma\lambda\theta} \left(\frac{-2\theta}{n}\right)^{-n\lambda/2} d\theta = \int_0^c e^{\gamma\lambda\theta} \theta^{-n\lambda/2} d\theta = \infty$$

for some $c > 0$. This means that $T_{\lambda,\gamma}$ is admissible if $\gamma = 0$ and $\lambda = 2/n$, or if $\gamma > 0$ and $\lambda \ge 2/n$. In particular, $2Y/(n+2)$ is admissible for estimating $E(T) = 2E(Y)/n = 2\sigma^2$, under the squared error loss. It is easy to see that $Y/(n+2)$ is then an admissible estimator of σ^2 under the squared error loss and the loss $(a - \sigma^2)^2/\sigma^4$. Hence $Y/(n+2)$ is minimax under the loss $(a - \sigma^2)^2/\sigma^4$.

Note that we cannot apply Corollary 4.3 directly since $\theta_+ = 0$. ∎

Example 4.21. Let $X_1, ..., X_n$ be i.i.d. from the Poisson distribution $P(\theta)$ with an unknown $\theta > 0$. The joint p.d.f. of X_i's w.r.t. the counting measure is $(x_1! \cdots x_n!)^{-1} e^{-n\theta} e^{n\bar{x} \log \theta}$. For $\eta = n \log \theta$, the conditions of Corollary 4.3 are satisfied with $T(X) = \bar{X}$. Since $E(T) = \theta$ and $\text{Var}(T) = \theta/n$, by Corollary 4.3, $\bar{X}$ is the unique minimax estimator of θ under the loss function $(a - \theta)^2/\theta$. ∎

4.3.3 Simultaneous estimation and shrinkage estimators

In this chapter (and most of Chapter 3) we have focused on the estimation of a real-valued ϑ. The problem of estimating a vector-valued ϑ under the decision theory approach is called *simultaneous estimation*. Many results for the case of a real-valued ϑ can be extended to simultaneous estimation in a straightforward manner.

Let ϑ be a p-vector of parameters (functions of θ) with range $\tilde{\Theta}$. A vector-valued estimator $T(X)$ can be viewed as a decision rule taking values in the action space $\mathbb{A} = \tilde{\Theta}$. Let $L(\theta, a)$ be a given nonnegative loss function on $\Theta \times \mathbb{A}$. A natural generalization of the squared error loss is

$$L(\theta, a) = \|a - \vartheta\|^2 = \sum_{i=1}^{p} (a_i - \vartheta_i)^2, \qquad (4.37)$$

where a_i and ϑ_i are the ith components of a and ϑ, respectively.

A vector-valued estimator T is called unbiased if and only if $E(T) = \vartheta$ for all $\theta \in \Theta$. If there is an unbiased estimator of ϑ, then ϑ is called estimable. It can be seen that the result in Theorem 3.1 extends to the case of vector-valued ϑ with any L strictly convex in a. If the loss function is given by (4.37) and T_i is a UMVUE of ϑ_i for each i, then $T = (T_1, ..., T_p)$ is a UMVUE of ϑ. If there is a sufficient and complete statistic $U(X)$ for θ, then by Theorem 2.5 (Rao-Blackwell theorem), T must be a function of $U(X)$ and is the unique best unbiased estimator of ϑ.

Example 4.22. Consider the general linear model (3.25) with assumption A1 and a full rank Z. Let $\vartheta = \beta$. An unbiased estimator of β is then the LSE $\hat{\beta}$. From the proof of Theorem 3.7, $\hat{\beta}$ is a function of the sufficient and complete statistic for $\theta = (\beta, \sigma^2)$. Hence, $\hat{\beta}$ is the unique best unbiased estimator of ϑ under any strictly convex loss function. In particular, $\hat{\beta}$ is the UMVUE of β under the loss function (4.37). ∎

Next, we consider Bayes estimators of ϑ, which is still defined to be Bayes actions considered as functions of X. Under the loss function (4.37), the Bayes estimator is still given by (4.4) with vector-valued $g(\theta) = \vartheta$.

Example 4.23. Let $X = (X_0, X_1, ..., X_k)$ have the multinomial distribution given in Example 2.7. Consider the estimation of the vector $\theta = (p_0, p_1, ..., p_k)$ under the loss function (4.37), and the Dirichlet prior for θ that has the Lebesgue p.d.f.

$$\frac{\Gamma(\alpha_0 + \cdots + \alpha_k)}{\Gamma(\alpha_0) \cdots \Gamma(\alpha_k)} p_0^{\alpha_0 - 1} \cdots p_k^{\alpha_k - 1} I_A(\theta), \qquad (4.38)$$

where α_j's are known positive constants and $A = \{\theta : 0 \le p_j, \sum_{j=0}^k p_j = 1\}$. It turns out that the Dirichlet prior is conjugate so that the posterior of θ given $X = x$ is also a Dirichlet distribution having the p.d.f. given by (4.38) with α_j replaced by $\alpha_j + x_j$, $j = 0, 1, ..., k$. Thus, the Bayes estimator of θ is $\delta = (\delta_0, \delta_1, ..., \delta_k)$ with

$$\delta_j(X) = \frac{\alpha_j + X_j}{\alpha_0 + \alpha_1 + \cdots + \alpha_k + n}, \qquad j = 0, 1, ..., k. \quad \blacksquare$$

After a suitable class of transformations is defined, the results in §4.2 for invariant estimators and MRIE's are still valid. This is illustrated by the following example.

Example 4.24. Let X be a sample with the Lebesgue p.d.f. $f(x - \theta)$, where f is a known Lebesgue p.d.f. on $\mathcal{R}^p$ with a finite second moment and $\theta \in \mathcal{R}^p$ is an unknown parameter. Consider the estimation of θ under the loss function (4.37). This problem is invariant under the location transformations $g(X) = X + c$, where $c \in \mathcal{R}^p$. Invariant estimators of θ are of the form $X + l$, $l \in \mathcal{R}^p$. It is easy to show that any invariant estimator has constant bias and risk (a generalization of Proposition 4.4) and the MRIE of θ is the unbiased invariant estimator. In particular, if f is the p.d.f. of $N_p(0, I_p)$, then the MRIE is X. $\quad \blacksquare$

The definition of minimax estimators applies without changes.

Example 4.25. Let X be a sample from $N_p(\theta, I_p)$ with an unknown $\theta \in \mathcal{R}^p$. Consider the estimation of θ under the loss function (4.37). A modification of the proof of Theorem 4.12 with independent priors for θ_i's shows that X is a minimax estimator of θ (exercise). $\quad \blacksquare$

Example 4.26. Consider Example 4.23. If we choose $\alpha_0 = \cdots = \alpha_k = \sqrt{n}/(k+1)$, then the Bayes estimator of θ in Example 4.23 has constant risk. Using the same argument in the proof of Theorem 4.11, we can show that this Bayes estimator is minimax. $\quad \blacksquare$

The previous results for simultaneous estimation are fairly straightforward generalizations of those for the case of a real-valued ϑ. Results for admissibility in simultaneous estimation, however, are quite different. A surprising result, due to Stein (1956), is that in estimating the vector mean $\theta = EX$ of a normally distributed p-vector X (Example 4.25), X is inadmissible under the loss function (4.37) when $p \geq 3$, although X is the UMVUE, MRIE (Example 4.24), and minimax estimator (Example 4.25). Since any estimator better than a minimax estimator is also minimax, there exist many (in fact, infinitely many) minimax estimators in Example 4.25 when $p \geq 3$, which is different from the case of $p = 1$ in which X is the unique admissible minimax estimator (Example 4.6 and Theorem 4.13).

We start with the simple case where X is from $N_p(\theta, I_p)$ with an unknown $\theta \in \mathcal{R}^p$. James and Stein (1961) proposed the following class of estimators of $\vartheta = \theta$ having smaller risks than X when the loss is given by (4.37) and $p \geq 3$:

$$\delta_c = X - \frac{p-2}{\|X-c\|^2}(X-c), \qquad (4.39)$$

where $c \in \mathcal{R}^p$ is fixed. The choice of c is discussed next and at the end of this section.

Before we prove that δ_c in (4.39) is better than X, we try to motivate δ_c from two viewpoints. First, suppose that it were thought a priori likely, though not certain, that $\theta = c$. Then we might first test a hypothesis $H_0 : \theta = c$ and estimate θ by c if H_0 is accepted and by X otherwise. The best rejection region has the form $\|X-c\|^2 > t$ for some constant $t > 0$ (see Chapter 6) so that we might estimate θ by

$$I_{(t,\infty)}(\|X-c\|^2)X + [1 - I_{(t,\infty)}(\|X-c\|^2)]c.$$

It can be seen that δ_c in (4.39) is a smoothed version of this estimator, since

$$\delta_c = \psi(\|X-c\|^2)X + [1 - \psi(\|X-c\|^2)]c \qquad (4.40)$$

for some function ψ. Any estimator having the form of the right-hand side of (4.40) shrinks the observations toward a given point c and, therefore, is called a *shrinkage estimator*.

Next, δ_c in (4.40) can be viewed as an empirical Bayes estimator (§4.1.2). In view of (2.25) in Example 2.25, a Bayes estimator of θ is of the form

$$\delta = (1-B)X + Bc,$$

where c is the prior mean of θ and B involves prior variances. If $1 - B$ is "estimated" by $\psi(\|X-c\|^2)$, then δ_c is an empirical Bayes estimator.

Theorem 4.15. Suppose that X is from $N_p(\theta, I_p)$ with $p \geq 3$. Then, under the loss function (4.37), the risks of the following estimators of θ,

$$\delta_{c,r} = X - \frac{r(p-2)}{\|X-c\|^2}(X-c), \qquad (4.41)$$

are given by

$$R_{\delta_{c,r}}(\theta) = p - (2r - r^2)(p-2)^2 E(\|X - c\|^{-2}), \qquad (4.42)$$

where $c \in \mathcal{R}^p$ and $r \in \mathcal{R}$ are known.
Proof. Let $Z = X - c$. Then

$$R_{\delta_{c,r}}(\theta) = E\|\delta_{c,r} - E(X)\|^2 = E\left\| \left[1 - \frac{r(p-2)}{\|Z\|^2}\right] Z - E(Z) \right\|^2.$$

Hence, we only need to show the case of $c = 0$. Let $h(\theta) = R_{\delta_{0,r}}(\theta)$, $g(\theta)$ be the right-hand side of (4.42) with $c = 0$, and $\pi_\alpha(\theta) = (2\pi\alpha)^{-p/2} e^{-\|\theta\|^2/(2\alpha)}$, which is the p.d.f. of $N_p(0, \alpha I_p)$. Note that the distribution of X can be viewed as the conditional distribution of X given $\boldsymbol{\theta} = \theta$, where $\boldsymbol{\theta}$ has the Lebesgue p.d.f. $\pi_\alpha(\theta)$. Then

$$\int_{\mathcal{R}^p} g(\theta)\pi_\alpha(\theta)d\theta = p - (2r - r^2)(p-2)^2 E[E(\|X\|^{-2}|\boldsymbol{\theta})]$$
$$= p - (2r - r^2)(p-2)^2 E(\|X\|^{-2})$$
$$= p - (2r - r^2)(p-2)/(\alpha + 1),$$

where the expectation in the second line of the previous expression is w.r.t. the joint distribution of $(X, \boldsymbol{\theta})$ and the last equality follows from the fact that the marginal distribution of X is $N_p(0, (\alpha+1)I_p)$, $\|X\|^2/(\alpha+1)$ has the chi-square distribution χ_p^2 and, therefore, $E(\|X\|^{-2}) = 1/[(p-2)(\alpha+1)]$. Let $B = 1/(\alpha + 1)$ and $\hat{B} = r(p-2)/\|X\|^2$. Then

$$\int_{\mathcal{R}^p} h(\theta)\pi_\alpha(\theta)d\theta = E\|(1 - \hat{B})X - \boldsymbol{\theta}\|^2$$
$$= E\{E[\|(1-\hat{B})X - \boldsymbol{\theta}\|^2|X]\}$$
$$= E\{E[\|\boldsymbol{\theta} - E(\boldsymbol{\theta}|X)\|^2|X]$$
$$\qquad + \|E(\boldsymbol{\theta}|X) - (1-\hat{B})X\|^2\}$$
$$= E\{p(1 - B) + (\hat{B} - B)^2\|X\|^2\}$$
$$= E\{p(1 - B) + B^2\|X\|^2$$
$$\qquad - 2Br(p-2) + r^2(p-2)^2\|X\|^{-2}\}$$
$$= p - (2r - r^2)(p-2)B,$$

where the fourth equality follows from the fact that the conditional distribution of $\boldsymbol{\theta}$ given X is $N_p((1-B)X, (1-B)I_p)$ and the last equality follows from $E\|X\|^{-2} = B/(p-2)$ and $E\|X\|^2 = p/B$. This proves

$$\int_{\mathcal{R}^p} g(\theta)\pi_\alpha(\theta)d\theta = \int_{\mathcal{R}^p} h(\theta)\pi_\alpha(\theta)d\theta, \qquad \alpha > 0. \qquad (4.43)$$

Note that $h(\theta)$ and $g(\theta)$ are expectations of functions of $\|X\|^2$, $\theta^\tau X$, and $\|\theta\|^2$. Make an orthogonal transformation from X to Y such that $Y_1 = \theta^\tau X/\|\theta\|$, $EY_j = 0$ for $j > 1$, and $\mathrm{Var}(Y) = I_p$. Then $h(\theta)$ and $g(\theta)$ are expectations of functions of Y_1, $\sum_{j=2}^p Y_j^2$, and $\|\theta\|^2$. Thus, both h and g are functions of $\|\theta\|^2$.

For the family of p.d.f.'s $\{\pi_\alpha(\theta) : \alpha > 0\}$, $\|\theta\|^2$ is a complete and sufficient "statistic". Hence, (4.43) and the fact that h and g are functions of $\|\theta\|^2$ imply that $h(\theta) = g(\theta)$ a.e. w.r.t. the Lebesgue measure. From Theorem 2.1, both h and g are continuous functions of $\|\theta\|^2$ and, therefore, $h(\theta) = g(\theta)$ for all $\theta \in \mathcal{R}^p$. This completes the proof. ∎

It follows from Theorem 4.15 that the risk of $\delta_{c,r}$ is smaller than that of X (for every value of θ) when $p \geq 3$ and $0 < r < 2$, since the risk of X is p under the loss function (4.37). From Example 4.6, X is admissible when $p = 1$. When $p = 2$, X is still admissible (Stein, 1956). But we have just shown that X is inadmissible when $p \geq 3$.

The James-Stein estimator δ_c in (4.39), which is a special case of (4.41) with $r = 1$, is better than any $\delta_{c,r}$ in (4.41) with $r \neq 1$, since the factor $2r - r^2$ takes on its maximum value 1 if and only if $r = 1$. To see that δ_c may have a substantial improvement over X in terms of risks, consider the special case where $\theta = c$. Since $\|X - c\|^2$ has the chi-square distribution χ_p^2 when $\theta = c$, $E\|X - c\|^{-2} = (p-2)^{-1}$ and the right-hand side of (4.42) equals 2. Thus, the ratio $R_X(\theta)/R_{\delta_c}(\theta)$ equals $p/2$ when $\theta = c$ and, therefore, can be substantially larger than 1 near $\theta = c$ when p is large.

Since X is minimax (Example 4.25), any shrinkage estimator of the form (4.41) is minimax provided that $p \geq 3$ and $0 < r < 2$.

Unfortunately, the James-Stein estimator with any c is also inadmissible. It is dominated by

$$\delta_c^+ = X - \min\left\{1, \frac{p-2}{\|X - c\|^2}\right\}(X - c); \tag{4.44}$$

see, for example, Lehmann (1983, Theorem 4.6.2). This estimator, however, is still inadmissible. An example of an admissible estimator of the form (4.40) is provided by Strawderman (1971); see also Lehmann (1983, p. 304). Although neither the James-Stein estimator δ_c nor δ_c^+ in (4.44) is admissible, it is found that no substantial improvements over δ_c^+ are possible (Efron and Morris, 1973).

To extend Theorem 4.15 to general $\mathrm{Var}(X)$, we consider the case where $\mathrm{Var}(X) = \sigma^2 D$ with an unknown $\sigma^2 > 0$ and a known positive definite matrix D. If σ^2 is known, then an extended James-Stein estimator is

$$\tilde{\delta}_{c,r} = X - \frac{r(p-2)\sigma^2}{\|D^{-1}(X - c)\|^2}D^{-1}(X - c). \tag{4.45}$$

One can show (exercise) that under the loss (4.37), the risk of $\tilde{\delta}_{c,r}$ is

$$\sigma^2 \left[\text{tr}(D) - (2r - r^2)(p - 2)^2 \sigma^2 E(\|D^{-1}(X - c)\|^{-2}) \right]. \qquad (4.46)$$

When σ^2 is unknown, we assume that there exists a statistic S_0^2 such that S_0^2 is independent of X and S_0^2/σ^2 has the chi-square distribution χ_m^2 (see Example 4.27). Replacing $r\sigma^2$ in (4.45) by $\hat{\sigma}^2 = tS_0^2$ with a constant $t > 0$ leads to the following extended James-Stein estimator:

$$\tilde{\delta}_c = X - \frac{(p - 2)\hat{\sigma}^2}{\|D^{-1}(X - c)\|^2} D^{-1}(X - c). \qquad (4.47)$$

By (4.46) and the independence of $\hat{\sigma}^2$ and X, the risk of $\tilde{\delta}_c$ (as an estimator of $\vartheta = EX$) is

$$\begin{aligned}
R_{\tilde{\delta}_c}(\theta) &= E\left[E(\|\tilde{\delta}_c - \vartheta\|^2 | \hat{\sigma}^2) \right] \\
&= E\left[E(\|\tilde{\delta}_{c,(\hat{\sigma}^2/\sigma^2)} - \vartheta\|^2 | \hat{\sigma}^2) \right] \\
&= \sigma^2 E\left\{ \text{tr}(D) - [2(\hat{\sigma}^2/\sigma^2) - (\hat{\sigma}^2/\sigma^2)^2](p - 2)^2 \sigma^2 \kappa(\theta) \right\} \\
&= \sigma^2 \left\{ \text{tr}(D) - [2E(\hat{\sigma}^2/\sigma^2) - E(\hat{\sigma}^2/\sigma^2)^2](p - 2)^2 \sigma^2 \kappa(\theta) \right\} \\
&= \sigma^2 \left\{ \text{tr}(D) - [2tm - t^2 m(m + 2)](p - 2)^2 \sigma^2 \kappa(\theta) \right\},
\end{aligned}$$

where $\theta = (\vartheta, \sigma^2)$ and $\kappa(\theta) = E(\|D^{-1}(X - c)\|^{-2})$. Since $2tm - t^2 m(m+2)$ is maximized at $t = 1/(m + 2)$, replacing t by $1/(m + 2)$ leads to

$$R_{\tilde{\delta}_c}(\theta) = \sigma^2 \left[\text{tr}(D) - m(m + 2)^{-1}(p - 2)^2 \sigma^2 E(\|D^{-1}(X - c)\|^{-2}) \right].$$

Hence, the risk of the extended James-Stein estimator in (4.47) is smaller than that of X for any fixed θ, when $p \geq 3$.

Example 4.27. Consider the general linear model (3.25) with assumption A1, $p \geq 3$, and a full rank Z, and the estimation of $\vartheta = \beta$ under the loss function (4.37). From Theorem 3.8, the LSE $\hat{\beta}$ is from $N(\beta, \sigma^2 D)$ with a known matrix $D = (Z^\tau Z)^{-1}$; $S_0^2 = SSR$ is independent of $\hat{\beta}$; and S_0^2/σ^2 has the chi-square distribution χ_{n-p}^2. Hence, from the previous discussion, the risk of the shrinkage estimator

$$\hat{\beta} - \frac{(p - 2)\hat{\sigma}^2}{\|Z^\tau Z(\hat{\beta} - c)\|^2} Z^\tau Z(\hat{\beta} - c)$$

is smaller than that of $\hat{\beta}$ for any β and σ^2, where $c \in \mathcal{R}^p$ is fixed and $\hat{\sigma}^2 = SSR/(n - p + 2)$. ∎

From the previous discussion, the James-Stein estimators improve X substantially when we shrink the observations toward a vector c that is near

$\vartheta = EX$. Of course, this cannot be done since ϑ is unknown. One may consider shrinking the observations toward the mean of the observations rather than a given point; that is, one may obtain a shrinkage estimator by replacing c in (4.39) or (4.47) by $\bar{X}J_p$, where $\bar{X} = p^{-1}\sum_{i=1}^{p} X_i$ and J_p is the p-vector of ones. However, we have to replace the factor $p - 2$ in (4.39) or (4.47) by $p - 3$. This leads to shrinkage estimators

$$X - \frac{p-3}{\|X - \bar{X}J_p\|^2}(X - \bar{X}J_p) \tag{4.48}$$

and

$$X - \frac{(p-3)\hat{\sigma}^2}{\|D^{-1}(X - \bar{X}J_p)\|^2}D^{-1}(X - \bar{X}J_p). \tag{4.49}$$

These estimators are better than X (and, hence, are minimax) when $p \geq 4$, under the loss function (4.37) (exercise).

The results discussed in this section for the simultaneous estimation of a vector of normal means can be extended to a wide variety of cases where the loss functions are not given by (4.37) (Brown, 1966). The results have also been extended to exponential families and to general location parameter families. For example, Berger (1976) studied the inadmissibility of generalized Bayes estimators of a location vector; Berger (1980) considered simultaneous estimation of gamma scale parameters; and Tsui (1981) investigated simultaneous estimation of several Poisson parameters. See Lehmann (1983, pp. 320-330) for some further references.

4.4 The Method of Maximum Likelihood

So far we have studied estimation methods in parametric families using the decision theory approach. The *maximum likelihood method* introduced next is the most popular method for deriving estimators in statistical inference that does not use any loss function.

4.4.1 The likelihood function and MLE's

To introduce the idea, let us consider an example.

Example 4.28. Let X be a single observation taking values from $\{0, 1, 2\}$ according to P_θ, where $\theta = \theta_0$ or θ_1 and the values of $P_{\theta_j}(\{i\})$ are given by the following table:

	$x = 0$	$x = 1$	$x = 2$
$\theta = \theta_0$	0.8	0.1	0.1
$\theta = \theta_1$	0.2	0.3	0.5

If $X = 0$ is observed, it is more plausible that it came from P_{θ_0}, since $P_{\theta_0}(\{0\})$ is much larger than $P_{\theta_1}(\{0\})$. We then estimate θ by θ_0. On the other hand, if $X = 1$ or 2, it is more plausible that it came from P_{θ_1}, although in this case the difference between the probabilities is not as large as that in the case of $X = 0$. This suggests the following estimator of θ:

$$T(X) = \begin{cases} \theta_0 & X = 0 \\ \theta_1 & X \neq 0. \end{cases} \quad \blacksquare$$

The idea in Example 4.28 can be easily extended to the case where P_θ is a discrete distribution and $\theta \in \Theta \subset \mathcal{R}^k$. If $X = x$ is observed, θ_1 is more plausible than θ_2 if and only if $P_{\theta_1}(\{x\}) > P_{\theta_2}(\{x\})$. We then estimate θ by a $\hat{\theta}$ that maximizes $P_\theta(\{x\})$ over $\theta \in \Theta$, if such a $\hat{\theta}$ exists. The word plausible rather than probable is used because θ is considered to be nonrandom and P_θ is not a distribution of θ. Under the Bayesian approach with a prior that is the discrete uniform distribution on $\{\theta_1, ..., \theta_m\}$, $P_\theta(\{x\})$ is proportional to the posterior probability and we can say that θ_1 is more probable than θ_2 if $P_{\theta_1}(\{x\}) > P_{\theta_2}(\{x\})$.

Note that $P_\theta(\{x\})$ in the previous discussion is the p.d.f. w.r.t. the counting measure. Hence, it is natural to extend the idea to the case of continuous (or arbitrary) X by using the p.d.f. of X w.r.t. some σ-finite measure on the range $\mathcal{X}$ of X. This leads to the following definition.

Definition 4.3. Let $X \in \mathcal{X}$ be a sample with a p.d.f. f_θ w.r.t. a σ-finite measure ν, where $\theta \in \Theta \subset \mathcal{R}^k$.
(i) For each $x \in \mathcal{X}$, $f_\theta(x)$ considered as a function of θ is called the *likelihood function* and denoted by $\ell(\theta)$.
(ii) Let $\bar{\Theta}$ be the closure of Θ. A $\hat{\theta} \in \bar{\Theta}$ satisfying $\ell(\hat{\theta}) = \max_{\theta \in \bar{\Theta}} \ell(\theta)$ is called a *maximum likelihood estimate* (MLE) of θ. If $\hat{\theta}$ is a Borel function of X a.e. ν, then $\hat{\theta}$ is called a *maximum likelihood estimator* (MLE) of θ.
(iii) Let g be a Borel function from Θ to $\mathcal{R}^p$, $p \leq k$. If $\hat{\theta}$ is an MLE of θ, then $\hat{\vartheta} = g(\hat{\theta})$ is defined to be an MLE of $\vartheta = g(\theta)$. $\quad \blacksquare$

Note that $\bar{\Theta}$ instead of Θ is used in the definition of an MLE. This is because a maximum of $\ell(\theta)$ may not exist when Θ is an open set (Examples 4.29 and 4.30). As an estimator, an MLE is defined a.e. ν. Part (iii) of Definition 4.3 is motivated by a fact given in Exercise 95 of §4.6.

If the parameter space Θ contains finitely many points, then $\bar{\Theta} = \Theta$ and an MLE can always be obtained by comparing finitely many values $\ell(\theta)$, $\theta \in \Theta$. If $\ell(\theta)$ is differentiable on Θ°, the interior of Θ, then possible candidates for MLE's are the values of $\theta \in \Theta^\circ$ satisfying

$$\frac{\partial \ell(\theta)}{\partial \theta} = 0, \tag{4.50}$$

which is called the *likelihood equation*. Note that θ's satisfying (4.50) may be local or global minima, local or global maxima, or simply stationary points. Also, extrema may occur at the boundary of Θ or when $\|\theta\| \to \infty$. Furthermore, if $\ell(\theta)$ is not always differentiable, then extrema may occur at nondifferentiable or discontinuity points of $\ell(\theta)$. Hence, it is important to analyze the entire likelihood function to find its maxima.

Since $\log x$ is a strictly increasing function and $\ell(\theta)$ can be assumed to be positive without loss of generality, $\hat{\theta}$ is an MLE if and only if it maximizes the log-likelihood function $\log \ell(\theta)$. It is often more convenient to work with $\log \ell(\theta)$ and the following analogue of (4.50) (which is called the log-likelihood equation or likelihood equation for simplicity):

$$\frac{\partial \log \ell(\theta)}{\partial \theta} = 0. \tag{4.51}$$

Example 4.29. Let $X_1, ..., X_n$ be i.i.d. binary random variables with $P(X_1 = 1) = p \in \Theta = (0, 1)$. When $(X_1, ..., X_n) = (x_1, ..., x_n)$ is observed, the likelihood function is

$$\ell(p) = \prod_{i=1}^{n} p^{x_i}(1-p)^{1-x_i} = p^{n\bar{x}}(1-p)^{n(1-\bar{x})},$$

where $\bar{x} = n^{-1}\sum_{i=1}^{n} x_i$. Note that $\bar{\Theta} = [0, 1]$ and $\Theta^\circ = \Theta$. The likelihood equation (4.51) reduces to

$$\frac{n\bar{x}}{p} - \frac{n(1-\bar{x})}{1-p} = 0.$$

If $0 < \bar{x} < 1$, then this equation has a unique solution $\bar{x}$. The second-order derivative of $\log \ell(p)$ is

$$-\frac{n\bar{x}}{p^2} - \frac{n(1-\bar{x})}{(1-p)^2},$$

which is always negative. Also, when p tends to 0 or 1 (the boundary of Θ), $\ell(p) \to 0$. Thus, $\bar{x}$ is the unique MLE of p.

When $\bar{x} = 0$, $\ell(p) = (1-p)^n$ is a strictly decreasing function of p and, therefore, its unique maximum is 0. Similarly, the MLE is 1 when $\bar{x} = 1$. Combining these results with the previous result, we conclude that the MLE of p is $\bar{x}$.

When $\bar{x} = 0$ or 1, a maximum of $\ell(p)$ does not exist on $\Theta = (0, 1)$, although $\sup_{p \in (0,1)} \ell(p) = 1$; the MLE takes a value outside of Θ and, hence, is not a reasonable estimator. However, if $p \in (0, 1)$, the probability that $\bar{x} = 0$ or 1 tends to 0 quickly as $n \to \infty$. ∎

Example 4.29 indicates that, for small n, a maximum of $\ell(\theta)$ may not exist on Θ and an MLE may be an unreasonable estimator; however, this is unlikely to occur when n is large. A rigorous result of this sort is given in §4.5.2, where we study asymptotic properties of MLE's.

Example 4.30. Let $X_1, ..., X_n$ be i.i.d. from $N(\mu, \sigma^2)$ with an unknown $\theta = (\mu, \sigma^2)$, where $n \geq 2$. Consider first the case where $\Theta = \mathcal{R} \times (0, \infty)$. When $(X_1, ..., X_n) = (x_1, ..., x_n)$ is observed, the log-likelihood function is

$$\log \ell(\theta) = -\frac{1}{2\sigma^2} \sum_{i=1}^{n} (x_i - \mu)^2 - \frac{n}{2} \log \sigma^2 - \frac{n}{2} \log(2\pi).$$

The likelihood equation (4.51) becomes

$$\frac{1}{\sigma^2} \sum_{i=1}^{n} (x_i - \mu) = 0 \qquad \text{and} \qquad \frac{1}{\sigma^4} \sum_{i=1}^{n} (x_i - \mu)^2 - \frac{n}{\sigma^2} = 0. \qquad (4.52)$$

Solving the first equation in (4.52) for μ, we obtain a unique solution $\bar{x} = n^{-1} \sum_{i=1}^{n} x_i$, and substituting $\bar{x}$ for μ in the second equation in (4.52), we obtain a unique solution $\hat{\sigma}^2 = n^{-1} \sum_{i=1}^{n} (x_i - \bar{x})^2$. To show that $\hat{\theta} = (\bar{x}, \hat{\sigma}^2)$ is an MLE, first note that Θ is an open set and $\ell(\theta)$ is differentiable everywhere; as θ tends to the boundary of Θ or $\|\theta\| \to \infty$, $\ell(\theta)$ tends to 0; and

$$\frac{\partial^2 \log \ell(\theta)}{\partial \theta \partial \theta^\tau} = - \begin{pmatrix} \frac{n}{\sigma^2} & \frac{1}{\sigma^4} \sum_{i=1}^{n} (x_i - \mu) \\ \frac{1}{\sigma^4} \sum_{i=1}^{n} (x_i - \mu) & \frac{1}{\sigma^6} \sum_{i=1}^{n} (x_i - \mu)^2 - \frac{n}{2\sigma^4} \end{pmatrix}$$

is negative definite when $\mu = \bar{x}$ and $\sigma^2 = \hat{\sigma}^2$. Hence $\hat{\theta}$ is the unique MLE. Sometimes we can avoid the calculation of the second-order derivatives. For instance, in this example we know that $\ell(\theta)$ is bounded and $\ell(\theta) \to 0$ as $\|\theta\| \to \infty$ or θ tends to the boundary of Θ; hence the unique solution to (4.52) must be the MLE. Another way to show that $\hat{\theta}$ is the MLE is indicated by the following discussion.

Consider next the case where $\Theta = (0, \infty) \times (0, \infty)$, i.e., μ is known to be positive. The likelihood function is differentiable on $\Theta^\circ = \Theta$ and $\bar{\Theta} = [0, \infty) \times [0, \infty)$. If $\bar{x} > 0$, then the same argument for the previous case can be used to show that $(\bar{x}, \hat{\sigma}^2)$ is the MLE. If $\bar{x} \leq 0$, then the first equation in (4.52) does not have a solution in Θ. However, the function $\log \ell(\theta) = \log \ell(\mu, \sigma^2)$ is strictly decreasing in μ for any fixed σ^2. Hence, a maximum of $\log \ell(\mu, \sigma^2)$ is $\mu = 0$, which does not depend on σ^2. Then, the MLE is $(0, \tilde{\sigma}^2)$, where $\tilde{\sigma}^2$ is the value maximizing $\log \ell(0, \sigma^2)$ over $\sigma^2 \geq 0$. Applying (4.51) to the function $\log \ell(0, \sigma^2)$ leads to $\tilde{\sigma}^2 = n^{-1} \sum_{i=1}^{n} x_i^2$. Thus, the MLE is

$$\hat{\theta} = \begin{cases} (\bar{x}, \hat{\sigma}^2) & \bar{x} > 0 \\ (0, \tilde{\sigma}^2) & \bar{x} \leq 0. \end{cases}$$

Again, the MLE in this case is not in Θ if $\bar{x} \leq 0$. One can show that a maximum of $\ell(\theta)$ does not exist on Θ when $\bar{x} \leq 0$. ∎

Example 4.31. Let $X_1, ..., X_n$ be i.i.d. from the uniform distribution on an interval $\mathcal{I}_\theta$ with an unknown θ. First, consider the case where $\mathcal{I}_\theta = (0, \theta)$ and $\theta > 0$. The likelihood function is $\ell(\theta) = \theta^{-n} I_{(x_{(n)}, \infty)}(\theta)$, which is not always differentiable. In this case $\Theta^\circ = (0, x_{(n)}) \cup (x_{(n)}, \infty)$. But, on $(0, x_{(n)})$, $\ell \equiv 0$ and on $(x_{(n)}, \infty)$, $\ell'(\theta) = -n\theta^{n-1} < 0$ for all θ. Hence, the method of using the likelihood equation is not applicable to this problem. Since $\ell(\theta)$ is strictly decreasing on $(x_{(n)}, \infty)$ and is 0 on $(0, x_{(n)})$, a unique maximum of $\ell(\theta)$ is $x_{(n)}$, which is a discontinuity point of $\ell(\theta)$. This shows that the MLE of θ is the largest order statistic $X_{(n)}$.

Next, consider the case where $\mathcal{I}_\theta = (\theta - \frac{1}{2}, \theta + \frac{1}{2})$ with $\theta \in \mathcal{R}$. The likelihood function is $\ell(\theta) = I_{(x_{(n)} - \frac{1}{2}, x_{(1)} + \frac{1}{2})}(\theta)$. Again, the method of using the likelihood equation is not applicable. However, it follows from Definition 4.3 that any statistic $T(X)$ satisfying $x_{(n)} - \frac{1}{2} \leq T(x) \leq x_{(1)} + \frac{1}{2}$ is an MLE of θ. This example indicates that MLE's may not be unique and can be unreasonable. ∎

Example 4.32. Let X be an observation from the hypergeometric distribution $HG(r, n, \theta - n)$ (Table 1.1, page 18) with known r, n, and an unknown $\theta = n+1, n+2, ...$ In this case, the likelihood function is defined on integers and the method of using the likelihood equation is certainly not applicable. Note that

$$\frac{\ell(\theta)}{\ell(\theta - 1)} = \frac{(\theta - r)(\theta - n)}{\theta(\theta - n - r + x)},$$

which is larger than 1 if and only if $\theta < rn/x$ and is smaller than 1 if and only if $\theta > rn/x$. Thus, $\ell(\theta)$ has a maximum $\theta = $ the integer part of rn/x, which is the MLE of θ. ∎

Example 4.33. Let $X_1, ..., X_n$ be i.i.d. from the gamma distribution $\Gamma(\alpha, \gamma)$ with unknown $\alpha > 0$ and $\gamma > 0$. The log-likelihood function is

$$\log \ell(\theta) = -n\alpha \log \gamma - n \log \Gamma(\alpha) + (\alpha - 1) \sum_{i=1}^{n} \log x_i - \frac{1}{\gamma} \sum_{i=1}^{n} x_i$$

and the likelihood equation (4.51) becomes

$$-n \log \gamma - \frac{n\Gamma'(\alpha)}{\Gamma(\alpha)} + \sum_{i=1}^{n} \log x_i = 0$$

and

$$-\frac{n\alpha}{\gamma} + \frac{1}{\gamma^2} \sum_{i=1}^{n} x_i = 0.$$

The second equation yields $\gamma = \bar{x}/\alpha$. Substituting $\gamma = \bar{x}/\alpha$ into the first equation we obtain that

$$\log \alpha - \frac{\Gamma'(\alpha)}{\Gamma(\alpha)} + \frac{1}{n} \sum_{i=1}^{n} \log x_i - \log \bar{x} = 0.$$

In this case, the likelihood equation does not have an explicit solution, although it can be shown (exercise) that a solution exists almost surely and it is the unique MLE. A numerical method has to be applied to compute the MLE for any given observations $x_1, ..., x_n$. ∎

These examples indicate that we need to use various methods to derive MLE's. In applications, MLE's typically do not have analytic forms and some numerical methods have to be used to compute MLE's. A commonly used numerical method is the Newton-Raphson iteration method, which repeatedly computes

$$\hat{\theta}^{(t+1)} = \hat{\theta}^{(t)} - \left[\left. \frac{\partial^2 \log \ell(\theta)}{\partial \theta \partial \theta^\tau} \right|_{\theta = \hat{\theta}^{(t)}} \right]^{-1} \left. \frac{\partial \log \ell(\theta)}{\partial \theta} \right|_{\theta = \hat{\theta}^{(t)}}, \qquad (4.53)$$

$t = 0, 1, ...$, where $\hat{\theta}^{(0)}$ is an initial value and $\partial^2 \log \ell(\theta)/\partial\theta\partial\theta^\tau$ is assumed of full rank for every $\theta \in \Theta$. If, at each iteration, we replace $\partial^2 \log \ell(\theta)/\partial\theta\partial\theta^\tau$ in (4.53) by its expected value $E[\partial^2 \log \ell(\theta)/\partial\theta\partial\theta^\tau]$, where the expectation is taken under P_θ, then the method is known as the Fisher-scoring method. If the iteration converges, then $\hat{\theta}^{(\infty)}$ or $\hat{\theta}^{(t)}$ with a sufficiently large t is a numerical approximation to a solution of the likelihood equation (4.51).

The following example shows that the MCMC methods discussed in §4.1.4 can also be useful in computing MLE's.

Example 4.34. Let X be a random k-vector from P_θ with the following p.d.f. w.r.t. a σ-finite measure ν:

$$f_\theta(x) = \int f_\theta(x, y) d\nu(y),$$

where $f_\theta(x, y)$ is a joint p.d.f. w.r.t. $\nu \times \nu$. This type of distribution is called a *mixture* distribution. Thus, the likelihood $\ell(\theta) = f_\theta(x)$ involves a k-dimensional integral. In many cases this integral has to be computed in order to compute an MLE of θ.

Let $\tilde{\ell}_m(\theta)$ be the MCMC approximation to $\ell(\theta)$ based on one of the MCMC methods described in §4.1.4 and a Markov chain of length m. Under the conditions of Theorem 4.4, $\tilde{\ell}_m(\theta) \to_{a.s.} \ell(\theta)$ for every fixed θ and x. Suppose that, for each m, there exists $\tilde{\theta}_m$ that maximizes $\tilde{\ell}_m(\theta)$ over $\theta \in \Theta$. Geyer (1994) studies the convergence of $\tilde{\theta}_m$ to an MLE. ∎

In terms of their mse's, MLE's are not necessarily better than UMVUE's or Bayes estimators. Also, MLE's are frequently inadmissible. This is not surprising, since MLE's are not derived under any given loss function. The main theoretical justification for MLE's is provided in the theory of asymptotic efficiency considered in §4.5.

4.4.2 MLE's in generalized linear models

Suppose that X has a distribution from a natural exponential family so that the likelihood function is

$$\ell(\eta) = \exp\{\eta^\tau T(x) - \zeta(\eta)\}h(x),$$

where $\eta \in \Xi$ is a vector of unknown parameters. The likelihood equation (4.51) is then

$$\frac{\partial \log \ell(\eta)}{\partial \eta} = T(x) - \frac{\partial \zeta(\eta)}{\partial \eta} = 0,$$

which has a unique solution $T(x) = \partial\zeta(\eta)/\partial\eta$, assuming that $T(x)$ is in the range of $\partial\zeta(\eta)/\partial\eta$. Note that

$$\frac{\partial^2 \log \ell(\eta)}{\partial \eta \partial \eta^\tau} = -\frac{\partial^2 \zeta(\eta)}{\partial \eta \partial \eta^\tau} = -\text{Var}(T) \qquad (4.54)$$

(see the proof of Proposition 3.2). Since $\text{Var}(T)$ is positive definite, $-\log \ell(\eta)$ is convex in η and $T(x)$ is the unique MLE of the parameter $\mu(\eta) = \partial\zeta(\eta)/\partial\eta$. By (4.54) again, the function $\mu(\eta)$ is one-to-one so that μ^{-1} exists. By Definition 4.3, the MLE of η is $\hat{\eta} = \mu^{-1}(T(x))$.

If the distribution of X is in a general exponential family and the likelihood function is

$$\ell(\theta) = \exp\{[\eta(\theta)]^\tau T(x) - \xi(\theta)\}h(x),$$

then the MLE of θ is $\hat{\theta} = \eta^{-1}(\hat{\eta})$, if η^{-1} exists and $\hat{\eta}$ is in the range of $\eta(\theta)$. Of course, $\hat{\theta}$ is also the solution of the likelihood equation

$$\frac{\partial \log \ell(\theta)}{\partial \theta} = \frac{\partial \eta(\theta)}{\partial \theta}T(x) - \frac{\partial \xi(\theta)}{\partial \theta} = 0.$$

The results for exponential families lead to an estimation method in a class of models that have very wide applications. These models are generalizations of the normal linear model (model (3.25) with assumption A1) discussed in §3.3.1-§3.3.2 and, therefore, are named *generalized linear models* (GLM).

A GLM has the following structure. The sample $X = (X_1, ..., X_n) \in \mathcal{R}^n$ has independent components and X_i has the p.d.f.

$$\exp\left\{\frac{\eta_i x_i - \zeta(\eta_i)}{\phi_i}\right\} h(x_i, \phi_i), \qquad i = 1, ..., n, \tag{4.55}$$

w.r.t. a σ-finite measure ν, where η_i and ϕ_i are unknown, $\phi_i > 0$,

$$\eta_i \in \Xi = \left\{\eta : \ 0 < \int h(x, \phi) e^{\eta x/\phi} d\nu(x) < \infty\right\} \subset \mathcal{R}$$

for all i, ζ and h are known functions, and $\zeta''(\eta) > 0$ is assumed for all $\eta \in \Xi^\circ$, the interior of Ξ. Note that the p.d.f. in (4.55) belongs to an exponential family if ϕ_i is known. As a consequence,

$$E(X_i) = \zeta'(\eta_i) \quad \text{and} \quad \text{Var}(X_i) = \phi_i \zeta''(\eta_i), \qquad i = 1, ..., n. \tag{4.56}$$

Define $\mu(\eta) = \zeta'(\eta)$. It is assumed that η_i is related to Z_i, the ith value of a p-vector of covariates (see (3.24)), through

$$g(\mu(\eta_i)) = \beta^\tau Z_i, \qquad i = 1, ..., n, \tag{4.57}$$

where β is a p-vector of unknown parameters and g, called a *link function*, is a known one-to-one, third-order continuously differentiable function on $\{\mu(\eta) : \eta \in \Xi^\circ\}$. If $\mu = g^{-1}$, then $\eta_i = \beta^\tau Z_i$ and g is called the *canonical* or *natural* link function. If g is not canonical, we assume that $\frac{d}{d\eta}(g \circ \mu)(\eta) \neq 0$ for all η.

In a GLM, the parameter of interest is β. We assume that the range of β is $B = \{\beta : (g \circ \mu)^{-1}(\beta^\tau z) \in \Xi^\circ \text{ for all } z \in \mathcal{Z}\}$, where $\mathcal{Z}$ is the range of Z_i's. ϕ_i's are called *dispersion* parameters and are considered to be nuisance parameters. It is often assumed that

$$\phi_i = \phi/t_i, \qquad i = 1, ..., n, \tag{4.58}$$

with an unknown $\phi > 0$ and known positive t_i's.

As we discussed earlier, the linear model (3.24) with $\varepsilon_i = N(0, \phi)$ is a special GLM. One can verify this by taking $g(\mu) \equiv \mu$ and $\zeta(\eta) = \eta^2/2$. The usefulness of the GLM is that it covers situations where the relationship between $E(X_i)$ and Z_i is nonlinear and/or X_i's are discrete (in which case the linear model (3.24) is clearly not appropriate). The following is an example.

Example 4.35. Let X_i's be independent discrete random variables taking values in $\{0, 1, ..., m\}$, where m is a known positive integer. First, suppose that X_i has the binomial distribution $Bi(p_i, m)$ with an unknown $p_i \in (0, 1)$, $i = 1, ..., n$. Let $\eta_i = \log \frac{p_i}{1-p_i}$ and $\zeta(\eta_i) = m \log(1 + e^{\eta_i})$. Then the p.d.f. of X_i (w.r.t. the counting measure) is given by (4.55) with $\phi_i = 1$,

$h(x_i, \phi_i) = \binom{m}{x_i}$, and $\Xi = \mathcal{R}$. Under (4.57) and the *logit* link (canonical link) $g(t) = m^{-1} \log \frac{t}{1-t}$,

$$E(X_i) = mp_i = \frac{me^{\eta_i}}{1 + e^{\eta_i}} = \frac{me^{\beta^\tau Z_i}}{1 + e^{\beta^\tau Z_i}}.$$

Another popular link in this problem is the *probit* link $g(t) = m^{-1}\Phi^{-1}(t)$, where Φ is the c.d.f. of the standard normal. Under the probit link, $E(X_i) = m\Phi(\beta^\tau Z_i)$.

The variance of X_i is $mp_i(1 - p_i)$ under the binomial distribution assumption. This assumption is often violated in applications, which results in an *over-dispersion*, i.e., the variance of X_i exceeds the nominal variance $mp_i(1 - p_i)$. Over-dispersion can arise in a number of ways, but the most common one is clustering in the population. Families, households, and litters are common instances of clustering. For example, suppose that $X_i = \sum_{j=1}^m X_{ij}$, where X_{ij} are binary random variables having a common distribution. If X_{ij}'s are independent, then X_i has a binomial distribution. However, if X_{ij}'s are from the same cluster (family or household), then they are often positively correlated. Suppose that the correlation coefficient (§1.3.2) between X_{ij} and X_{il}, $j \neq l$, is $\rho_i > 0$. Then

$$\text{Var}(X_i) = mp_i(1 - p_i) + m(m - 1)\rho_i p_i(1 - p_i) = \phi_i mp_i(1 - p_i),$$

where $\phi_i = 1 + (m - 1)\rho_i$ is the dispersion parameter. Of course, over-dispersion can occur only if $m > 1$ in this case.

This motivates the consideration of GLM (4.55)-(4.57) with dispersion parameters ϕ_i. If X_i has the p.d.f. (4.55) with $\zeta(\eta_i) = m \log(1 + e^{\eta_i})$, then

$$E(X_i) = \frac{me^{\eta_i}}{1 + e^{\eta_i}} \qquad \text{and} \qquad \text{Var}(X_i) = \phi_i \frac{me^{\eta_i}}{(1 + e^{\eta_i})^2},$$

which is exactly (4.56). Of course, the distribution of X_i is not binomial unless $\phi_i = 1$. ∎

We now derive an MLE of β in a GLM under assumption (4.58). Let $\theta = (\beta, \phi)$ and $\psi = (g \circ \mu)^{-1}$. Then the log-likelihood function is

$$\log \ell(\theta) = \sum_{i=1}^n \left[\log h(x_i, \phi/t_i) + \frac{\psi(\beta^\tau Z_i)x_i - \zeta(\psi(\beta^\tau Z_i))}{\phi/t_i} \right]$$

and the likelihood equation is

$$\frac{\partial \log \ell(\theta)}{\partial \beta} = \frac{1}{\phi} \sum_{i=1}^n \{[x_i - \mu(\psi(\beta^\tau Z_i))]\psi'(\beta^\tau Z_i)t_i Z_i\} = 0 \qquad (4.59)$$

and

$$\frac{\partial \log \ell(\theta)}{\partial \phi} = \sum_{i=1}^{n} \left\{ \frac{\partial \log h(x_i, \phi/t_i)}{\partial \phi} - \frac{t_i[\psi(\beta^\tau Z_i)x_i - \zeta(\psi(\beta^\tau Z_i))]}{\phi^2} \right\} = 0.$$

From the first equation, an MLE of β, if it exists, can be obtained without estimating ϕ. The second equation, however, is usually difficult to solve. Some other estimators of ϕ are suggested by various researchers; see, for example, McCullagh and Nelder (1989).

Suppose that there is a solution $\hat{\beta} \in B$ to equation (4.59). (The existence of $\hat{\beta}$ is studied in §4.5.2.) We now study whether $\hat{\beta}$ is an MLE of β. Let

$$M_n(\beta) = \sum_{i=1}^{n} [\psi'(\beta^\tau Z_i)]^2 \zeta''(\psi(\beta^\tau Z_i)) t_i Z_i Z_i^\tau \qquad (4.60)$$

and

$$R_n(\beta) = \sum_{i=1}^{n} [x_i - \mu(\psi(\beta^\tau Z_i))] \psi''(\beta^\tau Z_i) t_i Z_i Z_i^\tau. \qquad (4.61)$$

Then

$$\text{Var}\left(\frac{\partial \log \ell(\theta)}{\partial \beta}\right) = M_n(\beta)/\phi \qquad (4.62)$$

and

$$\frac{\partial^2 \log \ell(\theta)}{\partial \beta \partial \beta^\tau} = [R_n(\beta) - M_n(\beta)]/\phi. \qquad (4.63)$$

Consider first the simple case of canonical g. Then $\psi'' \equiv 0$ and $R_n \equiv 0$. If $M_n(\beta)$ is positive definite for all β, then $-\log \ell(\theta)$ is strictly convex in β for any fixed ϕ and, therefore, $\hat{\beta}$ is the unique MLE of β. For the case of noncanonical g, $R_n(\beta) \neq 0$ and $\hat{\beta}$ is not necessarily an MLE. If $R_n(\beta)$ is dominated by $M_n(\beta)$ (i.e., $[M_n(\beta)]^{-1/2} R_n(\beta) [M_n(\beta)]^{-1/2} \to 0$ in some sense), then $-\log \ell(\theta)$ is convex and $\hat{\beta}$ is an MLE for large n; see more details in the proof of Theorem 4.18 in §4.5.2.

Example 4.36. Consider the GLM (4.55) with $\zeta(\eta) = \eta^2/2$, $\eta \in \mathcal{R}$. If g in (4.57) is the canonical link, then the model is the same as (3.24) with independent ε_i's distributed as $N(0, \phi_i)$. If (4.58) holds with $t_i \equiv 1$, then (4.59) is exactly the same as equation (3.27). If Z is of full rank, then $M_n(\beta) = Z^\tau Z$ is positive definite. Thus, we have shown that the LSE $\hat{\beta}$ given by (3.28) is actually the unique MLE of β.

Suppose now that g is noncanonical but (4.58) still holds with $t_i \equiv 1$. Then the model reduces to the one with independent X_i's and

$$X_i = N\left(g^{-1}(\beta^\tau Z_i), \phi\right), \qquad i = 1, ..., n. \qquad (4.64)$$

This type of model is called a *nonlinear regression model* (with normal errors) and an MLE of β under this model is also called a nonlinear LSE, since maximizing the log-likelihood is equivalent to minimizing the sum of squares $\sum_{i=1}^{n}[X_i - g^{-1}(\beta^\tau Z_i)]^2$. Under certain conditions the matrix $R_n(\beta)$ is dominated by $M_n(\beta)$ and an MLE of β exists. More details can be found in §4.5.2. ∎

Example 4.37 (The Poisson model). Consider the GLM (4.55) with $\zeta(\eta) = e^\eta$, $\eta \in \mathcal{R}$. If $\phi_i \equiv 1$, then X_i has the Poisson distribution with mean e^{η_i}. Assume that (4.58) holds. Under the canonical link $g(t) = \log t$,

$$M_n(\beta) = \sum_{i=1}^{n} e^{\beta^\tau Z_i} t_i Z_i Z_i^\tau,$$

which is positive definite if $\inf_i e^{\beta^\tau Z_i} > 0$ and the matrix $(\sqrt{t_1}Z_1, ..., \sqrt{t_n}Z_n)$ is of full rank.

There is one noncanonical link that deserves attention. Suppose that we choose a link function so that $[\psi'(t)]^2 \zeta''(\psi(t)) \equiv 1$. Then $M_n(\beta) \equiv \sum_{i=1}^{n} t_i Z_i Z_i^\tau$ does not depend on β. In §4.5.2 it is shown that the asymptotic variance of the MLE $\hat{\beta}$ is $\phi[M_n(\beta)]^{-1}$. The fact that $M_n(\beta)$ does not depend on β makes the estimation of the asymptotic variance (and, thus, statistical inference) easy. Under the Poisson model, $\zeta''(t) = e^t$ and, therefore, we need to solve the differential equation $[\psi'(t)]^2 e^{\psi(t)} = 1$. A solution is $\psi(t) = 2\log(t/2)$, which gives the link function $g(\mu) = 2\sqrt{\mu}$. ∎

In a GLM, an MLE $\hat{\beta}$ usually does not have an analytic form. A numerical method such as the Newton-Raphson or the Fisher-scoring method has to be applied. Using the Newton-Raphson method, we have the following iteration procedure:

$$\hat{\beta}^{(t+1)} = \hat{\beta}^{(t)} - [R_n(\hat{\beta}^{(t)}) - M_n(\hat{\beta}^{(t)})]^{-1} s_n(\hat{\beta}^{(t)}), \qquad t = 0, 1, ...,$$

where $s_n(\beta) = \phi\partial \log \ell(\theta)/\partial\beta$. Note that $E[R_n(\beta)] = 0$ if β is the true parameter value and x_i is replaced by X_i. This means that the Fisher-scoring method uses the following iteration procedure:

$$\hat{\beta}^{(t+1)} = \hat{\beta}^{(t)} + [M_n(\hat{\beta}^{(t)})]^{-1} s_n(\hat{\beta}^{(t)}), \qquad t = 0, 1, ...$$

If the canonical link is used, then the two methods are identical.

4.4.3 Quasi-likelihoods and conditional likelihoods

We now introduce two variations of the method of using likelihoods.

Consider a GLM (4.55)-(4.57). Assumption (4.58) is often unrealistic in applications. If there is no restriction on ϕ_i's, however, there are too many parameters and an MLE of β may not exist. (Note that assumption (4.58) reduces n nuisance parameters to one.) One way to solve this problem is to assume that $\phi_i = \hbar(Z_i, \xi)$ for some known function $\hbar$ and unknown parameter vector ξ (which may include β as a subvector). Let $\theta = (\beta, \xi)$. Then we can try to solve the likelihood equation $\partial \log \ell(\theta)/\partial \theta = 0$ to obtain an MLE of β and/or ξ. We omit the details, which can be found, for example, in Smyth (1989).

Suppose that we do not impose any assumptions on ϕ_i's but still estimate β by solving

$$\tilde{s}_n(\beta) = \sum_{i=1}^{n} \{[x_i - \mu(\psi(\beta^\tau Z_i))]\psi'(\beta^\tau Z_i)t_i Z_i\} = 0. \qquad (4.65)$$

Note that (4.65) is not a likelihood equation unless (4.58) holds. In the special case of Example 4.36 where $X_i = N(\beta^\tau Z_i, \phi_i)$, $i = 1, ..., n$, a solution to (4.65) is simply an LSE of β whose properties are discussed at the end of §3.3.3. Estimating β by solving equation (4.65) is motivated by the following facts. First, if (4.58) does hold, then our estimate is an MLE. Second, if (4.58) is slightly violated, the performance of our estimate is still nearly the same as that of an MLE under assumption (4.58) (see the discussion of robustness at the end of §3.3.3). Finally, estimators obtained by solving (4.65) usually have good asymptotic properties. As a special case of a general result in §5.4, a solution to (4.65) is asymptotically normal under some regularity conditions.

In general, an equation such as (4.65) is called a *quasi-likelihood equation* if and only if it is a likelihood equation when certain assumptions hold. The "likelihood" corresponding to a quasi-likelihood equation is called *quasi-likelihood* and a maximum of the quasi-likelihood is then called a *maximum quasi-likelihood estimate* (MQLE). Thus, a solution to (4.65) is an MQLE.

Note that (4.65) is a likelihood equation if and only if both (4.55) and (4.58) hold. The LSE (§3.3) without normality assumption on X_i's is a simple example of an MQLE without (4.55). Without assumption (4.55), the model under consideration is usually nonparametric and, therefore, the MQLE's are studied in §5.4.

While the quasi-likelihoods are used to relax some assumptions in our models, the *conditional likelihoods* discussed next are used mainly in cases where MLE's are difficult to compute. We consider two cases. In the first case, $\theta = (\theta_1, \theta_2)$, θ_1 is the main parameter vector of interest, and θ_2 is a nuisance parameter vector. Suppose that there is a statistic $T_2(X)$ that is sufficient for θ_2 for each fixed θ_1. By the sufficiency, the conditional distribution of X given T_2 does not depend on θ_2. The likelihood function corresponding to the conditional p.d.f. of X given T_2 is called the conditional likelihood function. A conditional MLE of θ_1 can then be obtained by maximizing the conditional likelihood function. This method can be applied to the case where the dimension of θ is considerably larger than the dimension of θ_1 so that computing the unconditional MLE of θ is much more difficult than computing the conditional MLE of θ_1. Note that the conditional MLE's are usually different from the unconditional MLE's.

As a more specific example, suppose that X has a p.d.f. in an exponential family:

$$f_\theta(x) = \exp\{\theta_1^\tau T_1(x) + \theta_2^\tau T_2(x) - \zeta(\theta)\}h(x).$$

Then T_2 is sufficient for θ_2 for any given θ_1. Problems of this type are from comparisons of two binomial distributions or two Poisson distributions (Exercises 119-120).

The second case is when our sample $X = (X_1, ..., X_n)$ follows a first-order autoregressive time series model:

$$X_t - \mu = \rho(X_{t-1} - \mu) + \varepsilon_t, \qquad t = 2, ..., n,$$

where $\mu \in \mathcal{R}$ and $\rho \in (-1, 1)$ are unknown and ε_i's are i.i.d. from $N(0, \sigma^2)$ with an unknown $\sigma^2 > 0$. This model is often a satisfactory representation of the error time series in economic models, and is one of the simplest and most heavily used models in time series analysis (Fuller, 1996). Let $\theta = (\mu, \rho, \sigma^2)$. The log-likelihood function is

$$\log \ell(\theta) = -\frac{n}{2}\log(2\pi) - \frac{n}{2}\log\sigma^2 + \frac{1}{2}\log(1 - \rho^2)$$

$$- \frac{1}{2\sigma^2}\left\{(x_1 - \mu)^2(1 - \rho^2) + \sum_{t=2}^{n}[x_t - \mu - \rho(x_{t-1} - \mu)]^2\right\}.$$

The computation of the MLE is greatly simplified if we consider the conditional likelihood given $X_1 = x_1$:

$$\log \ell(\theta | x_1) = -\frac{n-1}{2}\log(2\pi) - \frac{n-1}{2}\log\sigma^2 - \frac{1}{2\sigma^2}\sum_{t=2}^{n}[x_t - \mu - \rho(x_{t-1} - \mu)]^2.$$

Let $(\bar{x}_{-1}, \bar{x}_0) = (n-1)^{-1}\sum_{t=2}^{n}(x_{t-1}, x_t)$. If

$$\hat{\rho} = \sum_{t=2}^{n}(x_t - \bar{x}_0)(x_{t-1} - \bar{x}_{-1}) \bigg/ \sum_{t=2}^{n}(x_{t-1} - \bar{x}_{-1})^2$$

is between -1 and 1, then it is the conditional MLE of ρ and the conditional MLE's of μ and σ^2 are, respectively,

$$\hat{\mu} = (\bar{x}_0 - \hat{\rho}\bar{x}_{-1})/(1 - \hat{\rho})$$

and

$$\hat{\sigma}^2 = \frac{1}{n-1} \sum_{t=2}^{n} [x_t - \bar{x}_0 - \hat{\rho}(x_{t-1} - \bar{x}_{-1})]^2.$$

Obviously, the result can be extended to the case where X follows a pth-order autoregressive time series model:

$$X_t - \mu = \rho_1(X_{t-1} - \mu) + \cdots + \rho_p(X_{t-p} - \mu) + \varepsilon_t, \qquad t = p+1, ..., n, \quad (4.66)$$

where ρ_j's are unknown parameters satisfying the constraint that the roots (which may be complex) of the polynomial $x^p - \rho_1 x^{p-1} - \cdots - \rho_p = 0$ are less than one in absolute value (exercise).

Some other likelihood based methods are introduced in §5.1.4. Although they can also be applied to parametric models, the methods in §5.1.4 are more useful in nonparametric models.

4.5 Asymptotically Efficient Estimation

In this section, we consider asymptotic optimality of point estimators in parametric models. We use the asymptotic mean squared error (amse, see §2.5.2) or its multivariate generalization to assess the performance of an estimator. Reasons for considering asymptotics have been discussed in §2.5.

We focus on estimators that are asymptotically normal, since this covers the majority of cases. Some cases of asymptotically nonnormal estimators are studied in Exercises 111-114 in §4.6.

4.5.1 Asymptotic optimality

Let $\{\hat{\theta}_n\}$ be a sequence of estimators of θ based on a sequence of samples $\{X = (X_1, ..., X_n) : n = 1, 2, ...\}$ whose distributions are in a parametric family indexed by θ. Suppose that as $n \to \infty$,

$$[V_n(\theta)]^{-1/2}(\hat{\theta}_n - \theta) \to_d N_k(0, I_k), \qquad (4.67)$$

where, for each n, $V_n(\theta)$ is a $k \times k$ positive definite matrix depending on θ. If θ is one-dimensional ($k = 1$), then $V_n(\theta)$ is the asymptotic variance as well as the amse of $\hat{\theta}_n$ (§2.5.2). When $k > 1$, $V_n(\theta)$ is called the *asymptotic covariance matrix* of $\hat{\theta}_n$ and can be used as a measure of asymptotic performance of estimators. If $\hat{\theta}_{jn}$ satisfies (4.67) with asymptotic covariance matrix $V_{jn}(\theta)$, $j = 1, 2$, and $V_{1n}(\theta) \leq V_{2n}(\theta)$ (in the sense that $V_{2n}(\theta) - V_{1n}(\theta)$ is nonnegative definite) for all $\theta \in \Theta$, then $\hat{\theta}_{1n}$ is said to be asymptotically more efficient than $\hat{\theta}_{2n}$. Of course, some sequences of estimators are not comparable under this criterion. Also, since the asymptotic covariance matrices are unique only in the limiting sense, we have to make our comparison based on their limits. When X_i's are i.i.d., $V_n(\theta)$ is usually of the form $n^{-\delta}V(\theta)$ for some $\delta > 0$ ($= 1$ in the majority of cases) and a positive definite matrix $V(\theta)$ that does not depend on n.

Note that (4.67) implies that $\hat{\theta}_n$ is an asymptotically unbiased estimator of θ. If $V_n(\theta) = \text{Var}(\hat{\theta}_n)$, then, under some regularity conditions, it follows from Theorem 3.3 that

$$V_n(\theta) \geq [I_n(\theta)]^{-1}, \tag{4.68}$$

where, for every n, $I_n(\theta)$ is the Fisher information matrix (see (3.5)) for X of size n. (Note that (4.68) holds if and only if $l^\tau V_n(\theta)l \geq l^\tau[I_n(\theta)]^{-1}l$ for every $l \in \mathcal{R}^k$.) Unfortunately, when $V_n(\theta)$ is an asymptotic covariance matrix, (4.68) may not hold (even in the limiting sense), even if the regularity conditions in Theorem 3.3 are satisfied.

Example 4.38 (Hodges). Let $X_1, ..., X_n$ be i.i.d. from $N(\theta, 1)$, $\theta \in \mathcal{R}$. Then $I_n(\theta) = n$. Define

$$\hat{\theta}_n = \begin{cases} \bar{X} & |\bar{X}| \geq n^{-1/4} \\ t\bar{X} & |\bar{X}| < n^{-1/4}, \end{cases}$$

where t is a fixed constant. By Proposition 3.2, all conditions in Theorem 3.3 are satisfied. It can be shown (exercise) that (4.67) holds with $V_n(\theta) = V(\theta)/n$, where $V(\theta) = 1$ if $\theta \neq 0$ and $V(\theta) = t^2$ if $\theta = 0$. If $t^2 < 1$, (4.68) does not hold when $\theta = 0$. ∎

However, the following result, due to Le Cam (1953), shows that (4.68) holds for i.i.d. X_i's except for θ in a set of Lebesgue measure 0.

Theorem 4.16. Let $X_1, ..., X_n$ be i.i.d. from a p.d.f. f_θ w.r.t. a σ-finite measure ν on $(\mathcal{R}, \mathcal{B})$, where $\theta \in \Theta$ and Θ is an open set in $\mathcal{R}^k$. Suppose that for every x in the range of X_1, $f_\theta(x)$ is twice continuously differentiable in θ and satisfies

$$\frac{\partial}{\partial \theta} \int \psi_\theta(x)d\nu = \int \frac{\partial}{\partial \theta}\psi_\theta(x)d\nu$$

for $\psi_\theta(x) = f_\theta(x)$ and $= \partial f_\theta(x)/\partial\theta$; the Fisher information matrix

$$I_1(\theta) = E\left\{\frac{\partial}{\partial \theta}\log f_\theta(X_1)\left[\frac{\partial}{\partial \theta}\log f_\theta(X_1)\right]^\tau\right\}$$

is positive definite; and for any given $\theta \in \Theta$, there exists a positive number c_θ and a positive function h_θ such that $E[h_\theta(X_1)] < \infty$ and

$$\sup_{\gamma:\|\gamma-\theta\|<c_\theta} \left\|\frac{\partial^2 \log f_\gamma(x)}{\partial\gamma\partial\gamma^\tau}\right\| \leq h_\theta(x) \tag{4.69}$$

for all x in the range of X_1, where $\|A\| = \sqrt{\mathrm{tr}(A^\tau A)}$ for any matrix A. If $\hat{\theta}_n$ is an estimator of θ (based on $X_1, ..., X_n$) and satisfies (4.67) with $V_n(\theta) = V(\theta)/n$, then there is a $\Theta_0 \subset \Theta$ with Lebesgue measure 0 such that (4.68) holds if $\theta \notin \Theta_0$.

Proof. We adopt the proof given by Bahadur (1964) and prove the case of univariate θ. The proof for multivariate θ is similar and can be found in Bahadur (1964). Let $x = (x_1, ..., x_n)$, $\theta_n = \theta + n^{-1/2} \in \Theta$, and

$$K_n(x, \theta) = [\log \ell(\theta_n) - \log \ell(\theta) + I_1(\theta)/2]/[I_1(\theta)]^{1/2}.$$

Under the assumed conditions, it can be shown (exercise) that

$$K_n(X, \theta) \to_d N(0, 1). \tag{4.70}$$

Let P_{θ_n} (or P_θ) be the distribution of X under the assumption that X_1 has the p.d.f. f_{θ_n} (or f_θ). Define $g_n(\theta) = |P_\theta(\hat{\theta}_n \leq \theta) - \frac{1}{2}|$. Let Φ denote the standard normal c.d.f. or its probability measure. By the dominated convergence theorem (Theorem 1.1(iii)), as $n \to \infty$,

$$\int g_n(\theta_n) d\Phi(\theta) = \int g_n(\theta) e^{n^{-1/2}\theta - (2n)^{-1}} d\Phi(\theta) \to 0,$$

since $g_n(\theta) \to 0$ under (4.67). By Theorem 1.8(ii) and (vi), there exists a sequence $\{n_k\}$ such that $g_{n_k}(\theta_{n_k}) \to_{a.s.} 0$ w.r.t. Φ. Since Φ is equivalent to the Lebesgue measure, we conclude that there is a $\Theta_0 \subset \Theta$ with Lebesgue measure 0 such that

$$\lim_{k \to \infty} g_{n_k}(\theta_{n_k}) = 0, \qquad \theta \notin \Theta_0. \tag{4.71}$$

Assume that $\theta \notin \Theta_0$. Then, for any $t > [I_1(\theta)]^{1/2}$,

$$
\begin{aligned}
P_{\theta_n}(K_n(X, \theta) \leq t) &= \int_{K_n(x,\theta) \leq t} \ell(\theta_n) d\nu \times \cdots \times d\nu \\
&= \int_{K_n(x,\theta) \leq t} \frac{\ell(\theta_n)}{\ell(\theta)} dP_\theta(x) \\
&= e^{-I_1(\theta)/2} \int_{K_n(x,\theta) \leq t} e^{[I_1(\theta)]^{1/2}K_n(x,\theta)} dP_\theta(x) \\
&= e^{-I_1(\theta)/2} \int_{-\infty}^{t} e^{[I_1(\theta)]^{1/2}z} dH_n(z) \\
&= e^{-I_1(\theta)/2} \int_{-\infty}^{t} e^{[I_1(\theta)]^{1/2}z} d\Phi(z) + o(1) \\
&= \Phi\left(t - [I_1(\theta)]^{1/2}\right) + o(1),
\end{aligned}
$$

where H_n denotes the distribution of $K_n(X, \theta)$ and the next to last equality follows from (4.70) and the dominated convergence theorem. This result and result (4.71) imply that there is a sequence $\{n_j\}$ such that for $j = 1, 2, \ldots,$

$$P_{\theta_{n_j}}(\hat{\theta}_{n_j} \leq \theta_{n_j}) < P_{\theta_{n_j}}(K_{n_j}(X, \theta) \leq t). \tag{4.72}$$

By the Neyman-Pearson lemma (Theorem 6.1 in §6.1.1), we conclude that (4.72) implies that for $j = 1, 2, \ldots,$

$$P_\theta(\hat{\theta}_{n_j} \leq \theta_{n_j}) < P_\theta(K_{n_j}(X, \theta) \leq t). \tag{4.73}$$

(The reader should come back to this after reading §6.1.1.) From (4.70) and (4.67) with $V_n(\theta) = V(\theta)/n$, (4.73) implies

$$\Phi\big([V(\theta)]^{-1/2}\big) \leq \Phi(t).$$

Hence $[V(\theta)]^{-1/2} \leq t$. Since $I_n(\theta) = nI_1(\theta)$ (Proposition 3.1(i)) and t is arbitrary but $> [I_1(\theta)]^{1/2}$, we conclude that (4.68) holds. ∎

Points at which (4.68) does not hold are called points of superefficiency. Motivated by the fact that the set of superefficiency points is of Lebesgue measure 0 under some regularity conditions, we have the following definition.

Definition 4.4. Assume that the Fisher information matrix $I_n(\theta)$ is well defined and positive definite for every n. A sequence of estimators $\{\hat{\theta}_n\}$ satisfying (4.67) is said to be *asymptotically efficient* or *asymptotically optimal* if and only if $V_n(\theta) = [I_n(\theta)]^{-1}$. ∎

Suppose that we are interested in estimating $\vartheta = g(\theta)$, where g is a differentiable function from Θ to $\mathcal{R}^p$, $1 \leq p \leq k$. If $\hat{\theta}_n$ satisfies (4.67), then, by Theorem 1.12(i), $\hat{\vartheta}_n = g(\hat{\theta}_n)$ is asymptotically distributed as $N_p(\vartheta, [\nabla g(\theta)]^\tau V_n(\theta) \nabla g(\theta))$. Thus, inequality (4.68) becomes

$$[\nabla g(\theta)]^\tau V_n(\theta) \nabla g(\theta) \geq [\tilde{I}_n(\vartheta)]^{-1},$$

where $\tilde{I}_n(\vartheta)$ is the Fisher information matrix about ϑ contained in X. If $p = k$ and g is one-to-one, then

$$[\tilde{I}_n(\vartheta)]^{-1} = [\nabla g(\theta)]^\tau [I_n(\theta)]^{-1} \nabla g(\theta)$$

and, therefore, $\hat{\vartheta}_n$ is asymptotically efficient if and only if $\hat{\theta}_n$ is asymptotically efficient. For this reason, in the case of $p < k$, $\hat{\vartheta}_n$ is considered to be asymptotically efficient if and only if $\hat{\theta}_n$ is asymptotically efficient, and we can focus on the estimation of θ only.

4.5.2 Asymptotic efficiency of MLE's and RLE's

We now show that under some regularity conditions, a root of the likelihood equation (RLE), which is a candidate for an MLE, is asymptotically efficient.

Theorem 4.17. Assume the conditions of Theorem 4.16.
(i) There is a sequence of estimators $\{\hat{\theta}_n\}$ such that

$$P\big(s_n(\hat{\theta}_n) = 0\big) \to 1 \qquad \text{and} \qquad \hat{\theta}_n \to_p \theta, \qquad (4.74)$$

where $s_n(\gamma) = \partial \log \ell(\gamma)/\partial\gamma$.
(ii) Any consistent sequence $\tilde{\theta}_n$ of RLE's is asymptotically efficient.
Proof. (i) Let $B_n(c) = \{\gamma : \|[I_n(\theta)]^{1/2}(\gamma - \theta)\| \le c\}$ for $c > 0$. Since Θ is open, for each $c > 0$, $B_n(c) \subset \Theta$ for sufficiently large n. Since $B_n(c)$ shrinks to $\{\theta\}$ as $n \to \infty$, the existence of $\hat{\theta}_n$ satisfying (4.74) is implied by the fact that for any $\epsilon > 0$, there exists $c > 0$ and $n_0 > 1$ such that

$$P\big(\log \ell(\gamma) - \log \ell(\theta) < 0 \quad \text{for all } \gamma \in \partial B_n(c)\big) \ge 1 - \epsilon, \qquad n \ge n_0, \quad (4.75)$$

where $\partial B_n(c)$ is the boundary of $B_n(c)$. (For a proof of the measurability of $\hat{\theta}_n$, see Serfling (1980, pp. 147-148).) For $\gamma \in \partial B_n(c)$, the Taylor expansion gives

$$\log \ell(\gamma) - \log \ell(\theta) = c\lambda^\tau [I_n(\theta)]^{-1/2} s_n(\theta) \qquad (4.76)$$
$$+ (c^2/2)\lambda^\tau [I_n(\theta)]^{-1/2} \nabla s_n(\gamma^*)[I_n(\theta)]^{-1/2}\lambda,$$

where $\lambda = [I_n(\theta)]^{1/2}(\gamma - \theta)/c$ satisfying $\|\lambda\| = 1$, $\nabla s_n(\gamma) = \partial s_n(\gamma)/\partial\gamma$, and γ^* lies between γ and θ. Note that

$$E\frac{\|\nabla s_n(\gamma^*) - \nabla s_n(\theta)\|}{n} \le E \max_{\gamma \in B_n(c)} \frac{\|\nabla s_n(\gamma) - \nabla s_n(\theta)\|}{n}$$

$$\le E \max_{\gamma \in B_n(c)} \left\| \frac{\partial^2 \log f_\gamma(X_1)}{\partial\gamma\partial\gamma^\tau} - \frac{\partial^2 \log f_\theta(X_1)}{\partial\theta\partial\theta^\tau} \right\|$$

$$\to 0, \qquad (4.77)$$

which follows from (a) $\partial^2 \log f_\gamma(x)/\partial\gamma\partial\gamma^\tau$ is continuous in a neighborhood of θ for any fixed x; (b) $B_n(c)$ shrinks to $\{\theta\}$; and (c) for sufficiently large n,

$$\max_{\gamma \in B_n(c)} \left\| \frac{\partial^2 \log f_\gamma(X_1)}{\partial\gamma\partial\gamma^\tau} - \frac{\partial^2 \log f_\theta(X_1)}{\partial\theta\partial\theta^\tau} \right\| \le 2h_\theta(X_1)$$

under condition (4.69). By the SLLN (Theorem 1.13) and Proposition 3.1, $n^{-1}\nabla s_n(\theta) \to_{a.s.} -I_1(\theta)$ (i.e., $\|n^{-1}\nabla s_n(\theta) + I_1(\theta)\| \to_{a.s.} 0$). These results, together with (4.76), imply that

$$\log \ell(\gamma) - \log \ell(\theta) = c\lambda^\tau [I_n(\theta)]^{-1/2} s_n(\theta) - [1 + o_p(1)]c^2/2. \qquad (4.78)$$

Note that $\max_\lambda \{\lambda^\tau [I_n(\theta)]^{-1/2} s_n(\theta)\} = \|[I_n(\theta)]^{-1/2} s_n(\theta)\|$. Hence, (4.75) follows from (4.78) and

$$
\begin{aligned}
P\big(\|[I_n(\theta)]^{-1/2} s_n(\theta)\| < c/4\big) &\geq 1 - (4/c)^2 E\|[I_n(\theta)]^{-1/2} s_n(\theta)\|^2 \\
&= 1 - k(4/c)^2 \\
&\geq 1 - \epsilon
\end{aligned}
$$

by choosing c sufficiently large. This completes the proof of (i).

(ii) Let $A_\epsilon = \{\gamma : \|\gamma - \theta\| \leq \epsilon\}$ for $\epsilon > 0$. Since Θ is open, $A_\epsilon \subset \Theta$ for sufficiently small ϵ. Let $\{\tilde{\theta}_n\}$ be a sequence of consistent RLE's, i.e., $P(s_n(\tilde{\theta}_n) = 0$ and $\tilde{\theta}_n \in A_\epsilon) \to 1$ for any $\epsilon > 0$. Hence, we can focus on the set on which $s_n(\tilde{\theta}_n) = 0$ and $\tilde{\theta}_n \in A_\epsilon$. Using the mean-value theorem for vector-valued functions, we obtain that

$$
-s_n(\theta) = \left[\int_0^1 \nabla s_n(\theta + t(\tilde{\theta}_n - \theta)) dt \right] (\tilde{\theta}_n - \theta).
$$

Note that

$$
\frac{1}{n} \left\| \int_0^1 \nabla s_n(\theta + t(\tilde{\theta}_n - \theta)) dt - \nabla s_n(\theta) \right\| \leq \max_{\gamma \in A_\epsilon} \frac{\|\nabla s_n(\gamma) - \nabla s_n(\theta)\|}{n}.
$$

Using the argument in proving (4.77) and the fact that $P(\tilde{\theta}_n \in A_\epsilon) \to 1$ for arbitrary $\epsilon > 0$, we obtain that

$$
\frac{1}{n} \left\| \int_0^1 \nabla s_n(\theta + t(\tilde{\theta}_n - \theta)) dt - \nabla s_n(\theta) \right\| \to_p 0.
$$

Since $n^{-1} \nabla s_n(\theta) \to_{a.s.} -I_1(\theta)$ and $I_n(\theta) = nI_1(\theta)$,

$$
-s_n(\theta) = -I_n(\theta)(\tilde{\theta}_n - \theta) + o_p\big(\|I_n(\theta)(\tilde{\theta}_n - \theta)\|\big).
$$

This and Slutsky's theorem (Theorem 1.11) imply that $\sqrt{n}(\tilde{\theta}_n - \theta)$ has the same asymptotic distribution as

$$
\sqrt{n}[I_n(\theta)]^{-1} s_n(\theta) = n^{-1/2}[I_1(\theta)]^{-1} s_n(\theta) \to_d N_k\big(0, [I_1(\theta)]^{-1}\big)
$$

by the CLT (Corollary 1.2), since $\text{Var}(s_n(\theta)) = I_n(\theta)$. ∎

Theorem 4.17(i) shows the asymptotic existence of a sequence of consistent RLE's, and Theorem 4.17(ii) shows the asymptotic efficiency of *any* sequence of consistent RLE's. However, for a given sequence of RLE's, its consistency has to be checked unless the RLE's are unique for sufficiently large n, in which case the consistency of the RLE's is guaranteed by Theorem 4.17(i).

RLE's are not necessarily MLE's. We still have to use the techniques discussed in §4.4 to check whether an RLE is an MLE. However, according to Theorem 4.17, when a sequence of RLE's is consistent, then it is asymptotically efficient and, therefore, we may not need to search for MLE's, if asymptotic efficiency is the only criterion to select estimators. The method of estimating θ by solving $s_n(\gamma) = 0$ over $\gamma \in \Theta$ is called *scoring* and the function $s_n(\gamma)$ is called the *score* function.

Example 4.39. Suppose that X_i has a distribution in a natural exponential family, i.e., the p.d.f. of X_i is

$$f_\eta(x_i) = \exp\{\eta^\tau T(x_i) - \zeta(\eta)\}h(x_i). \tag{4.79}$$

Since $\partial^2 \log f_\eta(x_i)/\partial\eta\partial\eta^\tau = -\partial^2\zeta(\eta)/\partial\eta\partial\eta^\tau$, condition (4.69) is satisfied. From Proposition 3.2, other conditions in Theorem 4.16 are also satisfied. For i.i.d. X_i's,

$$s_n(\eta) = \sum_{i=1}^{n}\left[T(X_i) - \frac{\partial\zeta(\eta)}{\partial\eta}\right].$$

If $\hat{\theta}_n = n^{-1}\sum_{i=1}^{n} T(X_i) \in \Theta$, the range of $\theta = g(\eta) = \partial\zeta(\eta)/\partial\eta$, then $\hat{\theta}_n$ is a unique RLE of θ, which is also a unique MLE of θ since $\partial^2\zeta(\eta)/\partial\eta\partial\eta^\tau = \text{Var}(T(X_i))$ is positive definite. Also, $\eta = g^{-1}(\theta)$ exists and a unique RLE (MLE) of η is $\hat{\eta}_n = g^{-1}(\hat{\theta}_n)$.

However, $\hat{\theta}_n$ may not be in Θ and the previous argument fails (e.g., Example 4.29). What Theorem 4.17 tells us in this case is that as $n \to \infty$, $P(\hat{\theta}_n \in \Theta) \to 1$ and, therefore, $\hat{\theta}_n$ (or $\hat{\eta}_n$) is the unique asymptotically efficient RLE (MLE) of θ (or η) in the limiting sense.

In an example like this we can directly show that $P(\hat{\theta}_n \in \Theta) \to 1$, using the fact that $\hat{\theta}_n \to_{a.s.} E[T(X_1)] = g(\eta)$ (the SLLN). ∎

The next theorem provides a similar result for the MLE or RLE in the GLM (§4.4.2).

Theorem 4.18. Consider the GLM (4.55)-(4.58) with t_i's in a fixed interval (t_0, t_∞), $0 < t_0 \le t_\infty < \infty$. Assume that the range of the unknown parameter β in (4.57) is an open subset of $\mathcal{R}^p$; at the true parameter value β, $0 < \inf_i \varphi(\beta^\tau Z_i) \le \sup_i \varphi(\beta^\tau Z_i) < \infty$, where $\varphi(t) = [\psi'(t)]^2\zeta''(\psi(t))$; as $n \to \infty$, $\max_{i \le n} Z_i^\tau(Z^\tau Z)^{-1}Z_i \to 0$ and $\lambda_-[Z^\tau Z] \to \infty$, where Z is the $n \times p$ matrix whose ith row is the vector Z_i and $\lambda_-[A]$ is the smallest eigenvalue of the matrix A.
(i) There is a unique sequence of estimators $\{\hat{\beta}_n\}$ such that

$$P\big(s_n(\hat{\beta}_n) = 0\big) \to 1 \quad \text{and} \quad \hat{\beta}_n \to_p \beta, \tag{4.80}$$

where $s_n(\gamma)$ is the score function defined to be the left-hand side of (4.59) with $\gamma = \beta$.

(ii) Let $I_n(\beta) = \text{Var}(s_n(\beta))$. Then

$$[I_n(\beta)]^{1/2}(\hat{\beta}_n - \beta) \to_d N_p(0, I_p). \tag{4.81}$$

(iii) If ϕ in (4.58) is known or the p.d.f. in (4.55) indexed by $\theta = (\beta, \phi)$ satisfies the conditions for f_θ in Theorem 4.16, then $\hat{\beta}_n$ is asymptotically efficient.

Proof. (i) The proof of the existence of $\hat{\beta}_n$ satisfying (4.80) is the same as that of Theorem 4.17(i) with $\theta = \beta$, except that we need to show

$$\max_{\gamma \in B_n(c)} \left\| [I_n(\beta)]^{-1/2} \nabla s_n(\gamma)[I_n(\beta)]^{-1/2} + I_p \right\| \to_p 0,$$

where $B_n(c) = \{\gamma : \|[I_n(\beta)]^{1/2}(\gamma - \beta)\| \leq c\}$. From (4.62) and (4.63), $I_n(\beta) = M_n(\beta)/\phi$ and $\nabla s_n(\gamma) = [R_n(\gamma) - M_n(\gamma)]/\phi$, where $M_n(\gamma)$ and $R_n(\gamma)$ are defined by (4.60)-(4.61) with $\gamma = \beta$. Hence, it suffices to show that for any $c > 0$,

$$\max_{\gamma \in B_n(c)} \left\| [M_n(\beta)]^{-1/2}[M_n(\gamma) - M_n(\beta)][M_n(\beta)]^{-1/2} \right\| \to 0 \tag{4.82}$$

and

$$\max_{\gamma \in B_n(c)} \left\| [M_n(\beta)]^{-1/2} R_n(\gamma)[M_n(\beta)]^{-1/2} \right\| \to_p 0. \tag{4.83}$$

The left-hand side of (4.82) is bounded by

$$\sqrt{p} \max_{\gamma \in B_n(c), i \leq n} \left| 1 - \varphi(\gamma^\tau Z_i)/\varphi(\beta^\tau Z_i) \right|,$$

which converges to 0 since φ is continuous and, for $\gamma \in B_n(c)$,

$$\begin{aligned}
|\gamma^\tau Z_i - \beta^\tau Z_i|^2 &= |(\gamma - \beta)^\tau [I_n(\beta)]^{1/2}[I_n(\beta)]^{-1/2} Z_i|^2 \\
&\leq \|[I_n(\beta)]^{1/2}(\gamma - \beta)\|^2 \|[I_n(\beta)]^{-1/2} Z_i\|^2 \\
&\leq c^2 \max_{i \leq n} Z_i^\tau [I_n(\beta)]^{-1} Z_i \\
&\leq c^2 \phi \left[t_0 \inf_i \varphi(\beta^\tau Z_i) \right]^{-1} \max_{i \leq n} Z_i^\tau (Z^\tau Z)^{-1} Z_i \\
&\to 0
\end{aligned}$$

under the assumed conditions. This proves (4.82).

Let $e_i = X_i - \mu(\psi(\beta^\tau Z_i))$,

$$U_n(\gamma) = \sum_{i=1}^n [\mu(\psi(\beta^\tau Z_i)) - \mu(\psi(\gamma^\tau Z_i))] \psi''(\gamma^\tau Z_i) t_i Z_i Z_i^\tau,$$

$$V_n(\gamma) = \sum_{i=1}^{n} e_i[\psi''(\gamma^\tau Z_i) - \psi''(\beta^\tau Z_i)]t_i Z_i Z_i^\tau,$$

and

$$W_n(\beta) = \sum_{i=1}^{n} e_i \psi''(\beta^\tau Z_i)t_i Z_i Z_i^\tau.$$

Then $R_n(\gamma) = U_n(\gamma) + V_n(\gamma) + W_n(\beta)$. Using the same argument as that in proving (4.82), we can show that

$$\max_{\gamma \in B_n(c)} \left\| [M_n(\beta)]^{-1/2} U_n(\gamma)[M_n(\beta)]^{-1/2} \right\| \to 0.$$

Note that $\left\| [M_n(\beta)]^{-1/2} V_n(\gamma)[M_n(\beta)]^{-1/2} \right\|$ is bounded by the product of

$$[M_n(\beta)]^{-1/2} \sum_{i=1}^{n} |e_i|t_i Z_i Z_i^\tau [M_n(\beta)]^{-1/2} = O_p(1)$$

and

$$\max_{\gamma \in B_n(c), i \leq n} \left| \psi''(\gamma^\tau Z_i) - \psi''(\beta^\tau Z_i) \right|,$$

which can be shown to be $o(1)$ using the same argument as that in proving (4.82). Hence,

$$\max_{\gamma \in B_n(c)} \left\| [M_n(\beta)]^{-1/2} V_n(\gamma)[M_n(\beta)]^{-1/2} \right\| \to_p 0$$

and (4.83) follows from

$$\left\| [M_n(\beta)]^{-1/2} W_n(\beta)[M_n(\beta)]^{-1/2} \right\| \to_p 0.$$

To show this result, we apply Theorem 1.14(ii). Since $E(e_i) = 0$ and e_i's are independent, it suffices to show that

$$\sum_{i=1}^{n} E \left| e_i \psi''(\beta^\tau Z_i)t_i Z_i^\tau [M_n(\beta)]^{-1} Z_i \right|^{1+\delta} \to 0 \qquad (4.84)$$

for some $\delta \in (0, 1)$. Note that $\sup_i E|e_i|^{1+\delta} < \infty$. Hence, there is a constant $C > 0$ such that the left-hand side of (4.84) is bounded by

$$C \sum_{i=1}^{n} \left| Z_i^\tau (Z^\tau Z)^{-1} Z_i \right|^{1+\delta} \leq pC \max_{i \leq n} \left| Z_i^\tau (Z^\tau Z)^{-1} Z_i \right|^\delta \to 0.$$

Hence, (4.84) follows from Theorem 1.14(ii). This proves (4.80). The uniqueness of $\hat{\beta}_n$ follows from (4.83) and the fact that $M_n(\gamma)$ is positive definite in a neighborhood of β. This completes the proof of (i).

(ii) The proof of (ii) is very similar to that of Theorem 4.17(ii). Using the results in the proof of (i) and Taylor's expansion, we can establish (exercise) that

$$[I_n(\beta)]^{1/2}(\hat{\beta}_n - \beta) = [I_n(\beta)]^{-1/2} s_n(\beta) + o_p(1). \tag{4.85}$$

Using the CLT (e.g., Corollary 1.3) and Theorem 1.9(iii), we can show (exercise) that

$$[I_n(\beta)]^{-1/2} s_n(\beta) \to_d N_p(0, I_p). \tag{4.86}$$

Result (4.81) follows from (4.85)-(4.86) and Slutsky's theorem.

(iii) The result is obvious if ϕ is known. When ϕ is unknown, it follows from (4.59) that

$$\frac{\partial}{\partial \phi} \left[\frac{\partial \log \ell(\theta)}{\partial \beta} \right] = -\frac{s_n(\beta)}{\phi}.$$

Since $E[s_n(\beta)] = 0$, the Fisher information about $\theta = (\beta, \phi)$ is

$$I_n(\beta, \phi) = -E \left[\frac{\partial^2 \log \ell(\theta)}{\partial \theta \partial \theta^\tau} \right] = \begin{pmatrix} I_n(\beta) & 0 \\ 0 & \tilde{I}_n(\phi) \end{pmatrix},$$

where $\tilde{I}_n(\phi)$ is the Fisher information about ϕ. The result then follows from (4.81) and the discussion in the end of §4.5.1. ∎

4.5.3 Other asymptotically efficient estimators

To study other asymptotically efficient estimators, we start with MRIE's in location-scale families. Since MLE's and RLE's are invariant (see Exercise 109 in §4.6), MRIE's are often asymptotically efficient; see, for example, Stone (1974).

Assume the conditions in Theorem 4.16 and let $s_n(\gamma)$ be the score function. Let $\hat{\theta}_n^{(0)}$ be an estimator of θ that may not be asymptotically efficient. The estimator

$$\hat{\theta}_n^{(1)} = \hat{\theta}_n^{(0)} - [\nabla s_n(\hat{\theta}_n^{(0)})]^{-1} s_n(\hat{\theta}_n^{(0)}) \tag{4.87}$$

is the first iteration in computing an MLE (or RLE) using the Newton-Raphson iteration method with $\hat{\theta}_n^{(0)}$ as the initial value (see (4.53)) and, therefore, is called the *one-step* MLE. Without any further iteration, $\hat{\theta}_n^{(1)}$ can be used as a numerical approximation to an MLE or RLE; and $\hat{\theta}_n^{(1)}$ is asymptotically efficient under some conditions, as the following result shows.

Theorem 4.19. Assume that the conditions in Theorem 4.16 hold and that $\hat{\theta}_n^{(0)}$ is $\sqrt{n}$-consistent for θ (Definition 2.10).

(i) The one-step MLE $\hat{\theta}_n^{(1)}$ is asymptotically efficient.

(ii) The one-step MLE obtained by replacing $\nabla s_n(\gamma)$ in (4.87) with its expected value, $-I_n(\gamma)$ (the Fisher-scoring method), is asymptotically efficient.

Proof. Since $\hat{\theta}_n^{(0)}$ is $\sqrt{n}$-consistent, we can focus on the event $\hat{\theta}_n^{(0)} \in A_\epsilon = \{\gamma : \|\gamma - \theta\| \le \epsilon\}$ for a sufficiently small ϵ such that $A_\epsilon \subset \Theta$. From the mean-value theorem,

$$s_n(\hat{\theta}_n^{(0)}) = s_n(\theta) + \left[\int_0^1 \nabla s_n\left(\theta + t(\hat{\theta}_n^{(0)} - \theta)\right) dt \right] (\hat{\theta}_n^{(0)} - \theta).$$

Substituting this into (4.87) we obtain that

$$\hat{\theta}_n^{(1)} - \theta = -[\nabla s_n(\hat{\theta}_n^{(0)})]^{-1} s_n(\theta) + [I_k - G_n(\hat{\theta}_n^{(0)})](\hat{\theta}_n^{(0)} - \theta),$$

where

$$G_n(\hat{\theta}_n^{(0)}) = [\nabla s_n(\hat{\theta}_n^{(0)})]^{-1} \int_0^1 \nabla s_n\left(\theta + t(\hat{\theta}_n^{(0)} - \theta)\right) dt.$$

From (4.77), $\|[I_n(\theta)]^{1/2}[\nabla s_n(\hat{\theta}_n^{(0)})]^{-1}[I_n(\theta)]^{1/2} + I_k\| \to_p 0$. Using an argument similar to those in the proofs of (4.77) and (4.82), we can show that $\|G_n(\hat{\theta}_n^{(0)}) - I_k\| \to_p 0$. These results and the fact that $\sqrt{n}(\hat{\theta}_n^{(0)} - \theta) = O_p(1)$ imply

$$\sqrt{n}(\hat{\theta}_n^{(1)} - \theta) = \sqrt{n}[I_n(\theta)]^{-1} s_n(\theta) + o_p(1).$$

This proves (i). The proof for (ii) is similar. ∎

Example 4.40. Let $X_1, ..., X_n$ be i.i.d. from the Weibull distribution $W(\theta, 1)$, where $\theta > 0$ is unknown. Note that

$$s_n(\theta) = \frac{n}{\theta} + \sum_{i=1}^n \log X_i - \sum_{i=1}^n X_i^\theta \log X_i$$

and

$$\nabla s_n(\theta) = -\frac{n}{\theta^2} - \sum_{i=1}^n X_i^\theta (\log X_i)^2.$$

Hence, the one-step MLE of θ is

$$\hat{\theta}_n^{(1)} = \hat{\theta}_n^{(0)} \left[1 + \frac{n + \hat{\theta}_n^{(0)}(\sum_{i=1}^n \log X_i - \sum_{i=1}^n X_i^{\hat{\theta}_n^{(0)}} \log X_i)}{n + (\hat{\theta}_n^{(0)})^2 \sum_{i=1}^n X_i^{\hat{\theta}_n^{(0)}} (\log X_i)^2} \right].$$

Usually one can use a moment estimator (§3.5.2) as the initial estimator $\hat{\theta}_n^{(0)}$. In this example, a moment estimator of θ is the solution of $\bar{X} = \Gamma(\theta^{-1} + 1)$. ∎

Results similar to that in Theorem 4.19 can be obtained in non-i.i.d. cases, for example, the GLM discussed in §4.4.2 (exercise); see also §5.4.

As we discussed in §4.1.3, Bayes estimators are usually consistent. The next result, due to Bickel and Yahav (1969) and Ibragimov and Has'minskii (1981), states that Bayes estimators are asymptotically efficient when X_i's are i.i.d.

Theorem 4.20. Assume the conditions of Theorem 4.16. Let $\pi(\gamma)$ be a prior p.d.f. (which may be improper) w.r.t. the Lebesgue measure on Θ and $p_n(\gamma)$ be the posterior p.d.f., given $X_1, ..., X_n$, $n = 1, 2, ...$ Assume that there exists an n_0 such that $p_{n_0}(\gamma)$ is continuous and positive for all $\gamma \in \Theta$, $\int p_{n_0}(\gamma)d\gamma = 1$ and $\int \|\gamma\|p_{n_0}(\gamma)d\gamma < \infty$. Suppose further that, for any $\epsilon > 0$, there exists a $\delta > 0$ such that

$$\lim_{n\to\infty} P\left(\sup_{\|\gamma-\theta\|\geq\epsilon} \frac{\log \ell(\gamma) - \log \ell(\theta)}{n} > -\delta\right) = 0 \qquad (4.88)$$

and

$$\lim_{n\to\infty} P\left(\sup_{\|\gamma-\theta\|\leq\delta} \frac{\|\nabla s_n(\gamma) - \nabla s_n(\theta)\|}{n} \geq \epsilon\right) = 0, \qquad (4.89)$$

where $\ell(\gamma)$ is the likelihood function and $s_n(\gamma)$ is the score function.
(i) Let $p_n^*(\gamma)$ be the posterior p.d.f. of $\sqrt{n}(\gamma - T_n)$, where $T_n = \theta + [I_n(\theta)]^{-1}s_n(\theta)$ and θ is the true parameter value, and let $\psi(\gamma)$ be the p.d.f. of $N_k(0, [I_1(\theta)]^{-1})$. Then

$$\int (1 + \|\gamma\|)|p_n^*(\gamma) - \psi(\gamma)|d\gamma \to_p 0. \qquad (4.90)$$

(ii) The Bayes estimator of θ under the squared error loss is asymptotically efficient. ∎

The proof of Theorem 4.20 is lengthy and is omitted; see Lehmann (1983, §6.7) for a proof of the case of univariate θ.

A number of conclusions can be drawn from Theorem 4.20. First, result (4.90) shows that the posterior p.d.f. is approximately normal with mean $\theta + [I_n(\theta)]^{-1}s_n(\theta)$ and covariance matrix $[I_n(\theta)]^{-1}$. This result is useful in Bayesian computation; see Berger (1985, §4.9.3). Second, (4.90) shows that the posterior distribution and its first-order moments converge to the degenerate distribution at θ and its first-order moments, which implies the consistency and asymptotic unbiasedness of Bayes estimators such as the posterior means. Third, the Bayes estimator under the squared error loss is asymptotically efficient, which provides an additional support for the early suggestion that the Bayesian approach is a useful method for generating estimators. Finally, the results hold regardless of the prior being used, indicating that the effect of the prior declines as n increases.

In addition to the regularity conditions in Theorem 4.16, Theorem 4.20 requires two more nontrivial regularity conditions, (4.88) and (4.89). Let us verify these conditions for natural exponential families (Example 4.39), i.e., X_i's are i.i.d. with p.d.f. (4.79). Since $\nabla s_n(\eta) = -n\partial^2\zeta(\eta)/\partial\eta\partial\eta^\tau$, (4.89) follows from the continuity of the second-order derivatives of ζ. To show (4.88), consider first the case of univariate η. Without loss of generality, we assume that $\gamma > \eta$. Note that

$$\frac{\log \ell(\gamma) - \log \ell(\eta)}{n} = \left[\bar{T} - \zeta'(\eta) + \zeta'(\eta) - \frac{\zeta(\gamma) - \zeta(\eta)}{\gamma - \eta}\right](\gamma - \eta), \quad (4.91)$$

where $\bar{T}$ is the average of $T(X_i)$'s. Since $\zeta(\gamma)$ is strictly convex, $\gamma > \eta$ implies $\zeta'(\eta) < [\zeta(\gamma) - \zeta(\eta)]/(\gamma - \eta)$. Also, $\bar{T} \to_{a.s.} \zeta'(\eta)$. Hence, with probability tending to 1, the factor in front of $(\gamma - \eta)$ on the right-hand side of (4.91) is negative. Then (4.88) holds with

$$\delta = \frac{\epsilon}{2} \inf_{\gamma \geq \eta + \epsilon} \left[\frac{\zeta(\gamma) - \zeta(\eta)}{\gamma - \eta} - \zeta'(\eta)\right].$$

To show how to extend this to multivariate η, consider the case of bivariate η. Let η_j, γ_j, and ξ_j be the jth components of η, γ, and $\bar{T} - \nabla\zeta(\eta)$, respectively. Assume $\gamma_1 > \eta_1$ and $\gamma_2 > \eta_2$. Let ζ'_j be the derivative of ζ w.r.t. the jth component of η. Then the left-hand side of (4.91) is the sum of

$$(\gamma_1 - \eta_1)\xi_1 - [\zeta(\eta_1, \gamma_2) - \zeta(\eta_1, \eta_2) - (\gamma_2 - \eta_2)\zeta'_2(\eta_1, \eta_2)]$$

and

$$(\gamma_2 - \eta_2)\xi_2 - [\zeta(\gamma_1, \gamma_2) - \zeta(\eta_1, \gamma_2) - (\gamma_1 - \eta_1)\zeta'_1(\eta_1, \gamma_2)],$$

where the last quantity is bounded by

$$(\gamma_2 - \eta_2)\xi_2 - [\zeta(\gamma_1, \gamma_2) - \zeta(\eta_1, \gamma_2) - (\gamma_1 - \eta_1)\zeta'_1(\eta_1, \gamma_2)],$$

since $\zeta'_1(\eta_1, \eta_2) \leq \zeta'_1(\eta_1, \gamma_2)$. The rest of the proof is the same as the case of univariate η.

When Bayes estimators have explicit forms under a specific prior, it is usually easy to prove the asymptotic efficiency of the Bayes estimators directly. For instance, in Example 4.7, the Bayes estimator of θ is

$$\frac{n\bar{X} + \gamma^{-1}}{n + \alpha - 1} = \bar{X} + \frac{\gamma^{-1} - (\alpha - 1)\bar{X}}{n + \alpha - 1} = \bar{X} + O\left(\frac{1}{n}\right) \quad \text{a.s.},$$

where $\bar{X}$ is the MLE of θ. Hence the Bayes estimator is asymptotically efficient by Slutsky's theorem. A similar result can be obtained for the Bayes estimator $\delta_t(X)$ in Example 4.7. Theorem 4.20, however, is useful in cases where Bayes estimators do not have explicit forms and/or the prior is not specified clearly. One such example is the problem in Example 4.40 (Exercises 153 and 154).

4.6 Exercises

1. Show that the priors in the following cases are conjugate priors:
   (a) $X_1, ..., X_n$ are i.i.d. from $N_k(\theta, I_k)$, $\theta \in \mathcal{R}^k$, and $\Pi = N_k(\mu_0, \Sigma_0)$ (Normal family);
   (b) $X_1, ..., X_n$ are i.i.d. from the binomial distribution $Bi(\theta, k)$, $\theta \in (0, 1)$, and $\Pi = B(\alpha, \beta)$ (Beta family);
   (c) $X_1, ..., X_n$ are i.i.d. from the uniform distribution $U(0, \theta)$, $\theta > 0$, and $\Pi = Pa(a, b)$ (Pareto family);
   (d) $X_1, ..., X_n$ are i.i.d. from the exponential distribution $E(0, \theta)$, $\theta > 0$, $\Pi = $ the inverse gamma distribution $\Gamma^{-1}(\alpha, \gamma)$ (a random variable Y has the inverse gamma distribution $\Gamma^{-1}(\alpha, \gamma)$ if and only if Y^{-1} has the gamma distribution $\Gamma(\alpha, \gamma)$);
   (e) X_1 is from the binomial distribution $Bi(p, \theta)$ with a known p, $\theta = 1, 2, ...$, and $\Pi = P(\lambda)$ (Poisson family).

2. In Exercise 1, find the posterior mean and variance for each case.

3. Let $X_1, ..., X_n$ be i.i.d. from the $N(\theta, 1)$ distribution and let the prior be the double exponential distribution $DE(0, 1)$. Obtain the posterior mean.

4. Let $X_1, ..., X_n$ be i.i.d. from the uniform distribution $U(0, \theta)$, where $\theta > 0$ is unknown. Let the prior of θ be the log-normal distribution $LN(\mu_0, \sigma_0^2)$, where $\mu_0 \in \mathcal{R}$ and $\sigma_0 > 0$ are known constants.
   (a) Find the posterior p.d.f. of $\vartheta = \log\theta$.
   (b) Find the rth posterior moment of θ.
   (c) Find a value that maximizes the posterior p.d.f. of θ.

5. Show that if $T(X)$ is a sufficient statistic for $\theta \in \Theta$, then the Bayes action $\delta(x)$ in (4.3) is a function of $T(x)$.

6. Let $\bar{X}$ be the sample mean of n i.i.d. observations from $N(\theta, \sigma^2)$ with a known $\sigma > 0$ and an unknown $\theta \in \mathcal{R}$. Let $\pi(\theta)$ be a prior p.d.f. w.r.t. a σ-finite measure on $\mathcal{R}$.
   (a) Show that the posterior mean of θ, given $\bar{X} = x$, is of the form
   $$\delta(x) = x + \frac{\sigma^2}{n} \frac{d\log(p(x))}{dx},$$
   where $p(x)$ is the marginal p.d.f. of $\bar{X}$, unconditional on θ.
   (b) Express the posterior variance of θ (given $\bar{X} = x$) as a function of the first two derivatives of $\log(p(x))$ w.r.t. x.
   (c) Find explicit expressions for $p(x)$ and $\delta(x)$ in (a) when the prior is $N(\mu_0, \sigma_0^2)$ with probability $1 - \epsilon$ and a point mass at μ_1 with probability ϵ, where μ_0, μ_1, and σ_0^2 are known constants.

7. Let $X_1, ..., X_n$ be i.i.d. binary random variables with $P(X_1 = 1) = p \in (0, 1)$. Find the Bayes action w.r.t. the uniform prior on $[0, 1]$ in the problem of estimating p under the loss $L(p, a) = (p-a)^2/[p(1-p)]$.

8. Consider the estimation of θ in Exercise 41 of §2.6 under the squared error loss. Suppose that the prior of θ is the uniform distribution $U(0, 1)$, the prior of j is $P(j = 1) = P(j = 2) = \frac{1}{2}$, and the joint prior of (θ, j) is the product probability of the two marginal priors. Show that the Bayes action is

$$\delta(x) = \frac{H(x)B(t+1) + G(t+1)}{H(x)B(t) + G(t)},$$

where $x = (x_1, ..., x_n)$ is the vector of observations, $t = x_1 + \cdots + x_n$, $B(t) = \int_0^1 \theta^t(1-\theta)^{n-t}d\theta$, $G(t) = \int_0^1 \theta^t e^{-n\theta}d\theta$, and $H(x)$ is a function of x with range $\{0, 1\}$.

9. Consider the estimation problem in Example 4.1 with the loss function $L(\theta, a) = w(\theta)[g(\theta) - a]^2$, where $w(\theta) \geq 0$ and $\int_\Theta w(\theta)[g(\theta)]^2 d\Pi < \infty$. Show that the Bayes action is

$$\delta(x) = \frac{\int_\Theta w(\theta)g(\theta)f_\theta(x)d\Pi}{\int_\Theta w(\theta)f_\theta(x)d\Pi}.$$

10. Let X be a sample from P_θ, $\theta \in \Theta \subset \mathcal{R}$. Consider the estimation of θ under the loss $L(|\theta - a|)$, where L is an increasing function on $[0, \infty)$. Let $\pi(\theta|x)$ be the posterior p.d.f. of θ given $X = x$. Suppose that $\pi(\theta|x)$ is symmetric about $\theta_0 \in \Theta$ and that $\pi(\theta|x)$ is nondecreasing for $\theta \leq \theta_0$ and nonincreasing for $\theta \geq \theta_0$. Show that δ satisfying $\pi(\delta|x) = \sup_{\theta \in \Theta} \pi(\theta|x)$ is a Bayes action, assuming that all integrals involved are finite.

11. Let X be a sample of size 1 from the geometric distribution $G(p)$ with an unknown $p \in (0, 1]$. Consider the estimation of p with $\mathbb{A} = [0, 1]$ and the loss function $L(p, a) = (p - a)^2/p$.
    (a) Show that δ is a Bayes action w.r.t. Π if and only if $\delta(x) = 1 - \int(1 - p)^x d\Pi(p)/\int(1 - p)^{x-1}d\Pi(p)$, $x = 1, 2, ...$.
    (b) Let δ_0 be a rule such that $\delta_0(1) = 1/2$ and $\delta_0(x) = 0$ for all $x > 1$. Show that δ_0 is a limit of Bayes actions.
    (c) Let δ_0 be a rule such that $\delta_0(x) = 0$ for all $x > 1$ and $\delta_0(1)$ is arbitrary. Show that δ_0 is a generalized Bayes action.

12. Let X be a single observation from $N(\mu, \sigma^2)$ with a known σ^2 and an unknown $\mu > 0$. Consider the estimation of μ under the squared error loss and the noninformative prior $\Pi = $ the Lebesgue measure on $(0, \infty)$. Show that the generalized Bayes action when $X = x$ is $\delta(x) = x + \sigma\Phi'(x/\sigma)/[1 - \Phi(-x/\sigma)]$, where Φ is the c.d.f. of the standard normal distribution and Φ' is its derivative.

13. Let X be a sample from P_θ having the p.d.f. $h(x)\exp\{\theta^\tau x - \zeta(\theta)\}$ w.r.t. ν. Let Π be the Lebesgue measure on $\Theta = \mathcal{R}^p$. Show that the generalized Bayes action under the loss $L(\theta, a) = \|E(X) - a\|^2$ is $\delta(x) = x$ when $X = x$.

14. Let $X_1, ..., X_n$ be i.i.d. random variables with the Lebesgue p.d.f. $\sqrt{2/\pi}e^{-(x-\theta)^2/2}I_{(\theta,\infty)}(x)$, where $\theta \in \mathcal{R}$ is unknown. Find the generalized Bayes action for estimating θ under the squared error loss, when the (improper) prior of θ is the Lebesgue measure on $\mathcal{R}$.

15. Let $X_1, ..., X_n$ be i.i.d. from $N(\mu, \sigma^2)$ and $\pi(\mu, \sigma^2) = \sigma^{-2}I_{(0,\infty)}(\sigma^2)$ be an improper prior for (μ, σ^2) w.r.t. the Lebesgue measure on $\mathcal{R}^2$.
    (a) Show that the posterior p.d.f. of (μ, σ^2) given $x = (x_1, ..., x_n)$ is $\pi(\mu, \sigma^2|x) = \pi_1(\mu|\sigma^2, x)\pi_2(\sigma^2|x)$, where $\pi_1(\mu|\sigma^2, x)$ is the p.d.f. of $N(\bar{x}, \sigma^2/n)$ and $\pi_2(\sigma^2|x)$ is the p.d.f. of the inverse gamma distribution $\Gamma^{-1}((n-1)/2, [\sum_{i=1}^n (x_i - \bar{x})^2/2]^{-1})$ (see Exercise 1(d)).
    (b) Show that the marginal posterior p.d.f. of μ given x is $f\left(\frac{\mu-\bar{x}}{\tau}\right)$, where $\tau^2 = \sum_{i=1}^n (x_i - \bar{x})^2/[n(n-1)]$ and f is the p.d.f. of the t-distribution t_{n-1}.
    (c) Obtain the generalized Bayes action for estimating μ/σ under the squared error loss.

16. Consider Example 3.13. Under the squared error loss and the prior with the improper Lebesgue density $\pi(\mu_1, ..., \mu_m, \sigma^2) = \sigma^{-2}$, obtain the generalized Bayes action for estimating $\theta = \sigma^{-2}\sum_{i=1}^m n_i(\mu_i - \bar{\mu})^2$, where $\bar{\mu} = n^{-1}\sum_{i=1}^m n_i\mu_i$.

17. Let X be a single observation from the Lebesgue p.d.f. $e^{-x+\theta}I_{(\theta,\infty)}(x)$, where $\theta > 0$ is an unknown parameter. Consider the estimation of
$$\vartheta = \begin{cases} j & \theta \in (j-1, j], \ j = 1, 2, 3, \\ 4 & \theta > 3 \end{cases}$$
under the loss $L(i, j)$, $1 \le i, j \le 4$, given by the following matrix:
$$\begin{pmatrix} 0 & 1 & 1 & 2 \\ 1 & 0 & 2 & 2 \\ 1 & 2 & 0 & 2 \\ 3 & 3 & 3 & 0 \end{pmatrix}.$$
When $X = 4$, find the Bayes action w.r.t. the prior with the Lebesgue p.d.f. $e^{-\theta}I_{(0,\infty)}(\theta)$.

18. (Bayesian hypothesis testing). Let X be a sample from P_θ, where $\theta \in \Theta$. Let $\Theta_0 \subset \Theta$ and $\Theta_1 = \Theta_0^c$, the complement of Θ_0. Consider the problem of testing $H_0 : \theta \in \Theta_0$ versus $H_1 : \theta \in \Theta_1$ under the loss
$$L(\theta, a_i) = \begin{cases} 0 & \theta \in \Theta_i \\ C_i & \theta \notin \Theta_i, \end{cases}$$

where $C_i > 0$ are known constants and $\{a_0, a_1\}$ is the action space. Let $\Pi_{\theta|x}$ be the posterior distribution of θ w.r.t. a prior distribution Π, given $X = x$. Show that the Bayes action $\delta(x) = a_1$ if and only if $\Pi_{\theta|x}(\Theta_1) \geq C_1/(C_0 + C_1)$.

19. In (b)-(d) of Exercise 1, assume that the parameters in priors are unknown. Using the method of moments, find empirical Bayes actions under the squared error loss.

20. In Example 4.5, assume that both μ_0 and σ_0^2 in the prior for μ are unknown. Let the second-stage joint prior for (μ_0, σ_0^2) be the product of $N(a, v^2)$ and the Lebesgue measure on $(0, \infty)$, where a and v are known. Under the squared error loss, obtain a formula for the hierarchical Bayes action in terms of a one-dimensional integral.

21. Let $X_1, ..., X_n$ be i.i.d. random variables from the uniform distribution $U(0, \theta)$, where $\theta > 0$ is unknown. Let $\pi(\theta) = ba^b\theta^{-(b+1)}I_{(a,\infty)}(\theta)$ be a prior p.d.f. w.r.t. the Lebesgue measure, where $b > 1$ is known but $a > 0$ is an unknown hyperparameter. Consider the estimation of θ under the squared error loss.
    (a) Show that the empirical Bayes method using the method of moments produces the empirical Bayes action $\delta(\hat{a})$, where $\delta(a) = \frac{b+n}{b+n-1} \max\{a, X_{(n)}\}$, $\hat{a} = \frac{2(b-1)}{bn} \sum_{i=1}^n X_i$, and $X_{(n)}$ is the largest order statistic.
    (b) Let $h(a) = a^{-1}I_{(0,\infty)}(a)$ be an improper Lebesgue prior density for a. Obtain explicitly the hierarchical generalized Bayes action.

22. Let X be a sample and $\delta(X)$ with any fixed $X = x \in A$ be a Bayes action, where δ is a measurable function and $\int_\Theta P_\theta(A)d\Pi = 1$. Show that $\delta(X)$ is a Bayes rule as defined in §2.3.2.

23. Let $X_1, ..., X_n$ be i.i.d. random variables with the Lebesgue p.d.f. $f_\theta(x) = \sqrt{2\theta/\pi}e^{-\theta x^2/2}I_{[0,\infty)}(x)$, where $\theta > 0$ is unknown. Let the prior of θ be the gamma distribution $\Gamma(\alpha, \gamma)$ with known α and γ. Find the Bayes estimator of $f_\theta(0)$ and its Bayes risk under the loss function $L(\theta, a) = (a - \theta)^2/\theta$.

24. Let X be a single observation from $N(0, \theta^2)$ and consider a prior p.d.f. $\pi_\xi(\theta) = c(\alpha, \mu, \tau)|\theta|^{-\alpha}e^{-(\theta^{-1}-\mu)^2/(2\tau^2)}$ w.r.t. the Lebesgue measure, where $\xi = (\alpha, \mu, \tau)$ is a vector of hyperparameters and $c(\alpha, \mu, \tau)$ ensures that $\pi_\xi(\theta)$ is a p.d.f.
    (a) Identify the constraints on the hyperparameters for $\pi_\xi(\theta)$ to be a proper prior.
    (b) Show that the posterior p.d.f. is $\pi_{\xi_*}(\theta)$ for given $X = x$ and identify ξ_*.

(c) Express the Bayes estimator of $|\theta|$ and its Bayes risk in terms of the function c and ξ_* and state any additional constraints needed on the hyperparameters.

25. Let $X_1, X_2, \ldots$ be i.i.d. from the exponential distribution $E(0,1)$. Suppose that we observe $T = X_1 + \cdots + X_\theta$, where θ is an unknown integer ≥ 1. Consider the estimation of θ under the loss function $L(\theta, a) = (\theta - a)^2/\theta$ and the geometric distribution $G(p)$ as the prior for θ, where $p \in (0,1)$ is known.
   (a) Show that the posterior expected loss is
   $$E[L(\theta, a)|T = t] = 1 + \xi - 2a + (1 - e^{-\xi})a^2/\xi,$$
   where $\xi = (1 - p)t$.
   (b) Find the Bayes estimator of θ and show that its posterior expected loss is $1 - \xi \sum_{m=1}^{\infty} e^{-m\xi}$.
   (c) Find the marginal distribution of $(1 - p)T$, unconditional on θ.
   (d) Obtain an explicit expression for the Bayes risk of the Bayes estimator in part (b).

26. Prove (ii) and (iii) of Theorem 4.2.

27. Let $X_1, \ldots, X_n$ be i.i.d. binary random variables with $P(X_1 = 1) = p \in (0,1)$.
   (a) Show that $\bar{X}$ is an admissible estimator of p under the loss function $(a - p)^2/[p(1 - p)]$.
   (b) Show that $\bar{X}$ is an admissible estimator of p under the squared error loss.

28. Let X be a sample (of size 1) from $N(\mu, 1)$. Consider the estimation of μ under the loss function $L(\mu, a) = |\mu - a|$. Show that X is an admissible estimator.

29. In Exercise 1, consider the posterior mean to be the Bayes estimator of the corresponding parameter in each case.
   (a) Show that the bias of the Bayes estimator converges to 0 if $n \to \infty$.
   (b) Show that the Bayes estimator is consistent.
   (c) Discuss whether the Bayes estimator is admissible.

30. Let $X_1, \ldots, X_n$ be i.i.d. binary random variables with $P(X_1 = 1) = p \in (0,1)$.
   (a) Obtain the Bayes estimator of $p(1 - p)$ w.r.t. $\Pi =$ the beta distribution $B(\alpha, \beta)$ with known α and β, under the squared error loss.
   (b) Compare the Bayes estimator in part (a) with the UMVUE of $p(1 - p)$.
   (c) Discuss the bias, consistency, and admissibility of the Bayes estimator in (a).

(d) Let $\pi(p) = [p(1 - p)]^{-1}I_{(0,1)}(p)$ be an improper Lebesgue prior density for p. Show that the posterior of p given X_i's is a p.d.f. provided that the sample mean $\bar{X} \in (0, 1)$.
(e) Under the squared error loss, find the generalized Bayes estimator of $p(1 - p)$ w.r.t. the improper prior in (d).

31. Let X be an observation from the negative binomial distribution $NB(p, r)$ with a known r and an unknown $p \in (0, 1)$.
    (a) Under the squared error loss, find the Bayes estimators of p and p^{-1} w.r.t. $\Pi =$ the beta distribution $B(\alpha, \beta)$ with known α and β.
    (b) Show that the Bayes estimators in (a) are consistent.

32. In Example 4.7, show that
    (a) $\bar{X}$ is the generalized Bayes estimator of θ w.r.t. the improper prior $\frac{d\Pi}{d\omega} = I_{(0,\infty)}(\omega)$ and is a limit of Bayes estimators (as $\alpha \to 1$ and $\gamma \to \infty$);
    (b) the Bayes estimator of θ with an $\alpha > 2$ is admissible;
    (c) the Bayes estimator of θ with an $\alpha < 2$ is inadmissible. (Does it contradict Theorem 4.2(i)?)

33. Consider Example 4.8. Show that the sample mean $\bar{X}$ is a generalized Bayes estimator of μ under the squared error loss and $\bar{X}$ is admissible using (a) Theorem 4.3 and (b) the result in Example 4.6.

34. Let X be an observation from the gamma distribution $\Gamma(\alpha, \theta)$ with a known α and an unknown $\theta > 0$. Show that $X/(\alpha+1)$ is an admissible estimator of θ under the squared error loss, using Theorem 4.3.

35. Let $X_1, ..., X_n$ be i.i.d. from the uniform distribution $U(\theta, \theta + 1)$, $\theta \in \mathcal{R}$. Consider the estimation of θ under the squared error loss.
    (a) Let $\pi(\theta)$ be a continuous and positive Lebesgue p.d.f. on $\mathcal{R}$. Derive the Bayes estimator w.r.t. the prior π and show that it is a consistent estimator of θ.
    (b) Show that $(X_{(1)} + X_{(n)} - 1)/2$ is an admissible estimator of θ and obtain its risk, where $X_{(j)}$ is the jth order statistic.

36. Consider the normal linear model $X = N_n(Z^\tau\beta, \sigma^2 I_n)$, where Z is an $n \times p$ known matrix of full rank, $p < n$, $\beta \in \mathcal{R}^p$, and $\sigma^2 > 0$.
    (a) Assume that σ^2 is known. Derive the posterior distribution of β when the prior distribution for β is $N_p(\beta_0, \sigma^2 V)$, where $\beta_0 \in \mathcal{R}^p$ is known and V is a known positive definite matrix, and find the Bayes estimator of $l^\tau\beta$ under the squared error loss, where $l \in \mathcal{R}^p$ is known.
    (b) Show that the Bayes estimator in (a) is admissible and consistent as $n \to \infty$, assuming that the minimum eigenvalue of $Z^\tau Z \to \infty$.
    (c) Repeat (a) and (b) when σ^2 is unknown and has the inverse gamma distribution $\Gamma^{-1}(\alpha, \gamma)$ (see Exercise 1(d)), where α and γ are known.

(d) In part (c), obtain Bayes estimators of σ^2 and $l^\tau \beta / \sigma$ and show that they are consistent under the condition in (b).

37. In Example 4.9, suppose that ε_{ij} has the Lebesgue p.d.f.

$$
\kappa(\delta) \sigma_i^{-1} \exp\left\{-c(\delta) |x/\sigma_i|^{2/(1+\delta)}\right\},
$$

where

$$
c(\delta) = \left[\frac{\Gamma\left(\frac{3(1+\delta)}{2}\right)}{\Gamma\left(\frac{1+\delta}{2}\right)}\right]^{\frac{1}{1+\delta}}, \qquad
\kappa(\delta) = \frac{\left[\Gamma\left(\frac{3(1+\delta)}{2}\right)\right]^{1/2}}{(1+\delta)\left[\Gamma\left(\frac{1+\delta}{2}\right)\right]^{3/2}},
$$

$-1 < \delta \leq 1$ and $\sigma_i > 0$.

(a) Assume that δ is known. Let $\omega_i = c(\delta)\sigma_i^{-2/(1+\delta)}$. Under the squared error loss and the same prior in Example 4.9, show that the Bayes estimator of σ_i^2 is

$$
q_i(\delta) \int \left[\frac{1}{\gamma} + \sum_{j=1}^{n_i} |x_{ij} - \beta^\tau Z_i|^{2/(1+\delta)}\right]^{1+\delta} f(\beta | x, \delta) d\beta,
$$

where $q_i(\delta) = [c(\delta)]^{1+\delta} \Gamma\left(\frac{1+\delta}{2} n_i + \alpha - \delta\right) / \Gamma\left(\frac{1+\delta}{2} n_i + \alpha + 1\right)$ and

$$
f(\beta | x, \delta) \propto \pi(\beta) \prod_{i=1}^{k} \left[\frac{1}{\gamma} + \sum_{j=1}^{n_i} |x_{ij} - \beta^\tau Z_i|^{2/(1+\delta)}\right]^{-\left(\alpha + 1 + \frac{1+\delta}{2} n_i\right)}.
$$

(b) Assume that δ has a prior p.d.f. $f(\delta)$ and that given δ, ω_i still has the same prior in (a). Derive a formula (similar to that in (a)) for the Bayes estimator of σ_i^2.

38. Suppose that we have observations

$$
X_{ij} = \mu_i + \varepsilon_{ij}, \quad i = 1, ..., k, \; j = 1, ..., m,
$$

where ε_{ij}'s are i.i.d. from $N(0, \sigma_\varepsilon^2)$, μ_i's are i.i.d. from $N(\mu, \sigma_\mu^2)$, and ε_{ij}'s and μ_i's are independent. Suppose that the distribution for σ_ε^2 is the inverse gamma distribution $\Gamma^{-1}(\alpha_1, \beta_1)$ (see Exercise 1(d)); the distribution for σ_μ^2 is the inverse gamma distribution $\Gamma^{-1}(\alpha_2, \beta_2)$; the distribution for μ is $N(\mu_0, \sigma_0^2)$; and σ_ε, σ_μ, and μ are independent. Describe a Gibbs sampler and obtain explicit forms of
(a) the distribution of μ, given X_{ij}'s, μ_i's, σ_ε^2, and σ_μ^2;
(b) the distribution of μ_i, given X_{ij}'s, μ, σ_ε^2, and σ_μ^2;
(c) the distribution of σ_ε^2, given X_{ij}'s, μ_i's, μ, and σ_μ^2;
(d) the distribution of σ_μ^2, given X_{ij}'s, μ_i's, μ, and σ_ε^2.

39. Prove (4.16).

40. Consider a Lebesgue p.d.f. $p(y) \propto (2+y)^{125}(1-y)^{38}y^{34}I_{(0,1)}(y)$. Generate Markov chains of length 10,000 and compute approximations to $\int yp(y)dy$, using the Metropolis kernel with $q(y, z)$ being the p.d.f. of $N(y, r^2)$, given y, where (a) $r = 0.001$; (b) $r = 0.05$; (c) $r = 0.12$.

41. Prove Proposition 4.4 for the cases of variance and risk.

42. In the proof of Theorem 4.5, show that if L is (strictly) convex and not monotone, then $E[L(T_0(x) - a)|D = d]$ is (strictly) convex and not monotone in a.

43. Prove part (iii) of Theorem 4.5.

44. Under the conditions of Theorem 4.5 and the loss function $L(\mu, a) = |\mu - a|$, show that $u_*(d)$ in Theorem 4.5 is any median (Exercise 92 in §2.6) of $T_0(X)$ under the conditional distribution of X given $D = d$ when $\mu = 0$.

45. Show that if there is a location invariant estimator T_0 of μ with finite mean, then $E_0[T(X)|D = d]$ is finite a.s. $\mathcal{P}$ for any location invariant estimator T.

46. Show (4.21) under the squared error loss.

47. In Exercise 14, find the MRIE of θ under the squared error loss.

48. In Example 4.12,
    (a) show that $X_{(1)} - \theta \log 2/n$ is an MRIE of μ under the absolute error loss $L(\mu - a) = |\mu - a|$;
    (b) show that $X_{(1)} - t$ is an MRIE under the loss function $L(\mu - a) = I_{(t,\infty)}(|\mu - a|)$.

49. In Example 4.13, show that T_* is also an MRIE of μ if the loss function is convex and even. (Hint: the distribution of $T_*(X)$ given D depends only on $X_{(n)} - X_{(1)}$ and is symmetric about 0 when $\mu = 0$.)

50. Let $X_1, ..., X_n$ be i.i.d. from the double exponential distribution $DE(\mu, 1)$ with an unknown $\mu \in \mathcal{R}$. Under the squared error loss, find the MRIE of μ. (Hint: for $x_1 < \cdots < x_n$ and $x_k < t < x_{k+1}$, $\sum_{i=1}^n |x_i - t| = \sum_{i=k+1}^n x_i - \sum_{i=1}^k x_i + (2k - n)t$.)

51. In Example 4.11, find the MRIE of μ under the loss function
$$L(\mu - a) = \begin{cases} -\alpha(\mu - a) & \mu < a \\ \beta(\mu - a) & \mu \geq a, \end{cases}$$
where α and β are positive constants. (Hint: show that if Y is a random variable with c.d.f. F, then $E[L(Y - u)]$ is minimized for any u satisfying $F(u) = \beta/(\alpha + \beta)$.)

52. Let T be a location invariant estimator of μ in a one-parameter location problem. Show that T is an MRIE under the squared error loss if and only if T is unbiased and $E[T(X)U(X)] = 0$ for any $U(X)$ satisfying $U(x_1 + c, ..., x_n + c) = U(x)$ for any c and $E[U(X)] = 0$ for any μ.

53. Assume the conditions in Theorem 4.6. Let T be a sufficient statistic for μ. Show that Pitman's estimator is a function of T.

54. Prove Proposition 4.5, Theorems 4.7 and 4.8, and Corollary 4.1.

55. Under the conditions of Theorem 4.8 and the loss function (4.24) with $p = 1$, show that $u_*(z)$ is any constant $c > 0$ satisfying

$$\int_0^c x dP_{x|z} = \int_c^\infty x dP_{x|z},$$

where $P_{x|z}$ is the conditional distribution of X given $Z = z$ when $\sigma = 1$.

56. In Example 4.15, show that the MRIE is $2^{(n+1)^{-1}} X_{(n)}$ when the loss is given by (4.24) with $p = 1$.

57. Let $X_1, ..., X_n$ be i.i.d. from the exponential distribution $E(0, \theta)$ with an unknown $\theta > 0$.
    (a) Find the MRIE of θ under the loss (4.24) with $p = 2$.
    (b) Find the MRIE of θ under the loss (4.24) with $p = 1$.
    (c) Find the MRIE of θ^2 under the loss (4.24) with $p = 2$.

58. Let $X_1, ..., X_n$ be i.i.d. with a Lebesgue p.d.f. $(2/\sigma)[1-(x/\sigma)]I_{(0,\sigma)}(x)$, where $\sigma > 0$ is an unknown scale parameter. Find Pitman's estimator of σ^h for $n = 2, 3$, and 4.

59. Let $X_1, ..., X_n$ be i.i.d. from the Pareto distribution $Pa(\alpha, \sigma)$, where $\sigma > 0$ is an unknown parameter and $\alpha > 2$ is known. Find the MRIE of σ under the loss function (4.24) with $p = 2$.

60. Assume that the sample X has a joint Lebesgue p.d.f. given by (4.25). Show that a loss function for the estimation of μ is invariant under the location-scale transformations $g_{c,r}(X) = (rX_1 + c, ..., rX_n + c)$, $r > 0$, $c \in \mathcal{R}$, if and only if it is of the form $L\left(\frac{a-\mu}{\sigma}\right)$.

61. Prove Proposition 4.6, Theorem 4.9, and Corollary 4.2.

62. Let $X_1, ..., X_n$ be i.i.d. from the exponential distribution $E(\mu, \sigma)$, where $\mu \in \mathcal{R}$ and $\sigma > 0$ are unknown.
    (a) Find the MRIE of σ under the loss (4.24) with $p = 1$ or 2.
    (b) Under the loss function $(a - \mu)^2/\sigma^2$, find the MRIE of μ.
    (c) Compute the bias of the MRIE of μ in (b).

63. Suppose that X and Y are two samples with p.d.f. given by (4.30).
    (a) Suppose that $\mu_x = \mu_y = 0$ and consider the estimation of $\eta = (\sigma_y/\sigma_x)^h$ with a fixed $h \neq 0$ under the loss $L(a/\eta)$. Show that the problem is invariant under the transformations $g(X,Y) = (rX, r'Y)$, $r > 0, r' > 0$. Generalize Proposition 4.5, Theorem 4.8, and Corollary 4.1 to the present problem.
    (b) Generalize the result in (a) to the case of unknown μ_x and μ_y under the transformations in (4.31).

64. Under the conditions of part (a) of the previous exercise and the loss function $(a - \eta)^2/\eta^2$, determine the MRIE of η in the following cases:
    (a) $m = n = 1$, X and Y are independent, X has the gamma distribution $\Gamma(\alpha_x, \gamma)$ with a known α_x and an unknown $\gamma = \sigma_x > 0$, and Y has the gamma distribution $\Gamma(\alpha_y, \gamma)$ with a known α_y and an unknown $\gamma = \sigma_y > 0$;
    (b) X is $N_m(0, \sigma_x^2 I_m)$, Y is $N_n(0, \sigma_y^2 I_n)$, and X and Y are independent;
    (c) X and Y are independent, the components of X are i.i.d. from the uniform distribution $U(0, \sigma_x)$, and the components of Y are i.i.d. from the uniform distribution $U(0, \sigma_y)$.

65. Let $X_1, ..., X_m$ and $Y_1, ..., Y_n$ be two independent samples, where X_i's are i.i.d. having the p.d.f. $\sigma_x^{-1} f\left(\frac{x - \mu_x}{\sigma_x}\right)$ with $\mu_x \in \mathcal{R}$ and $\sigma_x > 0$, and Y_i's are i.i.d. having the p.d.f. $\sigma_y^{-1} f\left(\frac{x - \mu_y}{\sigma_y}\right)$ with $\mu_y \in \mathcal{R}$ and $\sigma_y > 0$. Under the loss function $(a - \eta)^2/\eta^2$ and the transformations in (4.31), obtain the MRIE of $\eta = \sigma_y/\sigma_x$ when
    (a) f is the p.d.f. of $N(0, 1)$;
    (b) f is the p.d.f. of the exponential distribution $E(0, 1)$;
    (c) f is the p.d.f. of the uniform distribution $U\left(-\frac{1}{2}, \frac{1}{2}\right)$;
    (d) In (a)-(c), find the MRIE of $\Delta = \mu_y - \mu_x$ under the assumption that $\sigma_x = \sigma_y = \sigma$ and under the loss function $(a - \Delta)^2/\sigma^2$.

66. Consider the general linear model (3.25) under the assumption that ε_i's are i.i.d. with the p.d.f. $\sigma^{-1} f(x/\sigma)$, where f is a known Lebesgue p.d.f.
    (a) Show that the family of populations is invariant under the transformations in (4.32).
    (b) Show that the estimation of $l^\tau \beta$ with $l \in \mathcal{R}(Z)$ is invariant under the loss function $L\left(\frac{a - l^\tau \beta}{\sigma}\right)$.
    (c) Show that the LSE $l^\tau \hat{\beta}$ is an invariant estimator of $l^\tau \beta$, $l \in \mathcal{R}(Z)$.
    (d) Prove Theorem 4.10.

67. In Example 4.18, let T be a randomized estimator of p with probability $n/(n + 1)$ being $\bar{X}$ and probability $1/(n + 1)$ being $\frac{1}{2}$. Show that

T has a constant risk that is smaller than the maximum risk of $\bar{X}$.

68. Let X be a single sample from the geometric distribution $G(p)$ with an unknown $p \in (0, 1)$. Show that $I_{\{1\}}(X)$ is a minimax estimator of p under the loss function $(a - p)^2 / [p(1 - p)]$.

69. In Example 4.19, show that $\bar{X}$ is a minimax estimator of μ under the loss function $(a - \mu)^2 / \sigma^2$ when $\Theta = \mathcal{R} \times (0, \infty)$.

70. Let T be a minimax (or admissible) estimator of ϑ under the squared error loss. Show that $c_1 T + c_0$ is a minimax (or admissible) estimator of $c_1 \vartheta + c_0$ under the squared error loss, where c_1 and c_0 are constants.

71. Let X be a sample from P_θ with an unknown $\theta = (\theta_1, \theta_2)$, where $\theta_j \in \Theta_j$, $j = 1, 2$, and let Π_2 be a probability measure on Θ_2. Suppose that an estimator T_0 minimizes $\sup_{\theta_1 \in \Theta_1} \int R_T(\theta) d\Pi_2(\theta_2)$ over all estimators T and that $\sup_{\theta_1 \in \Theta_1} \int R_{T_0}(\theta) d\Pi_2(\theta_2) = \sup_{\theta_1 \in \Theta_1, \theta_2 \in \Theta_2} R_{T_0}(\theta)$. Show that T_0 is a minimax estimator.

72. Let $X_1, ..., X_m$ be i.i.d. from $N(\mu_x, \sigma_x^2)$ and $Y_1, ..., Y_n$ be i.i.d. from $N(\mu_y, \sigma_y^2)$. Assume that X_i's and Y_j's are independent. Consider the estimation of $\Delta = \mu_y - \mu_x$ under the squared error loss.
(a) Show that $\bar{Y} - \bar{X}$ is a minimax estimator of Δ when σ_x and σ_y are known, where $\bar{X}$ and $\bar{Y}$ are the sample means based on X_i's and Y_i's, respectively.
(b) Show that $\bar{Y} - \bar{X}$ is a minimax estimator of Δ when $\sigma_x \in (0, c_x]$ and $\sigma_y \in (0, c_y]$, where c_x and c_y are constants.

73. Consider the general linear model (3.25) with assumption A1 and the estimation of $l^\tau \beta$ under the squared error loss, where $l \in \mathcal{R}(Z)$. Show that the LSE $l^\tau \hat{\beta}$ is minimax if $\sigma^2 \in (0, c]$ with a constant c.

74. Let X be a random variable having the hypergeometric distribution $HG(r, \theta, N - \theta)$ (Table 1.1, page 18) with known N and r but an unknown θ. Consider the estimation of θ/N under the squared error loss.
(a) Show that the risk function of $T(X) = \alpha X/r + \beta$ is constant, where $\alpha = \{1 + \sqrt{(N - r)/[r(N - 1)]}\}^{-1}$ and $\beta = (1 - \alpha)/2$.
(b) Show that T in (a) is the minimax estimator of θ/N and the Bayes estimator w.r.t. the prior

$$\Pi(\{\theta\}) = \frac{\Gamma(2c)}{[\Gamma(c)]^2} \int_0^1 \binom{N}{\theta} t^{\theta + c - 1} (1 - t)^{N - \theta + c - 1} dt, \quad \theta = 1, 2, ...,$$

where $c = \beta/(\alpha/r - 1/N)$.

75. Let X be a single observation from $N(\mu, 1)$ and let μ have the improper Lebesgue prior density $\pi(\mu) = e^{\mu}$. Under the squared error loss, show that the generalized Bayes estimator of μ is $X + 1$, which is neither minimax nor admissible.

76. Let X be a random variable having the Poisson distribution $P(\theta)$ with an unknown $\theta > 0$. Consider the estimation of θ under the squared error loss.
    (a) Show that $\sup_\theta R_T(\theta) = \infty$ for any estimator $T = T(X)$.
    (b) Let $\Im = \{aX + b : a \in \mathcal{R}, b \in \mathcal{R}\}$. Show that 0 is a $\Im$-admissible estimator of θ.

77. Let $X_1, ..., X_n$ be i.i.d. from the exponential distribution $E(a, \theta)$ with a known θ and an unknown $a \in \mathcal{R}$. Under the squared error loss, show that $X_{(1)} - \theta/n$ is the unique minimax estimator of a.

78. Let $X_1, ..., X_n$ be i.i.d. from the uniform distribution $U(\mu - \frac{1}{2}, \mu + \frac{1}{2})$ with an unknown $\mu \in \mathcal{R}$. Under the squared error loss, show that $(X_{(1)} + X_{(n)})/2$ is the unique minimax estimator of μ.

79. Let $X_1, ..., X_n$ be i.i.d. from the double exponential distribution $DE(\mu, 1)$ with an unknown $\mu \in \mathcal{R}$. Under the squared error loss, find a minimax estimator of μ.

80. Consider Example 4.7. Show that $(n\bar{X} + b)/(n + 1)$ is an admissible estimator of θ under the squared error loss for any $b \geq 0$ and that $n\bar{X}/(n + 1)$ is a minimax estimator of θ under the loss function $L(\theta, a) = (a - \theta)^2/\theta^2$.

81. Let $X_1, ..., X_n$ be i.i.d. binary random variables with $P(X_1 = 1) = p \in (0, 1)$. Consider the estimation of p under the squared error loss. Using Theorem 4.14, show that $\bar{X}$ and $(\bar{X} + \gamma\lambda)/(1 + \lambda)$ with $\lambda > 0$ and $0 \leq \gamma \leq 1$ are admissible.

82. Let X be a single observation. Using Theorem 4.14, find values of α and β such that $\alpha X + \beta$ are admissible for estimating EX under the squared error loss when
    (a) X has the Poisson distribution $P(\theta)$ with an unknown $\theta > 0$;
    (b) X has the negative binomial distribution $NB(p, r)$ with a known r and an unknown $p \in (0, 1)$.

83. Let X be a single observation having the Lebesgue p.d.f. $\frac{1}{2}c(\theta)e^{\theta x - |x|}$, $|\theta| < 1$.
    (a) Show that $c(\theta) = 1 - \theta^2$.
    (b) Show that if $0 \leq \alpha \leq \frac{1}{2}$, then $\alpha X + \beta$ is admissible for estimating $E(X)$ under the squared error loss.

84. Let X be a single observation from the discrete p.d.f. $f_\theta(x)$
$= [x!(1 - e^{-\theta})]^{-1}\theta^x e^{-\theta} I_{\{1,2,...\}}(x)$, where $\theta > 0$ is unknown. Consider the estimation of $\vartheta = \theta/(1 - e^{-\theta})$ under the squared error loss.
(a) Show that the estimator X is admissible.
(b) Show that X is not minimax unless $\sup_\theta R_T(\theta) = \infty$ for any estimator $T = T(X)$.
(c) Find a loss function under which X is minimax and admissible.

85. In Example 4.23, find the UMVUE of $\theta = (p_1, ..., p_k)$ under the loss function (4.37).

86. Let X be a sample from P_θ, $\theta \in \Theta \subset \mathcal{R}^p$. Consider the estimation of θ under the loss $(\theta - a)^T Q(\theta - a)$, where $a \in \mathbb{A} = \Theta$ and Q is a known positive definite matrix. Show that the Bayes action is the posterior mean $E(\theta|X = x)$, assuming that all integrals involved are finite.

87. In Example 4.24, show that X is the MRIE of θ under the loss function (4.37), if
(a) $f(x - \theta) = \prod_{j=1}^p f_j(x_j - \theta_j)$, where each f_j is a known Lebesgue p.d.f. with mean 0;
(b) $f(x - \theta) = f(\|x - \theta\|)$ with $\int xf(\|x\|)dx = 0$.

88. Prove that X in Example 4.25 is a minimax estimator of θ under the loss function (4.37).

89. Let $X_1, ..., X_k$ be independent random variables, where X_i has the binomial distribution $Bi(p_i, n_i)$ with an unknown $p_i \in (0, 1)$ and a known n_i. For estimating $\theta = (p_1, ..., p_k)$ under the loss (4.37), find a minimax estimator of θ and determine whether it is admissible.

90. Show that the risk function in (4.42) tends to p as $\|\theta\| \to \infty$.

91. Suppose that X is $N_p(\theta, I_p)$. Consider the estimation of θ under the loss $(a - \theta)^T Q(a - \theta)$ with a positive definite $p \times p$ matrix Q. Show that the risk of the estimator
$$\delta^Q_{c,r} = X - \frac{r(p - 2)}{\|Q^{-1/2}(X - c)\|^2} Q^{-1}(X - c)$$
is equal to
$$\text{tr}(Q) - (2r - r^2)(p - 2)^2 E(\|Q^{-1/2}(X - c)\|^{-2}).$$

92. Show that under the loss (4.37), the risk of $\tilde{\delta}_{c,r}$ in (4.45) is given by (4.46).

93. Suppose that X is $N_p(\theta, V)$ with $p \geq 4$. Consider the estimation of θ under the loss function (4.37).
    (a) When $V = I_p$, show that the risk of the estimator in (4.48) is
    $p - (p-3)^2 E(\|X - \bar{X}J_p\|^{-2})$.
    (b) When $V = \sigma^2 D$ with an unknown $\sigma^2 > 0$ and a known matrix D, show that the risk function of the estimator in (4.49) is smaller than that of X for any θ and σ^2.

94. Let X be a sample from a p.d.f. f_θ and $T(X)$ be a sufficient statistic for θ. Show that if an MLE exists, it is a function of T but it may not be sufficient for θ.

95. Let $\{f_\theta : \theta \in \Theta\}$ be a family of p.d.f.'s w.r.t. a σ-finite measure, where $\Theta \subset \mathcal{R}^k$; h be a Borel function from Θ onto $\Lambda \subset \mathcal{R}^p$, $1 \leq p \leq k$; and let $\tilde{\ell}(\lambda) = \sup_{\theta:h(\theta)=\lambda} \ell(\theta)$ be the induced likelihood function for the transformed parameter λ. Show that if $\hat{\theta} \in \Theta$ is an MLE of θ, then $\hat{\lambda} = h(\hat{\theta})$ maximizes $\tilde{\ell}(\lambda)$.

96. Let $X_1, ..., X_n$ be i.i.d. with a p.d.f. f_θ. Find an MLE of θ in each of the following cases.
    (a) $f_\theta(x) = \theta^{-1} I_{\{1,...,\theta\}}(x)$, θ is an integer between 1 and θ_0.
    (b) $f_\theta(x) = e^{-(x-\theta)} I_{(\theta,\infty)}(x)$, $\theta > 0$.
    (c) $f_\theta(x) = \theta(1-x)^{\theta-1} I_{(0,1)}(x)$, $\theta > 1$.
    (d) $f_\theta(x) = \frac{\theta}{1-\theta} x^{(2\theta-1)/(1-\theta)} I_{(0,1)}(x)$, $\theta \in (\frac{1}{2}, 1)$.
    (e) $f_\theta(x) = 2^{-1} e^{-|x-\theta|}$, $\theta \in \mathcal{R}$.
    (f) $f_\theta(x) = \theta x^{-2} I_{(\theta,\infty)}(x)$, $\theta > 0$.
    (g) $f_\theta(x) = \theta^x (1-\theta)^{1-x} I_{\{0,1\}}(x)$, $\theta \in [\frac{1}{2}, \frac{3}{4}]$.
    (h) $f_\theta(x)$ is the p.d.f. of $N(\theta, \theta^2)$, $\theta \in \mathcal{R}$, $\theta \neq 0$.
    (i) $f_\theta(x)$ is the p.d.f. of the exponential distribution $E(\mu, \sigma)$, $\theta = (\mu, \sigma) \in \mathcal{R} \times (0, \infty)$.
    (j) $f_\theta(x)$ is the p.d.f. of the log-normal distribution $LN(\mu, \sigma^2)$, $\theta = (\mu, \sigma^2) \in \mathcal{R} \times (0, \infty)$.
    (k) $f_\theta(x) = I_{(0,1)}(x)$ if $\theta = 0$ and $f_\theta(x) = (2\sqrt{x})^{-1} I_{(0,1)}(x)$ if $\theta = 1$.
    (l) $f_\theta(x) = \beta^{-\alpha} \alpha x^{\alpha-1} I_{(0,\beta)}(x)$, $\theta = (\alpha, \beta) \in (0, \infty) \times (0, \infty)$.
    (m) $f_\theta(x) = \binom{\theta}{x} p^x (1-p)^{\theta-x} I_{\{0,1,...,\theta\}}(x)$, $\theta = 1, 2, ...$, where $p \in (0, 1)$ is known.
    (n) $f_\theta(x) = \frac{1}{2}(1 - \theta^2) e^{\theta x - |x|}$, $\theta \in (-1, 1)$.

97. In Exercise 14, obtain an MLE of θ when (a) $\theta \in \mathcal{R}$ and (b) $\theta \leq 0$.

98. Suppose that n observations are taken from $N(\mu, 1)$ with an unknown μ. Instead of recording all the observations, one records only whether the observation is less than 0. Find an MLE of μ.

99. Find an MLE of θ in Exercise 43 of §2.6.

100. Let $(Y_1, Z_1), ..., (Y_n, Z_n)$ be i.i.d. random 2-vectors such that Y_1 and Z_1 are independently distributed as the exponential distributions $E(0, \lambda)$ and $E(0, \mu)$, respectively, where $\lambda > 0$ and $\mu > 0$.
     (a) Find the MLE of (λ, μ).
     (b) Suppose that we only observe $X_i = \min\{Y_i, Z_i\}$ and $\Delta_i = 1$ if $X_i = Y_i$ and $\Delta_i = 0$ if $X_i = Z_i$. Find the MLE of (λ, μ).

101. In Example 4.33, show that almost surely the likelihood equation has a unique solution that is the MLE of $\theta = (\alpha, \gamma)$. Obtain iteration equation (4.53) for this example. Discuss how to apply the Fisher-scoring method in this example.

102. Let $X_1, ..., X_n$ be i.i.d. from the discrete p.d.f. in Exercise 84 with an unknown $\theta > 0$. Show that the likelihood equation has a unique root when the sample mean > 1. Show whether this root is an MLE of θ.

103. Let $X_1, ..., X_n$ be i.i.d. from the logistic distribution $LG(\mu, \sigma)$ (Table 1.2, page 20).
     (a) Show how to find an MLE of μ when $\mu \in \mathcal{R}$ and σ is known.
     (b) Show how to find an MLE of σ when $\sigma > 0$ and μ is known.

104. Let $(X_1, Y_1), ..., (X_n, Y_n)$ be i.i.d. from a two-dimensional normal distribution with $E(X_1) = E(Y_1) = 0$, $\text{Var}(X_1) = \text{Var}(Y_1) = 1$, and an unknown correlation coefficient $\rho \in (-1, 1)$. Show that the likelihood equation is a cubic in ρ and the probability that it has a unique root tends to 1 as $n \to \infty$.

105. Let $X_1, ..., X_n$ be i.i.d. from the Weibull distribution $W(\alpha, \theta)$ (Table 1.2, page 20) with unknown $\alpha > 0$ and $\theta > 0$. Show that the likelihood equation is equivalent to $h(\alpha) = n^{-1} \sum_{i=1}^{n} \log x_i$ and $\theta = n^{-1} \sum_{i=1}^{n} x_i^{\alpha}$, where $h(\alpha) = (\sum_{i=1}^{n} x_i^{\alpha})^{-1} \sum_{i=1}^{n} x_i^{\alpha} \log x_i - \alpha^{-1}$, and that the likelihood equation has a unique solution.

106. Consider the random effects model in Example 3.17. Assume that $\mu = 0$ and $n_i = n_0$ for all i. Provide a condition on X_{ij}'s under which a unique MLE of (σ_a^2, σ^2) exists and find this MLE.

107. Let $X_1, ..., X_n$ be i.i.d. with the p.d.f. $\theta f(\theta x)$, where f is a Lebesgue p.d.f. on $(0, \infty)$ or symmetric about 0, and $\theta > 0$ is an unknown parameter. Show that the likelihood equation has a unique root if $x f'(x)/f(x)$ is strictly decreasing for $x > 0$. Verify that this condition is satisfied if f is the p.d.f. of the Cauchy distribution $C(0, 1)$.

108. Let $X_1, ..., X_n$ be i.i.d. with the Lebesgue p.d.f. $f_\theta(x) = \theta f_1(x) + (1 - \theta) f_2(x)$, where f_j's are two different known Lebesgue p.d.f.'s and $\theta \in (0, 1)$ is unknown.

(a) Provide a necessary and sufficient condition for the likelihood equation to have a unique solution and show that if there is a solution, it is the MLE of θ.
(b) Derive the MLE of θ when the likelihood equation has no solution.

109. Consider the location family in §4.2.1 and the scale family in §4.2.2. In each case, show that an MLE or an RLE (root of the likelihood equation) of the parameter, if it exists, is invariant.

110. Let X be a sample from P_θ, $\theta \in \mathcal{R}$. Suppose that P_θ's have p.d.f.'s f_θ w.r.t. a common σ-finite measure and that $\{x : f_\theta(x) > 0\}$ does not depend on θ. Assume further that an estimator $\hat\theta$ of θ attains the Cramér-Rao lower bound and that the conditions in Theorem 3.3 hold for $\hat\theta$. Show that $\hat\theta$ is a unique MLE of θ.

111. Let X_{ij}, $j = 1, ..., r > 1$, $i = 1, ..., n$, be independently distributed as $N(\mu_i, \sigma^2)$. Find the MLE of $(\mu_1, ..., \mu_n, \sigma^2)$. Show that the MLE of σ^2 is not a consistent estimator (as $n \to \infty$).

112. Let $X_1, ..., X_n$ be i.i.d. from the uniform distribution $U(0, \theta)$, where $\theta > 0$ is unknown. Let $\hat\theta$ be the MLE of θ and T be the UMVUE.
(a) Obtain the ratio $\mathrm{mse}_T(\theta)/\mathrm{mse}_{\hat\theta}(\theta)$ and show that the MLE is inadmissible when $n \geq 2$.
(b) Let $Z_{a,\theta}$ be a random variable having the exponential distribution $E(a, \theta)$. Prove $n(\theta - \hat\theta) \to_d Z_{0,\theta}$ and $n(\theta - T) \to_d Z_{-\theta,\theta}$. Obtain the asymptotic relative efficiency of $\hat\theta$ w.r.t. T.

113. Let $X_1, ..., X_n$ be i.i.d. from the exponential distribution $E(a, \theta)$ with unknown a and θ. Obtain the asymptotic relative efficiency of the MLE of a (or θ) w.r.t. the UMVUE of a (or θ).

114. Let $X_1, ..., X_n$ be i.i.d. from the Pareto distribution $Pa(a, \theta)$ with unknown a and θ.
(a) Find the MLE of (a, θ).
(b) Find the asymptotic relative efficiency of the MLE of a w.r.t. the UMVUE of a.

115. In Exercises 40 and 41 of §2.6,
(a) obtain an MLE of (θ, j);
(b) show whether the MLE of j in part (a) is consistent;
(c) show that the MLE of θ is consistent and derive its nondegenerated asymptotic distribution.

116. In Example 4.36, obtain the MLE of β under the canonical link and assumption (4.58) but $t_i \not\equiv 1$.

117. Consider the GLM in Example 4.35 with $\phi_i \equiv 1$ and the canonical link. Assume that $\sum_{i=1}^n Z_i Z_i^\tau$ is positive definite for $n \geq n_0$. Show that the likelihood equation has at most one solution when $n \geq n_0$ and a solution exists with probability tending to 1.

118. Consider the linear model (3.25) with $\varepsilon = N_n(0, V)$, where V is an unknown positive definite matrix. Show that the LSE $\hat{\beta}$ defined by (3.29) is an MQLE and that $\hat{\beta}$ is an MLE if and only if one of (a)-(e) in Theorem 3.10 holds.

119. Let X_j be a random variable having the binomial distribution $Bi(p_j, n_j)$ with a known n_j and an unknown $p_j \in (0, 1)$, $j = 1, 2$. Assume that X_j's are independent. Obtain a conditional likelihood function of the odds ratio $\theta = \frac{p_1}{1-p_1} / \frac{p_2}{1-p_2}$, given $X_1 + X_2$.

120. Let X_1 and X_2 be independent from Poisson distributions $P(\mu_1)$ and $P(\mu_2)$, respectively. Suppose that we are interested in $\theta_1 = \mu_1/\mu_2$. Derive a conditional likelihood function of θ_1, using (a) $\theta_2 = \mu_1$; (b) $\theta_2 = \mu_1 + \mu_2$; and (c) $\theta_2 = \mu_1\mu_2$.

121. Assume model (4.66) with $p = 2$ and normally distributed i.i.d. ε_t's. Obtain the conditional likelihood given $(X_1, X_2) = (x_1, x_2)$.

122. Prove the claim in Example 4.38.

123. Prove (4.70). (Hint: Show, using the argument in proving (4.77), that $n^{-1}|\frac{\partial^2}{\partial \theta^2} \log \ell(\xi_n) - \frac{\partial^2}{\partial \theta^2} \log \ell(\theta)| = o_p(1)$ for any random variable ξ_n satisfying $|\xi_n - \theta| \leq |\theta - \theta_n|$.)

124. Let $X_1, ..., X_n$ be i.i.d. from $N(\mu, 1)$ truncated at two known points $\alpha < \beta$, i.e., the Lebesgue p.d.f. of X_i is

$$\{\sqrt{2\pi}[\Phi(\beta - \mu) - \Phi(\alpha - \mu)]\}^{-1} e^{-(x-\mu)^2/2} I_{(\alpha,\beta)}(x).$$

(a) Show that the sample mean $\bar{X}$ is asymptotically efficient for estimating $\theta = EX_1$.
(b) Show that $\bar{X}$ is the unique MLE of θ.

125. Let $X_1, ..., X_n$ be i.i.d. from the discrete p.d.f.

$$f_\theta(x) = [1 - (1 - \theta)^m]^{-1} \binom{m}{x} \theta^x (1 - \theta)^{m-x} I_{\{1,2,...,m\}}(x),$$

where $\theta \in (0, 1)$ is unknown and $m \geq 2$ is a known integer.
(a) When the sample mean $\bar{X} = m$, show that $\bar{X}/m$ is an MLE of θ.
(b) When $1 < \bar{X} < m$, show that the likelihood equation has at least one solution.
(c) Show that the regularity conditions of Theorem 4.16 are satisfied and find the asymptotic variance of a consistent RLE of θ.

126. In Exercise 96, check whether the regularity conditions of Theorem 4.16 are satisfied for cases (b), (c), (d), (e), (g), (h), (j) and (n). Obtain nondegenerated asymptotic distributions of RLE's for cases in which Theorem 4.17 can be applied.

127. Let $X_1, ..., X_n$ be i.i.d. random variables such that $\log X_i$ is $N(\theta, \theta)$ with an unknown $\theta > 0$.
(a) Obtain the likelihood equation and show that one of the solutions of the likelihood equation is the unique MLE of θ.
(b) Using Theorem 4.17, obtain the asymptotic distribution of the MLE of θ.

128. In Exercise 107 of §3.6, find the MLE's of α and β and obtain their nondegenerated asymptotic joint distribution.

129. In Example 4.30, show that the MLE (or RLE) of θ is asymptotically efficient by (a) applying Theorem 4.17 and (b) directly deriving the asymptotic distribution of the MLE.

130. In Example 4.23, show that there is a unique asymptotically efficient RLE of $\theta = (p_1, ..., p_k)$. Discuss whether this RLE is the MLE.

131. Let $X_1, ..., X_n$ be i.i.d. with $P(X_1 = 0) = 6\theta^2 - 4\theta + 1$, $P(X_1 = 1) = \theta - 2\theta^2$, and $P(X_1 = 2) = 3\theta - 4\theta^2$, where $\theta \in (0, \frac{1}{2})$ is unknown. Apply Theorem 4.17 to obtain the asymptotic distribution of an RLE of θ.

132. Let $X_1, ..., X_n$ be i.i.d. random variables from $N(\mu, 1)$, where $\mu \in \mathcal{R}$ is unknown. Let $\theta = P(X_1 \leq c)$, where c is a known constant. Find the asymptotic relative efficiency of the MLE of θ w.r.t. (a) the UMVUE of θ and (b) the estimator $n^{-1} \sum_{i=1}^{n} I_{(-\infty, c]}(X_i)$.

133. In Exercise 19 of §3.6, find the MLE's of θ and $\vartheta = P(Y_1 > 1)$ and find the asymptotic relative efficiency of the MLE of ϑ w.r.t. the UMVUE of ϑ in part (b).

134. Let $(X_1, Y_1), ..., (X_n, Y_n)$ be i.i.d. random 2-vectors. Suppose that both X_1 and Y_1 are binary, $P(X_1 = 1) = \frac{1}{2}$, $P(Y_1 = 1|X_1 = 0) = e^{-a\theta}$, and $P(Y_1 = 1|X_1 = 0) = e^{-b\theta}$, where $\theta > 0$ is unknown and $a > 0$ and $b > 0$ are known constants.
(a) Suppose that (X_i, Y_i), $i = 1, ..., n$, are observed. Find the MLE of θ and its nondegenerated asymptotic distribution.
(b) Suppose that only $Y_1, ..., Y_n$ are observed. Find the MLE of θ and its nondegenerated asymptotic distribution.
(c) Calculate the asymptotic relative efficiency of the MLE in (a) w.r.t. the MLE in (b). How much efficiency is lost in the special case of $a = b$?

135. In Exercise 110 of §3.6, derive
     (a) the MLE of (θ_1, θ_2);
     (b) a nondegenerated asymptotic distribution of the MLE of (θ_1, θ_2);
     (c) the asymptotic relative efficiencies of the MLE's w.r.t. the moment estimators in Exercise 110 of §3.6.

136. In Exercise 104, show that the RLE of ρ is asymptotically distributed as $N\big(\rho, (1 - \rho^2)^2 / [n(1 + \rho^2)]\big)$.

137. In Exercise 107, obtain a nondegenerated asymptotic distribution of the RLE of θ when f is the p.d.f. of the Cauchy distribution $C(0, 1)$.

138. Let $X_1, ..., X_n$ be i.i.d. from the logistic distribution $LG(\mu, \sigma)$ with unknown $\mu \in \mathcal{R}$ and $\sigma > 0$. Obtain a nondegenerated asymptotic distribution of the RLE of (μ, σ).

139. In Exercise 105, show that the conditions of Theorem 4.16 are satisfied.

140. Let $X_1, ..., X_n$ be i.i.d. binary random variables with $P(X_1 = 1) = p$, where $p \in (0, 1)$ is unknown. Let $\hat{\vartheta}_n$ be the MLE of $\vartheta = p(1 - p)$.
     (a) Show that $\hat{\vartheta}_n$ is asymptotically normal when $p \neq \frac{1}{2}$.
     (b) When $p = \frac{1}{2}$, derive a nondegenerated asymptotic distribution of $\hat{\vartheta}_n$ with an appropriate normalization.

141. Let $(X_1, Y_1), ..., (X_n, Y_n)$ be i.i.d. random 2-vectors satisfying $0 \leq X_1 \leq 1$, $0 \leq Y_1 \leq 1$, and

$$P(X_1 > x, Y_1 > y) = (1 - x)(1 - y)(1 - \max\{x, y\})^\theta$$

     for $0 \leq x \leq 1$, $0 \leq y \leq 1$, where $\theta \geq 0$ is unknown.
     (a) Obtain the likelihood function and the likelihood equation.
     (b) Show that an RLE of θ is asymptotically normal and derive its amse.

142. Assume the conditions in Theorem 4.16. Suppose that $\theta = (\theta_1, ..., \theta_k)$ and there is a positive integer $p < k$ such that $\partial \log \ell(\theta)/\partial \theta_i$ and $\partial \log \ell(\theta)/\partial \theta_j$ are uncorrelated whenever $i \leq p < j$. Show that the asymptotic distribution of the RLE of $(\theta_1, ..., \theta_p)$ is unaffected by whether $\theta_{p+1}, ..., \theta_k$ are known.

143. Let $X_1, ..., X_n$ be i.i.d. random p-vectors from $N_p(\mu, \Sigma)$ with unknown μ and Σ. Find the MLE's of μ and Σ and derive their nondegenerated asymptotic distributions.

144. Let $X_1, ..., X_n$ be i.i.d. bivariate normal random vectors with mean 0 and an unknown covariance matrix whose diagonal elements are

σ_1^2 and σ_2^2 and off-diagonal element is $\sigma_1\sigma_2\rho$. Let $\theta = (\sigma_1^2, \sigma_2^2, \rho)$. Obtain $I_n(\theta)$ and $[I_n(\theta)]^{-1}$ and derive a nondegenerated asymptotic distribution of the MLE of θ.

145. Let $X_1, ..., X_n$ be i.i.d. each with probability p as $N(\mu, \sigma^2)$ and probability $1 - p$ as $N(\eta, \tau^2)$, where $\theta = (\mu, \eta, \sigma^2, \tau^2, p)$ is unknown.
  (a) Show that the conditions in Theorem 4.16 are satisfied.
  (b) Show that the likelihood function is unbounded.
  (c) Show that an MLE may be inconsistent.

146. Let $X_1, ..., X_n$ and $Y_1, ..., Y_n$ be independently distributed as $N(\mu, \sigma^2)$ and $N(\mu, \tau^2)$, respectively, with unknown $\theta = (\mu, \sigma^2, \tau^2)$. Find the MLE of θ and show that it is asymptotically efficient.

147. Find a nondegenerated asymptotic distribution of the MLE of (σ_a^2, σ^2) in Exercise 106.

148. Under the conditions in Theorem 4.18, prove (4.85) and (4.86).

149. Assume linear model (3.25) with $\varepsilon = N_n(0, \sigma^2 I_n)$ and a full rank Z. Apply Theorem 4.18 to show that the LSE $\hat{\beta}$ is asymptotically efficient. Compare this result with that in Theorem 3.12.

150. Apply Theorem 4.18 to obtain the asymptotic distribution of the RLE of β in (a) Example 4.35 and (b) Example 4.37.

151. Let $X_1, ..., X_n$ be i.i.d. from the logistic distribution $LG(\mu, \sigma)$, $\mu \in \mathcal{R}$, $\sigma > 0$. Using Newton-Raphson and Fisher-scoring methods, find
  (a) one-step MLE's of μ when σ is known;
  (b) one-step MLE's of σ when μ is known;
  (c) one-step MLE's of (μ, σ).
  (d) Show how to obtain $\sqrt{n}$-consistent initial estimators in (a)-(c).

152. Under the GLM (4.55)-(4.58),
  (a) show how to obtain a one-step MLE of β, if an initial estimator $\hat{\beta}_n^{(0)}$ is available;
  (b) show that under the conditions in Theorem 4.18, the one-step MLE satisfies (4.81) if $\|[I_n(\beta)]^{1/2}(\hat{\beta}_n^{(0)} - \beta)\| = O_p(1)$.

153. In Example 4.40, show that the conditions in Theorem 4.20 concerning the likelihood function are satisfied.

154. Let $X_1, ..., X_n$ be i.i.d. from the logistic distribution $LG(\mu, \sigma)$ with unknown $\mu \in \mathcal{R}$ and $\sigma > 0$. Show that the conditions in Theorem 4.20 concerning the likelihood function are satisfied.

Chapter 5

Estimation in Nonparametric Models

Estimation methods studied in this chapter are useful for nonparametric models as well as for parametric models in which the parametric model assumptions might be violated (so that robust estimators are required) or the number of unknown parameters is exceptionally large. Some such methods have been introduced in Chapter 3; for example, the methods that produce UMVUE's in nonparametric models, the U- and V-statistics, the LSE's and BLUE's, the Horvitz-Thompson estimators, and the sample (central) moments.

The theoretical justification for estimators in nonparametric models, however, relies more on asymptotics than that in parametric models. This means that applications of nonparametric methods usually require large sample sizes. Also, estimators derived using parametric methods are asymptotically more efficient than those based on nonparametric methods when the parametric models are correct. Thus, to choose between a parametric method and a nonparametric method, we need to balance the advantage of requiring weaker model assumptions (robustness) against the drawback of losing efficiency, which results in requiring a larger sample size.

It is assumed in this chapter that a sample $X = (X_1, ..., X_n)$ is from a population in a nonparametric family, where X_i's are random vectors.

5.1 Distribution Estimators

In many applications the c.d.f.'s of X_i's are determined by a single c.d.f. F on $\mathcal{R}^d$; for example, X_i's are i.i.d. random d-vectors. In this section, we consider the estimation of F or $F(t)$ for several t's, under a nonparametric model in which very little is assumed about F.

5.1.1 Empirical c.d.f.'s in i.i.d. cases

For i.i.d. random variables $X_1, ..., X_n$, the empirical c.d.f. F_n is defined in (2.28). The definition of the empirical c.d.f. based on $X = (X_1, ..., X_n)$ in the case of $X_i \in \mathcal{R}^d$ is analogously given by

$$F_n(t) = \frac{1}{n} \sum_{i=1}^n I_{(-\infty, t]}(X_i), \qquad t \in \mathcal{R}^d, \tag{5.1}$$

where $(-\infty, a]$ denotes the set $(-\infty, a_1] \times \cdots \times (-\infty, a_d]$ for any $a = (a_1, ..., a_d) \in \mathcal{R}^d$. Similar to the case of $d = 1$ (Example 2.26), $F_n(t)$ as an estimator of $F(t)$ has the following properties. For any $t \in \mathcal{R}^d$, $nF_n(t)$ has the binomial distribution $Bi(F(t), n)$; $F_n(t)$ is unbiased with variance $F(t)[1 - F(t)]/n$; $F_n(t)$ is the UMVUE under some nonparametric models; and $F_n(t)$ is $\sqrt{n}$-consistent for $F(t)$. For any m fixed distinct points $t_1, ..., t_m$ in $\mathcal{R}^d$, it follows from the multivariate CLT (Corollary 1.2) and (5.1) that as $n \to \infty$,

$$\sqrt{n}\left[\left(F_n(t_1), ..., F_n(t_m)\right) - \left(F(t_1), ..., F(t_m)\right)\right] \to_d N_m(0, \Sigma), \tag{5.2}$$

where Σ is the $m \times m$ matrix whose (i, j)th element is

$$P\left(X_1 \in (-\infty, t_i] \cap (-\infty, t_j]\right) - F(t_i)F(t_j).$$

Note that these results hold without *any* assumption on F.

Considered as a function of t, F_n is a random element taking values in $\mathcal{F}$, the collection of all c.d.f.'s on $\mathcal{R}^d$. As $n \to \infty$, $\sqrt{n}(F_n - F)$ converges in some sense to a random element defined on some probability space. A detailed discussion of such a result is beyond our scope and can be found, for example, in Shorack and Wellner (1986). To discuss some global properties of F_n as an estimator of $F \in \mathcal{F}$, we need to define a closeness measure between the elements (c.d.f.'s) in $\mathcal{F}$.

Definition 5.1. Let $\mathcal{F}_0$ be a collection of c.d.f.'s on $\mathcal{R}^d$.
(i) A function ϱ from $\mathcal{F}_0 \times \mathcal{F}_0$ to $[0, \infty)$ is called a *distance* or *metric* on $\mathcal{F}_0$ if and only if for any G_j in $\mathcal{F}_0$, (a) $\varrho(G_1, G_2) = 0$ if and only if $G_1 = G_2$; (b) $\varrho(G_1, G_2) = \varrho(G_2, G_1)$; and (c) $\varrho(G_1, G_2) \leq \varrho(G_1, G_3) + \varrho(G_3, G_2)$.
(ii) Let $\mathcal{D} = \{c(G_1 - G_2) : c \in \mathcal{R}, \ G_j \in \mathcal{F}_0, \ j = 1, 2\}$. A function $\| \cdot \|$ from $\mathcal{D}$ to $[0, \infty)$ is called a *norm* on $\mathcal{D}$ if and only if (a) $\|\Delta\| = 0$ if and only if $\Delta = 0$; (b) $\|c\Delta\| = |c|\|\Delta\|$ for any $\Delta \in \mathcal{D}$ and $c \in \mathcal{R}$; and (c) $\|\Delta_1 + \Delta_2\| \leq \|\Delta_1\| + \|\Delta_2\|$ for any $\Delta_j \in \mathcal{D}, \ j = 1, 2$. ∎

Any norm $\|\cdot\|$ on $\boldsymbol{D}$ induces a distance given by $\varrho(G_1, G_2) = \|G_1 - G_2\|$. The most commonly used distance is the sup-norm distance ϱ_∞, i.e., the distance induced by the sup-norm

$$\|G_1 - G_2\|_\infty = \sup_{t \in \mathcal{R}^d} |G_1(t) - G_2(t)|, \quad G_j \in \boldsymbol{\mathcal{F}}. \tag{5.3}$$

The following result concerning the sup-norm distance between F_n and F is due to Dvoretzky, Kiefer, and Wolfowitz (1956).

Lemma 5.1. (DKW's inequality). Let F_n be the empirical c.d.f. based on i.i.d. $X_1, ..., X_n$ from a c.d.f. F on $\mathcal{R}^d$.
(i) When $d = 1$, there exists a positive constant C (not depending on F) such that

$$P\big(\varrho_\infty(F_n, F) > z\big) \le C e^{-2nz^2}, \quad z > 0, n = 1, 2, ...$$

(ii) When $d \ge 2$, for any $\epsilon > 0$, there exists a positive constant $C_{\epsilon,d}$ (not depending on F) such that

$$P\big(\varrho_\infty(F_n, F) > z\big) \le C_{\epsilon,d} e^{-(2-\epsilon)nz^2}, \quad z > 0, n = 1, 2, ... \quad \blacksquare$$

The proof of this lemma is omitted. The following results useful in statistics are direct consequences of Lemma 5.1.

Theorem 5.1. Let F_n be the empirical c.d.f. based on i.i.d. $X_1, ..., X_n$ from a c.d.f. F on $\mathcal{R}^d$. Then
(i) $\varrho_\infty(F_n, F) \to_{a.s.} 0$ as $n \to \infty$;
(ii) $E[\sqrt{n}\varrho_\infty(F_n, F)]^s = O(1)$ for any $s > 0$.
Proof. (i) From DKW's inequality,

$$\sum_{n=1}^{\infty} P\big(\varrho_\infty(F_n, F) > z\big) < \infty.$$

Hence, the result follows from Theorem 1.8(v).
(ii) Using DKW's inequality with $z = y^{1/s}/\sqrt{n}$ and the result in Exercise 55 of §1.6, we obtain that

$$E[\sqrt{n}\varrho_\infty(F_n, F)]^s = \int_0^\infty P\big(\sqrt{n}\varrho_\infty(F_n, F) > y^{1/s}\big) dy$$

$$\le C_{\epsilon,d} \int_0^\infty e^{-(2-\epsilon)y^{2/s}} dy$$

$$= O(1)$$

as long as $2 - \epsilon > 0$. $\blacksquare$

Theorem 5.1(i) means that $F_n(t) \to_{a.s.} F(t)$ uniformly in $t \in \mathcal{R}^d$, a result stronger than the strong consistency of $F_n(t)$ for every t. Theorem 5.1(ii) implies that $\sqrt{n}\varrho_\infty(F_n, F) = O_p(1)$, a result stronger than the $\sqrt{n}$-consistency of $F_n(t)$. These results hold without any condition on F.

Let $p \geq 1$ and $\mathcal{F}_p = \{G \in \mathcal{F} : \int \|t\|^p dG < \infty\}$, which is the subset of c.d.f.'s in $\mathcal{F}$ having finite pth moments. Mallows' distance between G_1 and G_2 in $\mathcal{F}_p$ is defined to be

$$\varrho_{M_p}(G_1, G_2) = \inf(E\|Y_1 - Y_2\|^p)^{1/p}, \tag{5.4}$$

where the infimum is taken over all pairs of Y_1 and Y_2 having c.d.f.'s G_1 and G_2, respectively. Let $\{G_j : j = 0, 1, 2, ...\} \subset \mathcal{F}_p$. Then $\varrho_{M_p}(G_j, G_0) \to 0$ as $j \to \infty$ if and only if $\int \|t\|^p dG_j \to \int \|t\|^p dG_0$ and $G_j(t) \to G_0(t)$ for every $t \in \mathcal{R}^d$ at which G_0 is continuous. It follows from Theorem 5.1 and the SLLN (Theorem 1.13) that $\varrho_{M_p}(F_n, F) \to_{a.s.} 0$ if $F \in \mathcal{F}_p$.

When $d = 1$, another useful distance for measuring the closeness between F_n and F is the L_p distance ϱ_{L_p} induced by the L_p-norm ($p \geq 1$)

$$\|G_1 - G_2\|_{L_p} = \left[\int |G_1(t) - G_2(t)|^p dt\right]^{1/p}, \quad G_j \in \mathcal{F}_1. \tag{5.5}$$

A result similar to Theorem 5.1 is given as follows.

Theorem 5.2. Let F_n be the empirical c.d.f. based on i.i.d. random variables $X_1, ..., X_n$ from a c.d.f. $F \in \mathcal{F}_1$. Then
(i) $\varrho_{L_p}(F_n, F) \to_{a.s.} 0$;
(ii) $E[\sqrt{n}\varrho_{L_p}(F_n, F)] = O(1)$ if $1 \leq p < 2$ and $\int \{F(t)[1 - F(t)]\}^{p/2} dt < \infty$, or $p \geq 2$.
Proof. (i) Since $[\varrho_{L_p}(F_n, F)]^p \leq [\varrho_\infty(F_n, F)]^{p-1}[\varrho_{L_1}(F_n, F)]$ and, by Theorem 5.1, $\varrho_\infty(F_n, F) \to_{a.s.} 0$, it suffices to show the result for $p = 1$. Let $Y_i = \int_{-\infty}^0 [I_{(-\infty, t]}(X_i) - F(t)] dt$. Then $Y_1, ..., Y_n$ are i.i.d. and

$$E|Y_i| \leq \int E|I_{(-\infty, t]}(X_i) - F(t)| dt = 2 \int F(t)[1 - F(t)] dt,$$

which is finite under the condition that $F \in \mathcal{F}_1$. By the SLLN,

$$\int_{-\infty}^0 [F_n(t) - F(t)] dt = \frac{1}{n} \sum_{i=1}^n Y_i \to_{a.s.} E(Y_1) = 0. \tag{5.6}$$

Since $[F_n(t) - F(t)]_- \leq F(t)$ and $\int_{-\infty}^0 F(t) dt < \infty$ (Exercise 55 in §1.6), it follows from Theorem 5.1 and the dominated convergence theorem that $\int_{-\infty}^0 [F_n(t) - F(t)]_- dt \to_{a.s.} 0$, which with (5.6) implies

$$\int_{-\infty}^0 |F_n(t) - F(t)| dt \to_{a.s.} 0. \tag{5.7}$$

The result follows since we can similarly show that (5.7) holds with $\int_{-\infty}^{0}$ replaced by $\int_{0}^{\infty}$.

(ii) When $1 \le p < 2$, the result follows from

$$
\begin{aligned}
E[\varrho_{L_p}(F_n, F)] &\le \left\{ \int E|F_n(t) - F(t)|^p dt \right\}^{1/p} \\
&\le \left\{ \int [E|F_n(t) - F(t)|^2]^{p/2} dt \right\}^{1/p} \\
&= n^{-1/2} \left\{ \int \{F(t)[1 - F(t)]\}^{p/2} dt \right\}^{1/p} \\
&= O(n^{-1/2}),
\end{aligned}
$$

where the two inequalities follow from Jensen's inequality. When $p \ge 2$,

$$
\begin{aligned}
E[\varrho_{L_p}(F_n, F)] &\le E\left\{ [\varrho_\infty(F_n, F)]^{1-2/p}[\varrho_{L_2}(F_n, F)]^{2/p} \right\} \\
&\le \left\{ E[\varrho_\infty(F_n, F)]^{(1-2/p)q} \right\}^{1/q} \left\{ E[\varrho_{L_2}(F_n, F)]^2 \right\}^{1/p} \\
&= \left\{ O(n^{-(1-2/p)q/2}) \right\}^{1/q} \left\{ E \int |F_n(t) - F(t)|^2 dt \right\}^{1/p} \\
&= O(n^{-(1-2/p)/2}) \left\{ \frac{1}{n} \int F(t)[1 - F(t)] dt \right\}^{1/p} \\
&= O(n^{-1/2}),
\end{aligned}
$$

where $\frac{1}{q} + \frac{1}{p} = 1$, the second inequality follows from Hölder's inequality (see (1.40) in §1.3.2), and the first equality follows from Theorem 5.1(ii). ∎

5.1.2 Empirical likelihoods

In §4.4 and §4.5, we have shown that the method of using likelihoods provides some asymptotically efficient estimators. We now introduce some likelihoods in nonparametric models. This not only provides another justification for the use of the empirical c.d.f. in (5.1), but also leads to a useful method of deriving estimators in various (possibly non-i.i.d.) cases, some of which are discussed later in this chapter.

Let $X_1, ..., X_n$ be i.i.d. with $F \in \mathcal{F}$ and P_G be the probability measure corresponding to $G \in \mathcal{F}$. Given $X_1 = x_1, ..., X_n = x_n$, the *nonparametric likelihood* function is defined to be the following functional from $\mathcal{F}$ to $[0, \infty)$:

$$
\ell(G) = \prod_{i=1}^{n} P_G(\{x_i\}), \qquad G \in \mathcal{F}. \tag{5.8}
$$

Apparently, $\ell(G) = 0$ if $P_G(\{x_i\}) = 0$ for at least one i. The following result, due to Kiefer and Wolfowitz (1956), shows that the empirical c.d.f. F_n is a nonparametric maximum likelihood estimator of F.

Theorem 5.3. Let $X_1, ..., X_n$ be i.i.d. with $F \in \mathcal{F}$ and $\ell(G)$ be defined by (5.8). Then F_n maximizes $\ell(G)$ over $G \in \mathcal{F}$.
Proof. We only need to consider $G \in \mathcal{F}$ such that $\ell(G) > 0$. Let $c \in (0, 1]$ and $\mathcal{F}(c)$ be the subset of $\mathcal{F}$ containing G's satisfying $p_i = P_G(\{x_i\}) > 0$, $i = 1, ..., n$, and $\sum_{i=1}^n p_i = c$. We now apply the Lagrange multiplier method to solve the problem of maximizing $\ell(G)$ over $G \in \mathcal{F}(c)$. Define

$$H(p_1, ..., p_n, \lambda) = \prod_{i=1}^n p_i + \lambda \left(\sum_{i=1}^n p_i - c \right),$$

where λ is the Lagrange multiplier. Set

$$\frac{\partial H}{\partial \lambda} = \sum_{i=1}^n p_i - c = 0, \qquad \frac{\partial H}{\partial p_j} = p_j^{-1} \prod_{i=1}^n p_i + \lambda = 0, \qquad j = 1, ..., n.$$

The solution is $p_i = c/n$, $i = 1, ..., n$, $\lambda = -(c/n)^{n-1}$. It can be shown (exercise) that this solution is a maximum of $H(p_1, ..., p_n, \lambda)$ over $p_i > 0$, $i = 1, ..., n$, $\sum_{i=1}^n p_i = c$. This shows that

$$\max_{G \in \mathcal{F}(c)} \ell(G) = (c/n)^n,$$

which is maximized at $c = 1$ for any fixed n. The result follows from $P_{F_n}(\{x_i\}) = n^{-1}$ for given $X_i = x_i$, $i = 1, ..., n$. ∎

From the proof of Theorem 5.3, F_n maximizes the likelihood $\ell(G)$ in (5.8) over $p_i > 0$, $i = 1, ..., n$, and $\sum_{i=1}^n p_i = 1$, where $p_i = P_G(\{x_i\})$. This method of deriving an estimator of F can be extended to various situations with some modifications of (5.8) and/or constraints on p_i's. Modifications of the likelihood in (5.8) are called *empirical likelihoods* (Owen, 1988, 2001; Qin and Lawless, 1994). An estimator obtained by maximizing an empirical likelihood is then called a *maximum empirical likelihood estimator* (MELE). We now discuss several applications of the method of empirical likelihoods.

Consider first the estimation of F with auxiliary information about F (and i.i.d. $X_1, ..., X_n$). For instance, suppose that there is a known Borel function u from $\mathcal{R}^d$ to $\mathcal{R}^s$ such that

$$\int u(x)dF = 0 \tag{5.9}$$

(e.g., some components of the mean of F are 0). It is then reasonable to expect that any estimate $\hat{F}$ of F has property (5.9), i.e., $\int u(x)d\hat{F} = 0$, which is not true for the empirical c.d.f. F_n in (5.1), since

$$\int u(x)dF_n = \frac{1}{n}\sum_{i=1}^{n} u(X_i) \neq 0$$

even if $E[u(X_1)] = 0$. Using the method of empirical likelihoods, a natural solution is to put another constraint in the process of maximizing the likelihood. That is, we maximize $\ell(G)$ in (5.8) subject to

$$p_i > 0, \quad i = 1, ..., n, \quad \sum_{i=1}^{n} p_i = 1, \quad \text{and} \quad \sum_{i=1}^{n} p_i u(x_i) = 0, \qquad (5.10)$$

where $p_i = P_G(\{x_i\})$. Using the Lagrange multiplier method and an argument similar to the proof of Theorem 5.3, it can be shown (exercise) that an MELE of F is

$$\hat{F}(t) = \sum_{i=1}^{n} \hat{p}_i I_{(-\infty,t]}(X_i), \qquad (5.11)$$

where the notation $(-\infty, t]$ is the same as that in (5.1),

$$\hat{p}_i = n^{-1}[1 + \lambda_n^\tau u(X_i)]^{-1}, \qquad i = 1, ..., n, \qquad (5.12)$$

and $\lambda_n \in \mathcal{R}^s$ is the Lagrange multiplier satisfying

$$\sum_{i=1}^{n} \hat{p}_i u(X_i) = \frac{1}{n}\sum_{i=1}^{n} \frac{u(X_i)}{1 + \lambda_n^\tau u(X_i)} = 0. \qquad (5.13)$$

Note that $\hat{F}$ reduces to F_n if $u \equiv 0$.

To see that (5.13) has a solution asymptotically, note that

$$\frac{\partial}{\partial\lambda}\left[\frac{1}{n}\sum_{i=1}^{n}\log\left(1 + \lambda^\tau u(X_i)\right)\right] = \frac{1}{n}\sum_{i=1}^{n}\frac{u(X_i)}{1 + \lambda^\tau u(X_i)}$$

and

$$\frac{\partial^2}{\partial\lambda\partial\lambda^\tau}\left[\frac{1}{n}\sum_{i=1}^{n}\log\left(1 + \lambda^\tau u(X_i)\right)\right] = -\frac{1}{n}\sum_{i=1}^{n}\frac{u(X_i)[u(X_i)]^\tau}{[1 + \lambda^\tau u(X_i)]^2},$$

which is negative definite if $\text{Var}(u(X_1))$ is positive definite. Also,

$$E\left\{\frac{\partial}{\partial\lambda}\left[\frac{1}{n}\sum_{i=1}^{n}\log\left(1 + \lambda^\tau u(X_i)\right)\right]\Big|_{\lambda=0}\right\} = E[u(X_1)] = 0.$$

Hence, using the same argument as in the proof of Theorem 4.18, we can show that there exists a unique sequence $\{\lambda_n(X)\}$ such that as $n \to \infty$,

$$P\left(\frac{1}{n}\sum_{i=1}^{n}\frac{u(X_i)}{1 + \lambda_n^\tau u(X_i)} = 0\right) \to 1 \quad \text{and} \quad \lambda_n \to_p 0. \qquad (5.14)$$

Theorem 5.4. Let $X_1, ..., X_n$ be i.i.d. with $F \in \mathcal{F}$, u be a Borel function on $\mathcal{R}^d$ satisfying (5.9), and $\hat{F}$ be given by (5.11)-(5.13). Suppose that $U = \mathrm{Var}(u(X_1))$ is positive definite. Then, for any m fixed distinct $t_1, ..., t_m$ in $\mathcal{R}^d$,

$$\sqrt{n}[(\hat{F}(t_1), ..., \hat{F}(t_m)) - (F(t_1), ..., F(t_m))] \to_d N_m(0, \Sigma_u), \qquad (5.15)$$

where

$$\Sigma_u = \Sigma - W^\tau U^{-1} W,$$

Σ is given in (5.2), $W = (W(t_1), ..., W(t_m))$, $W(t_j) = E[u(X_1)I_{(-\infty, t_j]}(X_1)]$, and the notation $(-\infty, t]$ is the same as that in (5.1).

Proof. We prove the case of $m = 1$. The case of $m \geq 2$ is left as an exercise. Let $\bar{u} = n^{-1} \sum_{i=1}^n u(X_i)$. It follows from (5.13), (5.14), and Taylor's expansion that

$$\bar{u} = \frac{1}{n} \sum_{i=1}^n u(X_i)[u(X_i)]^\tau \lambda_n [1 + o_p(1)].$$

By the SLLN and CLT,

$$U^{-1} \bar{u} = \lambda_n + o_p(n^{-1/2}).$$

Using Taylor's expansion and the SLLN again, we have

$$\frac{1}{n} \sum_{i=1}^n I_{(-\infty, t]}(X_i)(n\hat{p}_i - 1) = \frac{1}{n} \sum_{i=1}^n I_{(-\infty, t]}(X_i) \left[\frac{1}{1 + \lambda_n^\tau u(X_i)} - 1 \right]$$

$$= -\frac{1}{n} \sum_{i=1}^n I_{(-\infty, t]}(X_i) \lambda_n^\tau u(X_i) + o_p(n^{-1/2})$$

$$= -\lambda_n^\tau W(t) + o_p(n^{-1/2})$$

$$= -\bar{u}^\tau U^{-1} W(t) + o_p(n^{-1/2}).$$

Thus,

$$\hat{F}(t) - F(t) = F_n(t) - F(t) + \frac{1}{n} \sum_{i=1}^n I_{(-\infty, t]}(X_i)(n\hat{p}_i - 1)$$

$$= F_n(t) - F(t) - \bar{u}^\tau U^{-1} W(t) + o_p(n^{-1/2})$$

$$= \frac{1}{n} \sum_{i=1}^n \left\{ I_{(-\infty, t]}(X_i) - F(t) - [u(X_i)]^\tau U^{-1} W(t) \right\} + o_p(n^{-1/2}).$$

The result follows from the CLT and the fact that

$$\mathrm{Var}([W(t)]^\tau U^{-1} u(X_i)) = [W(t)]^\tau U^{-1} U U^{-1} W(t)$$

$$= [W(t)]^\tau U^{-1} W(t)$$

$$= E\{[W(t)]^\tau U^{-1} u(X_i) I_{(-\infty, t]}(X_i)\}$$

$$= \mathrm{Cov}(I_{(-\infty, t]}(X_i), [W(t)]^\tau U^{-1} u(X_i)). \quad \blacksquare$$

Comparing (5.15) with (5.2), we conclude that $\hat{F}$ is asymptotically more efficient than F_n.

Example 5.1 (Survey problems). An example of situations in which we have auxiliary information expressed as (5.9) is a survey problem (Example 2.3) where the population $\mathcal{P} = \{y_1, ..., y_N\}$ consists of two-dimensional y_j's, $y_j = (y_{1j}, y_{2j})$, and the population mean $\bar{Y}_2 = N^{-1} \sum_{j=1}^{N} y_{2j}$ is known. For example, suppose that y_{1j} is the current year's income of unit j in the population and y_{2j} is the last year's income. In many applications the population total or mean of y_{2j}'s is known, for example, from tax return records. Let $X_1, ..., X_n$ be a simple random sample (see Example 2.3) selected from $\mathcal{P}$ *with* replacement. Then X_i's are i.i.d. bivariate random vectors whose c.d.f. is

$$F(t) = \frac{1}{N} \sum_{j=1}^{N} I_{(-\infty, t]}(y_j), \qquad (5.16)$$

where the notation $(-\infty, t]$ is the same as that in (5.1). If $\bar{Y}_2$ is known, then it can be expressed as (5.9) with $u(x_1, x_2) = x_2 - \bar{Y}_2$. In survey problems X_i's are usually sampled *without* replacement so that $X_1, ..., X_n$ are not i.i.d. However, for a simple random sample without replacement, (5.8) can still be treated as an empirical likelihood, given X_i's. Note that F in (5.16) is the c.d.f. of X_i, regardless of whether X_i's are sampled with replacement.

If $X = (X_1, ..., X_n)$ is not a simple random sample, then the likelihood (5.8) has to be modified. Suppose that π_i is the probability that the ith unit is selected (see Theorem 3.15). Given $X = \{y_i, i \in s\}$, an empirical likelihood is

$$\ell(G) = \prod_{i \in s} [P_G(\{y_i\})]^{1/\pi_i} = \prod_{i \in s} p_i^{1/\pi_i}, \qquad (5.17)$$

where $p_i = P_G(\{y_i\})$. With the auxiliary information (5.9), an MELE of F in (5.16) can be obtained by maximizing $\ell(G)$ in (5.17) subject to (5.10). In this case F may not be the c.d.f. of X_i, but the c.d.f.'s of X_i's are determined by F and π_i's. It can be shown (exercise) that an MELE is given by (5.11) with

$$\hat{p}_i = \frac{1}{\pi_i[1 + \lambda_n^\tau u(y_i)]} \Big/ \sum_{i \in s} \frac{1}{\pi_i} \qquad (5.18)$$

and

$$\sum_{i \in s} \frac{u(y_i)}{\pi_i[1 + \lambda_n^\tau u(y_i)]} = 0. \qquad (5.19)$$

If π_i = a constant, then the MELE reduces to that in (5.11)-(5.13). If

$u(x) = 0$ (no auxiliary information), then the MELE is

$$\hat{F}(t) = \sum_{i \in \boldsymbol{S}} \frac{1}{\pi_i} I_{(-\infty, t]}(y_i) \bigg/ \sum_{i \in \boldsymbol{S}} \frac{1}{\pi_i},$$

which is a ratio of two Horvitz-Thompson estimators (§3.4.2). Some asymptotic properties of the MELE $\hat{F}$ can be found in Chen and Qin (1993). ∎

The second part of Example 5.1 shows how to use empirical likelihoods in a non-i.i.d. problem. Applications of empirical likelihoods in non-i.i.d. problems are usually straightforward extensions of those in i.i.d. cases. The following is another example.

Example 5.2 (Biased sampling). Biased sampling is often used in applications. Suppose that $n = n_1 + \cdots + n_k$, $k \geq 2$; X_i's are independent random variables; $X_1, ..., X_{n_1}$ are i.i.d. with F; and $X_{n_1 + \cdots + n_j + 1}, ..., X_{n_1 + \cdots + n_{j+1}}$ are i.i.d. with the c.d.f.

$$\int_{-\infty}^{t} w_{j+1}(s) dF(s) \bigg/ \int_{-\infty}^{\infty} w_{j+1}(s) dF(s),$$

$j = 1, ..., k - 1$, where w_j's are some nonnegative Borel functions. A simple example is that $X_1, ..., X_{n_1}$ are sampled from F and $X_{n_1+1}, ..., X_{n_1+n_2}$ are sampled from F but conditional on the fact that each sampled value exceeds a given value x_0 (i.e., $w_2(s) = I_{(x_0, \infty)}(s)$). For instance, X_i's are blood pressure measurements; $X_1, ..., X_{n_1}$ are sampled from ordinary people and $X_{n_1+1}, ..., X_{n_1+n_2}$ are sampled from patients whose blood pressures are higher than x_0. The name biased sampling comes from the fact that there is a bias in the selection of samples.

For simplicity we consider the case of $k = 2$, since the extension to $k \geq 3$ is straightforward. Denote w_2 by w. An empirical likelihood is

$$\ell(G) = \prod_{i=1}^{n_1} P_G(\{x_i\}) \prod_{i=n_1+1}^{n} \frac{w(x_i) P_G(\{x_i\})}{\int w(s) dG(s)}$$

$$= \left[\sum_{i=1}^{n} p_i w(x_i) \right]^{-n_2} \prod_{i=1}^{n} p_i \prod_{i=n_1+1}^{n} w(x_i), \qquad (5.20)$$

where $p_i = P_G(\{x_i\})$. An MELE of F can be obtained by maximizing the empirical likelihood (5.20) subject to $p_i > 0$, $i = 1, ..., n$, and $\sum_{i=1}^{n} p_i = 1$. Using the Lagrange multiplier method we can show (exercise) that an MELE $\hat{F}$ is given by (5.11) with

$$\hat{p}_i = [n_1 + n_2 w(X_i) / \hat{w}]^{-1}, \qquad i = 1, ..., n, \qquad (5.21)$$

where $\hat{w}$ satisfies

$$\hat{w} = \sum_{i=1}^{n} \frac{w(X_i)}{n_1 + n_2 w(X_i)/\hat{w}}.$$

An asymptotic result similar to that in Theorem 5.4 can be established (Vardi, 1985; Qin, 1993).

If the function w depends on an unknown parameter vector θ, then the method of profile empirical likelihood (see §5.1.4) can be applied. ∎

Our last example concerns an important application in survival analysis.

Example 5.3 (Censored data). Let $T_1, ..., T_n$ be *survival times* that are i.i.d. nonnegative random variables from a c.d.f. F, and $C_1, ..., C_n$ be i.i.d. nonnegative random variables independent of T_i's. In a variety of applications in biostatistics and life-time testing, we are only able to observe the smaller of T_i and C_i and an indicator of which variable is smaller:

$$X_i = \min\{T_i, C_i\}, \qquad \delta_i = I_{(0,C_i)}(T_i), \qquad i = 1, ..., n.$$

This is called a *random censorship model* and C_i's are called *censoring times*. We consider the estimation of the survival distribution F; see Kalbfleisch and Prentice (1980) for other problems involving censored data.

An MELE of F can be derived as follows. Let $x_{(1)} \leq \cdots \leq x_{(n)}$ be ordered values of X_i's and $\delta_{(i)}$ be the δ-value associated with $x_{(i)}$. Consider a c.d.f. G that assigns its mass to the points $x_{(1)}, ..., x_{(n)}$ and the interval $(x_{(n)}, \infty)$. Let $p_i = P_G(\{x_{(i)}\})$, $i = 1, ..., n$, and $p_{n+1} = 1 - G(x_{(n)})$. An MELE of F is then obtained by maximizing

$$\ell(G) = \prod_{i=1}^{n} p_i^{\delta_{(i)}} \left(\sum_{j=i+1}^{n+1} p_j \right)^{1-\delta_{(i)}} \tag{5.22}$$

subject to

$$p_i \geq 0, \quad i = 1, ..., n+1, \qquad \sum_{i=1}^{n+1} p_i = 1. \tag{5.23}$$

It can be shown (exercise) that an MELE is

$$\hat{F}(t) = \sum_{i=1}^{n+1} \hat{p}_i I_{(0,t]}(X_{(i)}), \tag{5.24}$$

where $X_{(0)} = 0$, $X_{(n+1)} = \infty$, $X_{(1)} \leq \cdots \leq X_{(n)}$ are order statistics, and

$$\hat{p}_i = \frac{\delta_{(i)}}{n-i+1} \prod_{j=1}^{i-1} \left(1 - \frac{\delta_{(j)}}{n-j+1} \right), \quad i = 1, ..., n, \quad \hat{p}_{n+1} = 1 - \sum_{j=1}^{n} \hat{p}_j.$$

The $\hat{F}$ in (5.24) can also be written as (exercise)

$$\hat{F}(t) = 1 - \prod_{X_{(i)} \leq t} \left(1 - \tfrac{\delta_{(i)}}{n-i+1}\right), \tag{5.25}$$

which is the well-known Kaplan-Meier (1958) *product-limit* estimator. Some asymptotic results for $\hat{F}$ in (5.25) can be found, for example, in Shorack and Wellner (1986). ∎

5.1.3 Density estimation

Suppose that $X_1, ..., X_n$ are i.i.d. random variables from F and that F is unknown but has a Lebesgue p.d.f. f. Estimation of F can be done by estimating f, which is called *density estimation*. Note that estimators of F derived in §5.1.1 and §5.1.2 do not have Lebesgue p.d.f.'s.

Since $f(t) = F'(t)$ a.e., a simple estimator of $f(t)$ is the difference quotient

$$f_n(t) = \frac{F_n(t + \lambda_n) - F_n(t - \lambda_n)}{2\lambda_n}, \qquad t \in \mathcal{R}, \tag{5.26}$$

where F_n is the empirical c.d.f. given by (2.28) or (5.1) with $d = 1$, and $\{\lambda_n\}$ is a sequence of positive constants. Since $2n\lambda_n f_n(t)$ has the binomial distribution $Bi(F(t + \lambda_n) - F(t - \lambda_n), n)$,

$$E[f_n(t)] \to f(t) \qquad \text{if } \lambda_n \to 0 \text{ as } n \to \infty$$

and

$$\mathrm{Var}\big(f_n(t)\big) \to 0 \qquad \text{if } \lambda_n \to 0 \text{ and } n\lambda_n \to \infty.$$

Thus, we should choose λ_n converging to 0 slower than n^{-1}. If we assume that $\lambda_n \to 0$, $n\lambda_n \to \infty$, and f is continuously differentiable at t, then it can be shown (exercise) that

$$\mathrm{mse}_{f_n(t)}(F) = \frac{f(t)}{2n\lambda_n} + o\left(\frac{1}{n\lambda_n}\right) + O(\lambda_n^2) \tag{5.27}$$

and, under the additional condition that $n\lambda_n^3 \to 0$,

$$\sqrt{n\lambda_n}[f_n(t) - f(t)] \to_d N\big(0, \tfrac{1}{2}f(t)\big). \tag{5.28}$$

A useful class of estimators is the class of *kernel density estimators* of the form

$$\hat{f}(t) = \frac{1}{n\lambda_n} \sum_{i=1}^{n} w\left(\tfrac{t-X_i}{\lambda_n}\right), \tag{5.29}$$

where w is a known Lebesgue p.d.f. on $\mathcal{R}$ and is called the kernel. If we choose $w(t) = \frac{1}{2}I_{[-1,1]}(t)$, then $\hat{f}(t)$ in (5.29) is essentially the same as the so-called histogram. The bias of $\hat{f}(t)$ in (5.29) is

$$E[\hat{f}(t)] - f(t) = \frac{1}{\lambda_n} \int w\left(\frac{t-z}{\lambda_n}\right) f(z)dz - f(t)$$

$$= \int w(y)[f(t - \lambda_n y) - f(t)]dy.$$

If f is bounded and continuous at t, then, by the dominated convergence theorem (Theorem 1.1(iii)), the bias of $\hat{f}(t)$ converges to 0 as $\lambda_n \to 0$; if f' is bounded and continuous at t and $\int |t|w(t)dt < \infty$, then the bias of $\hat{f}(t)$ is $O(\lambda_n)$. The variance of $\hat{f}(t)$ is

$$\text{Var}\left(\hat{f}(t)\right) = \frac{1}{n\lambda_n^2}\text{Var}\left(w\left(\frac{t-X_1}{\lambda_n}\right)\right)$$

$$= \frac{1}{n\lambda_n^2}\int \left[w\left(\frac{t-z}{\lambda_n}\right)\right]^2 f(z)dz$$

$$- \frac{1}{n}\left[\frac{1}{\lambda_n}\int w\left(\frac{t-z}{\lambda_n}\right)f(z)dz\right]^2$$

$$= \frac{1}{n\lambda_n}\int [w(y)]^2 f(t - \lambda_n y)dy + O\left(\frac{1}{n}\right)$$

$$= \frac{w_0 f(t)}{n\lambda_n} + o\left(\frac{1}{n\lambda_n}\right)$$

if f is bounded and continuous at t and $w_0 = \int [w(t)]^2 dt < \infty$. Hence, if $\lambda_n \to 0$, $n\lambda_n \to \infty$, and f' is bounded and continuous at t, then

$$\text{mse}_{\hat{f}(t)}(F) = \frac{w_0 f(t)}{n\lambda_n} + O(\lambda_n^2).$$

Using the CLT (Theorem 1.15), one can show (exercise) that if $\lambda_n \to 0$, $n\lambda_n \to \infty$, f is bounded and continuous at t, and $\int [w(t)]^{2+\delta} dt < \infty$ for some $\delta > 0$, then

$$\sqrt{n\lambda_n}\{\hat{f}(t) - E[\hat{f}(t)]\} \to_d N\left(0, w_0 f(t)\right). \tag{5.30}$$

Furthermore, if f' is bounded and continuous at t and $n\lambda_n^3 \to 0$, then

$$\sqrt{n\lambda_n}\{E[\hat{f}(t)] - f(t)\} = O\left(\sqrt{n\lambda_n}\lambda_n\right) \to 0$$

and, therefore, (5.30) holds with $E[\hat{f}(t)]$ replaced by $f(t)$.

Similar to the estimation of a c.d.f., we can also study global properties of f_n or $\hat{f}$ as an estimator of the density curve f, using a suitably defined

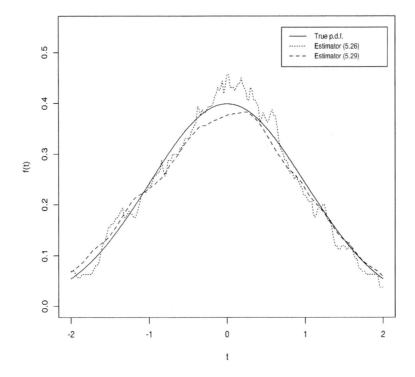

Figure 5.1: Density estimates in Example 5.4

distance between f and its density estimator. For example, we may study the convergence of $\sup_{t \in \mathcal{R}} |\hat{f}(t) - f(t)|$ or $\int |\hat{f}(t) - f(t)|^2 dt$. More details can be found, for example, in Silverman (1986).

Example 5.4. An i.i.d. sample of size $n = 200$ was generated from $N(0, 1)$. Density curve estimates (5.26) and (5.29) are plotted in Figure 5.1 with the curve of the true p.d.f. For the kernel density estimator (5.29), $w(t) = \frac{1}{2}e^{-|t|}$ is used and $\lambda_n = 0.4$. From Figure 5.1, it seems that the kernel estimate (5.29) is much better than the estimate (5.26). ∎

There are many other density estimation methods, for example, the nearest neighbor method (Stone, 1977), the smoothing splines (Wahba, 1990), and the method of empirical likelihoods described in §5.1.2 (see, e.g., Jones (1991)), which produces estimators of the form

$$\hat{f}(t) = \frac{1}{\lambda_n} \sum_{i=1}^{n} \hat{p}_i w \left(\frac{t - X_i}{\lambda_n} \right).$$

5.1.4 Semi-parametric methods

Suppose that the sample X is from a population in a family indexed by (θ, ξ), where θ is a parameter vector, i.e., $\theta \in \Theta \subset \mathcal{R}^k$ with a fixed positive integer k, but ξ is not vector-valued, e.g., ξ is a c.d.f. Such a model is often called a semi-parametric model, although it is nonparametric according to our definition in §2.1.2. A *semi-parametric method* refers to a statistical inference method that combines a parametric method and a nonparametric method in making an inference about the parametric component θ and the nonparametric component ξ. In the following, we consider two important examples of semi-parametric methods.

Partial likelihoods and proportional hazards models

The idea of *partial likelihood* (Cox, 1972) is similar to that of conditional likelihood introduced in §4.4.3. To illustrate this idea, we assume that X has a p.d.f. $f_{\theta,\xi}$ and ξ is also a vector-valued parameter. Suppose that X can be transformed into a sequence of pairs $(V_1, U_1), ..., (V_m, U_m)$ such that

$$f_{\theta,\xi}(x) = \left[\prod_{i=1}^{m} g_\theta(u_i|v_1, u_1, ..., u_{i-1}, v_i)\right]\left[\prod_{i=1}^{m} h_{\theta,\xi}(v_i|v_1, u_1, ..., v_{i-1}, u_{i-1})\right],$$

where $g_\theta(\cdot|v_1, u_1, ..., u_{i-1}, v_i)$ is the conditional p.d.f. of U_i given $V_1 = v_1, U_1 = u_1, ..., U_{i-1} = u_{i-1}, V_i = v_i$, which does not depend on ξ, and $h_{\theta,\xi}(\cdot|v_1, u_1, ..., v_{i-1}, u_{i-1})$ is the conditional p.d.f. of V_i given $V_1 = v_1, U_1 = u_1, ..., V_{i-1} = v_{i-1}, U_{i-1} = u_{i-1}$. The first product in the previous expression for $f_{\theta,\xi}(x)$ is called the partial likelihood for θ.

When ξ is a nonparametric component, the partial likelihood for θ can be similarly defined, in which case the full likelihood $f_{\theta,\xi}(x)$ should be replaced by a nonparametric likelihood or an empirical likelihood. As long as the conditional distributions of U_i given $V_1, U_1, ..., U_{i-1}, V_i$, $i = 1, ..., m$, are in a parametric family (indexed by θ), the partial likelihood is parametric.

A semi-parametric estimation method consists of a parametric method (typically the maximum likelihood method in §4.4) for estimating θ and a nonparametric method for estimating ξ.

To illustrate the application of the method of partial likelihoods, we consider the estimation of the c.d.f. of survival data in the random censorship model described in Example 5.3. Following the notation in Example 5.3, we assume that $\{T_1, ..., T_n\}$ (survival times) and $\{C_1, ..., C_n\}$ (censoring times) are two sets of independent nonnegative random variables and that $X_i = \min\{T_i, C_i\}$ and $\delta_i = I_{(0, C_i)}(T_i)$, $i = 1, ..., n$, are independent observations. In addition, we assume that there is a p-vector Z_i of covariate values associated with X_i and δ_i. The situation considered in Example 5.3

can be viewed as a special homogeneous case with $Z_i \equiv$ a constant.

The survival function when the covariate vector is equal to z is defined to be $S_z(t) = 1 - F_z(t)$, where F_z is the c.d.f. of the survival time T having the same distribution as T_i. Assume that $f_z(t) = F_z'(t)$ exists for all $t > 0$. The function $\lambda_z(t) = f_z(t)/S_z(t)$ is called the hazard function and the function $\Lambda_z(t) = \int_0^t \lambda_z(s)ds$ is called the cumulative hazard function, when the covariate vector is equal to z. A commonly adopted model for λ_z is the following proportional hazards model:

$$\lambda_z(t) = \lambda_0(t)\phi(\beta^\tau z), \tag{5.31}$$

where ϕ is a known function (typically $\phi(x) = e^x$), z is a value of the p-vector of covariates, $\beta \in \mathcal{R}^p$ is an unknown parameter vector, and $\lambda_0(t)$ is the unknown hazard function when the covariate vector is 0 and is referred to as the baseline hazard function. Under model (5.31),

$$1 - F_z(t) = \exp\{-\Lambda_z(t)\} = \exp\{\phi(\beta^\tau z)\Lambda_0(t)\}.$$

Thus, the estimation of the c.d.f. F_z or the survival function S_z can be done through the estimation of β, the parametric component of model (5.31), and Λ_0, the nonparametric component of model (5.31).

Consider first the estimation of β using the method of partial likelihoods. Suppose that there are l observed failures at times $T_{(1)} < \cdots < T_{(l)}$, where (i) is the label for the ith failure ordered according to the time to failure. (Note that a failure occurs when $\delta_i = 1$.) Suppose that there are m_i items censored at or after $T_{(i)}$ but before $T_{(i+1)}$ at times $T_{(i,1)}, ..., T_{(i,m_i)}$ (setting $T_{(0)} = 0$). Let $U_i = (i)$ and $V_i = (T_{(i)}, T_{(i-1,1)}, ..., T_{(i-1,m_{i-1})})$, $i = 1, ..., l$. Then the partial likelihood is

$$\prod_{i=1}^l P(U_i = (i)|V_1, U_1, ..., U_{i-1}, V_i).$$

Since $\lambda_z(t) = \lim_{\Delta>0, \Delta\to 0} \Delta^{-1} P_z(t \leq T < t + \Delta | T > t)$, where P_z denotes the probability measure of T when the covariate is equal to z,

$$P(U_i = (i)|V_1, U_1, ..., U_{i-1}, V_i) = \frac{\lambda_{Z_{(i)}}(t_i)}{\sum_{j \in R_i} \lambda_{Z_j}(t_i)} = \frac{\phi(\beta^\tau Z_{(i)})}{\sum_{j \in R_i} \phi(\beta^\tau Z_j)},$$

where t_i is the observed value of $T_{(i)}$, $R_i = \{j : X_j \geq t_i\}$ is called the risk set, and the last equality follows from assumption (5.31). This leads to the partial likelihood

$$\ell(\beta) = \prod_{i=1}^l \frac{\phi(\beta^\tau Z_{(i)})}{\sum_{j \in R_i} \phi(\beta^\tau Z_j)} = \prod_{i=1}^n \left[\frac{\phi(\beta^\tau Z_i)}{\sum_{j \in R_i} \phi(\beta^\tau Z_j)}\right]^{\delta_i},$$

which is a function of the parameter β, given the observed data. The maximum likelihood method introduced for parametric models in §4.4 can be applied to obtain a maximum partial likelihood estimator $\hat{\beta}$ of β. It is shown in Tsiatis (1981) that $\hat{\beta}$ is consistent for β and is asymptotically normal under some regularity conditions.

We now consider the estimation of Λ_0. First, assume that the covariate vector Z_i is random, (T_i, C_i, Z_i) are i.i.d., and T_i and C_i are conditionally independent given Z_i. Let (T, C, Z) be the random vector having the same distribution as (T_i, C_i, Z_i), $X = \min\{T, C\}$, and $\delta = I_{(0,C)}(T)$. Under assumption (5.31), it can be shown (exercise) that

$$Q(t) = P(X > t, \delta = 1) = \int\int_t^\infty \lambda_0(s)\phi(\beta^\tau z)H(s|z)dsdG(z), \qquad (5.32)$$

where $H(s|z) = P(X > s|Z = z)$ and G is the c.d.f. of Z. Then

$$\frac{dQ(t)}{dt} = -\lambda_0(t)\int \phi(\beta^\tau z)H(t|z)dG(z) \qquad (5.33)$$

and

$$\lambda_0(t) = -\frac{dQ(t)}{dt}\frac{1}{K(t)}, \qquad (5.34)$$

where $K(t) = E[\phi(\beta^\tau Z)I_{(t,\infty)}(X)]$ (exercise). Consequently,

$$\Lambda_0(t) = \int_0^t \lambda_0(s)ds = -\int_0^t \frac{dQ(s)}{K(s)}.$$

An estimator of Λ_0 can then be obtained by substituting Q and K in the previous expression by their estimators

$$\hat{Q}(t) = \frac{1}{n}\sum_{i=1}^n I_{\{X_i > t, \delta_i = 1\}}$$

and

$$\hat{K}(t) = \frac{1}{n}\sum_{i=1}^n \phi(\hat{\beta}^\tau Z_i)I_{(t,\infty)}(X_i). \qquad (5.35)$$

This estimator is known as Breslow's estimator. When $Z_1, ..., Z_n$ are non-random, we can still use Breslow estimator. Its asymptotic properties can be found, for example, in Fleming and Harrington (1991).

Profile likelihoods

Let $\ell(\theta, \xi)$ be a likelihood (or empirical likelihood), where θ and ξ are not necessarily vector-valued. It may be difficult to maximize the likelihood $\ell(\theta, \xi)$ simultaneously over θ and ξ. For each fixed θ, let $\xi(\theta)$ satisfy

$$\ell(\theta, \xi(\theta)) = \sup_\xi \ell(\theta, \xi).$$

The function
$$\ell_P(\theta) = \ell(\theta, \xi(\theta))$$
is called a *profile likelihood* function for θ. Suppose that $\hat{\theta}_P$ maximizes $\ell_P(\theta)$. Then $\hat{\theta}_P$ is called a maximum profile likelihood estimator of θ. Note that $\hat{\theta}_P$ may be different from an MLE of θ. Although this idea can be applied to parametric models, it is more useful in nonparametric models, especially when θ is a parametric component.

For example, consider the empirical likelihood in (5.8) subject to the constraints in (5.10). Sometimes it is more convenient to allow the function u in (5.10) to depend on an unknown parameter vector $\theta \in \mathcal{R}^k$, where $k \leq s$. This leads to the empirical likelihood $\ell(G)$ in (5.8) subject to (5.10) with $u(x)$ replaced by $\psi(x, \theta)$, where ψ is a known function from $\mathcal{R}^d \times \mathcal{R}^k$ to $\mathcal{R}^s$. Maximizing this empirical likelihood is equivalent to maximizing

$$\ell(p_1, ..., p_n, \omega, \lambda, \theta) = \prod_{i=1}^{n} p_i + \omega \left(1 - \sum_{i=1}^{n} p_i \right) + \sum_{i=1}^{n} p_i \lambda^\tau \psi(x_i, \theta),$$

where ω and λ are Lagrange multipliers. It follows from (5.12) and (5.13) that $\omega = n$, $\tilde{p}_i(\theta) = n^{-1}\{1 + [\lambda_n(\theta)]^\tau \psi(x_i, \theta)\}^{-1}$ with a $\lambda_n(\theta)$ satisfying

$$\frac{1}{n} \sum_{i=1}^{n} \frac{\psi(x_i, \theta)}{1 + [\lambda_n(\theta)]^\tau \psi(x_i, \theta)} = 0$$

maximize $\ell(p_1, ...p_n, \omega, \lambda, \theta)$ for any fixed θ. Substituting $\tilde{p}_i$ with $\sum_{i=1}^{n} \tilde{p}_i = 1$ into $\ell(p_1, ...p_n, \omega, \lambda, \theta)$ leads to the following profile empirical likelihood for θ:

$$\ell_P(\theta) = \prod_{i=1}^{n} \frac{1}{n\{1 + [\lambda_n(\theta)]^\tau \psi(x_i, \theta)\}}. \tag{5.36}$$

If $\hat{\theta}$ is a maximum of $\ell_P(\theta)$ in (5.36), then $\hat{\theta}$ is a maximum profile empirical likelihood estimator of θ and the corresponding estimator of p_i is $\tilde{p}_i(\hat{\theta})$. A result similar to Theorem 5.4 and a result on asymptotic normality of $\hat{\theta}$ are established in Qin and Lawless (1994), under some conditions on ψ.

Another example is the empirical likelihood (5.20) in the problem of biased sampling with a function $w(x) = w_\theta(x)$ depending on an unknown $\theta \in \mathcal{R}^k$. The profile empirical likelihood for θ is then

$$\ell_P(\theta) = \hat{w}_\theta^{-n_2} \prod_{i=1}^{n} \frac{1}{n_1 + n_2 w_\theta(x_i)/\hat{w}_\theta} \prod_{i=n_1+1}^{n} w_\theta(x_i),$$

where $\hat{w}_\theta$ satisfies

$$\hat{w}_\theta = \sum_{i=1}^{n} \frac{w_\theta(x_i)}{n_1 + n_2 w_\theta(x_i)/\hat{w}_\theta}.$$

Finally, we consider the problem of missing data. Assume that $X_1, ..., X_n$ are i.i.d. random variables from an unknown c.d.f. F and some X_i's are missing. Let $\delta_i = 1$ if X_i is observed and $\delta_i = 0$ if X_i is missing. Suppose that (X_i, δ_i) are i.i.d. Let

$$\pi(x) = P(\delta_i = 1 | X_i = x).$$

If X_i and δ_i are independent, i.e., $\pi(x) \equiv \pi$ does not depend on x, then the empirical c.d.f. based on observed data, i.e., the c.d.f. putting mass r^{-1} to each observed X_i, where r is the number of observed X_i's, is an unbiased and consistent estimator of F, provided that $\pi > 0$. On the other hand, if $\pi(x)$ depends on x, then the empirical c.d.f. based on observed data is a biased and inconsistent estimator of F. In fact, it can be shown (exercise) that the empirical c.d.f. based on observed data is an unbiased estimator of $P(X_i \leq x | \delta_i = 1)$, which is generally different from the unconditional probability $F(x) = P(X_i \leq x)$.

If both π and F are in parametric models, then we can apply the method of maximum likelihood. For example, if $\pi(x) = \pi_\theta(x)$ and $F(x) = F_\vartheta(x)$ has a p.d.f. f_ϑ, where θ and ϑ are vectors of unknown parameters, then a parametric likelihood of (θ, ϑ) is

$$\ell(\theta, \vartheta) = \prod_{i=1}^{n} [\pi_\theta(x_i) f_\vartheta(x_i)]^{\delta_i} (1 - \pi)^{1-\delta_i},$$

where $\pi = \int \pi_\theta(x) dF(x)$. Suppose now that $\pi(x) = \pi_\theta(x)$ is the parametric component and F is the nonparametric component. Then an empirical likelihood can be defined as

$$\ell(\theta, G) = \prod_{i=1}^{n} [\pi_\theta(x_i) P_G(\{x_i\})]^{\delta_i} (1 - \pi)^{1-\delta_i}$$

subject to $p_i \geq 0$, $\sum_{i=1}^{n} \delta_i p_i = 1$, $\sum_{i=1}^{n} \delta_i p_i [\pi_\theta(x_i) - \pi] = 0$, where $p_i = P_G(\{x_i\})$, $i = 1, ..., n$.

It can be shown (exercise) that the logarithm of the profile empirical likelihood for (θ, π) (with a Lagrange multiplier) is

$$\sum_{i=1}^{n} \left\{ \delta_i \log \left(\pi_\theta(x_i) \right) + (1 - \delta_i) \log(1 - \pi) - \delta_i \log \left(1 + \lambda[\pi_\theta(x_i) - \pi] \right) \right\}. \quad (5.37)$$

Under some regularity conditions, Qin, Leung, and Shao (2002) show that the estimators $\hat{\theta}$, $\hat{\pi}$, and $\hat{\lambda}$ obtained by maximizing the likelihood in (5.37) are consistent and asymptotically normal and that the empirical c.d.f. putting mass $\hat{p}_i = r^{-1}\{1 + \hat{\lambda}[\pi_{\hat{\theta}}(X_i) - \hat{\pi}]\}^{-1}$ to each observed X_i is consistent for F. The results are also extended to the case where a covariate vector Z_i associated with X_i is observed for all i.

5.2 Statistical Functionals

In many nonparametric problems, we are interested in estimating some characteristics (parameters) of the unknown population, not the entire population. We assume in this section that X_i's are i.i.d. from an unknown c.d.f. F on $\mathcal{R}^d$. Most characteristics of F can be written as $\mathrm{T}(F)$, where T is a functional from $\mathcal{F}$ to $\mathcal{R}^s$. If we estimate F by the empirical c.d.f. F_n in (5.1), then a natural estimator of $\mathrm{T}(F)$ is $\mathrm{T}(F_n)$, which is called a *statistical functional*.

Many commonly used statistics can be written as $\mathrm{T}(F_n)$ for some T. Two simple examples are given as follows. Let $\mathrm{T}(F) = \int \psi(x) dF(x)$ with an integrable function ψ, and $\mathrm{T}(F_n) = \int \psi(x) dF_n(x) = n^{-1} \sum_{i=1}^n \psi(X_i)$. The sample moments discussed in §3.5.2 are particular examples of this kind of statistical functional. For $d = 1$, let $\mathrm{T}(F) = F^{-1}(p) = \inf\{x : F(x) \geq p\}$, where $p \in (0, 1)$ is a fixed constant. $F^{-1}(p)$ is called the pth *quantile* of F. The statistical functional $\mathrm{T}(F_n) = F_n^{-1}(p)$ is called the pth *sample quantile*. More examples of statistical functionals are provided in §5.2.1 and §5.2.2.

In this section, we study asymptotic distributions of $\mathrm{T}(F_n)$. We focus on the case of real-valued T ($s = 1$), since the extension to the case of $s \geq 2$ is straightforward.

5.2.1 Differentiability and asymptotic normality

Note that $\mathrm{T}(F_n)$ is a function of the "statistic" F_n. In Theorem 1.12 (and §3.5.1) we have studied how to use Taylor's expansion to establish asymptotic normality of differentiable functions of statistics that are asymptotically normal. This leads to the approach of establishing asymptotic normality of $\mathrm{T}(F_n)$ by using some generalized Taylor expansions for functionals and using asymptotic properties of F_n given in §5.1.1.

First, we need a suitably defined differential of T. Several versions of differentials are given in the following definition.

Definition 5.2. Let T be a functional on $\mathcal{F}_0$, a collection of c.d.f.'s on $\mathcal{R}^d$, and let $\mathcal{D} = \{c(G_1 - G_2) : c \in \mathcal{R}, \ G_j \in \mathcal{F}_0, \ j = 1, 2\}$.
(i) A functional T on $\mathcal{F}_0$ is Gâteaux differentiable at $G \in \mathcal{F}_0$ if and only if there is a linear functional L_G on $\mathcal{D}$ (i.e., $\mathrm{L}_G(c_1\Delta_1 + c_2\Delta_2) = c_1\mathrm{L}_G(\Delta_1) + c_2\mathrm{L}_G(\Delta_2)$ for any $\Delta_j \in \mathcal{D}$ and $c_j \in \mathcal{R}$) such that $\Delta \in \mathcal{D}$ and $G + t\Delta \in \mathcal{F}_0$ imply

$$\lim_{t \to 0} \left[\frac{\mathrm{T}(G + t\Delta) - \mathrm{T}(G)}{t} - \mathrm{L}_G(\Delta) \right] = 0.$$

(ii) Let ϱ be a distance on $\mathcal{F}_0$ induced by a norm $\| \cdot \|$ on $\mathcal{D}$. A functional T on $\mathcal{F}_0$ is ϱ-Hadamard differentiable at $G \in \mathcal{F}_0$ if and only if there is a linear functional L_G on $\mathcal{D}$ such that for any sequence of numbers $t_j \to 0$ and $\{\Delta, \Delta_j, j = 1, 2, ...\} \subset \mathcal{D}$ satisfying $\|\Delta_j - \Delta\| \to 0$ and $G + t_j\Delta_j \in \mathcal{F}_0$,

$$\lim_{j \to \infty} \left[\frac{T(G + t_j\Delta_j) - T(G)}{t_j} - L_G(\Delta_j) \right] = 0.$$

(iii) Let ϱ be a distance on $\mathcal{F}_0$. A functional T on $\mathcal{F}_0$ is ϱ-Fréchet differentiable at $G \in \mathcal{F}_0$ if and only if there is a linear functional L_G on $\mathcal{D}$ such that for any sequence $\{G_j\}$ satisfying $G_j \in \mathcal{F}_0$ and $\varrho(G_j, G) \to 0$,

$$\lim_{j \to \infty} \frac{T(G_j) - T(G) - L_G(G_j - G)}{\varrho(G_j, G)} = 0. \quad \blacksquare$$

The functional L_G is called the *differential* of T at G. If we define $h(t) = T(G + t\Delta)$, then the Gâteaux differentiability is equivalent to the differentiability of the function $h(t)$ at $t = 0$, and $L_G(\Delta)$ is simply $h'(0)$. Let δ_x denote the d-dimensional c.d.f. degenerated at the point x and $\phi_G(x) = L_G(\delta_x - G)$. Then $\phi_F(x)$ is called the *influence function* of T at F, which is an important tool in robust statistics (see Hampel (1974)).

If T is Gâteaux differentiable at F, then we have the following expansion (taking $t = n^{-1/2}$ and $\Delta = \sqrt{n}(F_n - F)$):

$$\sqrt{n}[T(F_n) - T(F)] = L_F(\sqrt{n}(F_n - F)) + R_n. \tag{5.38}$$

Since L_F is linear,

$$L_F(\sqrt{n}(F_n - F)) = \frac{1}{\sqrt{n}} \sum_{i=1}^{n} \phi_F(X_i) \to_d N(0, \sigma_F^2) \tag{5.39}$$

by the CLT, provided that

$$E[\phi_F(X_1)] = 0 \quad \text{and} \quad \sigma_F^2 = E[\phi_F(X_1)]^2 < \infty \tag{5.40}$$

(which is usually true when ϕ_F is bounded or when F has some finite moments). By Slutsky's theorem and (5.39),

$$\sqrt{n}[T(F_n) - T(F)] \to_d N(0, \sigma_F^2) \tag{5.41}$$

if R_n in (5.38) is $o_p(1)$.

Unfortunately, Gâteaux differentiability is too weak to be useful in establishing $R_n = o_p(1)$ (or (5.41)). This is why we need other types of differentiability. Hadamard differentiability, which is also referred to as compact differentiability, is clearly stronger than Gâteaux differentiability but weaker than Fréchet differentiability (exercise). For a given functional

T, we can first find L_G by differentiating $h(t) = T(G + t\Delta)$ at $t = 0$ and then check whether T is ϱ-Hadamard (or ϱ-Fréchet) differentiable with a given ϱ. The most commonly used distances on $\mathcal{F}_0$ are the sup-norm distance ϱ_∞ and the L_p distance ϱ_{L_p}. Their corresponding norms are given by (5.3) and (5.5), respectively.

Theorem 5.5. Let $X_1, ..., X_n$ be i.i.d. from a c.d.f. F on $\mathcal{R}^d$.
(i) If T is ϱ_∞-Hadamard differentiable at F, then R_n in (5.38) is $o_p(1)$.
(ii) If T is ϱ-Fréchet differentiable at F with a distance ϱ satisfying

$$\sqrt{n}\varrho(F_n, F) = O_p(1), \tag{5.42}$$

then R_n in (5.38) is $o_p(1)$.
(iii) In either (i) or (ii), if (5.40) is also satisfied, then (5.41) holds.
Proof. Part (iii) follows directly from (i) or (ii). The proof of (i) involves some high-level mathematics and is omitted; see, for example, Fernholz (1983). We now prove (ii). From Definition 5.2(iii), for any $\epsilon > 0$, there is a $\delta > 0$ such that $|R_n| < \epsilon\sqrt{n}\varrho(F_n, F)$ whenever $\varrho(F_n, F) < \delta$. Then

$$P\left(|R_n| > \eta\right) \leq P\left(\sqrt{n}\varrho(F_n, F) > \eta/\epsilon\right) + P\left(\varrho(F_n, F) \geq \delta\right)$$

for any $\eta > 0$, which implies

$$\limsup_n P\left(|R_n| > \eta\right) \leq \limsup_n P\left(\sqrt{n}\varrho(F_n, F) > \eta/\epsilon\right).$$

The result follows from (5.42) and the fact that ϵ can be made arbitrarily small. ∎

Since ϱ-Fréchet differentiability implies ϱ-Hadamard differentiability, Theorem 5.5(ii) is useful when ϱ is not the sup-norm distance. There are functionals that are not ϱ_∞-Hadamard differentiable (and hence not ϱ_∞-Fréchet differentiable). For example, if $d = 1$ and $T(G) = g(\int x dG)$ with a differentiable function g, then T is not necessarily ϱ_∞-Hadamard differentiable, but is ϱ_{L_1}-Fréchet differentiable (exercise).

From Theorem 5.2, condition (5.42) holds for ϱ_{L_p} under the moment conditions on F given in Theorem 5.2.

Note that if ϱ and $\tilde{\varrho}$ are two distances on $\mathcal{F}_0$ satisfying $\tilde{\varrho}(G_1, G_2) \leq c\varrho(G_1, G_2)$ for a constant c and all $G_j \in \mathcal{F}_0$, then $\tilde{\varrho}$-Hadamard (Fréchet) differentiability implies ϱ-Hadamard (Fréchet) differentiability. This suggests the use of the distance $\varrho_{\infty+p} = \varrho_\infty + \varrho_{L_p}$, which also satisfies (5.42) under the moment conditions in Theorem 5.2. The distance $\varrho_{\infty+p}$ is useful in some cases (Theorem 5.6).

A ϱ_∞-Hadamard differentiable T having a bounded and continuous influence function ϕ_F is *robust* in Hampel's sense (see, e.g., Huber (1981)).

This is motivated by the fact that the asymptotic behavior of $T(F_n)$ is determined by that of $L_F(F_n - F)$, and a small change in the sample, i.e., small changes in all x_i's (rounding, grouping) or large changes in a few x_i's (gross errors, blunders), will result in a small change of $T(F_n)$ if and only if ϕ_F is bounded and continuous.

We now consider some examples. For the sample moments related to functionals of the form $T(G) = \int \psi(x)dG(x)$, it is clear that T is a linear functional. Any linear functional is trivially ϱ-Fréchet differentiable for any ϱ. Next, if F is one-dimensional and $F'(x) > 0$ for all x, then the quantile functional $T(G) = G^{-1}(p)$ is ϱ_∞-Hadamard differentiable at F (Fernholz, 1983). Hence, Theorem 5.5 applies to these functionals. But the asymptotic normality of sample quantiles can be established under weaker conditions, which are studied in §5.3.1.

Example 5.5 (Convolution functionals). Suppose that F is on $\mathcal{R}$ and for a fixed $z \in \mathcal{R}$,

$$T(G) = \int G(z - y)dG(y), \qquad G \in \mathcal{F}.$$

If X_1 and X_2 are i.i.d. with c.d.f. G, then $T(G)$ is the c.d.f. of $X_1 + X_2$ (Exercise 47 in §1.6), and is also called the convolution of G evaluated at z. For $t_j \to 0$ and $\|\Delta_j - \Delta\|_\infty \to 0$,

$$T(G + t_j\Delta_j) - T(G) = 2t_j \int \Delta_j(z - y)dG(y) + t_j^2 \int \Delta_j(z - y)d\Delta_j(y)$$

(for $\Delta = c_1 G_1 + c_2 G_2$, $G_j \in \mathcal{F}_0$, and $c_j \in \mathcal{R}$, $d\Delta$ denotes $c_1 dG_1 + c_2 dG_2$). Using Lemma 5.2, one can show (exercise) that

$$\int \Delta_j(z - y)d\Delta_j(y) = O(1). \tag{5.43}$$

Hence T is ϱ_∞-Hadamard differentiable at any $G \in \mathcal{F}$ with $L_G(\Delta) = 2\int \Delta(z-y)dG(y)$. The influence function, $\phi_F(x) = 2\int(\delta_x - F)(z-y)dF(y)$, is a bounded function and clearly satisfies (5.40). Thus, (5.41) holds. If F is continuous, then T is robust in Hampel's sense (exercise). ∎

Three important classes of statistical functionals, i.e., L-estimators, M-estimators, and rank statistics and R-estimators, are considered in §5.2.2.

Lemma 5.2. Let $\Delta \in \mathcal{D}$ and h be a continuous function on $\mathcal{R}$ such that $\int h(x)d\Delta(x)$ is finite. Then

$$\left| \int h(x)d\Delta(x) \right| \leq \|h\|_V \|\Delta\|_\infty,$$

where $\|h\|_V$ is the variation norm defined by

$$\|h\|_V = \lim_{a \to -\infty, b \to \infty} \left[\sup \sum_{j=1}^{m} |h(x_j) - h(x_{j-1})| \right]$$

with the supremum being taken over all partitions $a = x_0 < \cdots < x_m = b$ of the interval $[a, b]$. ∎

The proof of Lemma 5.2 can be found in Natanson (1961, p. 232).

The differentials in Definition 5.2 are first-order differentials. For some functionals, we can also consider their second-order differentials, which provides a way of defining the order of the asymptotic biases via expansion (2.37).

Definition 5.3. Let T be a functional on $\mathcal{F}_0$ and ϱ be a distance on $\mathcal{F}_0$.
(i) T is second-order ϱ-Hadamard differentiable at $G \in \mathcal{F}_0$ if and only if there is a functional Q_G on $\mathcal{D}$ such that for any sequence of numbers $t_j \to 0$ and $\{\Delta, \Delta_j, j = 1, 2, ...\} \subset \mathcal{D}$ satisfying $\|\Delta_j - \Delta\| \to 0$ and $G + t_j \Delta_j \in \mathcal{F}_0$,

$$\lim_{j \to \infty} \frac{T(G + t_j \Delta_j) - T(G) - Q_G(t_j \Delta_j)}{t_j^2} = 0,$$

where $Q_G(\Delta) = \int \int \psi_G(x, y) d(G + \Delta)(x) d(G + \Delta)(y)$ for a function ψ_G satisfying $\psi_G(x, y) = \psi_G(y, x)$, $\int \int \psi_G(x, y) dG(x) dG(y) = 0$, and $\mathcal{D}$ and $\| \cdot \|$ are the same as those in Definition 5.2(ii).
(ii) T is second-order ϱ-Fréchet differentiable at $G \in \mathcal{F}_0$ if and only if, for any sequence $\{G_j\}$ satisfying $G_j \in \mathcal{F}_0$ and $\varrho(G_j, G) \to 0$,

$$\lim_{j \to \infty} \frac{T(G_j) - T(G) - Q_G(G_j - G)}{[\varrho(G_j, G)]^2} = 0,$$

where Q_G is the same as that in (i). ∎

For a second-order differentiable T, we have the following expansion:

$$n[T(F_n) - T(F)] = nV_n + R_n, \qquad (5.44)$$

where

$$V_n = Q_F(F_n - F) = \int \int \psi_F(x, y) dF_n(x) dF_n(y) = \frac{1}{n^2} \sum_{j=1}^{n} \sum_{i=1}^{n} \psi_F(X_i, X_j)$$

is a "V-statistic" (§3.5.3) whose asymptotic properties are given by Theorem 3.16. If R_n in (5.44) is $o_p(1)$, then the asymptotic behavior of $T(F_n) - T(F)$ is the same as that of V_n.

Proposition 5.1. Let $X_1, ..., X_n$ be i.i.d. from F.

(i) If T is second-order ϱ_∞-Hadamard differentiable at F, then R_n in (5.44) is $o_p(1)$.

(ii) If T is second-order ϱ-Fréchet differentiable at F with a distance ϱ satisfying (5.42), then R_n in (5.44) is $o_p(1)$.   ∎

Combining Proposition 5.1 with Theorem 3.16, we conclude that if

$$\zeta_1 = \text{Var}\left(\int \psi_F(X_1, y)dF(y)\right) > 0,$$

then (5.41) holds with $\sigma_F^2 = 4\zeta_1$ and $\text{amse}_{T(F_n)}(P) = \sigma_F^2/n$; if $\zeta_1 = 0$, then

$$n[\text{T}(F_n) - \text{T}(F)] \to_d \sum_{j=1}^{\infty} \lambda_j \chi_{1j}^2$$

and $\text{amse}_{T(F_n)}(P) = \{2\text{Var}(\psi_F(X_1, X_2)) + [E\psi_F(X_1, X_1)]^2\}/n^2$. In any case, expansion (2.37) holds and the n^{-1} order asymptotic bias of $T(F_n)$ is $E\psi_F(X_1, X_1)/n$.

If T is also first-order differentiable, then it can be shown (exercise) that

$$\phi_F(x) = 2\int \psi_F(x, y)dF(y). \tag{5.45}$$

Then $\zeta_1 = 4^{-1}\text{Var}(\phi_F(X_1))$ and $\zeta_1 = 0$ corresponds to the case of $\phi_F(x) \equiv 0$. However, second-order ϱ-Hadamard (Fréchet) differentiability does not imply first-order ϱ-Hadamard (Fréchet) differentiability (exercise).

The technique in this section can be applied to non-i.i.d. X_i's when the c.d.f.'s of X_i's are determined by an unknown c.d.f. F, provided that results similar to (5.39) and (5.42) (with F_n replaced by some other estimator $\hat{F}$) can be established.

5.2.2 L-, M-, and R-estimators and rank statistics

Three large classes of statistical functionals based on i.i.d. X_i's are studied in this section.

L-estimators

Let $J(t)$ be a Borel function on $[0, 1]$. An *L-functional* is defined as

$$\text{T}(G) = \int xJ(G(x))dG(x), \qquad G \in \mathcal{F}_0, \tag{5.46}$$

where $\mathcal{F}_0$ contains all c.d.f.'s on $\mathcal{R}$ for which T is well defined. For $X_1, ..., X_n$ i.i.d. from $F \in \mathcal{F}_0$, $\text{T}(F_n)$ is called an *L-estimator* of $\text{T}(F)$.

Example 5.6. The following are some examples of commonly used L-estimators.

(i) When $J \equiv 1$, $T(F_n) = \bar{X}$, the sample mean.

(ii) When $J(t) = 4t - 2$, $T(F_n)$ is proportional to Gini's mean difference.

(iii) When $J(t) = (\beta - \alpha)^{-1} I_{(\alpha,\beta)}(t)$ for some constants $\alpha < \beta$, $T(F_n)$ is called the trimmed sample mean. ∎

For an L-functional T, it can be shown (exercise) that

$$T(G) - T(F) = \int \phi_F(x) d(G - F)(x) + R(G, F), \qquad (5.47)$$

where

$$\phi_F(x) = -\int (\delta_x - F)(y) J(F(y)) dy, \qquad (5.48)$$

$$R(G, F) = -\int W_G(x)[G(x) - F(x)] dx,$$

and

$$W_G(x) = \begin{cases} [G(x) - F(x)]^{-1} \int_{F(x)}^{G(x)} J(t) dt - J(F(x)) & G(x) \neq F(x) \\ 0 & G(x) = F(x). \end{cases}$$

A sufficient condition for (5.40) in this case is that F has a finite variance (exercise). However, (5.40) is also satisfied if ϕ_F is bounded. The differentiability of T can be verified under some conditions on J.

Theorem 5.6. Let T be an L-functional defined by (5.46).

(i) Suppose that J is bounded, $J(t) = 0$ when $t \in [0, \alpha] \cup [\beta, 1]$ for some constants $\alpha < \beta$, and that the set $D = \{x : J \text{ is discontinuous at } F(x)\}$ has Lebesgue measure 0. Then T is ϱ_∞-Fréchet differentiable at F with the influence function ϕ_F given by (5.48), and ϕ_F is bounded and continuous and satisfies (5.40).

(ii) Suppose that J is bounded, the set D in (i) has Lebesgue measure 0, and J is continuous on $[0, \alpha] \cup [\beta, 1]$ for some constants $\alpha < \beta$. Then T is $\varrho_{\infty+1}$-Fréchet differentiable at F.

(iii) Suppose that $|J(t) - J(s)| \leq C|t - s|^{p-1}$, where $C > 0$ and $p > 1$ are some constants. Then T is ϱ_{L_p}-Fréchet differentiable at F.

(iv) If, in addition to the conditions in part (i), J' is continuous on $[\alpha, \beta]$, then T is second-order ϱ_∞-Fréchet differentiable at F with

$$\psi_F(x, y) = \phi_F(x) + \phi_F(y) - \int (\delta_x - F)(z)(\delta_y - F)(z) J'(F(z)) dz.$$

(v) Suppose that J' is continuous on $[0, 1]$. Then T is second-order ϱ_{L_2}-Fréchet differentiable at F with the same ψ_F given in (iv).

Proof. We prove (i)-(iii). The proofs for (iv) and (v) are similar and are left to the reader.

(i) Let $G_j \in \mathcal{F}$ and $\varrho_\infty(G_j, F) \to 0$. Let c and d be two constants such that $F(c) > \beta$ and $F(d) < \alpha$. Then, for sufficiently large j, $G_j(x) \in [0, \alpha] \cup [\beta, 1]$ if $x > c$ or $x < d$. Hence, for sufficiently large j,

$$|R(G_j, F)| = \left| \int_d^c W_{G_j}(x)(G_j - F)(x) dx \right|$$

$$\leq \varrho_\infty(G_j, F) \int_d^c |W_{G_j}(x)| dx.$$

Since J is continuous at $F(x)$ when $x \notin D$ and D has Lebesgue measure 0, $W_{G_j}(x) \to 0$ a.e. Lebesgue. By the dominated convergence theorem, $\int_d^c |W_{G_j}(x)| dx \to 0$. This proves that T is ϱ_∞-Fréchet differentiable. The assertions on ϕ_F can be proved by noting that

$$\phi_F(x) = -\int_d^c (\delta_x - F)(y) J(F(y)) dy.$$

(ii) From the proof of (i), we only need to show that

$$\left| \int_A W_{G_j}(x)(G_j - F)(x) dx \right| \Big/ \varrho_{\infty+1}(G_j, F) \to 0, \tag{5.49}$$

where $A = \{x : F(x) \leq \alpha \text{ or } F(x) > \beta\}$. The quantity on the left-hand side of (5.49) is bounded by $\sup_{x \in A} |W_{G_j}(x)|$, which converges to 0 under the continuity assumption of J on $[0, \alpha] \cup [\beta, 1]$. Hence (5.49) follows.

(iii) The result follows from

$$|R(G, F)| \leq C \int |G(x) - F(x)|^p dx = O\left([\varrho_{L_p}(G, F)]^p\right)$$

and the fact that $p > 1$. ∎

An L-estimator with $J(t) = 0$ when $t \in [0, \alpha] \cup [\beta, 1]$ is called a trimmed L-estimator. Theorem 5.6(i) shows that trimmed L-estimators satisfy (5.41) and are robust in Hampel's sense. In cases (ii) and (iii) of Theorem 5.6, (5.41) holds if $\text{Var}(X_1) < \infty$, but $T(F_n)$ may not be robust in Hampel's sense. It can be shown (exercise) that one or several of (i)-(v) of Theorem 5.6 can be applied to each of the L-estimators in Example 5.6.

M-estimators

Let $\rho(x, t)$ be a Borel function on $\mathcal{R}^d \times \mathcal{R}$ and Θ be an open subset of $\mathcal{R}$. An *M-functional* is defined to be a solution of

$$\int \rho(x, T(G)) dG(x) = \min_{t \in \Theta} \int \rho(x, t) dG(x), \qquad G \in \mathcal{F}_0, \tag{5.50}$$

where $\mathcal{F}_0$ contains all c.d.f.'s on $\mathcal{R}^d$ for which the integrals in (5.50) are well defined. For $X_1, ..., X_n$ i.i.d. from $F \in \mathcal{F}_0$, $T(F_n)$ is called an *M-estimator* of $T(F)$. Assume that $\psi(x, t) = \partial\rho(x, t)/\partial t$ exists a.e. and

$$\lambda_G(t) = \int \psi(x, t)dG(x) = \frac{\partial}{\partial t}\int \rho(x, t)dG(x). \qquad (5.51)$$

Then $\lambda_G(T(G)) = 0$.

Example 5.7. The following are some examples of M-estimators.
(i) If $\rho(x, t) = (x - t)^2/2$, then $\psi(x, t) = t - x$; $T(G) = \int x dG(x)$ is the mean functional; and $T(F_n) = \bar{X}$ is the sample mean.
(ii) If $\rho(x, t) = |x - t|^p/p$, where $p \in [1, 2)$, then

$$\psi(x, t) = \begin{cases} |x - t|^{p-1} & x \le t \\ -|x - t|^{p-1} & x > t. \end{cases}$$

When $p = 1$, $T(F_n)$ is the sample median. When $1 < p < 2$, $T(F_n)$ is called the pth least absolute deviations estimator or the minimum L_p distance estimator.
(iii) Let $\mathcal{F}_0 = \{f_\theta : \theta \in \Theta\}$ be a parametric family of p.d.f.'s with $\Theta \subset \mathcal{R}$ and $\rho(x, t) = -\log f_t(x)$. Then $T(F_n)$ is an MLE. This indicates that M-estimators are extensions of MLE's in parametric models.
(iv) Let $C > 0$ be a constant. Huber (1964) considers

$$\rho(x, t) = \begin{cases} \frac{1}{2}(x - t)^2 & |x - t| \le C \\ \frac{1}{2}C^2 & |x - t| > C \end{cases}$$

with

$$\psi(x, t) = \begin{cases} t - x & |x - t| \le C \\ 0 & |x - t| > C. \end{cases}$$

The corresponding $T(F_n)$ is a type of trimmed sample mean.
(v) Let $C > 0$ be a constant. Huber (1964) considers

$$\rho(x, t) = \begin{cases} \frac{1}{2}(x - t)^2 & |x - t| \le C \\ C|x - t| - \frac{1}{2}C^2 & |x - t| > C \end{cases}$$

with

$$\psi(x, t) = \begin{cases} C & t - x > C \\ t - x & |x - t| \le C \\ -C & t - x < -C. \end{cases}$$

The corresponding $T(F_n)$ is a type of Winsorized sample mean.
(vi) Hampel (1974) considers $\psi(x, t) = \psi_0(t - x)$ with $\psi_0(s) = -\psi_0(-s)$ and

$$\psi_0(s) = \begin{cases} s & 0 \le s \le a \\ a & a < s \le b \\ \frac{a(c-s)}{c-b} & b < s \le c \\ 0 & s > c, \end{cases}$$

where $0 < a < b < c$ are constants. A smoothed version of ψ_0 is

$$\psi_1(s) = \begin{cases} \sin(as) & 0 \le s < \pi/a \\ 0 & s > \pi/a. \end{cases} \quad \blacksquare$$

For bounded and continuous ψ, the following result shows that T is ϱ_∞-Hadamard differentiable with a bounded and continuous influence function and, hence, $T(F_n)$ satisfies (5.41) and is robust in Hampel's sense.

Theorem 5.7. Let T be an M-functional defined by (5.50). Assume that ψ is a bounded and continuous function on $\mathcal{R}^d \times \mathcal{R}$ and that $\lambda_F(t)$ is continuously differentiable at $T(F)$ and $\lambda_F'(T(F)) \ne 0$. Then T is ϱ_∞-Hadamard differentiable at F with

$$\phi_F(x) = -\psi(x, T(F))/\lambda_F'(T(F)).$$

Proof. Let $t_j \to 0$, $\Delta_j \in \mathcal{D}$, $\|\Delta_j - \Delta\|_\infty \to 0$, and $G_j = F + t_j\Delta_j \in \mathcal{F}$. Since $\lambda_G(T(G)) = 0$,

$$|\lambda_F(T(G_j)) - \lambda_F(T(F))| = \left| t_j \int \psi(x, T(G_j))d\Delta_j(x) \right| \to 0$$

by $\|\Delta_j - \Delta\|_\infty \to 0$ and the boundedness of ψ. Note that $\lambda_F'(T(F)) \ne 0$. Hence, the inverse of $\lambda_F(t)$ exists and is continuous in a neighborhood of $0 = \lambda_F(T(F))$. Therefore,

$$T(G_j) - T(F) \to 0. \tag{5.52}$$

Let $h_F(T(F)) = \lambda_F'(T(F))$, $h_F(t) = [\lambda_F(t) - \lambda_F(T(F))]/[t - T(F)]$ if $t \ne T(F)$,

$$R_{1j} = \int \psi(x, T(F))d\Delta_j(x)\left[\frac{1}{\lambda_F'(T(F))} - \frac{1}{h_F(T(G_j))}\right],$$

$$R_{2j} = \frac{1}{h_F(T(G_j))} \int [\psi(x, T(G_j)) - \psi(x, T(F))]d\Delta_j(x),$$

and

$$L_F(\Delta) = -\frac{1}{\lambda_F'(T(F))} \int \psi(x, T(F))d\Delta(x), \qquad \Delta \in \mathcal{D}.$$

Then

$$T(G_j) - T(F) = -L_F(t_j\Delta_j) + t_j(R_{1j} - R_{2j}).$$

By (5.52), $\|\Delta_j - \Delta\|_\infty \to 0$, and the boundedness of ψ, $R_{j1} \to 0$. The result then follows from $R_{2j} \to 0$, which follows from $\|\Delta_j - \Delta\|_\infty \to 0$ and the boundedness and continuity of ψ (exercise). $\blacksquare$

Some ψ functions in Example 5.7 satisfy the conditions in Theorem 5.7 (exercise). Under more conditions on ψ, it can be shown that an M-functional is ϱ_∞-Fréchet differentiable at F (Clarke, 1986; Shao, 1993). Some M-estimators that satisfy (5.41) but are not differentiable functionals are studied in §5.4.

Rank statistics and R-estimators

Assume that $X_1, ..., X_n$ are i.i.d. from a c.d.f. F on $\mathcal{R}$. The *rank* of X_i among $X_1, ..., X_n$, denoted by R_i, is defined to be the number of X_j's satisfying $X_j \leq X_i$, $i = 1, ..., n$. The rank of $|X_i|$ among $|X_1|, ..., |X_n|$ is similarly defined and denoted by $\tilde{R}_i$. A statistic that is a function of R_i's or $\tilde{R}_i$'s is called a *rank statistic*. For $G \in \mathcal{F}$, let

$$\tilde{G}(x) = G(x) - G((-x)-), \qquad x > 0,$$

where $g(x-)$ denotes the left limit of the function g at x. Define a functional T by

$$T(G) = \int_0^\infty J(\tilde{G}(x))dG(x), \qquad G \in \mathcal{F}, \tag{5.53}$$

where J is a function on $[0, 1]$ with a bounded derivative J'. Then

$$T(F_n) = \int_0^\infty J(\tilde{F}_n(x))dF_n(x) = \frac{1}{n}\sum_{i=1}^n J\left(\frac{\tilde{R}_i}{n}\right) I_{(0,\infty)}(X_i)$$

is a (one-sample) signed rank statistic. If $J(t) = t$, then $T(F_n)$ is the well-known Wilcoxon signed rank test statistic (§6.5.1).

Statistics based on ranks (or signed ranks) are robust against changes in values of x_i's, but may not provide efficient inference procedures, since the values of x_i's are discarded after ranks (or signed ranks) are determined.

It can be shown (exercise) that T in (5.53) is ϱ_∞-Hadamard differentiable at F with the differential

$$L_F(\Delta) = \int_0^\infty J'(\tilde{F}(x))\tilde{\Delta}(x)dF(x) + \int_0^\infty J(\tilde{F}(x))d\Delta(x), \tag{5.54}$$

where $\Delta \in \mathcal{D}$ and $\tilde{\Delta}(x) = \Delta(x) - \Delta((-x)-)$.

These results can be extended to the case where $X_1, ..., X_n$ are i.i.d. from a c.d.f. F on $\mathcal{R}^2$. For any c.d.f. G on $\mathcal{R}^2$, let J be a function on $[0, 1]$ with $J(1 - t) = -J(t)$ and a bounded J',

$$\bar{G}(y) = [G(y, \infty) + G(\infty, y)]/2, \qquad y \in \mathcal{R},$$

and

$$T(G) = \int J(\bar{G}(y))dG(y, \infty). \tag{5.55}$$

Let $X_i = (Y_i, Z_i)$, R_i be the rank of Y_i, and U_i be the number of Z_j's satisfying $Z_j \leq Y_i$, $i = 1, ..., n$. Then

$$T(F_n) = \int J(\bar{F}_n(y))dF_n(y, \infty) = \frac{1}{n}\sum_{i=1}^{n} J\left(\frac{R_i + U_i}{2n}\right)$$

is called a two-sample linear rank statistic. It can be shown (exercise) that T in (5.55) is ϱ_∞-Hadamard differentiable at F with the differential

$$L_F(\Delta) = \int J'(\bar{F}(y))\bar{\Delta}(y)dF(y, \infty) + \int J(\bar{F}(y))d\Delta(y, \infty), \qquad (5.56)$$

where $\bar{\Delta}(y) = [\Delta(y, \infty) + \Delta(\infty, y)]/2$.

Rank statistics (one-sample or two-sample) are asymptotically normal and robust in Hampel's sense (exercise). These results are useful in testing hypotheses (§6.5).

Let F be a continuous c.d.f. on $\mathcal{R}$ symmetric about an unknown parameter $\theta \in \mathcal{R}$. An estimator of θ closely related to a rank statistic can be derived as follows. Let X_i be i.i.d. from F and $W_i = (X_i, 2t - X_i)$ with a fixed $t \in \mathcal{R}$. The functional T in (5.55) evaluated at the c.d.f. of W_i is equal to

$$\lambda_F(t) = \int J\left(\frac{F(x) + 1 - F(2t-x)}{2}\right) dF(x). \qquad (5.57)$$

If J is strictly increasing and F is strictly increasing in a neighborhood of θ, then $\lambda_F(t) = 0$ if and only if $t = \theta$ (exercise). For $G \in \mathcal{F}$, define T(G) to be a solution of

$$\int J\left(\frac{G(x) + 1 - G(2T(G)-x)}{2}\right) dG(x) = 0. \qquad (5.58)$$

T(F_n) is called an *R-estimator* of T$(F) = \theta$. When $J(t) = t - \frac{1}{2}$ (which is related to the Wilcoxon signed rank test), T(F_n) is the well-known Hodges-Lehmann estimator and is equal to any value between the two middle points of the values $(X_i + X_j)/2$, $i = 1, ..., n$, $j = 1, ..., n$.

Theorem 5.8. Let T be the functional defined by (5.58). Suppose that F is continuous and symmetric about θ, the derivatives F' and J' exist, and J' is bounded. Then T is ϱ_∞-Hadamard differentiable at F with the influence function

$$\phi_F(x) = \frac{J(F(x))}{\int J'(F(x))F'(x)dF(x)}.$$

Proof. Since F is symmetric about θ, $F(x) + F(2\theta - x) = 1$. Under the assumed conditions, $\lambda_F(t)$ is continuous and $\int J'(F(x))F'(x)dF(x) = -\lambda'_F(\theta) \neq 0$ (exercise). Hence, the inverse of λ_F exists and is continuous at $0 = \lambda_F(\theta)$. Suppose that $t_j \to 0$, $\Delta_j \in \mathcal{D}$, $\|\Delta_j - \Delta\|_\infty \to 0$, and $G_j = F + t_j \Delta_j \in \mathcal{F}$. Then

$$\int [J(G_j(x,t)) - J(F(x,t))]dG_j(x) \to 0$$

uniformly in t, where $G(x,t) = [G(x) + 1 - G(2t - x)]/2$, and

$$\int J(F(x,t))d(G_j - F)(x) = \int (F - G_j)(x)J'(F(x,t))dF(x,t) \to 0$$

uniformly in t. Let $\lambda_G(t)$ be defined by (5.57) with F replaced by G. Then

$$\lambda_{G_j}(t) - \lambda_F(t) \to 0$$

uniformly in t. Thus, $\lambda_F(\mathrm{T}(G_j)) \to 0$, which implies

$$\mathrm{T}(G_j) \to \mathrm{T}(F) = \theta. \qquad (5.59)$$

Let $\xi_G(t) = \int J(F(x,t))dG(x)$, $h_F(t) = [\lambda_F(t) - \lambda_F(\theta)]/(t - \theta)$ if $t \neq \theta$, and $h_F(\theta) = \lambda'_F(\theta)$. Then $\mathrm{T}(G_j) - \mathrm{T}(F) - \int \phi_F(x)d(G_j - F)(x)$ is equal to

$$\xi_{G_j}(\theta)\left[\frac{1}{\lambda'_F(\theta)} - \frac{1}{h_F(\mathrm{T}(G_j))}\right] + \frac{\lambda_F(\mathrm{T}(G_j)) - \xi_{G_j}(\theta)}{h_F(\mathrm{T}(G_j))}. \qquad (5.60)$$

Note that

$$\xi_{G_j}(\theta) = \int J(F(x))dG_j(x) = t_j \int J(F(x))d\Delta_j(x).$$

By (5.59), Lemma 5.2, and $\|\Delta_j - \Delta\|_\infty \to 0$, the first term in (5.60) is $o(t_j)$. The second term in (5.60) is the sum of

$$-\frac{t_j}{h_F(\mathrm{T}(G_j))}\int [J(F(x,\mathrm{T}(G_j))) - J(F(x))]d\Delta_j(x) \qquad (5.61)$$

and

$$\frac{1}{h_F(\mathrm{T}(G_j))}\int [J(F(x,\mathrm{T}(G_j))) - J(G_j(x,\mathrm{T}(G_j)))]dG_j(x). \qquad (5.62)$$

From the continuity of J and F, the quantity in (5.61) is $o(t_j)$. Similarly, the quantity in (5.62) is equal to

$$\frac{1}{h_F(\mathrm{T}(G_j))}\int [J(F(x,\mathrm{T}(G_j))) - J(G_j(x,\mathrm{T}(G_j)))]dF(x) + o(t_j). \qquad (5.63)$$

From Taylor's expansion, (5.59), and $\|\Delta_j - \Delta\|_\infty \to 0$, the quantity in (5.63) is equal to

$$\frac{t_j}{h_F(\mathrm{T}(G_j))}\int J'(F(x))\Delta(x,\theta)dF(x) + o(t_j). \qquad (5.64)$$

Since $J(1-t) = -J(t)$, the integral in (5.64) is 0. This proves that the second term in (5.60) is $o(t_j)$ and thus the result. ∎

It is clear that the influence function ϕ_F for an R-estimator is bounded and continuous if J and F are continuous. Thus, R-estimators satisfy (5.41) and are robust in Hampel's sense.

Example 5.8. Let $J(t) = t - \frac{1}{2}$. Then $\text{T}(F_n)$ is the Hodges-Lehmann estimator. From Theorem 5.8, $\phi_F(x) = [F(x) - \frac{1}{2}]/\gamma$, where $\gamma = \int F'(x)dF(x)$. Since $F(X_1)$ has a uniform distribution on $[0,1]$, $\phi_F(X_1)$ has mean 0 and variance $(12\gamma^2)^{-1}$. Thus, $\sqrt{n}[\text{T}(F_n) - \text{T}(F)] \to_d N(0, (12\gamma^2)^{-1})$. ∎

5.3 Linear Functions of Order Statistics

In this section, we study statistics that are linear functions of order statistics $X_{(1)} \leq \cdots \leq X_{(n)}$ based on independent random variables $X_1, ..., X_n$ (in §5.3.1 and §5.3.2, $X_1, ..., X_n$ are assumed i.i.d.). Order statistics, first introduced in Example 2.9, are usually sufficient and often complete (or minimal sufficient) for nonparametric families (Examples 2.12 and 2.14).

L-estimators defined in §5.2.2 are in fact linear functions of order statistics. If T is given by (5.46), then

$$\text{T}(F_n) = \int xJ(F_n(x))dF_n(x) = \frac{1}{n}\sum_{i=1}^{n} J\left(\tfrac{i}{n}\right) X_{(i)}, \qquad (5.65)$$

since $F_n(X_{(i)}) = i/n$, $i = 1, ..., n$. If J is a smooth function, such as those given in Example 5.6 or those satisfying the conditions in Theorem 5.6, the corresponding L-estimator is often called a smooth L-estimator. Asymptotic properties of smooth L-estimators can be obtained using Theorem 5.6 and the results in §5.2.1. Results on L-estimators that are slightly different from that in (5.65) can be found in Serfling (1980, Chapter 8).

In §5.3.1, we consider another useful class of linear functions of order statistics, the sample quantiles described in the beginning of §5.2. In §5.3.2, we study robust linear functions of order statistics (in Hampel's sense) and their relative efficiencies w.r.t. the sample mean $\bar{X}$, an efficient but nonrobust estimator. In §5.3.3, extensions to linear models are discussed.

5.3.1 Sample quantiles

Recall that $G^{-1}(p)$ is defined to be $\inf\{x : G(x) \geq p\}$ for any c.d.f. G on $\mathcal{R}$, where $p \in (0, 1)$ is a fixed constant. For i.i.d. $X_1, ..., X_n$ from F, let $\theta_p = F^{-1}(p)$ and $\hat{\theta}_p = F_n^{-1}(p)$ denote the pth quantile of F and the pth sample quantile, respectively. Then

$$\hat{\theta}_p = c_{np}X_{(m_p)} + (1 - c_{np})X_{(m_p+1)}, \tag{5.66}$$

where m_p is the integer part of np, $c_{np} = 1$ if np is an integer, and $c_{np} = 0$ if np is not an integer. Thus, $\hat{\theta}_p$ is a linear function of order statistics.

Note that $F(\theta_p-) \leq p \leq F(\theta_p)$. If F is not flat in a neighborhood of θ_p, then $F(\theta_p - \epsilon) < p < F(\theta_p + \epsilon)$ for any $\epsilon > 0$.

Theorem 5.9. Let $X_1, ..., X_n$ be i.i.d. random variables from a c.d.f. F satisfying $F(\theta_p - \epsilon) < p < F(\theta_p + \epsilon)$ for any $\epsilon > 0$. Then, for every $\epsilon > 0$ and $n = 1, 2, ...,$

$$P\big(|\hat{\theta}_p - \theta_p| > \epsilon\big) \leq 2Ce^{-2n\delta_\epsilon^2}, \tag{5.67}$$

where δ_ϵ is the smaller of $F(\theta_p + \epsilon) - p$ and $p - F(\theta_p - \epsilon)$ and C is the same constant in Lemma 5.1(i).
Proof. Let $\epsilon > 0$ be fixed. Note that $G(x) \geq t$ if and only if $x \geq G^{-1}(t)$ for any c.d.f. G on $\mathcal{R}$ (exercise). Hence

$$\begin{aligned}
P\big(\hat{\theta}_p > \theta_p + \epsilon\big) &= P\big(p > F_n(\theta_p + \epsilon)\big) \\
&= P\big(F(\theta_p + \epsilon) - F_n(\theta_p + \epsilon) > F(\theta_p + \epsilon) - p\big) \\
&\leq P\big(\varrho_\infty(F_n, F) > \delta_\epsilon\big) \\
&\leq Ce^{-2n\delta_\epsilon^2},
\end{aligned}$$

where the last inequality follows from DKW's inequality (Lemma 5.1(i)). Similarly,

$$P\big(\hat{\theta}_p < \theta_p - \epsilon\big) \leq Ce^{-2n\delta_\epsilon^2}.$$

This proves (5.67). ∎

Result (5.67) implies that $\hat{\theta}_p$ is strongly consistent for θ_p (exercise) and that $\hat{\theta}_p$ is $\sqrt{n}$-consistent for θ_p if $F'(\theta_p-)$ and $F'(\theta_p+)$ (the left and right derivatives of F at θ_p) exist (exercise).

The exact distribution of $\hat{\theta}_p$ can be obtained as follows. Since $nF_n(t)$ has the binomial distribution $Bi(F(t), n)$ for any $t \in \mathcal{R}$,

$$\begin{aligned}
P\big(\hat{\theta}_p \leq t\big) &= P\big(F_n(t) \geq p\big) \\
&= \sum_{i=m_p}^{n} \binom{n}{i} [F(t)]^i [1 - F(t)]^{n-i}, \tag{5.68}
\end{aligned}$$

where m_p is given in (5.66). If F has a Lebesgue p.d.f. f, then $\hat{\theta}_p$ has the Lebesgue p.d.f.

$$\varphi_n(t) = n\binom{n-1}{m_p-1} [F(t)]^{m_p-1} [1 - F(t)]^{n-m_p} f(t). \tag{5.69}$$

The following result provides an asymptotic distribution for $\sqrt{n}(\hat{\theta}_p - \theta_p)$.

Theorem 5.10. Let $X_1, ..., X_n$ be i.i.d. random variables from F.
(i) $P(\sqrt{n}(\hat{\theta}_p - \theta_p) \leq 0) \to \Phi(0) = \frac{1}{2}$, where Φ is the c.d.f. of the standard normal.
(ii) If F is continuous at θ_p and there exists $F'(\theta_p-) > 0$, then

$$P(\sqrt{n}(\hat{\theta}_p - \theta_p) \leq t) \to \Phi(t/\sigma_F^-), \qquad t < 0,$$

where $\sigma_F^- = \sqrt{p(1-p)}/F'(\theta_p-)$.
(iii) If F is continuous at θ_p and there exists $F'(\theta_p+) > 0$, then

$$P(\sqrt{n}(\hat{\theta}_p - \theta_p) \leq t) \to \Phi(t/\sigma_F^+), \qquad t > 0,$$

where $\sigma_F^+ = \sqrt{p(1-p)}/F'(\theta_p+)$.
(iv) If $F'(\theta_p)$ exists and is positive, then

$$\sqrt{n}(\hat{\theta}_p - \theta_p) \to_d N(0, \sigma_F^2), \qquad (5.70)$$

where $\sigma_F = \sqrt{p(1-p)}/F'(\theta_p)$.
Proof. The proof of (i) is left as an exercise. Part (iv) is a direct consequence of (i)-(iii) and the proofs of (ii) and (iii) are similar. Thus, we only give a proof for (iii).

Let $t > 0$, $p_{nt} = F(\theta_p + t\sigma_F^+ n^{-1/2})$, $c_{nt} = \sqrt{n}(p_{nt} - p)/\sqrt{p_{nt}(1 - p_{nt})}$, and $Z_{nt} = [B_n(p_{nt}) - np_{nt}]/\sqrt{np_{nt}(1 - p_{nt})}$, where $B_n(q)$ denotes a random variable having the binomial distribution $Bi(q, n)$. Then

$$P(\hat{\theta}_p \leq \theta_p + t\sigma_F^+ n^{-1/2}) = P(p \leq F_n(\theta_p + t\sigma_F^+ n^{-1/2}))$$
$$= P(Z_{nt} \geq -c_{nt}).$$

Under the assumed conditions on F, $p_{nt} \to p$ and $c_{nt} \to t$. Hence, the result follows from

$$P(Z_{nt} < -c_{nt}) - \Phi(-c_{nt}) \to 0.$$

But this follows from the CLT (Example 1.33) and Pólya's theorem (Proposition 1.16). ∎

If both $F'(\theta_p-)$ and $F'(\theta_p+)$ exist and are positive, but $F'(\theta_p-) \neq F'(\theta_p+)$, then the asymptotic distribution of $\sqrt{n}(\hat{\theta}_p - \theta_p)$ has the c.d.f. $\Phi(t/\sigma_F^-)I_{(-\infty,0)}(t) + \Phi(t/\sigma_F^+)I_{[0,\infty)}(t)$, a mixture of two normal distributions. An example of such a case when $p = \frac{1}{2}$ is

$$F(x) = xI_{[0,\frac{1}{2})}(x) + (2x - \tfrac{1}{2})I_{[\frac{1}{2},\frac{3}{4})}(x) + I_{[\frac{3}{4},\infty)}(x).$$

When $F'(\theta_p-) = F'(\theta_p+) = F'(\theta_p) > 0$, (5.70) shows that the asymptotic distribution of $\sqrt{n}(\hat{\theta}_p-\theta_p)$ is the same as that of $\sqrt{n}[F_n(\theta_p)-F(\theta_p)]/F'(\theta_p)$ (see (5.2)). The following result reveals a stronger relationship between sample quantiles and the empirical c.d.f.

Theorem 5.11 (Bahadur's representation). Let $X_1, ..., X_n$ be i.i.d. random variables from F. Suppose that $F'(\theta_p)$ exists and is positive. Then

$$\hat{\theta}_p = \theta_p + \frac{F(\theta_p) - F_n(\theta_p)}{F'(\theta_p)} + o_p\left(\frac{1}{\sqrt{n}}\right). \tag{5.71}$$

Proof. Let $t \in \mathcal{R}$, $\theta_{nt} = \theta_p + tn^{-1/2}$, $Z_n(t) = \sqrt{n}[F(\theta_{nt})-F_n(\theta_{nt})]/F'(\theta_p)$, and $U_n(t) = \sqrt{n}[F(\theta_{nt}) - F_n(\hat{\theta}_p)]/F'(\theta_p)$. It can be shown (exercise) that

$$Z_n(t) - Z_n(0) = o_p(1). \tag{5.72}$$

Note that $|p - F_n(\hat{\theta}_p)| \leq n^{-1}$. Then

$$\begin{aligned}
U_n(t) &= \sqrt{n}[F(\theta_{nt}) - p + p - F_n(\hat{\theta}_p)]/F'(\theta_p) \\
&= \sqrt{n}[F(\theta_{nt}) - p]/F'(\theta_p) + O(n^{-1/2}) \\
&\to t.
\end{aligned} \tag{5.73}$$

Let $\xi_n = \sqrt{n}(\hat{\theta}_p - \theta_p)$. Then, for any $t \in \mathcal{R}$ and $\epsilon > 0$,

$$\begin{aligned}
P(\xi_n \leq t, Z_n(0) \geq t + \epsilon) &= P(Z_n(t) \leq U_n(t), Z_n(0) \geq t + \epsilon) \\
&\leq P(|Z_n(t) - Z_n(0)| \geq \epsilon/2) \\
&\quad + P(|U_n(t) - t| \geq \epsilon/2) \\
&\to 0
\end{aligned} \tag{5.74}$$

by (5.72) and (5.73). Similarly,

$$P(\xi_n \geq t + \epsilon, Z_n(0) \leq t) \to 0. \tag{5.75}$$

It follows from the result in Exercise 128 of §1.6 that

$$\xi_n - Z_n(0) = o_p(1),$$

which is the same as (5.71). ∎

If F has a positive Lebesgue p.d.f., then $\hat{\theta}_p$ viewed as a statistical functional (§5.2) is ϱ_∞-Hadamard differentiable at F (Fernholz, 1983) with the influence function

$$\phi_F(x) = [F(\theta_p) - I_{(-\infty,\theta_p]}(x)]/F'(\theta_p).$$

This implies result (5.71). Note that ϕ_F is bounded and is continuous except when $x = \theta_p$.

Corollary 5.1. Let $X_1, ..., X_n$ be i.i.d. random variables from F having positive derivatives at θ_{p_j}, where $0 < p_1 < \cdots < p_m < 1$ are fixed constants. Then
$$\sqrt{n}[(\hat{\theta}_{p_1}, ..., \hat{\theta}_{p_m}) - (\theta_{p_1}, ..., \theta_{p_m})] \to_d N_m(0, D),$$
where D is the $m \times m$ symmetric matrix whose (i, j)th element is
$$p_i(1 - p_j)/[F'(\theta_{p_i})F'(\theta_{p_j})], \qquad i \leq j. \quad \blacksquare$$

The proof of this corollary is left to the reader.

Example 5.9 (Interquartile range). One application of Corollary 5.1 is the derivation of the asymptotic distribution of the *interquartile range* $\hat{\theta}_{0.75} - \hat{\theta}_{0.25}$. The interquartile range is used as a measure of the variability among X_i's. It can be shown (exercise) that
$$\sqrt{n}[(\hat{\theta}_{0.75} - \hat{\theta}_{0.25}) - (\theta_{0.75} - \theta_{0.25})] \to_d N(0, \sigma_F^2)$$
with
$$\sigma_F^2 = \frac{3}{16[F'(\theta_{0.75})]^2} + \frac{3}{16[F'(\theta_{0.25})]^2} - \frac{1}{8F'(\theta_{0.75})F'(\theta_{0.25})}. \quad \blacksquare$$

There are some applications of using extreme order statistics such as $X_{(1)}$ and $X_{(n)}$. One example is given in Example 2.34. Some other examples and references can be found in Serfling (1980, pp. 89-91).

5.3.2 Robustness and efficiency

Let F be a c.d.f. on $\mathcal{R}$ symmetric about $\theta \in \mathcal{R}$ with $F'(\theta) > 0$. Then $\theta = \theta_{0.5}$ and is called the *median* of F. If F has a finite mean, then θ is also equal to the mean. In this section, we consider the estimation of θ based on i.i.d. X_i's from F.

If F is normal, it has been shown in previous chapters that the sample mean $\bar{X}$ is the UMVUE, MRIE, and MLE of θ, and is asymptotically efficient. On the other hand, if F is the c.d.f. of the Cauchy distribution $C(\theta, 1)$, it follows from Exercise 78 in §1.6 that $\bar{X}$ has the same distribution as X_1, i.e., $\bar{X}$ is as variable as X_1, and is inconsistent as an estimator of θ.

Why does $\bar{X}$ perform so differently? An important difference between the normal and Cauchy p.d.f.'s is that the former tends to 0 at the rate

$e^{-x^2/2}$ as $|x| \to \infty$, whereas the latter tends to 0 at the much slower rate x^{-2}, which results in $\int |x| dF(x) = \infty$. The poor performance of $\bar{X}$ in the Cauchy case is due to the high probability of getting extreme observations and the fact that $\bar{X}$ is sensitive to large changes in a few of the X_i's. (Note that $\bar{X}$ is not robust in Hampel's sense, since the functional $\int x dG(x)$ has an unbounded influence function at F.) This suggests the use of a robust estimator that discards some extreme observations. The *sample median*, which is defined to be the 50%th sample quantile $\hat{\theta}_{0.5}$ described in §5.3.1, is insensitive to the behavior of F as $|x| \to \infty$.

Since both the sample mean and the sample median can be used to estimate θ, a natural question is when is one better than the other, using a criterion such as the amse. Unfortunately, a general answer does not exist, since the asymptotic relative efficiency between these two estimators depends on the unknown distribution F. If F does not have a finite variance, then $\text{Var}(\bar{X}) = \infty$ and $\bar{X}$ may be inconsistent. In such a case the sample median is certainly preferred, since $\hat{\theta}_{0.5}$ is consistent and asymptotically normal as long as $F'(\theta) > 0$, and may have a finite variance (Exercise 60). The following example, which compares the sample mean and median in some cases, shows that the sample median can be better even if $\text{Var}(X_1) < \infty$.

Example 5.10. Suppose that $\text{Var}(X_1) < \infty$. Then, by the CLT,

$$\sqrt{n}(\bar{X} - \theta) \to_d N(0, \text{Var}(X_1)).$$

By Theorem 5.10(iv),

$$\sqrt{n}(\hat{\theta}_{0.5} - \theta) \to_d N(0, [2F'(\theta)]^{-2}).$$

Hence, the asymptotic relative efficiency of $\hat{\theta}_{0.5}$ w.r.t. $\bar{X}$ is

$$e(F) = 4[F'(\theta)]^2 \text{Var}(X_1).$$

(i) If F is the c.d.f. of $N(\theta, \sigma^2)$, then $\text{Var}(X_1) = \sigma^2$, $F'(\theta) = (\sqrt{2\pi}\sigma)^{-1}$, and $e(F) = 2/\pi = 0.637$.
(ii) If F is the c.d.f. of the logistic distribution $LG(\theta, \sigma)$, then $\text{Var}(X_1) = \sigma^2\pi^2/3$, $F'(\theta) = (4\sigma)^{-1}$, and $e(F) = \pi^2/12 = 0.822$.
(iii) If $F(x) = F_0(x - \theta)$ and F_0 is the c.d.f. of the t-distribution t_ν with $\nu \geq 3$, then $\text{Var}(X_1) = \nu/(\nu - 2)$, $F'(\theta) = \Gamma(\frac{\nu+1}{2})/[\sqrt{\nu\pi}\Gamma(\frac{\nu}{2})]$, $e(F) = 1.62$ when $\nu = 3$, $e(F) = 1.12$ when $\nu = 4$, and $e(F) = 0.96$ when $\nu = 5$.
(iv) If F is the c.d.f. of the double exponential distribution $DE(\theta, \sigma)$, then $F'(\theta) = (2\sigma)^{-1}$ and $e(F) = 2$.
(v) Consider the Tukey model

$$F(x) = (1 - \epsilon)\Phi\left(\frac{x-\theta}{\sigma}\right) + \epsilon\Phi\left(\frac{x-\theta}{\tau\sigma}\right), \tag{5.76}$$

where $\sigma > 0$, $\tau > 0$, and $0 < \epsilon < 1$. Then $\text{Var}(X_1) = (1 - \epsilon)\sigma^2 + \epsilon\tau^2\sigma^2$, $F'(\theta) = (1 - \epsilon + \epsilon/\tau)/(\sqrt{2\pi}\sigma)$, and $e(F) = 2(1 - \epsilon + \epsilon\tau^2)(1 - \epsilon + \epsilon/\tau)^2/\pi$. Note that $\lim_{\epsilon \to 0} e(F) = 2/\pi$ and $\lim_{\tau \to \infty} e(F) = \infty$. $\blacksquare$

Since the sample median uses at most two actual values of x_i's, it may go too far in discarding observations, which results in a possible loss of efficiency. The trimmed sample mean introduced in Example 5.6(iii) is a natural compromise between the sample mean and median. Since F is symmetric, we consider $\beta = 1 - \alpha$ in the trimmed mean, which results in the following L-estimator:

$$\bar{X}_\alpha = \frac{1}{n - 2m_\alpha} \sum_{j=m_\alpha+1}^{n-m_\alpha} X_{(j)}, \tag{5.77}$$

where m_α is the integer part of $n\alpha$ and $\alpha \in (0, \frac{1}{2})$. The estimator in (5.77) is called the α-trimmed sample mean. It discards the m_α smallest and m_α largest observations. The sample mean and median can be viewed as two extreme cases of $\bar{X}_\alpha$ as $\alpha \to 0$ and $\frac{1}{2}$, respectively.

It follows from Theorem 5.6 that if $F(x) = F_0(x - \theta)$, where F_0 is symmetric about 0 and has a Lebesgue p.d.f. positive in the range of X_1, then

$$\sqrt{n}(\bar{X}_\alpha - \theta) \to_d N(0, \sigma_\alpha^2), \tag{5.78}$$

where

$$\sigma_\alpha^2 = \frac{2}{(1 - 2\alpha)^2} \left\{ \int_0^{F_0^{-1}(1-\alpha)} x^2 dF_0(x) + \alpha[F_0^{-1}(1 - \alpha)]^2 \right\}.$$

Lehmann (1983, §5.4) provides various values of the asymptotic relative efficiency $e_{\bar{X}_\alpha, \bar{X}}(F) = \text{Var}(X_1)/\sigma_\alpha^2$. For instance, when $F(x) = F_0(x - \theta)$ and F_0 is the c.d.f. of the t-distribution t_3, $e_{\bar{X}_\alpha, \bar{X}}(F) = 1.70$, 1.91, and 1.97 for $\alpha = 0.05$, 0.125, and 0.25, respectively; when F is given by (5.76) with $\tau = 3$ and $\epsilon = 0.05$, $e_{\bar{X}_\alpha, \bar{X}}(F) = 1.20$, 1.19, and 1.09 for $\alpha = 0.05$, 0.125, and 0.25, respectively; when F is given by (5.76) with $\tau = 3$ and $\epsilon = 0.01$, $e_{\bar{X}_\alpha, \bar{X}}(F) = 1.04$, 0.98, and 0.89 for $\alpha = 0.05$, 0.125, and 0.25, respectively.

Robustness and efficiency of other L-estimators can be discussed similarly. For an L-estimator $\text{T}(F_n)$ with T given by (5.46), if the conditions in one of (i)-(iii) of Theorem 5.6 are satisfied, then (5.41) holds with

$$\sigma_F^2 = \int_{-\infty}^{\infty} \int_{-\infty}^{\infty} J(F(x))J(F(y))[F(\min\{x, y\}) - F(x)F(y)]dxdy, \tag{5.79}$$

provided that $\sigma_F^2 < \infty$ (exercise). If F is symmetric about θ and J is symmetric about $\frac{1}{2}$, then $\text{T}(F) = \theta$ (exercise) and, therefore, the asymptotic relative efficiency of $\text{T}(F_n)$ w.r.t. $\bar{X}$ is $\text{Var}(X_1)/\sigma_F^2$.

5.3.3 L-estimators in linear models

In this section, we extend L-estimators to the following linear model:

$$X_i = \beta^\tau Z_i + \varepsilon_i, \qquad i = 1, ..., n, \tag{5.80}$$

with i.i.d. ε_i's having an unknown c.d.f. F_0 and a full rank Z whose ith row is the vector Z_i. Note that the c.d.f. of X_i is $F_0(x - \beta^\tau Z_i)$. Instead of assuming $E(\varepsilon_i) = 0$ (as we did in Chapter 3), we assume that

$$\int xJ(F_0(x))dF_0(x) = 0, \tag{5.81}$$

where J is a Borel function on $[0, 1]$ (the same as that in (5.46)). Note that (5.81) may hold without any assumption on the existence of $E(\varepsilon_i)$. For instance, (5.81) holds if F_0 is symmetric about 0 and J is symmetric about $\frac{1}{2}$ (Exercise 69).

Since X_i's are not identically distributed, the use of the order statistics and the empirical c.d.f. based on $X_1, ..., X_n$ may not be appropriate. Instead, we consider the ordered values of residuals $r_i = X_i - Z_i^\tau \hat{\beta}$, $i = 1, ..., n$, and some empirical c.d.f.'s based on residuals, where $\hat{\beta} = (Z^\tau Z)^{-1} Z^\tau X$ is the LSE of β (§3.3.1).

To illustrate the idea, let us start with the case where β and Z_i are univariate. First, assume that $Z_i \geq 0$ for all i (or $Z_i \leq 0$ for all i). Let $\hat{F}_0$ be the c.d.f. putting mass $Z_i / \sum_{i=1}^n Z_i$ at r_i, $i = 1, ..., n$. An L-estimator of β is defined to be

$$\hat{\beta}_L = \hat{\beta} + \int xJ(\hat{F}_0(x))d\hat{F}_0(x) \sum_{i=1}^n Z_i \Big/ \sum_{i=1}^n Z_i^2.$$

When $J(t) = (1 - 2\alpha)^{-1} I_{(\alpha, 1-\alpha)}(t)$ with an $\alpha \in (0, \frac{1}{2})$, $\hat{\beta}_L$ is similar to the α-trimmed sample mean in the i.i.d. case.

If not all Z_i's have the same sign, we can define L-estimators as follows. Let $Z_i^+ = \max\{Z_i, 0\}$ and $Z_i^- = Z_i^+ - Z_i$. Let $\hat{F}_0^\pm$ be the c.d.f. putting mass $Z_i^\pm / \sum_{i=1}^n Z_i^\pm$ at r_i, $i = 1, ..., n$. An L-estimator of β is defined to be

$$\hat{\beta}_L = \hat{\beta} + \int xJ(\hat{F}_0^+(x))d\hat{F}_0^+(x) \sum_{i=1}^n Z_i^+ \Big/ \sum_{i=1}^n Z_i^2$$

$$- \int xJ(\hat{F}_0^-(x))d\hat{F}_0^-(x) \sum_{i=1}^n Z_i^- \Big/ \sum_{i=1}^n Z_i^2.$$

For a general p-vector Z_i, let z_{ij} be the jth component of Z_i, $j = 1, ..., p$. Let $z_{ij}^+ = \max\{z_{ij}, 0\}$, $z_{ij}^- = z_{ij}^+ - z_{ij}$, and $\hat{F}_{0j}^\pm$ be the c.d.f. putting mass

$z_{ij}^{\pm} / \sum_{i=1}^{n} z_{ij}^{\pm}$ at r_i, $i = 1, ..., n$. For any j, if $z_{ij} \geq 0$ for all i (or $z_{ij} \leq 0$ for all i), then we set $\hat{F}_{0j}^{+} \equiv 0$ (or $\hat{F}_{0j}^{-} \equiv 0$). An L-estimator of β is defined to be

$$\hat{\beta}_L = \hat{\beta} + (Z^{\tau} Z)^{-1}(A^+ - A^-), \tag{5.82}$$

where

$$A^{\pm} = \left(\int x J(\hat{F}_{01}^{\pm}(x)) d\hat{F}_{01}^{\pm}(x) \sum_{i=1}^{n} z_{i1}^{\pm}, ..., \int x J(\hat{F}_{0p}^{\pm}(x)) d\hat{F}_{0p}^{\pm}(x) \sum_{i=1}^{n} z_{ip}^{\pm} \right).$$

Obviously, $\hat{\beta}_L$ in (5.82) reduces to the previously defined $\hat{\beta}_L$ when β and Z_i are univariate.

Theorem 5.12. Assume model (5.80) with i.i.d. ε_i's from a c.d.f. F_0 satisfying (5.81) for a given J. Suppose that F_0 has a uniformly continuous, positive, and bounded derivative on the range of ε_1. Suppose further that the conditions on Z_i's in Theorem 3.12 are satisfied.
(i) If the function J is continuous on (α_1, α_2) and equals 0 on $[0, \alpha_1] \cup [\alpha_2, 1]$, where $0 < \alpha_1 < \alpha_2 < 1$ are constants, then

$$\sigma_{F_0}^{-1}(Z^{\tau} Z)^{1/2}(\hat{\beta}_L - \beta) \to_d N_p(0, I_p), \tag{5.83}$$

where $\sigma_{F_0}^2$ is given by (5.79) with $F = F_0$.
(ii) Result (5.83) also holds if J' is bounded on $[0, 1]$, $E|\varepsilon_1| < \infty$, and $\sigma_{F_0}^2$ is finite. ∎

The proof of this theorem can be found in Bickel (1973). Robustness and efficiency comparisons between the LSE $\hat{\beta}$ and L-estimators $\hat{\beta}_L$ can be made in a way similar to those in §5.3.2.

5.4 Generalized Estimating Equations

The method of *generalized estimating equations* (GEE) is a powerful and general method of deriving point estimators, which includes many previously described methods as special cases. In §5.4.1, we begin with a description of this method and, to motivate the idea, we discuss its relationship with other methods that have been studied. Consistency and asymptotic normality of estimators derived from generalized estimating equations are studied in §5.4.2 and §5.4.3.

Throughout this section, we assume that $X_1, ..., X_n$ are independent (not necessarily identically distributed) random vectors, where the dimension of X_i is d_i, $i = 1, ..., n$ ($\sup_i d_i < \infty$), and that we are interested in estimating θ, a k-vector of unknown parameters related to the unknown population.

5.4.1 The GEE method and its relationship with others

The sample mean and, more generally, the LSE in linear models are solutions of equations of the form

$$\sum_{i=1}^{n}(X_i - \gamma^\tau Z_i)Z_i = 0.$$

Also, MLE's (or RLE's) in §4.4 and, more generally, M-estimators in §5.2.2 are solutions to equations of the form

$$\sum_{i=1}^{n}\psi(X_i, \gamma) = 0.$$

This leads to the following general estimation method. Let $\Theta \subset \mathcal{R}^k$ be the range of θ, ψ_i be a Borel function from $\mathcal{R}^{d_i} \times \Theta$ to $\mathcal{R}^k$, $i = 1, ..., n$, and

$$s_n(\gamma) = \sum_{i=1}^{n}\psi_i(X_i, \gamma), \qquad \gamma \in \Theta. \tag{5.84}$$

If θ is estimated by $\hat{\theta} \in \Theta$ satisfying $s_n(\hat{\theta}) = 0$, then $\hat{\theta}$ is called a GEE estimator. The equation $s_n(\gamma) = 0$ is called a GEE. Apparently, the LSE's, RLE's, MQLE's, and M-estimators are special cases of GEE estimators.

Usually GEE's are chosen so that

$$E[s_n(\theta)] = \sum_{i=1}^{n}E[\psi_i(X_i, \theta)] = 0, \tag{5.85}$$

where the expectation E may be replaced by an asymptotic expectation defined in §2.5.2 if the exact expectation does not exist. If this is true, then $\hat{\theta}$ is motivated by the fact that $s_n(\hat{\theta}) = 0$ is a sample analogue of $E[s_n(\theta)] = 0$.

To motivate the idea, let us study the relationship between the GEE method and other methods that have been introduced.

M-estimators

The M-estimators defined in §5.2.2 for univariate $\theta = \mathrm{T}(F)$ in the i.i.d. case are special cases of GEE estimators. Huber (1981) also considers regression M-estimators in the linear model (5.80). A regression M-estimator of β is defined as a solution to the GEE

$$\sum_{i=1}^{n}\psi(X_i - \gamma^\tau Z_i)Z_i = 0,$$

where ψ is one of the functions given in Example 5.7.

LSE's in linear and nonlinear regression models

Suppose that
$$X_i = f(Z_i, \theta) + \varepsilon_i, \qquad i = 1, ..., n, \tag{5.86}$$

where Z_i's are the same as those in (5.80), θ is an unknown k-vector of parameters, f is a known function, and ε_i's are independent random variables. Model (5.86) is the same as model (5.80) if f is linear in θ and is called a *nonlinear regression model* otherwise. Note that model (4.64) is a special case of model (5.86). The LSE under model (5.86) is any point in Θ minimizing $\sum_{i=1}^{n} [X_i - f(Z_i, \gamma)]^2$ over $\gamma \in \Theta$. If f is differentiable, then the LSE is a solution to the GEE

$$\sum_{i=1}^{n} [X_i - f(Z_i, \gamma)] \frac{\partial f(Z_i, \gamma)}{\partial \gamma} = 0.$$

Quasi-likelihoods

This is a continuation of the discussion of the quasi-likelihoods introduced in §4.4.3. Assume first that X_i's are univariate ($d_i \equiv 1$). If X_i's follow a GLM, i.e., X_i has the p.d.f. in (4.55) and (4.57) holds, and if (4.58) holds, then the likelihood equation (4.59) can be written as

$$\sum_{i=1}^{n} \frac{x_i - \mu_i(\gamma)}{v_i(\gamma)} G_i(\gamma) = 0, \tag{5.87}$$

where $\mu_i(\gamma) = \mu(\psi(\gamma^\tau Z_i))$, $G_i(\gamma) = \partial \mu_i(\gamma)/\partial \gamma$, $v_i(\gamma) = \text{Var}(X_i)/\phi$, and we have used the following fact:

$$\psi'(t) = (\mu^{-1})'(g^{-1}(t))(g^{-1})'(t) = (g^{-1})'(t)/\zeta''(\psi(t)).$$

Equation (5.87) is a quasi-likelihood equation if either X_i does not have the p.d.f. in (4.55) or (4.58) does not hold. Note that this generalizes the discussion in §4.4.3. If X_i does not have the p.d.f. in (4.55), then the problem is often nonparametric. Let $s_n(\gamma)$ be the left-hand side of (5.87). Then $s_n(\gamma) = 0$ is a GEE and $E[s_n(\beta)] = 0$ is satisfied as long as the first condition in (4.56), $E(X_i) = \mu_i(\beta)$, is satisfied.

For general d_i's, let $X_i = (X_{i1}, ..., X_{id_i})$, $i = 1, ..., n$, where each X_{it} satisfies (4.56) and (4.57), i.e.,

$$E(X_{it}) = \mu(\eta_{it}) = g^{-1}(\beta^\tau Z_{it}) \quad \text{and} \quad \text{Var}(X_{it}) = \phi_i \mu'(\eta_{it}),$$

and Z_{it}'s are k-vector values of covariates. In biostatistics and life-time testing problems, components of X_i are *repeated measurements* at different times from subject i and are called *longitudinal data*. Although X_i's are assumed independent, X_{it}'s are likely to be dependent for each i. Let R_i be the $d_i \times d_i$ correlation matrix whose (t, l)th element is the correlation coefficient between X_{it} and X_{il}. Then

$$\text{Var}(X_i) = \phi_i [D_i(\beta)]^{1/2} R_i [D_i(\beta)]^{1/2}, \tag{5.88}$$

where $D_i(\gamma)$ is the $d_i \times d_i$ diagonal matrix with the tth diagonal element $(g^{-1})'(\gamma^\tau Z_{it})$. If R_i's in (5.88) are known, then an extension of (5.87) to the multivariate x_i's is

$$\sum_{i=1}^{n} G_i(\gamma) \{[D_i(\gamma)]^{1/2} R_i [D_i(\gamma)]^{1/2}\}^{-1} [x_i - \mu_i(\gamma)] = 0, \tag{5.89}$$

where $\mu_i(\gamma) = (\mu(\psi(\gamma^\tau Z_{i1})), ..., \mu(\psi(\gamma^\tau Z_{id_i})))$ and $G_i(\gamma) = \partial \mu_i(\gamma)/\partial \gamma$. In most applications, R_i is unknown and its form is hard to model. Let $\tilde{R}_i$ be a known correlation matrix (called a *working correlation matrix*). Replacing R_i in (5.89) by $\tilde{R}_i$ leads to the quasi-likelihood equation

$$\sum_{i=1}^{n} G_i(\gamma) \{[D_i(\gamma)]^{1/2} \tilde{R}_i [D_i(\gamma)]^{1/2}\}^{-1} [x_i - \mu_i(\gamma)] = 0. \tag{5.90}$$

For example, we may assume that the components of X_i are independent and take $\tilde{R}_i = I_{d_i}$. Although the working correlation matrix $\tilde{R}_i$ may not be the same as the true unknown correlation matrix R_i, an MQLE obtained from (5.90) is still consistent and asymptotically normal (§5.4.2 and §5.4.3). Of course, MQLE's are asymptotically more efficient if $\tilde{R}_i$ is closer to R_i. Even if $\tilde{R}_i = R_i$ and $\phi_i \equiv \phi$, (5.90) is still a quasi-likelihood equation, since the covariance matrix of X_i cannot determine the distribution of X_i unless X_i is normal.

Since an $\tilde{R}_i$ closer to R_i results in a better MQLE, sometimes it is suggested to replace $\tilde{R}_i$ in (5.90) by $\hat{R}_i$, an estimator of R_i (Liang and Zeger, 1986). The resulting equation is called a *pseudo-likelihood equation*. As long as $\max_{i \leq n} \|\hat{R}_i - U_i\| \to_p 0$ as $n \to \infty$, where $\|A\| = \sqrt{\text{tr}(A^\tau A)}$ for a matrix A and U_i is a correlation matrix (not necessarily the same as R_i), $i = 1, ..., n$, MQLE's are consistent and asymptotically normal.

Empirical likelihoods

The previous discussion shows that the GEE method coincides with the method of deriving M-estimators, LSE's, MLE's, or MQLE's. The following discussion indicates that the GEE method is also closely related to the method of empirical likelihoods introduced in §5.1.4.

Assume that X_i's are i.i.d. from a c.d.f. F on $\mathcal{R}^d$ and $\psi_i = \psi$ for all i. Then condition (5.85) reduces to $E[\psi(X_1, \theta)] = 0$. Hence, we can consider the empirical likelihood

$$\ell(G) = \prod_{i=1}^{n} P_G(\{x_i\}), \qquad G \in \mathcal{F}$$

subject to

$$p_i \geq 0, \quad \sum_{i=1}^{n} p_i = 1, \quad \text{and} \quad \sum_{i=1}^{n} p_i \psi(x_i, \theta) = 0, \tag{5.91}$$

where $p_i = P_G(\{x_i\})$. However, in this case the dimension of the function ψ is the same as the dimension of the parameter θ and, hence, the last equation in (5.91) does not impose any restriction on p_i's. Then, it follows from Theorem 5.3 that $(p_1, ..., p_n) = (n^{-1}, ..., n^{-1})$ maximizes $\ell(G)$ for any fixed θ. Substituting $p_i = n^{-1}$ into the last equation in (5.91) leads to

$$\frac{1}{n} \sum_{i=1}^{n} \psi(x_i, \theta) = 0.$$

That is, any MELE $\hat{\theta}$ of θ is a GEE estimator.

5.4.2 Consistency of GEE estimators

We now study under what conditions (besides (5.85)) GEE estimators are consistent. For each n, let $\hat{\theta}_n$ be a GEE estimator, i.e., $s_n(\hat{\theta}_n) = 0$, where $s_n(\gamma)$ is defined by (5.84).

First, Theorem 5.7 and its proof can be extended to multivariate T in a straightforward manner. Hence, we have the following result.

Proposition 5.2. Suppose that $X_1, ..., X_n$ are i.i.d. from F and $\psi_i \equiv \psi$, a bounded and continuous function from $\mathcal{R}^d \times \Theta$ to $\mathcal{R}^k$. Let $\Psi(t) = \int \psi(x, t) dF(x)$. Suppose that $\Psi(\theta) = 0$ and $\partial \Psi(t)/\partial t$ exists and is of full rank at $t = \theta$. Then $\hat{\theta}_n \to_p \theta$. ∎

For unbounded ψ in the i.i.d. case, the following result and its proof can be found in Qin and Lawless (1994).

Proposition 5.3. Suppose that $X_1, ..., X_n$ are i.i.d. from F and $\psi_i \equiv \psi$. Assume that $\varphi(x, \gamma) = \partial \psi(x, \gamma)/\partial \gamma$ exists in N_θ, a neighborhood of θ, and is continuous at θ; there is a function $h(x)$ such that $\sup_{\gamma \in N_\theta} \|\varphi(x, \gamma)\| \leq h(x)$, $\sup_{\gamma \in N_\theta} \|\psi(x, \gamma)\|^3 \leq h(x)$, and $E[h(X_1)] < \infty$; $E[\varphi(X_1, \theta)]$ is of full rank; $E\{\psi(X_1, \theta)[\psi(X_1, \theta)]^\tau\}$ is positive definite; and (5.85) holds. Then, there exists a sequence of random vectors $\{\hat{\theta}_n\}$ such that

$$P\left(s_n(\hat{\theta}_n) = 0\right) \to 1 \quad \text{and} \quad \hat{\theta}_n \to_p \theta. \quad \blacksquare \tag{5.92}$$

Next, we consider non-i.i.d. X_i's.

Proposition 5.4. Suppose that $X_1, ..., X_n$ are independent and θ is univariate. Assume that $\psi_i(x, \gamma)$ is real-valued and nonincreasing in γ for all i; there is a $\delta > 0$ such that $\sup_i E|\psi_i(X_i, \gamma)|^{1+\delta} < \infty$ for any γ in N_θ, a neighborhood of θ (this condition can be replaced by $E|\psi(X_1, \gamma)| < \infty$ for any γ in N_θ when X_i's are i.i.d. and $\psi_i \equiv \psi$); $\psi_i(x, \gamma)$ are continuous in N_θ; (5.85) holds; and

$$\limsup_n E[\Psi_n(\theta + \epsilon)] < 0 < \liminf_n E[\Psi_n(\theta - \epsilon)] \tag{5.93}$$

for any $\epsilon > 0$, where $\Psi_n(\gamma) = n^{-1}s_n(\gamma)$. Then, there exists a sequence of random variables $\{\hat{\theta}_n\}$ such that (5.92) holds. Furthermore, any sequence $\{\hat{\theta}_n\}$ satisfying $s_n(\hat{\theta}_n) = 0$ satisfies (5.92).
Proof. Since ψ_i's are nonincreasing, the functions $\Psi_n(\gamma)$ and $E[\Psi_n(\gamma)]$ are nonincreasing. Let $\epsilon > 0$ be fixed so that $\theta \pm \epsilon \in N_\theta$. Under the assumed conditions,

$$\Psi_n(\theta \pm \epsilon) - E[\Psi_n(\theta \pm \epsilon)] \to_p 0$$

(Theorem 1.14(ii)). By condition (5.93),

$$P\big(\Psi_n(\theta + \epsilon) < 0 < \Psi_n(\theta - \epsilon)\big) \to 1.$$

The rest of the proof is left as an exercise.   ∎

To establish the next result, we need the following lemma. First, we need the following concept. A sequence of functions $\{g_i\}$ from $\mathcal{R}^k$ to $\mathcal{R}^k$ is called *equicontinuous* on an open set $\mathcal{O} \subset \mathcal{R}^k$ if and only if, for any $\epsilon > 0$, there is a $\delta_\epsilon > 0$ such that $\sup_i \|g_i(t) - g_i(s)\| < \epsilon$ whenever $t \in \mathcal{O}$, $s \in \mathcal{O}$, and $\|t - s\| < \delta_\epsilon$. Since a continuous function on a compact set is uniformly continuous, functions such as $g_i(\gamma) = g(t_i, \gamma)$ form an equicontinuous sequence on $\mathcal{O}$ if t_i's vary in a compact set containing $\mathcal{O}$ and $g(t, \gamma)$ is a continuous function in (t, γ).

Lemma 5.3. Suppose that Θ is a compact subset of $\mathcal{R}^k$. Let $h_i(X_i) = \sup_{\gamma \in \Theta} \|\psi_i(X_i, \gamma)\|$, $i = 1, 2, ...$ Suppose that $\sup_i E|h_i(X_i)|^{1+\delta} < \infty$ and $\sup_i E\|X_i\|^\delta < \infty$ for some $\delta > 0$ (this condition can be replaced by $E|h(X_1)| < \infty$ when X_i's are i.i.d. and $\psi_i \equiv \psi$). Suppose further that for any $c > 0$ and sequence $\{x_i\}$ satisfying $\|x_i\| \leq c$, the sequence of functions $\{g_i(\gamma) = \psi_i(x_i, \gamma)\}$ is equicontinuous on any open subset of Θ. Then

$$\sup_{\gamma \in \Theta} \left\| \frac{1}{n} \sum_{i=1}^n \{\psi_i(X_i, \gamma) - E[\psi_i(X_i, \gamma)]\} \right\| \to_p 0.$$

Proof. Since we only need to consider components of ψ_i's, without loss of generality we can assume that ψ_i's are functions from $\mathcal{R}^{d_i} \times \Theta$ to $\mathcal{R}$. For any $c > 0$,

$$\sup_n E\left[\frac{1}{n}\sum_{i=1}^n h_i(X_i)I_{(c,\infty)}(\|X_i\|)\right] \leq \sup_i E[h_i(X_i)I_{(c,\infty)}(\|X_i\|)].$$

Let $c_0 = \sup_i E|h_i(X_i)|^{1+\delta}$ and $c_1 = \sup_i E\|X_i\|^\delta$. By Hölder's inequality,

$$E[h_i(X_i)I_{(c,\infty)}(\|X_i\|)] \leq \left[E|h_i(X_i)|^{1+\delta}\right]^{1/(1+\delta)}[P(\|X_i\| > c)]^{\delta/(1+\delta)}$$
$$\leq c_0^{1/(1+\delta)}c_1^{\delta/(1+\delta)}c^{-\delta^2/(1+\delta)}$$

for all i. For $\epsilon > 0$ and $\tilde{\epsilon} > 0$, choose a c such that $c_0^{1/(1+\delta)}c_1^{\delta/(1+\delta)}c^{-\delta^2/(1+\delta)} < \epsilon\tilde{\epsilon}/2$. Then, for any $\mathcal{O} \subset \Theta$, the probability

$$P\left(\frac{1}{n}\sum_{i=1}^n \left\{\sup_{\gamma \in \mathcal{O}} \psi_i(X_i, \gamma) - \inf_{\gamma \in \mathcal{O}} \psi_i(X_i, \gamma)\right\} I_{(c,\infty)}(\|X_i\|) > \frac{\epsilon}{2}\right) \quad (5.94)$$

is bounded by $\tilde{\epsilon}$ (exercise). From the equicontinuity of $\{\psi_i(x_i, \gamma)\}$, there is a $\delta_\epsilon > 0$ such that

$$\frac{1}{n}\sum_{i=1}^n \left\{\sup_{\gamma \in \mathcal{O}_\epsilon} \psi_i(X_i, \gamma) - \inf_{\gamma \in \mathcal{O}_\epsilon} \psi_i(X_i, \gamma)\right\} I_{[0,c]}(\|X_i\|) < \frac{\epsilon}{2}$$

for sufficiently large n, where $\mathcal{O}_\epsilon$ denotes any open ball in $\mathcal{R}^k$ with radius less than δ_ϵ. These results, together with Theorem 1.14(ii) and the fact that $\|\psi_i(X_i, \gamma)\| \leq h_i(X_i)$, imply that

$$P\left(\frac{1}{n}\sum_{i=1}^n \left\{\sup_{\gamma \in \mathcal{O}_\epsilon} \psi_i(X_i, \gamma) - E\left[\inf_{\gamma \in \mathcal{O}_\epsilon} \psi_i(X_i, \gamma)\right]\right\} > \epsilon\right) \to 0. \quad (5.95)$$

Let $H_n(\gamma) = n^{-1}\sum_{i=1}^n \{\psi_i(X_i, \gamma) - E[\psi_i(X_i, \gamma)]\}$. Then

$$\sup_{\gamma \in \mathcal{O}_\epsilon} H_n(\gamma) \leq \frac{1}{n}\sum_{i=1}^n \left\{\sup_{\gamma \in \mathcal{O}_\epsilon} \psi_i(X_i, \gamma) - E\left[\inf_{\gamma \in \mathcal{O}_\epsilon} \psi_i(X_i, \gamma)\right]\right\},$$

which with (5.95) implies that

$$P\big(H_n(\gamma) > \epsilon \text{ for all } \gamma \in \mathcal{O}_\epsilon\big) = P\left(\sup_{\gamma \in \mathcal{O}_\epsilon} H_n(\gamma) > \epsilon\right) \to 0.$$

Similarly we can show that

$$P\big(H_n(\gamma) < -\epsilon \text{ for all } \gamma \in \mathcal{O}_\epsilon\big) \to 0.$$

Since Θ is compact, there exists m_ϵ open balls $\mathcal{O}_{\epsilon,j}$ such that $\Theta \subset \cup \mathcal{O}_{\epsilon,j}$. Then, the result follows from

$$P\left(\sup_{\gamma \in \Theta} |H_n(\gamma)| > \epsilon\right) \leq \sum_{j=1}^{m_\epsilon} P\left(\sup_{\gamma \in \mathcal{O}_{\epsilon,j}} |H_n(\gamma)| > \epsilon\right) \to 0. \quad \blacksquare$$

Example 5.11. Consider the quasi-likelihood equation (5.90). Let $\{\tilde{R}_i\}$ be a sequence of working correlation matrices and

$$\psi_i(x_i, \gamma) = G_i(\gamma)\{[D_i(\gamma)]^{1/2}\tilde{R}_i[D_i(\gamma)]^{1/2}\}^{-1}[x_i - \mu_i(\gamma)]. \tag{5.96}$$

It can be shown (exercise) that ψ_i's satisfy the conditions of Lemma 5.3 if Θ is compact and $\sup_i \|Z_i\| < \infty$. $\quad \blacksquare$

Proposition 5.5. Assume (5.85) and the conditions in Lemma 5.3 (with Θ replaced by any compact subset of the parameter space). Suppose that the functions $\Delta_n(\gamma) = E[n^{-1}s_n(\gamma)]$ have the property that $\lim_{n\to\infty} \Delta_n(\gamma) = 0$ if and only if $\gamma = \theta$. (If Δ_n converges to a function Δ, then this condition and (5.85) imply that Δ has a unique 0 at θ.) Suppose that $\{\hat{\theta}_n\}$ is a sequence of GEE estimators and that $\hat{\theta}_n = O_p(1)$. Then $\hat{\theta}_n \to_p \theta$.
Proof. First, assume that Θ is a compact subset of $\mathcal{R}^k$. From Lemma 5.3 and $s_n(\hat{\theta}_n) = 0$, $\Delta_n(\hat{\theta}_n) \to_p 0$. By Theorem 1.8(vi), there is a subsequence $\{n_i\}$ such that

$$\Delta_{n_i}(\hat{\theta}_{n_i}) \to_{a.s.} 0. \tag{5.97}$$

Let $x_1, x_2, \ldots$ be a fixed sequence such that (5.97) holds and let θ_0 be a limit point of $\{\hat{\theta}_n\}$. Since Θ is compact, $\theta_0 \in \Theta$ and there is a subsequence $\{m_j\} \subset \{n_i\}$ such that $\hat{\theta}_{m_j} \to \theta_0$. Using the argument in the proof of Lemma 5.3, it can be shown (exercise) that $\{\Delta_n(\gamma)\}$ is equicontinuous on any open subset of Θ. Then

$$\Delta_{m_j}(\hat{\theta}_{m_j}) - \Delta_{m_j}(\theta_0) \to 0,$$

which with (5.97) implies $\Delta_{m_j}(\theta_0) \to 0$. Under the assumed condition, $\theta_0 = \theta$. Since this is true for any limit point of $\{\hat{\theta}_n\}$, $\hat{\theta}_n \to_p \theta$.

Next, consider a general Θ. For any $\epsilon > 0$, there is an $M_\epsilon > 0$ such that $P(\|\hat{\theta}_n\| \leq M_\epsilon) > 1 - \epsilon$. The result follows from the previous proof by considering the closure of $\Theta \cap \{\gamma : \|\gamma\| \leq M_\epsilon\}$ as the parameter space. $\quad \blacksquare$

Condition $\hat{\theta}_n = O_p(1)$ in Proposition 5.5 is obviously necessary for the consistency of $\hat{\theta}_n$. It has to be checked in any particular problem.

If a GEE is a likelihood equation under some conditions, then we can often show, using an argument similar to the proof of Theorem 4.17 or 4.18, that there exists a consistent sequence of GEE estimators.

Proposition 5.6. Suppose that $s_n(\gamma) = \partial \log \ell_n(\gamma)/\partial \gamma$ for some function ℓ_n; $D_n(\theta) = \text{Var}(s_n(\theta)) \to 0$; $\varphi_i(x,\gamma) = \partial \psi_i(x,\gamma)/\partial \gamma$ exists and the sequence of functions $\{\varphi_{ij}, i = 1, 2, ...\}$ satisfies the conditions in Lemma 5.3 with Θ replaced by a compact neighborhood of θ, where φ_{ij} is the jth row of φ_i, $j = 1, ..., k$; $-\liminf_n [D_n(\theta)]^{1/2} E[\nabla s_n(\theta)][D_n(\theta)]^{1/2}$ is positive definite, where $\nabla s_n(\gamma) = \partial s_n(\gamma)/\partial \gamma$; and (5.85) holds. Then, there exists a sequence of estimators $\{\hat{\theta}_n\}$ satisfying (5.92). ∎

The proof of Proposition 5.6 is similar to that of Theorem 4.17 or Theorem 4.18 and is left as an exercise.

Example 5.12. Consider the quasi-likelihood equation (5.90) with $\tilde{R}_i = I_{d_i}$ for all i. Then the GEE is a likelihood equation under a GLM (§4.4.2) assumption. It can be shown (exercise) that the conditions of Proposition 5.6 are satisfied if $\sup_i \|Z_i\| < \infty$. ∎

5.4.3 Asymptotic normality of GEE estimators

Asymptotic normality of a consistent sequence of GEE estimators can be established under some conditions. We first consider the special case where θ is univariate and $X_1, ..., X_n$ are i.i.d.

Theorem 5.13. Let $X_1, ..., X_n$ be i.i.d. from F, $\psi_i \equiv \psi$, and $\theta \in \mathcal{R}$. Suppose that $\Psi(\gamma) = \int \psi(x,\gamma)dF(x) = 0$ if and only if $\gamma = \theta$, $\Psi'(\theta)$ exists and $\Psi'(\theta) \neq 0$.
(i) Assume that $\psi(x,\gamma)$ is nonincreasing in γ and that $\int [\psi(x,\gamma)]^2 dF(x)$ is finite for γ in a neighborhood of θ and is continuous at θ. Then, any sequence of GEE estimators (M-estimators) $\{\hat{\theta}_n\}$ satisfies

$$\sqrt{n}(\hat{\theta}_n - \theta) \to_d N(0, \sigma_F^2), \qquad (5.98)$$

where

$$\sigma_F^2 = \int [\psi(x,\theta)]^2 dF(x)/[\Psi'(\theta)]^2.$$

(ii) Assume that $\int [\psi(x,\theta)]^2 dF(x) < \infty$, $\psi(x,\gamma)$ is continuous in x, and $\lim_{\gamma \to \theta} \|\psi(\cdot,\gamma) - \psi(\cdot,\theta)\|_V = 0$, where $\|\cdot\|_V$ is the variation norm defined in Lemma 5.2. Then, any consistent sequence of GEE estimators $\{\hat{\theta}_n\}$ satisfies (5.98).
Proof. (i) Let $\Psi_n(\gamma) = n^{-1}s_n(\gamma)$. Since Ψ_n is nonincreasing,

$$P(\Psi_n(t) < 0) \leq P(\hat{\theta}_n \leq t) \leq P(\Psi_n(t) \leq 0)$$

for any $t \in \mathcal{R}$. Then, (5.98) follows from

$$\lim_{n \to \infty} P(\Psi_n(t_n) < 0) = \lim_{n \to \infty} P(\Psi_n(t_n) \leq 0) = \Phi(t)$$

for all $t \in \mathcal{R}$, where $t_n = \theta + t\sigma_F n^{-1/2}$. Let $s_{t,n}^2 = \mathrm{Var}(\psi(X_1, t_n))$ and $Y_{ni} = [\psi(X_i, t_n) - \Psi(t_n)]/s_{t,n}$. Then, it suffices to show that

$$\lim_{n \to \infty} P\left(\frac{1}{\sqrt{n}} \sum_{i=1}^n Y_{ni} \leq -\frac{\sqrt{n}\Psi(t_n)}{s_{t,n}}\right) = \Phi(t)$$

for all t. Under the assumed conditions, $\sqrt{n}\Psi(t_n) \to \Psi'(\theta)t\sigma_F$ and $s_{t,n} \to -\Psi'(\theta)\sigma_F$. Hence, it suffices to show that

$$\frac{1}{\sqrt{n}} \sum_{i=1}^n Y_{ni} \to_d N(0, 1).$$

Note that $Y_{n1}, ..., Y_{nn}$ are i.i.d. random variables. Hence we can apply Lindeberg's CLT (Theorem 1.15). In this case, Lindeberg's condition (1.92) is implied by

$$\lim_{n \to \infty} \int_{|\psi(x, t_n)| > \sqrt{n}\epsilon} [\psi(x, t_n)]^2 dF(x) = 0$$

for any $\epsilon > 0$. For any $\eta > 0$, $\psi(x, \theta + \eta) \leq \psi(x, t_n) \leq \psi(x, \theta - \eta)$ for all x and sufficiently large n. Let $u(x) = \max\{|\psi(x, \theta - \eta)|, |\psi(x, \theta + \eta)|\}$. Then

$$\int_{|\psi(x, t_n)| > \sqrt{n}\epsilon} [\psi(x, t_n)]^2 dF(x) \leq \int_{u(x) > \sqrt{n}\epsilon} [u(x)]^2 dF(x),$$

which converges to 0 since $\int [\psi(x, \gamma)]^2 dF(x)$ is finite for γ in a neighborhood of θ. This proves (i).

(ii) Let $\phi_F(x) = -\psi(x, \theta)/\Psi'(\theta)$. Following the proof of Theorem 5.7, we have

$$\sqrt{n}(\hat{\theta}_n - \theta) = \frac{1}{\sqrt{n}} \sum_{i=1}^n \phi_F(X_i) + R_{1n} - R_{2n},$$

where

$$R_{1n} = \frac{1}{\sqrt{n}} \sum_{i=1}^n \psi(X_i, \theta) \left[\frac{1}{\Psi'(\theta)} - \frac{1}{h_F(\hat{\theta}_n)}\right],$$

$$R_{2n} = \frac{\sqrt{n}}{h_F(\hat{\theta}_n)} \int [\psi(x, \hat{\theta}_n) - \psi(x, \theta)] d(F_n - F)(x),$$

and h_F is defined in the proof of Theorem 5.7 with $\Psi = \lambda_F$. By the CLT and the consistency of $\hat{\theta}_n$, $R_{1n} = o_p(1)$. Hence, the result follows if we can show that $R_{2n} = o_p(1)$. By Lemma 5.2,

$$|R_{2n}| \leq \sqrt{n}|h_F(\hat{\theta}_n)|^{-1} \varrho_\infty(F_n, F) \|\psi(\cdot, \hat{\theta}_n) - \psi(\cdot, \theta)\|_V.$$

The result follows from the assumed condition on ψ and the fact that $\sqrt{n}\varrho_\infty(F_n, F) = O_p(1)$ (Theorem 5.1). ∎

Note that the result in Theorem 5.13 coincides with the result in Theorem 5.7 and (5.41).

Example 5.13. Consider the M-estimators given in Example 5.7 based on i.i.d. random variables $X_1, ..., X_n$. If ψ is bounded and continuous, then Theorem 5.7 applies and (5.98) holds. For case (ii), $\psi(x, \gamma)$ is not bounded but is nondecreasing in γ ($-\psi(x, \gamma)$ is nonincreasing in γ). Hence Theorem 5.13 can be applied to this case.

Consider Huber's ψ given in Example 5.7(v). Assume that F is differentiable in neighborhoods of $\theta - C$ and $\theta + C$. Then

$$\Psi(\gamma) = \int_{\gamma-C}^{\gamma+C} (\gamma - x) dF(x) - CF(\gamma - C) + C[1 - F(\gamma + C)]$$

is differentiable at θ (exercise); $\Psi(\theta) = 0$ if F is symmetric about θ (exercise); and

$$\int [\psi(x, \gamma)]^2 dF(x) = \int_{\gamma-C}^{\gamma+C} (\gamma - x)^2 dF(x) + C^2 F(\gamma - C) + C^2[1 - F(\gamma + C)]$$

is continuous at θ (exercise). Therefore, (5.98) holds with

$$\sigma_F^2 = \frac{\int_{\theta-C}^{\theta+C} (\theta - x)^2 dF(x) + C^2 F(\theta - C) + C^2[1 - F(\theta + C)]}{[F(\theta + C) - F(\theta - C)]^2}$$

(exercise). Note that Huber's M-estimator is robust in Hampel's sense. Asymptotic relative efficiency of $\hat{\theta}_n$ w.r.t. the sample mean $\bar{X}$ can be obtained (exercise). ∎

The next result is for general θ and independent X_i's.

Theorem 5.14. Suppose that $\varphi_i(x, \gamma) = \partial \psi_i(x, \gamma)/\partial \gamma$ exists and the sequence of functions $\{\varphi_{ij}, i = 1, 2, ...\}$ satisfies the conditions in Lemma 5.3 with Θ replaced by a compact neighborhood of θ, where φ_{ij} is the jth row of φ_i; $\sup_i E\|\psi_i(X_i, \theta)\|^{2+\delta} < \infty$ for some $\delta > 0$ (this condition can be replaced by $E\|\psi(X_1, \theta)\|^2 < \infty$ if X_i's are i.i.d. and $\psi_i \equiv \psi$); $E[\psi_i(X_i, \theta)] = 0$; $\liminf_n \lambda_-[n^{-1}\text{Var}(s_n(\theta))] > 0$ and $\liminf_n \lambda_-[n^{-1}M_n(\theta)] > 0$, where $M_n(\theta) = -E[\nabla s_n(\theta)]$ and $\lambda_-[A]$ is the smallest eigenvalue of the matrix A. If $\{\hat{\theta}_n\}$ is a consistent sequence of GEE estimators, then

$$V_n^{-1/2}(\hat{\theta}_n - \theta) \to_d N_k(0, I_k), \tag{5.99}$$

where

$$V_n = [M_n(\theta)]^{-1} \text{Var}(s_n(\theta))[M_n(\theta)]^{-1}. \tag{5.100}$$

Proof. The proof is similar to that of Theorem 4.17. By the consistency of $\hat{\theta}_n$, we can focus on the event $\{\hat{\theta}_n \in A_\epsilon\}$, where $A_\epsilon = \{\gamma : \|\gamma - \theta\| \le \epsilon\}$ with a given $\epsilon > 0$. For sufficiently small ϵ, it can be shown (exercise) that

$$\max_{\gamma \in A_\epsilon} \frac{\|\nabla s_n(\gamma) - \nabla s_n(\theta)\|}{n} = o_p(1), \tag{5.101}$$

using an argument similar to the proof of Lemma 5.3. From the mean-value theorem and $s_n(\hat{\theta}_n) = 0$,

$$-s_n(\theta) = \left[\int_0^1 \nabla s_n \big(\theta + t(\hat{\theta}_n - \theta)\big) dt \right] (\hat{\theta}_n - \theta).$$

It follows from (5.101) that

$$\frac{1}{n} \left\| \int_0^1 \nabla s_n \big(\theta + t(\hat{\theta}_n - \theta)\big) dt - \nabla s_n(\theta) \right\| = o_p(1).$$

Also, by Theorem 1.14(ii),

$$n^{-1} \|\nabla s_n(\theta) + M_n(\theta)\| = o_p(1).$$

This and $\liminf_n \lambda_- [n^{-1} M_n(\theta)] > 0$ imply

$$[M_n(\theta)]^{-1} s_n(\theta) = [1 + o_p(1)](\hat{\theta}_n - \theta).$$

The result follows if we can show that

$$V_n^{-1/2} [M_n(\theta)]^{-1} s_n(\theta) \to_d N_k(0, I_k). \tag{5.102}$$

For any nonzero $l \in \mathcal{R}^k$,

$$\frac{1}{(l^\tau V_n l)^{1+\delta/2}} \sum_{i=1}^n E|l^\tau [M_n(\theta)]^{-1} \psi_i(X_i, \theta)|^{2+\delta} \to 0, \tag{5.103}$$

since $\liminf_n \lambda_- [n^{-1} \mathrm{Var}(s_n(\theta))] > 0$ and $\sup_i E\|\psi_i(X_i, \theta)\|^{2+\delta} < \infty$ (exercise). Applying the CLT (Theorem 1.15) with Liapounov's condition (5.103), we obtain that

$$l^\tau [M_n(\theta)]^{-1} s_n(\theta) / \sqrt{l^\tau V_n l} \to_d N(0, 1) \tag{5.104}$$

for any l, which implies (5.102) (exercise). ∎

Asymptotic normality of GEE estimators can be established under various other conditions; see, for example, Serfling (1980, Chapter 7) and He and Shao (1996).

If X_i's are i.i.d. and $\psi_i \equiv \psi$, the asymptotic covariance matrix in (5.100) reduces to

$$V_n = n^{-1}\{E[\varphi(X_1, \theta)]\}^{-1} E\{\psi(X_1, \theta)[\psi(X_1, \theta)]^\tau\}\{E[\varphi(X_1, \theta)]\}^{-1},$$

where $\varphi(x, \gamma) = \partial\psi(x, \gamma)/\partial\gamma$. When θ is univariate, V_n further reduces to

$$V_n = n^{-1} E[\psi(X_1, \theta)]^2/\{E[\varphi(X_1, \theta)]\}^2.$$

Under the conditions of Theorem 5.14,

$$E[\varphi(X_1, \theta)] = \int \frac{\partial\psi(x, \theta)}{\partial\theta} dF(x) = \frac{\partial}{\partial\theta} \int \psi(x, \theta) dF(x).$$

Hence, the result in Theorem 5.14 coincides with that in Theorem 5.13.

Example 5.14. Consider the quasi-likelihood equation in (5.90) and ψ_i in (5.96). If $\sup_i \|Z_i\| < \infty$, then ψ_i satisfies the conditions in Theorem 5.14 (exercise). Let $\hat{V}_n(\gamma) = [D_i(\gamma)]^{1/2}\tilde{R}_i[D_i(\gamma)]^{1/2}$. Then

$$\text{Var}(s_n(\theta)) = \sum_{i=1}^n G_i(\theta)[\tilde{V}_n(\theta)]^{-1}\text{Var}(X_i)[\tilde{V}_n(\theta)]^{-1}[G_i(\theta)]^\tau$$

and

$$M_n(\theta) = \sum_{i=1}^n G_i(\theta)[\tilde{V}_n(\theta)]^{-1}[G_i(\theta)]^\tau.$$

If $\tilde{R}_i = R_i$ (the true correlation matrix) for all i, then

$$\text{Var}(s_n(\theta)) = \sum_{i=1}^n \phi_i G_i(\theta)[\tilde{V}_n(\theta)]^{-1}[G_i(\theta)]^\tau.$$

If, in addition, $\phi_i \equiv \phi$, then

$$V_n = [M_n(\theta)]^{-1}\text{Var}(s_n(\theta))[M_n(\theta)]^{-1} = \phi[M_n(\theta)]^{-1}. \quad \blacksquare$$

5.5 Variance Estimation

In statistical inference the accuracy of a point estimator is usually assessed by its mse or amse. If the bias or asymptotic bias of an estimator is (asymptotically) negligible w.r.t. its mse or amse, then assessing the mse or amse is equivalent to assessing variance or asymptotic variance. Since variances and asymptotic variances usually depend on the unknown population, we have to estimate them in order to report accuracies of point estimators. Variance estimation is an important part of statistical inference, not only for assessing accuracy, but also for constructing inference procedures studied in Chapters 6 and 7. See also the discussion at the end of §2.5.1.

Let θ be a parameter of interest and $\hat{\theta}_n$ be its estimator. Suppose that, as the sample size $n \to \infty$,

$$V_n^{-1/2}(\hat{\theta}_n - \theta) \to_d N_k(0, I_k), \qquad (5.105)$$

where V_n is the covariance matrix or an asymptotic covariance matrix of $\hat{\theta}_n$. An essential asymptotic requirement in variance estimation is the consistency of variance estimators according to the following definition. See also (3.60) and Exercise 116 in §3.6.

Definition 5.4. Let $\{V_n\}$ be a sequence of $k \times k$ positive definite matrices and $\hat{V}_n$ be a positive definite matrix estimator of V_n for each n. Then $\{\hat{V}_n\}$ or $\hat{V}_n$ is said to be consistent for V_n (or strongly consistent for V_n) if and only if

$$\|V_n^{-1/2}\hat{V}_n V_n^{-1/2} - I_k\| \to_p 0 \qquad (5.106)$$

(or (5.106) holds with $\to_p$ replaced by $\to_{a.s.}$). ∎

Note that (5.106) is different from $\|\hat{V}_n - V_n\| \to_p 0$, because $\|V_n\| \to 0$ in most applications. It can be shown (Exercise 93) that (5.106) holds if and only if $l_n^\tau \hat{V}_n l_n / l_n^\tau V_n l_n \to_p 1$ for any sequence of nonzero vectors $\{l_n\} \subset \mathcal{R}^k$. If (5.105) and (5.106) hold, then

$$\hat{V}_n^{-1/2}(\hat{\theta}_n - \theta) \to_d N_k(0, I_k)$$

(exercise), a result useful for asymptotic inference discussed in Chapters 6 and 7.

If the unknown population is in a parametric family indexed by θ, then V_n is a function of θ, say $V_n = V_n(\theta)$, and it is natural to estimate $V_n(\theta)$ by $V_n(\hat{\theta}_n)$. Consistency of $V_n(\hat{\theta}_n)$ according to Definition 5.4 can usually be directly established. Thus, variance estimation in parametric problems is usually simple. In a nonparametric problem, V_n may depend on unknown quantities other than θ and, thus, variance estimation is much more complex.

We introduce three commonly used variance estimation methods in this section, the substitution method, the jackknife, and the bootstrap.

5.5.1 The substitution method

Suppose that we can obtain a formula for the covariance or asymptotic covariance matrix V_n in (5.105). Then a direct method of variance estimation is to substitute unknown quantities in the variance formula by some estimators. To illustrate, consider the simplest case where $X_1, ..., X_n$ are i.i.d. random d-vectors with $E\|X_1\|^2 < \infty$, $\theta = g(\mu)$, $\mu = EX_1$, $\hat{\theta}_n = g(\bar{X})$, and g is a function from $\mathcal{R}^d$ to $\mathcal{R}^k$. Suppose that g is differentiable at μ. Then, by the CLT and Theorem 1.12(i), (5.105) holds with

$$V_n = [\nabla g(\mu)]^\tau \text{Var}(X_1) \nabla g(\mu)/n, \tag{5.107}$$

which depends on unknown quantities μ and $\text{Var}(X_1)$. A substitution estimator of V_n is

$$\hat{V}_n = [\nabla g(\bar{X})]^\tau S^2 \nabla g(\bar{X})/n, \tag{5.108}$$

where

$$S^2 = \frac{1}{n-1} \sum_{i=1}^n (X_i - \bar{X})(X_i - \bar{X})^\tau$$

is the *sample covariance matrix*, an extension of the sample variance to the multivariate X_i's.

By the SLLN, $\bar{X} \to_{a.s.} \mu$ and $S^2 \to_{a.s.} \text{Var}(X_1)$. Hence, $\hat{V}_n$ in (5.108) is strongly consistent for V_n in (5.107), provided that $\nabla g(\mu) \neq 0$ and ∇g is continuous at μ.

Example 5.15. Let $Y_1, ..., Y_n$ be i.i.d. random variables with finite $\mu_y = EY_1$, $\sigma_y^2 = \text{Var}(Y_1)$, $\gamma_y = EY_1^3$, and $\kappa_y = EY_1^4$. Consider the estimation of $\theta = (\mu_y, \sigma_y^2)$. Let $\hat{\theta}_n = (\bar{X}, \hat{\sigma}_y^2)$, where $\hat{\sigma}_y^2 = n^{-1} \sum_{i=1}^n (Y_i - \bar{Y})^2$. If $X_i = (Y_i, Y_i^2)$, then $\hat{\theta}_n = g(\bar{X})$ with $g(x) = (x_1, x_2 - x_1^2)$. Hence, (5.105) holds with

$$\text{Var}(X_1) = \begin{pmatrix} \sigma_y^2 & \gamma_y - \mu_y(\sigma_y^2 + \mu_y^2) \\ \gamma_y - \mu_y(\sigma_y^2 + \mu_y^2) & \kappa_y - (\sigma_y^2 + \mu_y^2)^2 \end{pmatrix}$$

and

$$\nabla g(x) = \begin{pmatrix} 1 & 0 \\ -2x_1 & 1 \end{pmatrix}.$$

The estimator $\hat{V}_n$ in (5.108) is strongly consistent, since $\nabla g(x)$ is obviously a continuous function. ∎

Similar results can be obtained for problems in Examples 3.21 and 3.23 and Exercises 100 and 101 in §3.6.

A key step in the previous discussion is the derivation of formula (5.107) for the asymptotic covariance matrix of $\hat{\theta}_n = g(\bar{X})$ via Taylor's expansion (Theorem 1.12) and the CLT. Thus, the idea can be applied to the case where $\hat{\theta}_n = T(F_n)$, a differentiable statistical functional.

We still consider i.i.d. random d-vectors $X_1, ..., X_n$ from F. Suppose that T is a vector-valued functional whose components are ϱ-Hadamard differentiable at F, where ϱ is either ϱ_∞ or a distance satisfying (5.42). Let ϕ_F be the vector of influence functions of components of T. If the components of ϕ_F satisfy (5.40), then (5.105) holds with $\theta = T(F)$, $\hat{\theta}_n = T(F_n)$, $F_n = $ the empirical c.d.f. in (5.1), and

$$V_n = \frac{\text{Var}(\phi_F(X_1))}{n} = \frac{1}{n}\int \phi_F(x)[\phi_F(x)]^\tau dF(x). \qquad (5.109)$$

Formula (5.109) leads to a natural substitution variance estimator

$$\hat{V}_n = \frac{1}{n}\int \phi_{F_n}(x)[\phi_{F_n}(x)]^\tau dF_n(x) = \frac{1}{n^2}\sum_{i=1}^n \phi_{F_n}(X_i)[\phi_{F_n}(X_i)]^\tau, \quad (5.110)$$

provided that $\phi_{F_n}(x)$ is well defined, i.e., the components of T are Gâteaux differentiable at F_n for sufficiently large n. Under some more conditions on ϕ_{F_n} we can establish the consistency of $\hat{V}_n$ in (5.110).

Theorem 5.15. Let $X_1, ..., X_n$ be i.i.d. random d-vectors from F, T be a vector-valued functional whose components are Gâteaux differentiable at F and F_n, and ϕ_F be the vector of influence functions of components of T. Suppose that $\sup_{\|x\|\le c}\|\phi_{F_n}(x) - \phi_F(x)\| = o_p(1)$ for any $c > 0$ and that there exist a constant $c_0 > 0$ and a function $h(x) \ge 0$ such that $E[h(X_1)] < \infty$ and $P(\|\phi_{F_n}(x)\|^2 \le h(x)$ for all $\|x\| \ge c_0) \to 1$. Then $\hat{V}_n$ in (5.110) is consistent for V_n in (5.109).
Proof. Let $\zeta(x) = \phi_F(x)[\phi_F(x)]^\tau$ and $\zeta_n(x) = \phi_{F_n}(x)[\phi_{F_n}(x)]^\tau$. By the SLLN,

$$\frac{1}{n}\sum_{i=1}^n \zeta(X_i) \to_{a.s.} \int \zeta(x)dF(x).$$

Hence the result follows from

$$\left\|\frac{1}{n}\sum_{i=1}^n [\zeta_n(X_i) - \zeta(X_i)]\right\| = o_p(1).$$

Using the assumed conditions and the argument in the proof of Lemma 5.3, we can show that for any $\epsilon > 0$, there is a $c > 0$ such that

$$P\left(\frac{1}{n}\sum_{i=1}^n \|\zeta_n(X_i) - \zeta(X_i)\|I_{(c,\infty)}(\|X_i\|) > \frac{\epsilon}{2}\right) \le \epsilon$$

and

$$P\left(\frac{1}{n}\sum_{i=1}^n \|\zeta_n(X_i) - \zeta(X_i)\|I_{[0,c]}(\|X_i\|) > \frac{\epsilon}{2}\right) \le \epsilon$$

for sufficiently large n. This completes the proof.   ∎

Example 5.16. Consider the L-functional defined in (5.46) and the L-estimator $\hat{\theta}_n = \mathtt{T}(F_n)$. Theorem 5.6 shows that $\mathtt{T}$ is Hadamard differentiable at F under some conditions on J. It can be shown (exercise) that $\mathtt{T}$ is Gâteaux differentiable at F_n with $\phi_{F_n}(x)$ given by (5.48) (with F replaced by F_n). Then the difference $\phi_{F_n}(x) - \phi_F(x)$ is equal to

$$\int (F_n - F)(y) J(F_n(y)) dy + \int (F - \delta_x)(y) [J(F_n(y)) - J(F(y))] dy.$$

One can show (exercise) that the conditions in Theorem 5.15 are satisfied if the conditions in Theorem 5.6(i) or (ii) (with $E|X_1| < \infty$) hold. ∎

Substitution variance estimators for M-estimators and U-statistics can also be derived (exercises).

The substitution method can clearly be applied to non-i.i.d. cases. For example, the LSE $\hat{\beta}$ in linear model (3.25) with a full rank Z and i.i.d. ε_i's has $\mathrm{Var}(\hat{\beta}) = \sigma^2 (Z^\tau Z)^{-1}$, where $\sigma^2 = \mathrm{Var}(\varepsilon_1)$. A consistent substitution estimator of $\mathrm{Var}(\hat{\beta})$ can be obtained by replacing σ^2 in the formula of $\mathrm{Var}(\hat{\beta})$ by a consistent estimator of σ^2 such as $SSR/(n - p)$ (see (3.35)).

We now consider variance estimation for the GEE estimators described in §5.4.1. By Theorem 5.14, the asymptotic covariance matrix of the GEE estimator $\hat{\theta}_n$ is given by (5.100), where

$$\mathrm{Var}(s_n(\theta)) = \sum_{i=1}^n E\{\psi_i(X_i, \theta)[\psi_i(X_i, \theta)]^\tau\},$$

$$M_n(\theta) = \sum_{i=1}^n E[\varphi_i(X_i, \theta)],$$

and $\varphi_i(x, \gamma) = \partial \psi_i(x, \gamma)/\partial \gamma$. Substituting θ by $\hat{\theta}_n$ and the expectations by their empirical analogues, we obtain the substitution estimator $\hat{V}_n = \hat{M}_n^{-1} \widehat{\mathrm{Var}}(s_n) \hat{M}_n^{-1}$, where

$$\widehat{\mathrm{Var}}(s_n) = \sum_{i=1}^n \psi_i(X_i, \hat{\theta}_n)[\psi_i(X_i, \hat{\theta}_n)]^\tau$$

and

$$\hat{M}_n = \sum_{i=1}^n \varphi_i(X_i, \hat{\theta}_n).$$

The proof of the following result is left as an exercise.

Theorem 5.16. Let $X_1, ..., X_n$ be independent and $\{\hat{\theta}_n\}$ be a consistent sequence of GEE estimators. Assume the conditions in Theorem 5.14. Suppose further that the sequence of functions $\{h_{ij}, i = 1, 2, ...\}$ satisfies the conditions in Lemma 5.3 with Θ replaced by a compact neighborhood of θ, where $h_{ij}(x, \gamma)$ is the jth row of $\psi_i(x, \gamma)[\psi_i(x, \gamma)]^\tau$, $j = 1, ..., k$. Let V_n be given by (5.100). Then $\hat{V}_n = \hat{M}_n^{-1}\widehat{\text{Var}}(s_n)\hat{M}_n^{-1}$ is consistent for V_n. ∎

5.5.2 The jackknife

Applying the substitution method requires the derivation of a formula for the covariance matrix or asymptotic covariance matrix of a point estimator. There are variance estimation methods that can be used without actually deriving such a formula (only the existence of the covariance matrix or asymptotic covariance matrix is assumed), at the expense of requiring a large number of computations. These methods are called *resampling* methods, *replication* methods, or *data reuse* methods. The *jackknife* method introduced here and the *bootstrap* method in §5.5.3 are the most popular resampling methods.

The jackknife method was proposed by Quenouille (1949) and Tukey (1958). Let $\hat{\theta}_n$ be a vector-valued estimator based on independent X_i's, where each X_i is a random d_i-vector and $\sup_i d_i < \infty$. Let $\hat{\theta}_{-i}$ be the same estimator but based on $X_1, ..., X_{i-1}, X_{i+1}, ..., X_n$, $i = 1, ..., n$. Note that $\hat{\theta}_{-i}$ also depends on n but the subscript n is omitted for simplicity. Since $\hat{\theta}_n$ and $\hat{\theta}_{-1}, ..., \hat{\theta}_{-n}$ are estimators of the same quantity, the "sample covariance matrix"

$$\frac{1}{n-1} \sum_{i=1}^{n} \left(\hat{\theta}_{-i} - \bar{\theta}_n\right) \left(\hat{\theta}_{-i} - \bar{\theta}_n\right)^\tau \tag{5.111}$$

can be used as a measure of the variation of $\hat{\theta}_n$, where $\bar{\theta}_n$ is the average of $\hat{\theta}_{-i}$'s.

There are two major differences between the quantity in (5.111) and the sample covariance matrix S^2 previously discussed. First, $\hat{\theta}_{-i}$'s are not independent. Second, $\hat{\theta}_{-i} - \hat{\theta}_{-j}$ usually converges to 0 at a fast rate (such as n^{-1}). Hence, to estimate the asymptotic covariance matrix of $\hat{\theta}_n$, the quantity in (5.111) should be multiplied by a correction factor c_n. If $\hat{\theta}_n = \bar{X}$ ($d_i \equiv d$), then $\hat{\theta}_{-i} = (n-1)^{-1}(\bar{X} - X_i)$ and the quantity in (5.111) reduces to

$$\frac{1}{(n-1)^3} \sum_{i=1}^{n} \left(X_i - \bar{X}\right) \left(X_i - \bar{X}\right)^\tau = \frac{1}{(n-1)^2} S^2,$$

where S^2 is the sample covariance matrix. Thus, the correction factor c_n is $(n-1)^2/n$ for the case of $\hat{\theta}_n = \bar{X}$ since, by the SLLN, S^2/n is strongly consistent for $\text{Var}(\bar{X})$.

It turns out that the same correction factor works for many other estimators. This leads to the following *jackknife variance estimator* for $\hat{\theta}_n$:

$$\hat{V}_J = \frac{n-1}{n} \sum_{i=1}^{n} \left(\hat{\theta}_{-i} - \bar{\theta}_n\right) \left(\hat{\theta}_{-i} - \bar{\theta}_n\right)^\tau. \qquad (5.112)$$

Theorem 5.17. Let $X_1, ..., X_n$ be i.i.d. random d-vectors from F with finite $\mu = E(X_1)$ and $\text{Var}(X_1)$, and let $\hat{\theta}_n = g(\bar{X})$. Suppose that ∇g is continuous at μ and $\nabla g(\mu) \neq 0$. Then the jackknife variance estimator $\hat{V}_J$ in (5.112) is strongly consistent for V_n in (5.107).

Proof. We prove the case where g is real-valued. The proof of the general case is left to the reader. Let $\bar{X}_{-i}$ be the sample mean based on $X_1, ..., X_{i-1}, X_{i+1}, ..., X_n$. From the mean-value theorem, we have

$$\begin{aligned}
\hat{\theta}_{-i} - \hat{\theta}_n &= g(\bar{X}_{-i}) - g(\bar{X}) \\
&= [\nabla g(\xi_{n,i})]^\tau (\bar{X}_{-i} - \bar{X}) \\
&= [\nabla g(\bar{X})]^\tau (\bar{X}_{-i} - \bar{X}) + R_{n,i},
\end{aligned}$$

where $R_{n,i} = \left[\nabla g(\xi_{n,i}) - \nabla g(\bar{X})\right]^\tau (\bar{X}_{-i} - \bar{X})$ and $\xi_{n,i}$ is a point on the line segment between $\bar{X}_{-i}$ and $\bar{X}$. From $\bar{X}_{-i} - \bar{X} = (n-1)^{-1}(\bar{X} - X_i)$, it follows that $\sum_{i=1}^{n}(\bar{X}_{-i} - \bar{X}) = 0$ and

$$\frac{1}{n}\sum_{i=1}^{n}(\hat{\theta}_{-i} - \hat{\theta}_n) = \frac{1}{n}\sum_{i=1}^{n} R_{n,i} = \bar{R}_n.$$

From the definition of the jackknife estimator in (5.112),

$$\hat{V}_J = A_n + B_n + 2C_n,$$

where

$$A_n = \frac{n-1}{n}[\nabla g(\bar{X})]^\tau \sum_{i=1}^{n}(\bar{X}_{-i} - \bar{X})(\bar{X}_{-i} - \bar{X})^\tau \nabla g(\bar{X}),$$

$$B_n = \frac{n-1}{n}\sum_{i=1}^{n}(R_{n,i} - \bar{R}_n)^2,$$

and

$$C_n = \frac{n-1}{n}\sum_{i=1}^{n}(R_{n,i} - \bar{R}_n)[\nabla g(\bar{X})]^\tau(\bar{X}_{-i} - \bar{X}).$$

By $\bar{X}_{-i} - \bar{X} = (n-1)^{-1}(\bar{X} - X_i)$, the SLLN, and the continuity of ∇g at μ,

$$A_n/V_n \to_{a.s.} 1.$$

Also,

$$(n-1)\sum_{i=1}^{n}\|\bar{X}_{-i}-\bar{X}\|^2 = \frac{1}{n-1}\sum_{i=1}^{n}\|X_i-\bar{X}\|^2 = O(1) \quad \text{a.s.} \qquad (5.113)$$

Hence

$$\max_{i\leq n}\|\bar{X}_{-i}-\bar{X}\|^2 \to_{a.s.} 0,$$

which, together with the continuity of ∇g at μ and $\|\xi_{n,i}-\bar{X}\|\leq\|\bar{X}_{-i}-\bar{X}\|$, implies that

$$u_n = \max_{i\leq n}\|\nabla g(\xi_{n,i})-\nabla g(\bar{X})\| \to_{a.s.} 0.$$

From (5.107) and (5.113), $\sum_{i=1}^{n}\|\bar{X}_{-i}-\bar{X}\|^2/V_n = O(1)$ a.s. Hence

$$\frac{B_n}{V_n} \leq \frac{n-1}{V_n n}\sum_{i=1}^{n}R_{n,i}^2 \leq \frac{u_n}{V_n}\sum_{i=1}^{n}\|\bar{X}_{-i}-\bar{X}\|^2 \to_{a.s.} 0.$$

By the Cauchy-Schwarz inequality, $(C_n/V_n)^2 \leq (A_n/V_n)(B_n/V_n) \to_{a.s.} 0$. This proves the result. ∎

A key step in the proof of Theorem 5.17 is that $\hat{\theta}_{-i}-\hat{\theta}_n$ can be approximated by $[\nabla g(\bar{X})]^\tau(\bar{X}_{-i}-\bar{X})$ and the contributions of the remainders, $R_{n,1},...,R_{n,n}$, are sufficiently small, i.e., $B_n/V_n \to_{a.s.} 0$. This indicates that the jackknife estimator (5.112) is consistent for $\hat{\theta}_n$ that can be well approximated by some linear statistic. In fact, the jackknife estimator (5.112) has been shown to be consistent when $\hat{\theta}_n$ is a U-statistic (Arvesen, 1969) or a statistical functional that is Hadamard differentiable and *continuously* Gâteaux differentiable at F (which includes certain types of L-estimators and M-estimators). More details can be found in Shao and Tu (1995, Chapter 2).

The jackknife method can be applied to non-i.i.d. problems. A detailed discussion of the use of the jackknife method in survey problems can be found in Shao and Tu (1995, Chapter 6). We now consider the jackknife variance estimator for the LSE $\hat{\beta}$ in linear model (3.25). For simplicity, assume that Z is of full rank. Assume also that ε_i's are independent with $E(\varepsilon_i)=0$ and $\text{Var}(\varepsilon_i)=\sigma_i^2$. Then

$$\text{Var}(\hat{\beta}) = (Z^\tau Z)^{-1}\sum_{i=1}^{n}\sigma_i^2 Z_i Z_i^\tau (Z^\tau Z)^{-1}.$$

Let $\hat{\beta}_{-i}$ be the LSE of β based on the data with the ith pair (X_i, Z_i) deleted. Using the fact that $(A+cc^\tau)^{-1} = A^{-1} - A^{-1}cc^\tau A^{-1}/(1+c^\tau A^{-1}c)$ for a matrix A and a vector c, we can show that (exercise)

$$\hat{\beta}_{-i} = \hat{\beta} - r_i Z_i/(1-h_i), \qquad (5.114)$$

where $r_i = X_i - Z_i^\tau \hat\beta$ is the ith residual and $h_i = Z_i^\tau (Z^\tau Z)^{-1} Z_i$. Hence

$$\hat V_J = \frac{n-1}{n} (Z^\tau Z)^{-1} \left[\sum_{i=1}^n \frac{r_i^2 Z_i Z_i^\tau}{(1-h_i)^2} - \frac{1}{n} \sum_{i=1}^n \frac{r_i Z_i}{1-h_i} \sum_{i=1}^n \frac{r_i Z_i^\tau}{1-h_i} \right] (Z^\tau Z)^{-1}.$$

Wu (1986) proposed the following weighted jackknife variance estimator that improves $\hat V_J$:

$$\hat V_{WJ} = \sum_{i=1}^n (1-h_i) \left(\hat\beta_{-i} - \hat\beta \right) \left(\hat\beta_{-i} - \hat\beta \right)^\tau = (Z^\tau Z)^{-1} \sum_{i=1}^n \frac{r_i^2 Z_i Z_i^\tau}{1-h_i} (Z^\tau Z)^{-1}.$$

Theorem 5.18. Assume the conditions in Theorem 3.12 and that ε_i's are independent. Then both $\hat V_J$ and $\hat V_{WJ}$ are consistent for $\mathrm{Var}(\hat\beta)$.
Proof. Let $l_n \in \mathcal{R}^p$, $n = 1, 2, ...$, be nonzero vectors and $l_i = l_n^\tau (Z^\tau Z)^{-1} Z_i$. Since $\max_{i \le n} h_i \to 0$, the result for $\hat V_{WJ}$ follows from

$$\sum_{i=1}^n l_i^2 r_i^2 \Big/ \sum_{i=1}^n l_i^2 \sigma_i^2 \to_p 1 \tag{5.115}$$

(see Exercise 93). By the WLLN (Theorem 1.14(ii)) and $\max_{i \le n} h_i \to 0$,

$$\sum_{i=1}^n l_i^2 \varepsilon_i^2 \Big/ \sum_{i=1}^n l_i^2 \sigma_i^2 \to_p 1.$$

Note that $r_i = \varepsilon_i + Z_i^\tau (\beta - \hat\beta)$ and

$$\max_{i \le n} [Z_i^\tau (\beta - \hat\beta)]^2 \le \|Z(\beta - \hat\beta)\|^2 \max_{i \le n} h_i = o_p(1).$$

Hence (5.115) holds.

The consistency of $\hat V_J$ follows from (5.115) and

$$\frac{n-1}{n^2} \left(\sum_{i=1}^n \frac{l_i r_i}{1-h_i} \right)^2 \Big/ \sum_{i=1}^n l_i^2 \sigma_i^2 = o_p(1). \tag{5.116}$$

The proof of (5.116) is left as an exercise. ∎

Finally, let us consider the jackknife estimators for GEE estimators in §5.4.1. Under the conditions of Proposition 5.5 or 5.6, it can be shown that

$$\max_{i \le n} \|\hat\theta_{-i} - \hat\theta\| = o_p(1), \tag{5.117}$$

where $\hat\theta_{-i}$ is a root of $s_{ni}(\gamma) = 0$ and

$$s_{ni}(\gamma) = \sum_{j \ne i, j \le n} \psi_j(X_j, \gamma).$$

Assume that $\psi_i(x, \gamma)$ is continuously differentiable w.r.t. γ in a neighborhood of θ. Using Taylor's expansion and the fact that $s_{ni}(\hat{\theta}_{-i}) = 0$ and $s_n(\hat{\theta}_n) = 0$, we obtain that

$$\psi_i(X_i, \hat{\theta}_{-i}) = \left[\int_0^1 \nabla s_n(\hat{\theta}_n + t(\hat{\theta}_{-i} - \hat{\theta}_n))dt\right](\hat{\theta}_{-i} - \hat{\theta}_n).$$

Following the proof of Theorem 5.14, we obtain that

$$\hat{V}_J = [M_n(\theta)]^{-1} \sum_{i=1}^n \psi_i(X_i, \hat{\theta}_{-i})[\psi_i(X_i, \hat{\theta}_{-i})]^\tau [M_n(\theta)]^{-1} + R_n,$$

where R_n satisfies $\|V_n^{-1/2} R_n V_n^{-1/2}\| = o_p(1)$ for V_n in (5.100). Under the conditions of Theorem 5.16, it follows from (5.117) that $\hat{V}_J$ is consistent.

If $\hat{\theta}_n$ is computed using an iteration method, then the computation of $\hat{V}_J$ requires n additional iteration processes. We may use the idea of a one-step MLE to reduce the amount of computation. For each i, let

$$\hat{\theta}_{-i} = \hat{\theta}_n - [\nabla s_{ni}(\hat{\theta}_n)]^{-1} s_{ni}(\hat{\theta}_n), \tag{5.118}$$

which is the result from the first iteration when the Newton-Raphson method is applied in computing a root of $s_{ni}(\gamma) = 0$ and $\hat{\theta}_n$ is used as the initial point. Note that $\hat{\theta}_{-i}$'s in (5.118) satisfy (5.117) (exercise). If the jackknife variance estimator is based on $\hat{\theta}_{-i}$'s in (5.118), then

$$\hat{V}_J = [M_n(\theta)]^{-1} \sum_{i=1}^n \psi_i(X_i, \hat{\theta}_n)[\psi_i(X_i, \hat{\theta}_n)]^\tau [M_n(\theta)]^{-1} + \tilde{R}_n,$$

where $\tilde{R}_n$ satisfies $\|V_n^{-1/2} \tilde{R}_n V_n^{-1/2}\| = o_p(1)$. These results are summarized in the following theorem.

Theorem 5.19. Assume the conditions in Theorems 5.14 and 5.16. Assume further that $\hat{\theta}_{-i}$'s are given by (5.118) or GEE estimators satisfying (5.117). Then the jackknife variance estimator $\hat{V}_J$ is consistent for V_n given in (5.100). ∎

5.5.3 The bootstrap

The basic idea of the bootstrap method can be described as follows. Suppose that P is a population or model that generates the sample X and that we need to estimate $\text{Var}(\hat{\theta})$, where $\hat{\theta} = \hat{\theta}(X)$ is an estimator, a statistic based on X. Suppose further that the unknown population P is estimated by $\hat{P}$, based on the sample X. Let X^* be a sample (called a *bootstrap* *sample*) taken from the estimated population $\hat{P}$ using the same or a similar sampling procedure used to obtain X, and let $\hat{\theta}^* = \hat{\theta}(X^*)$, which is the same as $\hat{\theta}$ but with X replaced by X^*. If we believe that $P = \hat{P}$ (i.e., we have a perfect estimate of the population), then $\text{Var}(\hat{\theta}) = \text{Var}_*(\hat{\theta}^*)$, where Var_* is the conditional variance w.r.t. the randomness in generating X^*, given X. In general, $P \neq \hat{P}$ and, therefore, $\text{Var}(\hat{\theta}) \neq \text{Var}_*(\hat{\theta}^*)$. But $\hat{V}_B = \text{Var}_*(\hat{\theta}^*)$ is an empirical analogue of $\text{Var}(\hat{\theta})$ and can be used as an estimate of $\text{Var}(\hat{\theta})$.

In a few cases, an explicit form of $\hat{V}_B = \text{Var}_*(\hat{\theta}^*)$ can be obtained. First, consider i.i.d. $X_1, ..., X_n$ from a c.d.f. F on $\mathcal{R}^d$. The population is determined by F. Suppose that we estimate F by the empirical c.d.f. F_n in (5.1) and that $X_1^*, ..., X_n^*$ are i.i.d. from F_n. For $\hat{\theta} = \bar{X}$, its bootstrap analogue is $\hat{\theta}^* = \bar{X}^*$, the average of X_i^*'s. Then

$$\hat{V}_B = \text{Var}_*(\bar{X}^*) = \frac{1}{n^2} \sum_{i=1}^{n} (X_i - \bar{X})(X_i - \bar{X})^\tau = \frac{n-1}{n^2} S^2,$$

where S^2 is the sample covariance matrix. In this case $\hat{V}_B = \text{Var}_*(\bar{X}^*)$ is a strongly consistent estimator for $\text{Var}(\bar{X})$. Next, consider i.i.d. random variables $X_1, ..., X_n$ from a c.d.f. F on $\mathcal{R}$ and $\hat{\theta} = F_n^{-1}(\frac{1}{2})$, the sample median. Suppose that $n = 2l - 1$ for an integer l. Let $X_1^*, ..., X_n^*$ be i.i.d. from F_n and $\hat{\theta}^*$ be the sample median based on $X_1^*, ..., X_n^*$. Then

$$\hat{V}_B = \text{Var}_*(\hat{\theta}^*) = \sum_{j=1}^{n} p_j \left(X_{(j)} - \sum_{i=1}^{n} p_i X_{(i)} \right)^2,$$

where $X_{(1)} \leq \cdots \leq X_{(n)}$ are order statistics and $p_j = P(\hat{\theta}^* = X_{(j)}|X)$. It can be shown (exercise) that

$$p_j = \sum_{t=0}^{l-1} \binom{n}{t} \frac{(j-1)^t (n-j+1)^{n-t} - j^t (n-j)^{n-t}}{n^n}. \tag{5.119}$$

However, in most cases $\hat{V}_B$ does not have a simple explicit form. When P is known, the Monte Carlo method described in §4.1.4 can be used to approximate $\text{Var}(\hat{\theta})$. That is, we draw repeatedly new data sets from P and then use the sample covariance matrix based on the values of $\hat{\theta}$ computed from new data sets as a numerical approximation to $\text{Var}(\hat{\theta})$. This idea can be used to approximate $\hat{V}_B$, since $\hat{P}$ is a known population. That is, we can draw m bootstrap data sets $X^{*1}, ..., X^{*m}$ independently from $\hat{P}$ (conditioned on X), compute $\hat{\theta}^{*j} = \hat{\theta}(X^{*j})$, $j = 1, ..., m$, and approximate $\hat{V}_B$ by

$$\hat{V}_B^m = \frac{1}{m} \sum_{j=1}^{m} \left(\hat{\theta}^{*j} - \bar{\theta}^* \right) \left(\hat{\theta}^{*j} - \bar{\theta}^* \right)^\tau,$$

where $\bar{\theta}^*$ is the average of $\hat{\theta}^{*j}$'s. Since each X^{*j} is a data set generated from $\hat{P}$, $\hat{V}_B^m$ is a resampling estimator. From the SLLN, as $m \to \infty$, $\hat{V}_B^m \to_{a.s.} \hat{V}_B$, conditioned on X. Both $\hat{V}_B$ and its Monte Carlo approximation $\hat{V}_B^m$ are called *bootstrap variance estimators* for $\hat{\theta}$. $\hat{V}_B^m$ is more useful in practical applications, whereas in theoretical studies, we usually focus on $\hat{V}_B$.

The consistency of the bootstrap variance estimator $\hat{V}_B$ is a much more complicated problem than that of the jackknife variance estimator in §5.5.2. Some examples can be found in Shao and Tu (1995, §3.2.2).

The bootstrap method can also be applied to estimate quantities other than $\text{Var}(\hat{\theta})$. For example, let $K(t) = P(\hat{\theta} \le t)$ be the c.d.f. of a real-valued estimator $\hat{\theta}$. From the previous discussion, a bootstrap estimator of $K(t)$ is the conditional probability $P(\hat{\theta}^* \le t|X)$, which can be approximated by the Monte Carlo approximation $m^{-1} \sum_{j=1}^m I_{(-\infty,t]}(\hat{\theta}^{*j})$. An important application of bootstrap distribution estimators in problems of constructing confidence sets is studied in §7.4. Here, we study the use of a bootstrap distribution estimator to form a consistent estimator of the asymptotic variance of a real-valued estimator $\hat{\theta}$.

Suppose that

$$\sqrt{n}(\hat{\theta} - \theta) \to_d N(0, v), \qquad (5.120)$$

where v is unknown. Let $H_n(t)$ be the c.d.f. of $\sqrt{n}(\hat{\theta} - \theta)$ and

$$\hat{H}_B(t) = P(\sqrt{n}(\hat{\theta}^* - \hat{\theta}) \le t|X) \qquad (5.121)$$

be a bootstrap estimator of $H_n(t)$. If

$$\hat{H}_B(t) - H_n(t) \to_p 0$$

for any t, then, by (5.120),

$$\hat{H}_B(t) - \Phi\left(t/\sqrt{v}\right) \to_p 0,$$

which implies (Exercise 112) that

$$\hat{H}_B^{-1}(\alpha) \to_p \sqrt{v}z_\alpha$$

for any $\alpha \in (0, 1)$, where $z_\alpha = \Phi^{-1}(\alpha)$. Then, for $\alpha \ne \frac{1}{2}$,

$$\hat{H}_B^{-1}(1 - \alpha) - \hat{H}_B^{-1}(\alpha) \to_p \sqrt{v}(z_{1-\alpha} - z_\alpha).$$

Therefore, a consistent estimator of v/n, the asymptotic variance of $\hat{\theta}$, is

$$\tilde{V}_B = \frac{1}{n}\left[\frac{\hat{H}_B^{-1}(1 - \alpha) - \hat{H}_B^{-1}(\alpha)}{z_{1-\alpha} - z_\alpha}\right]^2.$$

The following result gives some conditions under which $\hat{H}_B(t) - H_n(t) \to_p 0$. The proof of part (i) is omitted. The proof of part (ii) is given in Exercises 113-115 in §5.6.

Theorem 5.20. Suppose that $X_1, ..., X_n$ are i.i.d. from a c.d.f. F on $\mathcal{R}^d$. Let $\hat{\theta} = \mathrm{T}(F_n)$, where T is a real-valued functional, $\hat{\theta}^* = \mathrm{T}(F_n^*)$, where F_n^* is the empirical c.d.f. based on a bootstrap sample $X_1^*, ..., X_n^*$ i.i.d. from F_n, and let $\hat{H}_B$ be given by (5.121).
(i) If T is ϱ_∞-Hadamard differentiable at F and (5.40) holds, then

$$\varrho_\infty(\hat{H}_B, H_n) \to_p 0. \tag{5.122}$$

(ii) If $d = 1$ and T is ϱ_{L_p}-Fréchet differentiable at F ($\int \{F(t)[1 - F(t)]\}^{p/2} dt < \infty$ if $1 \le p < 2$) and (5.40) holds, then (5.122) holds. ∎

Applications of the bootstrap method to non-i.i.d. cases can be found, for example, in Efron and Tibshirani (1993), Hall (1992), and Shao and Tu (1995).

5.6 Exercises

1. Let ϱ_∞ be the sup-norm distance. Find an example of a sequence $\{G_n\}$ of c.d.f.'s satisfying $G_n \to_w G$ for a c.d.f. G, but $\varrho_\infty(G_n, G)$ does not converge to 0.

2. Let $X_1, ..., X_n$ be i.i.d. random d-vectors with c.d.f. F and F_n be the empirical c.d.f. defined by (5.1). Show that for any $t > 0$ and $\epsilon > 0$, there is a $C_{\epsilon,d}$ such that for all $n = 1, 2, ...$,

$$P\left(\sup_{m \ge n} \varrho_\infty(F_m, F) > t\right) \le \frac{C_{\epsilon,d} e^{-(2-\epsilon)t^2 n}}{1 - e^{-(2-\epsilon)t^2}}.$$

3. Show that ϱ_{M_p} defined by (5.4) is a distance on $\mathcal{F}_p$, $p \ge 1$.

4. Show that $\|\cdot\|_{L_p}$ in (5.5) is a norm for any $p \ge 1$.

5. Let $\mathcal{F}_1$ be the collection of c.d.f.'s on $\mathcal{R}$ with finite means.
   (a) Show that $\varrho_{M_1}(G_1, G_2) = \int_0^1 |G_1^{-1}(z) - G_2^{-1}(z)| dz$, where $G^{-1}(z) = \inf\{t : G(t) \ge z\}$ for any $G \in \mathcal{F}$.
   (b) Show that $\varrho_{M_1}(G_1, G_2) = \varrho_{L_1}(G_1, G_2)$.

6. Find an example of a sequence $\{G_j\} \subset \mathcal{F}$ for which
   (a) $\lim_{j \to \infty} \varrho_\infty(G_j, G_0) = 0$ but $\varrho_{M_2}(G_j, G_0)$ does not converge to 0;
   (b) $\lim_{j \to \infty} \varrho_{M_2}(G_j, G_0) = 0$ but $\varrho_\infty(G_j, G_0)$ does not converge to 0.

7. Repeat the previous exercise with ϱ_{M_2} replaced by ϱ_{L_2}.

8. Let X be a random variable having c.d.f. F. Show that
   (a) $E|X|^2 < \infty$ implies $\int \{F(t)[1 - F(t)]\}^{p/2} dt < \infty$ for $p \in (1, 2)$;
   (b) $E|X|^{2+\delta} < \infty$ with some $\delta > 0$ implies $\int \{F(t)[1 - F(t)]\}^{1/2} dt < \infty$.

9. For any one-dimensional $G_j \in \mathcal{F}_1$, $j = 1, 2$, show that $\varrho_{L_1}(G_1, G_2) \geq |\int x dG_1 - \int x dG_2|$.

10. In the proof of Theorem 5.3, show that $p_i = c/n$, $i = 1, ..., n$, $\lambda = -(c/n)^{n-1}$ is a maximum of the function $H(p_1, ..., p_n, \lambda)$ over $p_i > 0$, $i = 1, ..., n$, $\sum_{i=1}^{n} p_i = c$.

11. Show that (5.11)-(5.13) is a solution to the problem of maximizing $\ell(G)$ in (5.8) subject to (5.10).

12. In the proof of Theorem 5.4, prove the case of $m \geq 2$.

13. Show that a maximum of $\ell(G)$ in (5.17) subject to (5.10) is given by (5.11) with $\hat{p}_i$ defined by (5.18) and (5.19).

14. In Example 5.2, show that an MELE is given by (5.11) with $\hat{p}_i$'s given by (5.21).

15. In Example 5.3, show that
    (a) maximizing (5.22) subject to (5.23) is equivalent to maximizing

$$\prod_{i=1}^{n} q_i^{\delta_{(i)}} (1 - q_i)^{n-i+1-\delta_{(i)}},$$

where $q_i = p_i / \sum_{j=i}^{n+1} p_j$, $i = 1, ..., n$;
    (b) $\hat{F}$ given by (5.24) maximizes (5.22) subject to (5.23); (Hint: use part (a) and the fact that $p_i = q_i \prod_{j=1}^{i-1}(1 - q_j)$.)
    (c) $\hat{F}$ given by (5.25) is the same as that in (5.24);
    (d) if $\delta_i = 1$ for all i (no censoring), then $\hat{F}$ in (5.25) is the same as the empirical c.d.f. in (5.1).

16. Let f_n be given by (5.26).
    (a) Show that f_n is a Lebesgue p.d.f. on $\mathcal{R}$.
    (b) Suppose that f is continuously differentiable at t, $\lambda_n \to 0$, and $n\lambda_n \to \infty$. Show that (5.27) holds.
    (c) Under $n\lambda_n^3 \to 0$ and the conditions of (b), show that (5.28) holds.
    (d) Suppose that f is continuous on $[a, b]$, $-\infty < a < b < \infty$, $\lambda_n \to 0$, and $n\lambda_n \to \infty$. Show that $\int_a^b f_n(t)dt \to_p \int_a^b f(t)dt$.

17. Let $\hat{f}$ be given by (5.29).
   (a) Show that $\hat{f}$ is a Lebesgue p.d.f. on $\mathcal{R}$.
   (b) Prove (5.30) under the condition that $\lambda_n \to 0$, $n\lambda_n \to \infty$, and f is bounded and continuous at t and $\int [w(t)]^{2+\delta} dt < \infty$ for some $\delta > 0$. (Hint: check Liapounov's condition and apply Theorem 1.15.)
   (c) Suppose that $\lambda_n \to 0$, $n\lambda_n \to \infty$, and f is bounded and continuous on $[a, b]$, $-\infty < a < b < \infty$. Show that $\int_a^b \hat{f}(t)dt \to_p \int_a^b f(t)dt$.

18. Prove (5.32)-(5.34) under the conditions described in §5.1.4.

19. Show that $\hat{K}(t)$ in (5.35) is a consistent estimator of $K(t)$ in (5.34), assuming that $\hat{\beta} \to_p \beta$, ϕ is a continuous function on $\mathcal{R}$, (X_i, Z_i)'s are i.i.d., and $\|Z_i\| \le c$ for a constant $c > 0$.

20. Let $\ell(\theta, \xi)$ be a likelihood. Show that a maximum profile likelihood estimator $\hat{\theta}$ of θ is an MLE if $\xi(\theta)$, the maximum of $\sup_\xi \ell(\theta, \xi)$ for a fixed θ, does not depend on θ.

21. Let $X_1, ..., X_n$ be i.i.d. from $N(\mu, \sigma^2)$. Derive the profile likelihood function for μ or σ^2. Discuss in each case whether the maximum profile likelihood estimator is the same as the MLE.

22. Derive the profile empirical likelihoods in (5.36) and (5.37).

23. Let $X_1, ..., X_n$ be i.i.d. random variables from a c.d.f. F and let $\pi(x) = P(\delta_i = 1 | X_i = x)$, where $\delta_i = 1$ if X_i is observed and $\delta_i = 0$ if X_i is missing. Assume that $0 < \pi = \int \pi(x) dF(x) < 1$.
   (a) Let $F_1(x) = P(X_i \le x | \delta_i = 1)$. Show that F and F_1 are the same if and only if $\pi(x) \equiv \pi$.
   (b) Let $\hat{F}$ be the c.d.f. putting mass r^{-1} to each observed X_i, where r is the number of observed X_i's. Show that $\hat{F}(x)$ is unbiased and consistent for $F_1(x)$, $x \in \mathcal{R}$.
   (c) When $\pi(x) \equiv \pi$, show that $\hat{F}(x)$ in part (b) is unbiased and consistent for $F(x)$, $x \in \mathcal{R}$. When $\pi(x)$ is not constant, show that $\hat{F}(x)$ is biased and inconsistent for $F(x)$ for some $x \in \mathcal{R}$.

24. Show that ϱ-Fréchet differentiability implies ϱ-Hadamard differentiability.

25. Suppose that a functional T is Gâteaux differentiable at F with a continuous differential L_F in the sense that $\varrho_\infty(\Delta_j, \Delta) \to 0$ implies $L_F(\Delta_j) \to L_F(\Delta)$. Show that ϕ_F is bounded.

26. Suppose that a functional T is Gâteaux differentiable at F with a bounded and continuous influence function ϕ_F. Show that the differential L_F is continuous in the sense described in the previous exercise.

27. Let $T(G) = g(\int x dG)$ be a functional defined on $\mathcal{F}_1$, the collection of one-dimensional c.d.f.'s with finite means.
    (a) Find a differentiable function g for which the functional T is not ϱ_∞-Hadamard differentiable at F.
    (b) Show that if g is a differentiable function, then T is ϱ_{L_1}-Fréchet differentiable at F. (Hint: use the result in Exercise 9.)

28. In Example 5.5, show that (5.43) holds. (Hint: for $\Delta = c(G_1 - G_2)$, show that $\|\Delta\|_V \leq |c|(\|G_1\|_V + \|G_2\|_V) = 2|c|$.)

29. In Example 5.5, show that ϕ_F is continuous if F is continuous.

30. In Example 5.5, show that T is not ϱ_∞-Fréchet differentiable at F.

31. Prove Proposition 5.1(ii).

32. Suppose that T is first-order and second-order ϱ-Hadamard differentiable at F. Prove (5.45).

33. Find an example of a second-order ϱ-Fréchet differentiable functional T that is not first-order ϱ-Hadamard differentiable.

34. Prove (5.47) and that (5.40) is satisfied for an L-functional if F has a finite variance.

35. Prove (iv) and (v) of Theorem 5.6.

36. Discuss which of (i)-(v) in Theorem 5.6 can be applied to each of the L-estimators in Example 5.6.

37. Obtain explicit forms of the influence functions for L-estimators in Example 5.6. Discuss which of them are bounded and continuous.

38. Provide an example in which the L-functional T given by (5.46) is not ϱ_∞-Hadamard differentiable at F. (Hint: consider an untrimmed J.)

39. Discuss which M-functionals defined in (i)-(vi) of Example 5.7 satisfy the conditions of Theorem 5.7.

40. In the proof of Theorem 5.7, show that $R_{2j} \to 0$.

41. Show that the second equality in (5.51) holds when ψ is Borel and bounded.

42. Show that the functional T in (5.53) is ϱ_∞-Hadamard differentiable at F with the differential given by (5.54). Obtain the influence function ϕ_F and show that it is bounded and continuous if F is continuous.

43. Show that the functional T in (5.55) is ϱ_∞-Hadamard differentiable at F with the differential given by (5.56). Obtain the influence function ϕ_F and show that it is bounded and continuous if $F(y, \infty)$ and $F(\infty, z)$ are continuous.

44. Let F be a continuous c.d.f. on $\mathcal{R}$. Suppose that F is symmetric about θ and is strictly increasing in a neighborhood of θ. Show that $\lambda_F(t) = 0$ if and only if $t = \theta$, where $\lambda_F(t)$ is defined by (5.57) with a strictly increasing J satisfying $J(1 - t) = -J(t)$.

45. Show that $\lambda_F(t)$ in (5.57) is differentiable at θ and $\lambda'_F(\theta)$ is equal to $-\int J'(F(x))F'(x)dF(x)$.

46. Let $T(F_n)$ be an R-estimator satisfying the conditions in Theorem 5.8. Show that (5.41) holds with

$$\sigma_F^2 = \int_0^1 [J(t)]^2 dt \bigg/ \left[\int_{-\infty}^\infty J'(F(x))F'(x)dF(x)\right]^2.$$

47. Calculate the asymptotic relative efficiency of the Hodges-Lehmann estimator in Example 5.8 w.r.t. the sample mean based on an i.i.d. sample from F when
    (a) F is the c.d.f. of $N(\mu, \sigma^2)$;
    (b) F is the c.d.f. of the logistic distribution $LG(\mu, \sigma)$;
    (c) F is the c.d.f. of the double exponential distribution $DE(\mu, \sigma)$;
    (d) $F(x) = F_0(x - \theta)$, where $F_0(x)$ is the c.d.f. of the t-distribution t_ν with $\nu \geq 3$.

48. Let G be a c.d.f. on $\mathcal{R}$. Show that $G(x) \geq t$ if and only if $x \geq G^{-1}(t)$.

49. Show that (5.67) implies that $\hat{\theta}_p$ is strongly consistent for θ_p and is $\sqrt{n}$-consistent for θ_p if $F'(\theta_p-)$ and $F'(\theta_p+)$ exist and are positive.

50. Under the condition of Theorem 5.9, show that, for $\rho_\epsilon = e^{-2\delta_\epsilon^2}$,

$$P\left(\sup_{m \geq n} |\hat{\theta}_p - \theta_p| > \epsilon\right) \leq \frac{2C\rho_\epsilon^n}{1 - \rho_\epsilon}, \qquad n = 1, 2, ...$$

51. Prove that $\varphi_n(t)$ in (5.69) is the Lebesgue p.d.f. of the pth sample quantile $\hat{\theta}_p$ when F has the Lebesgue p.d.f. f by
    (a) differentiating the c.d.f. of $\hat{\theta}_p$ in (5.68);
    (b) using result (5.66) and the result in Example 2.9.

52. Let $X_1, ..., X_n$ be i.i.d. random variables from F with a finite mean. Show that $\hat{\theta}_p$ has a finite jth moment for sufficiently large n, $j = 1, 2,...$.

53. Prove Theorem 5.10(i).

54. Suppose that a c.d.f. F has a Lebesgue p.d.f. f. Using the p.d.f. in (5.69) and Scheffé's theorem (Proposition 1.18), prove part (iv) of Theorem 5.10.

55. Let $\{k_n\}$ be a sequence of integers satisfying $k_n/n = p + o(n^{-1/2})$ with $p \in (0,1)$, and let $X_1, ..., X_n$ be i.i.d. random variables from a c.d.f. F with $F'(\theta_p) > 0$. Show that
$$\sqrt{n}(X_{(k_n)} - \theta_p) \to_d N(0, p(1-p)/[F'(\theta_p)]^2).$$

56. In the proof of Theorem 5.11, prove (5.72), (5.75), and inequality (5.74).

57. Prove Corollary 5.1.

58. Prove the claim in Example 5.9.

59. Let $T(G) = G^{-1}(p)$ be the pth quantile functional. Suppose that F has a positive derivative F' in a neighborhood of $\theta = F^{-1}(p)$. Show that T is Gâteaux differentiable at F and obtain the influence function.

60. Let $X_1, ..., X_n$ be i.i.d. from the Cauchy distribution $C(0,1)$.
    (a) Show that $E(X_{(j)})^2 < \infty$ if and only if $3 \le j \le n - 2$.
    (b) Show that $E(\hat{\theta}_{0.5})^2 < \infty$ for $n \ge 5$.

61. Suppose that F is the c.d.f. of the uniform distribution $U(\theta - \frac{1}{2}, \theta + \frac{1}{2})$, $\theta \in \mathcal{R}$. Obtain the asymptotic relative efficiency of the sample median w.r.t. the sample mean, based on an i.i.d. sample of size n from F.

62. Suppose that $F(x) = F_0(x - \theta)$ and F_0 is the c.d.f. of the Cauchy distribution $C(0,1)$ truncated at c and $-c$, i.e., F_0 has the Lebesgue p.d.f. $(1 + x^2)^{-1}I_{(-c,c)}(x)/\int_{-c}^{c}(1 + x^2)^{-1}dt$. Obtain the asymptotic relative efficiency of the sample median w.r.t. the sample mean, based on an i.i.d. sample of size n from F.

63. Let $X_1, ..., X_n$ be i.i.d. with the c.d.f. $(1-\epsilon)\Phi\left(\frac{x-\mu}{\sigma}\right) + \epsilon D\left(\frac{x-\mu}{\sigma}\right)$, where $\epsilon \in (0,1)$ is a known constant, Φ is the c.d.f. of the standard normal distribution, D is the c.d.f. of the double exponential distribution $D(0,1)$, and $\mu \in \mathcal{R}$ and $\sigma > 0$ are unknown parameters. Consider the estimation of μ. Obtain the asymptotic relative efficiency of the sample mean w.r.t. the sample median.

64. Let $X_1, ..., X_n$ be i.i.d. with the Lebesgue p.d.f. $2^{-1}(1 - \theta^2)e^{\theta x - |x|}$, where $\theta \in (-1, 1)$ is unknown.
    (a) Show that the median of the distribution of X_1 is given by $m(\theta) =$

$(1-\theta)^{-1}\log(1+\theta)$ when $\theta \geq 0$ and $m(\theta) = -m(-\theta)$ when $\theta < 0$.

(b) Show that the mean of the distribution of X_1 is $\mu(\theta) = 2\theta/(1-\theta^2)$.

(c) Show that the inverse functions of $m(\theta)$ and $\mu(\theta)$ exist. Obtain the asymptotic relative efficiency of $m^{-1}(\hat{m})$ w.r.t. $\mu^{-1}(\bar{X})$, where $\hat{m}$ is the sample median and $\bar{X}$ is the sample mean.

(e) Is $\mu^{-1}(\bar{X})$ in (d) asymptotically efficient in estimating θ?

65. Show that $\bar{X}_\alpha$ in (5.77) is the L-estimator corresponding to the J function given in Example 5.6(iii) with $\beta = 1 - \alpha$.

66. Let $X_1, ..., X_n$ be i.i.d. random variables from F, where F is symmetric about θ.

(a) Show that $X_{(j)} - \theta$ and $\theta - X_{(n-j+1)}$ have the same distribution.

(b) Show that $\sum_{j=1}^{n} w_j X_{(j)}$ has a c.d.f. symmetric about θ, if w_i's are constants satisfying $\sum_{i=1}^{n} w_i = 1$ and $w_j = w_{n-j+1}$ for all j.

(c) Show that the trimmed sample mean $\bar{X}_\alpha$ has a c.d.f. symmetric about θ.

67. Under the conditions in one of (i)-(iii) of Theorem 5.6, show that (5.41) holds for $T(F_n)$ with σ_F^2 given by (5.79), if $\sigma_F^2 < \infty$.

68. Prove (5.78) under the assumed conditions.

69. For the functional T given by (5.46), show that $T(F) = \theta$ if F is symmetric about θ and J is symmetric about $\frac{1}{2}$.

70. Obtain the asymptotic relative efficiency of the trimmed sample mean $\bar{X}_\alpha$ w.r.t. the sample mean, based on an i.i.d. sample of size n from the double exponential distribution $DE(\theta, 1)$, where $\theta \in \mathcal{R}$ is unknown.

71. Obtain the asymptotic relative efficiency of the trimmed sample mean $\bar{X}_\alpha$ w.r.t. the sample median, based on an i.i.d. sample of size n from the Cauchy distribution $C(\theta, 1)$, where $\theta \in \mathcal{R}$ is unknown.

72. Consider the α-trimmed sample mean defined in (5.77). Show that σ_α^2 in (5.78) is the same as σ_F^2 in (5.79) with $J(t) = (1-2\alpha)^{-1}I_{(\alpha,1-\alpha)}(t)$, when $F(x) = F_0(x-\theta)$ and F_0 is symmetric about 0.

73. For σ_α^2 in (5.78), show that

(a) if $F_0'(0)$ exists and is positive, then $\lim_{\alpha \to \frac{1}{2}} \sigma_\alpha^2 = 1/[2F_0'(0)]^2$;

(b) if $\sigma^2 = \int x^2 dF_0(x) < \infty$, then $\lim_{\alpha \to 0} \sigma_\alpha^2 = \sigma^2$.

74. Show that if $J \equiv 1$, then σ_F^2 in (5.79) is equal to the variance of the c.d.f. F.

75. Calculate σ_F^2 in (5.79) with $J(t) = 4t - 2$ and F being the double exponential distribution $DE(\theta, 1)$, $\theta \in \mathcal{R}$.

76. Consider the simple linear model in Example 3.12 with positive t_i's. Derive the L-estimator of β defined by (5.82) with a J symmetric about $\frac{1}{2}$ and compare it with the LSE of β.

77. Consider the one-way ANOVA model in Example 3.13. Derive the L-estimator of β defined by (5.82) when (a) J is symmetric about $\frac{1}{2}$ and (b) $J(t) = (1 - 2\alpha)^{-1} I_{(\alpha, 1-\alpha)}(t)$. Compare these L-estimators with the LSE of β.

78. Show that the method of moments in §3.5.2 is a special case of the GEE method.

79. Complete the proof of Proposition 5.4.

80. In the proof of Lemma 5.3, show that the probability in (5.94) is bounded by ϵ.

81. In Example 5.11, show that ψ_i's satisfy the conditions of Lemma 5.3 if Θ is compact and $\sup_i \|Z_i\| < \infty$.

82. In the proof of Proposition 5.5, show that $\{\Delta_n(\gamma)\}$ is equicontinuous on any open subset of Θ.

83. Prove Proposition 5.6.

84. Prove the claim in Example 5.12.

85. Prove the claims in Example 5.13.

86. For Huber's M-estimator discussed in Example 5.13, obtain a formula for $e(F)$, the asymptotic relative efficiency of $\hat{\theta}_n$ w.r.t. $\bar{X}$, when F is given by (5.76). Show that $\lim_{\tau \to \infty} e(F) = \infty$. Find the value of $e(F)$ when $\epsilon = 0$, $\sigma = 1$, and $C = 1.5$.

87. Consider the ψ function in Example 5.7(ii). Show that under some conditions on F, ψ satisfies the conditions given in Theorem 5.13(i) or (ii). Obtain σ_F^2 in (5.98) in this case.

88. In the proof of Theorem 5.14, show that
    (a) (5.101) holds;
    (b) (5.103) holds;
    (c) (5.104) implies (5.102). (Hint: use Theorem 1.9(iii).)

89. Prove the claim in Example 5.14, assuming some necessary moment conditions.

90. Derive the asymptotic distribution of the MQLE (the GEE estimator based on (5.90)), assuming that $X_i = (X_{i1}, ..., X_{id_i})$, $E(X_{it}) = me^{\eta_i}/(1 + e^{\eta_i})$, $\mathrm{Var}(X_{it}) = m\phi_i e^{\eta_i}/(1 + e^{\eta_i})^2$, and (4.57) holds with $g(t) = \log \frac{t}{1-t}$.

91. Repeat the previous exercise under the assumption that $E(X_{it}) = e^{\eta_i}$, $\text{Var}(X_{it}) = \phi_i e^{\eta_i}$, and (4.57) holds with $g(t) = \log t$ or $g(t) = 2\sqrt{t}$.

92. In Theorem 5.14, show that result (5.99) still holds if $\tilde{R}_i$ is replaced by an estimator $\hat{R}_i$ satisfying $\max_{i \leq n} \|\hat{R}_i - U_i\| = o_p(1)$, where U_i's are correlation matrices.

93. Show that (5.106) holds if and only if one of the following holds:
    (a) $\lambda_- \to_p 1$ and $\lambda_+ \to_p 1$, where λ_- and λ_+ are respectively the smallest and largest eigenvalues of $V_n^{-1/2} \hat{V}_n V_n^{-1/2}$.
    (b) $l_n^\tau \hat{V}_n l_n / l_n^\tau V_n l_n \to_p 1$, where $\{l_n\}$ is any sequence of nonzero vectors in $\mathcal{R}^k$.

94. Show that (5.105) and (5.106) imply $\hat{V}_n^{-1/2}(\hat{\theta}_n - \theta) \to_d N_k(0, I_k)$.

95. Suppose that $X_1, ..., X_n$ are independent (not necessarily identically distributed) random d-vectors with $E(X_i) = \mu$ for all i. Suppose also that $\sup_i E\|X_i\|^{2+\delta} < \infty$ for some $\delta > 0$. Let $\mu = E(X_1)$, $\theta = g(\mu)$, and $\hat{\theta}_n = g(\bar{X})$. Show that
    (a) (5.105) holds with $V_n = n^{-2}[\nabla g(\mu)]^\tau \sum_{i=1}^n \text{Var}(X_i) \nabla g(\mu)$;
    (b) $\hat{V}_n$ in (5.108) is consistent for V_n in part (a).

96. Consider the ratio estimator in Example 3.21. Derive the estimator $\hat{V}_n$ given by (5.108) and show that $\hat{V}_n$ is consistent for the asymptotic variance of the ratio estimator.

97. Derive a consistent variance estimator for $\hat{R}(t)$ in Example 3.23.

98. Prove the claims in Example 5.16.

99. Let $\sigma_{F_n}^2$ be given by (5.79) with F replaced by the empirical c.d.f. F_n.
    (a) Show that $\sigma_{F_n}^2/n$ is the same as $\hat{V}_n$ in (5.110) for an L-estimator with influence function ϕ_F.
    (b) Show directly (without using Theorem 5.15) $\sigma_{F_n}^2 \to_{a.s.} \sigma_F^2$ in (5.79), under the conditions in Theorem 5.6(i) or (ii) (with $EX_1^2 < \infty$).

100. Derive a consistent variance estimator for a U-statistic satisfying the conditions in Theorem 3.5(i).

101. Derive a consistent variance estimator for Huber's M-estimator discussed in Example 5.13.

102. Assume the conditions in Theorem 5.8. Let $r \in (0, \frac{1}{2})$.
    (a) Show that $n^r \lambda_F(\mathrm{T}(F_n) + n^{-r}) \to_p \lambda_F(\mathrm{T}(F))$.
    (b) Show that $n^r[\lambda_{F_n}(\mathrm{T}(F_n) + n^{-r}) - \lambda_F(\mathrm{T}(F_n) + n^{-r})] \to_p 0$.
    (c) Derive a consistent estimator of the asymptotic variance of $\mathrm{T}(F_n)$, using the results in (a) and (b).

103. Prove Theorem 5.16.

104. Let $X_1, ..., X_n$ be random variables and $\hat{\theta} = \bar{X}^2$. Show that the jackknife estimator in (5.112) equals $\frac{4\bar{X}^2\hat{c}_2}{n-1} - \frac{4\bar{X}\hat{c}_3}{(n-1)^2} + \frac{\hat{c}_4 - \hat{c}_2^2}{(n-1)^3}$, where $\hat{c}_j$'s are the sample central moments defined by (3.52).

105. Prove Theorem 5.17 for the case where g is from $\mathcal{R}^d$ to $\mathcal{R}^k$ and $k \geq 2$.

106. Prove (5.114).

107. In the proof of Theorem 5.18, prove (5.116).

108. Show that $\hat{\theta}_{-i}$'s in (5.118) satisfy (5.117), under the conditions of Theorem 5.14.

109. Prove Theorem 5.19.

110. Prove (5.119).

111. Let $X_1, ..., X_n$ be random variables and $\hat{\theta} = \bar{X}^2$. Show that the bootstrap variance estimator based on i.i.d. X_i^*'s from F_n is equal to $\hat{V}_B = \frac{4\bar{X}^2\hat{c}_2}{n} + \frac{4\bar{X}\hat{c}_3}{n^2} + \frac{\hat{c}_4}{n^3}$, where $\hat{c}_j$'s are the sample central moments defined by (3.52).

112. Let $G, G_1, G_2,...,$ be c.d.f.'s on $\mathcal{R}$. Suppose that $\varrho_\infty(G_j, G) \to 0$ as $j \to \infty$ and $G'(x)$ exists and is positive for all $x \in \mathcal{R}$. Show that $G_j^{-1}(p) \to G^{-1}(p)$ for any $p \in (0, 1)$.

113. Let $X_1, ..., X_n$ be i.i.d. from a c.d.f. F on $\mathcal{R}^d$ with a finite $\text{Var}(X_1)$. Let $X_1^*, ..., X_n^*$ be i.i.d. from the empirical c.d.f. F_n. Show that for almost all given sequences $X_1, X_2, ...,$ $\sqrt{n}(\bar{X}^* - \bar{X}) \to_d N(0, \text{Var}(X_1))$. (Hint: verify Lindeberg's condition.)

114. Let $X_1, ..., X_n$ be i.i.d. from a c.d.f. F on $\mathcal{R}^d$, $X_1^*, ..., X_n^*$ be i.i.d. from the empirical c.d.f. F_n, and let F_n^* be the empirical c.d.f. based on X_i^*'s. Using DKW's inequality (Lemma 5.1), show that
(a) $\varrho_\infty(F_n^*, F) \to_{a.s.} 0$;
(b) $\varrho_\infty(F_n^*, F) = O_p(n^{-1/2})$;
(c) $\varrho_{L_p}(F_n^*, F) = O_p(n^{-1/2})$, under the condition in Theorem 5.20(ii).

115. Using the results from the previous two exercises, prove Theorem 5.20(ii).

116. Under the conditions in Theorem 5.11, establish a Bahadur's representation for the bootstrap sample quantile $\hat{\theta}_p^*$.

Chapter 6

Hypothesis Tests

A general theory of testing hypotheses is presented in this chapter. Let X be a sample from a population P in $\mathcal{P}$, a family of populations. Based on the observed X, we test a given hypothesis $H_0 : P \in \mathcal{P}_0$ versus $H_1 : P \in \mathcal{P}_1$, where $\mathcal{P}_0$ and $\mathcal{P}_1$ are two disjoint subsets of $\mathcal{P}$ and $\mathcal{P}_0 \cup \mathcal{P}_1 = \mathcal{P}$. Notational conventions and basic concepts (such as two types of errors, significance levels, and sizes) given in Example 2.20 and §2.4.2 are used in this chapter.

6.1 UMP Tests

A test for a hypothesis is a statistic $T(X)$ taking values in $[0, 1]$. When $X = x$ is observed, we reject H_0 with probability $T(x)$ and accept H_0 with probability $1 - T(x)$. If $T(X) = 1$ or 0 a.s. $\mathcal{P}$, then $T(X)$ is a nonrandomized test. Otherwise $T(X)$ is a randomized test. For a given test $T(X)$, the *power function* of $T(X)$ is defined to be

$$\beta_T(P) = E[T(X)], \qquad P \in \mathcal{P}, \qquad (6.1)$$

which is the type I error probability of $T(X)$ when $P \in \mathcal{P}_0$ and one minus the type II error probability of $T(X)$ when $P \in \mathcal{P}_1$.

As we discussed in §2.4.2, with a sample of a fixed size, we are not able to minimize two error probabilities simultaneously. Our approach involves maximizing the power $\beta_T(P)$ over all $P \in \mathcal{P}_1$ (i.e., minimizing the type II error probability) and over all tests T satisfying

$$\sup_{P \in \mathcal{P}_0} \beta_T(P) \leq \alpha, \qquad (6.2)$$

where $\alpha \in [0, 1]$ is a given level of significance. Recall that the left-hand side of (6.2) is defined to be the size of T.

Definition 6.1. A test T_* of size α is a *uniformly most powerful* (UMP) test if and only if $\beta_{T_*}(P) \geq \beta_T(P)$ for all $P \in \mathcal{P}_1$ and T of level α.   ∎

If $U(X)$ is a sufficient statistic for $P \in \mathcal{P}$, then for any test $T(X)$, $E(T|U)$ has the same power function as T and, therefore, to find a UMP test we may consider tests that are functions of U only.

The existence and characteristics of UMP tests are studied in this section.

6.1.1 The Neyman-Pearson lemma

A hypothesis H_0 (or H_1) is said to be *simple* if and only if $\mathcal{P}_0$ (or $\mathcal{P}_1$) contains exactly one population. The following useful result, which has already been used once in the proof of Theorem 4.16, provides the form of UMP tests when *both* H_0 and H_1 are simple.

Theorem 6.1 (Neyman-Pearson lemma). Suppose that $\mathcal{P}_0 = \{P_0\}$ and $\mathcal{P}_1 = \{P_1\}$. Let f_j be the p.d.f. of P_j w.r.t. a σ-finite measure ν (e.g., $\nu = P_0 + P_1$), $j = 0, 1$.
(i) (Existence of a UMP test). For every α, there exists a UMP test of size α, which is equal to

$$T_*(X) = \begin{cases} 1 & f_1(X) > cf_0(X) \\ \gamma & f_1(X) = cf_0(X) \\ 0 & f_1(X) < cf_0(X), \end{cases} \tag{6.3}$$

where $\gamma \in [0, 1]$ and $c \geq 0$ are some constants chosen so that $E[T_*(X)] = \alpha$ when $P = P_0$ ($c = \infty$ is allowed).
(ii) (Uniqueness). If T_{**} is a UMP test of size α, then

$$T_{**}(X) = \begin{cases} 1 & f_1(X) > cf_0(X) \\ 0 & f_1(X) < cf_0(X) \end{cases} \quad \text{a.s. } \mathcal{P}. \tag{6.4}$$

Proof. The proof for the case of $\alpha = 0$ or 1 is left as an exercise. Assume now that $0 < \alpha < 1$.
(i) We first show that there exist γ and c such that $E_0[T_*(X)] = \alpha$, where E_j is the expectation w.r.t. P_j. Let $\gamma(t) = P_0(f_1(X) > tf_0(X))$. Then $\gamma(t)$ is nonincreasing, $\gamma(-\infty) = 1$, and $\gamma(\infty) = 0$ (why?). Thus, there exists a $c \in (0, \infty)$ such that $\gamma(c) \leq \alpha \leq \gamma(c-)$. Set

$$\gamma = \begin{cases} \frac{\alpha - \gamma(c)}{\gamma(c-) - \gamma(c)} & \gamma(c-) \neq \gamma(c) \\ 0 & \gamma(c-) = \gamma(c). \end{cases}$$

Note that $\gamma(c-) - \gamma(c) = P(f_1(X) = cf_0(X))$. Then

$$E_0[T_*(X)] = P_0\big(f_1(X) > cf_0(X)\big) + \gamma P_0\big(f_1(X) = cf_0(X)\big) = \alpha.$$

Next, we show that T_* in (6.3) is a UMP test. Suppose that $T(X)$ is a test satisfying $E_0[T(X)] \leq \alpha$. If $T_*(x) - T(x) > 0$, then $T_*(x) > 0$ and, therefore, $f_1(x) \geq c f_0(x)$. If $T_*(x) - T(x) < 0$, then $T_*(x) < 1$ and, therefore, $f_1(x) \leq c f_0(x)$. In any case, $[T_*(x) - T(x)][f_1(x) - c f_0(x)] \geq 0$ and, therefore,

$$\int [T_*(x) - T(x)][f_1(x) - c f_0(x)] d\nu \geq 0,$$

i.e.,

$$\int [T_*(x) - T(x)] f_1(x) d\nu \geq c \int [T_*(x) - T(x)] f_0(x) d\nu. \qquad (6.5)$$

The left-hand side of (6.5) is $E_1[T_*(X)] - E_1[T(X)]$ and the right-hand side of (6.5) is $c\{E_0[T_*(X)] - E_0[T(X)]\} = c\{\alpha - E_0[T(X)]\} \geq 0$. This proves the result in (i).

(ii) Let $T_{**}(X)$ be a UMP test of size α. Define

$$A = \{x : T_*(x) \neq T_{**}(x), \quad f_1(x) \neq c f_0(x)\}.$$

Then $[T_*(x) - T_{**}(x)][f_1(x) - c f_0(x)] > 0$ when $x \in A$ and $= 0$ when $x \in A^c$, and

$$\int [T_*(x) - T_{**}(x)][f_1(x) - c f_0(x)] d\nu = 0,$$

since both T_* and T_{**} are UMP tests of size α. By Proposition 1.6(ii), $\nu(A) = 0$. This proves (6.4). ∎

Theorem 6.1 shows that when both H_0 and H_1 are simple, there exists a UMP test that can be determined by (6.4) uniquely (a.s. $\mathcal{P}$) except on the set $B = \{x : f_1(x) = c f_0(x)\}$. If $\nu(B) = 0$, then we have a unique nonrandomized UMP test; otherwise UMP tests are randomized on the set B and the randomization is necessary for UMP tests to have the given size α; furthermore, we can always choose a UMP test that is constant on B.

Example 6.1. Suppose that X is a sample of size 1, $\mathcal{P}_0 = \{P_0\}$, and $\mathcal{P}_1 = \{P_1\}$, where P_0 is $N(0,1)$ and P_1 is the double exponential distribution $DE(0,2)$ with the p.d.f. $4^{-1} e^{-|x|/2}$. Since $P(f_1(X) = c f_0(X)) = 0$, there is a unique nonrandomized UMP test. From (6.3), the UMP test $T_*(x) = 1$ if and only if $\frac{\pi}{8} e^{x^2 - |x|} > c^2$ for some $c > 0$, which is equivalent to $|x| > t$ or $|x| < 1 - t$ for some $t > \frac{1}{2}$. Suppose that $\alpha < \frac{1}{3}$. To determine t, we use

$$\alpha = E_0[T_*(X)] = P_0(|X| > t) + P_0(|X| < 1 - t). \qquad (6.6)$$

If $t \leq 1$, then $P_0(|X| > t) \geq P_0(|X| > 1) = 0.3374 > \alpha$. Hence t should be larger than 1 and (6.6) becomes

$$\alpha = P_0(|X| > t) = \Phi(-t) + 1 - \Phi(t).$$

Thus, $t = \Phi^{-1}(1 - \alpha/2)$ and $T_*(X) = I_{(t,\infty)}(|X|)$. Note that it is not necessary to find out what c is.

Intuitively, the reason why the UMP test in this example rejects H_0 when $|X|$ is large is that the probability of getting a large $|X|$ is much higher under H_1 (i.e., P is the double exponential distribution $DE(0,2)$).

The power of T_* when $P \in \mathcal{P}_1$ is

$$E_1[T_*(X)] = P_1(|X| > t) = 1 - \frac{1}{4}\int_{-t}^{t} e^{-|x|/2}dx = e^{-t/2}. \quad \blacksquare$$

Example 6.2. Let $X_1, ..., X_n$ be i.i.d. binary random variables with $p = P(X_1 = 1)$. Suppose that $H_0 : p = p_0$ and $H_1 : p = p_1$, where $0 < p_0 < p_1 < 1$. By Theorem 6.1, a UMP test of size α is

$$T_*(Y) = \begin{cases} 1 & \lambda(Y) > c \\ \gamma & \lambda(Y) = c \\ 0 & \lambda(Y) < c, \end{cases}$$

where $Y = \sum_{i=1}^{n} X_i$ and

$$\lambda(Y) = \left(\frac{p_1}{p_0}\right)^Y \left(\frac{1-p_1}{1-p_0}\right)^{n-Y}.$$

Since $\lambda(Y)$ is increasing in Y, there is an integer $m > 0$ such that

$$T_*(Y) = \begin{cases} 1 & Y > m \\ \gamma & Y = m \\ 0 & Y < m, \end{cases}$$

where m and γ satisfy $\alpha = E_0[T_*(Y)] = P_0(Y > m) + \gamma P_0(Y = m)$. Since Y has the binomial distribution $Bi(p, n)$, we can determine m and γ from

$$\alpha = \sum_{j=m+1}^{n} \binom{n}{j} p_0^j (1-p_0)^{n-j} + \gamma \binom{n}{m} p_0^m (1-p_0)^{n-m}. \quad (6.7)$$

Unless

$$\alpha = \sum_{j=m+1}^{n} \binom{n}{j} p_0^j (1-p_0)^{n-j}$$

for some integer m, in which case we can choose $\gamma = 0$, the UMP test T_* is a randomized test. $\quad \blacksquare$

An interesting phenomenon in Example 6.2 is that the UMP test T_* does not depend on p_1. In such a case, T_* is in fact a UMP test for testing $H_0 : p = p_0$ versus $H_1 : p > p_0$.

Lemma 6.1. Suppose that there is a test T_* of size α such that for every $P_1 \in \mathcal{P}_1$, T_* is UMP for testing H_0 versus the hypothesis $P = P_1$. Then T_* is UMP for testing H_0 versus H_1.

Proof. For any test T of level α, T is also of level α for testing H_0 versus the hypothesis $P = P_1$ with any $P_1 \in \mathcal{P}_1$. Hence $\beta_{T*}(P_1) \geq \beta_T(P_1)$. ∎

We conclude this section with the following generalized Neyman-Pearson lemma. Its proof is left to the reader. Other extensions of the Neyman-Pearson lemma can be found in Exercises 8 and 9 in §6.6.

Proposition 6.1. Let $f_1, ..., f_{m+1}$ be Borel functions on $\mathcal{R}^p$ that are integrable w.r.t. a σ-finite measure ν. For given constants $t_1, ..., t_m$, let $\mathcal{T}$ be the class of Borel functions ϕ (from $\mathcal{R}^p$ to $[0, 1]$) satisfying

$$\int \phi f_i d\nu \leq t_i, \quad i = 1, ..., m, \tag{6.8}$$

and $\mathcal{T}_0$ be the set of ϕ's in $\mathcal{T}$ satisfying (6.8) with all inequalities replaced by equalities. If there are constants $c_1, ..., c_m$ such that

$$\phi_*(x) = \begin{cases} 1 & f_{m+1}(x) > c_1 f_1(x) + \cdots + c_m f_m(x) \\ 0 & f_{m+1}(x) < c_1 f_1(x) + \cdots + c_m f_m(x) \end{cases} \tag{6.9}$$

is a member of $\mathcal{T}_0$, then ϕ_* maximizes $\int \phi f_{m+1} d\nu$ over $\phi \in \mathcal{T}_0$. If $c_i \geq 0$ for all i, then ϕ_* maximizes $\int \phi f_{m+1} d\nu$ over $\phi \in \mathcal{T}$. ∎

The existence of constants c_i's in (6.9) is considered in the following lemma whose proof can be found in Lehmann (1986, pp. 97-99).

Lemma 6.2. Let $f_1, ..., f_m$ and ν be given by Proposition 6.1. Then the set $M = \{(\int \phi f_1 d\nu, ..., \int \phi f_m d\nu) : \phi$ is from $\mathcal{R}^p$ to $[0, 1]\}$ is convex and closed. If $(t_1, ..., t_m)$ is an interior point of M, then there exist constants $c_1, ..., c_m$ such that the function defined by (6.9) is in $\mathcal{T}_0$. ∎

6.1.2 Monotone likelihood ratio

The case where both H_0 and H_1 are simple is mainly of theoretical interest. If a hypothesis is not simple, it is called composite. As we discussed in §6.1.1, UMP tests for composite H_1 exist in the problem discussed in Example 6.2. We now extend this result to a class of parametric problems in which the likelihood functions have a special property.

Definition 6.2. Suppose that the distribution of X is in $\mathcal{P} = \{P_\theta : \theta \in \Theta\}$, a parametric family indexed by a real-valued θ, and that $\mathcal{P}$ is dominated by a σ-finite measure ν. Let $f_\theta = dP_\theta/d\nu$. The family $\mathcal{P}$ is said to have *monotone likelihood ratio* in $Y(X)$ (a real-valued statistic) if and only if, for any $\theta_1 < \theta_2$, $f_{\theta_2}(x)/f_{\theta_1}(x)$ is a nondecreasing function of $Y(x)$ for values x at which at least one of $f_{\theta_1}(x)$ and $f_{\theta_2}(x)$ is positive. ∎

The following lemma states a useful result for a family with monotone likelihood ratio.

Lemma 6.3. Suppose that the distribution of X is in a parametric family $\mathcal{P}$ indexed by a real-valued θ and that $\mathcal{P}$ has monotone likelihood ratio in $Y(X)$. If ψ is a nondecreasing function of Y, then $g(\theta) = E[\psi(Y)]$ is a nondecreasing function of θ.
Proof. Let $\theta_1 < \theta_2$, $A = \{x : f_{\theta_1}(x) > f_{\theta_2}(x)\}$, $a = \sup_{x \in A} \psi(Y(x))$, $B = \{x : f_{\theta_1}(x) < f_{\theta_2}(x)\}$, and $b = \inf_{x \in B} \psi(Y(x))$. Since $\mathcal{P}$ has monotone likelihood ratio in $Y(X)$ and ψ is nondecreasing in Y, $b \geq a$. Then the result follows from

$$g(\theta_2) - g(\theta_1) = \int \psi(Y(x))(f_{\theta_2} - f_{\theta_1})(x)d\nu$$

$$\geq a \int_A (f_{\theta_2} - f_{\theta_1})(x)d\nu + b \int_B (f_{\theta_2} - f_{\theta_1})(x)d\nu$$

$$= (b - a) \int_B (f_{\theta_2} - f_{\theta_1})(x)d\nu$$

$$\geq 0. \quad ∎$$

Before discussing UMP tests in families with monotone likelihood ratio, let us consider some examples of such families.

Example 6.3. Let θ be real-valued and $\eta(\theta)$ be a nondecreasing function of θ. Then the one-parameter exponential family with

$$f_\theta(x) = \exp\{\eta(\theta)Y(x) - \xi(\theta)\}h(x) \tag{6.10}$$

has monotone likelihood ratio in $Y(X)$. From Tables 1.1-1.2 (§1.3.1), this includes the binomial family $\{Bi(\theta, r)\}$, the Poisson family $\{P(\theta)\}$, the negative binomial family $\{NB(\theta, r)\}$, the log-distribution family $\{L(\theta)\}$, the normal family $\{N(\theta, c^2)\}$ or $\{N(c, \theta)\}$, the exponential family $\{E(c, \theta)\}$, the gamma family $\{\Gamma(\theta, c)\}$ or $\{\Gamma(c, \theta)\}$, the beta family $\{B(\theta, c)\}$ or $\{B(c, \theta)\}$, and the double exponential family $\{DE(c, \theta)\}$, where r or c is known. ∎

Example 6.4. Let $X_1, ..., X_n$ be i.i.d. from the uniform distribution on $(0, \theta)$, where $\theta > 0$. The Lebesgue p.d.f. of $X = (X_1, ..., X_n)$ is $f_\theta(x) = \theta^{-n} I_{(0,\theta)}(x_{(n)})$, where $x_{(n)}$ is the value of the largest order statistic $X_{(n)}$. For $\theta_1 < \theta_2$,

$$\frac{f_{\theta_2}(x)}{f_{\theta_1}(x)} = \frac{\theta_1^n \, I_{(0,\theta_2)}(x_{(n)})}{\theta_2^n \, I_{(0,\theta_1)}(x_{(n)})},$$

which is a nondecreasing function of $x_{(n)}$ for x's at which at least one of $f_{\theta_1}(x)$ and $f_{\theta_2}(x)$ is positive, i.e., $x_{(n)} < \theta_2$. Hence the family of distributions of X has monotone likelihood ratio in $X_{(n)}$. $\blacksquare$

Example 6.5. The following families have monotone likelihood ratio:
(a) the double exponential distribution family $\{DE(\theta, c)\}$ with a known c;
(b) the exponential distribution family $\{E(\theta, c)\}$ with a known c;
(c) the logistic distribution family $\{LG(\theta, c)\}$ with a known c;
(d) the uniform distribution family $\{U(\theta, \theta + 1)\}$;
(e) the hypergeometric distribution family $\{HG(r, \theta, N - \theta)\}$ with known r and N (Table 1.1, page 18).

An example of a family that does not have monotone likelihood ratio is the Cauchy distribution family $\{C(\theta, c)\}$ with a known c. $\blacksquare$

Hypotheses of the form $H_0 : \theta \leq \theta_0$ (or $H_0 : \theta \geq \theta_0$) versus $H_1 : \theta > \theta_0$ (or $H_1 : \theta < \theta_0$) are called *one-sided* hypotheses for any given constant θ_0. The following result provides UMP tests for testing one-sided hypotheses when the distribution of X is in a parametric family with monotone likelihood ratio.

Theorem 6.2. Suppose that X has a distribution in $\mathcal{P} = \{P_\theta : \theta \in \Theta\}$ ($\Theta \subset \mathcal{R}$) that has monotone likelihood ratio in $Y(X)$. Consider the problem of testing $H_0 : \theta \leq \theta_0$ versus $H_1 : \theta > \theta_0$, where θ_0 is a given constant.
(i) There exists a UMP test of size α, which is given by

$$
T_*(X) = \begin{cases} 1 & Y(X) > c \\ \gamma & Y(X) = c \\ 0 & Y(X) < c, \end{cases} \tag{6.11}
$$

where c and γ are determined by $\beta_{T_*}(\theta_0) = \alpha$, and $\beta_T(\theta) = E[T(X)]$ is the power function of a test T.
(ii) $\beta_{T_*}(\theta)$ is strictly increasing for all θ's for which $0 < \beta_{T_*}(\theta) < 1$.
(iii) For any $\theta < \theta_0$, T_* minimizes $\beta_T(\theta)$ (the type I error probability of T) among all tests T satisfying $\beta_T(\theta_0) = \alpha$.
(iv) Assume that $P_\theta(f_\theta(X) = cf_{\theta_0}(X)) = 0$ for any $\theta > \theta_0$ and $c \geq 0$, where f_θ is the p.d.f. of P_θ. If T is a test with $\beta_T(\theta_0) = \beta_{T_*}(\theta_0)$, then for any $\theta > \theta_0$, either $\beta_T(\theta) < \beta_{T_*}(\theta)$ or $T = T_*$ a.s. P_θ.
(v) For any fixed θ_1, T_* is UMP for testing $H_0 : \theta \leq \theta_1$ versus $H_1 : \theta > \theta_1$, with size $\beta_{T_*}(\theta_1)$.
Proof. (i) Consider the hypotheses $\theta = \theta_0$ versus $\theta = \theta_1$ with any $\theta_1 > \theta_0$. From Theorem 6.1, a UMP test is given by (6.3) with $f_j = $ the p.d.f. of P_{θ_j}, $j = 0, 1$. Since $\mathcal{P}$ has monotone likelihood ratio in $Y(X)$, this UMP test can be chosen to be the same as T_* in (6.11) with possibly different c and γ satisfying $\beta_{T_*}(\theta_0) = \alpha$. Since T_* does not depend on θ_1, it follows from

Lemma 6.1 that T_* is UMP for testing the hypothesis $\theta = \theta_0$ versus H_1.

Note that if T_* is UMP for testing $\theta = \theta_0$ versus H_1, then it is UMP for testing H_0 versus H_1, provided that $\beta_{T_*}(\theta) \leq \alpha$ for all $\theta \leq \theta_0$, i.e., the size of T_* is α. But this follows from Lemma 6.3, i.e., $\beta_{T_*}(\theta)$ is nondecreasing in θ. This proves (i).
(ii) See Exercise 2 in §6.6.
(iii) The result can be proved using Theorem 6.1 with all inequalities reversed.
(iv) The proof for (iv) is left as an exercise.
(v) The proof for (v) is similar to that of (i).   ∎

By reversing inequalities throughout, we can obtain UMP tests for testing $H_0 : \theta \geq \theta_0$ versus $H_1 : \theta < \theta_0$.

A major application of Theorem 6.2 is to problems with one-parameter exponential families.

Corollary 6.1. Suppose that X has the p.d.f. given by (6.10) w.r.t. a σ-finite measure, where η is a strictly monotone function of θ. If η is increasing, then T_* given by (6.11) is UMP for testing $H_0 : \theta \leq \theta_0$ versus $H_1 : \theta > \theta_0$, where γ and c are determined by $\beta_{T_*}(\theta_0) = \alpha$. If η is decreasing or $H_0 : \theta \geq \theta_0$ ($H_1 : \theta < \theta_0$), the result is still valid by reversing inequalities in (6.11).   ∎

Example 6.6. Let $X_1, ..., X_n$ be i.i.d. from the $N(\mu, \sigma^2)$ distribution with an unknown $\mu \in \mathcal{R}$ and a known σ^2. Consider $H_0 : \mu \leq \mu_0$ versus $H_1 : \mu > \mu_0$, where μ_0 is a fixed constant. The p.d.f. of $X = (X_1, ..., X_n)$ is of the form (6.10) with $Y(X) = \bar{X}$ and $\eta(\mu) = n\mu/\sigma^2$. By Corollary 6.1 and the fact that $\bar{X}$ is $N(\mu, \sigma^2/n)$, the UMP test is $T_*(X) = I_{(c_\alpha, \infty)}(\bar{X})$, where $c_\alpha = \sigma z_{1-\alpha}/\sqrt{n} + \mu_0$ and $z_a = \Phi^{-1}(a)$ (see also Example 2.28).   ∎

To derive a UMP test for testing $H_0 : \theta \leq \theta_0$ versus $H_1 : \theta > \theta_0$ when X has the p.d.f. (6.10), it is essential to know the distribution of $Y(X)$. Typically, a nonrandomized test can be obtained if the distribution of Y is continuous; otherwise UMP tests are randomized.

Example 6.7. Let $X_1, ..., X_n$ be i.i.d. binary random variables with $p = P(X_1 = 1)$. The p.d.f. of $X = (X_1, ..., X_n)$ is of the form (6.10) with $Y = \sum_{i=1}^n X_i$ and $\eta(p) = \log \frac{p}{1-p}$. Note that $\eta(p)$ is a strictly increasing function of p. By Corollary 6.1, a UMP test for $H_0 : p \leq p_0$ versus $H_1 : p > p_0$ is given by (6.11), where c and γ are determined by (6.7) with $c = m$.   ∎

Example 6.8. Let $X_1, ..., X_n$ be i.i.d. random variables from the Poisson distribution $P(\theta)$ with an unknown $\theta > 0$. The p.d.f. of $X = (X_1, ..., X_n)$

is of the form (6.10) with $Y(X) = \sum_{i=1}^{n} X_i$ and $\eta(\theta) = \log \theta$. Note that Y has the Poisson distribution $P(n\theta)$. By Corollary 6.1, a UMP test for $H_0 : \theta \leq \theta_0$ versus $H_1 : \theta > \theta_0$ is given by (6.11) with c and γ satisfying

$$\alpha = \sum_{j=c+1}^{\infty} \frac{e^{n\theta_0}(n\theta_0)^j}{j!} + \gamma \frac{e^{n\theta_0}(n\theta_0)^c}{c!}. \quad \blacksquare$$

Example 6.9. Let $X_1, ..., X_n$ be i.i.d. random variables from the uniform distribution $U(0, \theta)$, $\theta > 0$. Consider the hypotheses $H_0 : \theta \leq \theta_0$ and $H_1 : \theta > \theta_0$. Since the p.d.f. of $X = (X_1, ..., X_n)$ is in a family with monotone likelihood ratio in $Y(X) = X_{(n)}$ (Example 6.4), by Theorem 6.2, a UMP test is of the form (6.11). Since $X_{(n)}$ has the Lebesgue p.d.f. $n\theta^{-n}x^{n-1}I_{(0,\theta)}(x)$, the UMP test in (6.11) is nonrandomized and

$$\alpha = \beta_{T_*}(\theta_0) = \frac{n}{\theta_0^n} \int_c^{\theta_0} x^{n-1}dx = 1 - \frac{c^n}{\theta_0^n}.$$

Hence $c = \theta_0(1 - \alpha)^{1/n}$. The power function of T_* when $\theta > \theta_0$ is

$$\beta_{T_*}(\theta) = \frac{n}{\theta^n} \int_c^{\theta} x^{n-1}dx = 1 - \frac{\theta_0^n(1 - \alpha)}{\theta^n}.$$

In this problem, however, UMP tests are not unique. (Note that the condition $P_\theta(f_\theta(X) = cf_{\theta_0}(X)) = 0$ in Theorem 6.2(iv) is not satisfied.) It can be shown (exercise) that the following test is also UMP with size α:

$$T(X) = \begin{cases} 1 & X_{(n)} > \theta_0 \\ \alpha & X_{(n)} \leq \theta_0. \end{cases} \quad \blacksquare$$

6.1.3 UMP tests for two-sided hypotheses

The following hypotheses are called two-sided hypotheses:

$$H_0 : \theta \leq \theta_1 \text{ or } \theta \geq \theta_2 \quad \text{versus} \quad H_1 : \theta_1 < \theta < \theta_2, \quad (6.12)$$

$$H_0 : \theta_1 \leq \theta \leq \theta_2 \quad \text{versus} \quad H_1 : \theta < \theta_1 \text{ or } \theta > \theta_2, \quad (6.13)$$

$$H_0 : \theta = \theta_0 \quad \text{versus} \quad H_1 : \theta \neq \theta_0, \quad (6.14)$$

where θ_0, θ_1, and θ_2 are given constants and $\theta_1 < \theta_2$.

Theorem 6.3. Suppose that X has the p.d.f. given by (6.10) w.r.t. a σ-finite measure, where η is a strictly increasing function of θ.
(i) For testing hypotheses (6.12), a UMP test of size α is

$$T_*(X) = \begin{cases} 1 & c_1 < Y(X) < c_2 \\ \gamma_i & Y(X) = c_i, \ i = 1, 2 \\ 0 & Y(X) < c_1 \text{ or } Y(X) > c_2, \end{cases} \quad (6.15)$$

where c_i's and γ_i's are determined by

$$\beta_{T_*}(\theta_1) = \beta_{T_*}(\theta_2) = \alpha. \qquad (6.16)$$

(ii) The test defined by (6.15) minimizes $\beta_T(\theta)$ over all $\theta < \theta_1$, $\theta > \theta_2$, and T satisfying $\beta_T(\theta_1) = \beta_T(\theta_2) = \alpha$.

(iii) If T_* and T_{**} are two tests satisfying (6.15) and $\beta_{T_*}(\theta_1) = \beta_{T_{**}}(\theta_1)$ and if the region $\{T_{**} = 1\}$ is to the right of $\{T_* = 1\}$, then $\beta_{T_*}(\theta) < \beta_{T_{**}}(\theta)$ for $\theta > \theta_1$ and $\beta_{T_*}(\theta) > \beta_{T_{**}}(\theta)$ for $\theta < \theta_1$. If both T_* and T_{**} satisfy (6.15) and (6.16), then $T_* = T_{**}$ a.s. $\mathcal{P}$.

Proof. (i) The distribution of Y has a p.d.f.

$$g_\theta(y) = \exp\{\eta(\theta)y - \xi(\theta)\} \qquad (6.17)$$

(Theorem 2.1). Since Y is sufficient for θ, we only need to consider tests of the form $T(Y)$. Let $\theta_1 < \theta_3 < \theta_2$. Consider the problem of testing $\theta = \theta_1$ or $\theta = \theta_2$ versus $\theta = \theta_3$. Clearly, (α, α) is an interior point of the set of all points $(\beta_T(\theta_1), \beta_T(\theta_2))$ as T ranges over all tests of the form $T(Y)$. By (6.17) and Lemma 6.2, there are constants $\tilde{c}_1$ and $\tilde{c}_2$ such that

$$T_*(Y) = \begin{cases} 1 & a_1 e^{b_1 Y} + a_2 e^{b_2 Y} < 1 \\ 0 & a_1 e^{b_1 Y} + a_2 e^{b_2 Y} > 1 \end{cases}$$

satisfies (6.16), where $a_i = \tilde{c}_i e^{\xi(\theta_3) - \xi(\theta_i)}$ and $b_i = \eta(\theta_i) - \eta(\theta_3)$, $i = 1, 2$. Clearly a_i's cannot both be ≤ 0. If one of the a_i's is ≤ 0 and the other is > 0, then $a_1 e^{b_1 Y} + a_2 e^{b_2 Y}$ is strictly monotone (since $b_1 < 0 < b_2$) and T_* or $1 - T_*$ is of the form (6.11), which has a strictly monotone power function (Theorem 6.2) and, therefore, cannot satisfy (6.16). Thus, both a_i's are positive. Then, T_* is of the form (6.15) (since $b_1 < 0 < b_2$) and it follows from Proposition 6.1 that T_* is UMP for testing $\theta = \theta_1$ or $\theta = \theta_2$ versus $\theta = \theta_3$. Since T_* does not depend on θ_3, it follows from Lemma 6.1 that T_* is UMP for testing $\theta = \theta_1$ or $\theta = \theta_2$ versus H_1.

To show that T_* is a UMP test of size α for testing H_0 versus H_1, it remains to show that $\beta_{T_*}(\theta) \leq \alpha$ for $\theta \leq \theta_1$ or $\theta \geq \theta_2$. But this follows from part (ii) of the theorem by comparing T_* with the test $T(Y) \equiv \alpha$.

(ii) The proof is similar to that in (i) and is left as an exercise.

(iii) The first claim in (iii) follows from Lemma 6.4, since the function $T_{**} - T_*$ has a single change of sign. The second claim in (iii) follows from the first claim. ∎

Lemma 6.4. Suppose that X has a p.d.f. in $\{f_\theta(x) : \theta \in \Theta\}$, a parametric family of p.d.f.'s w.r.t. a single σ-finite measure ν on $\mathcal{R}$, where $\Theta \subset \mathcal{R}$. Suppose that this family has monotone likelihood ratio in X. Let ψ be a function with a single change of sign.

(i) There exists $\theta_0 \in \Theta$ such that $E_\theta[\psi(X)] \leq 0$ for $\theta < \theta_0$ and $E_\theta[\psi(X)] \geq 0$

for $\theta > \theta_0$, where E_θ is the expectation w.r.t. f_θ.

(ii) Suppose that $f_\theta(x) > 0$ for all x and θ, that $f_{\theta_1}(x)/f_\theta(x)$ is strictly increasing in x for $\theta < \theta_1$, and that $\nu(\{x : \psi(x) \neq 0\}) > 0$. If $E_{\theta_0}[\psi(X)] = 0$, then $E_\theta[\psi(X)] < 0$ for $\theta < \theta_0$ and $E_\theta[\psi(X)] > 0$ for $\theta > \theta_0$.

Proof. (i) Suppose that there is an $x_0 \in \mathcal{R}$ such that $\psi(x) \leq 0$ for $x < x_0$ and $\psi(x) \geq 0$ for $x > x_0$. Let $\theta_1 < \theta_2$. We first show that $E_{\theta_1}[\psi(X)] > 0$ implies $E_{\theta_2}[\psi(X)] \geq 0$. If $f_{\theta_2}(x_0)/f_{\theta_1}(x_0) = \infty$, then $f_{\theta_1}(x) = 0$ for $x \geq x_0$ and, therefore, $E_{\theta_1}[\psi(X)] \leq 0$. Hence $f_{\theta_2}(x_0)/f_{\theta_1}(x_0) = c < \infty$. Then $\psi(x) \geq 0$ on the set $A = \{x : f_{\theta_1}(x) = 0 \text{ and } f_{\theta_2}(x) > 0\}$. Thus,

$$
\begin{aligned}
E_{\theta_2}[\psi(X)] &\geq \int_{A^c} \psi \frac{f_{\theta_2}}{f_{\theta_1}} f_{\theta_1} d\nu \\
&\geq \int_{x < x_0} c\psi f_{\theta_1} d\nu + \int_{x \geq x_0} c\psi f_{\theta_1} d\nu \quad (6.18) \\
&= cE_{\theta_1}[\psi(X)].
\end{aligned}
$$

The result follows by letting $\theta_0 = \inf\{\theta : E_\theta[\psi(X)] > 0\}$.

(ii) Under the assumed conditions, $f_{\theta_2}(x_0)/f_{\theta_1}(x_0) = c < \infty$. The result follows from the proof in (i) with θ_1 replaced by θ_0 and the fact that $\geq$ should be replaced by $>$ in (6.18) under the assumed conditions. ∎

Part (iii) of Theorem 6.3 shows that the c_i's and γ_i's are uniquely determined by (6.15) and (6.16). It also indicates how to select the c_i's and γ_i's. One can start with some trial values $c_1^{(0)}$ and $\gamma_1^{(0)}$, find $c_2^{(0)}$ and $\gamma_2^{(0)}$ such that $\beta_{T_*}(\theta_1) = \alpha$, and compute $\beta_{T_*}(\theta_2)$. If $\beta_{T_*}(\theta_2) < \alpha$, by Theorem 6.3(iii), the correct rejection region $\{T_* = 1\}$ is to the right of the one chosen so that one should try $c_1^{(1)} > c_1^{(0)}$ or $c_1^{(1)} = c_1^{(0)}$ and $\gamma_1^{(1)} < \gamma_1^{(0)}$; the converse holds if $\beta_{T_*}(\theta_2) > \alpha$.

Example 6.10. Let $X_1, ..., X_n$ be i.i.d. from $N(\theta, 1)$. By Theorem 6.3, a UMP test for testing (6.12) is $T_*(X) = I_{(c_1, c_2)}(\bar{X})$, where c_i's are determined by

$$
\Phi\big(\sqrt{n}(c_2 - \theta_1)\big) - \Phi\big(\sqrt{n}(c_1 - \theta_1)\big) = \alpha
$$

and

$$
\Phi\big(\sqrt{n}(c_2 - \theta_2)\big) - \Phi\big(\sqrt{n}(c_1 - \theta_2)\big) = \alpha. \quad ∎
$$

When the distribution of X is not given by (6.10), UMP tests for hypotheses (6.12) exist in some cases (see Exercises 17 and 26). Unfortunately, a UMP test does not exist in general for testing hypotheses (6.13) or (6.14) (Exercises 28 and 29). A key reason for this phenomenon is that UMP tests for testing one-sided hypotheses do not have level α for testing (6.12); but they are of level α for testing (6.13) or (6.14) and there does not exist a single test more powerful than all tests that are UMP for testing one-sided hypotheses.

6.2 UMP Unbiased Tests

When a UMP test does not exist, we may use the same approach used in estimation problems, i.e., imposing a reasonable restriction on the tests to be considered and finding optimal tests within the class of tests under the restriction. Two such types of restrictions in estimation problems are unbiasedness and invariance. We consider *unbiased tests* in this section. The class of *invariant tests* is studied in §6.3.

6.2.1 Unbiasedness, similarity, and Neyman structure

A UMP test T of size α has the property that

$$\beta_T(P) \leq \alpha, \quad P \in \mathcal{P}_0 \qquad \text{and} \qquad \beta_T(P) \geq \alpha, \quad P \in \mathcal{P}_1. \qquad (6.19)$$

This means that T is at least as good as the silly test $T \equiv \alpha$. Thus, we have the following definition.

Definition 6.3. Let α be a given level of significance. A test T for $H_0 : P \in \mathcal{P}_0$ versus $H_1 : P \in \mathcal{P}_1$ is said to be unbiased of level α if and only if (6.19) holds. A test of size α is called a *uniformly most powerful unbiased* (UMPU) test if and only if it is UMP within the class of unbiased tests of level α. ∎

   Since a UMP test is UMPU, the discussion of unbiasedness of tests is useful only when a UMP test does not exist. In a large class of problems for which a UMP test does not exist, there do exist UMPU tests.

   Suppose that U is a sufficient statistic for $P \in \mathcal{P}$. Then, similar to the search for a UMP test, we need to consider functions of U only in order to find a UMPU test, since, for any unbiased test $T(X)$, $E(T|U)$ is unbiased and has the same power function as T.

   Throughout this section, we consider the following hypotheses:

$$H_0 : \theta \in \Theta_0 \qquad \text{versus} \qquad H_1 : \theta \in \Theta_1, \qquad (6.20)$$

where $\theta = \theta(P)$ is a functional from $\mathcal{P}$ onto Θ and Θ_0 and Θ_1 are two disjoint Borel sets with $\Theta_0 \cup \Theta_1 = \Theta$. Note that $\mathcal{P}_j = \{P : \theta \in \Theta_j\}$, $j = 0, 1$. For instance, $X_1, ..., X_n$ are i.i.d. from F but we are interested in testing $H_0 : \theta \leq 0$ versus $H_1 : \theta > 0$, where $\theta = EX_1$ or the median of F.

Definition 6.4. Consider the hypotheses specified by (6.20). Let α be a given level of significance and let $\bar{\Theta}_{01}$ be the common boundary of Θ_0 and Θ_1, i.e., the set of points θ that are points or limit points of both Θ_0 and Θ_1. A test T is *similar* on $\bar{\Theta}_{01}$ if and only if

$$\beta_T(P) = \alpha, \qquad \theta \in \bar{\Theta}_{01}. \quad ∎ \qquad (6.21)$$

It is more convenient to work with (6.21) than to work with (6.19) when the hypotheses are given by (6.20). Thus, the following lemma is useful. For a given test T, the power function $\beta_T(P)$ is said to be continuous in θ if and only if for any $\{\theta_j : j = 0, 1, 2, ...\} \subset \Theta$, $\theta_j \to \theta_0$ implies $\beta_T(P_j) \to \beta_T(P_0)$, where $P_j \in \mathcal{P}$ satisfying $\theta(P_j) = \theta_j$, $j = 0, 1, ...$. Note that if β_T is a function of θ, then this continuity property is simply the continuity of $\beta_T(\theta)$.

Lemma 6.5. Consider hypotheses (6.20). Suppose that, for every T, $\beta_T(P)$ is continuous in θ. If T_* is uniformly most powerful among all tests satisfying (6.21) and has size α, then T_* is a UMPU test.
Proof. Under the continuity assumption on β_T, the class of tests satisfying (6.21) contains the class of tests satisfying (6.19). Since T_* is uniformly at least as powerful as the test $T \equiv \alpha$, T_* is unbiased. Hence, T_* is a UMPU test. ∎

Using Lemma 6.5, we can derive a UMPU test for testing hypotheses given by (6.13) or (6.14), when X has the p.d.f. (6.10) in a one-parameter exponential family. (Note that a UMP test does not exist in these cases.) We do not provide the details here, since the results for one-parameter exponential families are special cases of those in §6.2.2 for multiparameter exponential families. To prepare for the discussion in §6.2.2, we introduce the following result that simplifies (6.21) when there is a statistic sufficient and complete for $P \in \bar{\mathcal{P}} = \{P : \theta(P) \in \bar{\Theta}_{01}\}$.

Let $U(X)$ be a sufficient statistic for $P \in \bar{\mathcal{P}}$ and let $\bar{\mathcal{P}}_U$ be the family of distributions of U as P ranges over $\bar{\mathcal{P}}$. If T is a test satisfying

$$E[T(X)|U] = \alpha \qquad \text{a.s. } \bar{\mathcal{P}}_U, \qquad (6.22)$$

then

$$E[T(X)] = E\{E[T(X)|U]\} = \alpha \qquad P \in \bar{\mathcal{P}},$$

i.e., T is similar on $\bar{\Theta}_{01}$. A test satisfying (6.22) is said to have *Neyman structure* w.r.t. U. If all tests similar on $\bar{\Theta}_{01}$ have Neyman structure w.r.t. U, then working with (6.21) is the same as working with (6.22).

Lemma 6.6. Let $U(X)$ be a sufficient statistic for $P \in \bar{\mathcal{P}}$. Then a necessary and sufficient condition for all tests similar on $\bar{\Theta}_{01}$ to have Neyman structure w.r.t. U is that U is boundedly complete for $P \in \bar{\mathcal{P}}$.
Proof. (i) Suppose first that U is boundedly complete for $P \in \bar{\mathcal{P}}$. Let $T(X)$ be a test similar on $\bar{\Theta}_{01}$. Then $E[T(X) - \alpha] = 0$ for all $P \in \bar{\mathcal{P}}$. From the boundedness of $T(X)$, $E[T(X)|U]$ is bounded (Proposition 1.10). Since $E\{E[T(X)|U] - \alpha\} = E[T(X) - \alpha] = 0$ for all $P \in \bar{\mathcal{P}}$, (6.22) holds.
(ii) Suppose now that U is not boundedly complete for $P \in \bar{\mathcal{P}}$. Then there is a function h such that $|h(u)| \leq C$, $E[h(U)] = 0$ for all $P \in \bar{\mathcal{P}}$, and $h(U) \neq 0$ with positive probability for some $P \in \bar{\mathcal{P}}$. Let $T(X) = \alpha + ch(U)$,

where $c = \min\{\alpha, 1 - \alpha\}/C$. The result follows from the fact that T is a test similar on $\bar{\Theta}_{01}$ but does not have Neyman structure w.r.t. U. ∎

6.2.2 UMPU tests in exponential families

Suppose that the distribution of X is in a multiparameter natural exponential family (§2.1.3) with the following p.d.f. w.r.t. a σ-finite measure:

$$f_{\theta,\varphi}(x) = \exp\left\{\theta Y(x) + \varphi^\tau U(x) - \zeta(\theta,\varphi)\right\}, \tag{6.23}$$

where θ is a real-valued parameter, φ is a vector-valued parameter, and Y (real-valued) and U (vector-valued) are statistics. It follows from Theorem 2.1(i) that the p.d.f. of (Y, U) (w.r.t. a σ-finite measure) is in a natural exponential family of the form $\exp\left\{\theta y + \varphi^\tau u - \zeta(\theta, \varphi)\right\}$ and, given $U = u$, the p.d.f. of the conditional distribution of Y (w.r.t. a σ-finite measure ν_u) is in a natural exponential family of the form $\exp\left\{\theta y - \zeta_u(\theta)\right\}$.

Theorem 6.4. Suppose that the distribution of X is in a multiparameter natural exponential family given by (6.23).
(i) For testing $H_0 : \theta \leq \theta_0$ versus $H_1 : \theta > \theta_0$, a UMPU test of size α is

$$T_*(Y, U) = \begin{cases} 1 & Y > c(U) \\ \gamma(U) & Y = c(U) \\ 0 & Y < c(U), \end{cases} \tag{6.24}$$

where $c(u)$ and $\gamma(u)$ are Borel functions determined by

$$E_{\theta_0}[T_*(Y, U)|U = u] = \alpha \tag{6.25}$$

for every u, and E_{θ_0} is the expectation w.r.t. $f_{\theta_0,\varphi}$.
(ii) For testing hypotheses (6.12), a UMPU test of size α is

$$T_*(Y, U) = \begin{cases} 1 & c_1(U) < Y < c_2(U) \\ \gamma_i(U) & Y = c_i(U), \ i = 1, 2, \\ 0 & Y < c_1(U) \text{ or } Y > c_2(U), \end{cases} \tag{6.26}$$

where $c_i(u)$'s and $\gamma_i(u)$'s are Borel functions determined by

$$E_{\theta_1}[T_*(Y, U)|U = u] = E_{\theta_2}[T_*(Y, U)|U = u] = \alpha \tag{6.27}$$

for every u.
(iii) For testing hypotheses (6.13), a UMPU test of size α is

$$T_*(Y, U) = \begin{cases} 1 & Y < c_1(U) \text{ or } Y > c_2(U) \\ \gamma_i(U) & Y = c_i(U), \ i = 1, 2, \\ 0 & c_1(U) < Y < c_2(U), \end{cases} \tag{6.28}$$

where $c_i(u)$'s and $\gamma_i(u)$'s are Borel functions determined by (6.27) for every u.

(iv) For testing hypotheses (6.14), a UMPU test of size α is given by (6.28), where $c_i(u)$'s and $\gamma_i(u)$'s are Borel functions determined by (6.25) and

$$E_{\theta_0}[T_*(Y,U)Y|U=u] = \alpha E_{\theta_0}(Y|U=u) \tag{6.29}$$

for every u.

Proof. Since (Y,U) is sufficient for (θ,φ), we only need to consider tests that are functions of (Y,U). Hypotheses in (i)-(iv) are of the form (6.20) with $\bar{\Theta}_{01} = \{(\theta,\varphi) : \theta = \theta_0\}$ or $= \{(\theta,\varphi) : \theta = \theta_i, i = 1,2\}$. In case (i) or (iv), U is sufficient and complete for $P \in \bar{\mathcal{P}}$ and, hence, Lemma 6.6 applies. In case (ii) or (iii), applying Lemma 6.6 to each $\{(\theta,\varphi) : \theta = \theta_i\}$ also shows that working with (6.21) is the same as working with (6.22). By Theorem 2.1, the power functions of all tests are continuous and, hence, Lemma 6.5 applies. Thus, for (i)-(iii), we only need to show that T_* is UMP among all tests T satisfying (6.25) (for part (i)) or (6.27) (for part (ii) or (iii)) with T_* replaced by T. For (iv), any unbiased T should satisfy (6.25) with T_* replaced by T and

$$\frac{\partial}{\partial\theta} E_{\theta,\varphi}[T(Y,U)] = 0, \qquad \theta \in \bar{\Theta}_{01}. \tag{6.30}$$

By Theorem 2.1, the differentiation can be carried out under the expectation sign. Hence, one can show (exercise) that (6.30) is equivalent to

$$E_{\theta,\varphi}[T(Y,U)Y - \alpha Y] = 0, \qquad \theta \in \bar{\Theta}_{01}. \tag{6.31}$$

Using the argument in the proof of Lemma 6.6, one can show (exercise) that (6.31) is equivalent to (6.29) with T_* replaced by T. Hence, to prove (iv) we only need to show that T_* is UMP among all tests T satisfying (6.25) and (6.29) with T_* replaced by T.

Note that the power function of any test $T(Y,U)$ is

$$\beta_T(\theta,\varphi) = \int \left[\int T(y,u) dP_{Y|U=u}(y) \right] dP_U(u).$$

Thus, it suffices to show that for every fixed u and $\theta \in \Theta_1$, T_* maximizes

$$\int T(y,u) dP_{Y|U=u}(y)$$

over all T subject to the given side conditions. Since $P_{Y|U=u}$ is in a one-parameter exponential family, the results in (i) and (ii) follow from Corollary 6.1 and Theorem 6.3, respectively. The result in (iii) follows from Theorem 6.3(ii) by considering $1 - T_*$ with T_* given by (6.15). To prove the result in (iv), it suffices to show that if Y has the p.d.f. given by (6.10) and if U is treated as a constant in (6.25), (6.28), and (6.29), T_* in (6.28) is UMP subject to conditions (6.25) and (6.29). We now omit U in the following proof for (iv), which is very similar to the proof of Theorem 6.3. First, $(\alpha, \alpha E_{\theta_0}(Y))$ is an interior point of the set of points $(E_{\theta_0}[T(Y)], E_{\theta_0}[T(Y)Y])$ as T ranges over all tests of the form $T(Y)$ (exercise). By Lemma 6.2 and Proposition 6.1, for testing $\theta = \theta_0$ versus $\theta = \theta_1$, the UMP test is equal to 1 when

$$(k_1 + k_2 y)e^{\theta_0 y} < C(\theta_0, \theta_1)e^{\theta_1 y}, \tag{6.32}$$

where k_i's and $C(\theta_0, \theta_1)$ are constants. Note that (6.32) is equivalent to

$$a_1 + a_2 y < e^{by}$$

for some constants a_1, a_2, and b. This region is either one-sided or the outside of an interval. By Theorem 6.2(ii), a one-sided test has a strictly monotone power function and therefore cannot satisfy (6.29). Thus, this test must have the form (6.28). Since T_* in (6.28) does not depend on θ_1, by Lemma 6.1, it is UMP over all tests satisfying (6.25) and (6.29); in particular, the test $\equiv \alpha$. Thus, T_* is UMPU.

Finally, it can be shown that all the c- and γ-functions in (i)-(iv) are Borel functions (see Lehmann (1986, p. 149)). ∎

Example 6.11. A problem arising in many different contexts is the comparison of two treatments. If the observations are integer-valued, the problem often reduces to testing the equality of two Poisson distributions (e.g., a comparison of the radioactivity of two substances or the car accident rate in two cities) or two binomial distributions (when the observation is the number of successes in a sequence of trials for each treatment).

Consider first the Poisson problem in which X_1 and X_2 are independently distributed as the Poisson distributions $P(\lambda_1)$ and $P(\lambda_2)$, respectively. The p.d.f. of $X = (X_1, X_2)$ is

$$\frac{e^{-(\lambda_1+\lambda_2)}}{x_1!x_2!} \exp\{x_2 \log(\lambda_2/\lambda_1) + (x_1 + x_2)\log\lambda_1\} \tag{6.33}$$

w.r.t. the counting measure on $\{(i, j) : i = 0, 1, 2, ..., j = 0, 1, 2, ...\}$. Let $\theta = \log(\lambda_2/\lambda_1)$. Then hypotheses such as $\lambda_1 = \lambda_2$ and $\lambda_1 \geq \lambda_2$ are equivalent to $\theta = 0$ and $\theta \leq 0$, respectively. The p.d.f. in (6.33) is of the form (6.23) with $\varphi = \log\lambda_1$, $Y = X_2$, and $U = X_1 + X_2$. Thus, Theorem 6.4 applies. To obtain various tests in Theorem 6.4, it is enough to derive the conditional distribution of $Y = X_2$ given $U = X_1 + X_2 = u$. Using the fact that $X_1 + X_2$ has the Poisson distribution $P(\lambda_1 + \lambda_2)$, one can show that

$$P(Y = y | U = u) = \binom{u}{y} p^y (1 - p)^{u-y} I_{\{0,1,...,u\}}(y), \quad u = 0, 1, 2, ...,$$

where $p = \lambda_2/(\lambda_1 + \lambda_2) = e^\theta/(1 + e^\theta)$. This is the binomial distribution $Bi(p, u)$. On the boundary set $\bar{\Theta}_{01}$, $\theta = \theta_j$ (a known value) and the distribution $P_{Y|U=u}$ is known.

The previous result can obviously be extended to the case where two independent samples, $X_{i1}, ..., X_{in_i}$, $i = 1, 2$, are i.i.d. from the Poisson distributions $P(\lambda_i)$, $i = 1, 2$, respectively.

Consider next the binomial problem in which X_j, $j = 1, 2$, are independently distributed as the binomial distributions $Bi(p_j, n_j)$, $j = 1, 2$, respectively, where n_j's are known but p_j's are unknown. The p.d.f. of $X = (X_1, X_2)$ is

$$\binom{n_1}{x_1}\binom{n_2}{x_2}(1 - p_1)^{n_1}(1 - p_2)^{n_2}\exp\left\{x_2 \log \frac{p_2(1-p_1)}{p_1(1-p_2)} + (x_1 + x_2)\log \frac{p_1}{(1-p_1)}\right\}$$

w.r.t. the counting measure on $\{(i, j) : i = 0, 1, ..., n_1, j = 0, 1, ..., n_2\}$. This p.d.f. is of the form (6.23) with $\theta = \log \frac{p_2(1-p_1)}{p_1(1-p_2)}$, $Y = X_2$, and $U = X_1 + X_2$. Thus, Theorem 6.4 applies. Note that hypotheses such as $p_1 = p_2$ and $p_1 \geq p_2$ are equivalent to $\theta = 0$ and $\theta \leq 0$, respectively. Using the joint distribution of (X_1, X_2), one can show (exercise) that

$$P(Y = y|U = u) = K_u(\theta)\binom{n_1}{u - y}\binom{n_2}{y}e^{\theta y}I_A(y), \quad u = 0, 1, ..., n_1 + n_2,$$

where $A = \{y : y = 0, 1, ..., \min\{u, n_2\}, u - y \leq n_1\}$ and

$$K_u(\theta) = \left[\sum_{y \in A}\binom{n_1}{u - y}\binom{n_2}{y}e^{\theta y}\right]^{-1}. \tag{6.34}$$

If $\theta = 0$, this distribution reduces to a known distribution: the hypergeometric distribution $HG(u, n_2, n_1)$ (Table 1.1, page 18). ∎

Example 6.12 (2×2 contingency tables). Let A and B be two different events in a probability space related to a random experiment. Suppose that n independent trials of the experiment are carried out and that we observe the frequencies of the occurrence of the events $A \cap B$, $A \cap B^c$, $A^c \cap B$, and $A^c \cap B^c$. The results can be summarized in the following 2×2 *contingency table*:

	A	A^c	Total
B	X_{11}	X_{12}	n_1
B^c	X_{21}	X_{22}	n_2
Total	m_1	m_2	n

The distribution of $X = (X_{11}, X_{12}, X_{21}, X_{22})$ is multinomial (Example 2.7) with probabilities p_{11}, p_{12}, p_{21}, and p_{22}, where $p_{ij} = E(X_{ij})/n$. Thus, the p.d.f. of X is

$$\frac{n!}{x_{11}!x_{12}!x_{21}!x_{22}!} p_{22}^n \exp\left\{ x_{11} \log \tfrac{p_{11}}{p_{22}} + x_{12} \log \tfrac{p_{12}}{p_{22}} + x_{21} \log \tfrac{p_{21}}{p_{22}} \right\}$$

w.r.t. the counting measure on the range of X. This p.d.f. is clearly of the form (6.23). By Theorem 6.4, we can derive UMPU tests for any parameter of the form

$$\theta = a_0 \log \tfrac{p_{11}}{p_{22}} + a_1 \log \tfrac{p_{12}}{p_{22}} + a_2 \log \tfrac{p_{21}}{p_{22}},$$

where a_i's are given constants. In particular, testing independence of A and B is equivalent to the hypotheses $H_0 : \theta = 0$ versus $H_1 : \theta \neq 0$ when $a_0 = 1$ and $a_1 = a_2 = -1$ (exercise).

For hypotheses concerning θ with $a_0 = 1$ and $a_1 = a_2 = -1$, the p.d.f. of X can be written as (6.23) with $Y = X_{11}$ and $U = (X_{11} + X_{12}, X_{11} + X_{21})$. A direct calculation shows that $P(Y = y | X_{11} + X_{12} = n_1, X_{11} + X_{21} = m_1)$ is equal to

$$K_{m_1}(\theta) \binom{n_1}{y} \binom{n_2}{m_1 - y} e^{\theta(m_1 - y)} I_A(y),$$

where $A = \{y : y = 0, 1, ..., \min\{m_1, n_1\}, m_1 - y \leq n_2\}$ and $K_u(\theta)$ is given by (6.34). This distribution is known when $\theta = \theta_j$ is known. In particular, for testing independence of A and B, $\theta = 0$ implies that $P_{Y|U=u}$ is the hypergeometric distribution $HG(m_1, n_1, n_2)$, and the UMPU test in Theorem 6.4(iv) is also known as Fisher's exact test.

Suppose that X_{ij}'s in the 2×2 contingency table are from two binomial distributions, i.e., X_{i1} is from the binomial distribution $Bi(p_i, n_i)$, $X_{i2} = n_i - X_{i1}$, $i = 1, 2$, and that X_{i1}'s are independent. Then the UMPU test for independence of A and B previously derived is exactly the same as the UMPU test for $p_1 = p_2$ given in Example 6.11. The only difference is that n_i's are fixed for testing the equality of two binomial distributions, whereas n_i's are random for testing independence of A and B. This is also true for the general $r \times c$ contingency tables considered in §6.4.3. ∎

6.2.3 UMPU tests in normal families

An important application of Theorem 6.4 to problems with continuous distributions in exponential families is the derivation of UMPU tests in normal families. The results presented here are the basic justifications for tests in elementary textbooks concerning parameters in normal families.

We start with the following lemma, which is useful especially when X is from a population in a normal family.

Lemma 6.7. Suppose that X has the p.d.f. (6.23) and that $V(Y, U)$ is a statistic independent of U when $\theta = \theta_j$, where θ_j's are known values given in the hypotheses in (i)-(iv) of Theorem 6.4.

(i) If $V(y, u)$ is increasing in y for each u, then the UMPU tests in (i)-(iii) of Theorem 6.4 are equivalent to those given by (6.24)-(6.28) with Y and (Y, U) replaced by V and with $c_i(U)$ and $\gamma_i(U)$ replaced by constants c_i and γ_i, respectively.

(ii) If there are Borel functions $a(u) > 0$ and $b(u)$ such that $V(y, u) = a(u)y + b(u)$, then the UMPU test in Theorem 6.4(iv) is equivalent to that given by (6.25), (6.28), and (6.29) with Y and (Y, U) replaced by V and with $c_i(U)$ and $\gamma_i(U)$ replaced by constants c_i and γ_i, respectively.

Proof. (i) Since V is increasing in y, $Y > c_i(u)$ is equivalent to $V > d_i(u)$ for some d_i. The result follows from the fact that V is independent of U so that d_i's and γ_i's do not depend on u when Y is replaced by V.

(ii) Since $V = a(U)Y + b(U)$, the UMPU test in Theorem 6.4(iv) is the same as

$$T_*(V, U) = \begin{cases} 1 & V < c_1(U) \text{ or } V > c_2(U) \\ \gamma_i(U) & V = c_i(U), \ i = 1, 2, \\ 0 & c_1(U) < V < c_2(U), \end{cases} \tag{6.35}$$

subject to $E_{\theta_0}[T_*(V, U)|U = u] = \alpha$ and

$$E_{\theta_0}\left[T_*(V, U)\frac{V - b(U)}{a(U)}\bigg|U\right] = \alpha E_{\theta_0}\left[\frac{V - b(U)}{a(U)}\bigg|U\right]. \tag{6.36}$$

Under $E_{\theta_0}[T_*(V, U)|U = u] = \alpha$, (6.36) is the same as $E_{\theta_0}[T_*(V, U)V|U] = \alpha E_{\theta_0}(V|U)$. Since V and U are independent when $\theta = \theta_0$, $c_i(u)$'s and $\gamma_i(u)$'s do not depend on u and, therefore, T_* in (6.35) does not depend on U. ∎

If the conditions of Lemma 6.7 are satisfied, then UMPU tests can be derived by working with the distribution of V instead of $P_{Y|U=u}$. In exponential families, a $V(Y, U)$ independent of U can often be found by applying Basu's theorem (Theorem 2.4).

When we consider normal families, γ_i's can be chosen to be 0 since the c.d.f. of Y given $U = u$ or the c.d.f. of V is continuous.

One-sample problems

Let $X_1, ..., X_n$ be i.i.d. from $N(\mu, \sigma^2)$ with unknown $\mu \in \mathcal{R}$ and $\sigma^2 > 0$, where $n \geq 2$. The joint p.d.f. of $X = (X_1, ..., X_n)$ is

$$\frac{1}{(2\pi\sigma^2)^{n/2}} \exp\left\{-\frac{1}{2\sigma^2}\sum_{i=1}^{n} x_i^2 + \frac{\mu}{\sigma^2}\sum_{i=1}^{n} x_i - \frac{n\mu^2}{2\sigma^2}\right\}.$$

Consider first hypotheses concerning σ^2. The p.d.f. of X has the form (6.23) with $\theta = -(2\sigma^2)^{-1}$, $\varphi = n\mu/\sigma^2$, $Y = \sum_{i=1}^n X_i^2$, and $U = \bar{X}$. By Basu's theorem, $V = (n-1)S^2$ is independent of $U = \bar{X}$ (Example 2.18), where S^2 is the sample variance. Also,

$$\sum_{i=1}^n X_i^2 = (n-1)S^2 + n\bar{X}^2,$$

i.e., $V = Y - nU^2$. Hence the conditions of Lemma 6.7 are satisfied. Since V/σ^2 has the chi-square distribution χ_{n-1}^2 (Example 2.18), values of c_i's for hypotheses in (i)-(iii) of Theorem 6.4 are related to quantiles of χ_{n-1}^2. For testing $H_0 : \theta = \theta_0$ versus $H_1 : \theta \neq \theta_0$ (which is equivalent to testing $H_0 : \sigma^2 = \sigma_0^2$ versus $H_1 : \sigma^2 \neq \sigma_0^2$), $d_i = c_i/\sigma_0^2$, $i = 1, 2$, are determined by

$$\int_{d_1}^{d_2} f_{n-1}(v)dv = 1 - \alpha \quad \text{and} \quad \int_{d_1}^{d_2} v f_{n-1}(v)dv = (n-1)(1-\alpha),$$

where f_m is the Lebesgue p.d.f. of the chi-square distribution χ_m^2. Since $v f_{n-1}(v) = (n-1)f_{n+1}(v)$, d_1 and d_2 are determined by

$$\int_{d_1}^{d_2} f_{n-1}(v)dv = \int_{d_1}^{d_2} f_{n+1}(v)dv = 1 - \alpha.$$

If $n - 1 \approx n + 1$, then d_1 and d_2 are nearly the $(\alpha/2)$th and $(1 - \alpha/2)$th quantiles of χ_{n-1}^2, respectively, in which case the UMPU test in Theorem 6.4(iv) is the same as the "equal-tailed" chi-square test for H_0 in elementary textbooks.

Consider next hypotheses concerning μ. The p.d.f. of X has the form (6.23) with $Y = \bar{X}$, $U = \sum_{i=1}^n (X_i - \mu_0)^2$, $\theta = n(\mu - \mu_0)/\sigma^2$, and $\varphi = -(2\sigma^2)^{-1}$. For testing hypotheses $H_0 : \mu \leq \mu_0$ versus $H_1 : \mu > \mu_0$, we take V to be $t(X) = \sqrt{n}(\bar{X} - \mu_0)/S$. By Basu's theorem, $t(X)$ is independent of U when $\mu = \mu_0$. Hence it satisfies the conditions in Lemma 6.7(i). From Examples 1.16 and 2.18, $t(X)$ has the t-distribution t_{n-1} when $\mu = \mu_0$. Thus, $c(U)$ in Theorem 6.4(i) is the $(1 - \alpha)$th quantile of t_{n-1}. For the two-sided hypotheses $H_0 : \mu = \mu_0$ versus $H_1 : \mu \neq \mu_0$, the statistic $V = (\bar{X} - \mu_0)/\sqrt{U}$ satisfies the conditions in Lemma 6.7(ii) and has a distribution symmetric about 0 when $\mu = \mu_0$. Then the UMPU test in Theorem 6.4(iv) rejects H_0 when $|V| > d$, where d satisfies $P(|V| > d) = \alpha$ when $\mu = \mu_0$. Since

$$t(X) = \sqrt{(n-1)n}V(X)/\sqrt{1 - n[V(X)]^2},$$

the UMPU test rejects H_0 if and only if $|t(X)| > t_{n-1,\alpha/2}$, where $t_{n-1,\alpha}$ is the $(1 - \alpha)$th quantile of the t-distribution t_{n-1}. The UMPU tests derived here are the so-called one-sample t-tests in elementary textbooks.

The power function of a one-sample t-test is related to the noncentral t-distribution introduced in §1.3.1 (see Exercise 36).

Two-sample problems

The problem of comparing the parameters of two normal distributions arises in the comparison of two treatments, products, and so on (see also Example 6.11). Suppose that we have two independent samples, $X_{i1}, ..., X_{in_i}$, $i = 1, 2$, i.i.d. from $N(\mu_i, \sigma_i^2)$, $i = 1, 2$, respectively, where $n_i \geq 2$. The joint p.d.f. of X_{ij}'s is

$$C(\mu_1, \mu_2, \sigma_1^2, \sigma_2^2) \exp\left\{ -\sum_{i=1}^{2} \frac{1}{2\sigma_i^2} \sum_{j=1}^{n_i} x_{ij}^2 + \sum_{i=1}^{2} \frac{n_i \mu_i}{\sigma_i^2} \bar{x}_i \right\},$$

where $\bar{x}_i$ is the sample mean based on $x_{i1}, ..., x_{in_i}$ and $C(\cdot)$ is a known function.

Consider first the hypothesis $H_0 : \sigma_2^2/\sigma_1^2 \leq \Delta_0$ or $H_0 : \sigma_2^2/\sigma_1^2 = \Delta_0$. The p.d.f. of X_{ij}'s is of the form (6.23) with

$$\theta = \frac{1}{2\Delta_0 \sigma_1^2} - \frac{1}{2\sigma_2^2}, \qquad \varphi = \left(-\frac{1}{2\sigma_1^2}, \frac{n_1 \mu_1}{\sigma_1^2}, \frac{n_2 \mu_2}{\sigma_2^2} \right),$$

$$Y = \sum_{j=1}^{n_2} X_{2j}^2, \qquad U = \left(\sum_{j=1}^{n_1} X_{1j}^2 + \frac{1}{\Delta_0} \sum_{j=1}^{n_2} X_{2j}^2, \bar{X}_1, \bar{X}_2 \right).$$

To apply Lemma 6.7, consider

$$V = \frac{(n_2 - 1)S_2^2/\Delta_0}{(n_1 - 1)S_1^2 + (n_2 - 1)S_2^2/\Delta_0} = \frac{(Y - n_2 U_3)/\Delta_0}{U_1 - n_1 U_2 - n_2 U_3/\Delta_0},$$

where S_i^2 is the sample variance based on $X_{i1}, ..., X_{in_i}$ and U_j is the jth component of U. By Basu's theorem, V and U are independent when $\theta = 0$ ($\sigma_2^2 = \Delta_0 \sigma_1^2$). Since V is increasing and linear in Y, the conditions of Lemma 6.7 are satisfied. Thus, a UMPU test rejects $H_0 : \theta \leq 0$ (which is equivalent to $H_0 : \sigma_2^2/\sigma_1^2 \leq \Delta_0$) when $V > c_0$, where c_0 satisfies $P(V > c_0) = \alpha$ when $\theta = 0$; and a UMPU test rejects $H_0 : \theta = 0$ (which is equivalent to $H_0 : \sigma_2^2/\sigma_1^2 = \Delta_0$) when $V < c_1$ or $V > c_2$, where c_i's satisfy $P(c_1 < V < c_2) = 1 - \alpha$ and $E[VT_*(V)] = \alpha E(V)$ when $\theta = 0$. Note that

$$V = \frac{(n_2 - 1)F}{n_1 - 1 + (n_2 - 1)F} \qquad \text{with} \qquad F = \frac{S_2^2/\Delta_0}{S_1^2}.$$

It follows from Example 1.16 that F has the F-distribution F_{n_2-1, n_1-1} (Table 1.2, page 20) when $\theta = 0$. Since V is a strictly increasing function of F, a UMPU test rejects $H_0 : \theta \leq 0$ when $F > F_{n_2-1, n_1-1, \alpha}$, where $F_{a,b,\alpha}$ is the $(1 - \alpha)$th quantile of the F-distribution $F_{a,b}$. This is the F-test in elementary textbooks.

When $\theta = 0$, V has the beta distribution $B((n_2 - 1)/2, (n_1 - 1)/2)$ and $E(V) = (n_2 - 1)/(n_1 + n_2 - 2)$ (Table 1.2). Then, $E[VT_*(V)] = \alpha E(V)$ when $\theta = 0$ is the same as

$$\frac{(1 - \alpha)(n_2 - 1)}{n_1 + n_2 - 2} = \int_{c_1}^{c_2} v f_{(n_2-1)/2,(n_1-1)/2}(v)dv,$$

where $f_{a,b}$ is the p.d.f. of the beta distribution $B(a, b)$. Using the fact that $v f_{(n_2-1)/2,(n_1-1)/2}(v) = (n_1 + n_2 - 2)^{-1}(n_2 - 1)f_{(n_2+1)/2,(n_1-1)/2}(v)$, we conclude that a UMPU test rejects $H_0 : \theta = 0$ when $V < c_1$ or $V > c_2$, where c_1 and c_2 are determined by

$$1 - \alpha = \int_{c_1}^{c_2} v f_{(n_2-1)/2,(n_1-1)/2}(v)dv = \int_{c_1}^{c_2} f_{(n_2+1)/2,(n_1-1)/2}(v)dv.$$

If $n_2 - 1 \approx n_2 + 1$ (i.e., n_2 is large), then this UMPU test can be approximated by the F-test that rejects $H_0 : \theta = 0$ if and only if $F < F_{n_2-1,n_1-1,1-\alpha/2}$ or $F > F_{n_2-1,n_1-1,\alpha/2}$.

Consider next the hypothesis $H_0 : \mu_1 \geq \mu_2$ or $H_0 : \mu_1 = \mu_2$. If $\sigma_1^2 \neq \sigma_2^2$, the problem is the so-called Behrens-Fisher problem and is not accessible by the method introduced in this section. We now assume that $\sigma_1^2 = \sigma_2^2 = \sigma^2$ but σ^2 is unknown. The p.d.f. of X_{ij}'s is then

$$C(\mu_1, \mu_2, \sigma^2) \exp \left\{ -\frac{1}{2\sigma^2} \sum_{i=1}^{2} \sum_{j=1}^{n_i} x_{ij}^2 + \frac{n_1\mu_1}{\sigma^2}\bar{x}_1 + \frac{n_2\mu_2}{\sigma^2}\bar{x}_2 \right\},$$

which is of the form (6.23) with

$$\theta = \frac{\mu_2 - \mu_1}{(n_1^{-1} + n_2^{-1})\sigma^2}, \qquad \varphi = \left(\frac{n_1\mu_1 + n_2\mu_2}{(n_1 + n_2)\sigma^2}, -\frac{1}{2\sigma^2} \right),$$

$$Y = \bar{X}_2 - \bar{X}_1, \qquad U = \left(n_1\bar{X}_1 + n_2\bar{X}_2, \sum_{i=1}^{2} \sum_{j=1}^{n_i} X_{ij}^2 \right).$$

For testing $H_0 : \theta \leq 0$ (i.e., $\mu_1 \geq \mu_2$) versus $H_1 : \theta > 0$, we consider V in Lemma 6.7 to be

$$t(X) = \frac{(\bar{X}_2 - \bar{X}_1)/\sqrt{n_1^{-1} + n_2^{-1}}}{\sqrt{[(n_1 - 1)S_1^2 + (n_2 - 1)S_2^2]/(n_1 + n_2 - 2)}}. \tag{6.37}$$

When $\theta = 0$, $t(X)$ is independent of U (Basu's theorem) and satisfies the conditions in Lemma 6.7(i); the numerator and the denominator of $t(X)$ (after division by σ) are independently distributed as $N(0, 1)$ and the chi-square distribution $\chi^2_{n_1+n_2-2}$, respectively. Hence $t(X)$ has the t-distribution $t_{n_1+n_2-2}$ and a UMPU test rejects H_0 when $t(X) > t_{n_1+n_2-2,\alpha}$, where $t_{n_1+n_2-2,\alpha}$ is the $(1-\alpha)$th quantile of the t-distribution $t_{n_1+n_2-2}$. This is the so-called (one-sided) two-sample t-test.

For testing $H_0 : \theta = 0$ (i.e., $\mu_1 = \mu_2$) versus $H_1 : \theta \neq 0$, it follows from a similar argument used in the derivation of the (two-sided) one-sample t-test that a UMPU test rejects H_0 when $|t(X)| > t_{n_1+n_2-2,\alpha/2}$ (exercise). This is the (two-sided) two-sample t-test.

The power function of a two-sample t-test is related to a noncentral t-distribution.

Normal linear models

Consider linear model (3.25) with assumption A1, i.e.,

$$X = (X_1, ..., X_n) \text{ is } N_n(Z\beta, \sigma^2 I_n), \tag{6.38}$$

where β is a p-vector of unknown parameters, Z is the $n \times p$ matrix whose ith row is the vector Z_i, Z_i's are the values of a p-vector of deterministic covariates, and $\sigma^2 > 0$ is an unknown parameter. Assume that $n > p$ and the rank of Z is $r \leq p$. Let $l \in \mathcal{R}(Z)$ (the linear space generated by the rows of Z) and θ_0 be a fixed constant. We consider the hypotheses

$$H_0 : l^\tau \beta \leq \theta_0 \qquad \text{versus} \qquad H_1 : l^\tau \beta > \theta_0 \tag{6.39}$$

or

$$H_0 : l^\tau \beta = \theta_0 \qquad \text{versus} \qquad H_1 : l^\tau \beta \neq \theta_0. \tag{6.40}$$

Since $H = Z(Z^\tau Z)^- Z^\tau$ is a projection matrix of rank r, there exists an $n \times n$ orthogonal matrix Γ such that

$$\Gamma = (\Gamma_1 \ \Gamma_2) \qquad \text{and} \qquad H\Gamma = (\Gamma_1 \ 0), \tag{6.41}$$

where Γ_1 is $n \times r$ and Γ_2 is $n \times (n-r)$. Let $Y_j = \Gamma_j^\tau X$, $j = 1, 2$. Consider the transformation $(Y_1, Y_2) = \Gamma^\tau X$. Since $\Gamma^\tau \Gamma = I_n$ and X is $N_n(Z\beta, \sigma^2 I_n)$, (Y_1, Y_2) is $N_n(\Gamma^\tau Z\beta, \sigma^2 I_n)$. It follows from (6.41) that

$$E(Y_2) = E(\Gamma_2^\tau X) = \Gamma_2^\tau Z\beta = \Gamma_2^\tau HZ\beta = 0.$$

Let $\eta = \Gamma_1^\tau Z\beta = E(Y_1)$. Then the p.d.f. of (Y_1, Y_2) is

$$\frac{1}{(2\pi\sigma^2)^{n/2}} \exp\left\{ \frac{\eta^\tau Y_1}{\sigma^2} - \frac{\|Y_1\|^2 + \|Y_2\|^2}{2\sigma^2} - \frac{\|\eta\|^2}{2\sigma^2} \right\}. \tag{6.42}$$

Since l in (6.39) or (6.40) is in $\mathcal{R}(Z)$, there exists $\lambda \in \mathcal{R}^n$ such that $l = Z^\tau \lambda$. Then

$$l^\tau \hat{\beta} = \lambda^\tau HX = \lambda^\tau \Gamma\Gamma^\tau HX = \lambda^\tau \Gamma_1\Gamma_1^\tau X = \lambda^\tau \Gamma_1 Y_1, \tag{6.43}$$

where $\hat{\beta}$ is the LSE defined by (3.27). By (6.43) and Theorem 3.6(ii),

$$E(l^\tau \hat{\beta}) = l^\tau \beta = \lambda^\tau \Gamma_1 E(Y_1) = a^\tau \eta,$$

where $a = \Gamma_1^\tau \lambda$. Let $\eta = (\eta_1, ..., \eta_r)$ and $a = (a_1, ..., a_r)$. Without loss of generality, we assume that $a_1 \neq 0$. Then the p.d.f. in (6.42) is of the form (6.23) with

$$\theta = \frac{a^\tau \eta}{a_1 \sigma^2}, \qquad \varphi = \left(-\frac{1}{2\sigma^2}, \frac{\eta_2}{\sigma^2}, ..., \frac{\eta_r}{\sigma^2} \right),$$

$$Y = Y_{11}, \qquad U = \left(\|Y_1\|^2 + \|Y_2\|^2, Y_{12} - \frac{a_2 Y_{11}}{a_1}, ..., Y_{1r} - \frac{a_r Y_{11}}{a_1} \right),$$

where Y_{1j} is the jth component of Y_1. By Basu's theorem,

$$t(X) = \frac{\sqrt{n-r}(a^\tau Y_1 - \theta_0)}{\|Y_2\| \|a\|}$$

is independent of U when $a^\tau \eta = l^\tau \beta = \theta_0$. Note that $\|Y_2\|^2 = SSR$ in (3.35) and $\|a\|^2 = \lambda^\tau \Gamma_1 \Gamma_1^\tau \lambda = \lambda^\tau H \lambda = l^\tau (Z^\tau Z)^- l$. Hence, by (6.43),

$$t(X) = \frac{l^\tau \hat{\beta} - \theta_0}{\sqrt{l^\tau (Z^\tau Z)^- l} \sqrt{SSR/(n-r)}},$$

which has the t-distribution t_{n-r} (Theorem 3.8). Using the same arguments in deriving the one-sample or two-sample t-test, we obtain that a UMPU test for the hypotheses in (6.39) rejects H_0 when $t(X) > t_{n-r,\alpha}$, and that a UMPU test for the hypotheses in (6.40) rejects H_0 when $|t(X)| > t_{n-r,\alpha/2}$.

Testing for independence in the bivariate normal family

Suppose that $X_1, ..., X_n$ are i.i.d. from a bivariate normal distribution, i.e., the p.d.f. of $X = (X_1, ..., X_n)$ is

$$\frac{1}{(2\pi\sigma_1\sigma_2\sqrt{1-\rho^2})^n} \exp\left\{ -\frac{\|Y_1 - \mu_1\|^2}{2\sigma_1^2(1-\rho^2)} + \frac{\rho(Y_1-\mu_1)^\tau(Y_2-\mu_2)}{\sigma_1\sigma_2(1-\rho^2)} - \frac{\|Y_2-\mu_2\|^2}{2\sigma_2^2(1-\rho^2)} \right\}, \quad (6.44)$$

where $Y_j = (X_{1j}, ..., X_{nj})$ and X_{ij} is the jth component of X_i, $j = 1, 2$.

Testing for independence of the two components of X_1 (or Y_1 and Y_2) is equivalent to testing $H_0 : \rho = 0$ versus $H_1 : \rho \neq 0$. In some cases, one may also be interested in the one-sided hypotheses $H_0 : \rho \leq 0$ versus $H_1 : \rho > 0$. It can be shown (exercise) that the p.d.f. in (6.44) is of the form (6.23) with $\theta = \frac{\rho}{\sigma_1\sigma_2(1-\rho^2)}$ and

$$Y = \sum_{i=1}^n X_{i1} X_{i2}, \qquad U = \left(\sum_{i=1}^n X_{i1}^2, \sum_{i=1}^n X_{i2}^2, \sum_{i=1}^n X_{i1}, \sum_{i=1}^n X_{i2} \right).$$

The hypothesis $\rho \leq 0$ is equivalent to $\theta \leq 0$. The sample correlation coefficient is

$$R = \sum_{i=1}^{n}(X_{i1} - \bar{X}_1)(X_{i2} - \bar{X}_2) \Big/ \left[\sum_{i=1}^{n}(X_{i1} - \bar{X}_1)^2 \sum_{i=1}^{n}(X_{i2} - \bar{X}_2)^2\right]^{1/2},$$

where $\bar{X}_j$ is the sample mean of $X_{1j}, ..., X_{nj}$ and is independent of U when $\rho = 0$ (Basu's theorem), $j = 1, 2$. To apply Lemma 6.7, we consider

$$V = \sqrt{n-2}R/\sqrt{1-R^2}. \tag{6.45}$$

It can be shown (exercise) that R is linear in Y and that V has the t-distribution t_{n-2} when $\rho = 0$. Hence, a UMPU test for $H_0 : \rho \leq 0$ versus $H_1 : \rho > 0$ rejects H_0 when $V > t_{n-2,\alpha}$ and a UMPU test for $H_0 : \rho = 0$ versus $H_1 : \rho \neq 0$ rejects H_0 when $|V| > t_{n-2,\alpha/2}$, where $t_{n-2,\alpha}$ is the $(1-\alpha)$th quantile of the t-distribution t_{n-2}.

6.3 UMP Invariant Tests

In the previous section the unbiasedness principle is considered to derive an optimal test within the class of unbiased tests when a UMP test does not exist. In this section, we study the same problem with unbiasedness replaced by invariance under a given group of transformations. The principles of unbiasedness and invariance often complement each other in that each is successful in cases where the other is not.

6.3.1 Invariance and UMPI tests

The invariance principle considered here is similar to that introduced in §2.3.2 (Definition 2.9) and in §4.2. Although a hypothesis testing problem can be treated as a particular statistical decision problem (see, e.g., Example 2.20), in the following definition we define invariant tests without using any loss function which is a basic element in statistical decision theory. However, the reader is encouraged to compare Definition 2.9 with the following definition.

Definition 6.5. Let X be a sample from $P \in \mathcal{P}$ and $\mathcal{G}$ be a group (Definition 2.9(i)) of one-to-one transformations of X.
(i) We say that the problem of testing $H_0 : P \in \mathcal{P}_0$ versus $H_1 : P \in \mathcal{P}_1$ is *invariant* under $\mathcal{G}$ if and only if both $\mathcal{P}_0$ and $\mathcal{P}_1$ are invariant under $\mathcal{G}$ in the sense of Definition 2.9(ii).
(ii) In an invariant testing problem, a test $T(X)$ is said to be *invariant*

under $\mathcal{G}$ if and only if

$$T(g(x)) = T(x) \qquad \text{for all } x \text{ and } g. \tag{6.46}$$

(iii) A test of size α is said to be a *uniformly most powerful invariant* (UMPI) test if and only if it is UMP within the class of level α tests that are invariant under $\mathcal{G}$.

(iv) A statistic $M(X)$ is said to be *maximal invariant* under $\mathcal{G}$ if and only if (6.46) holds with T replaced by M and

$$M(x_1) = M(x_2) \qquad \text{implies } x_1 = g(x_2) \text{ for some } g \in \mathcal{G}. \quad \blacksquare \tag{6.47}$$

The following result indicates that invariance reduces the data X to a maximal invariant statistic $M(X)$ whose distribution may depend only on a functional of P that shrinks $\mathcal{P}$.

Proposition 6.2. Let $M(X)$ be maximal invariant under $\mathcal{G}$.
(i) A test $T(X)$ is invariant under $\mathcal{G}$ if and only if there is a function h such that $T(x) = h(M(x))$ for all x.
(ii) Suppose that there is a functional $\theta(P)$ on $\mathcal{P}$ satisfying $\theta(\bar{g}(P)) = \theta(P)$ for all $g \in \mathcal{G}$ and $P \in \mathcal{P}$ and

$$\theta(P_1) = \theta(P_2) \qquad \text{implies } P_1 = \bar{g}(P_2) \text{ for some } g \in \mathcal{G}$$

(i.e., $\theta(P)$ is "maximal invariant"), where $\bar{g}(P_X) = P_{g(X)}$ is given in Definition 2.9(ii). Then the distribution of $M(X)$ depends only on $\theta(P)$.
Proof. (i) If $T(x) = h(M(x))$ for all x, then $T(g(x)) = h(M(g(x))) = h(M(x)) = T(x)$ so that T is invariant. If T is invariant and if $M(x_1) = M(x_2)$, then $x_1 = g(x_2)$ for some g and $T(x_1) = T(g(x_2)) = T(x_2)$. Hence T is a function of M.
(ii) Suppose that $\theta(P_1) = \theta(P_2)$. Then $P_2 = \bar{g}(P_1)$ for some $g \in \mathcal{G}$ and for any event B in the range of $M(X)$,

$$\begin{aligned}
P_2\big(M(X) \in B\big) &= \bar{g}(P_1)\big(M(X) \in B\big) \\
&= P_1\big(M(g(X)) \in B\big) \\
&= P_1\big(M(X) \in B\big).
\end{aligned}$$

Hence the distribution of $M(X)$ depends only on $\theta(P)$. $\blacksquare$

In applications, maximal invariants $M(X)$ and $\theta = \theta(P)$ are frequently real-valued. If the hypotheses of interest can be expressed in terms of θ, then there may exist a test UMP among those depending only on $M(X)$ (e.g., when the distribution of $M(X)$ is in a parametric family having monotone likelihood ratio). Such a test is then a UMPI test.

Example 6.13 (Location-scale families). Suppose that X has the Lebesgue p.d.f. $f_{i,\mu}(x) = f_i(x_1 - \mu, ..., x_n - \mu)$, where $n \geq 2$, $\mu \in \mathcal{R}$ is unknown, and f_i, $i = 0, 1$, are known Lebesgue p.d.f.'s. We consider the problem of testing

$$H_0 : X \text{ is from } f_{0,\mu} \qquad \text{versus} \qquad H_1 : X \text{ is from } f_{1,\mu}. \qquad (6.48)$$

Consider $\mathcal{G} = \{g_c : c \in \mathcal{R}\}$ with $g_c(x) = (x_1 + c, ..., x_n + c)$. For any $g_c \in \mathcal{G}$, it induces a transformation $\bar{g}_c(f_{i,\mu}) = f_{i,\mu+c}$ and the problem of testing H_0 versus H_1 in (6.48) is invariant under $\mathcal{G}$.

We now show that a maximal invariant under $\mathcal{G}$ is $D(X) = (D_1, ..., D_{n-1})$ $= (X_1 - X_n, ..., X_{n-1} - X_n)$. First, it is easy to see that $D(X)$ is invariant under $\mathcal{G}$. Let $x = (x_1, ..., x_n)$ and $y = (y_1, ..., y_n)$ be two points in the range of X. Suppose that $x_i - x_n = y_i - y_n$ for $i = 1, ..., n - 1$. Putting $c = y_n - x_n$, we have $y_i = x_i + c$ for all i. Hence, $D(X)$ is maximal invariant under $\mathcal{G}$.

By Proposition 1.8, D has the p.d.f. $\int f_i(d_1 + t, ..., d_{n-1} + t, t)dt$ under H_i, $i = 0, 1$, which does not depend on μ. In fact, in this case Proposition 6.2 applies with $M(X) = D(X)$ and $\theta(f_{i,\mu}) = i$. If we consider tests that are functions of $D(X)$, then the problem of testing the hypotheses in (6.48) becomes one of testing a simple hypothesis versus a simple hypothesis. By Theorem 6.1, the test UMP among functions of $D(X)$, which is then the UMPI test, rejects H_0 in (6.48) when

$$\frac{\int f_1(d_1 + t, ..., d_{n-1} + t, t)dt}{\int f_0(d_1 + t, ..., d_{n-1} + t, t)dt} = \frac{\int f_1(x_1 + t, ..., x_n + t)dt}{\int f_0(x_1 + t, ..., x_n + t)dt} > c,$$

where c is determined by the size of the UMPI test.

The previous result can be extended to the case of a location-scale family where the p.d.f. of X is one of $f_{i,\mu,\sigma} = \frac{1}{\sigma^n} f_i\left(\frac{x_1-\mu}{\sigma}, ..., \frac{x_n-\mu}{\sigma}\right)$, $i = 0, 1$, $f_{i,\mu,\sigma}$ is symmetric about μ, the hypotheses of interest are given by (6.48) with $f_{i,\mu}$ replaced by $f_{i,\mu,\sigma}$, and $\mathcal{G} = \{g_{c,r} : c \in \mathcal{R}, r \neq 0\}$ with $g_{c,r}(x) = (rx_1 + c, ..., rx_n + c)$. When $n \geq 3$, it can be shown that a maximal invariant under $\mathcal{G}$ is $W(X) = (W_1, ..., W_{n-2})$, where $W_i = (X_i - X_n)/(X_{n-1} - X_n)$, and that the p.d.f. of W does not depend on (μ, σ). A UMPI test can then be derived (exercise). ∎

The next example considers finding a maximal invariant in a problem that is not a location-scale family problem.

Example 6.14. Let $\mathcal{G}$ be the set of $n!$ permutations of the components of $x \in \mathcal{R}^n$. Then a maximal invariant is the vector of order statistics. This is because a permutation of the components of x does not change the values of these components and two x's with the same set of ordered components can be obtained from each other through a permutation of coordinates.

Suppose that $\mathcal{P}$ contains continuous c.d.f.'s on $\mathcal{R}^n$. Let $\mathcal{G}$ be the class of all transformations of the form $g(x) = (\psi(x_1), ..., \psi(x_n))$, where ψ is continuous and strictly increasing. For $x = (x_1, ..., x_n)$, let $R(x) = (R_1, ..., R_n)$ be the vector of ranks (§5.2.2), i.e., $x_i = x_{(R_i)}$, where $x_{(j)}$ is the jth smallest value of x_i's. Clearly, $R(g(x)) = R(x)$ for any $g \in \mathcal{G}$. For any x and y in $\mathcal{R}^n$ with $R(x) = R(y)$, define $\psi(t)$ to be linear between $x_{(j)}$ and $x_{(j+1)}$, $j = 1, ..., n-1$, $\psi(t) = t + (y_{(1)} - x_{(1)})$ for $t \le x_{(1)}$, and $\psi(t) = t + (y_{(n)} - x_{(n)})$ for $t \ge x_{(n)}$. Then $\psi(x_i) = \psi(y_i)$, $i = 1, ..., n$. This shows that the vector of rank statistics is maximal invariant. ∎

When there is a sufficient statistic $U(X)$, it is convenient first to reduce the data to $U(X)$ before applying invariance. If there is a test $T(U)$ UMP among all invariant tests depending only on U, one would like to conclude that $T(U)$ is a UMPI test. Unfortunately, this may not be true in general, since it is not clear that for any invariant test based on X there is an equivalent invariant test based only on $U(X)$. The following result provides a sufficient condition under which it is enough to consider invariant tests depending only on $U(X)$. Its proof is omitted and can be found in Lehmann (1986, pp. 297-302).

Proposition 6.3. Let $\mathcal{G}$ be a group of transformations on $\mathcal{X}$ (the range of X) and $(\mathcal{G}, \mathcal{B}_\mathcal{G}, \lambda)$ be a measure space with a σ-finite λ. Suppose that the testing problem under consideration is invariant under $\mathcal{G}$, that for any set $A \in \mathcal{B}_\mathcal{X}$, the set of points (x, g) for which $g(x) \in A$ is in $\sigma(\mathcal{B}_\mathcal{X} \times \mathcal{B}_\mathcal{G})$, and that $\lambda(B) = 0$ implies $\lambda(\{h \circ g : h \in B\}) = 0$ for all $g \in \mathcal{G}$. Suppose further that there is a statistic $U(X)$ sufficient for $P \in \mathcal{P}$ and that $U(x_1) = U(x_2)$ implies $U(g(x_1)) = U(g(x_2))$ for all $g \in \mathcal{G}$ so that $\mathcal{G}$ induces a group $\mathcal{G}_U$ of transformations on the range of U through $g_U(U(x)) = U(g(x))$. Then, for any test $T(X)$ invariant under $\mathcal{G}$, there exists a test based on $U(X)$ that is invariant under $\mathcal{G}$ (and $\mathcal{G}_U$) and has the same power function as $T(X)$. ∎

In many problems $g(x) = \psi(x, g)$, where g ranges over a set $\mathcal{G}$ in $\mathcal{R}^m$ and ψ is a Borel function on $\mathcal{R}^{n+m}$. Then the measurability condition in Proposition 6.3 is satisfied by choosing $\mathcal{B}_\mathcal{G}$ to be the Borel σ-field on $\mathcal{G}$. In such cases it is usually not difficult to find a measure λ satisfying the condition in Proposition 6.3.

Example 6.15. Let $X_1, ..., X_n$ be i.i.d. from $N(\mu, \sigma^2)$ with unknown $\mu \in \mathcal{R}$ and $\sigma^2 > 0$. The problem of testing $H_0 : \sigma^2 \ge \sigma_0^2$ versus $H_1 : \sigma^2 < \sigma_0^2$ is invariant under $\mathcal{G} = \{g_c : c \in \mathcal{R}\}$ with $g_c(x) = (x_1 + c, ..., x_n + c)$. It can be shown (exercise) that $\mathcal{G}$ and the sufficient statistic $U = (\bar{X}, S^2)$ satisfy the conditions in Proposition 6.3 with $\mathcal{G}_U = \{h_c : c \in \mathcal{R}\}$ and $h_c(u_1, u_2) = (u_1 + c, u_2)$, and that S^2 is maximal invariant under $\mathcal{G}_U$. It follows from Proposition 6.3, Corollary 6.1, and the fact that $(n-1)S^2/\sigma_0^2$

has the chi-square distribution χ^2_{n-1} when $\sigma^2 = \sigma^2_0$ that a UMPI test of size α rejects H_0 when $(n-1)S^2/\sigma^2_0 \leq \chi^2_{n-1,1-\alpha}$, where $\chi^2_{n-1,\alpha}$ is the $(1-\alpha)$th quantile of the chi-square distribution χ^2_{n-1}. This test coincides with the UMPU test given in §6.2.3. ∎

Example 6.16. Let $X_{i1}, ..., X_{in_i}$, $i = 1, 2$, be two independent samples i.i.d. from $N(\mu_i, \sigma^2_i)$, $i = 1, 2$, respectively. The problem of testing $H_0 : \sigma^2_2/\sigma^2_1 \leq \Delta_0$ versus $H_1 : \sigma^2_2/\sigma^2_1 > \Delta_0$ is invariant under

$$\mathcal{G} = \{g_{c_1,c_2,r} : c_i \in \mathcal{R}, i = 1, 2, r > 0\}$$

with

$$g_{c_1,c_2,r}(x_1, x_2) = (rx_{11} + c_1, ..., rx_{1n_1} + c_1, rx_{21} + c_2, ..., rx_{2n_2} + c_2).$$

It can be shown (exercise) that the sufficient statistic $U = (\bar{X}_1, \bar{X}_2, S^2_1, S^2_2)$ and $\mathcal{G}$ satisfy the conditions in Proposition 6.3 with

$$\mathcal{G}_U = \{h_{c_1,c_2,r} : c_i \in \mathcal{R}, i = 1, 2, r > 0\}$$

and

$$h_{c_1,c_2,r}(u_1, u_2, u_3, u_4) = (ru_1 + c_1, ru_2 + c_2, ru_3, ru_4).$$

A maximal invariant under $\mathcal{G}_U$ is S^2/S_1. Let $\Delta = \sigma^2_2/\sigma^2_1$. Then $(S^2_2/S^2_1)/\Delta$ has an F-distribution and, therefore, $V = S^2_2/S^2_1$ has a Lebesgue p.d.f. of the form

$$f_\Delta(v) = C(\Delta)v^{(n_2-3)/2}[\Delta + (n_2-1)v/(n_1-1)]^{-(n_1+n_2-2)/2}I_{(0,\infty)}(v),$$

where $C(\Delta)$ is a known function of Δ. It can be shown (exercise) that the family $\{f_\Delta : \Delta > 0\}$ has monotone likelihood ratio in V so that a UMPI test of size α rejects H_0 when $V > F_{n_2-1,n_1-1,\alpha}$, where $F_{a,b,\alpha}$ is the $(1-\alpha)$th quantile of the F-distribution $F_{a,b}$. Again, this UMPI test coincides with the UMPU test given in §6.2.3. ∎

The following result shows that, in Examples 6.15 and 6.16, the fact that UMPI tests are the same as the UMPU tests is not a simple coincidence.

Proposition 6.4. Consider a testing problem invariant under $\mathcal{G}$. If there exists a UMPI test of size α, then it is unbiased. If there also exists a UMPU test of size α that is invariant under $\mathcal{G}$, then the two tests have the same power function on $P \in \mathcal{P}_1$. If either the UMPI test or the UMPU test is unique a.s. $\mathcal{P}$, then the two tests are equal a.s. $\mathcal{P}$.
Proof. We only need to prove that a UMPI test of size α is unbiased. This follows from the fact that the test $T \equiv \alpha$ is invariant under $\mathcal{G}$. ∎

The next example shows an application of invariance in a situation where a UMPU test may not exist.

Example 6.17. Let $X_1, ..., X_n$ be i.i.d. from $N(\mu, \sigma^2)$ with unknown μ and σ^2. Let $\theta = (\mu - u)/\sigma$, where u is a known constant. Consider the problem of testing $H_0 : \theta \leq \theta_0$ versus $H_1 : \theta > \theta_0$. Note that H_0 is the same as $P(X_1 \leq u) \geq p_0$ for a known constant $p_0 = \Phi(-\theta_0)$. Without loss of generality, we consider the case of $u = 0$.

The problem is invariant under $\mathcal{G} = \{g_r : r > 0\}$ with $g_r(x) = rx$. By Proposition 6.3, we can consider tests that are functions of the sufficient statistic $(\bar{X}, S^2)$ only. A maximal invariant under $\mathcal{G}$ is $t(X) = \sqrt{n}\bar{X}/S$. To find a UMPI test, it remains to find a test UMP among all tests that are functions of $t(X)$.

From the discussion in §1.3.1, $t(X)$ has the noncentral t-distribution $t_{n-1}(\sqrt{n}\theta)$. Let $f_\theta(t)$ be the Lebesgue p.d.f. of $t(X)$, i.e., f_θ is given by (1.32) with n replaced by $n - 1$ and $\delta = \sqrt{n}\theta$. It can be shown (exercise) that the family of p.d.f.'s, $\{f_\theta(t) : \theta \in \mathcal{R}\}$, has monotone likelihood ratio in t. Hence, by Theorem 6.2, a UMPI test of size α rejects H_0 when $t(X) > c$, where c is the $(1 - \alpha)$th quantile of $t_{n-1}(\sqrt{n}\theta_0)$.

In some problems, we may have to apply both unbiasedness and invariance principles. For instance, suppose that in the current problem we would like to test $H_0 : \theta = \theta_0$ versus $H_1 : \theta \neq \theta_0$. The problem is still invariant under $\mathcal{G}$. Following the previous discussion, we only need to consider tests that are functions of $t(X)$. But a test UMP among functions of $t(X)$ does not exist in this case. A test UMP among all unbiased tests of level α that are functions of $t(X)$ rejects H_0 when $t(X) < c_1$ or $t(X) > c_2$, where c_1 and c_2 are determined by

$$\int_{c_1}^{c_2} f_{\theta_0}(t)dt = 1 - \alpha \qquad \text{and} \qquad \frac{d}{d\theta}\left[\int_{c_1}^{c_2} f_\theta(t)dt\right]\Bigg|_{\theta=\theta_0} = 0$$

(see Exercise 26). This test is then UMP among all tests that are invariant and unbiased of level α. Whether it is also UMPU without the restriction to invariant tests is an open problem. ∎

6.3.2 UMPI tests in normal linear models

Consider normal linear model (6.38):

$$X = N_n(Z\beta, \sigma^2 I_n),$$

where β is a p-vector of unknown parameters, $\sigma^2 > 0$ is unknown, and Z is a fixed $n \times p$ matrix of rank $r \leq p < n$. In §6.2.3, UMPU tests for testing

(6.39) or (6.40) are derived. A frequently encountered problem in practice is to test

$$H_0 : L\beta = 0 \qquad \text{versus} \qquad H_1 : L\beta \neq 0, \qquad (6.49)$$

where L is an $s \times p$ matrix of rank $s \leq r$ and all rows of L are in $\mathcal{R}(Z)$. However, a UMPU test for (6.49) does not exist if $s > 1$. We now derive a UMPI test for testing (6.49). We use without proof the following result from linear algebra: there exists an orthogonal matrix Γ such that (6.49) is equivalent to

$$H_0 : \eta_1 = 0 \qquad \text{versus} \qquad H_1 : \eta_1 \neq 0, \qquad (6.50)$$

where η_1 is the s-vector containing the first s components of η, η is the r-vector containing the first r components of $\Gamma Z\beta$, and the last $n - r$ components of $\Gamma Z\beta$ are 0's. Let $Y = \Gamma X$. Then $Y = N_n((\eta, 0), \sigma^2 I_n)$ with the p.d.f. given by (6.42). Let $Y = (Y_1, Y_2)$, where Y_1 is an r-vector, and let $Y_1 = (Y_{11}, Y_{12})$, where Y_{11} is an s-vector. Define

$$\mathcal{G} = \{g_{\Lambda, c, \gamma} : c \in \mathcal{R}^{r-s}, \ \gamma > 0, \ \Lambda \text{ is an } s \times s \text{ orthogonal matrix}\}$$

with

$$g_{\Lambda, c, \gamma}(Y) = \gamma(\Lambda Y_{11}, Y_{12} + c, Y_2).$$

Testing (6.50) is invariant under $\mathcal{G}$. By Proposition 6.3, we can restrict our attention to the sufficient statistic $U = (Y_1, \|Y_2\|^2)$. The statistic

$$M(U) = \|Y_{11}\|^2 / \|Y_2\|^2 \qquad (6.51)$$

is invariant under $\mathcal{G}_U$, the group of transformations on the range of U defined by $\tilde{g}_{\Lambda, c, \gamma}(U(Y)) = U(g_{\Lambda, c, \gamma}(Y))$. We now show that $M(U)$ is maximal invariant under $\mathcal{G}_U$. Let $l_i \in \mathcal{R}^s$, $l_i \neq 0$, and $t_i \in (0, \infty)$, $i = 1, 2$. If $\|l_1\|^2 / t_1^2 = \|l_2\|^2 / t_2^2$, then $t_1 = \gamma t_2$ with $\gamma = \|l_1\| / \|l_2\|$. Since $l_1 / \|l_1\|$ and $l_2 / \|l_2\|$ are two points having the same distance from the origin, there exists orthogonal matrix Λ such that $l_1 / \|l_1\| = \Lambda l_2 / \|l_2\|$, i.e., $l_1 = \gamma \Lambda l_2$. This proves that if $M(u^{(1)}) = M(u^{(2)})$ with $u^{(j)} = (y_{11}^{(j)}, y_{12}^{(j)}, t_j^2)$, then $y_{11}^{(1)} = \gamma \Lambda y_{11}^{(2)}$ and $t_1 = \gamma t_2$ for some $\gamma > 0$ and orthogonal matrix Λ and, therefore, $u^{(1)} = \tilde{g}_{\Lambda, c, \gamma}(u^{(2)})$ with $c = \gamma^{-1} y_{12}^{(1)} - y_{12}^{(2)}$. Thus, $M(U)$ is maximal invariant under $\mathcal{G}_U$.

It can be shown (exercise) that $W = M(U)(n - r)/s$ has the noncentral F-distribution $F_{s, n-r}(\theta)$ with $\theta = \|\eta_1\|^2 / \sigma^2$ (see §1.3.1). Let $f_\theta(w)$ be the Lebesgue p.d.f. of W, i.e., f_θ is given by (1.33) with $n_1 = s$, $n_2 = n - r$, and $\delta = \theta$. Note that under H_0, $\theta = 0$ and f_θ reduces to the p.d.f. of the central F-distribution $F_{s, n-r}$ (Table 1.2, page 20). Also, it can be shown (exercise) that the ratio $f_{\theta_1}(w) / f_0(w)$ is an increasing function of w for any given $\theta_1 > 0$. By Theorem 6.1, a UMPI test of size α for testing $H_0 : \theta = 0$

versus $H_1 : \theta = \theta_1$ rejects H_0 when $W > F_{s,n-r,\alpha}$, where $F_{s,n-r,\alpha}$ is the $(1 - \alpha)$th quantile of the F-distribution $F_{s,n-r}$. Since this test does not depend on θ_1, by Lemma 6.1, it is also a UMPI test of size α for testing $H_0 : \theta = 0$ versus $H_1 : \theta > 0$, which is equivalent to testing (6.50).

In applications it is not convenient to carry out the test by finding explicitly the orthogonal matrix Γ. Hence, we now express the statistic W in terms of X. Since $Y = \Gamma X$ and $E(Y) = \Gamma E(X) = \Gamma Z \beta$,

$$\|Y_1 - \eta\|^2 + \|Y_2\|^2 = \|X - Z\beta\|^2$$

and, therefore,

$$\min_{\eta} \|Y_1 - \eta\|^2 + \|Y_2\|^2 = \min_{\beta} \|X - Z\beta\|^2,$$

which is the same as

$$\|Y_2\|^2 = \|X - Z\hat\beta\|^2 = SSR,$$

where $\hat\beta$ is the LSE defined by (3.27). Similarly,

$$\|Y_{11}\|^2 + \|Y_2\|^2 = \min_{\beta:L\beta=0} \|X - Z\beta\|^2.$$

If we define $\hat\beta_{H_0}$ to be a solution of

$$\|X - Z\hat\beta_{H_0}\|^2 = \min_{\beta:L\beta=0} \|X - Z\beta\|^2,$$

which is called the LSE of β under H_0 or the LSE of β subject to $L\beta = 0$, then

$$W = \frac{(\|X - Z\hat\beta_{H_0}\|^2 - \|X - Z\hat\beta\|^2)/s}{\|X - Z\hat\beta\|^2/(n - r)}. \tag{6.52}$$

Thus, the UMPI test for (6.49) can be used without finding Γ.

When $s = 1$, the UMPI test derived here is the same as the UMPU test for (6.40) given in §6.2.3.

Example 6.18. Consider the one-way ANOVA model in Example 3.13:

$$X_{ij} = N(\mu_i, \sigma^2), \qquad j = 1, ..., n_i, \ i = 1, ..., m,$$

and X_{ij}'s are independent. A common testing problem in applications is the test for homogeneity of means, i.e.,

$$H_0 : \mu_1 = \cdots = \mu_m \qquad \text{versus} \qquad H_1 : \mu_i \neq \mu_k \text{ for some } i \neq k. \tag{6.53}$$

One can easily find a matrix L for which (6.53) is equivalent to (6.49). But it is not necessary to find such a matrix in order to compute the statistic W that defines the UMPI test. Note that the LSE of $(\mu_1, ..., \mu_m)$ is $(\bar{X}_1, ..., \bar{X}_m)$, where $\bar{X}_i$ is the sample mean based on $X_{i1}, ..., X_{in_i}$, and the LSE under H_0 is simply $\bar{X}$, the sample mean based on all X_{ij}'s. Thus,

$$SSR = \|X - Z\hat{\beta}\|^2 = \sum_{i=1}^{m} \sum_{j=1}^{n_i} (X_{ij} - \bar{X}_i)^2,$$

$$SST = \|X - Z\hat{\beta}_{H_0}\|^2 = \sum_{i=1}^{m} \sum_{j=1}^{n_i} (X_{ij} - \bar{X})^2,$$

and

$$SSA = SST - SSR = \sum_{i=1}^{m} n_i(\bar{X}_i - \bar{X})^2.$$

Then

$$W = \frac{SSA/(m-1)}{SSR/(n-m)},$$

where $n = \sum_{i=1}^{m} n_i$. The name ANOVA comes from the fact that the UMPI test is carried out by comparing two sources of variation: the variation within each group of observations (measured by SSR) and the variation among m groups (measured by SSA), and that $SSA + SSR = SST$ is the total variation in the data set.

In this case, the distribution of W can also be derived using Cochran's theorem (Theorem 1.5). See Exercise 75. ■

Example 6.19. Consider the two-way balanced ANOVA model in Example 3.14:

$$X_{ijk} = N(\mu_{ij}, \sigma^2), \qquad i = 1, ..., a, \ j = 1, ..., b, \ k = 1, ..., c,$$

where $\mu_{ij} = \mu + \alpha_i + \beta_j + \gamma_{ij}$, $\sum_{i=1}^{a} \alpha_i = \sum_{j=1}^{b} \beta_j = \sum_{i=1}^{a} \gamma_{ij} = \sum_{j=1}^{b} \gamma_{ij} = 0$, and X_{ijk}'s are independent. Typically the following hypotheses are of interest:

$$H_0 : \alpha_i = 0 \text{ for all } i \qquad \text{versus} \qquad H_1 : \alpha_i \neq 0 \text{ for some } i, \qquad (6.54)$$

$$H_0 : \beta_j = 0 \text{ for all } j \qquad \text{versus} \qquad H_1 : \beta_j \neq 0 \text{ for some } j, \qquad (6.55)$$

and

$$H_0 : \gamma_{ij} = 0 \text{ for all } i, j \qquad \text{versus} \qquad H_1 : \gamma_{ij} \neq 0 \text{ for some } i, j. \qquad (6.56)$$

In applications, α_i's are effects of a factor A (a variable taking finitely many values), β_j's are effects of a factor B, and γ_{ij}'s are effects of the interaction of factors A and B. Hence, testing hypotheses in (6.54), (6.55), and (6.56) are the same as testing effects of factor A, of factor B, and of the interaction between A and B, respectively.

The LSE's of μ, α_i, β_j, and γ_{ij} are given by (Example 3.14) $\hat{\mu} = \bar{X}_{...}$, $\hat{\alpha}_i = \bar{X}_{i..} - \bar{X}_{...}$, $\hat{\beta}_j = \bar{X}_{.j.} - \bar{X}_{...}$, $\hat{\gamma}_{ij} = \bar{X}_{ij.} - \bar{X}_{i..} - \bar{X}_{.j.} + \bar{X}_{...}$, and a dot is used to denote averaging over the indicated subscript. Let

$$SSR = \sum_{i=1}^{a}\sum_{j=1}^{b}\sum_{k=1}^{c}(X_{ijk} - \bar{X}_{ij.})^2,$$

$$SSA = bc\sum_{i=1}^{a}(\bar{X}_{i..} - \bar{X}_{...})^2,$$

$$SSB = ac\sum_{j=1}^{b}(\bar{X}_{.j.} - \bar{X}_{...})^2,$$

and

$$SSC = c\sum_{i=1}^{a}\sum_{j=1}^{b}(\bar{X}_{ij.} - \bar{X}_{i..} - \bar{X}_{.j.} + \bar{X}_{...})^2.$$

Then, one can show (exercise) that for testing (6.54), (6.55), and (6.56), the statistics W in (6.52) (for the UMPI tests) are, respectively,

$$\frac{SSA/(a-1)}{SSR/[(c-1)ab]}, \quad \frac{SSB/(b-1)}{SSR/[(c-1)ab]}, \quad \text{and} \quad \frac{SSC/[(a-1)(b-1)]}{SSR/[(c-1)ab]}. \quad \blacksquare$$

We end this section with a discussion of testing for random effects in the following balanced one-way random effects model (Example 3.17):

$$X_{ij} = \mu + A_i + e_{ij}, \qquad i = 1, ..., a, j = 1, ..., b, \tag{6.57}$$

where μ is an unknown parameter, A_i's are i.i.d. random effects from $N(0, \sigma_a^2)$, e_{ij}'s are i.i.d. measurement errors from $N(0, \sigma^2)$, and A_i's and e_{ij}'s are independent. Consider the problem of testing

$$H_0 : \sigma_a^2/\sigma^2 \le \Delta_0 \qquad \text{versus} \qquad H_1 : \sigma_a^2/\sigma^2 > \Delta_0 \tag{6.58}$$

for a given Δ_0. When Δ_0 is small, hypothesis H_0 in (6.58) means that the random effects are negligible relative to the measurement variation.

Let $(Y_{i1}, ..., Y_{ib}) = \Gamma(X_{i1}, ..., X_{ib})$, where Γ is a $b \times b$ orthogonal matrix whose elements in the first row are all equal to $1/\sqrt{b}$. Then

$$Y_{i1} = \sqrt{b}\bar{X}_{i.} = \sqrt{b}(\mu + A_i + \bar{e}_{i.}), \qquad i = 1, ..., a,$$

are i.i.d. from $N(\sqrt{b}\mu, \sigma^2 + b\sigma_a^2)$, Y_{ij}, $i = 1, ..., a, j = 2, ..., b$, are i.i.d. from $N(0, \sigma^2)$, and Y_{ij}'s are independent. The reason why $E(Y_{ij}) = 0$ when $j > 1$ is because row j of Γ is orthogonal to the first row of Γ.

Let Λ be an $a \times a$ orthogonal matrix whose elements in the first row are all equal to $1/\sqrt{a}$ and $(U_{11}, ..., U_{a1}) = \Lambda(Y_{11}, ..., Y_{a1})$. Then $U_{11} = \sqrt{a}\bar{Y}_{\cdot 1}$ is $N(\sqrt{ab}\mu, \sigma^2 + b\sigma_a^2)$, U_{i1}, $i = 2, ..., a$, are from $N(0, \sigma^2 + b\sigma_a^2)$, and U_{i1}'s are independent. Let $U_{ij} = Y_{ij}$ for $j = 2, ..., b$, $i = 1, ..., a$.

The problem of testing (6.58) is invariant under the group of transformations that transform U_{11} to $rU_{11} + c$ and U_{ij} to rU_{ij}, $(i, j) \neq (1, 1)$, where $r > 0$ and $c \in \mathcal{R}$. It can be shown (exercise) that the maximal invariant under this group of transformations is SSA/SSR, where

$$SSA = \sum_{i=2}^{a} U_{i1}^2 \quad \text{and} \quad SSR = \sum_{i=1}^{a}\sum_{j=2}^{b} U_{ij}^2.$$

Note that H_0 in (6.58) is equivalent to $(\sigma^2 + b\sigma_a^2)/\sigma^2 \leq 1 + b\Delta_0$. Also, $SSA/(\sigma^2 + b\sigma_a^2)$ has the chi-square distribution χ_{a-1}^2 and SSR/σ^2 has the chi-square distribution $\chi_{a(b-1)}^2$. Hence, the p.d.f. of the statistic

$$W = \frac{1}{1 + b\Delta_0} \frac{SSA/(a-1)}{SSR/[a(b-1)]}$$

is in a parametric family (indexed by the parameter $(\sigma^2 + b\sigma_a^2)/\sigma^2$) with monotone likelihood ratio in W. Thus, a UMPI test of size α for testing (6.58) rejects H_0 when $W > F_{a-1,a(b-1),\alpha}$, where $F_{a-1,a(b-1),\alpha}$ is the $(1 - \alpha)$th quantile of the F-distribution $F_{a-1,a(b-1)}$.

It remains to express W in terms of X_{ij}'s. Note that

$$SSR = \sum_{i=1}^{a}\sum_{j=2}^{b} Y_{ij}^2 = \sum_{i=1}^{a}\left(\sum_{j=1}^{b} e_{ij}^2 - b\bar{e}_{i\cdot}^2\right) = \sum_{i=1}^{a}\sum_{j=1}^{b}(X_{ij} - \bar{X}_{i\cdot})^2$$

and

$$SSA = \sum_{i=1}^{a} U_{i1}^2 - U_{11}^2 = \sum_{i=1}^{a} Y_{i1}^2 - a\bar{Y}_{\cdot 1}^2 = b\sum_{i=1}^{a}(\bar{X}_{i\cdot} - \bar{X}_{\cdot\cdot})^2.$$

The SSR and SSA derived here are the same as those in Example 6.18 when $n_i = b$ for all i and $m = a$. It can also be seen that if $\Delta_0 = 0$, then testing (6.58) is equivalent to testing $H_0 : \sigma_a^2 = 0$ versus $H_1 : \sigma_a^2 > 0$ and the derived UMPI test is exactly the same as that in Example 6.18, although the testing problems are different in these two cases.

Extensions to balanced two-way random effects models can be found in Lehmann (1986, §7.12).

6.4 Tests in Parametric Models

A UMP, UMPU, or UMPI test often does not exist in a particular problem. In the rest of this chapter, we study some methods for constructing tests that have intuitive appeal and frequently coincide with optimal tests (UMP or UMPU tests) when optimal tests do exist. We consider tests in parametric models in this section, whereas tests in nonparametric models are studied in §6.5.

When the hypothesis H_0 is not simple, it is often difficult or even impossible to obtain a test that has exactly a given size α, since it is hard to find a population P that maximizes the power function of the test over all $P \in \mathcal{P}_0$. In such cases a common approach is to find tests having asymptotic significance level α (Definition 2.13). This involves finding the limit of the power of a test at $P \in \mathcal{P}_0$, which is studied in this section and §6.5.

Throughout this section, we assume that a sample X is from $P \in \mathcal{P} = \{P_\theta : \theta \in \Theta\}$, $\Theta \subset \mathcal{R}^k$, $f_\theta = \frac{dP_\theta}{d\nu}$ exists w.r.t. a σ-finite measure ν for all θ, and the testing problem is

$$H_0 : \theta \in \Theta_0 \qquad \text{versus} \qquad H_1 : \theta \in \Theta_1, \tag{6.59}$$

where $\Theta_0 \cup \Theta_1 = \Theta$ and $\Theta_0 \cap \Theta_1 = \emptyset$.

6.4.1 Likelihood ratio tests

When both H_0 and H_1 are simple (i.e., both $\Theta_0 = \{\theta_0\}$ and $\Theta_1 = \{\theta_1\}$ are single-point sets), Theorem 6.1 applies and a UMP test rejects H_0 when

$$\frac{f_{\theta_1}(X)}{f_{\theta_0}(X)} > c_0 \tag{6.60}$$

for some $c_0 > 0$, which is equivalent to (exercise)

$$\frac{f_{\theta_0}(X)}{\max\{f_{\theta_0}(X), f_{\theta_1}(X)\}} < c \tag{6.61}$$

for some $c \in (0, 1]$. The following definition is a natural extension of this idea.

Definition 6.6. Let $\ell(\theta) = f_\theta(X)$ be the likelihood function. For testing (6.59), a *likelihood ratio* (LR) test is any test that rejects H_0 if and only if $\lambda(X) < c$, where $c \in [0, 1]$ and $\lambda(X)$ is the likelihood ratio defined by

$$\lambda(X) = \frac{\sup\limits_{\theta \in \Theta_0} \ell(\theta)}{\sup\limits_{\theta \in \Theta} \ell(\theta)}. \quad \blacksquare$$

If $\lambda(X)$ is well defined, then $\lambda(X) \leq 1$. The rationale behind LR tests is that when H_0 is true, $\lambda(X)$ tends to be close to 1, whereas when H_1 is true, $\lambda(X)$ tends to be away from 1. If there is a sufficient statistic, then $\lambda(X)$ depends only on the sufficient statistic. LR tests are as widely applicable as MLE's in §4.4 and, in fact, they are closely related to MLE's. If $\hat{\theta}$ is an MLE of θ and $\hat{\theta}_0$ is an MLE of θ subject to $\theta \in \Theta_0$ (i.e., Θ_0 is treated as the parameter space), then

$$\lambda(X) = \ell(\hat{\theta}_0)/\ell(\hat{\theta}).$$

For a given $\alpha \in (0, 1)$, if there exists a $c_\alpha \in [0, 1]$ such that

$$\sup_{\theta \in \Theta_0} P_\theta(\lambda(X) < c_\alpha) = \alpha, \tag{6.62}$$

then an LR test of size α can be obtained. Even when the c.d.f. of $\lambda(X)$ is continuous or randomized LR tests are introduced, it is still possible that a c_α satisfying (6.62) does not exist.

When a UMP or UMPU test exists, an LR test is often the same as this optimal test. For a real-valued θ, we have the following result.

Proposition 6.5. Suppose that X has the p.d.f. given by (6.10) w.r.t. a σ-finite measure ν, where η is a strictly increasing function of θ.
(i) For testing $H_0 : \theta \leq \theta_0$ versus $H_1 : \theta > \theta_0$, there is an LR test whose rejection region is the same as that of the UMP test T_* given by (6.11).
(ii) For testing the hypotheses in (6.12), there is an LR test whose rejection region is the same as that of the UMP test T_* given by (6.15).
(iii) For testing the hypotheses in (6.13) or (6.14), there is an LR test whose rejection region is equivalent to $Y(X) < c_1$ or $Y(X) > c_2$ for some constants c_1 and c_2.
Proof. (i) Let $\hat{\theta}$ be the MLE of θ. Note that $\ell(\theta)$ is increasing when $\theta \leq \hat{\theta}$ and decreasing when $\theta > \hat{\theta}$. Thus,

$$\lambda(X) = \begin{cases} 1 & \hat{\theta} \leq \theta_0 \\ \frac{\ell(\theta_0)}{\ell(\hat{\theta})} & \hat{\theta} > \theta_0. \end{cases}$$

Then $\lambda(X) < c$ is the same as $\hat{\theta} > \theta_0$ and $\ell(\theta_0)/\ell(\hat{\theta}) < c$. From the discussion in §4.4.2, $\hat{\theta}$ is a strictly increasing function of Y. It can be shown that $\log \ell(\hat{\theta}) - \log \ell(\theta_0)$ is strictly increasing in Y when $\hat{\theta} > \theta_0$ and strictly decreasing in Y when $\hat{\theta} < \theta_0$ (exercise). Hence, for any $d \in \mathcal{R}$, $\hat{\theta} > \theta_0$ and $\ell(\theta_0)/\ell(\hat{\theta}) < c$ is equivalent to $Y > d$ for some $c \in (0, 1)$.
(ii) The proof is similar to that in (i). Note that

$$\lambda(X) = \begin{cases} 1 & \hat{\theta} < \theta_1 \text{ or } \hat{\theta} > \theta_2 \\ \frac{\max\{\ell(\theta_1), \ell(\theta_2)\}}{\ell(\hat{\theta})} & \theta_1 \leq \hat{\theta} \leq \theta_2. \end{cases}$$

Hence $\lambda(X) < c$ is equivalent to $c_1 < Y < c_2$.
(iii) The proof for (iii) is left as an exercise. ∎

Proposition 6.5 can be applied to problems concerning one-parameter exponential families such as the binomial, Poisson, negative binomial, and normal (with one parameter known) families. The following example shows that the same result holds in a situation where Proposition 6.5 is not applicable.

Example 6.20. Consider the testing problem $H_0 : \theta = \theta_0$ versus $H_1 : \theta \neq \theta_0$ based on i.i.d. $X_1, ..., X_n$ from the uniform distribution $U(0, \theta)$. We now show that the UMP test with rejection region $X_{(n)} > \theta_0$ or $X_{(n)} \leq \theta_0 \alpha^{1/n}$ given in Exercise 19(c) is an LR test. Note that $\ell(\theta) = \theta^{-n} I_{(X_{(n)}, \infty)}(\theta)$. Hence

$$\lambda(X) = \begin{cases} (X_{(n)}/\theta_0)^n & X_{(n)} \leq \theta_0 \\ 0 & X_{(n)} > \theta_0 \end{cases}$$

and $\lambda(X) < c$ is equivalent to $X_{(n)} > \theta_0$ or $X_{(n)}/\theta_0 < c^{1/n}$. Taking $c = \alpha$ ensures that the LR test has size α. ∎

More examples of this kind can be found in §6.6. The next example considers multivariate θ.

Example 6.21. Consider normal linear model (6.38) and the hypotheses in (6.49). The likelihood function in this problem is

$$\ell(\theta) = \left(\frac{1}{2\pi\sigma^2}\right)^{n/2} \exp\left\{-\frac{1}{2\sigma^2}\|X - Z\beta\|^2\right\},$$

where $\theta = (\beta, \sigma^2)$. Let $\hat{\beta}$ be the LSE defined by (3.27). Since $\|X - Z\beta\|^2 \geq \|X - Z\hat{\beta}\|^2$ for any β,

$$\ell(\theta) \leq \left(\frac{1}{2\pi\sigma^2}\right)^{n/2} \exp\left\{-\frac{1}{2\sigma^2}\|X - Z\hat{\beta}\|^2\right\}.$$

Treating the right-hand side of the previous expression as a function of σ^2, it is easy to show that it has a maximum at $\sigma^2 = \hat{\sigma}^2 = \|X - Z\hat{\beta}\|^2/n$ and, therefore,

$$\sup_{\theta \in \Theta} \ell(\theta) = (2\pi\hat{\sigma}^2)^{-n/2} e^{-n/2}.$$

Similarly, let $\hat{\beta}_{H_0}$ be the LSE under H_0 and $\hat{\sigma}^2_{H_0} = \|X - Z\hat{\beta}_{H_0}\|^2/n$. Then

$$\sup_{\theta \in \Theta_0} \ell(\theta) = (2\pi\hat{\sigma}^2_{H_0})^{-n/2} e^{-n/2}.$$

Thus,

$$\lambda(X) = (\hat{\sigma}^2/\hat{\sigma}^2_{H_0})^{n/2} = \left(\frac{\|X - Z\hat{\beta}\|^2}{\|X - Z\hat{\beta}_{H_0}\|^2}\right)^{n/2} = \left(\frac{sW}{n-r} + 1\right)^{-n/2},$$

where W is given in (6.52). This shows that LR tests are the same as the UMPI tests derived in §6.3.2.

The one-sample or two-sample two-sided t-tests derived in §6.2.3 are special cases of LR tests. For a one-sample problem, we define $\beta = \mu$ and $Z = J_n$, the n-vector of ones. Note that $\hat{\beta} = \bar{X}$, $\hat{\sigma}^2 = (n-1)S^2/n$, $\hat{\beta}_{H_0}^2 = 0$ ($H_0 : \beta = 0$), and $\hat{\sigma}_{H_0}^2 = \|X\|^2/n = (n-1)S^2/n + \bar{X}^2$. Hence

$$\lambda(X) = \left[1 + \frac{n\bar{X}^2}{(n-1)S^2} \right]^{-n/2} = \left(1 + \frac{[t(X)]^2}{n-1} \right)^{-n/2},$$

where $t(X) = \sqrt{n}\bar{X}/S$ has the t-distribution t_{n-1} under H_0. Thus, $\lambda(X) < c$ is equivalent to $|t(X)| > c_0$, which is the rejection region of a one-sample two-sided t-test.

For a two-sample problem, we let $n = n_1 + n_2$, $\beta = (\mu_1, \mu_2)$, and

$$Z = \begin{pmatrix} J_{n_1} & 0 \\ 0 & J_{n_2} \end{pmatrix}.$$

Testing $H_0 : \mu_1 = \mu_2$ versus $H_1 : \mu_1 \neq \mu_2$ is the same as testing (6.49) with $L = (\ 1 \ \ -1 \ )$. Since $\hat{\beta}_{H_0} = \bar{X}$ and $\hat{\beta} = (\bar{X}_1, \bar{X}_2)$, where $\bar{X}_1$ and $\bar{X}_2$ are the sample means based on $X_1, ..., X_{n_1}$ and $X_{n_1+1}, ..., X_n$, respectively, we have

$$n\hat{\sigma}^2 = \sum_{i=1}^{n_1}(X_i - \bar{X}_1)^2 + \sum_{i=n_1+1}^{n} (X_i - \bar{X}_2)^2 = (n_1 - 1)S_1^2 + (n_2 - 1)S_2^2$$

and

$$n\hat{\sigma}_{H_0}^2 = (n-1)S^2 = n^{-1}n_1 n_2(\bar{X}_1 - \bar{X}_2)^2 + (n_1 - 1)S_1^2 + (n_2 - 1)S_2^2.$$

Therefore, $\lambda(X) < c$ is equivalent to $|t(X)| > c_0$, where $t(X)$ is given by (6.37), and LR tests are the same as the two-sample two-sided t-tests in §6.2.3. ∎

6.4.2 Asymptotic tests based on likelihoods

As we can see from Proposition 6.5 and the previous examples, an LR test is often equivalent to a test based on a statistic $Y(X)$ whose distribution under H_0 can be used to determine the rejection region of the LR test with size α. When this technique fails, it is difficult or even impossible to find an LR test with size α, even if the c.d.f. of $\lambda(X)$ is continuous. The following result shows that in the i.i.d. case we can obtain the asymptotic distribution (under H_0) of the likelihood ratio $\lambda(X)$ so that an LR test having asymptotic significance level α can be obtained. Assume that Θ_0 is determined by

$$H_0 : \theta = g(\vartheta), \qquad (6.63)$$

where ϑ is a $(k - r)$-vector of unknown parameters and g is a continuously differentiable function from $\mathcal{R}^{k-r}$ to $\mathcal{R}^k$ with a full rank $\partial g(\vartheta)/\partial \vartheta$. For example, if $\Theta = \mathcal{R}^2$ and $\Theta_0 = \{(\theta_1, \theta_2) \in \Theta : \theta_1 = 0\}$, then $\vartheta = \theta_2$, $g_1(\vartheta) = 0$, and $g_2(\vartheta) = \vartheta$.

Theorem 6.5. Assume the conditions in Theorem 4.16. Suppose that H_0 is determined by (6.63). Under H_0, $-2 \log \lambda_n \to_d \chi_r^2$, where $\lambda_n = \lambda(X)$ and χ_r^2 is a random variable having the chi-square distribution χ_r^2. Consequently, the LR test with rejection region $\lambda_n < e^{-\chi_{r,\alpha}^2/2}$ has asymptotic significance level α, where $\chi_{r,\alpha}^2$ is the $(1 - \alpha)$th quantile of the chi-square distribution χ_r^2.

Proof. Without loss of generality, we assume that there exist an MLE $\hat{\theta}$ and an MLE $\hat{\vartheta}$ under H_0 such that

$$\lambda_n = \frac{\sup_{\theta \in \Theta_0} \ell(\theta)}{\sup_{\theta \in \Theta} \ell(\theta)} = \frac{\ell(g(\hat{\vartheta}))}{\ell(\hat{\theta})}.$$

Following the proof of Theorem 4.17 in §4.5.2, we can obtain that

$$\sqrt{n} I_1(\theta)(\hat{\theta} - \theta) = n^{-1/2} s_n(\theta) + o_p(1),$$

where $s_n(\theta) = \partial \log \ell(\theta)/\partial \theta$ and $I_1(\theta)$ is the Fisher information about θ contained in X_1, and that

$$2[\log \ell(\hat{\theta}) - \log \ell(\theta)] = n(\hat{\theta} - \theta)^\tau I_1(\theta)(\hat{\theta} - \theta) + o_p(1).$$

Then

$$2[\log \ell(\hat{\theta}) - \log \ell(\theta)] = n^{-1}[s_n(\theta)]^\tau [I_1(\theta)]^{-1} s_n(\theta) + o_p(1).$$

Similarly, under H_0,

$$2[\log \ell(g(\hat{\vartheta})) - \log \ell(g(\vartheta))] = n^{-1}[\tilde{s}_n(\vartheta)]^\tau [\tilde{I}_1(\vartheta)]^{-1} \tilde{s}_n(\vartheta) + o_p(1),$$

where $\tilde{s}_n(\vartheta) = \partial \log \ell(g(\vartheta))/\partial \vartheta = D(\vartheta) s_n(g(\vartheta))$, $D(\vartheta) = \partial g(\vartheta)/\partial \vartheta$, and $\tilde{I}_1(\vartheta)$ is the Fisher information about ϑ (under H_0) contained in X_1. Combining these results, we obtain that

$$-2 \log \lambda_n = 2[\log \ell(\hat{\theta}) - \log \ell(g(\hat{\vartheta}))]$$
$$= n^{-1}[s_n(g(\vartheta))]^\tau B(\vartheta) s_n(g(\vartheta)) + o_p(1)$$

under H_0, where

$$B(\vartheta) = [I_1(g(\vartheta))]^{-1} - [D(\vartheta)]^\tau [\tilde{I}_1(\vartheta)]^{-1} D(\vartheta).$$

By the CLT, $n^{-1/2}[I_1(\theta)]^{-1/2}s_n(\theta) \to_d Z$, where $Z = N_k(0, I_k)$. Then, it follows from Theorem 1.10(iii) that, under H_0,

$$-2\log\lambda_n \to_d Z^\tau[I_1(g(\vartheta))]^{1/2}B(\vartheta)[I_1(g(\vartheta))]^{1/2}Z.$$

Let $D = D(\vartheta)$, $B = B(\vartheta)$, $A = I_1(g(\vartheta))$, and $C = \tilde{I}_1(\vartheta)$. Then

$$
\begin{aligned}
(A^{1/2}BA^{1/2})^2 &= A^{1/2}BABA^{1/2} \\
&= A^{1/2}(A^{-1} - D^\tau C^{-1}D)A(A^{-1} - D^\tau C^{-1}D)A^{1/2} \\
&= (I_k - A^{1/2}D^\tau C^{-1}DA^{1/2})(I_k - A^{1/2}D^\tau C^{-1}DA^{1/2}) \\
&= I_k - 2A^{1/2}D^\tau C^{-1}DA^{1/2} + A^{1/2}D^\tau C^{-1}DAD^\tau C^{-1}DA^{1/2} \\
&= I_k - A^{1/2}D^\tau C^{-1}DA^{1/2} \\
&= A^{1/2}BA^{1/2},
\end{aligned}
$$

where the fourth equality follows from the fact that $C = DAD^\tau$. This shows that $A^{1/2}BA^{1/2}$ is a projection matrix. The rank of $A^{1/2}BA^{1/2}$ is

$$
\begin{aligned}
\text{tr}(A^{1/2}BA^{1/2}) &= \text{tr}(I_k - D^\tau C^{-1}DA) \\
&= k - \text{tr}(C^{-1}DAD^\tau) \\
&= k - \text{tr}(C^{-1}C) \\
&= k - (k - r) \\
&= r.
\end{aligned}
$$

Thus, by Exercise 51 in §1.6, $Z^\tau[I_1(g(\vartheta))]^{1/2}B(\vartheta)[I_1(g(\vartheta))]^{1/2}Z = \chi_r^2$. ∎

As an example, Theorem 6.5 can be applied to testing problems in Example 4.33 where the exact rejection region of the LR test of size α is difficult to obtain but the likelihood ratio λ_n can be calculated numerically.

Tests whose rejection regions are constructed using asymptotic theory (so that these tests have asymptotic significance level α) are called *asymptotic tests*, which are useful when a test of exact size α is difficult to find. There are two popular asymptotic tests based on likelihoods that are asymptotically equivalent to LR tests. Note that the hypothesis in (6.63) is equivalent to a set of $r \le k$ equations:

$$H_0 : R(\theta) = 0, \tag{6.64}$$

where $R(\theta)$ is a continuously differentiable function from $\mathcal{R}^k$ to $\mathcal{R}^r$. Wald (1943) introduced a test that rejects H_0 when the value of

$$W_n = [R(\hat{\theta})]^\tau \{[C(\hat{\theta})]^\tau [I_n(\hat{\theta})]^{-1} C(\hat{\theta})\}^{-1} R(\hat{\theta})$$

is large, where $C(\theta) = \partial R(\theta)/\partial\theta$, $I_n(\theta)$ is the Fisher information matrix based on $X_1, ..., X_n$, and $\hat{\theta}$ is an MLE or RLE of θ. For testing $H_0 : \theta = \theta_0$

with a known θ_0, $R(\theta) = \theta - \theta_0$ and W_n simplifies to

$$W_n = (\hat{\theta} - \theta_0)^\tau I_n(\hat{\theta})(\hat{\theta} - \theta_0).$$

Rao (1947) introduced a *score* test that rejects H_0 when the value of

$$R_n = [s_n(\tilde{\theta})]^\tau [I_n(\tilde{\theta})]^{-1} s_n(\tilde{\theta})$$

is large, where $s_n(\theta) = \partial \log \ell(\theta)/\partial \theta$ is the score function and $\tilde{\theta}$ is an MLE or RLE of θ under H_0 in (6.64).

Theorem 6.6. Assume the conditions in Theorem 4.16.
(i) Under H_0 given by (6.64), $W_n \to_d \chi_r^2$ and, therefore, the test rejects H_0 if and only if $W_n > \chi_{r,\alpha}^2$ has asymptotic significance level α, where $\chi_{r,\alpha}^2$ is the $(1 - \alpha)$th quantile of the chi-square distribution χ_r^2.
(ii) The result in (i) still holds if W_n is replaced by R_n.
Proof. (i) Using Theorems 1.12 and 4.17,

$$\sqrt{n}[R(\hat{\theta}) - R(\theta)] \to_d N_r\left(0, [C(\theta)]^\tau [I_1(\theta)]^{-1} C(\theta)\right),$$

where $I_1(\theta)$ is the Fisher information about θ contained in X_1. Under H_0, $R(\theta) = 0$ and, therefore,

$$n[R(\hat{\theta})]^\tau \{[C(\theta)]^\tau [I_1(\theta)]^{-1} C(\theta)\}^{-1} R(\hat{\theta}) \to_d \chi_r^2$$

(Theorem 1.10). Then the result follows from Slutsky's theorem (Theorem 1.11) and the fact that $\hat{\theta} \to_p \theta$ and $I_1(\theta)$ and $C(\theta)$ are continuous at θ.
(ii) From the Lagrange multiplier, $\tilde{\theta}$ satisfies

$$s_n(\tilde{\theta}) + C(\tilde{\theta})\lambda_n = 0 \qquad \text{and} \qquad R(\tilde{\theta}) = 0.$$

Using Taylor's expansion, one can show (exercise) that under H_0,

$$[C(\theta)]^\tau (\tilde{\theta} - \theta) = o_p(n^{-1/2}) \tag{6.65}$$

and

$$s_n(\theta) - I_n(\theta)(\tilde{\theta} - \theta) + C(\theta)\lambda_n = o_p(n^{1/2}), \tag{6.66}$$

where $I_n(\theta) = nI_1(\theta)$. Multiplying $[C(\theta)]^\tau [I_n(\theta)]^{-1}$ to the left-hand side of (6.66) and using (6.65), we obtain that

$$[C(\theta)]^\tau [I_n(\theta)]^{-1} C(\theta)\lambda_n = -[C(\theta)]^\tau [I_n(\theta)]^{-1} s_n(\theta) + o_p(n^{-1/2}), \tag{6.67}$$

which implies

$$\lambda_n^\tau [C(\theta)]^\tau [I_n(\theta)]^{-1} C(\theta)\lambda_n \to_d \chi_r^2 \tag{6.68}$$

(exercise). Then the result follows from (6.68) and the fact that $C(\tilde{\theta})\lambda_n = -s_n(\tilde{\theta})$, $I_n(\theta) = nI_1(\theta)$, and $I_1(\theta)$ is continuous at θ. ∎

Thus, Wald's tests, Rao's score tests, and LR tests are asymptotically equivalent. Note that Wald's test requires computing $\hat{\theta}$, not $\tilde{\theta} = g(\hat{\vartheta})$, whereas Rao's score test requires computing $\tilde{\theta}$, not $\hat{\theta}$. On the other hand, an LR test requires computing both $\hat{\theta}$ and $\tilde{\theta}$ (or solving two maximization problems). Hence, one may choose one of these tests that is easy to compute in a particular application.

The results in Theorems 6.5 and 6.6 can be extended to non-i.i.d. situations (e.g., the GLM in §4.4.2). We state without proof the following result.

Theorem 6.7. Assume the conditions in Theorem 4.18. Consider the problem of testing H_0 in (6.64) (or equivalently, (6.63)) with $\theta = (\beta, \phi)$. Then the results in Theorems 6.5 and 6.6 still hold. $\blacksquare$

Example 6.22. Consider the GLM (4.55)-(4.58) with t_i's in a fixed interval (t_0, t_∞), $0 < t_0 \le t_\infty < \infty$. Then the Fisher information matrix

$$I_n(\theta) = \begin{pmatrix} \phi^{-1} M_n(\beta) & 0 \\ 0 & \tilde{I}_n(\beta, \phi) \end{pmatrix},$$

where $M_n(\beta)$ is given by (4.60) and $\tilde{I}_n(\beta, \phi)$ is the Fisher information about ϕ.

Consider the problem of testing $H_0 : \beta = \beta_0$ versus $H_1 : \beta \neq \beta_0$, where β_0 is a fixed vector. Then $R(\beta, \phi) = \beta - \beta_0$. Let $(\hat{\beta}, \hat{\phi})$ be the MLE (or RLE) of (β, ϕ). Then, Wald's test is based on

$$W_n = \hat{\phi}^{-1}(\hat{\beta} - \beta_0)^\tau M_n(\hat{\beta})(\hat{\beta} - \beta_0)$$

and Rao's score test is based on

$$R_n = \tilde{\phi}[\tilde{s}_n(\beta_0)]^\tau [M_n(\beta_0)]^{-1} \tilde{s}_n(\beta_0),$$

where $\tilde{s}_n(\beta)$ is given by (4.65) and $\tilde{\phi}$ is a solution of $\partial \log \ell(\beta_0, \phi)/\partial \phi = 0$. It follows from Theorem 4.18 that both W_n and R_n are asymptotically distributed as χ_p^2 under H_0. By Slutsky's theorem, we may replace $\hat{\phi}$ or $\tilde{\phi}$ by any consistent estimator of ϕ. $\blacksquare$

Wald's tests, Rao's score tests, and LR tests are typically consistent according to Definition 2.13(iii). They are also Chernoff-consistent (Definition 2.13(iv)) if α is chosen to be $\alpha_n \to 0$ and $\chi_{r,\alpha_n}^2 = o(n)$ as $n \to \infty$ (exercise). Other asymptotic optimality properties of these tests are discussed in Wald (1943); see also Serfling (1980, Chapter 10).

6.4.3 χ^2-tests

A test that is related to the asymptotic tests described in §6.4.2 is the so-called χ^2-test for testing cell probabilities in a multinomial distribution. Consider a sequence of n independent trials with k possible outcomes for each trial. Let $p_j > 0$ be the cell probability of occurrence of the jth outcome in any given trial and X_j be the number of occurrences of the jth outcome in n trials. Then $X = (X_1, ..., X_k)$ has the multinomial distribution (Example 2.7) with the parameter $\boldsymbol{p} = (p_1, ..., p_k)$. Let $\xi_i = (0, ..., 0, 1, 0, ..., 0)$, where the single nonzero component 1 is located in the jth position if the ith trial yields the jth outcome. Then $\xi_1, ..., \xi_n$ are i.i.d. and $X/n = \bar{\xi} = \sum_{i=1}^{n} \xi_i/n$. By the CLT,

$$Z_n(\boldsymbol{p}) = \sqrt{n} \left(\tfrac{X}{n} - \boldsymbol{p} \right) = \sqrt{n}(\bar{\xi} - \boldsymbol{p}) \to_d N_k(0, \Sigma), \qquad (6.69)$$

where $\Sigma = \mathrm{Var}(X/\sqrt{n})$ is a symmetric $k \times k$ matrix whose ith diagonal element is $p_i(1 - p_i)$ and (i, j)th off-diagonal element is $-p_i p_j$.

Consider the problem of testing

$$H_0 : \boldsymbol{p} = \boldsymbol{p}_0 \qquad \text{versus} \qquad H_1 : \boldsymbol{p} \neq \boldsymbol{p}_0, \qquad (6.70)$$

where $\boldsymbol{p}_0 = (p_{01}, ..., p_{0k})$ is a known vector of cell probabilities. A popular test for (6.70) is based on the following χ^2-statistic:

$$\chi^2 = \sum_{j=1}^{k} \frac{(X_j - np_{0j})^2}{np_{0j}} = \|D(\boldsymbol{p}_0)Z_n(\boldsymbol{p}_0)\|^2, \qquad (6.71)$$

where $Z_n(\boldsymbol{p})$ is given by (6.69) and $D(c)$ with $c = (c_1, ..., c_k)$ is the $k \times k$ diagonal matrix whose jth diagonal element is $c_j^{-1/2}$. Another popular test is based on the following modified χ^2-statistic:

$$\tilde{\chi}^2 = \sum_{j=1}^{k} \frac{(X_j - np_{0j})^2}{X_j} = \|D(X/n)Z_n(\boldsymbol{p}_0)\|^2. \qquad (6.72)$$

Note that X/n is an unbiased estimator of $\boldsymbol{p}$.

Theorem 6.8. Let $\phi = (\sqrt{p_1}, ..., \sqrt{p_k})$ and Λ be a $k \times k$ projection matrix. (i) If $\Lambda\phi = a\phi$, then

$$[Z_n(\boldsymbol{p})]^\tau D(\boldsymbol{p})\Lambda D(\boldsymbol{p})Z_n(\boldsymbol{p}) \to_d \chi_r^2,$$

where χ_r^2 has the chi-square distribution χ_r^2 with $r = \mathrm{tr}(\Lambda) - a$. (ii) The same result holds if $D(\boldsymbol{p})$ in (i) is replaced by $D(X/n)$.

Proof. (i) Let $D = D(\boldsymbol{p})$, $Z_n = Z_n(\boldsymbol{p})$, and $Z = N_k(0, I_k)$. From (6.69) and Theorem 1.10,

$$Z_n^\tau D\Lambda D Z_n \to_d Z^\tau A Z \qquad \text{with} \quad A = \Sigma^{1/2} D\Lambda D\Sigma^{1/2}.$$

From Exercise 51 in §1.6, the result in (i) follows if we can show that $A^2 = A$ (i.e., A is a projection matrix) and $r = \text{tr}(A)$. Since Λ is a projection matrix and $\Lambda\phi = a\phi$, a must be either 0 or 1. Note that $D\Sigma D = I_k - \phi\phi^\tau$. Then

$$
\begin{aligned}
A^3 &= \Sigma^{1/2} D\Lambda D\Sigma D\Lambda D\Sigma D\Lambda D\Sigma^{1/2} \\
&= \Sigma^{1/2} D(\Lambda - a\phi\phi^\tau)(\Lambda - a\phi\phi^\tau)\Lambda D\Sigma^{1/2} \\
&= \Sigma^{1/2} D(\Lambda - 2a\phi\phi^\tau + a^2\phi\phi^\tau)\Lambda D\Sigma^{1/2} \\
&= \Sigma^{1/2} D(\Lambda - a\phi\phi^\tau)\Lambda D\Sigma^{1/2} \\
&= \Sigma^{1/2} D\Lambda D\Sigma D\Lambda D\Sigma^{1/2} \\
&= A^2,
\end{aligned}
$$

which implies that the eigenvalues of A must be 0 or 1. Therefore, $A^2 = A$. Also,

$$\text{tr}(A) = \text{tr}[\Lambda(D\Sigma D)] = \text{tr}(\Lambda - a\phi\phi^\tau) = \text{tr}(\Lambda) - a.$$

(ii) The result in (ii) follows from the result in (i) and $X/n \to_p \boldsymbol{p}$. ∎

Note that the χ^2-statistic in (6.71) and the modified χ^2-statistic in (6.72) are special cases of the statistics in Theorem 6.8(i) and (ii), respectively, with $\Lambda = I_k$ satisfying $\Lambda\phi = \phi$. Hence, a test of asymptotic significance level α for testing (6.70) rejects H_0 when $\chi^2 > \chi^2_{k-1,\alpha}$ (or $\tilde{\chi}^2 > \chi^2_{k-1,\alpha}$), where $\chi^2_{k-1,\alpha}$ is the $(1 - \alpha)$th quantile of χ^2_{k-1}. These tests are called (asymptotic) χ^2-tests.

Example 6.23 (Goodness of fit tests). Let $Y_1, ..., Y_n$ be i.i.d. from F. Consider the problem of testing

$$H_0 : F = F_0 \qquad \text{versus} \qquad H_1 : F \neq F_0, \tag{6.73}$$

where F_0 is a known c.d.f. For instance, $F_0 = N(0, 1)$. One way to test (6.73) is to partition the range of Y_1 into k disjoint events $A_1, ..., A_k$ and test (6.70) with $p_j = P_F(A_j)$ and $p_{0j} = P_{F_0}(A_j)$, $j = 1, ..., k$. Let X_j be the number of Y_i's in A_j, $j = 1, ..., k$. Based on X_j's, the χ^2-tests discussed previously can be applied to this problem and they are called *goodness of fit* tests. ∎

In the goodness of fit tests discussed in Example 6.23, F_0 in H_0 is known so that p_{0j}'s can be computed. In some cases, we need to test the following hypotheses that are slightly different from those in (6.73):

$$H_0 : F = F_\theta \qquad \text{versus} \qquad H_1 : F \neq F_\theta, \tag{6.74}$$

where θ is an unknown parameter in $\Theta \subset \mathcal{R}^s$. For example, $F_\theta = N(\mu, \sigma^2)$, $\theta = (\mu, \sigma^2)$. If we still try to test (6.70) with $p_j = P_{F_\theta}(A_j)$, $j = 1, ..., k$, the result in Example 6.23 is not applicable since p is unknown under H_0. A generalized χ^2-test for (6.74) can be obtained using the following result. Let $p(\theta) = (p_1(\theta), ..., p_k(\theta))$ be a k-vector of known functions of $\theta \in \Theta \subset \mathcal{R}^s$, where $s < k$. Consider the testing problem

$$H_0 : p = p(\theta) \qquad \text{versus} \qquad H_1 : p \neq p(\theta). \tag{6.75}$$

Note that (6.70) is the special case of (6.75) with $s = 0$, i.e., θ is known. Let $\hat{\theta}$ be an MLE of θ under H_0. Then, by Theorem 6.5, the LR test that rejects H_0 when $-2 \log \lambda_n > \chi^2_{k-s-1,\alpha}$ has asymptotic significance level α, where $\chi^2_{k-s-1,\alpha}$ is the $(1-\alpha)$th quantile of χ^2_{k-s-1} and

$$\lambda_n = \prod_{j=1}^{k} [p_j(\hat{\theta})]^{X_j} \Big/ (X_j/n)^{X_j}.$$

Using the fact that $p_j(\hat{\theta})/(X_j/n) \to_p 1$ under H_0 and

$$\log(1 + x) = x - x^2/2 + o(|x|^2) \qquad \text{as } |x| \to 0,$$

we obtain that

$$-2 \log \lambda_n = -2 \sum_{j=1}^{k} X_j \log \left(1 + \frac{p_j(\hat{\theta})}{X_j/n} - 1 \right)$$

$$= -2 \sum_{j=1}^{k} X_j \left(\frac{p_j(\hat{\theta})}{X_j/n} - 1 \right) + \sum_{j=1}^{k} X_j \left(\frac{p_j(\hat{\theta})}{X_j/n} - 1 \right)^2 + o_p(1)$$

$$= \sum_{j=1}^{k} \frac{[X_j - np_j(\hat{\theta})]^2}{X_j} + o_p(1)$$

$$= \sum_{j=1}^{k} \frac{[X_j - np_j(\hat{\theta})]^2}{np_j(\hat{\theta})} + o_p(1),$$

where the third equality follows from $\sum_{j=1}^{k} p_j(\hat{\theta}) = \sum_{j=1}^{k} X_j/n = 1$. Define the generalized χ^2-statistics χ^2 and $\tilde{\chi}^2$ to be the χ^2 and $\tilde{\chi}^2$ in (6.71) and (6.72), respectively, with p_{0j}'s replaced by $p_j(\hat{\theta})$'s. We then have the following result.

Theorem 6.9. Under H_0 given by (6.75), the generalized χ^2-statistics converge in distribution to χ^2_{k-s-1}. The χ^2-test with rejection region $\chi^2 > \chi^2_{k-s-1,\alpha}$ (or $\tilde{\chi}^2 > \chi^2_{k-s-1,\alpha}$) has asymptotic significance level α, where $\chi^2_{k-s-1,\alpha}$ is the $(1-\alpha)$th quantile of χ^2_{k-s-1}. ∎

Theorem 6.9 can be applied to derive a goodness of fit test for hypotheses (6.74). However, one has to formulate (6.75) and compute an MLE of θ under $H_0 : \boldsymbol{p} = \boldsymbol{p}(\theta)$, which is different from an MLE under $H_0 : F = F_\theta$ unless (6.74) and (6.75) are the same; see Moore and Spruill (1975). The next example is the main application of Theorem 6.9.

Example 6.24 ($r \times c$ contingency tables). The following $r \times c$ contingency table is a natural extension of the 2×2 contingency table considered in Example 6.12:

	A_1	A_2	$\cdots$	A_c	Total
B_1	X_{11}	X_{12}	$\cdots$	X_{1c}	n_1
B_2	X_{21}	X_{22}	$\cdots$	X_{2c}	n_2
$\cdots$	$\cdots$	$\cdots$	$\cdots$	$\cdots$	$\cdots$
B_r	X_{r1}	X_{r2}	$\cdots$	X_{rc}	n_r
Total	m_1	m_2	$\cdots$	m_c	n

where A_i's are disjoint events with $A_1 \cup \cdots \cup A_c = \Omega$ (the sample space of a random experiment), B_i's are disjoint events with $B_1 \cup \cdots \cup B_r = \Omega$, and X_{ij} is the observed frequency of the outcomes in $A_j \cap B_i$. Similar to the case of the 2×2 contingency table discussed in Example 6.12, there are two important applications in this problem. We first consider testing independence of $\{A_j : j = 1, ..., c\}$ and $\{B_i : i = 1, ..., r\}$ with hypotheses

$$H_0 : p_{ij} = p_{i\cdot}p_{\cdot j} \quad \text{for all } i, j \qquad \text{versus} \qquad H_1 : p_{ij} \neq p_{i\cdot}p_{\cdot j} \quad \text{for some } i, j,$$

where $p_{ij} = P(A_j \cap B_i) = E(X_{ij})/n$, $p_{i\cdot} = P(B_i)$, and $p_{\cdot j} = P(A_j)$, $i = 1, ..., r$, $j = 1, ..., c$. In this case, $X = (X_{ij}, i = 1, ..., r, j = 1, ..., c)$ has the multinomial distribution with parameters p_{ij}, $i = 1, ..., r$, $j = 1, ..., c$. Under H_0, MLE's of $p_{i\cdot}$ and $p_{\cdot j}$ are $\bar{X}_{i\cdot} = n_i/n$ and $\bar{X}_{\cdot j} = m_j/n$, respectively, $i = 1, ..., r$, $j = 1, ..., c$ (exercise). By Theorem 6.9, the χ^2-test rejects H_0 when $\chi^2 > \chi^2_{(r-1)(c-1),\alpha}$, where

$$\chi^2 = \sum_{i=1}^{r}\sum_{j=1}^{c} \frac{(X_{ij} - n\bar{X}_{i\cdot}\bar{X}_{\cdot j})^2}{n\bar{X}_{i\cdot}\bar{X}_{\cdot j}} \tag{6.76}$$

and $\chi^2_{(r-1)(c-1),\alpha}$ is the $(1 - \alpha)$th quantile of the chi-square distribution $\chi^2_{(r-1)(c-1)}$ (exercise). One can also obtain the modified χ^2-test by replacing $n\bar{X}_{i\cdot}\bar{X}_{\cdot j}$ by X_{ij} in the denominator of each term of the sum in (6.76).

Next, suppose that $(X_{1j}, ..., X_{rj})$, $j = 1, ..., c$, are c independent random vectors having the multinomial distributions with parameters $(p_{1j}, ..., p_{rj})$, $j = 1, ..., c$, respectively. Consider the problem of testing whether c multinomial distributions are the same, i.e.,

$$H_0 : p_{ij} = p_{i1} \quad \text{for all } i, j \qquad \text{versus} \qquad H_1 : p_{ij} \neq p_{i1} \quad \text{for some } i, j.$$

It turns out that the rejection region of the χ^2-test given in Theorem 6.9 is still $\chi^2 > \chi^2_{(r-1)(c-1),\alpha}$ with χ^2 given by (6.76) (exercise).

One can also obtain the LR test in this problem. When $r = c = 2$, the LR test is equivalent to Fisher's exact test given in Example 6.12, which is a UMPU test. When $r > 2$ or $c > 2$, however, a UMPU test does not exist in this problem. ■

6.4.4 Bayes tests

An LR test actually compares $\sup_{\theta \in \Theta_0} \ell(\theta)$ with $\sup_{\theta \in \Theta_1} \ell(\theta)$ for testing (6.59). Instead of comparing two maximum values, one may compare two averages such as $\hat{\pi}_j = \int_{\Theta_j} \ell(\theta) d\Pi(\theta) / \int_\Theta \ell(\theta) d\Pi(\theta)$, $j = 0, 1$, where $\Pi(\theta)$ is a c.d.f. on Θ, and reject H_0 when $\hat{\pi}_1 > \hat{\pi}_0$. If Π is treated as a prior c.d.f., then $\hat{\pi}_j$ is the posterior probability of Θ_j, and this test is a particular Bayes action (see Exercise 18 in §4.6) and is called a Bayes test.

In Bayesian analysis, one often considers the *Bayes factor* defined to be

$$\beta = \frac{\text{posterior odds ratio}}{\text{prior odds ratio}} = \frac{\hat{\pi}_0/\hat{\pi}_1}{\pi_0/\pi_1},$$

where $\pi_j = \Pi(\Theta_j)$ is the prior probability of Θ_j.

Clearly, if there is a statistic sufficient for θ, then the Bayes test and Bayes factor depend only on the sufficient statistic.

Consider the special case where $\Theta_0 = \{\theta_0\}$ and $\Theta_1 = \{\theta_1\}$ are simple hypotheses. For given $X = x$,

$$\hat{\pi}_j = \frac{\pi_j f_{\theta_j}(x)}{\pi_0 f_{\theta_0}(x) + \pi_1 f_{\theta_1}(x)}.$$

Rejecting H_0 when $\hat{\pi}_1 > \hat{\pi}_0$ is the same as rejecting H_0 when

$$\frac{f_{\theta_1}(x)}{f_{\theta_0}(x)} > \frac{\pi_0}{\pi_1}. \tag{6.77}$$

This is equivalent to the UMP test T_* in (6.3) (Theorem 6.1) with $c = \pi_0/\pi_1$ and $\gamma = 0$. The Bayes factor in this case is

$$\beta = \frac{\hat{\pi}_0 \pi_1}{\hat{\pi}_1 \pi_0} = \frac{f_{\theta_0}(x)}{f_{\theta_1}(x)}.$$

Thus, the UMP test T_* in (6.3) is equivalent to the test that rejects H_0 when the Bayes factor is small. Note that the rejection region given by (6.77) depends on prior probabilities, whereas the Bayes factor does not.

When either Θ_0 or Θ_1 is not simple, however, Bayes factors also depend on the prior Π.

If Π is an improper prior, the Bayes test is still defined as long as the posterior probabilities $\hat{\pi}_j$ are finite. However, the Bayes factor may not be well defined when Π is improper.

Example 6.25. Let $X_1, ..., X_n$ be i.i.d. from $N(\mu, \sigma^2)$ with an unknown $\mu \in \mathcal{R}$ and a known $\sigma^2 > 0$. Let the prior of μ be $N(\xi, \tau^2)$. Then the posterior of μ is $N(\mu_*(x), c^2)$, where

$$\mu_*(x) = \frac{\sigma^2}{n\tau^2 + \sigma^2}\xi + \frac{n\tau^2}{n\tau^2 + \sigma^2}\bar{x} \quad \text{and} \quad c^2 = \frac{\tau^2\sigma^2}{n\tau^2 + \sigma^2}$$

(see Example 2.25). Consider first the problem of testing $H_0 : \mu \leq \mu_0$ versus $H_1 : \mu > \mu_0$. Let Φ be the c.d.f. of the standard normal. Then the posterior probability of Θ_0 and the Bayes factor are, respectively,

$$\hat{\pi}_0 = \Phi\left(\frac{\mu_0 - \mu_*(x)}{c}\right) \quad \text{and} \quad \beta = \frac{\Phi\left(\frac{\mu_0 - \mu_*(x)}{c}\right)\Phi\left(\frac{\xi - \mu_0}{\tau}\right)}{\Phi\left(\frac{\mu_*(x) - \mu_0}{c}\right)\Phi\left(\frac{\mu_0 - \xi}{\tau}\right)}.$$

It is interesting to see that if we let $\tau \to \infty$, which is the same as considering the improper prior $\Pi =$ the Lebesgue measure on $\mathcal{R}$, then

$$\hat{\pi}_0 \to \Phi\left(\frac{\mu_0 - \bar{x}}{\sigma/\sqrt{n}}\right),$$

which is exactly the p-value $\hat{\alpha}(x)$ derived in Example 2.29.

Consider next the problem of testing $H_0 : \mu = \mu_0$ versus $H_1 : \mu \neq \mu_0$. In this case the prior c.d.f. cannot be continuous at μ_0. We consider $\Pi(\mu) = \pi_0 I_{[\mu_0, \infty)}(\mu) + (1 - \pi_0)\Phi\left(\frac{\mu - \xi}{\tau}\right)$. Let $\ell(\mu)$ be the likelihood function based on $\bar{x}$. Then

$$m_1(x) = \int_{\mu \neq \mu_0} \ell(\mu) d\Phi\left(\frac{\mu - \xi}{\tau}\right) = \frac{1}{\sqrt{\tau^2 + \sigma^2/n}}\Phi'\left(\frac{\bar{x} - \xi}{\sqrt{\tau^2 + \sigma^2/n}}\right),$$

where $\Phi'(t)$ is the p.d.f. of the standard normal distribution, and

$$\hat{\pi}_0 = \frac{\pi_0 \ell(\mu_0)}{\pi_0 \ell(\mu_0) + (1 - \pi_0)m_1(x)} = \left(1 + \frac{1 - \pi_0}{\pi_0 \beta}\right)^{-1},$$

where

$$\beta = \frac{\ell(\mu_0)}{m_1(x)} = \frac{\sqrt{n\tau^2 + \sigma^2}\Phi'\left(\frac{\bar{x} - \mu_0}{\sigma/\sqrt{n}}\right)}{\sigma\Phi'\left(\frac{\bar{x} - \xi}{\sqrt{\tau^2 + \sigma^2/n}}\right)}$$

is the Bayes factor. ∎

More discussions about Bayesian hypothesis tests can be found in Berger (1985, §4.3.3).

6.5   Tests in Nonparametric Models

In a nonparametric problem, a UMP, UMPU, or UMPI test usually does not exist. In this section we study some nonparametric tests that have size α, limiting size α, or asymptotic significance level α. Consistency (Definition 2.13) of these nonparametric tests is also discussed.

Nonparametric tests are derived using some intuitively appealing ideas. They are commonly referred to as *distribution-free* tests, since almost no assumption is imposed on the population under consideration. But a nonparametric test may not be as good as a parametric test (in terms of its power) when the parametric model is correct. This is very similar to the case where we consider parametric estimation methods versus nonparametric estimation methods.

6.5.1   Sign, permutation, and rank tests

Three popular classes of nonparametric tests are introduced here. The first one is the class of *sign tests*. Let $X_1, ..., X_n$ be i.i.d. random variables from F, u be a fixed constant, and $p = F(u)$. Consider the problem of testing $H_0 : p \leq p_0$ versus $H_1 : p > p_0$, or testing $H_0 : p = p_0$ versus $H_1 : p \neq p_0$, where p_0 is a fixed constant in $(0, 1)$. Let

$$\Delta_i = \begin{cases} 1 & X_i - u \leq 0 \\ 0 & X_i - u > 0, \end{cases} \qquad i = 1, ..., n.$$

Then $\Delta_1, ..., \Delta_n$ are i.i.d. binary random variables with $p = P(\Delta_i = 1)$. For testing $H_0 : p \leq p_0$ versus $H_1 : p > p_0$, it follows from Corollary 6.1 that the test

$$T_*(Y) = \begin{cases} 1 & Y > m \\ \gamma & Y = m \\ 0 & Y < m \end{cases} \qquad (6.78)$$

is of size α and UMP among tests based on Δ_i's, where $Y = \sum_{i=1}^n \Delta_i$ and m and γ satisfy (6.7). Although T_* is of size α, we cannot conclude immediately that T_* is a UMP test, since $\Delta_1, ..., \Delta_n$ may not be sufficient for F. However, it can be shown that T_* is in fact a UMP test (Lehmann, 1986, pp. 106-107) in this particular case. Note that no assumption is imposed on F.

For testing $H_0 : p = p_0$ versus $H_1 : p \neq p_0$, it follows from Theorem 6.4 that the test

$$T_*(Y) = \begin{cases} 1 & Y < c_1 \text{ or } Y > c_2 \\ \gamma_i & Y = c_i, \ i = 1, 2, \\ 0 & c_1 < Y < c_2 \end{cases} \qquad (6.79)$$

is of size α and UMP among unbiased tests based on Δ_i's, where γ and c_i's are chosen so that $E(T_*) = \alpha$ and $E(T_*Y) = \alpha n p_0$ when $p = p_0$. This test is in fact a UMPU test (Lehmann, 1986, p. 166).

Since Y is equal to the number of nonnegative signs of $(u - X_i)$'s, tests based on T_* in (6.78) or (6.79) are called sign tests. One can easily extend the sign tests to the case where $p = P(X_1 \in B)$ with any fixed event B. Another extension is to the case where we observe i.i.d. $(X_1, Y_1), ..., (X_n, Y_n)$ (matched pairs). By using $\Delta_i = X_i - Y_i - u$, one can obtain sign tests for hypotheses concerning $P(X_1 - Y_1 \le u)$.

Next, we introduce the class of *permutation tests*. Let $X_{i1}, ..., X_{in_i}$, $i = 1, 2$, be two independent samples i.i.d. from F_i, $i = 1, 2$, respectively, where F_i's are c.d.f.'s on $\mathcal{R}$. In §6.2.3, we showed that the two-sample t-tests are UMPU tests for testing hypotheses concerning the means of F_i's, under the assumption that F_i's are normal with the same variance. Such types of testing problems arise from the comparison of two treatments. Suppose now we remove the normality assumption and replace it by a much weaker assumption that F_i's are in the nonparametric family $\mathcal{F}$ containing all continuous c.d.f.'s on $\mathcal{R}$. Consider the problem of testing

$$H_0 : F_1 = F_2 \qquad \text{versus} \qquad H_1 : F_1 \ne F_2, \tag{6.80}$$

which is the same as testing the equality of the means of F_i's when F_i's are normal with the same variance.

Let $X = (X_{ij}, j = 1, ..., n_i, i = 1, 2)$, $n = n_1 + n_2$, and α be a given significance level. A test $T(X)$ satisfying

$$\frac{1}{n!} \sum_{z \in \pi(x)} T(z) = \alpha \tag{6.81}$$

is called a permutation test, where $\pi(x)$ is the set of $n!$ points obtained from $x \in \mathcal{R}^n$ by permuting the components of x. Permutation tests are of size α (exercise). Under the assumption that $F_1(x) = F_2(x - \theta)$ and $F_1 \in \mathcal{F}$ containing all c.d.f.'s having Lebesgue p.d.f.'s that are continuous a.e., which is still much weaker than the assumption that F_i's are normal with the same variance, the class of permutation tests of size α is exactly the same as the class of unbiased tests of size α; see, for example, Lehmann (1986, p. 231).

Unfortunately, a test UMP among all permutation tests of size α does not exist. In applications, we usually choose a Lebesgue p.d.f. h and define a permutation test

$$T(X) = \begin{cases} 1 & h(X) > h_m \\ \gamma & h(X) = h_m \\ 0 & h(X) < h_m, \end{cases} \tag{6.82}$$

where h_m is the $(m + 1)$th largest value of the set $\{h(z) : z \in \pi(x)\}$, m is the integer part of $\alpha n!$, and $\gamma = \alpha n! - m$. This permutation test is optimal in some sense (Lehmann, 1986, §5.11).

While the class of permutation tests is motivated by the unbiasedness principle, the third class of tests introduced here is motivated by the invariance principle.

Consider first the one-sample problem in which $X_1, ..., X_n$ are i.i.d. random variables from a continuous c.d.f. F and we would like to test

$$H_0 : F \text{ is symmetric about } 0 \quad \text{versus} \quad H_1 : F \text{ is not symmetric about } 0.$$

Let $\mathcal{G}$ be the class of transformations $g(x) = (\psi(x_1), ..., \psi(x_n))$, where ψ is continuous, odd, and strictly increasing. Let $\tilde{R}(X)$ be the vector of ranks of $|X_i|$'s and $R_+(X)$ (or $R_-(X)$) be the subvector of $\tilde{R}(X)$ containing ranks corresponding to positive (or negative) X_i's. It can be shown (exercise) that (R_+, R_-) is maximal invariant under $\mathcal{G}$. Furthermore, sufficiency permits a reduction from R_+ and R_- to R_+^o, the vector of ordered components of R_+. A test based on R_+^o is called a (one-sample) *signed rank test*.

Similar to the case of permutation tests, there is no UMP test within the class of signed rank tests. A common choice is the signed rank test that rejects H_0 when $W(R_+^o)$ is too large or too small, where

$$W(R_+^o) = J(R_{+1}^o/n) + \cdots + J(R_{+n_*}^o/n), \qquad (6.83)$$

J is a continuous and strictly increasing function on $[0, 1]$, R_{+i}^o is the ith component of R_+^o, and n_* is the number of positive X_i's. This is motivated by the fact that H_0 is unlikely to be true if W in (6.83) is too large or too small. Note that W/n is equal to $T(F_n)$ with T given by (5.53) and $J(t) = t$, and the test based on W in (6.83) is the well-known one-sample Wilcoxon signed rank test.

Under H_0, $P(R_+^o = y) = 2^{-n}$ for each $y \in \mathcal{Y}$ containing 2^n n_*-tuples $y = (y_1, ..., y_{n_*})$ satisfying $1 \le y_1 < \cdots < y_{n_*} \le n$. Then, the following signed rank test is of size α:

$$T(X) = \begin{cases} 1 & W(R_+^o) < c_1 \text{ or } W(R_+^o) > c_2 \\ \gamma & W(R_+^o) = c_i, i = 1, 2 \\ 0 & c_1 < W(R_+^o) < c_2, \end{cases} \qquad (6.84)$$

where c_1 and c_2 are the $(m + 1)$th smallest and largest values of the set $\{W(y) : y \in \mathcal{Y}\}$, m is the integer part of $\alpha 2^n/2$, and $\gamma = \alpha 2^n/2 - m$.

Consider next the two-sample problem of testing (6.80) based on two independent samples, $X_{i1}, ..., X_{in_i}$, $i = 1, 2$, i.i.d. from F_i, $i = 1, 2$, respectively. Let $\mathcal{G}$ be the class of transformations $g(x) = (\psi(x_{ij}), j = 1, ..., n_i, i =$

1, 2), where ψ is continuous and strictly increasing. Let $R(X)$ be the vector of ranks of all X_{ij}'s. In Example 6.14, we showed that R is maximal invariant under $\mathcal{G}$. Again, sufficiency permits a reduction from R to R_1^o, the vector of ordered values of the ranks of $X_{11}, ..., X_{1n_1}$. A test for (6.80) based on R_1^o is called a two-sample *rank test*. Under H_0, $P(R_1^o = y) = \binom{n}{n_1}^{-1}$ for each $y \in \mathcal{Y}$ containing $\binom{n}{n_1}$ n_1-tuples $y = (y_1, ..., y_{n_1})$ satisfying $1 \le y_1 < \cdots < y_{n_1} \le n$. Let $R_1^o = (R_{11}^o, ..., R_{1n_1}^o)$. Then a commonly used two-sample rank test is given by (6.83)-(6.84) with R_{+i}^o, n_*, and 2^n replaced by R_{1i}^o, n_1, and $\binom{n}{n_1}$, respectively. When $n_1 = n_2$, the statistic W/n is equal to $T(F_n)$ with T given by (5.55). When $J(t) = t - \frac{1}{2}$, this reduces to the well-known two-sample Wilcoxon rank test.

A common feature of the permutation and rank tests previously introduced is that tests of size α can be obtained for each fixed sample size n, but the computation involved in determining the rejection regions $\{T(X) = 1\}$ may be cumbersome if n is large. Thus, one may consider approximations to permutation and rank tests when n is large. Permutation tests can often be approximated by the two-sample t-tests derived in §6.2.3 (Lehmann, 1986, §5.13). Using the results in §5.2.2, we now derive one-sample signed rank tests having limiting size α (Definition 2.13(ii)), which can be viewed as signed rank tests of size approximately α when n is large.

From the discussion in §5.2.2, $W/n = T(F_n)$ with a ϱ_∞-Hadamard differentiable functional T given by (5.53) and, by Theorem 5.5,

$$\sqrt{n}[W/n - T(F)] \to_d N(0, \sigma_F^2),$$

where $\sigma_F^2 = E[\phi_F(X_1)]^2$,

$$\phi_F(x) = \int_0^\infty J'(\tilde{F}(y))(\tilde{\delta}_x - \tilde{F})(y)dF(y) + J(\tilde{F}(x)) - T(F)$$

(see (5.54)), and $\tilde{\delta}_x$ denotes the c.d.f. degenerated at x. Since F is continuous, $\tilde{F}(x) = F(x) - F(-x)$. Under H_0, $F(x) = 1 - F(-x)$. Hence, σ_F^2 under H_0 is equal to $v_1 + v_2 + 2v_{12}$, where

$$v_1 = \text{Var}\big(J(\tilde{F}(X_1))\big) = \frac{1}{2}\int_0^\infty [J(\tilde{F}(x))]^2 d\tilde{F}(x),$$

$$v_2 = \text{Var}\left(\int_0^\infty J'(\tilde{F}(y))(\tilde{\delta}_{X_1} - \tilde{F})(y)dF(y)\right)$$

$$= E\int_0^\infty \int_0^\infty J'(\tilde{F}(y))J'(\tilde{F}(z))(\tilde{\delta}_{X_1} - \tilde{F})(y)(\tilde{\delta}_{X_1} - \tilde{F})(z)dF(y)dF(z)$$

$$= \frac{1}{4}\int_0^\infty \int_0^\infty J'(\tilde{F}(y))J'(\tilde{F}(z))[\tilde{F}(\min\{y, z\}) - \tilde{F}(y)\tilde{F}(z)]d\tilde{F}(y)d\tilde{F}(z)$$

$$= \frac{1}{2}\int_{0<z<y<\infty} J'(\tilde{F}(y))J'(\tilde{F}(z))\tilde{F}(z)[1 - \tilde{F}(y)]d\tilde{F}(y)d\tilde{F}(z),$$

and

$$v_{12} = \text{Cov}\left(J(\tilde{F}(X_1)), \int_0^\infty J'(\tilde{F}(y))(\tilde{\delta}_{X_1} - \tilde{F})(y)dF(y)\right)$$

$$= E\int_0^\infty J(\tilde{F}(X_1))J'(\tilde{F}(y))(\tilde{\delta}_{X_1} - \tilde{F})(y)dF(y)$$

$$= \int_{-\infty}^\infty \int_0^\infty J(\tilde{F}(x))J'(\tilde{F}(y))(\delta_{|x|} - \tilde{F})(y)dF(y)dF(x)$$

$$= \frac{1}{2}\int_0^\infty \int_0^\infty J(\tilde{F}(x))J'(\tilde{F}(y))(\delta_x - \tilde{F})(y)d\tilde{F}(y)d\tilde{F}(x).$$

Note that under H_0, the distribution of W is completely known. Indeed, letting $s = \tilde{F}(y)$ and $t = \tilde{F}(z)$, we conclude that $\sigma_F^2 = v_1 + v_2 + 2v_{12}$ and

$$\mathrm{T}(F) = \int_0^\infty J(\tilde{F}(x))dF(x) = \frac{1}{2}\int_0^1 J(s)ds$$

do not depend on F. Hence, a signed rank test T that rejects H_0 when

$$\sqrt{n}|W/n - t_0| > \sigma_0 z_{1-\alpha/2}, \tag{6.85}$$

where $z_a = \Phi^{-1}(a)$ and $t_0 = \mathrm{T}(F)$ and $\sigma_0^2 = \sigma_F^2$ under H_0 are known constants, has the property that

$$\sup_{P\in\mathcal{P}_0} \beta_T(P) = \sup_{P\in\mathcal{P}_0} P\left(\sqrt{n}|W/n - t_0| > \sigma_0 z_{1-\alpha/2}\right)$$

$$= P_W\left(\sqrt{n}|W/n - t_0| > \sigma_0 z_{1-\alpha/2}\right)$$

$$\to \alpha,$$

i.e., T has limiting size α.

Two-sample rank tests having limiting size α can be similarly derived (exercise).

6.5.2 Kolmogorov-Smirnov and Cramér-von Mises tests

In this section we introduce two types of tests for hypotheses concerning continuous c.d.f.'s on $\mathcal{R}$. Let $X_1, ..., X_n$ be i.i.d. random variables from a continuous c.d.f. F. Suppose that we would like to test hypotheses (6.73), i.e., $H_0 : F = F_0$ versus $H_1 : F \neq F_0$ with a fixed F_0. Let F_n be the empirical c.d.f. and

$$D_n(F) = \sup_{x\in\mathcal{R}} |F_n(x) - F(x)|, \tag{6.86}$$

which is in fact the distance $\varrho_\infty(F_n, F)$. Intuitively, $D_n(F_0)$ should be small if H_0 is true. From the results in §5.1.1, we know that $D_n(F_0) \to_{a.s.} 0$ if and only if H_0 is true. The statistic $D_n(F_0)$ is called the Kolmogorov-Smirnov statistic. Tests with rejection region $D_n(F_0) > c$ are called Kolmogorov-Smirnov tests.

In some cases we would like to test "one-sided" hypotheses $H_0 : F = F_0$ versus $H_1 : F \geq F_0$, $F \neq F_0$, or $H_0 : F = F_0$ versus $H_1 : F \leq F_0$, $F \neq F_0$. The corresponding Kolmogorov-Smirnov statistic is $D_n^+(F_0)$ or $D_n^-(F_0)$, where

$$D_n^+(F) = \sup_{x \in \mathcal{R}}[F_n(x) - F(x)] \qquad (6.87)$$

and

$$D_n^-(F) = \sup_{x \in \mathcal{R}}[F(x) - F_n(x)].$$

The rejection regions of one-sided Kolmogorov-Smirnov tests are, respectively, $D_n^+(F_0) > c$ and $D_n^-(F_0) > c$.

Let $X_{(1)} < \cdots < X_{(n)}$ be the order statistics and define $X_{(0)} = -\infty$ and $X_{(n+1)} = \infty$. Since $F_n(x) = i/n$ when $X_{(i)} \leq x < X_{(i+1)}$, $i = 0, 1, ..., n$,

$$\begin{aligned}
D_n^+(F) &= \max_{0 \leq i \leq n} \sup_{X_{(i)} \leq x < X_{(i+1)}} \left[\frac{i}{n} - F(x) \right] \\
&= \max_{0 \leq i \leq n} \left[\frac{i}{n} - \inf_{X_{(i)} \leq x < X_{(i+1)}} F(x) \right] \\
&= \max_{0 \leq i \leq n} \left[\frac{i}{n} - F(X_{(i)}) \right].
\end{aligned}$$

When F is continuous, $F(X_{(i)})$ is the ith order statistic of a sample of size n from the uniform distribution $U(0, 1)$ irrespective of what F is. Therefore, the distribution of $D_n^+(F)$ does not depend on F, if we restrict our attention to continuous c.d.f.'s on $\mathcal{R}$. The distribution of $D_n^-(F)$ is the same as that of $D_n^+(F)$ because of symmetry (exercise). Since

$$D_n(F) = \max\{D_n^+(F), D_n^-(F)\},$$

the distribution of $D_n(F)$ does not depend on F. This means that the distributions of Kolmogorov-Smirnov statistics are known under H_0.

Theorem 6.10. Let $D_n(F)$ and $D_n^+(F)$ be defined by (6.86) and (6.87), respectively, for a continuous c.d.f. F on $\mathcal{R}$.
(i) For any fixed n,

$$P\big(D_n^+(F) \leq t\big) = \begin{cases} 0 & t \leq 0 \\ n! \displaystyle\prod_{i=1}^{n} \int_{\max\{0, \frac{n-i+1}{n} - t\}}^{u_{n-i+2}} du_1 \cdots du_n & 0 < t < 1 \\ 1 & t \geq 1 \end{cases}$$

and

$$P\big(D_n(F) \le t\big) = \begin{cases} 0 & t \le \frac{1}{2n} \\ n! \prod_{i=1}^{n} \int_{\max\{0, \frac{n-i+1}{n} - t\}}^{\min\{u_{n-i+2}, \frac{n-i}{n} + t\}} du_1 \cdots du_n & \frac{1}{2n} < t < 1 \\ 1 & t \ge 1, \end{cases}$$

where $u_{n+1} = 1$.

(ii) For $t > 0$,

$$\lim_{n \to \infty} P\big(\sqrt{n} D_n^+(F) \le t\big) = 1 - e^{-2t^2}$$

and

$$\lim_{n \to \infty} P\big(\sqrt{n} D_n(F) \le t\big) = 1 - 2 \sum_{j=1}^{\infty} (-1)^{j-1} e^{-2j^2 t^2}. \quad \blacksquare$$

The proof of Theorem 6.10(i) is left as an exercise. The proof of Theorem 6.10(ii) can be found in Kolmogorov (1933) and Smirnov (1944).

When n is not large, Kolmogorov-Smirnov tests of size α can be obtained using the results in Theorem 6.10(i). When n is large, using the results in Theorem 6.10(i) is not convenient. We can obtain Kolmogorov-Smirnov tests of limiting size α using the results in Theorem 6.10(ii).

Another test for $H_0 : F = F_0$ versus $H_1 : F \ne F_0$ is the Cramér-von Mises test, which rejects H_0 when $C_n(F_0) > c$, where

$$C_n(F) = \int [F_n(x) - F(x)]^2 dF(x) \tag{6.88}$$

is another measure of disparity between F_n and F. Similar to $D_n(F)$, the distribution of $C_n(F)$ does not depend on F (exercise). Hence, a Cramér-von Mises test of size α can be obtained. When n is large, it is more convenient to use a Cramér-von Mises test of limiting size α. Note that $C_n(F_0)$ is actually a V-statistic (§3.5.3) with kernel

$$h(x_1, x_2) = \int [\delta_{x_1}(y) - F_0(y)][\delta_{x_2}(y) - F_0(y)] dF_0(y)$$

and

$$h_1(x_1) = E[h(x_1, X_2)] = \int [\delta_{x_1}(y) - F_0(y)][F(y) - F_0(y)] dF_0(y),$$

where δ_x denotes the c.d.f. degenerated at x. It follows from Theorem 3.16 that if H_1 is true, $C_n(F_0)$ is asymptotically normal, whereas if H_0 is true, $h_1(x_1) \equiv 0$ and

$$nC_n(F_0) \to_d \sum_{j=1}^{\infty} \lambda_j \chi_{1j}^2,$$

where χ^2_{1j}'s are i.i.d. from the chi-square distribution χ^2_1 and λ_j's are constants. In this case, Durbin (1973) showed that $\lambda_j = j^{-2}\pi^{-2}$.

For testing (6.73), it is worthwhile to compare the goodness of fit test introduced in Example 6.23 with the Kolmogorov-Smirnov test (or Cramér-von Mises test). The former requires a partition of the range of observations and may lose information through partitioning, whereas the latter requires that F be continuous and univariate; the latter is of size α (or limiting size α), whereas the former is only of asymptotic significance level α; and the former can be modified to allow estimation of unknown parameters under H_0 (i.e., hypotheses (6.74)), whereas the latter does not have this flexibility. Note that goodness of fit tests are nonparametric in nature, although χ^2-tests are derived from a parametric model.

Kolmogorov-Smirnov tests can be extended to two-sample problems to test hypotheses in (6.80). Let $X_{i1}, ..., X_{in_i}$, $i = 1, 2$, be two independent samples i.i.d. from F_i on $\mathcal{R}$, $i = 1, 2$, and let F_{in_i} be the empirical c.d.f. based on $X_{i1}, ..., X_{in_i}$. A Kolmogorov-Smirnov test rejects H_0 when $D_{n_1,n_2} > c$, where

$$D_{n_1,n_2} = \sup_{x \in \mathcal{R}} |F_{1n_1}(x) - F_{2n_2}(x)|.$$

A Kolmogorov-Smirnov test of limiting size α can be obtained using

$$\lim_{n_1,n_2 \to \infty} P\big(\sqrt{n_1 n_2/(n_1 + n_2)} D_{n_1,n_2} \le t\big) = \sum_{j=-\infty}^{\infty} (-1)^{j-1} e^{-2j^2 t^2}, \quad t > 0.$$

6.5.3 Empirical likelihood ratio tests

The method of likelihood ratio is useful in deriving tests under parametric models. In nonparametric problems, we now introduce a similar method based on the empirical likelihoods introduced in §5.1.2 and §5.1.4.

Suppose that a sample X is from a population determined by a c.d.f. $F \in \mathcal{F}$, where $\mathcal{F}$ is a class of c.d.f.'s on $\mathcal{R}^d$. Consider the problem of testing

$$H_0 : \mathrm{T}(F) = t_0 \qquad \text{versus} \qquad H_1 : \mathrm{T}(F) \neq t_0, \qquad (6.89)$$

where T is a functional from $\mathcal{F}$ to $\mathcal{R}^k$ and t_0 is a fixed vector in $\mathcal{R}^k$. Let $\ell(G)$, $G \in \mathcal{F}$, be a given empirical likelihood, $\hat{F}$ be an MELE of F, and $\hat{F}_{H_0}$ be an MELE of F under H_0, i.e., $\hat{F}_{H_0}$ is an MELE of F subject to $\mathrm{T}(F) = t_0$. Then the empirical likelihood ratio is defined as

$$\lambda_n(X) = \ell(\hat{F}_{H_0})/\ell(\hat{F}).$$

A test with rejection region $\lambda_n(X) < c$ is called an *empirical likelihood ratio test*.

As a specific example, consider the following empirical likelihood (or nonparametric likelihood) when $X = (X_1, ..., X_n)$ with i.i.d. X_i's:

$$\ell(G) = \prod_{i=1}^{n} p_i \quad \text{subject to} \quad p_i \geq 0, \ \sum_{i=1}^{n} p_i = 1,$$

where $p_i = P_G(\{x_i\})$, $i = 1, ..., n$. Suppose that $T(G) = \int u(x) dG(x)$ with a known function $u(x)$ from $\mathcal{R}^d$ to $\mathcal{R}^r$. Then $\hat{F} = F_n$; H_0 in (6.89) with $t_0 = 0$ is the same as the case where assumption (5.9) holds; $\hat{F}_{H_0}$ is the MELE given by (5.11); and the empirical likelihood ratio is

$$\lambda_n(X) = n^n \prod_{i=1}^{n} \hat{p}_i, \tag{6.90}$$

where $\hat{p}_i$ is given by (5.12). An empirical likelihood ratio test with asymptotic significance level α can be obtained using the following result.

Theorem 6.11. Assume the conditions in Theorem 5.4. Under the hypothesis H_0 in (6.89) with $t_0 = 0$ (i.e., (5.9) holds),

$$-2 \log \lambda_n \to_d \chi_r^2,$$

where $\lambda_n = \lambda_n(X)$ is given by (6.90) and χ_r^2 has the chi-square distribution χ_r^2. ∎

The proof of this result can be found in Owen (1988, 1990). In fact, the result in Theorem 6.11 holds for some other functionals T such as the median functional.

We can also derive tests based on the profile empirical likelihoods discussed in §5.4.1. Consider an empirical likelihood

$$\ell(G) = \prod_{i=1}^{n} p_i \quad \text{subject to} \quad p_i \geq 0, \ \sum_{i=1}^{n} p_i = 1, \ \sum_{i=1}^{n} p_i \psi(x_i, \theta) = 0,$$

where θ is a k-vector of unknown parameters and ψ is a known function. Let $\theta = (\vartheta, \varphi)$, where ϑ is an r-vector and φ is a $(k - r)$-vector. Suppose that we would like to test

$$H_0 : \vartheta = \vartheta_0 \quad \text{versus} \quad H_1 : \vartheta \neq \vartheta_0,$$

where ϑ_0 is a fixed r-vector. Let $\hat{\theta}$ be a maximum of the profile empirical likelihood $\ell_P(\theta)$ given by (5.36) and let $\hat{\varphi}$ be a maximum of $\ell_P(\varphi) = \ell_P(\vartheta_0, \varphi)$. Then a profile empirical likelihood ratio test rejects H_0 when $\lambda_n(X) < c$, where

$$\lambda_n(X) = \prod_{i=1}^{n} \frac{1 + [\xi_n(\hat{\theta})]^\tau \psi(x_i, \hat{\theta})}{1 + [\zeta_n(\vartheta_0, \hat{\varphi})]^\tau \psi(x_i, \vartheta_0, \hat{\varphi})}, \tag{6.91}$$

$\hat{\theta}$ and $\hat{\varphi}$ are maximum profile empirical likelihood estimators, $\xi_n(\hat{\theta})$ satisfies

$$\sum_{i=1}^{n} \frac{\psi(x_i, \hat{\theta})}{1 + [\xi_n(\hat{\theta})]^\tau \psi(x_i, \hat{\theta})} = 0,$$

and $\zeta_n(\vartheta_0, \hat{\varphi})$ satisfies

$$\sum_{i=1}^{n} \frac{\psi(x_i, \vartheta_0, \hat{\varphi})}{1 + [\zeta_n(\vartheta_0, \hat{\varphi})]^\tau \psi(x_i, \vartheta_0, \hat{\varphi})} = 0.$$

From the discussion in §5.4.1, $\hat{\theta}$ is a solution of the GEE $\sum_{i=1}^{n} \psi(X_i, \theta) = 0$ when the dimension of ψ is k. Under some regularity conditions (e.g., the conditions in Proposition 5.3), Qin and Lawless (1994) showed that the result in Theorem 6.11 holds with $\lambda_n(X)$ given by (6.91). Thus, a profile empirical likelihood ratio test with asymptotic significance level α can be obtained.

Example 6.26. Let $Y_1, ..., Y_n$ be i.i.d. random 2-vectors from F. Consider the problem of testing $H_0 : \mu_1 = \mu_2$ versus $H_1 : \mu_1 \neq \mu_2$, where $(\mu_1, \mu_2) = E(Y_1)$. Let $Y_i = (Y_{i1}, Y_{i2})$, $X_{i1} = Y_{i1} - Y_{i2}$, $X_{i2} = Y_{i1} + Y_{i2}$, and $X_i = (X_{i1}, X_{i2})$, $i = 1, ..., n$. Then $X_1, ..., X_n$ are i.i.d. with $E(X_1) = \theta = (\vartheta, \varphi)$, where $\vartheta = \mu_1 - \mu_2$ and $\varphi = \mu_1 + \mu_2$. The hypotheses of interest becomes $H_0 : \vartheta = 0$ versus $H_1 : \vartheta \neq 0$.

To apply the profile empirical likelihood method, we define $\psi(x, \theta) = x - \theta$, $x \in \mathcal{R}^2$. Note that a solution of the GEE $\sum_{i=1}^{n}(X_i - \theta) = 0$ is the sample mean $\hat{\theta} = \bar{X}$. The profile empirical likelihood ratio is then given by

$$\lambda_n(X) = \prod_{i=1}^{n} \frac{1 + [\xi_n(\bar{X})]^\tau (X_i - \bar{X})}{1 + [\zeta_n(0, \hat{\varphi})]^\tau [X_i - (0, \hat{\varphi})]},$$

where $\xi_n(\bar{X})$, $\zeta_n(0, \hat{\varphi})$, and $\hat{\varphi}$ satisfy

$$\sum_{i=1}^{n} \frac{X_i - \bar{X}}{1 + [\xi_n(\bar{X})]^\tau (X_i - \bar{X})} = 0,$$

$$\sum_{i=1}^{n} \frac{X_i - (0, \hat{\varphi})}{1 + [\zeta_n(0, \hat{\varphi})]^\tau [X_i - (0, \hat{\varphi})]} = 0,$$

and $\ell_P(0, \hat{\varphi}) = \max_\varphi \ell_P(0, \varphi)$ with

$$\ell_P(0, \varphi) = \prod_{i=1}^{n} \frac{1}{n\{1 + [\xi_n(0, \varphi)]^\tau [X_i - (0, \varphi)]\}}. \quad \blacksquare$$

Empirical likelihood ratio tests or profile empirical likelihood ratio tests in various other problems can be found, for example, in Owen (1988, 1990, 2001), Chen and Qin (1993), Qin (1993), and Qin and Lawless (1994).

6.5.4 Asymptotic tests

We now introduce a simple method of constructing asymptotic tests (i.e., tests with asymptotic significance level α). This method works for almost all problems (parametric or nonparametric) in which the hypotheses being tested are $H_0 : \theta = \theta_0$ versus $H_1 : \theta \neq \theta_0$, where θ is a vector of parameters, and an asymptotically normally distributed estimator of θ can be found. However, this simple method may not provide the best or even nearly best solution to the problem, especially when there are different asymptotically normally distributed estimators of θ.

Let X be a sample of size n from a population P and $\hat{\theta}_n$ be an estimator of θ, a k-vector of parameters related to P. Suppose that under H_0,

$$V_n^{-1/2}(\hat{\theta}_n - \theta) \to_d N_k(0, I_k), \tag{6.92}$$

where V_n is the asymptotic covariance matrix of $\hat{\theta}_n$. If V_n is known when $\theta = \theta_0$, then a test with rejection region

$$(\hat{\theta}_n - \theta_0)^\tau V_n^{-1}(\hat{\theta}_n - \theta_0) > \chi^2_{k,\alpha} \tag{6.93}$$

has asymptotic significance level α, where $\chi^2_{k,\alpha}$ is the $(1-\alpha)$th quantile of the chi-squared distribution χ^2_k. If the distribution of $\hat{\theta}_n$ does not depend on the unknown population P under H_0 and (6.92) holds, then a test with rejection region (6.93) has limiting size α.

If V_n in (6.93) depends on the unknown population P even if H_0 is true ($\theta = \theta_0$), then we have to replace V_n in (6.93) by an estimator $\hat{V}_n$. If, under H_0, $\hat{V}_n$ is consistent according to Definition 5.4, then the test having rejection region (6.93) with V_n replaced by $\hat{V}_n$ has asymptotic significance level α. Variance estimation methods introduced in §5.5 can be used to construct a consistent estimator $\hat{V}_n$.

In some cases result (6.92) holds for any P. Then, the following result shows that the test having rejection region (6.93) is asymptotically correct (§2.5.3), i.e., it is a consistent asymptotic test (Definition 2.13).

Theorem 6.12. Assume that (6.92) holds for any P and that $\lambda_+[V_n] \to 0$, where $\lambda_+[V_n]$ is the largest eigenvalue of V_n.
(i) The test having rejection region (6.93) (with a known V_n or V_n replaced by an estimator $\hat{V}_n$ that is consistent for any P) is consistent.
(ii) If we choose $\alpha = \alpha_n \to 0$ as $n \to \infty$ and $\chi^2_{k,1-\alpha_n}\lambda_+[V_n] = o(1)$, then the test in (i) is Chernoff-consistent.
Proof. The proof of (ii) is left as an exercise. We only prove (i) for the case where V_n is known. Let $Z_n = V_n^{-1/2}(\hat{\theta}_n - \theta)$ and $l_n = V_n^{-1/2}(\theta - \theta_0)$. Then $\|Z_n\| = O_p(1)$ and $\|l_n\| = \|V_n^{-1/2}(\theta - \theta_0)\| \to \infty$ when $\theta \neq \theta_0$. The

result follows from the fact that when $\theta \neq \theta_0$,

$$
\begin{aligned}
(\hat{\theta}_n - \theta_0)^\tau V_n^{-1}(\hat{\theta}_n - \theta_0) &= \|Z_n\|^2 + \|l_n\|^2 + 2l_n^\tau Z_n \\
&\geq \|Z_n\|^2 + \|l_n\|^2 - 2\|l_n\|\|Z_n\| \\
&= O_p(1) + \|l_n\|^2[1 - o_p(1)]
\end{aligned}
$$

and, therefore,

$$
P\left((\hat{\theta}_n - \theta_0)^\tau V_n^{-1}(\hat{\theta}_n - \theta_0) > \chi_{k,\alpha}^2\right) \to 1. \quad \blacksquare
$$

Example 6.27. Let $X_1, ..., X_n$ be i.i.d. random variables from a symmetric c.d.f. F having finite variance and positive F'. Consider the problem of testing $H_0 : F$ is symmetric about 0 versus $H_1 : F$ is not symmetric about 0. Under H_0, there are many estimators satisfying (6.92). We consider the following five estimators:

(1) $\hat{\theta}_n = \bar{X}$ and $\theta = E(X_1)$;

(2) $\hat{\theta}_n = \hat{\theta}_{0.5}$ (the sample median) and $\theta = F^{-1}(\frac{1}{2})$ (the median of F);

(3) $\hat{\theta}_n = \bar{X}_a$ (the a-trimmed sample mean defined by (5.77)) and $\theta = T(F)$, where T is given by (5.46) with $J(t) = (1 - 2a)^{-1}I_{(a,1-a)}(t)$, $a \in (0, \frac{1}{2})$;

(4) $\hat{\theta}_n =$ the Hodges-Lehmann estimator (Example 5.8) and $\theta = F^{-1}(\frac{1}{2})$;

(5) $\hat{\theta}_n = W/n - \frac{1}{2}$, where W is given by (6.83) with $J(t) = t$, and $\theta = T(F) - \frac{1}{2}$ with T given by (5.53).

Although the θ's in (1)-(5) are different in general, in all cases $\theta = 0$ is equivalent to that H_0 holds.

For $\bar{X}$, it follows from the CLT that (6.92) holds with $V_n = \sigma^2/n$ for any F, where $\sigma^2 = \text{Var}(X_1)$. From the SLLN, S^2/n is a consistent estimator of V_n for any F. Thus, the test having rejection region (6.93) with $\hat{\theta}_n = \bar{X}$ and V_n replaced by S^2/n is asymptotically correct. This test is asymptotically equivalent to the one-sample t-test derived in §6.2.3.

From Theorem 5.10, $\hat{\theta}_{0.5}$ satisfies (6.92) with $V_n = 4^{-1}[F'(\theta)]^{-2}n^{-1}$ for any F. A consistent estimator of V_n can be obtained using the bootstrap method considered in §5.5.3. Another consistent estimator of V_n can be obtained using Woodruff's interval introduced in §7.4 (see Exercise 86 in §7.6). The test having rejection region (6.93) with $\hat{\theta}_n = \hat{\theta}_{0.5}$ and V_n replaced by a consistent estimator is asymptotically correct.

It follows from the discussion in §5.3.2 that $\bar{X}_a$ satisfies (6.92) for any F. A consistent estimator of V_n can be obtained using formula (5.110) or the jackknife method in §5.5.2. The test having rejection region (6.93) with $\hat{\theta}_n = \bar{X}_a$ and V_n replaced by a consistent estimator is asymptotically correct.

From Example 5.8, the Hodges-Lehmann estimator satisfies (6.92) for any F and $V_n = 12^{-1}\gamma^{-2}n^{-1}$ under H_0, where $\gamma = \int F'(x)dF(x)$. A consistent estimator of V_n under H_0 can be obtained using the result in Exercise 102 in §5.6. The test having rejection region (6.93) with $\hat{\theta}_n = $ the Hodges-Lehmann estimator and V_n replaced by a consistent estimator is asymptotically correct.

Note that all tests discussed so far are not of limiting size α, since the distributions of $\hat{\theta}_n$ are still unknown under H_0.

The test having rejection region (6.93) with $\hat{\theta}_n = W/n - \frac{1}{2}$ and $V_n = (12n)^{-1}$ is equivalent to the one-sample Wilcoxon signed rank test and is shown to have limiting size α (§6.5.1). Also, (6.92) is satisfied for any F (§5.2.2). Although Theorem 6.12 is not applicable, a modified proof of Theorem 6.12 can be used to show the consistency of this test (exercise).

It is not clear which one of the five tests discussed here is to be preferred in general.

The results for $\hat{\theta}_n$ in (1)-(3) and (5) still hold for testing $H_0 : \theta = 0$ versus $H_1 : \theta \neq 0$ without the assumption that F is symmetric. ∎

An example of asymptotic tests for one-sided hypotheses is given in Exercise 123. Most tests in §6.1-§6.4 derived under parametric models are asymptotically correct even when the parametric model assumptions are removed. Some examples are given in Exercises 121-123.

Finally, a study of asymptotic efficiencies of various tests can be found, for example, in Serfling (1980, Chapter 10).

6.6 Exercises

1. Prove Theorem 6.1 for the case of $\alpha = 0$ or 1.

2. Assume the conditions in Theorem 6.1. Let $\beta(P)$ be the power function of a UMP test of size $\alpha \in (0, 1)$. Show that $\alpha < \beta(P_1)$ unless $P_0 = P_1$.

3. Let T_* be given by (6.3) with $c = c(\alpha)$ for an $\alpha > 0$.
   (a) Show that if $\alpha_1 < \alpha_2$, then $c(\alpha_1) \geq c(\alpha_2)$.
   (b) Show that if $\alpha_1 < \alpha_2$, then the type II error probability of T_* of size α_1 is larger than that of T_* of size α_2.

4. Let H_0 and H_1 be simple and let $\alpha \in (0, 1)$. Suppose that T_* is a UMP test of size α for testing H_0 versus H_1 and that $\beta < 1$, where β is the power of T_* when H_1 is true. Show that $1 - T_*$ is a UMP test of size $1 - \beta$ for testing H_1 versus H_0.

5. Let X be a sample of size 1 from a Lebesgue p.d.f. f_θ. Find a UMP test of size $\alpha \in (0, \frac{1}{2})$ for $H_0 : \theta = \theta_0$ versus $H_1 : \theta = \theta_1$ when

(a) $f_\theta(x) = 2\theta^{-2}(\theta - x)I_{(0,\theta)}(x)$, $\theta_0 < \theta_1$;

(b) $f_\theta(x) = 2[\theta x + (1 - \theta)(1 - x)]I_{(0,1)}(x)$, $0 \leq \theta_1 < \theta_0 \leq 1$;

(c) f_{θ_0} is the p.d.f. of $N(0,1)$ and f_{θ_1} is the p.d.f. of the Cauchy distribution $C(0,1)$;

(d) $f_{\theta_0}(x) = 4xI_{(0,\frac{1}{2})}(x) + 4(1 - x)I_{(\frac{1}{2},1)}(x)$ and $f_{\theta_1}(x) = I_{(0,1)}(x)$;

(e) f_θ is the p.d.f. of the Cauchy distribution $C(\theta,1)$ and $\theta_i = i$;

(f) $f_{\theta_0}(x) = e^{-x}I_{(0,\infty)}(x)$ and $f_{\theta_1}(x) = 2^{-1}x^2 e^{-x}I_{(0,\infty)}(x)$.

6. Let $X_1, ..., X_n$ be i.i.d. from a Lebesgue p.d.f. f_θ. Find a UMP test of size α for $H_0 : \theta = \theta_0$ versus $H_1 : \theta = \theta_1$ in the following cases:

(a) $f_\theta(x) = e^{-(x-\theta)}I_{(\theta,\infty)}(x)$, $\theta_0 < \theta_1$;

(b) $f_\theta(x) = \theta x^{-2}I_{(\theta,\infty)}(x)$, $\theta_0 \neq \theta_1$.

7. Prove Proposition 6.1.

8. Let $X \in \mathcal{R}^n$ be a sample with a p.d.f. f w.r.t. a σ-finite measure ν. Consider the problem of testing $H_0 : f = f_\theta$ versus $H_1 : f = g$, where $\theta \in \Theta$, $f_\theta(x)$ is Borel on $(\mathcal{R}^n \times \Theta, \sigma(\mathcal{B}^n \times \mathcal{F}))$, and $(\Theta, \mathcal{F}, \Lambda)$ is a probability space. Let $c > 0$ be a constant and

$$\phi_*(x) = \begin{cases} 1 & g(x) \geq c \int_\Theta f_\theta(x)d\Lambda \\ 0 & g(x) < c \int_\Theta f_\theta(x)d\Lambda. \end{cases}$$

Suppose that $\int \phi_*(x)f_\theta(x)d\nu = \sup_{\theta \in \Theta} \int \phi_*(x)f_\theta(x)d\nu = \alpha$ for any $\theta \in \Theta'$ with $\Lambda(\Theta') = 1$. Show that ϕ_* is a UMP test of size α.

9. Let f_0 and f_1 be Lebesgue integrable functions on $\mathcal{R}$ and ϕ_* be the indicator function of the set $\{x: f_0(x) < 0\} \cup \{x: f_0(x) = 0, f_1(x) \geq 0\}$. Show that ϕ_* maximizes $\int \phi(x)f_1(x)dx$ overall Borel functions ϕ on $\mathcal{R}$ satisfying $0 \leq \phi(x) \leq 1$ and $\int \phi(x)f_0(x)dx = \int \phi_*(x)f_0(x)dx$.

10. Let F_1 and F_2 be two c.d.f.'s on $\mathcal{R}$. Show that $F_1(x) \leq F_2(x)$ for all x if and only if $\int g(x)dF_2(x) \leq \int g(x)dF_1(x)$ for any nondecreasing function g.

11. Prove the claims in Example 6.5.

12. Show that the family $\{f_\theta : \theta \in \mathcal{R}\}$ has monotone likelihood ratio, where $f_\theta(x) = c(\theta)h(x)I_{(a(\theta),b(\theta))}(x)$, h is a positive Lebesgue integrable function, and a and b are nondecreasing functions of θ.

13. Prove part (iv) and part (v) of Theorem 6.2.

14. Let $X_1, ..., X_n$ be i.i.d. from a Lebesgue p.d.f. f_θ, $\theta \in \Theta \subset \mathcal{R}$. Find a UMP test of size α for testing $H_0 : \theta \leq \theta_0$ versus $H_1 : \theta > \theta_0$ when

(a) $f_\theta(x) = \theta^{-1}e^{-x/\theta}I_{(0,\infty)}(x)$, $\theta > 0$;

(b) $f_\theta(x) = \theta^{-1}x^{\theta-1}I_{(0,1)}(x)$, $\theta > 0$;

(c) $f_\theta(x)$ is the p.d.f. of $N(1,\theta)$;

(d) $f_\theta(x) = \theta^{-c}cx^{c-1}e^{-(x/\theta)^c}I_{(0,\infty)}(x)$, $\theta > 0$, where $c > 0$ is known.

15. Suppose that the distribution of X is in a family with monotone likelihood ratio in $Y(X)$, where $Y(X)$ has a continuous distribution. Consider the hypotheses $H_0 : \theta \leq \theta_0$ versus $H_1 : \theta > \theta_0$. Show that the p-value (§2.4.2) of the UMP test is given by $P_{\theta_0}(Y \geq y)$, where y is the observed value of Y.

16. Let $X_1, ..., X_m$ be i.i.d. from $N(\mu_x, \sigma_x^2)$ and $Y_1, ..., Y_n$ be i.i.d. from $N(\mu_y, \sigma_y^2)$. Suppose that X_i's and Y_j's are independent.
    (a) When $\sigma_x = \sigma_y = 1$, find a UMP test of size α for testing $H_0 : \mu_x \leq \mu_y$ versus $H_1 : \mu_x > \mu_y$. (Hint: see Lehmann (1986, §3.9).)
    (b) When μ_x and μ_y are known, find a UMP test of size α for testing $H_0 : \sigma_x \leq \sigma_y$ versus $H_1 : \sigma_x > \sigma_y$. (Hint: see Lehmann (1986, §3.9).)

17. Let F and G be two known c.d.f.'s on $\mathcal{R}$ and X be a single observation from the c.d.f. $\theta F(x) + (1 - \theta)G(x)$, where $\theta \in [0, 1]$ is unknown.
    (a) Find a UMP test of size α for testing $H_0: \theta \leq \theta_0$ versus $H_1: \theta > \theta_0$.
    (b) Show that the test $T_*(X) \equiv \alpha$ is a UMP test of size α for testing $H_0 : \theta \leq \theta_1$ or $\theta \geq \theta_2$ versus $H_1 : \theta_1 < \theta < \theta_2$.

18. Let $X_1, ..., X_n$ be i.i.d. from the uniform distribution $U(\theta, \theta + 1)$, $\theta \in \mathcal{R}$. Suppose that $n \geq 2$.
    (a) Find the joint distribution of $X_{(1)}$ and $X_{(n)}$.
    (b) Show that a UMP test of size α for testing $H_0 : \theta \leq 0$ versus $H_1 : \theta > 0$ is of the form

$$T_*(X_{(1)}, X_{(n)}) = \begin{cases} 0 & X_{(1)} < 1 - \alpha^{1/n}, \ X_{(n)} < 1 \\ 1 & \text{otherwise.} \end{cases}$$

(c) Does the family of all possible distributions of $(X_{(1)}, X_{(n)})$ have monotone likelihood ratio? (Hint: see Lehmann (1986, p. 115).)

19. Suppose that $X_1, ..., X_n$ are i.i.d. from the discrete uniform distribution $DU(1, ..., \theta)$ (Table 1.1, page 18) with an unknown $\theta = 1, 2, ...$
    (a) Consider $H_0 : \theta \leq \theta_0$ versus $H_1 : \theta > \theta_0$. Show that

$$T_*(X) = \begin{cases} 1 & X_{(n)} > \theta_0 \\ \alpha & X_{(n)} \leq \theta_0 \end{cases}$$

is a UMP test of size α.
    (b) Consider $H_0 : \theta = \theta_0$ versus $H_1 : \theta \neq \theta_0$. Show that

$$T_*(X) = \begin{cases} 1 & X_{(n)} > \theta_0 \text{ or } X_{(n)} \leq \theta_0 \alpha^{1/n} \\ 0 & \text{otherwise} \end{cases}$$

is a UMP test of size α.
    (c) Show that the results in (a) and (b) still hold if the discrete uniform distribution is replaced by the uniform distribution $U(0, \theta)$, $\theta > 0$.

20. Let $X_1, ..., X_n$ be i.i.d. from the exponential distribution $E(a, \theta)$, $a \in \mathcal{R}$, $\theta > 0$.

    (a) Derive a UMP test of size α for testing $H_0 : a = a_0$ versus $H_1 : a \neq a_0$, when θ is known.

    (b) For testing $H_0 : a = a_0$ versus $H_1 : a = a_1 < \theta_0$, show that any UMP test T_* of size α satisfies $\beta_{T_*}(a_1) = 1 - (1 - \alpha)e^{-n(a_0 - a_1)/\theta}$.

    (c) For testing $H_0 : a = a_0$ versus $H_1 : a = a_1 < \theta_0$, show that the power of any size α test that rejects H_0 when $Y \leq c_1$ or $Y \geq c_2$ is the same as that in part (b), where $Y = (X_{(1)} - a_0)/\sum_{i=1}^{n}(X_i - X_{(1)})$.

    (d) Show that the test in part (c) is a UMP test of size α for testing $H_0 : a = a_0$ versus $H_1 : a \neq a_0$.

    (e) Derive a UMP test of size α for testing $H_0 : \theta = \theta_0, a = a_0$ versus $H_1 : \theta < \theta_0, a < a_0$.

21. Let $X_1, ..., X_n$ be i.i.d. from the Pareto distribution $Pa(a, \theta)$, $\theta > 0$, $a > 0$.

    (a) Derive a UMP test of size α for testing $H_0 : a = a_0$ versus $H_1 : a \neq a_0$ when θ is known.

    (b) Derive a UMP test of size α for testing $H_0 : a = a_0, \theta = \theta_0$ versus $H_1 : \theta > \theta_0, a < a_0$.

22. In Exercise 19(a) of §3.6, derive a UMP test of size $\alpha \in (0, 1)$ for testing $H_0 : \theta \leq \theta_0$ versus $H_1 : \theta > \theta_0$, where θ_0 is known and $\theta_0 > (1 - \alpha)^{-1/n}$.

23. In Exercise 55 of §2.6, derive a UMP test of size α for testing $H_0 : \theta \geq \theta_0$ versus $H_1 : \theta < \theta_0$ based on data $X_1, ..., X_n$, where $\theta_0 > 0$ is a fixed value.

24. Prove part (ii) of Theorem 6.3.

25. Consider Example 6.10. Suppose that $\theta_2 = -\theta_1$. Show that $c_2 = -c_1$ and discuss how to find the value of c_2.

26. Suppose that the distribution of X is in a family of p.d.f.'s indexed by a real-valued parameter θ; there is a real-valued sufficient statistic $U(X)$ such that $f_{\theta_2}(u)/f_{\theta_1}(u)$ is strictly increasing in u for $\theta_1 < \theta_2$, where $f_\theta(u)$ is the Lebesgue p.d.f. of $U(X)$ and is continuous in u for each θ; and that for all $\theta_1 < \theta_2 < \theta_3$ and $u_1 < u_2 < u_3$,

$$\begin{vmatrix} f_{\theta_1}(u_1) & f_{\theta_1}(u_2) & f_{\theta_1}(u_3) \\ f_{\theta_2}(u_1) & f_{\theta_2}(u_2) & f_{\theta_2}(u_3) \\ f_{\theta_3}(u_1) & f_{\theta_3}(u_2) & f_{\theta_3}(u_3) \end{vmatrix} > 0.$$

Show that the conclusions of Theorem 6.3 remain valid.

27. (p-values). Suppose that X has a distribution P_θ, where $\theta \in \mathcal{R}$ is unknown. Consider a family of nonrandomized level α tests for $H_0 : \theta = \theta_0$ (or $\theta \le \theta_0$) with rejection region C_α such that $P_{\theta_0}(X \in C_\alpha) = \alpha$ for all $0 < \alpha < 1$ and $C_{\alpha_1} = \cap_{\alpha > \alpha_1} C_\alpha$ for all $0 < \alpha_1 < 1$.
    (a) Show that the p-value is $\hat{\alpha}(x) = \inf\{\alpha : x \in C_\alpha\}$.
    (b) Show that when $\theta = \theta_0$, $\hat{\alpha}(X)$ has the uniform distribution $U(0, 1)$.
    (c) If the tests with rejection regions C_α are unbiased of level α, show that under H_1, $P_\theta(\hat{\alpha}(X) \le \alpha) \ge \alpha$.

28. Suppose that X has the p.d.f. (6.10). Consider hypotheses (6.13) or (6.14). Show that a UMP test does not exist. (Hint: this follows from a consideration of the UMP tests for the one-sided hypotheses $H_0 : \theta \ge \theta_1$ and $H_0 : \theta \le \theta_2$.)

29. Consider Exercise 17 with $H_0 : \theta \in [\theta_1, \theta_2]$ versus $H_1 : \theta \notin [\theta_1, \theta_2]$, where $0 < \theta_1 \le \theta_2 < 1$.
    (a) Show that a UMP test does not exist.
    (b) Obtain a UMPU test of size α.

30. In the proof of Theorem 6.4, show that
    (a) (6.30) is equivalent to (6.31);
    (b) (6.31) is equivalent to (6.29) with T_* replaced by T;
    (c) when $0 < \alpha < 1$, $(\alpha, \alpha E_{\theta_0}(Y))$ is an interior point of the set of points $(E_{\theta_0}(T), E_{\theta_0}(TY))$ as T ranges over all tests of the form $T = T(Y)$;
    (d) the UMPU tests are unique a.s. $\mathcal{P}$ if attention is restricted to tests depending on (Y, U) and (Y, U) has a continuous c.d.f.

31. Consider the decision problem in Example 2.20 with the 0-1 loss. Show that if a UMPU test of size α exists and is unique (in the sense that decision rules that are equivalent in terms of the risk are treated the same), then it is admissible.

32. Let $X_1, ..., X_n$ be i.i.d. binary random variables with $p = P(X_1 = 1)$.
    (a) Determine the c_i's and γ_i's in (6.15) and (6.16) for testing $H_0 : p \le 0.2$ or $p \ge 0.7$ when $\alpha = 0.1$ and $n = 15$. Find the power of the UMP test (6.15) when $p = 0.4$.
    (b) Derive a UMPU test of size α for $H_0 : p = p_0$ versus $H_1 : p \ne p_0$ when $n = 10$, $\alpha = 0.05$, and $p_0 = 0.4$.

33. Suppose that X has the Poisson distribution $P(\theta)$ with an unknown $\theta > 0$. Show that (6.29) reduces to

$$\sum_{x=c_1+1}^{c_2-1} \frac{\theta_0^{x-1} e^{-\theta_0}}{(x-1)!} + \sum_{i=1}^{2} (1 - \gamma_i) \frac{\theta_0^{c_i-1} e^{-\theta_0}}{(c_i - 1)!} = 1 - \alpha,$$

provided that $c_1 > 1$.

34. Let X be a random variable from the geometric distribution $G(p)$. Find a UMPU test of size α for $H_0 : p = p_0$ versus $H_1 : p \neq p_0$.

35. In Exercise 33 of §2.6, derive a UMPU test of size $\alpha \in (0, 1)$ for testing $H_0 : p \leq \frac{1}{2}$ versus $H_1 : p > \frac{1}{2}$.

36. Let $X_1, ..., X_n$ be i.i.d. from $N(\mu, \sigma^2)$ with unknown μ and σ^2.
    (a) Show how the power of the one-sample t-test depends on a non-central t-distribution.
    (b) Show that the power of the one-sample t-test is an increasing function of μ/σ in the one-sided case ($H_0 : \mu \leq \mu_0$ versus $H_1 : \mu > \mu_0$), and of $|\mu|/\sigma$ in the two-sided case ($H_0 : \mu = \mu_0$ versus $H_1 : \mu \neq \mu_0$).

37. Let $X_1, ..., X_n$ be i.i.d. from the gamma distribution $\Gamma(\theta, \gamma)$ with unknown θ and γ.
    (a) For testing $H_0 : \theta \leq \theta_0$ versus $H_1 : \theta > \theta_0$ and $H_0 : \theta = \theta_0$ versus $H_1 : \theta \neq \theta_0$, show that there exist UMPU tests whose rejection regions are based on $V = \prod_{i=1}^{n}(X_i/\bar{X})$.
    (b) For testing $H_0 : \gamma \leq \gamma_0$ versus $H_1 : \gamma > \gamma_0$, show that a UMPU test rejects H_0 when $\sum_{i=1}^{n} X_i > C(\prod_{i=1}^{n} X_i)$ for some function C.

38. Let X_1 and X_2 be independently distributed as the Poisson distributions $P(\lambda_1)$ and $P(\lambda_2)$, respectively.
    (a) Find a UMPU test of size α for testing $H_0 : \lambda_1 \geq \lambda_2$ versus $H_1 : \lambda_1 < \lambda_2$.
    (b) Calculate the power of the UMPU test in (a) when $\alpha = 0.1$, $(\lambda_1, \lambda_2) = (0.1, 0.2)$, (1,2), (10,20), and (0.1,0.4).

39. Consider the binomial problem in Example 6.11.
    (a) Prove the claim about $P(Y = y|U = u)$.
    (b) Find a UMPU test of size α for testing $H_0 : p_1 \geq p_2$ versus $H_1 : p_1 < p_2$.
    (c) Repeat (b) for $H_0 : p_1 = p_2$ versus $H_1 : p_1 \neq p_2$.

40. Let X_1 and X_2 be independently distributed as the negative binomial distributions $NB(p_1, n_1)$ and $NB(p_2, n_2)$, respectively, where n_i's are known and p_i's are unknown.
    (a) Show that there exists a UMPU test of size α for testing $H_0 : p_1 \leq p_2$ versus $H_1 : p_1 > p_2$.
    (b) Determine the conditional distribution $P_{Y|U=u}$ in Theorem 6.4 when $n_1 = n_2 = 1$.

41. Let (X_0, X_1, X_2) be a random vector having a multinomial distribution (Example 2.7) with $k = 2$, $p_0 = 1 - p_1 - p_2$, and unknown $p_1 \in (0, 1)$ and $p_2 \in (0, 1)$. Derive a UMPU test of size α for testing $H_0 : p_0 = p^2, p_1 = 2p(1-p), p_2 = (1-p)^2$ versus $H_1 : H_0$ is not true, where $p \in (0, 1)$ is unknown.

42. Consider Example 6.12.

    (a) Show that A and B are independent if and only if $\log \frac{p_{11}}{p_{22}} = \log \frac{p_{12}}{p_{22}} + \log \frac{p_{21}}{p_{22}}$.

    (b) Derive a UMPU test of size α for testing $H_0 : P(A) = P(B)$ versus $H_1 : P(A) \neq P(B)$.

43. Let X_1 and X_2 be independently distributed according to p.d.f.'s given by (6.10) with ξ, η, θ, Y, and h replaced by ξ_i, η_i, θ_i, Y_i, and h_i, $i = 1, 2$, respectively. Show that there exists a UMPU test of size α for testing

    (a) $H_0 : \eta_2(\theta_2) - \eta_1(\theta_1) \leq \eta_0$ versus $H_1 : \eta_2(\theta_2) - \eta_1(\theta_1) > \eta_0$;

    (b) $H_0 : \eta_2(\theta_2) + \eta_1(\theta_1) \leq \eta_0$ versus $H_1 : \eta_2(\theta_2) + \eta_1(\theta_1) > \eta_0$.

44. Let X_j, $j = 1, 2, 3$, be independent from the Poisson distributions $P(\lambda_j)$, $j = 1, 2, 3$, respectively. Show that there exists a UMPU test of size α for testing $H_0 : \lambda_1 \lambda_2 \leq \lambda_3^2$ versus $H_1 : \lambda_1 \lambda_2 > \lambda_3^2$.

45. Let X_{ij}, $i = 1, 2$, $j = 1, 2$, be independent from the Poisson distributions $P(\lambda_i p_{ij})$, where $\lambda_i > 0$, $0 < p_{ij} < 1$, and $p_{i1} + p_{i2} = 0$. Derive a UMPU test of size α for testing $H_0 : p_{11} \leq p_{21}$ versus $H_1 : p_{11} > p_{21}$.

46. Let X_{ij} be independent random variables satisfying $P(X_{ij} = 0) = \theta_i$, $P(X_{ij} = k) = (1 - \theta_i)(1 - p_i)^{j-1} p_i$, $k = 1, 2, ...$, where $0 < \theta_i < 1$ and $0 < p_i < 1$, $j = 1, ..., n_i$ and $i = 1, 2$. Derive a UMPU test of size α for testing $H_0 : p_1 \leq p_2$ versus $H_1 : p_1 > p_2$.

47. Let $X_{11}, ..., X_{1n_1}$ and $X_{21}, ..., X_{2n_2}$ be two independent samples i.i.d. from the gamma distributions $\Gamma(\theta_1, \gamma_1)$ and $\Gamma(\theta_2, \gamma_2)$, respectively.

    (a) Assume that γ_1 and γ_2 are known. For testing $H_0 : \theta_1 \leq \theta_2$ versus $H_1 : \theta_1 > \theta_2$ and $H_0 : \theta_1 = \theta_2$ versus $H_1 : \theta_1 \neq \theta_2$, show that there exist UMPU tests and that the rejection regions can be determined by using beta distributions.

    (b) If γ_i's are unknown in (a), show that there exist UMPU tests and describe their general forms.

    (c) Assume that $\theta_1 = \theta_2$ (unknown). For testing $H_0 : \gamma_1 \leq \gamma_2$ versus $H_1 : \gamma_1 > \gamma_2$ and $H_0 : \gamma_1 = \gamma_2$ versus $H_1 : \gamma_1 \neq \gamma_2$, show that there exist UMPU tests and describe their general forms.

48. Let N be a random variable with the following discrete p.d.f.:

    $$P(N = n) = C(\lambda) a(n) \lambda^n I_{\{0,1,2,...\}}(n),$$

    where $\lambda > 0$ is unknown and a and C are known functions. Suppose that given $N = n$, $X_1, ..., X_n$ are i.i.d. from the p.d.f. given in (6.10). Show that, based on $(N, X_1, ..., X_N)$, there exists a UMPU test of size α for $H_0 : \eta(\theta) \leq \eta_0$ versus $H_1 : \eta(\theta) > \eta_0$.

49. Let $X_{i1}, ..., X_{in_i}$, $i = 1, 2$, be two independent samples i.i.d. from $N(\mu_i, \sigma^2)$, respectively, $n_i \geq 2$. Show that a UMPU test of size α for $H_0 : \mu_1 = \mu_2$ versus $H_1 : \mu_1 \neq \mu_2$ rejects H_0 when $|t(X)| > t_{n_1+n_2-1, \alpha/2}$, where $t(X)$ is given by (6.37) and $t_{n_1+n_2-1, \alpha}$ is the $(1 - \alpha)$th quantile of the t-distribution $t_{n_1+n_2-1}$. Derive the power function of this test.

50. In the two-sample problem discussed in §6.2.3, show that when $n_1 = n_2$, a UMPU test of size α for testing $H_0 : \sigma_2^2 = \Delta_0 \sigma_1^2$ versus $H_1 : \sigma_2^2 \neq \Delta_0 \sigma_1^2$ rejects H_0 when

$$\max \left\{ \frac{S_2^2}{\Delta_0 S_1^2}, \frac{\Delta_0 S_1^2}{S_2^2} \right\} > \frac{1 - c}{c},$$

where $\int_0^c f_{(n_1-1)/2, (n_1-1)/2}(v) dv = \alpha/2$ and $f_{a,b}$ is the p.d.f. of the beta distribution $B(a, b)$.

51. Suppose that $X_i = \beta_0 + \beta_1 t_i + \varepsilon_i$, where t_i's are fixed constants that are not all the same, ε_i's are i.i.d. from $N(0, \sigma^2)$, and β_0, β_1, and σ^2 are unknown parameters. Derive a UMPU test of size α for testing
    (a) $H_0 : \beta_0 \leq \theta_0$ versus $H_1 : \beta_0 > \theta_0$;
    (b) $H_0 : \beta_0 = \theta_0$ versus $H_1 : \beta_0 \neq \theta_0$;
    (c) $H_0 : \beta_1 \leq \theta_0$ versus $H_1 : \beta_1 > \theta_0$;
    (d) $H_0 : \beta_1 = \theta_0$ versus $H_1 : \beta_1 \neq \theta_0$.

52. In the previous exercise, derive the power function in each of (a)-(d) in terms of a noncentral t-distribution.

53. Consider the normal linear model in §6.2.3 (i.e., model (3.25) with $\varepsilon = N_n(0, \sigma^2 I_n)$). For testing $H_0 : \sigma^2 \leq \sigma_0^2$ versus $H_1 : \sigma^2 > \sigma_0^2$ and $H_0 : \sigma^2 = \sigma_0^2$ versus $H_1 : \sigma^2 \neq \sigma_0^2$, show that UMPU tests of size α are functions of SSR and their rejection regions can be determined using chi-square distributions.

54. In the problem of testing for independence in the bivariate normal family, show that
    (a) the p.d.f. in (6.44) is of the form (6.23) and identify φ;
    (b) the sample correlation coefficient R is independent of U when $\rho = 0$;
    (c) R is linear in Y, and V in (6.45) has the t-distribution t_{n-2} when $\rho = 0$.

55. Let $X_1, ..., X_n$ be i.i.d. bivariate normal with the p.d.f. in (6.44) and let $S_j^2 = \sum_{i=1}^n (X_{ij} - \bar{X}_j)^2$ and $S_{12} = \sum_{i=1}^n (X_{i1} - \bar{X}_1)(X_{i2} - \bar{X}_2)$.
    (a) Show that a UMPU test for testing $H_0 : \sigma_2/\sigma_1 = \Delta_0$ versus $H_1 : \sigma_2/\sigma_1 \neq \Delta_0$ rejects H_0 when

$$R = |\Delta_0^2 S_1^2 - S_2^2| / \sqrt{(\Delta_0^2 S_1^2 + S_2^2)^2 - 4\Delta_0^2 S_{12}^2} > c.$$

(b) Find the p.d.f. of R in (a) when $\sigma_2/\sigma_1 = \Delta_0$.
(c) Assume that $\sigma_1 = \sigma_2$. Show that a UMPU test for $H_0 : \mu_1 = \mu_2$ versus $H_1 : \mu_1 \neq \mu_2$ rejects H_0 when

$$V = |\bar{X}_2 - \bar{X}_1| / \sqrt{S_1^2 + S_2^2 - 2S_{12}} > c.$$

(d) Find the p.d.f. of V in (c) when $\mu_1 = \mu_2$.

56. Let $(X_1, Y_1), ..., (X_n, Y_n)$ be i.i.d. random 2-vectors having the bivariate normal distribution with $EX_1 = EY_1 = 0$, $\text{Var}(X_1) = \sigma_x^2$, $\text{Var}(Y_1) = \sigma_y^2$, and $\text{Cov}(X_1, Y_1) = \rho\sigma_x\sigma_y$, where $\sigma_x > 0$, $\sigma_y > 0$, and $\rho \in [0, 1)$ are unknown. Derive the form and exact distribution of a UMPU test of size α for testing $H_0 : \rho = 0$ versus $H_1 : \rho > 0$.

57. Let $X_1, ..., X_n$ be i.i.d. from the exponential distribution $E(a, \theta)$ with unknown a and θ. Let $V = 2\sum_{i=1}^{n}(X_i - X_{(1)})$, where $X_{(1)}$ is the smallest order statistic.
(a) For testing $H_0 : \theta = 1$ versus $H_1 : \theta \neq 1$, show that a UMPU test of size α rejects H_0 when $V < c_1$ or $V > c_2$, where c_i's are determined by

$$\int_{c_1}^{c_2} f_{2n-2}(v)dv = \int_{c_1}^{c_2} f_{2n}(v)dv = 1 - \alpha,$$

and $f_m(v)$ is the p.d.f. of the chi-square distribution χ_m^2.
(b) For testing $H_0 : a = 0$ versus $H_1 : a \neq 0$, show that a UMPU test of size α rejects H_0 when $X_{(1)} < 0$ or $2nX_{(1)}/V > c$, where c is determined by

$$(n-1)\int_0^c (1+v)^{-n}dv = 1 - \alpha.$$

58. Let $X_1, ..., X_n$ be i.i.d. random variables from the uniform distribution $U(\theta, \vartheta)$, $-\infty < \theta < \vartheta < \infty$.
(a) Show that the conditional distribution of $X_{(1)}$ given $X_{(n)} = x$ is the distribution of the minimum of a sample of size $n - 1$ from the uniform distribution $U(\theta, x)$.
(b) Find a UMPU test of size α for testing $H_0 : \theta \leq 0$ versus $H_1 : \theta > 0$.

59. Let $X_1, ..., X_n$ be independent random variables having the binomial distributions $Bi(p_i, k_i)$, $i = 1, ..., n$, respectively, where $p_i = e^{a+bt_i}/(1+e^{a+bt_i})$, $(a, b) \in \mathcal{R}^2$ is unknown, and t_i's are known covariate values that are not all the same. Derive the UMPU test of size α for testing (a) $H_0 : a \geq 0$ versus $H_1 : a < 0$; (b) $H_0 : b \geq 0$ versus $H_1 : b < 0$.

60. In the previous exercise, derive approximations to the UMPU tests by considering the limiting distributions of the test statistics.

61. Let $\mathcal{X} = \{x \in \mathcal{R}^n : \text{all components of } x \text{ are nonzero}\}$ and $\mathcal{G}$ be the group of transformations $g(x) = (cx_1, ..., cx_n)$, $c > 0$. Show that a maximal invariant under $\mathcal{G}$ is $(\text{sgn}(x_n), x_1/x_n, ..., x_{n-1}/x_n)$, where $\text{sgn}(x)$ is 1 or -1 as x is positive or negative.

62. Let $X_1, ..., X_n$ be i.i.d. with a Lebesgue p.d.f. $\sigma^{-1} f(x/\sigma)$ and f_i, $i = 0, 1$, be two known Lebesgue p.d.f.'s on $\mathcal{R}$ that are either 0 for $x < 0$ or symmetric about 0. Consider $H_0 : f = f_0$ versus $H_1 : f = f_1$ and $\mathcal{G} = \{g_r : r > 0\}$ with $g_r(x) = rx$.
    (a) Show that a UMPI test rejects H_0 when
$$\frac{\int_0^\infty v^{n-1} f_1(vX_1) \cdots f_1(vX_n) dv}{\int_0^\infty v^{n-1} f_0(vX_1) \cdots f_0(vX_n) dv} > c.$$
    (b) Show that if $f_0 = N(0, 1)$ and $f_1(x) = e^{-|x|}/2$, then the UMPI test in (a) rejects H_0 when $(\sum_{i=1}^n X_i^2)^{1/2} / \sum_{i=1}^n |X_i| > c$.
    (c) Show that if $f_0(x) = I_{(0,1)}(x)$ and $f_1(x) = 2x I_{(0,1)}(x)$, then the UMPI test in (a) rejects H_0 when $X_{(n)}/(\prod_{i=1}^n X_i)^{1/n} < c$.
    (d) Find the value of c in part (c) when the UMPI test is of size α.

63. Consider the location-scale family problem (with unknown parameters μ and σ) in Example 6.13.
    (a) Show that W is maximal invariant under the given $\mathcal{G}$.
    (b) Show that Proposition 6.2 applies and find the form of the functional $\theta(f_{i,\mu,\sigma})$.
    (c) Derive the p.d.f. of $W(X)$ under H_i, $i = 0, 1$.
    (d) Obtain a UMPI test.

64. In Example 6.13, find the rejection region of the UMPI test when $X_1, ..., X_n$ are i.i.d. and
    (a) $f_{0,\mu,\sigma}$ is $N(\mu, \sigma^2)$ and $f_{1,\mu,\sigma}$ is the p.d.f. of the uniform distribution $U(\mu - \frac{1}{2}\sigma, \mu + \frac{1}{2}\sigma)$;
    (b) $f_{0,\mu,\sigma}$ is $N(\mu, \sigma^2)$ and $f_{1,\mu,\sigma}$ is the p.d.f. of the exponential distribution $E(\mu, \sigma)$;
    (c) $f_{0,\mu,\sigma}$ is the p.d.f. of $U(\mu - \frac{1}{2}\sigma, \mu + \frac{1}{2}\sigma)$ and $f_{1,\mu,\sigma}$ is the p.d.f. of $E(\mu, \sigma)$;
    (d) $f_{0,\mu}$ is $N(\mu, 1)$ and $f_{1,\mu}(x) = \exp\{-e^{x-\mu} + x - \mu\}$.

65. Prove the claims in Example 6.15.

66. Let $X_1, ..., X_n$ be i.i.d. from $N(\mu, \sigma^2)$ with unknown μ and σ^2. Consider the problem of testing $H_0 : \mu = 0$ versus $H_1 : \mu \neq 0$ and the group of transformations $g_c(X_i) = cX_i$, $c \neq 0$.

(a) Show that the testing problem is invariant under $\mathcal{G}$.
(b) Show that the one-sample two-sided t-test in §6.2.3 is a UMPI test.

67. Prove the claims in Example 6.16.

68. Consider Example 6.16 with H_0 and H_1 replaced by $H_0 : \mu_1 = \mu_2$ and $H_1 : \mu_1 \neq \mu_2$, and with $\mathcal{G}$ changed to $\{g_{c_1,c_2,r} : c_1 = c_2 \in \mathcal{R}, r \neq 0\}$.
(a) Show that the testing problem is invariant under $\mathcal{G}$.
(b) Show that the two-sample two-sided t-test in §6.2.3 is a UMPI test.

69. Show that the UMPU tests in Exercise 37(a) and Exercise 47(a) are also UMPI tests under $\mathcal{G} = \{g_r : r > 0\}$ with $g_r(x) = rx$.

70. In Example 6.17, show that $t(X)$ has the noncentral t-distribution $t_{n-1}(\sqrt{n}\theta)$; the family $\{f_\theta(t) : \theta \in \mathcal{R}\}$ has monotone likelihood ratio in t; and that for testing $H_0 : \theta = \theta_0$ versus $H_1 : \theta \neq \theta_0$, a test that is UMP among all level α unbiased tests based on $t(X)$ rejects H_0 when $t(X) < c_1$ or $t(X) > c_2$. (Hint: consider Exercise 26.)

71. Let X_1 and X_2 be independently distributed as the exponential distributions $E(0, \theta_i)$, $i = 1, 2$, respectively. Define $\theta = \theta_1/\theta_2$.
(a) For testing $H_0 : \theta \leq 1$ versus $\theta > 1$, show that the problem is invariant under the group of transformations $g_c(x_1, x_2) = (cx_1, cx_2)$, $c > 0$, and that a UMPI test of size α rejects H_0 when $X_2/X_1 > (1 - \alpha)/\alpha$.
(b) For testing $H_0 : \theta = 1$ versus $\theta \neq 1$, show that the problem is invariant under the group of transformations in (a) and $g(x_1, x_2) = (x_2, x_1)$, and that a UMPI test of size α rejects H_0 when $X_1/X_2 > (2 - \alpha)/\alpha$ and $X_2/X_1 > (2 - \alpha)/\alpha$.

72. Let $X_1, ..., X_m$ and $Y_1, ..., Y_n$ be two independent samples i.i.d. from the exponential distributions $E(a_1, \theta_1)$ and $E(a_2, \theta_2)$, respectively. Let $g_{r,c,d}(x, y) = (rx_1 + c, ..., rx_m + c, ry_1 + d, ..., ry_n + d)$ and let $\mathcal{G} = \{g_{r,c,d} : r > 0, c \in \mathcal{R}, d \in \mathcal{R}\}$.
(a) Show that a UMPI test of size α for testing $H_0 : \theta_1/\theta_2 \geq \Delta_0$ versus $H_1 : \theta_1/\theta_2 < \Delta_0$ rejects H_0 when $\sum_{i=1}^n (Y_i - Y_{(1)}) > c \sum_{i=1}^m (X_i - X_{(1)})$ for some constant c.
(b) Find the value of c in (a).
(c) Show that the UMPI test in (a) is also a UMPU test.

73. Let $M(U)$ be given by (6.51) and $W = M(U)(n - r)/s$.
(a) Show that W has the noncentral F-distribution $F_{s,n-r}(\theta)$.
(b) Show that $f_{\theta_1}(w)/f_0(w)$ is an increasing function of w for any given $\theta_1 > 0$.

74. Consider normal linear model (6.38). Show that
    (a) the UMPI test derived in §6.3.2 for testing (6.49) is the same as the UMPU test for (6.40) given in §6.2.3 when $s = 1$ and $\theta_0 = 0$;
    (b) the test with the rejection region $W > F_{s,n-r,\alpha}$ is a UMPI test of size α for testing $H_0 : L\beta = \theta_0$ versus $H_1 : L\beta \neq \theta_0$, where W is given by (6.52), θ_0 is a fixed constant, L is the same as that in (6.49), and $F_{s,n-r,\alpha}$ is the $(1 - \alpha)$th quantile of the F-distribution $F_{s,n-r}$.

75. In Examples 6.18-6.19,
    (a) prove the claim in Example 6.19;
    (b) derive the distribution of W by applying Cochran's theorem.

76. (Two-way additive model). Assume that X_{ij}'s are independent and
$$X_{ij} = N(\mu_{ij}, \sigma^2), \qquad i = 1, ..., a, \ j = 1, ..., b,$$
where $\mu_{ij} = \mu + \alpha_i + \beta_j$ and $\sum_{i=1}^{a} \alpha_i = \sum_{j=1}^{b} \beta_j = 0$. Derive the forms of the UMPI tests in §6.3.2 for testing (6.54) and (6.55).

77. (Three-way additive model). Assume that X_{ijk}'s are independent and
$$X_{ijk} = N(\mu_{ijk}, \sigma^2), \qquad i = 1, ..., a, \ j = 1, ..., b, \ k = 1, ..., c,$$
where $\mu_{ijk} = \mu + \alpha_i + \beta_j + \gamma_k$ and $\sum_{i=1}^{a} \alpha_i = \sum_{j=1}^{b} \beta_j = \sum_{k=1}^{c} \gamma_k = 0$. Derive the UMPI test based on the W in (6.52) for testing $H_0 : \alpha_i = 0$ for all i versus $H_1 : \alpha_i \neq 0$ for some i.

78. Let $X_1, ..., X_m$ and $Y_1, ..., Y_n$ be independently normally distributed with a common unknown variance σ^2 and means
$$E(X_i) = \mu_x + \beta_x(u_i - \bar{u}), \qquad E(Y_j) = \mu_y + \beta_y(v_j - \bar{v}),$$
where u_i's and v_j's are known constants, $\bar{u} = m^{-1} \sum_{i=1}^{m} u_i$, $\bar{v} = n^{-1} \sum_{i=1}^{n} v_i$, and μ_x, μ_y, β_x, and β_y are unknown. Derive the UMPI test based on the W in (6.52) for testing
    (a) $H_0 : \beta_x = \beta_y$ versus $H_1 : \beta_x \neq \beta_y$;
    (b) $H_0 : \beta_x = \beta_y$ and $\mu_x = \mu_y$ versus $H_1 : \beta_x \neq \beta_y$ or $\mu_x \neq \mu_y$.

79. Let $(X_1, Y_1), ..., (X_n, Y_n)$ be i.i.d. from a bivariate normal distribution with unknown means, variances, and correlation coefficient ρ.
    (a) Show that the problem of testing $H_0 : \rho \leq \rho_0$ versus $H_1 : \rho > \rho_0$ is invariant under $\mathcal{G}$ containing transformations $rX_i + c$, $sY_i + d$, $i = 1, ..., n$, where $r > 0$, $s > 0$, $c \in \mathcal{R}$, and $d \in \mathcal{R}$. Show that a UMPI test rejects H_0 when $R > c$, where R is the sample correlation coefficient given in (6.45). (Hint: see Lehmann (1986, p. 340).)
    (b) Show that the problem of testing $H_0 : \rho = 0$ versus $H_1 : \rho \neq 0$ is invariant in addition (to the transformations in (a)) under the transformation $g(X_i, Y_i) = (X_i, -Y_i)$, $i = 1, ..., n$. Show that a UMPI test rejects H_0 when $|R| > c$.

80. Under the random effects model (6.57), show that
    (a) SSA/SSR is maximal invariant under the group of transformations described in §6.3.2;
    (b) the UMPI test for (6.58) derived in §6.3.2 is also a UMPU test.

81. Show that (6.60) is equivalent to (6.61).

82. In Proposition 6.5,
    (a) show that $\log \ell(\hat{\theta}) - \log \ell(\theta_0)$ is strictly increasing (or decreasing) in Y when $\hat{\theta} > \theta_0$ (or $\hat{\theta} < \theta_0$);
    (b) prove part (iii).

83. In Exercises 40 and 41 of §2.6, consider $H_0 : j = 1$ versus $H_1 : j = 2$.
    (a) Derive the likelihood ratio $\lambda(X)$.
    (b) Obtain an LR test of size α in Exercise 40 of §2.6.

84. In Exercise 17, derive the likelihood ratio $\lambda(X)$ when (a) $H_0 : \theta \leq \theta_0$; (b) $H_0 : \theta_1 \leq \theta \leq \theta_2$; and (c) $H_0 : \theta \leq \theta_1$ or $\theta \geq \theta_2$.

85. Let $X_1, ..., X_n$ be i.i.d. from the discrete uniform distribution on $\{1, ..., \theta\}$, where θ is an integer ≥ 2. Find a level α LR test for
    (a) $H_0 : \theta \leq \theta_0$ versus $H_1 : \theta > \theta_0$, where θ_0 is a known integer ≥ 2;
    (b) $H_0 : \theta = \theta_0$ versus $H_1 : \theta \neq \theta_0$.

86. Let X be a sample of size 1 from the p.d.f. $2\theta^{-2}(\theta - x)I_{(0,\theta)}(x)$, where $\theta > 0$ is unknown. Find an LR test of size α for testing $H_0 : \theta = \theta_0$ versus $H_1 : \theta \neq \theta_0$.

87. Let $X_1, ..., X_n$ be i.i.d. from the exponential distribution $E(a, \theta)$.
    (a) Suppose that θ is known. Find an LR test of size α for testing $H_0 : a \leq a_0$ versus $H_1 : a > a_0$.
    (b) Suppose that θ is known. Find an LR test of size α for testing $H_0 : a = a_0$ versus $H_1 : a \neq a_0$.
    (c) Repeat part (a) for the case where θ is also unknown.
    (d) When both θ and a are unknown, find an LR test of size α for testing $H_0 : \theta = \theta_0$ versus $H_1 : \theta \neq \theta_0$.
    (e) When $a > 0$ and $\theta > 0$ are unknown, find an LR test of size α for testing $H_0 : a = \theta$ versus $H_1 : a \neq \theta$.

88. Let $X_1, ..., X_n$ be i.i.d. from the Pareto distribution $Pa(\gamma, \theta)$, where $\theta > 0$ and $\gamma > 0$ are unknown. Show that an LR test for $H_0 : \theta = 1$ versus $H_1 : \theta \neq 1$ rejects H_0 when $Y < c_1$ or $Y > c_2$, where $Y = \log\left(\prod_{i=1}^{n} X_i/X_{(1)}^n\right)$ and c_1 and c_2 are positive constants. Find values of c_1 and c_2 so that this LR test has size α.

89. Let $X_{i1}, ..., X_{in_i}$, $i = 1, 2$, be two independent samples i.i.d. from the uniform distributions $U(0, \theta_i)$, $i = 1, 2$, respectively, where $\theta_1 > 0$

and $\theta_2 > 0$ are unknown.

(a) Find an LR test of size α for testing $H_0 : \theta_1 = \theta_2$ versus $H_1 : \theta_1 \neq \theta_2$.

(b) Derive the limit distribution of $-2 \log \lambda$, where λ is the likelihood ratio in part (a).

90. Let $X_{i1}, ..., X_{in_i}$, $i = 1, 2$, be two independent samples i.i.d. from $N(\mu_i, \sigma_i^2)$, $i = 1, 2$, respectively, where μ_i's and σ_i^2's are unknown. For testing $H_0 : \sigma_2^2/\sigma_1^2 = \Delta_0$ versus $H_1 : \sigma_2^2/\sigma_1^2 \neq \Delta_0$, derive an LR test of size α and compare it with the UMPU test derived in §6.2.3.

91. Let $(X_{11}, X_{12}), ..., (X_{n1}, X_{n2})$ be i.i.d. from a bivariate normal distribution with unknown mean and covariance matrix. For testing $H_0 : \rho = 0$ versus $H_1 : \rho \neq 0$, where ρ is the correlation coefficient, show that the test rejecting H_0 when $|W| > 0$ is an LR test, where

$$W = \sum_{i=1}^{n}(X_{i1} - \bar{X}_1)(X_{i2} - \bar{X}_2) \Big/ \left[\sum_{i=1}^{n}(X_{i1} - \bar{X}_1)^2 + \sum_{i=1}^{n}(X_{i2} - \bar{X}_2)^2 \right].$$

Find the distribution of W.

92. Let X_1 and X_2 be independently distributed as the Poisson distributions $P(\lambda_1)$ and $P(\lambda_2)$, respectively. Find an LR test of significance level α for testing

(a) $H_0 : \lambda_1 = \lambda_2$ versus $H_1 : \lambda_1 \neq \lambda_2$;

(b) $H_0 : \lambda_1 \geq \lambda_2$ versus $H_1 : \lambda_1 < \lambda_2$. (Is this test a UMPU test?)

93. Let X_1 and X_2 be independently distributed as the binomial distributions $Bi(p_1, n_1)$ and $Bi(p_2, n_2)$, respectively, where n_i's are known and p_i's are unknown. Find an LR test of significance level α for testing

(a) $H_0 : p_1 = p_2$ versus $H_1 : p_1 \neq p_2$;

(b) $H_0 : p_1 \geq p_2$ versus $H_1 : p_1 < p_2$. (Is this test a UMPU test?)

94. Let X_1 and X_2 be independently distributed as the negative binomial distributions $NB(p_1, n_1)$ and $NB(p_2, n_2)$, respectively, where n_i's are known and p_i's are unknown. Find an LR test of significance level α for testing

(a) $H_0 : p_1 = p_2$ versus $H_1 : p_1 \neq p_2$;

(b) $H_0 : p_1 \leq p_2$ versus $H_1 : p_1 > p_2$.

95. Let X_1 and X_2 be independently distributed as the exponential distributions $E(0, \theta_i)$, $i = 1, 2$, respectively. Define $\theta = \theta_1/\theta_2$. Find an LR test of size α for testing

(a) $H_0 : \theta = 1$ versus $H_1 : \theta \neq 1$;

(b) $H_0 : \theta \leq 1$ versus $H_1 : \theta > 1$.

96. Let $X_{i1}, ..., X_{in_i}$, $i = 1, 2$, be independently distributed as the beta distributions with p.d.f.'s $\theta_i x^{\theta_i - 1} I_{(0,1)}(x)$, $i = 1, 2$, respectively. For testing $H_0 : \theta_1 = \theta_2$ versus $H_1 : \theta_1 \neq \theta_2$, find the forms of the LR test, Wald's test, and Rao's score test.

97. In the proof of Theorem 6.6(ii), show that (6.65) and (6.66) hold and that (6.67) implies (6.68).

98. Let $X_1, ..., X_n$ be i.i.d. from $N(\mu, \sigma^2)$.
    (a) Suppose that $\sigma^2 = \gamma \mu^2$ with unknown $\gamma > 0$ and $\mu \in \mathcal{R}$. Find an LR test for testing $H_0 : \gamma = 1$ versus $H_1 : \gamma \neq 1$.
    (b) In the testing problem in (a), find the forms of W_n for Wald's test and R_n for Rao's score test, and discuss whether Theorems 6.5 and 6.6 can be applied.
    (c) Repeat (a) and (b) when $\sigma^2 = \gamma \mu$ with unknown $\gamma > 0$ and $\mu > 0$.

99. Suppose that $X_1, ..., X_n$ are i.i.d. from the Weibull distribution with p.d.f. $\theta^{-1} \gamma x^{\gamma - 1} e^{-x^\gamma / \theta} I_{(0,\infty)}(x)$, where $\gamma > 0$ and $\theta > 0$ are unknown. Consider the problem of testing $H_0 : \gamma = 1$ versus $H_1 : \gamma \neq 1$.
    (a) Find an LR test and discuss whether Theorem 6.5 can be applied.
    (b) Find the forms of W_n for Wald's test and R_n for Rao's score test.

100. Suppose that $X = (X_1, ..., X_k)$ has the multinomial distribution with the parameter $p = (p_1, ..., p_k)$. Consider the problem of testing (6.70). Find the forms of W_n for Wald's test and R_n for Rao's score test.

101. In Example 6.12, consider testing $H_0 : P(A) = P(B)$ versus $H_1 : P(A) \neq P(B)$.
    (a) Derive the likelihood ratio λ_n and the limiting distribution of $-2 \log \lambda_n$ under H_0.
    (b) Find the forms of W_n for Wald's test and R_n for Rao's score test.

102. Prove the claims in Example 6.24.

103. Consider testing independence in the $r \times c$ contingency table problem in Example 6.24. Find the forms of W_n for Wald's test and R_n for Rao's score test.

104. Under the conditions of Theorems 6.5 and 6.6, show that Wald's tests are Chernoff-consistent (Definition 2.13) if α is chosen to be $\alpha_n \to 0$ and $\chi^2_{r,\alpha_n} = o(n)$ as $n \to \infty$, where $\chi^2_{r,\alpha}$ is the $(1 - \alpha)$th quantile of χ^2_r.

105. Let $X_1, ..., X_n$ be i.i.d. binary random variables with $\theta = P(X_1 = 1)$.
    (a) Let the prior $\Pi(\theta)$ be the c.d.f. of the beta distribution $B(a, b)$. Find the Bayes factor and the Bayes test for $H_0 : \theta \leq \theta_0$ versus $H_1 : \theta > \theta_0$.

(b) Let the prior c.d.f. be $\pi_0 I_{[\theta_0,\infty)}(\theta) + (1 - \pi_0)\Pi(\theta)$, where Π is the same as that in (a). Find the Bayes factor and the Bayes test for $H_0 : \theta = \theta_0$ versus $H_1 : \theta \neq \theta_0$.

106. Let $X_1, ..., X_n$ be i.i.d. from the Poisson distribution $P(\theta)$.
   (a) Let the prior c.d.f. be $\Pi(\theta) = (1 - e^{-\theta})I_{(0,\infty)}(\theta)$. Find the Bayes factor and the Bayes test for $H_0 : \theta \leq \theta_0$ versus $H_1 : \theta > \theta_0$.
   (b) Let the prior c.d.f. be $\pi_0 I_{[\theta_0,\infty)}(\theta) + (1 - \pi_0)\Pi(\theta)$, where Π is the same as that in (a). Find the Bayes factor and the Bayes test for $H_0 : \theta = \theta_0$ versus $H_1 : \theta \neq \theta_0$.

107. Let X_i, $i = 1, 2$, be independent observations from the gamma distributions $\Gamma(a, \gamma_1)$ and $\Gamma(a, \gamma_2)$, respectively, where $a > 0$ is known and $\gamma_i > 0$, $i = 1, 2$, are unknown. Find the Bayes factor and the Bayes test for $H_0 : \gamma_1 = \gamma_2$ versus $H_1 : \gamma_1 \neq \gamma_2$ under the prior c.d.f. $\Pi = \pi_0\Pi_0 + (1 - \pi_0)\Pi_1$, where $\Pi_0(x_1, x_2) = G(\min\{x_1, x_2\})$, $\Pi_1(x_1, x_2) = G(x_1)G(x_2)$, $G(x)$ is the c.d.f. of a known gamma distribution, and π_0 is a known constant.

108. Find a condition under which the UMPI test given in Example 6.17 is better than the sign test given by (6.78) in terms of their power functions under H_1.

109. For testing (6.80), show that a test T satisfying (6.81) is of size α and that the test in (6.82) satisfies (6.81).

110. Let $\mathcal{G}$ be the class of transformations $g(x) = (\psi(x_1), ..., \psi(x_n))$, where ψ is continuous, odd, and strictly increasing. Let $\tilde{R}$ be the vector of ranks of $|x_1|, ..., |x_n|$ and R_+ (or R_-) be the subvector of $\tilde{R}$ containing ranks corresponding to positive (or negative) x_i's. Show that (R_+, R_-) is maximal invariant under $\mathcal{G}$. (Hint: see Example 6.14.)

111. Under H_0, obtain the distribution of W in (6.83) for the one-sample Wilcoxon signed rank test when $n = 3$ or 4.

112. For the one-sample Wilcoxon signed rank test, show that t_0 and σ_0^2 in (6.85) are equal to $\frac{1}{4}$ and $\frac{1}{12}$, respectively.

113. Using the results in §5.2.2, derive a two-sample rank test for testing (6.80) that has limiting size α.

114. Prove Theorem 6.10(i) and show that $D_n^-(F)$ and $D_n^+(F)$ have the same distribution.

115. Show that the one-sided and two-sided Kolmogorov-Smirnov tests are consistent according to Definition 2.13.

116. Let $C_n(F)$ be given by (6.88) for any continuous c.d.f. F on $\mathcal{R}$. Show that the distribution of $C_n(F)$ does not vary with F.

117. Show that the Cramér-von Mises tests are consistent.

118. In Example 6.27, show that the one-sample Wilcoxon signed rank test is consistent.

119. Let $X_1, ..., X_n$ be i.i.d. from a c.d.f. F on $\mathcal{R}^d$ and $\theta = E(X_1)$.
(a) Derive the empirical likelihood ratio for testing $H_0 : \theta = \theta_0$ versus $H_1 : \theta \neq \theta_0$.
(b) Let $\theta = (\vartheta, \varphi)$. Derive the profile empirical likelihood ratio for testing $H_0 : \vartheta = \vartheta_0$ versus $H_1 : \vartheta \neq \vartheta_0$.

120. Prove Theorem 6.12(ii).

121. Let $X_{i1}, ..., X_{in_i}$, $i = 1, 2$, be two independent samples i.i.d. from F_i on $\mathcal{R}$, $i = 1, 2$, respectively, and let $\mu_i = E(X_i)$.
(a) Show that the two-sample t-test derived in §6.2.3 for testing $H_0 : \mu_1 = \mu_2$ versus $H_1 : \mu_1 \neq \mu_2$ has asymptotic significance level α and is consistent, if $n_1 \to \infty$, $n_1/n_2 \to c \in (0, 1)$, and $\sigma_1^2 = \sigma_2^2$.
(b) Derive a consistent asymptotic test for testing $H_0 : \mu_1/\mu_2 = \Delta_0$ versus $H_1 : \mu_1/\mu_2 \neq \Delta_0$, assuming that $\mu_2 \neq 0$.

122. Consider the general linear model (3.25) with i.i.d. ε_i's having $E(\varepsilon_i) = 0$ and $E(\varepsilon_i^2) = \sigma^2$.
(a) Under the conditions of Theorem 3.12, derive a consistent asymptotic test based on the LSE $l^\tau \hat\beta$ for testing $H_0 : l^\tau \beta = \theta_0$ versus $H_1 : l^\tau \beta \neq \theta_0$, where $l \in \mathcal{R}(Z)$.
(b) Show that the LR test in Example 6.21 has asymptotic significance level α and is consistent.

123. Let $\hat\theta_n$ be an estimator of a real-valued parameter θ such that (6.92) holds for any θ and let $\hat V_n$ be a consistent estimator of V_n. Suppose that $V_n \to 0$.
(a) Show that the test with rejection region $\hat V_n^{-1/2}(\hat\theta_n - \theta_0) > z_{1-\alpha}$ is a consistent asymptotic test for testing $H_0 : \theta \leq \theta_0$ versus $H_1 : \theta > \theta_0$.
(b) Apply the result in (a) to show that the one-sample one-sided t-test in §6.2.3 is a consistent asymptotic test.

124. Let $X_1, ..., X_n$ be i.i.d. from the gamma distribution $\Gamma(\theta, \gamma)$, where $\theta > 0$ and $\gamma > 0$ are unknown. Let $T_n = n \sum_{i=1}^n X_i^2 / (\sum_{i=1}^n X_i)^2$. Show how to use T_n to obtain an asymptotically correct test for $H_0 : \theta = 1$ versus $H_1 : \theta \neq 1$.

Chapter 7

Confidence Sets

Various methods of constructing confidence sets are introduced in this chapter, along with studies of properties of confidence sets. Throughout this chapter $X = (X_1, ..., X_n)$ denotes a sample from a population $P \in \mathcal{P}$; $\theta = \theta(P)$ denotes a functional from $\mathcal{P}$ to $\Theta \subset \mathcal{R}^k$ for a fixed integer k; and $C(X)$ denotes a *confidence set* for θ, a set in $\mathcal{B}_\Theta$ (the class of Borel sets on Θ) depending only on X. We adopt the basic concepts of confidence sets introduced in §2.4.3. In particular, $\inf_{P \in \mathcal{P}} P(\theta \in C(X))$ is the confidence coefficient of $C(X)$ and, if the confidence coefficient of $C(X)$ is $\geq 1 - \alpha$ for fixed $\alpha \in (0, 1)$, then we say that $C(X)$ has significance level $1 - \alpha$ or $C(X)$ is a level $1 - \alpha$ confidence set.

7.1 Construction of Confidence Sets

In this section, we introduce some basic methods for constructing confidence sets that have a given significance level (or confidence coefficient) for any fixed n. Properties and comparisons of confidence sets are given in §7.2.

7.1.1 Pivotal quantities

Perhaps the most popular method of constructing confidence sets is the use of pivotal quantities defined as follows.

Definition 7.1. A known Borel function $\Re$ of (X, θ) is called a *pivotal quantity* if and only if the distribution of $\Re(X, \theta)$ does not depend on P. ∎

Note that a pivotal quantity depends on P through $\theta = \theta(P)$. A pivotal quantity is usually not a statistic, although its distribution is known.

With a pivotal quantity $\Re(X, \theta)$, a level $1 - \alpha$ confidence set for any given $\alpha \in (0, 1)$ can be obtained as follows. First, find two constants c_1 and c_2 such that

$$P(c_1 \leq \Re(X, \theta) \leq c_2) \geq 1 - \alpha. \tag{7.1}$$

Next, define

$$C(X) = \{\theta \in \Theta : c_1 \leq \Re(X, \theta) \leq c_2\}. \tag{7.2}$$

Then $C(X)$ is a level $1 - \alpha$ confidence set, since

$$
\begin{aligned}
\inf_{P \in \mathcal{P}} P\big(\theta \in C(X)\big) &= \inf_{P \in \mathcal{P}} P\big(c_1 \leq \Re(X, \theta) \leq c_2\big) \\
&= P\big(c_1 \leq \Re(X, \theta) \leq c_2\big) \\
&\geq 1 - \alpha.
\end{aligned}
$$

Note that the confidence coefficient of $C(X)$ may not be $1 - \alpha$. If $\Re(X, \theta)$ has a continuous c.d.f., then we can choose c_i's such that the equality in (7.1) holds and, therefore, the confidence set $C(X)$ has confidence coefficient $1 - \alpha$.

In a given problem, there may not exist any pivotal quantity, or there may be many different pivotal quantities. When there are many pivotal quantities, one has to choose one based on some principles or criteria, which are discussed in §7.2. For example, pivotal quantities based on sufficient statistics are certainly preferred. In many cases we also have to choose c_i's in (7.1) based on some criteria.

When $\Re(X, \theta)$ and c_i's are chosen, we need to compute the confidence set $C(X)$ in (7.2). This can be done by inverting $c_1 \leq \Re(X, \theta) \leq c_2$. For example, if θ is real-valued and $\Re(X, \theta)$ is monotone in θ when X is fixed, then $C(X) = \{\theta : \underline{\theta}(X) \leq \theta \leq \bar{\theta}(X)\}$ for some $\underline{\theta}(X) < \bar{\theta}(X)$, i.e., $C(X)$ is an interval (finite or infinite); if $\Re(X, \theta)$ is not monotone, then $C(X)$ may be a union of several intervals. For real-valued θ, a confidence interval rather than a complex set such as a union of several intervals is generally preferred since it is simple and the result is easy to interpret. When θ is multivariate, inverting $c_1 \leq \Re(X, \theta) \leq c_2$ may be complicated. In most cases where explicit forms of $C(X)$ do not exist, $C(X)$ can still be obtained numerically.

Example 7.1 (Location-scale families). Suppose that $X_1, ..., X_n$ are i.i.d. with a Lebesgue p.d.f. $\frac{1}{\sigma} f\left(\frac{x - \mu}{\sigma}\right)$, where $\mu \in \mathcal{R}$, $\sigma > 0$, and f is a known Lebesgue p.d.f.

Consider first the case where σ is known and $\theta = \mu$. For any fixed i, $X_i - \mu$ is a pivotal quantity. Also, $\bar{X} - \mu$ is a pivotal quantity, since any function of independent pivotal quantities is pivotal. In many cases $\bar{X} - \mu$ is preferred. Let c_1 and c_2 be constants such that $P(c_1 \leq \bar{X} - \mu \leq c_2) = 1 - \alpha$.

Then $C(X)$ in (7.2) is

$$C(X) = \{\mu : c_1 \leq \bar{X} - \mu \leq c_2\} = \{\mu : \bar{X} - c_2 \leq \mu \leq \bar{X} - c_1\},$$

i.e., $C(X)$ is the interval $[\bar{X} - c_2, \bar{X} - c_1] \subset \mathcal{R} = \Theta$. This interval has confidence coefficient $1 - \alpha$. The choice of c_i's is not unique. Some criteria discussed in §7.2 can be applied to choose c_i's. One particular choice (not necessarily the best choice) frequently used by practitioners is $c_1 = -c_2$. The resulting $C(X)$ is symmetric about $\bar{X}$ and is also an *equal-tail* confidence interval (a confidence interval $[\underline{\theta}, \bar{\theta}]$ is equal-tail if $P(\theta < \underline{\theta}) = P(\theta > \bar{\theta})$) if the distribution of $\bar{X}$ is symmetric about μ. Note that the confidence interval in Example 2.31 is a special case of the intervals considered here.

Consider next the case where μ is known and $\theta = \sigma$. The following quantities are pivotal: $(X_i - \mu)/\sigma$, $i = 1, ..., n$, $\prod_{i=1}^{n}(X_i - \mu)/\sigma$, $(\bar{X} - \mu)/\sigma$, and S/σ, where S^2 is the sample variance. Consider the confidence set (7.2) with $\mathcal{R} = S/\sigma$. Let c_1 and c_2 be chosen such that $P(c_1 \leq S/\sigma \leq c_2) = 1 - \alpha$. If both c_i's are positive, then

$$C(X) = \{\sigma : S/c_2 \leq \sigma \leq S/c_1\} = [S/c_2, S/c_1]$$

is a finite interval. Similarly, if $c_1 = 0$ $(0 < c_2 < \infty)$ or $c_2 = \infty$ $(0 < c_1 < \infty)$, then $C(X) = [S/c_2, \infty)$ or $(0, S/c_1]$.

When $\theta = \sigma$ and μ is also unknown, S/σ is still a pivotal quantity and, hence, confidence intervals of σ based on S are still valid. Note that $(\bar{X} - \mu)/\sigma$ and $\prod_{i=1}^{n}(X_i - \mu)/\sigma$ are not pivotal when μ is unknown.

Finally, we consider the case where both μ and σ are unknown and $\theta = \mu$. There are still many different pivotal quantities, but the most commonly used pivotal quantity is $t(X) = \sqrt{n}(\bar{X} - \mu)/S$. The distribution of $t(X)$ does not depend on (μ, σ). When f is normal, $t(X)$ has the t-distribution t_{n-1}. The pivotal quantity $t(X)$ is often called a studentized statistic or t-statistic, although $t(X)$ is not a statistic and $t(X)$ does not have a t-distribution when f is not normal. A confidence interval for μ based on $t(X)$ is of the form

$$\{\mu : c_1 \leq \sqrt{n}(\bar{X} - \mu)/S \leq c_2\} = [\bar{X} - c_2 S/\sqrt{n}, \bar{X} - c_1 S/\sqrt{n}],$$

where c_i's are chosen so that $P(c_1 \leq t(X) \leq c_2) = 1 - \alpha$. ∎

Example 7.2. Let $X_1, ..., X_n$ be i.i.d. random variables from the uniform distribution $U(0, \theta)$. Consider the problem of finding a confidence set for θ. Note that the family $\mathcal{P}$ in this case is a scale family so that the results in Example 7.1 can be used. But a better confidence interval can be obtained based on the sufficient and complete statistic $X_{(n)}$ for which $X_{(n)}/\theta$ is a pivotal quantity (Example 7.13). Note that $X_{(n)}/\theta$ has the Lebesgue p.d.f.

$nx^{n-1}I_{(0,1)}(x)$. Hence c_i's in (7.1) should satisfy $c_2^n - c_1^n = 1 - \alpha$. The resulting confidence interval for θ is $[c_2^{-1}X_{(n)}, c_1^{-1}X_{(n)}]$. Choices of c_i's are discussed in Example 7.13. ∎

Example 7.3 (Fieller's interval). Let (X_{i1}, X_{i2}), $i = 1, ..., n$, be i.i.d. bivariate normal with unknown $\mu_j = E(X_{1j})$, $\sigma_j^2 = \mathrm{Var}(X_{1j})$, $j = 1, 2$, and $\sigma_{12} = \mathrm{Cov}(X_{11}, X_{12})$. Let $\theta = \mu_2/\mu_1$ be the parameter of interest ($\mu_1 \neq 0$). Define $Y_i(\theta) = X_{i2} - \theta X_{i1}$. Then $Y_1(\theta), ..., Y_n(\theta)$ are i.i.d. from $N(0, \sigma_2^2 - 2\theta\sigma_{12} + \theta^2\sigma_1^2)$. Let

$$S^2(\theta) = \frac{1}{n-1} \sum_{i=1}^{n} [Y_i(\theta) - \bar{Y}(\theta)]^2 = S_2^2 - 2\theta S_{12} + \theta^2 S_1^2,$$

where $\bar{Y}(\theta)$ is the average of $Y_i(\theta)$'s and S_i^2 and S_{12} are sample variances and covariance based on X_{ij}'s. It follows from Examples 1.16 and 2.18 that $\sqrt{n}\bar{Y}(\theta)/S(\theta)$ has the t-distribution t_{n-1} and, therefore, is a pivotal quantity. Let $t_{n-1,\alpha}$ be the $(1 - \alpha)$th quantile of the t-distribution t_{n-1}. Then

$$C(X) = \{\theta : n[\bar{Y}(\theta)]^2/S^2(\theta) \leq t_{n-1,\alpha/2}^2\}$$

is a confidence set for θ with confidence coefficient $1 - \alpha$. Note that $n[\bar{Y}(\theta)]^2 = t_{n-1,\alpha/2}^2 S^2(\theta)$ defines a parabola in θ. Depending on the roots of the parabola, $C(X)$ can be a finite interval, the complement of a finite interval, or the whole real line (exercise). ∎

Example 7.4. Consider the normal linear model $X = N_n(Z\beta, \sigma^2 I_n)$, where $\theta = \beta$ is a p-vector of unknown parameters and Z is a known $n \times p$ matrix of full rank. A pivotal quantity is

$$\Re(X, \beta) = \frac{(\hat{\beta} - \beta)^\tau Z^\tau Z(\hat{\beta} - \beta)/p}{\|X - Z\hat{\beta}\|^2/(n-p)},$$

where $\hat{\beta}$ is the LSE of β. By Theorem 3.8 and Example 1.16, $\Re(X, \beta)$ has the F-distribution $F_{p,n-p}$. We can then obtain a confidence set

$$C(X) = \{\beta : c_1 \leq \Re(X, \beta) \leq c_2\}.$$

Note that $\{\beta : \Re(X, \beta) < c\}$ is the interior of an ellipsoid in $\mathcal{R}^p$. ∎

The following result indicates that in many problems, there exist pivotal quantities.

Proposition 7.1. Let $T(X) = (T_1(X), ..., T_s(X))$ and $T_1, ..., T_s$ be independent statistics. Suppose that each T_i has a continuous c.d.f. $F_{T_i,\theta}$ indexed by θ. Then $\Re(X, \theta) = \prod_{i=1}^{s} F_{T_i,\theta}(T_i(X))$ is a pivotal quantity.

Proof. The result follows from the fact that $F_{T_i,\theta}(T_i)$'s are i.i.d. from the uniform distribution $U(0,1)$. ∎

When $X_1, ..., X_n$ are i.i.d. from a parametric family indexed by θ, the simplest way to apply Proposition 7.1 is to take $T(X) = X$. However, the resulting pivotal quantity may not be the best pivotal quantity. For instance, the pivotal quantity in Example 7.2 is a function of the one obtained by applying Proposition 7.1 with $T(X) = X_{(n)}$ ($s = 1$), which is better than the one obtained by using $T(X) = X$ (Example 7.13).

The result in Proposition 7.1 holds even when P is in a nonparametric family, but in a nonparametric problem, it may be difficult to find a statistic T whose c.d.f. is indexed by θ, the parameter vector of interest.

When θ and T in Proposition 7.1 are real-valued, we can use the following result to construct confidence intervals for θ even when the c.d.f. of T is not continuous.

Theorem 7.1. Suppose that P is in a parametric family indexed by a real-valued θ. Let $T(X)$ be a real-valued statistic with c.d.f. $F_{T,\theta}(t)$ and let α_1 and α_2 be fixed positive constants such that $\alpha_1 + \alpha_2 = \alpha < \frac{1}{2}$.
(i) Suppose that $F_{T,\theta}(t)$ and $F_{T,\theta}(t-)$ are nonincreasing in θ for each fixed t. Define

$$\bar\theta = \sup\{\theta : F_{T,\theta}(T) \geq \alpha_1\} \qquad \text{and} \qquad \underline\theta = \inf\{\theta : F_{T,\theta}(T-) \leq 1 - \alpha_2\}.$$

Then $[\underline\theta(T), \bar\theta(T)]$ is a level $1 - \alpha$ confidence interval for θ.
(ii) If $F_{T,\theta}(t)$ and $F_{T,\theta}(t-)$ are nondecreasing in θ for each t, then the same result holds with

$$\underline\theta = \inf\{\theta : F_{T,\theta}(T) \geq \alpha_1\} \qquad \text{and} \qquad \bar\theta = \sup\{\theta : F_{T,\theta}(T-) \leq 1 - \alpha_2\}.$$

(iii) If $F_{T,\theta}$ is a continuous c.d.f. for any θ, then $F_{T,\theta}(T)$ is a pivotal quantity and the confidence interval in (i) or (ii) has confidence coefficient $1 - \alpha$.
Proof. We only need to prove (i). Under the given condition, $\theta > \bar\theta$ implies $F_{T,\theta}(T) < \alpha_1$ and $\theta < \underline\theta$ implies $F_{T,\theta}(T-) > 1 - \alpha_2$. Hence,

$$P(\underline\theta \leq \theta \leq \bar\theta) \geq 1 - P(F_{T,\theta}(T) < \alpha_1) - P(F_{T,\theta}(T-) > 1 - \alpha_2).$$

The result follows from

$$P(F_{T,\theta}(T) < \alpha_1) \leq \alpha_1 \quad \text{and} \quad P(F_{T,\theta}(T-) > 1 - \alpha_2) \leq \alpha_2. \tag{7.3}$$

The proof of (7.3) is left as an exercise. ∎

When the parametric family in Theorem 7.1 has monotone likelihood ratio in $T(X)$, it follows from Lemma 6.3 that the condition in Theorem 7.1(i) holds; in fact, it follows from Exercise 2 in §6.6 that $F_{T,\theta}(t)$ is strictly decreasing for any t at which $0 < F_{T,\theta}(t) < 1$. If $F_{T,\theta}(t)$ is also continuous in θ, $\lim_{\theta \to \theta_-} F_{T,\theta}(t) > \alpha_1$, and $\lim_{\theta \to \theta_+} F_{T,\theta}(t) < \alpha_1$, where θ_- and θ_+ are the two ends of the parameter space, then $\bar{\theta}$ is the unique solution of $F_{T,\theta}(t) = \alpha_1$. A similar conclusion can be drawn for $\underline{\theta}$.

Theorem 7.1 can be applied to obtain the confidence interval for θ in Example 7.2 (exercise). The following example concerns a discrete $F_{T,\theta}$.

Example 7.5. Let $X_1, ..., X_n$ be i.i.d. random variables from the Poisson distribution $P(\theta)$ with an unknown $\theta > 0$ and $T(X) = \sum_{i=1}^{n} X_i$. Note that T is sufficient and complete for θ and has the Poisson distribution $P(n\theta)$. Thus,

$$F_{T,\theta}(t) = \sum_{j=0}^{t} \frac{e^{-n\theta}(n\theta)^j}{j!}, \qquad t = 0, 1, 2, ...$$

Since the Poisson family has monotone likelihood ratio in T and $0 < F_{T,\theta}(t) < 1$ for any t, $F_{T,\theta}(t)$ is strictly decreasing in θ. Also, $F_{T,\theta}(t)$ is continuous in θ and $F_{T,\theta}(t)$ tends to 1 and 0 as θ tends to 0 and ∞, respectively. Thus, Theorem 7.1 applies and $\bar{\theta}$ is the unique solution of $F_{T,\theta}(T) = \alpha_1$. Since $F_{T,\theta}(t-) = F_{T,\theta}(t-1)$ for $t > 0$, $\underline{\theta}$ is the unique solution of $F_{T,\theta}(t-1) = 1 - \alpha_2$ when $T = t > 0$ and $\underline{\theta} = 0$ when $T = 0$. In fact, in this case explicit forms of $\underline{\theta}$ and $\bar{\theta}$ can be obtained from

$$\frac{1}{\Gamma(t)} \int_\lambda^\infty x^{t-1} e^{-x} dx = \sum_{j=0}^{t-1} \frac{e^{-\lambda} \lambda^j}{j!}, \qquad t = 1, 2, ...$$

Using this equality, it can be shown (exercise) that

$$\bar{\theta} = (2n)^{-1} \chi^2_{2(T+1), 1-\alpha_1} \qquad \text{and} \qquad \underline{\theta} = (2n)^{-1} \chi^2_{2T, \alpha_2}, \qquad (7.4)$$

where $\chi^2_{r,\alpha}$ is the $(1 - \alpha)$th quantile of the chi-square distribution χ^2_r and $\chi^2_{0,\alpha}$ is defined to be 0. ∎

So far we have considered examples for parametric problems. In a non-parametric problem, a pivotal quantity may not exist and we have to consider approximate pivotal quantities (§7.3 and §7.4). The following is an example of a nonparametric problem in which there exist pivotal quantities.

Example 7.6. Let $X_1, ..., X_n$ be i.i.d. random variables from $F \in \mathcal{F}$ containing all continuous and symmetric distributions on $\mathcal{R}$. Suppose that F is symmetric about θ. Let $\tilde{R}(\theta)$ be the vector of ranks of $|X_i - \theta|$'s and $R_+(\theta)$ be the subvector of $\tilde{R}(\theta)$ containing ranks corresponding to positive $(X_i - \theta)$'s. Then, any real-valued Borel function of $R_+(\theta)$ is a pivotal quantity (see the discussion in §6.5.1). Various confidence sets can be constructed using these pivotal quantities. More details can be found in Example 7.10. ∎

7.1.2 Inverting acceptance regions of tests

Another popular method of constructing confidence sets is to use a close relationship between confidence sets and hypothesis tests. For any test T, the set $\{x : T(x) \neq 1\}$ is called the *acceptance region*. Note that this terminology is not precise when T is a randomized test.

Theorem 7.2. For each $\theta_0 \in \Theta$, let T_{θ_0} be a test for $H_0 : \theta = \theta_0$ (versus some H_1) with significance level α and acceptance region $A(\theta_0)$. For each x in the range of X, define

$$C(x) = \{\theta : x \in A(\theta)\}.$$

Then $C(X)$ is a level $1 - \alpha$ confidence set for θ. If T_{θ_0} is nonrandomized and has size α for every θ_0, then $C(X)$ has confidence coefficient $1 - \alpha$.
Proof. We prove the first assertion only. The proof for the second assertion is similar. Under the given condition,

$$\sup_{\theta=\theta_0} P\big(X \notin A(\theta_0)\big) = \sup_{\theta=\theta_0} P(T_{\theta_0} = 1) \leq \alpha,$$

which is the same as

$$1 - \alpha \leq \inf_{\theta=\theta_0} P\big(X \in A(\theta_0)\big) = \inf_{\theta=\theta_0} P\big(\theta_0 \in C(X)\big).$$

Since this holds for all θ_0, the result follows from

$$\inf_{P \in \mathcal{P}} P\big(\theta \in C(X)\big) = \inf_{\theta_0 \in \Theta} \inf_{\theta=\theta_0} P\big(\theta_0 \in C(X)\big) \geq 1 - \alpha. \quad \blacksquare$$

The converse of Theorem 7.2 is partially true, which is stated in the next result whose proof is left as an exercise.

Proposition 7.2. Let $C(X)$ be a confidence set for θ with significance level (or confidence coefficient) $1 - \alpha$. For any $\theta_0 \in \Theta$, define a region $A(\theta_0) = \{x : \theta_0 \in C(x)\}$. Then the test $T(X) = 1 - I_{A(\theta_0)}(X)$ has significance level α for testing $H_0 : \theta = \theta_0$ versus some H_1. $\quad \blacksquare$

In general, $C(X)$ in Theorem 7.2 can be determined numerically, if it does not have an explicit form. Theorem 7.2 can be best illustrated in the case where θ is real-valued and $A(\theta) = \{Y : a(\theta) \leq Y \leq b(\theta)\}$ for a real-valued statistic $Y(X)$ and some nondecreasing functions $a(\theta)$ and $b(\theta)$. When we observe $Y = y$, $C(X)$ is an interval with limits $\underline{\theta}$ and $\overline{\theta}$, which are the θ-values at which the horizontal line $Y = y$ intersects the curves $Y = b(\theta)$ and $Y = a(\theta)$ (Figure 7.1), respectively. If $y = b(\theta)$ (or $y = a(\theta)$) has no solution or more than one solution, $\underline{\theta} = \inf\{\theta : y \leq b(\theta)\}$

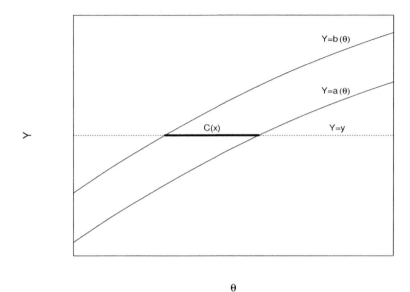

Figure 7.1: A confidence interval obtained by inverting $A(\theta) = [a(\theta), b(\theta)]$

(or $\bar{\theta} = \sup\{\theta : a(\theta) \leq y\}$). $C(X)$ does not include $\underline{\theta}$ (or $\bar{\theta}$) if and only if at $\underline{\theta}$ (or $\bar{\theta}$), $b(\theta)$ (or $a(\theta)$) is only left-continuous (or right-continuous).

Example 7.7. Suppose that X has the following p.d.f. in a one-parameter exponential family: $f_\theta(x) = \exp\{\eta(\theta)Y(x) - \xi(\theta)\}h(x)$, where θ is real-valued and $\eta(\theta)$ is nondecreasing in θ. First, we apply Theorem 7.2 with $H_0 : \theta = \theta_0$ and $H_1 : \theta > \theta_0$. By Theorem 6.2, the acceptance region of the UMP test of size α given by (6.11) is $A(\theta_0) = \{x : Y(x) \leq c(\theta_0)\}$, where $c(\theta_0) = c$ in (6.11). It can be shown (exercise) that $c(\theta)$ is nondecreasing in θ. Inverting $A(\theta)$ according to Figure 7.1 with $b(\theta) = c(\theta)$ and $a(\theta)$ ignored, we obtain $C(X) = [\underline{\theta}(X), \infty)$ or $(\underline{\theta}(X), \infty)$, a one-sided confidence interval for θ with significance level $1 - \alpha$. ($\underline{\theta}(X)$ is a called a lower confidence bound for θ in §2.4.3.) When the c.d.f. of $Y(X)$ is continuous, $C(X)$ has confidence coefficient $1 - \alpha$.

In the previous derivation, if $H_0 : \theta = \theta_0$ and $H_1 : \theta < \theta_0$ are considered, then $C(X) = \{\theta : Y(X) \geq c(\theta)\}$ and is of the form $(-\infty, \bar{\theta}(X)]$ or $(-\infty, \bar{\theta}(X))$. ($\bar{\theta}(X)$ is called an upper confidence bound for θ.)

Consider next $H_0 : \theta = \theta_0$ and $H_1 : \theta \neq \theta_0$. By Theorem 6.4, the acceptance region of the UMPU test of size α defined in (6.28) is given by $A(\theta_0) = \{x : c_1(\theta_0) \leq Y(x) \leq c_2(\theta_0)\}$, where $c_i(\theta)$ are nondecreasing (exercise). A confidence interval can be obtained by inverting $A(\theta)$ according to Figure 7.1 with $a(\theta) = c_1(\theta)$ and $b(\theta) = c_2(\theta)$.

Let us consider a specific example in which $X_1, ..., X_n$ are i.i.d. binary random variables with $p = P(X_i = 1)$. Note that $Y(X) = \sum_{i=1}^{n} X_i$. Suppose that we need a lower confidence bound for p so that we consider $H_0 : p = p_0$ and $H_1 : p > p_0$. From Example 6.2, the acceptance region of a UMP test of size $\alpha \in (0, 1)$ is $A(p_0) = \{y : y \le m(p_0)\}$, where $m(p_0)$ is an integer between 0 and n such that

$$\sum_{j=m(p_0)+1}^{n} \binom{n}{j} p_0^j (1 - p_0)^{n-j} \le \alpha < \sum_{j=m(p_0)}^{n} \binom{n}{j} p_0^j (1 - p_0)^{n-j}.$$

Thus, $m(p)$ is an integer-valued, nondecreasing step-function of p. Define

$$\underline{p} = \inf\{p : m(p) \ge y\} = \inf \left\{ p : \sum_{j=y}^{n} \binom{n}{j} p^j (1 - p)^{n-j} \ge \alpha \right\}. \qquad (7.5)$$

Then a level $1 - \alpha$ confidence interval for p is $(\underline{p}, 1]$ (exercise). One can compare this confidence interval with the one obtained by applying Theorem 7.1 (exercise). See also Example 7.16.   ∎

Example 7.8. Suppose that X has the following p.d.f. in a multiparameter exponential family: $f_{\theta,\varphi}(x) = \exp\{\theta Y(x) + \varphi^\tau U(x) - \zeta(\theta, \varphi)\}$. By Theorem 6.4, the acceptance region of a UMPU test of size α for testing $H_0 : \theta = \theta_0$ versus $H_1 : \theta > \theta_0$ or $H_0 : \theta = \theta_0$ versus $H_1 : \theta \ne \theta_0$ is

$$A(\theta_0) = \{(y, u) : y \le c_2(u, \theta_0)\}$$

or

$$A(\theta_0) = \{(y, u) : c_1(u, \theta_0) \le y \le c_2(u, \theta_0)\},$$

where $c_i(u, \theta)$, $i = 1, 2$, are nondecreasing functions of θ. Confidence intervals for θ can then be obtained by inverting $A(\theta)$ according to Figure 7.1 with $b(\theta) = c_2(u, \theta)$ and $a(\theta) = c_1(u, \theta)$ or $a(\theta) \equiv -\infty$, for any observed u.

Consider more specifically the case where X_1 and X_2 are independently distributed as the Poisson distributions $P(\lambda_1)$ and $P(\lambda_2)$, respectively, and we need a lower confidence bound for the ratio $\rho = \lambda_2/\lambda_1$. From Example 6.11, a UMPU test of size α for testing $H_0 : \rho = \rho_0$ versus $H_1 : \rho > \rho_0$ has the acceptance region $A(\rho_0) = \{(y, u) : y \le c(u, \rho_0)\}$, where $c(u, \rho_0)$ is determined by the conditional distribution of $Y = X_2$ given $U = X_1 + X_2 = u$. Since the conditional distribution of Y given $U = u$ is the binomial distribution $Bi(\rho/(1 + \rho), u)$, we can use the result in Example 7.7, i.e., $c(u, \rho)$ is the same as $m(p)$ in Example 7.7 with $n = u$ and $p = \rho/(1 + \rho)$. Then a level $1 - \alpha$ lower confidence bound for p is $\underline{p}$ given by (7.5) with $n = u$. Since $\rho = p/(1 - p)$ is a strictly increasing function of p, a level $1 - \alpha$ lower confidence bound for ρ is $\underline{p}/(1 - \underline{p})$.   ∎

Example 7.9. Consider the normal linear model $X = N_n(Z\beta, \sigma^2 I_n)$ and the problem of constructing a confidence set for $\theta = L\beta$, where L is an $s \times p$ matrix of rank s and all rows of L are in $\mathcal{R}(Z)$. It follows from the discussion in §6.3.2 and Exercise 74 in §6.6 that a nonrandomized UMPI test of size α for $H_0 : \theta = \theta_0$ versus $H_1 : \theta \neq \theta_0$ has the acceptance region

$$A(\theta_0) = \{X : W(X, \theta_0) \leq c_\alpha\},$$

where c_α is the $(1 - \alpha)$th quantile of the F-distribution $F_{s,n-r}$,

$$W(X, \theta) = \frac{[\|X - Z\hat{\beta}(\theta)\|^2 - \|X - Z\hat{\beta}\|^2]/s}{\|X - Z\hat{\beta}\|^2/(n-r)},$$

r is the rank of Z, $r \geq s$, $\hat{\beta}$ is the LSE of β and, for each fixed θ, $\hat{\beta}(\theta)$ is a solution of

$$\|X - Z\hat{\beta}(\theta)\|^2 = \min_{\beta:L\beta=\theta} \|X - Z\beta\|^2.$$

Inverting $A(\theta)$, we obtain the following confidence set for θ with confidence coefficient $1-\alpha$: $C(X) = \{\theta : W(X, \theta) \leq c_\alpha\}$, which forms a closed ellipsoid in $\mathcal{R}^s$. ∎

The last example concerns inverting the acceptance regions of tests in a nonparametric problem.

Example 7.10. Consider the problem in Example 7.6. We now derive a confidence interval for θ by inverting the acceptance regions of the signed rank tests given by (6.84). Note that testing whether the c.d.f. of X_i is symmetric about θ is equivalent to testing whether the c.d.f. of $X_i - \theta$ is symmetric about 0. Let c_i's be given by (6.84), W be given by (6.83), and, for each θ, let $R_+^o(\theta)$ be the vector of ordered components of $R_+(\theta)$ described in Example 7.6. A level $1 - \alpha$ confidence set for θ is

$$C(X) = \{\theta : c_1 \leq W(R_+^o(\theta)) \leq c_2\}.$$

The region $C(X)$ can be computed numerically for any observed X. From the discussion in Example 7.6, $W(R_+^o(\theta))$ is a pivotal quantity and, therefore, $C(X)$ is the same as the confidence set obtained by using a pivotal quantity. ∎

7.1.3 The Bayesian approach

In Bayesian analysis, analogues to confidence sets are called *credible sets*. Consider a sample X from a population in a parametric family indexed by

$\theta \in \Theta \subset \mathcal{R}^k$ and dominated by a σ-finite measure. Let $f_\theta(x)$ be the p.d.f. of X and $\pi(\theta)$ be a prior p.d.f. w.r.t. a σ-finite measure λ on $(\Theta, \mathcal{B}_\Theta)$. Let

$$p_x(\theta) = f_\theta(x)\pi(\theta)/m(x)$$

be the posterior p.d.f. w.r.t. λ, where x is the observed X and $m(x) = \int_\Theta f_\theta(x)\pi(\theta)d\lambda$. For any $\alpha \in (0,1)$, a level $1 - \alpha$ credible set for θ is any $C \in \mathcal{B}_\Theta$ with

$$P_{\theta|x}(\theta \in C) = \int_C p_x(\theta)d\lambda \geq 1 - \alpha. \tag{7.6}$$

A level $1 - \alpha$ *highest posterior density* (HPD) credible set for θ is defined to be the event

$$C(x) = \{\theta : p_x(\theta) \geq c_\alpha\}, \tag{7.7}$$

where c_α is chosen so that $\int_{C(x)} p_x(\theta)d\lambda \geq 1 - \alpha$. When $p_x(\theta)$ has a continuous c.d.f., we can replace $\geq$ in (7.6) and (7.7) by $=$. An HPD credible set is often an interval with the shortest length among all credible intervals of the same level (Exercise 40).

The Bayesian credible sets and the confidence sets we have discussed so far very different in terms of their meanings and interpretations, although sometimes they look similar. In a credible set, x is fixed and θ is considered random and the probability statement in (7.6) is w.r.t. the posterior probability $P_{\theta|x}$. On the other hand, in a confidence set θ is nonrandom (although unknown) but X is considered random, and the significance level is w.r.t. $P(\theta \in C(X))$, the probability related to the distribution of X. The set $C(X)$ in (7.7) is not necessarily a confidence set with significance level $1 - \alpha$.

When $\pi(\theta)$ is constant, which is usually an improper prior, the HPD credible set $C(x)$ in (7.7) is related to the idea of maximizing likelihood (a non-Bayesian approach introduced in §4.4; see also §7.3.2), since $p_x(\theta) = f_\theta(x)/m(x)$ is proportional to $f_\theta(x) = \ell(\theta)$, the likelihood function. In such a case $C(X)$ may be a confidence set with significance level $1 - \alpha$.

Example 7.11. Let $X_1, ..., X_n$ be i.i.d. as $N(\theta, \sigma^2)$ with an unknown $\theta \in \mathcal{R}$ and a known σ^2. Let $\pi(\theta)$ be the p.d.f. of $N(\mu_0, \sigma_0^2)$ with known μ_0 and σ_0^2. Then, $p_x(\theta)$ is the p.d.f. of $N(\mu_*(x), c^2)$ (Example 2.25), where $\mu_*(x)$ and c^2 are given by (2.25), and the HPD credible set in (7.7) is

$$C(x) = \left\{\theta : e^{-[\theta - \mu_*(x)]^2/(2c^2)} \geq c_\alpha\sqrt{2\pi}c\right\}$$
$$= \left\{\theta : |\theta - \mu_*(x)| \leq \sqrt{2}c[-\log(c_\alpha\sqrt{2\pi}c)]^{1/2}\right\}.$$

Let Φ be the standard normal c.d.f. The quantity $\sqrt{2}c[-\log(c_\alpha\sqrt{2\pi}c)]^{1/2}$ must be $cz_{1-\alpha/2}$, where $z_a = \Phi^{-1}(a)$, since it is chosen so that $P_{\theta|x}(C(x)) =$

$1 - \alpha$ and $P_{\theta|x} = N(\mu_*(x), c^2)$. Therefore,

$$C(x) = [\mu_*(x) - cz_{1-\alpha/2}, \ \mu_*(x) + cz_{1-\alpha/2}].$$

If we let $\sigma_0^2 \to \infty$, which is equivalent to taking the Lebesgue measure as the (improper) prior, then $\mu_*(x) = \bar{x}$, $c^2 = \sigma^2/n$, and

$$C(x) = [\bar{x} - \sigma z_{1-\alpha/2}/\sqrt{n}, \ \bar{x} + \sigma z_{1-\alpha/2}/\sqrt{n}],$$

which is the same as the confidence interval in Example 2.31 for θ with confidence coefficient $1 - \alpha$. Although the Bayesian credible set coincides with the classical confidence interval, which is frequently the case when a noninformative prior is used, their interpretations are still different. ∎

More details about Bayesian credible sets can be found, for example, in Berger (1985, §4.3).

7.1.4 Prediction sets

In some problems the quantity of interest is the future (or unobserved) value of a random variable ξ. An inference procedure about a random quantity instead of an unknown nonrandom parameter is called *prediction*. If the distribution of ξ is known, then a level $1 - \alpha$ *prediction set* for ξ is any event C satisfying $P_\xi(\xi \in C) \geq 1 - \alpha$. In applications, however, the distribution of ξ is usually unknown.

Suppose that the distribution of ξ is related to the distribution of a sample X from which prediction will be made. For instance, $X = (X_1, ..., X_n)$ is the observed sample and $\xi = X_{n+1}$ is to be predicted, where $X_1, ..., X_n, X_{n+1}$ are i.i.d. random variables. A set $C(X)$ depending only on the sample X is said to be a level $1 - \alpha$ *prediction set* for ξ if

$$\inf_{P \in \mathcal{P}} P\big(\xi \in C(X)\big) \geq 1 - \alpha,$$

where P is the joint distribution of (ξ, X) and $\mathcal{P}$ contains all possible P.

Note that prediction sets are very similar and closely related to confidence sets. Hence, some methods for constructing confidence sets can be applied to obtained prediction sets. For example, if $\Re(X, \xi)$ is a pivotal quantity in the sense that its distribution does not depend on P, then a prediction set can be obtained by inverting $c_1 \leq \Re(X, \xi) \leq c_2$. The following example illustrates this idea.

Example 7.12. Many prediction problems encountered in practice can be formulated as follows. The variable ξ to be predicted is related to a vector-valued covariate ζ (called predictor) according to $E(\xi|\zeta) = \zeta^\tau \beta$, where β is a p-vector of unknown parameters. Suppose that at $\zeta = Z_i$, we observe $\xi = X_i$, $i = 1, ..., n$, and X_i's are independent. Based on $(X_1, Z_1), ..., (X_n, Z_n)$, we would like to construct a prediction set for the value of $\xi = X_0$ when $\zeta = Z_0 \in \mathcal{R}(Z)$, where Z is the $n \times p$ matrix whose ith row is the vector Z_i. The Z_i's are either fixed or random observations (in the latter case all probabilities and expectations given in the following discussion are conditional on $Z_0, Z_1, ..., Z_n$).

Assume further that $X = (X_1, ..., X_n) = N_n(Z\beta, \sigma^2 I_n)$ follows a normal linear model and is independent of $X_0 = N(Z_0^\tau \beta, \sigma^2)$. Let $\hat\beta$ be the LSE of β, $\hat\sigma^2 = \|X - Z\hat\beta\|^2/(n - r)$, and $\|Z_0\|_Z^2 = Z_0^\tau(Z^\tau Z)^- Z_0$, where r is the rank of Z. Then

$$\Re(X, X_0) = \frac{X_0 - Z_0^\tau \hat\beta}{\hat\sigma\sqrt{1 + \|Z_0\|_Z^2}}$$

has the t-distribution t_{n-r} and, therefore, is a pivotal quantity. This is because X_0 and $Z_0^\tau \hat\beta$ are independently normal,

$$E(X_0 - Z_0^\tau \hat\beta) = 0, \qquad \text{Var}(X_0 - Z_0^\tau \hat\beta) = \sigma^2(1 + \|Z_0\|_Z^2),$$

$(n - r)\hat\sigma^2$ has the chi-square distribution χ_{n-r}^2, and X_0, $Z_0^\tau \hat\beta$, and $\hat\sigma^2$ are independent (Theorem 3.8). A level $1 - \alpha$ prediction interval for X_0 is then

$$\left[Z_0^\tau \hat\beta - t_{n-r,\alpha/2}\hat\sigma\sqrt{1 + \|Z_0\|_Z^2}, \ Z_0^\tau \hat\beta + t_{n-r,\alpha/2}\hat\sigma\sqrt{1 + \|Z_0\|_Z^2} \right], \quad (7.8)$$

where $t_{n-r,\alpha}$ is the $(1 - \alpha)$th quantile of the t-distribution t_{n-r}.

To compare prediction sets with confidence sets, let us consider a confidence interval for $E(X_0) = Z_0^\tau \beta$. Using the pivotal quantity

$$\Re(X, Z_0^\tau \beta) = \frac{Z_0^\tau \beta - Z_0^\tau \hat\beta}{\hat\sigma\|Z_0\|_Z},$$

we obtain the following confidence interval for $Z_0^\tau \beta$ with confidence coefficient $1 - \alpha$:

$$\left[Z_0^\tau \hat\beta - t_{n-r,\alpha/2}\hat\sigma\|Z_0\|_Z, \ Z_0^\tau \hat\beta + t_{n-r,\alpha/2}\hat\sigma\|Z_0\|_Z \right]. \quad (7.9)$$

Since a random variable is more variable than its average (an unknown parameter), the prediction interval (7.8) is always longer than the confidence interval (7.9), although each of them covers the quantity of interest with probability $1 - \alpha$. In fact, when $\|Z_0\|_Z^2 \to 0$ as $n \to \infty$, the length of the confidence interval (7.9) tends to 0 a.s., whereas the length of the prediction interval (7.8) tends to a positive constant a.s. ∎

Because of the similarity between confidence sets and prediction sets, in the rest of this chapter we do not discuss prediction sets in detail. Some examples are given in Exercises 30 and 31.

7.2 Properties of Confidence Sets

In this section, we study some properties of confidence sets and introduce several criteria for comparing them.

7.2.1 Lengths of confidence intervals

For confidence intervals of a real-valued θ with the same confidence coefficient, an apparent measure of their performance is the interval length. Shorter confidence intervals are preferred, since they are more informative. In most problems, however, shortest-length confidence intervals do not exist. A common approach is to consider a reasonable class of confidence intervals (with the same confidence coefficient) and then find a confidence interval with the shortest length within the class.

When confidence intervals are constructed by using pivotal quantities or by inverting acceptance regions of tests, choosing a reasonable class of confidence intervals amounts to selecting good pivotal quantities or tests. Functions of sufficient statistics should be used, when sufficient statistics exist. In many problems pivotal quantities or tests are related to some point estimators of θ. For example, in a location family problem (Example 7.1), a confidence interval for $\theta = \mu$ is often of the form $[\hat{\theta} - c, \hat{\theta} + c]$, where $\hat{\theta}$ is an estimator of θ and c is a constant. In such a case a more accurate estimator of θ should intuitively result in a better confidence interval. For instance, when $X_1, ..., X_n$ are i.i.d. $N(\mu, 1)$, it can be shown (exercise) that the interval $[\bar{X} - c_1, \bar{X} + c_1]$ is better than the interval $[X_1 - c_2, X_1 + c_2]$ in terms of their lengths, where c_i's are chosen so that these confidence intervals have confidence coefficient $1 - \alpha$. However, we cannot have the same conclusion when X_i's are from the Cauchy distribution $C(\mu, 1)$ (Exercise 32). The following is another example.

Example 7.13. Let $X_1, ..., X_n$ be i.i.d. from the uniform distribution $U(0, \theta)$ with an unknown $\theta > 0$. A confidence interval for θ of the form $[b^{-1}X_{(n)}, a^{-1}X_{(n)}]$ is derived in Example 7.2, where a and b are constants chosen so that this confidence interval has confidence coefficient $1 - \alpha$. Another confidence interval obtained by applying Proposition 7.1 with $T = X$ is of the form $[b_1^{-1}\tilde{X}, a_1^{-1}\tilde{X}]$, where $\tilde{X} = (\prod_{i=1}^n X_i)^{1/n}$. We now argue that when n is large enough, the former has a shorter length than the latter. Note that $\sqrt{n}(\tilde{X} - \theta)/\theta \to_d N(0, 1)$. Thus,

$$P\left(\left(1 + \tfrac{d}{\sqrt{n}}\right)^{-1}\tilde{X} \leq \theta \leq \left(1 + \tfrac{c}{\sqrt{n}}\right)^{-1}\tilde{X}\right) = P\left(\tfrac{c}{\sqrt{n}} \leq \tfrac{\tilde{X}-\theta}{\theta} \leq \tfrac{d}{\sqrt{n}}\right) \to 1 - \alpha$$

for some constants c and d. This means that $a_1 \approx 1 + c/\sqrt{n}$, $b_1 \approx 1 + d/\sqrt{n}$, and the length of $[b_1^{-1}\tilde{X}, a_1^{-1}\tilde{X}]$ converges to 0 a.s. at the rate $n^{-1/2}$. On the other hand,

$$P\left(\left(1+\tfrac{d}{n}\right)^{-1}X_{(n)} \leq \theta \leq \left(1+\tfrac{c}{n}\right)^{-1}X_{(n)}\right) = P\left(\tfrac{c}{n} \leq \tfrac{X_{(n)}-\theta}{\theta} \leq \tfrac{d}{n}\right) \to 1-\alpha$$

for some constants c and d, since $n(X_{(n)} - \theta)/\theta$ has a known limiting distribution (Example 2.34). This means that the length of $[b^{-1}X_{(n)}, a^{-1}X_{(n)}]$ converges to 0 a.s. at the rate n^{-1} and, therefore, $[b^{-1}X_{(n)}, a^{-1}X_{(n)}]$ is shorter than $[b_1^{-1}\tilde{X}, a_1^{-1}\tilde{X}]$ for sufficiently large n a.s.

Similarly, one can show that the confidence interval based on the pivotal quantity $\bar{X}/\theta$ is not as good as $[b^{-1}X_{(n)}, a^{-1}X_{(n)}]$ in terms of their lengths.

Thus, it is reasonable to consider the class of confidence intervals of the form $[b^{-1}X_{(n)}, a^{-1}X_{(n)}]$ subject to $P(b^{-1}X_{(n)} \leq \theta \leq a^{-1}X_{(n)}) = 1-\alpha$. The shortest-length interval within this class can be derived as follows. Note that $X_{(n)}/\theta$ has the Lebesgue p.d.f. $nx^{n-1}I_{(0,1)}(x)$. Hence

$$1 - \alpha = P(b^{-1}X_{(n)} \leq \theta \leq a^{-1}X_{(n)}) = \int_a^b nx^{n-1}dx = b^n - a^n.$$

This implies that $1 \geq b > a \geq 0$ and $\frac{da}{db} = (\frac{b}{a})^{n-1}$. Since the length of the interval $[b^{-1}X_{(n)}, a^{-1}X_{(n)}]$ is $\psi(a,b) = X_{(n)}(a^{-1} - b^{-1})$,

$$\frac{d\psi}{db} = X_{(n)}\left(\frac{1}{b^2} - \frac{1}{a^2}\frac{da}{db}\right) = X_{(n)}\frac{a^{n+1} - b^{n+1}}{b^2 a^{n+1}} < 0.$$

Hence the minimum occurs at $b = 1$ $(a = \alpha^{1/n})$. This shows that the shortest-length interval is $[X_{(n)}, \alpha^{-1/n}X_{(n)}]$. ∎

As Example 7.13 indicates, once a reasonable class of confidence intervals is chosen (using some good estimators, pivotal quantities, or tests), we may find the shortest-length confidence interval within the class by directly analyzing the lengths of the intervals. For a large class of problems, the following result can be used.

Theorem 7.3. Let θ be a real-valued parameter and $T(X)$ be a real-valued statistic.
(i) Let $U(X)$ be a positive statistic. Suppose that $(T - \theta)/U$ is a pivotal quantity having a Lebesgue p.d.f. f that is *unimodal* at $x_0 \in \mathcal{R}$ in the sense that $f(x)$ is nondecreasing for $x \leq x_0$ and $f(x)$ is nonincreasing for $x \geq x_0$. Consider the following class of confidence intervals for θ:

$$\mathcal{C} = \left\{[T - bU, T - aU] : a \in \mathcal{R}, b \in \mathcal{R}, \int_a^b f(x)dx = 1 - \alpha\right\}. \quad (7.10)$$

If $[T - b_*U, T - a_*U] \in \mathcal{C}$, $f(a_*) = f(b_*) > 0$, and $a_* \leq x_0 \leq b_*$, then the interval $[T - b_*U, T - a_*U]$ has the shortest length within $\mathcal{C}$.

(ii) Suppose that $T > 0$, $\theta > 0$, T/θ is a pivotal quantity having a Lebesgue p.d.f. f, and that $x^2 f(x)$ is unimodal at x_0. Consider the following class of confidence intervals for θ:

$$\mathcal{C} = \left\{ [b^{-1}T, a^{-1}T] : a > 0, b > 0, \int_a^b f(x)dx = 1 - \alpha \right\}. \qquad (7.11)$$

If $[b_*^{-1}T, a_*^{-1}T] \in \mathcal{C}$, $a_*^2 f(a_*) = b_*^2 f(b_*) > 0$, and $a_* \leq x_0 \leq b_*$, then the interval $[b_*^{-1}T, a_*^{-1}T]$ has the shortest length within $\mathcal{C}$.

Proof. We prove (i) only. The proof of (ii) is left as an exercise. Note that the length of an interval in $\mathcal{C}$ is $(b - a)U$. Thus, it suffices to show that if $a < b$ and $b - a < b_* - a_*$, then $\int_a^b f(x)dx < 1 - \alpha$. Assume that $a < b$, $b - a < b_* - a_*$, and $a \leq a_*$ (the proof for $a > a_*$ is similar).

If $b \leq a_*$, then $a \leq b \leq a_* \leq x_0$ and

$$\int_a^b f(x)dx \leq f(a_*)(b - a) < f(a_*)(b_* - a_*) \leq \int_{a_*}^{b_*} f(x)dx = 1 - \alpha,$$

where the first inequality follows from the unimodality of f, the strict inequality follows from $b - a < b_* - a_*$ and $f(a_*) > 0$, and the last inequality follows from the unimodality of f and the fact that $f(a_*) = f(b_*)$.

If $b > a_*$, then $a \leq a_* < b < b_*$. By the unimodality of f,

$$\int_a^{a_*} f(x)dx \leq f(a_*)(a_* - a) \qquad \text{and} \qquad \int_b^{b_*} f(x)dx \geq f(b_*)(b_* - b).$$

Then

$$\int_a^b f(x)dx = \int_{a_*}^{b_*} f(x)dx + \int_a^{a_*} f(x)dx - \int_b^{b_*} f(x)dx$$

$$= 1 - \alpha + \int_a^{a_*} f(x)dx - \int_b^{b_*} f(x)dx$$

$$\leq 1 - \alpha + f(a_*)(a_* - a) - f(b_*)(b_* - b)$$

$$= 1 - \alpha + f(a_*)[(a_* - a) - (b_* - b)]$$

$$= 1 - \alpha + f(a_*)[(b - a) - (b_* - a_*)]$$

$$< 1 - \alpha. \quad \blacksquare$$

Example 7.14. Let $X_1, ..., X_n$ be i.i.d. from $N(\mu, \sigma^2)$ with unknown μ and σ^2. Confidence intervals for $\theta = \mu$ using the pivotal quantity $\sqrt{n}(\bar{X} - \mu)/S$ form the class $\mathcal{C}$ in (7.10) with f being the p.d.f. of the t-distribution t_{n-1}, which is unimodal at $x_0 = 0$. Hence, we can apply Theorem 7.3(i). Since f is symmetric about 0, $f(a_*) = f(b_*)$ implies $a_* = -b_*$ (exercise). Therefore, the equal-tail confidence interval

$$\left[\bar{X} - t_{n-1,\alpha/2} S/\sqrt{n}, \bar{X} + t_{n-1,\alpha/2} S/\sqrt{n} \right] \qquad (7.12)$$

has the shortest length within $\mathcal{C}$.

If $\theta = \mu$ and σ^2 is known, then we can replace S by σ and f by the standard normal p.d.f. (i.e., use the pivotal quantity $\sqrt{n}(\bar{X} - \mu)/\sigma$ instead of $\sqrt{n}(\bar{X} - \mu)/S$). The resulting confidence interval is

$$\left[\bar{X} - \Phi^{-1}(1 - \alpha/2)\sigma/\sqrt{n}, \bar{X} + \Phi^{-1}(1 - \alpha/2)\sigma/\sqrt{n} \right], \qquad (7.13)$$

which is the shortest interval of the form $[\bar{X} - b, \bar{X} - a]$ with confidence coefficient $1-\alpha$. The difference in length of the intervals in (7.12) and (7.13) is a random variable so that we cannot tell which one is better in general. But the expected length of the interval (7.13) is always shorter than that of the interval (7.12) (exercise). This again shows the importance of picking the right pivotal quantity.

Consider next confidence intervals for $\theta = \sigma^2$ using the pivotal quantity $(n - 1)S^2/\sigma^2$, which form the class $\mathcal{C}$ in (7.11) with f being the p.d.f. of the chi-square distribution χ_{n-1}^2. Note that $x^2 f(x)$ is unimodal, but not symmetric. By Theorem 7.3(ii), the shortest-length interval within $\mathcal{C}$ is

$$[b_*^{-1}(n - 1)S^2, a_*^{-1}(n - 1)S^2], \qquad (7.14)$$

where a_* and b_* are solutions of $a_*^2 f(a_*) = b_*^2 f(b_*)$ and $\int_{a_*}^{b_*} f(x)dx = 1 - \alpha$. Numerical values of a_* and b_* can be obtained (Tate and Klett, 1959). Note that this interval is not equal-tail.

If $\theta = \sigma^2$ and μ is known, then a better pivotal quantity is T/σ^2, where $T = \sum_{i=1}^{n}(X_i - \mu)^2$. One can show (exercise) that if we replace $(n-1)S^2$ by T and f by the p.d.f. of the chi-square distribution χ_n^2, then the resulting interval has shorter expected length than that of the interval in (7.14).

Suppose that we need a confidence interval for $\theta = \sigma$ when μ is unknown. Consider the class of confidence intervals

$$\left[b^{-1/2}\sqrt{n - 1}\, S, a^{-1/2}\sqrt{n - 1}\, S \right]$$

with $\int_a^b f(x)dx = 1 - \alpha$ and f being the p.d.f. of χ_{n-1}^2. The shortest-length interval, however, is not the one with the endpoints equal to the square roots of the endpoints of the interval (7.14) (Exercise 36(c)). ∎

Note that Theorem 7.3(ii) cannot be applied to obtain the result in Example 7.13 unless $n = 1$, since the p.d.f. of $X_{(n)}/\theta$ is strictly increasing when $n > 1$. A result similar to Theorem 7.3, which can be applied to Example 7.13, is given in Exercise 38.

The result in Theorem 7.3 can also be applied to justify the idea of HPD credible sets in Bayesian analysis (Exercise 40).

If a confidence interval has the shortest length within a class of confidence intervals, then its expected length is also the shortest within the same class, provided that its expected length is finite. In a problem where a shortest-length confidence interval does not exist, we may have to use the expected length as the criterion in comparing confidence intervals. For instance, the expected length of the interval in (7.13) is always shorter than that of the interval in (7.12), whereas the probability that the interval in (7.12) is shorter than the interval in (7.13) is positive for any fixed n. Another example is the interval $[X_{(n)}, \alpha^{-1/n} X_{(n)}]$ in Example 7.13. Although we are not able to say that this interval has the shortest length among all confidence intervals for θ with confidence coefficient $1 - \alpha$, we can show that it has the shortest expected length, using the results in Theorems 7.4 and 7.6 (§7.2.2).

For one-sided confidence intervals (confidence bounds) of a real-valued θ, their lengths may be infinity. We can use the distance between the confidence bound and θ as a criterion in comparing confidence bounds, which is equivalent to comparing the tightness of confidence bounds. Let $\underline{\theta}_j$, $j = 1, 2$, be two lower confidence bounds for θ with the same confidence coefficient. If $\underline{\theta}_1 - \theta \geq \underline{\theta}_2 - \theta$ is always true, then $\underline{\theta}_1 \geq \underline{\theta}_2$ and $\underline{\theta}_1$ is tighter (more informative) than $\underline{\theta}_2$. Again, since $\underline{\theta}_j$ are random, we may have to consider $E(\underline{\theta}_j - \theta)$ and choose $\underline{\theta}_1$ if $E(\underline{\theta}_1) \geq E(\underline{\theta}_2)$. As a specific example, consider i.i.d. $X_1, ..., X_n$ from $N(\theta, 1)$. If we use the pivotal quantity $\bar{X} - \mu$, then $\underline{\theta}_1 = \bar{X} - \Phi^{-1}(1 - \alpha)/\sqrt{n}$. If we use the pivotal quantity $X_1 - \mu$, then $\underline{\theta}_2 = X_1 - \Phi^{-1}(1 - \alpha)$. Clearly $E(\underline{\theta}_1) \geq E(\underline{\theta}_2)$. Although $\underline{\theta}_1$ is intuitively preferred, $\underline{\theta}_1 < \underline{\theta}_2$ with a positive probability for any fixed $n > 1$.

Some ideas discussed previously can be extended to the comparison of confidence sets for multivariate θ. For bounded confidence sets in $\mathcal{R}^k$, for example, we may consider their volumes (Lebesgue measures). However, in multivariate cases it is difficult to compare the volumes of confidence sets with different shapes. Some results about expected volumes of confidence sets are given in Theorem 7.6.

7.2.2 UMA and UMAU confidence sets

For a confidence set obtained by inverting the acceptance regions of some UMP or UMPU tests, it is expected that the confidence set inherits some optimality property.

Definition 7.2. Let $\theta \in \Theta$ be an unknown parameter and Θ' be a subset of Θ that does not contain the true parameter value θ. A confidence set $C(X)$ for θ with confidence coefficient $1 - \alpha$ is said to be Θ'-*uniformly most accurate* (UMA) if and only if for any other confidence set $C_1(X)$ with significance level $1 - \alpha$,

$$P\big(\theta' \in C(X)\big) \leq P\big(\theta' \in C_1(X)\big) \qquad \text{for all } \theta' \in \Theta'. \tag{7.15}$$

$C(X)$ is UMA if and only if it is Θ'-UMA with $\Theta' = \{\theta\}^c$. ∎

The probabilities in (7.15) are probabilities of covering false values. Intuitively, confidence sets with small probabilities of covering wrong parameter values are preferred. The reason why we sometimes need to consider a Θ' different from $\{\theta\}^c$ (the set containing all false values) is that for some confidence sets, such as one-sided confidence intervals, we do not need to worry about the probabilities of covering some false values. For example, if we consider a lower confidence bound for a real-valued θ, we are asserting that θ is larger than a certain value and we only need to worry about covering values of θ that are too small. Thus, $\Theta' = \{\theta' \in \Theta : \theta' < \theta\}$. A similar discussion leads to the consideration of $\Theta' = \{\theta' \in \Theta : \theta' > \theta\}$ for upper confidence bounds.

Theorem 7.4. Let $C(X)$ be a confidence set for θ obtained by inverting the acceptance regions of nonrandomized tests T_{θ_0} for testing $H_0 : \theta = \theta_0$ versus $H_1 : \theta \in \Theta_{\theta_0}$. Suppose that for each θ_0, T_{θ_0} is UMP of size α. Then $C(X)$ is Θ'-UMA with confidence coefficient $1 - \alpha$, where $\Theta' = \{\theta' : \theta \in \Theta_{\theta'}\}$.
Proof. The fact that $C(X)$ has confidence coefficient $1 - \alpha$ follows from Theorem 7.2. Let $C_1(X)$ be another confidence set with significance level $1 - \alpha$. By Proposition 7.2, the test $T_{1\theta_0}(X) = 1 - I_{A_1(\theta_0)}(X)$ with $A_1(\theta_0) = \{x : \theta_0 \in C_1(x)\}$ has significance level α for testing $H_0 : \theta = \theta_0$ versus $H_1 : \theta \in \Theta_{\theta_0}$. For any $\theta' \in \Theta'$, $\theta \in \Theta_{\theta'}$ and, hence, the population P is in the family defined by $H_1 : \theta \in \Theta_{\theta'}$. Thus,

$$\begin{aligned} P\big(\theta' \in C(X)\big) &= 1 - P\big(T_{\theta'}(X) = 1\big) \\ &\leq 1 - P\big(T_{1\theta'}(X) = 1\big) \\ &= P\big(\theta' \in C_1(X)\big), \end{aligned}$$

where the first equality follows from the fact that $T_{\theta'}$ is nonrandomized and the inequality follows from the fact that $T_{\theta'}$ is UMP. ∎

Theorem 7.4 can be applied to construct UMA confidence bounds in problems where the population is in a one-parameter parametric family with monotone likelihood ratio so that UMP tests exist (Theorem 6.2). It can also be applied to a few cases to construct two-sided UMA confidence intervals. For example, the confidence interval $[X_{(n)}, \alpha^{-1/n} X_{(n)}]$ in Example 7.13 is UMA (exercise).

As we discussed in §6.2, in many problems there are UMPU tests but not UMP tests. This leads to the following definition.

Definition 7.3. Let $\theta \in \Theta$ be an unknown parameter, Θ' be a subset of Θ that does not contain the true parameter value θ, and $1 - \alpha$ be a given significance level.

(i) A level $1 - \alpha$ confidence set $C(X)$ is said to be Θ'-*unbiased* (unbiased when $\Theta' = \{\theta\}^c$) if and only if $P(\theta' \in C(X)) \leq 1 - \alpha$ for all $\theta' \in \Theta'$.

(ii) Let $C(X)$ be a Θ'-unbiased confidence set with confidence coefficient $1 - \alpha$. If (7.15) holds for any other Θ'-unbiased confidence set $C_1(X)$ with significance level $1 - \alpha$, then $C(X)$ is Θ'-*uniformly most accurate unbiased* (UMAU). $C(X)$ is UMAU if and only if it is Θ'-UMAU with $\Theta' = \{\theta\}^c$.   ∎

Theorem 7.5. Let $C(X)$ be a confidence set for θ obtained by inverting the acceptance regions of nonrandomized tests T_{θ_0} for testing $H_0 : \theta = \theta_0$ versus $H_1 : \theta \in \Theta_{\theta_0}$. If T_{θ_0} is unbiased of size α for each θ_0, then $C(X)$ is Θ'-unbiased with confidence coefficient $1 - \alpha$, where $\Theta' = \{\theta' : \theta \in \Theta_{\theta'}\}$; if T_{θ_0} is also UMPU for each θ_0, then $C(X)$ is Θ'-UMAU.   ∎

The proof of Theorem 7.5 is very similar to that of Theorem 7.4.

It follows from Theorem 7.5 and the results in §6.2 that the confidence intervals in (7.12), (7.13), and (7.14) are UMAU, since they can be obtained by inverting acceptance regions of UMPU tests (Exercise 23).

Example 7.15. Consider the normal linear model in Example 7.9 and the parameter $\theta = l^\tau \beta$, where $l \in \mathcal{R}(Z)$. From §6.2.3, the nonrandomized test with acceptance region

$$A(\theta_0) = \left\{ X : l^\tau \hat{\beta} - \theta_0 > t_{n-r,\alpha} \sqrt{l^\tau (Z^\tau Z)^- l SSR/(n-r)} \right\}$$

is UMPU with size α for testing $H_0 : \theta = \theta_0$ versus $H_1 : \theta < \theta_0$, where $\hat{\beta}$ is the LSE of β and $t_{n-r,\alpha}$ is the $(1-\alpha)$th quantile of the t-distribution t_{n-r}. Inverting $A(\theta)$ we obtain the following Θ'-UMAU upper confidence bound with confidence coefficient $1 - \alpha$ and $\Theta' = (\theta, \infty)$:

$$\bar{\theta} = l^\tau \hat{\beta} - t_{n-r,\alpha} \sqrt{l^\tau (Z^\tau Z)^- l SSR/(n-r)}.$$

A UMAU confidence interval for θ can be similarly obtained.

If $\theta = L\beta$ with L described in Example 7.9 and $s > 1$, then θ is multivariate. It can be shown that the confidence set derived in Example 7.9 is unbiased (exercise), but it may not be UMAU.   ∎

The volume of a confidence set $C(X)$ for $\theta \in \mathcal{R}^k$ when $X = x$ is defined to be $\mathrm{vol}(C(x)) = \int_{C(x)} d\theta'$, which is the Lebesgue measure of the set $C(x)$ and may be infinite. In particular, if θ is real-valued and $C(X) = [\underline{\theta}(X), \overline{\theta}(X)]$ is a confidence interval, then $\mathrm{vol}(C(x))$ is simply the length of $C(x)$. The next result reveals a relationship between the expected volume (length) and the probability of covering a false value of a confidence set (interval).

Theorem 7.6 (Pratt's theorem). Let X be a sample from P and $C(X)$ be a confidence set for $\theta \in \mathcal{R}^k$. Suppose that $\text{vol}(C(x)) = \int_{C(x)} d\theta'$ is finite a.s. P. Then the expected volume of $C(X)$ is

$$E[\text{vol}(C(X))] = \int_{\theta \neq \theta'} P(\theta' \in C(X)) d\theta'. \qquad (7.16)$$

Proof. By Fubini's theorem,

$$E[\text{vol}(C(X))] = \int \text{vol}(C(X)) dP$$

$$= \int \left[\int_{C(x)} d\theta' \right] dP(x)$$

$$= \int \int_{\theta' \in C(x)} d\theta' dP(x)$$

$$= \int \left[\int_{\theta' \in C(x)} dP(x) \right] d\theta'$$

$$= \int P(\theta' \in C(X)) d\theta'$$

$$= \int_{\theta \neq \theta'} P(\theta' \in C(X)) d\theta'.$$

This proves the result. ∎

It follows from Theorem 7.6 that if $C(X)$ is UMA (or UMAU) with confidence coefficient $1 - \alpha$, then it has the smallest expected volume among all confidence sets (or all unbiased confidence sets) with significance level $1 - \alpha$. For example, the confidence interval (7.13) in Example 7.14 (when σ^2 is known) or $[X_{(n)}, \alpha^{-1/n} X_{(n)}]$ in Example 7.13 has the shortest expected length among all confidence intervals with significance level $1 - \alpha$; the confidence interval (7.12) or (7.14) has the shortest expected length among all unbiased confidence intervals with significance level $1 - \alpha$.

7.2.3 Randomized confidence sets

Applications of Theorems 7.4 and 7.5 require that $C(X)$ be obtained by inverting acceptance regions of nonrandomized tests. Thus, these results cannot be directly applied to discrete problems. In fact, in discrete problems inverting acceptance regions of randomized tests may not lead to a confidence set with a given confidence coefficient. Note that randomization is used in hypothesis testing to obtain tests with a given size. Thus, the same idea can be applied to confidence sets, i.e., we may consider *randomized* confidence sets.

Suppose that we invert acceptance regions of randomized tests T_{θ_0} that reject $H_0 : \theta = \theta_0$ with probability $T_{\theta_0}(x)$ when $X = x$. Let U be a random variable that is independent of X and has the uniform distribution $U(0, 1)$. Then the test $\tilde{T}_{\theta_0}(X, U) = I_{(U,1]}(T_{\theta_0})$ has the same power function as T_{θ_0} and is "nonrandomized" if U is viewed as part of the sample. Let

$$A_U(\theta_0) = \{(x, U) : U \geq T_{\theta_0}(x)\}$$

be the acceptance region of $\tilde{T}_{\theta_0}(X, U)$. If T_{θ_0} has size α for all θ_0, then inverting $A_U(\theta)$ we obtain a confidence set

$$C(X, U) = \{\theta : (X, U) \in A_U(\theta)\}$$

having confidence coefficient $1 - \alpha$, since

$$P(\theta \in C(X, U)) = E[P(U \geq T_\theta(X)|X)] = E[1 - T_\theta(X)].$$

If T_{θ_0} is UMP (or UMPU) for each θ_0, then $C(X, U)$ is UMA (or UMAU). However, $C(X, U)$ is a randomized confidence set since it is still random when we observe $X = x$.

When T_{θ_0} is a function of an integer-valued statistic, we can use the method in the following example to derive $C(X, U)$.

Example 7.16. Let $X_1, ..., X_n$ be i.i.d. binary random variables with $p = P(X_i = 1)$. The confidence coefficient of $(\underline{p}, 1]$ may not be $1 - \alpha$, where $\underline{p}$ is given by (7.5).

From Example 6.2 and the previous discussion, a randomized UMP test for testing $H_0 : p = p_0$ versus $H_1 : p > p_0$ can be constructed based on $Y = \sum_{i=1}^n X_i$ and U, a random variable that is independent of Y and has the uniform distribution $U(0, 1)$. Since Y is integer-valued and $U \in (0, 1)$, $W = Y + U$ is equivalent to (Y, U). It can be shown (exercise) that W has the following Lebesgue p.d.f.:

$$f_p(w) = \binom{n}{[w]} p^{[w]}(1 - p)^{n-[w]} I_{(0,n+1)}(w), \tag{7.17}$$

where $[w]$ is the integer part of w, and that the family $\{f_p : p \in (0, 1)\}$ has monotone likelihood ratio in W. It follows from Theorem 6.2 that the test $\tilde{T}_{p_0}(Y, U) = I_{(c(p_0),n+1)}(W)$ is UMP of size α for testing $H_0 : p = p_0$ versus $H_1 : p > p_0$, where $\alpha = \int_{c(p_0)}^{n+1} f_{p_0}(w)dw$. Since $\int_W^{n+1} f_p(w)dw$ is increasing in p (Lemma 6.3), inverting the acceptance regions of $\tilde{T}_p(Y, U)$ leads to $C(X, U) = \left\{ p : \int_W^{n+1} f_p(w)dw \geq \alpha \right\} = [\underline{p}_1, 1]$, where $\underline{p}_1$ is the solution of

$$\int_{Y+U}^{n+1} \binom{n}{[w]} p^{[w]}(1 - p)^{n-[w]} dw = \alpha.$$

$(\underline{p}_1 = 0$ if $Y = 0$ and $U < 1 - \alpha$; $\underline{p}_1 = 1$ if $Y = 1$ and $U > 1 - \alpha)$. The lower confidence bound $\underline{p}_1$ has confidence coefficient $1 - \alpha$ and is Θ'-UMA with $\Theta' = (0, p)$. ∎

Using a randomized confidence set, we can achieve the purpose of obtaining a confidence set with a given confidence coefficient as well as some optimality properties such as UMA, UMAU, or shortest expected length. On the other hand, randomization may not be desired in practical problems.

7.2.4 Invariant confidence sets

Let $C(X)$ be a confidence set for θ and g be a one-to-one transformation of X. The invariance principle requires that $C(x)$ change in a specified way when x is transformed to $g(x)$, where $x \in \mathcal{X}$ and $\mathcal{X}$ is the range of X.

Definition 7.4. Let $\mathcal{G}$ be a group of one-to-one transformations of X such that $\mathcal{P}$ is invariant under $\mathcal{G}$ (Definition 2.9). Let $\theta = \theta(P)$ be a parameter with range Θ. Assume that $\tilde{g}(\theta) = \theta(P_{g(X)})$ is well defined for any $g \in \mathcal{G}$, i.e., $\tilde{g}$ is a transformation on Θ induced by g ($\tilde{g} = \bar{g}$ given in Definition 2.9 if $\mathcal{P}$ is indexed by θ).
(i) A confidence set $C(X)$ is *invariant* under $\mathcal{G}$ if and only if $\theta \in C(x)$ is equivalent to $\tilde{g}(\theta) \in C(g(x))$ for every $x \in \mathcal{X}$, $\theta \in \Theta$, and $g \in \mathcal{G}$.
(ii) $C(X)$ is Θ'-*uniformly most accurate invariant* (UMAI) with confidence coefficient $1 - \alpha$ if and only if $C(X)$ is invariant with confidence coefficient $1 - \alpha$ and (7.15) holds for any other invariant confidence set $C_1(X)$ with significance level $1 - \alpha$. $C(X)$ is UMAI if and only if it is Θ'-UMAI with $\Theta' = \{\theta\}^c$. ∎

Example 7.17. Consider the confidence intervals in Example 7.14. Let $\mathcal{G} = \{g_{r,c} : r > 0, c \in \mathcal{R}\}$ with $g_{r,c}(x) = (rx_1 + c, ..., rx_n + c)$. Let $\theta = \mu$. Then $\bar{g}_{r,c}(\mu, \sigma^2) = (r\mu + c, r^2\sigma^2)$ and $\tilde{g}(\mu) = r\mu + c$. Clearly, confidence interval (7.12) is invariant under $\mathcal{G}$.

When σ^2 is known, the family $\mathcal{P}$ is not invariant under $\mathcal{G}$ and we consider $\mathcal{G}_1 = \{g_{1,c} : c \in \mathcal{R}\}$. Then both confidence intervals (7.12) and (7.13) are invariant under $\mathcal{G}_1$.

Suppose now that $\theta = \sigma^2$. For $g_{r,c} \in \mathcal{G}$, $\tilde{g}(\sigma^2) = r^2\sigma^2$. Hence confidence interval (7.14) is invariant under $\mathcal{G}$. ∎

If a confidence set $C(X)$ is UMA and invariant, then it is UMAI. If $C(X)$ is UMAU and invariant, it is not so obvious whether it is UMAI, since a UMAI confidence set (if it exists) is not necessarily unbiased. The following result may be used to construct a UMAI confidence set.

Theorem 7.7. Suppose that for each $\theta_0 \in \Theta$, $A(\theta_0)$ is the acceptance region of a nonrandomized UMPI test of size α for $H_0 : \theta = \theta_0$ versus $H_1 :$ $\theta \in \Theta_{\theta_0}$ under $\mathcal{G}_{\theta_0}$ and that for any θ_0 and $g \in \mathcal{G}_{\theta_0}$, $\tilde{g}$, the transformation on Θ induced by g, is well defined. If $C(X) = \{\theta : x \in A(\theta)\}$ is invariant under $\mathcal{G}$, the smallest group containing $\cup_{\theta \in \Theta} \mathcal{G}_\theta$, then it is Θ'-UMAI with confidence coefficient $1 - \alpha$, where $\Theta' = \{\theta' : \theta \in \Theta_{\theta'}\}$. ∎

The proofs of Theorem 7.7 and the following result are given as exercises.

Proposition 7.3. Let $\mathcal{P}$ be a parametric family indexed by θ and $\mathcal{G}$ be a group of transformations such that $\tilde{g}$ is well defined by $P_{\tilde{g}(\theta)} = P_{g(X)}$. Suppose that, for any θ, $\theta' \in \Theta$, there is a $g \in \mathcal{G}$ such that $\tilde{g}(\theta) = \theta'$. Then, for any invariant confidence set $C(X)$, $P(\theta \in C(X))$ is a constant. ∎

Example 7.18. Let $X_1, ..., X_n$ be i.i.d. from $N(\mu, \sigma^2)$ with unknown μ and σ^2. Consider the problem of setting a lower confidence bound for $\theta = \mu/\sigma$ and $\mathcal{G} = \{g_r : r > 0\}$ with $g_r(x) = rx$. From Example 6.17, a nonrandomized UMPI test of size α for $H_0 : \theta = \theta_0$ versus $H_1 : \theta > \theta_0$ has the acceptance region $A(\theta_0) = \{x : t(x) \leq c(\theta_0)\}$, where $t(X) = \sqrt{n}\bar{X}/S$ and $c(\theta)$ is the $(1-\alpha)$th quantile of the noncentral t-distribution $t_{n-1}(\sqrt{n}\theta)$. Applying Theorem 7.7 with $\mathcal{G}_{\theta_0} = \mathcal{G}$ for all θ_0, one can show (exercise) that the solution of $\int_{t(x)}^{\infty} f_\theta(u)du = \alpha$ is a Θ'-UMAI lower confidence bound for θ with confidence coefficient $1 - \alpha$, where f_θ is the Lebesgue p.d.f. of the noncentral t-distribution $t_{n-1}(\sqrt{n}\theta)$ and $\Theta' = (-\infty, \theta)$. ∎

Example 7.19. Consider again the confidence intervals in Example 7.14. In Example 7.17, confidence interval (7.12) is shown to be invariant under $\mathcal{G} = \{g_{r,c} : r > 0, c \in \mathcal{R}\}$ with $g_{r,c}(x) = (rx_1 + c, ..., rx_n + c)$. Although confidence interval (7.12) is UMAU, it is not obvious whether it is UMAI. This interval can be obtained by inverting $A(\mu_0) = \{x : |\bar{X} - \mu_0| \leq t_{n-1,\alpha/2}S/\sqrt{n}\}$, which is the acceptance region of a nonrandomized test UMP among unbiased and invariant tests of size α for $H_0 : \mu = \mu_0$ versus $H_1 : \mu \neq \mu_0$, under $\mathcal{G}_{\mu_0} = \{h_{r,\mu_0} : r > 0\}$ with $h_{r,\mu_0}(x) = (r(x_1 - \mu_0) + \mu_0, ..., r(x_n - \mu_0) + \mu_0)$ (exercise). Note that the testing problem $H_0 : \mu = \mu_0$ versus $H_1 : \mu \neq \mu_0$ is not invariant under $\mathcal{G}$. Since $\mathcal{G}$ is the smallest group containing $\cup_{\mu_0 \in \mathcal{R}} \mathcal{G}_{\mu_0}$ (exercise), by Theorem 7.7, interval (7.12) is UMA among unbiased and invariant confidence intervals with confidence coefficient $1 - \alpha$, under $\mathcal{G}$.

Using similar arguments one can show (exercise) that confidence intervals (7.13) and (7.14) are UMA among unbiased and invariant confidence intervals with confidence coefficient $1 - \alpha$, under $\mathcal{G}_1$ (in Example 7.17) and $\mathcal{G}$, respectively. ∎

When UMPI tests are randomized, one can construct randomized UMAI confidence sets, using the techniques introduced in Theorem 7.7 and §7.2.3.

7.3 Asymptotic Confidence Sets

In some problems, especially in nonparametric problems, it is difficult to find a reasonable confidence set with a given confidence coefficient or significance level $1 - \alpha$. A common approach is to find a confidence set whose confidence coefficient or significance level is nearly $1 - \alpha$ when the sample size n is large. A confidence set $C(X)$ for θ has asymptotic significance level $1 - \alpha$ if $\liminf_n P(\theta \in C(X)) \geq 1 - \alpha$ for any $P \in \mathcal{P}$ (Definition 2.14). If $\lim_{n \to \infty} P(\theta \in C(X)) = 1 - \alpha$ for any $P \in \mathcal{P}$, then $C(X)$ is a $1 - \alpha$ *asymptotically correct* confidence set. Note that asymptotic correctness is not the same as having limiting confidence coefficient $1 - \alpha$ (Definition 2.14).

7.3.1 Asymptotically pivotal quantities

A known Borel function of (X, θ), $\Re_n(X, \theta)$, is said to be *asymptotically pivotal* if and only if the limiting distribution of $\Re_n(X, \theta)$ does not depend on P. Like a pivotal quantity in constructing confidence sets (§7.1.1) with a given confidence coefficient or significance level, an asymptotically pivotal quantity can be used in constructing asymptotically correct confidence sets.

Most asymptotically pivotal quantities are of the form $\hat{V}_n^{-1/2}(\hat{\theta}_n - \theta)$, where $\hat{\theta}_n$ is an estimator of θ that is asymptotically normal, i.e.,

$$V_n^{-1/2}(\hat{\theta}_n - \theta) \to_d N_k(0, I_k), \tag{7.18}$$

and $\hat{V}_n$ is an estimator of the asymptotic covariance matrix V_n and is consistent according to Definition 5.4. The resulting $1 - \alpha$ asymptotically correct confidence sets are of the form

$$C(X) = \{\theta : \|\hat{V}_n^{-1/2}(\hat{\theta}_n - \theta)\|^2 \leq \chi_{k,\alpha}^2\}, \tag{7.19}$$

where $\chi_{k,\alpha}^2$ is the $(1 - \alpha)$th quantile of the chi-square distribution χ_k^2. If θ is real-valued ($k = 1$), then $C(X)$ in (7.19) is a $1 - \alpha$ asymptotically correct confidence interval. When $k > 1$, $C(X)$ in (7.19) is an ellipsoid.

Example 7.20 (Functions of means). Suppose that $X_1, ..., X_n$ are i.i.d. random vectors having a c.d.f. F on $\mathcal{R}^d$ and that the unknown parameter of interest is $\theta = g(\mu)$, where $\mu = E(X_1)$ and g is a known differentiable function from $\mathcal{R}^d$ to $\mathcal{R}^k$, $k \leq d$. From the CLT, Theorem 1.12, and the result in §5.5.1, (7.18) holds with $\hat{\theta}_n = g(\bar{X})$ and $\hat{V}_n$ given by (5.108). Thus, $C(X)$ in (7.19) is a $1 - \alpha$ asymptotically correct confidence set for θ. ∎

Example 7.21 (Statistical functionals). Suppose that $X_1, ..., X_n$ are i.i.d. random vectors having a c.d.f. F on $\mathcal{R}^d$ and that the unknown parameter of interest is $\theta = T(F)$, where T is a k-vector-valued functional. Let F_n be the empirical c.d.f. defined by (5.1) and $\hat{\theta}_n = T(F_n)$. Suppose that each component of T is ϱ_∞-Hadamard differentiable with an influence function satisfying (5.40) and that the conditions in Theorem 5.15 hold. Then, by Theorems 5.5 and 5.15 and the discussions in §5.2.1, (7.18) holds with $\hat{V}_n$ given by (5.110) and $C(X)$ in (7.19) is a $1 - \alpha$ asymptotically correct confidence set for θ. ∎

Example 7.22 (Linear models). Consider linear model (3.25): $X = Z\beta + \varepsilon$, where ε has i.i.d. components with mean 0 and variance σ^2. Assume that Z is of full rank and that the conditions in Theorem 3.12 hold. It follows from Theorem 1.9(iii) and Theorem 3.12 that (7.18) holds with $\hat{\theta}_n = \hat{\beta}$ and $\hat{V}_n = (n - p)^{-1} SSR(Z^\tau Z)^{-1}$ (see §5.5.1). Thus, a $1 - \alpha$ asymptotically correct confidence set for β is

$$C(X) = \{\beta : (\hat{\beta} - \beta)^\tau (Z^\tau Z)(\hat{\beta} - \beta) \le \chi^2_{p,\alpha} SSR/(n - p)\}.$$

Note that this confidence set is different from the one in Example 7.9 derived under the normality assumption on ε. ∎

The problems in the previous three examples are nonparametric. The method of using asymptotically pivotal quantities can also be applied to parametric problems. Note that in a parametric problem where the unknown parameter θ is multivariate, a confidence set for θ with a given confidence coefficient may be difficult or impossible to obtain.

Typically, in a given problem there exist many different asymptotically pivotal quantities that lead to different $1 - \alpha$ asymptotically correct confidence sets for θ. Intuitively, if two asymptotic confidence sets are constructed using (7.18) with two different estimators, $\hat{\theta}_{1n}$ and $\hat{\theta}_{2n}$, and if $\hat{\theta}_{1n}$ is asymptotically more efficient than $\hat{\theta}_{2n}$ (§4.5.1), then the confidence set based on $\hat{\theta}_{1n}$ should be better than the one based on $\hat{\theta}_{2n}$ in some sense. This is formally stated in the following result.

Proposition 7.4. Let $C_j(X)$, $j = 1, 2$, be the confidence sets given in (7.19) with $\hat{\theta}_n = \hat{\theta}_{jn}$ and $\hat{V}_n = \hat{V}_{jn}$, $j = 1, 2$, respectively. Suppose that for each j, (7.18) holds for $\hat{\theta}_{jn}$ and $\hat{V}_{jn}$ is consistent for V_{jn}, the asymptotic covariance matrix of $\hat{\theta}_{jn}$. If $\text{Det}(V_{1n}) < \text{Det}(V_{2n})$ for sufficiently large n, where $\text{Det}(A)$ is the determinant of A, then

$$P\big(\text{vol}(C_1(X)) < \text{vol}(C_2(X))\big) \to 1.$$

Proof. The result follows from the consistency of $\hat{V}_{jn}$ and the fact that the volume of the ellipsoid $C(X)$ defined by (7.19) is equal to

$$\text{vol}(C(X)) = \frac{\pi^{k/2}(\chi^2_{k,\alpha})^{k/2}[\text{Det}(\hat{V}_n)]^{1/2}}{\Gamma(1 + k/2)}. \quad ∎$$

If $\hat{\theta}_{1n}$ is asymptotically more efficient than $\hat{\theta}_{2n}$ (§4.5.1), then $\text{Det}(V_{1n}) \leq \text{Det}(V_{2n})$. Hence, Proposition 7.4 indicates that a more efficient estimator of θ results in a better confidence set of the form (7.19) in terms of volume. If $\hat{\theta}_n$ is asymptotically efficient (optimal in the sense of having the smallest asymptotic covariance matrix; see Definition 4.4), then the confidence set $C(X)$ in (7.19) is asymptotically optimal (in terms of volume) among the confidence sets of the form (7.19).

Asymptotically correct confidence sets for θ can also be constructed by inverting acceptance regions of asymptotic tests for testing $H_0 : \theta = \theta_0$ versus some H_1. If asymptotic tests are constructed using asymptotically pivotal quantities (see §6.4.2, §6.4.3, and §6.5.4), the resulting confidence sets are almost the same as those based on asymptotically pivotal quantities.

7.3.2 Confidence sets based on likelihoods

As we discussed in §7.3.1, a $1 - \alpha$ asymptotically correct confidence set is asymptotically optimal in some sense if it is based on an asymptotically efficient point estimator. In parametric problems, it is shown in §4.5 that MLE's or RLE's are asymptotically efficient. Thus, in this section we study more closely the asymptotic confidence sets based on MLE's and RLE's or, more generally, based on likelihoods.

Consider the case where $\mathcal{P} = \{P_\theta : \theta \in \Theta\}$ is a parametric family dominated by a σ-finite measure, where $\Theta \subset \mathcal{R}^k$. For convenience, we consider $\theta = (\vartheta, \varphi)$ and confidence sets for ϑ with dimension r. Let $\ell(\theta)$ be the likelihood function based on the observation $X = x$. The acceptance region of the LR test defined in §6.4.1 with $\Theta_0 = \{\theta : \vartheta = \vartheta_0\}$ is

$$A(\vartheta_0) = \{x : \ell(\vartheta_0, \hat{\varphi}_{\vartheta_0}) \geq e^{-c_\alpha/2}\ell(\hat{\theta})\},$$

where $\ell(\hat{\theta}) = \sup_{\theta \in \Theta} \ell(\theta)$, $\ell(\vartheta, \hat{\varphi}_\vartheta) = \sup_\varphi \ell(\vartheta, \varphi)$, and c_α is a constant related to the significance level α. Under the conditions of Theorem 6.5, if c_α is chosen to be $\chi^2_{r,\alpha}$, the $(1-\alpha)$th quantile of the chi-square distribution χ^2_r, then

$$C(X) = \{\vartheta : \ell(\vartheta, \hat{\varphi}_\vartheta) \geq e^{-c_\alpha/2}\ell(\hat{\theta})\} \qquad (7.20)$$

is a $1 - \alpha$ asymptotically correct confidence set. Note that this confidence set and the one given by (7.19) are generally different.

In many cases $-\ell(\vartheta, \varphi)$ is a convex function of ϑ and, therefore, the set defined by (7.20) is a bounded set in $\mathcal{R}^k$; in particular, $C(X)$ in (7.20) is a bounded interval when $k = 1$.

In §6.4.2 we discussed two asymptotic tests closely related to the LR test: Wald's test and Rao's score test. When $\Theta_0 = \{\theta : \vartheta = \vartheta_0\}$, Wald's test has acceptance region

$$A(\vartheta_0) = \{x : (\hat{\vartheta} - \vartheta_0)^\tau \{C^\tau[I_n(\hat{\theta})]^{-1}C\}^{-1}(\hat{\vartheta} - \vartheta_0) \le \chi^2_{r,\alpha}\}, \qquad (7.21)$$

where $\hat{\theta} = (\hat{\vartheta}, \hat{\varphi})$ is an MLE or RLE of $\theta = (\vartheta, \varphi)$, $I_n(\theta)$ is the Fisher information matrix based on X, $C^\tau = (\ I_r\ 0\ )$, and 0 is an $r \times (k - r)$ matrix of 0's. By Theorem 4.17 or 4.18, the confidence set obtained by inverting $A(\vartheta)$ in (7.21) is the same as that in (7.19) with $\theta = \vartheta$ and $\hat{V}_n = C^\tau[I_n(\hat{\theta})]^{-1}C$.

When $\Theta_0 = \{\theta : \vartheta = \vartheta_0\}$, Rao's score test has acceptance region

$$A(\vartheta_0) = \{x : [s_n(\vartheta_0, \hat{\varphi}_{\vartheta_0})]^\tau [I_n(\vartheta_0, \hat{\varphi}_{\vartheta_0})]^{-1} s_n(\vartheta_0, \hat{\varphi}_{\vartheta_0}) \le \chi^2_{r,\alpha}\}, \qquad (7.22)$$

where $s_n(\theta) = \partial \log \ell(\theta)/\partial\theta$ and $\hat{\varphi}_\vartheta$ is defined in (7.20). The confidence set obtained by inverting $A(\vartheta)$ in (7.22) is also $1 - \alpha$ asymptotically correct. To illustrate these likelihood-based confidence sets and their differences, we consider the following two examples.

Example 7.23. Let $X_1, ..., X_n$ be i.i.d. binary random variables with $p = P(X_i = 1)$. Since confidence sets for p with a given confidence coefficient are usually randomized (§7.2.3), asymptotically correct confidence sets may be considered when n is large.

The likelihood ratio for testing $H_0 : p = p_0$ versus $H_1 : p \ne p_0$ is

$$\lambda(Y) = p_0^Y(1 - p_0)^{n-Y}/\hat{p}^Y(1 - \hat{p})^{n-Y},$$

where $Y = \sum_{i=1}^n X_i$ and $\hat{p} = Y/n$ is the MLE of p. The confidence set (7.20) is then equal to

$$C_1(X) = \{p : p^Y(1 - p)^{n-Y} \ge e^{-c_\alpha/2}\hat{p}^Y(1 - \hat{p})^{n-Y}\}.$$

When $0 < Y < n$, $-p^Y(1-p)^{n-Y}$ is strictly convex and equals 0 if $p = 0$ or 1 and, hence, $C_1(X) = [\underline{p}, \overline{p}]$ with $0 < \underline{p} < \overline{p} < 1$. When $Y = 0$, $(1 - p)^n$ is strictly decreasing and, therefore, $C_1(X) = (0, \overline{p}]$ with $0 < \overline{p} < 1$. Similarly, when $Y = n$, $C_1(X) = [\underline{p}, 1)$ with $0 < \underline{p} < 1$.

The confidence set obtained by inverting acceptance regions of Wald's tests is simply

$$C_2(X) = [\hat{p} - z_{1-\alpha/2}\sqrt{\hat{p}(1 - \hat{p})/n},\ \hat{p} + z_{1-\alpha/2}\sqrt{\hat{p}(1 - \hat{p})/n}\,],$$

since $I_n(p) = n/[p(1 - p)]$ and $(\chi^2_{1,\alpha})^{1/2} = z_{1-\alpha/2}$, the $(1 - \alpha/2)$th quantile of $N(0, 1)$. Note that

$$s_n(p) = \frac{Y}{p} - \frac{n - Y}{1 - p} = \frac{Y - pn}{p(1 - p)}$$

and

$$[s_n(p)]^2[I_n(p)]^{-1} = \frac{(Y - pn)^2}{p^2(1-p)^2}\frac{p(1-p)}{n} = \frac{n(\hat{p} - p)^2}{p(1-p)}.$$

Hence, the confidence set obtained by inverting acceptance regions of Rao's score tests is

$$C_3(X) = \{p : n(\hat{p} - p)^2 \leq p(1-p)\chi_{1,\alpha}^2\}.$$

It can be shown (exercise) that $C_3(X) = [p_-, p_+]$ with

$$p_{\pm} = \frac{2Y + \chi_{1,\alpha}^2 \pm \sqrt{\chi_{1,\alpha}^2[4n\hat{p}(1-\hat{p}) + \chi_{1,\alpha}^2]}}{2(n + \chi_{1,\alpha}^2)}. \quad \blacksquare$$

Example 7.24. Let $X_1, ..., X_n$ be i.i.d. from $N(\mu, \varphi)$ with unknown $\theta = (\mu, \varphi)$. Consider the problem of constructing a $1 - \alpha$ asymptotically correct confidence set for θ. The log-likelihood function is

$$\log \ell(\theta) = -\frac{1}{2\varphi}\sum_{i=1}^{n}(X_i - \mu)^2 - \frac{n}{2}\log\varphi - \frac{n}{2}\log(2\pi).$$

Since $(\bar{X}, \hat{\varphi})$ is the MLE of θ, where $\hat{\varphi} = (n-1)S^2/n$, the confidence set based on LR tests is

$$C_1(X) = \left\{\theta : \frac{1}{\varphi}\sum_{i=1}^{n}(X_i - \mu)^2 + n\log\varphi \leq \chi_{2,\alpha}^2 + n + n\log\hat{\varphi}\right\}.$$

Note that

$$s_n(\theta) = \left(\frac{n(\bar{X} - \mu)}{\varphi}, \frac{1}{2\varphi^2}\sum_{i=1}^{n}(X_i - \mu)^2 - \frac{n}{2\varphi}\right)$$

and

$$I_n(\theta) = \begin{pmatrix} \frac{n}{\varphi} & 0 \\ 0 & \frac{n}{2\varphi^2} \end{pmatrix}.$$

Hence, the confidence set based on Wald's tests is

$$C_2(X) = \left\{\theta : \frac{(\bar{X} - \mu)^2}{\hat{\varphi}} + \frac{(\hat{\varphi} - \varphi)^2}{2\hat{\varphi}^2} \leq \frac{\chi_{2,\alpha}^2}{n}\right\},$$

which is an ellipsoid in $\mathcal{R}^2$, and the confidence set based on Rao's score tests is

$$C_3(X) = \left\{\theta : \frac{(\bar{X} - \mu)^2}{\varphi} + \frac{1}{2}\left[\frac{1}{n\varphi}\sum_{i=1}^{n}(X_i - \mu)^2 - 1\right]^2 \leq \frac{\chi_{2,\alpha}^2}{n}\right\}.$$

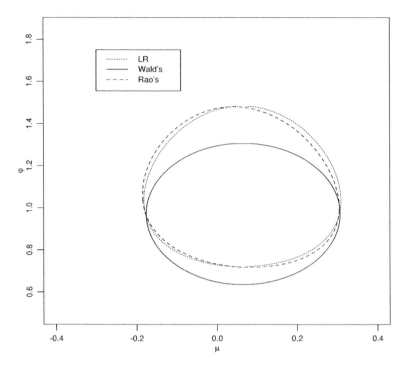

Figure 7.2: Confidence sets obtained by inverting LR, Wald's, and Rao's score tests in Example 7.24

In general, $C_j(X)$, $j = 1, 2, 3$, are different. An example of these three confidence sets is given in Figure 7.2, where $n = 100$, $\mu = 0$, and $\varphi = 1$.

Consider now the construction of a confidence set for μ. It can be shown (exercise) that the confidence set based on Wald's tests is defined by $C_2(X)$ with φ replaced by $\hat{\varphi}$, whereas the confidence sets based on LR tests and Rao's score tests are defined by $C_1(X)$ and $C_3(X)$, respectively, with φ replaced by $n^{-1} \sum_{i=1}^{n} (X_i - \mu)^2$. ∎

In nonparametric problems, asymptotic confidence sets can be obtained by inverting acceptance regions of empirical likelihood ratio tests or profile empirical likelihood ratio tests (§6.5.3). We consider the following problem as an example. Let $X_1, ..., X_n$ be i.i.d. from F and $\theta = (\vartheta, \varphi)$ be a k-vector of unknown parameters defined by $E[\psi(X_1, \theta)] = 0$, where ψ is a known function. Using the empirical likelihood

$$\ell(G) = \prod_{i=1}^{n} p_i \quad \text{subject to} \quad p_i \geq 0, \ \sum_{i=1}^{n} p_i = 1, \ \sum_{i=1}^{n} p_i \psi(x_i, \theta) = 0,$$

we can obtain a confidence set for ϑ by inverting acceptance regions of the profile empirical likelihood ratio tests based on the ratio $\lambda_n(X)$ in (6.91). This leads to the confidence set defined by

$$C(X) = \left\{ \vartheta : \prod_{i=1}^{n} \frac{1 + [\xi_n(\hat{\theta})]^\tau \psi(x_i, \hat{\theta})}{1 + [\zeta_n(\vartheta, \hat{\varphi})]^\tau \psi(x_i, \vartheta, \hat{\varphi})} \geq e^{-\chi^2_{r,\alpha}/2} \right\}, \qquad (7.23)$$

where the notation is the same as that in (6.91) and $\chi^2_{r,\alpha}$ is the $(1 - \alpha)$th quantile of the chi-square distribution χ^2_r with $r = $ the dimension of ϑ. By Theorem 6.11, this confidence set is $1 - \alpha$ asymptotically correct. Inverting the function of ϑ in (7.23) may be complicated, but $C(X)$ can usually be obtained numerically. More discussions about confidence sets based on empirical likelihoods can be found in Owen (1988, 1990, 2001), Chen and Qin (1993), Qin (1993), and Qin and Lawless (1994).

7.3.3 Confidence intervals for quantiles

Let $X_1, ..., X_n$ be i.i.d. from a continuous c.d.f. F on $\mathcal{R}$ and let $\theta = F^{-1}(p)$ be the pth quantile of F, $0 < p < 1$. The general methods we previously discussed can be applied to obtain a confidence set for θ, but we introduce here a method that works especially for quantile problems.

In fact, for any given α, it is possible to derive a confidence interval (or bound) for θ with confidence coefficient $1 - \alpha$ (Exercise 84), but the numerical computation of such a confidence interval may be cumbersome. We focus on asymptotic confidence intervals for θ. Our result is based on the following result due to Bahadur (1966). Its proof is omitted.

Theorem 7.8. Let $X_1, ..., X_n$ be i.i.d. from a continuous c.d.f. F on $\mathcal{R}$ that is twice differentiable at $\theta = F^{-1}(p)$, $0 < p < 1$, with $F'(\theta) > 0$. Let $\{k_n\}$ be a sequence of integers satisfying $1 \leq k_n \leq n$ and $k_n/n = p + o((\log n)^\delta/\sqrt{n})$ for some $\delta > 0$. Let F_n be the empirical c.d.f. defined in (5.1). Then

$$X_{(k_n)} = \theta + \frac{(k_n/n) - F_n(\theta)}{F'(\theta)} + O\left(\frac{(\log n)^{(1+\delta)/2}}{n^{3/4}} \right) \quad \text{a.s.} \quad \blacksquare$$

The result in Theorem 7.8 is a refinement of the Bahadur representation in Theorem 5.11. The following corollary of Theorem 7.8 is useful in statistics. Let $\hat{\theta}_n = F_n^{-1}(p)$ be the sample pth quantile.

Corollary 7.1. Assume the conditions in Theorem 7.8 and $k_n/n = p + cn^{-1/2} + o(n^{-1/2})$ with a constant c. Then

$$\sqrt{n}(X_{(k_n)} - \hat{\theta}_n) \rightarrow_{a.s.} c/F'(\theta). \quad \blacksquare$$

The proof of Corollary 7.1 is left as an exercise. Using Corollary 7.1, we can obtain a confidence interval for θ with limiting confidence coefficient $1 - \alpha$ (Definition 2.14) for any given $\alpha \in (0, \frac{1}{2})$.

Corollary 7.2. Assume the conditions in Theorem 7.8. Let $\{k_{1n}\}$ and $\{k_{2n}\}$ be two sequences of integers satisfying $1 \le k_{1n} < k_{2n} \le n$,

$$k_{1n}/n = p - z_{1-\alpha/2}\sqrt{p(1-p)/n} + o(n^{-1/2}),$$

and

$$k_{2n}/n = p + z_{1-\alpha/2}\sqrt{p(1-p)/n} + o(n^{-1/2}),$$

where $z_a = \Phi^{-1}(a)$. Then the confidence interval $C(X) = [X_{(k_{1n})}, X_{(k_{2n})}]$ has the property that $P(\theta \in C(X))$ does not depend on P and

$$\lim_{n \to \infty} \inf_{P \in \mathcal{P}} P\big(\theta \in C(X)\big) = \lim_{n \to \infty} P\big(\theta \in C(X)\big) = 1 - \alpha. \qquad (7.24)$$

Furthermore,

$$\text{the length of } C(X) = \frac{2z_{1-\alpha/2}\sqrt{p(1-p)}}{F'(\theta)\sqrt{n}} + o\left(\frac{1}{\sqrt{n}}\right) \quad \text{a.s.}$$

Proof. Note that $P(X_{(k_{1n})} \le \theta \le X_{(k_{2n})}) = P(U_{(k_{1n})} \le p \le U_{(k_{2n})})$, where $U_{(k)}$ is the kth order statistic based on a sample $U_1, ..., U_n$ i.i.d. from the uniform distribution $U(0,1)$ (Exercise 84). Hence, $P(\theta \in C(X))$ does not depend on P and the first equality in (7.24) holds.

By Corollary 7.1, Theorem 5.10, and Slutsky's theorem,

$$
\begin{aligned}
P(X_{(k_{1n})} > \theta) &= P\left(\hat{\theta}_n - z_{1-\alpha/2}\frac{\sqrt{p(1-p)}}{F'(\theta)\sqrt{n}} + o_p(n^{-1/2}) > \theta\right) \\
&= P\left(\frac{\sqrt{n}(\hat{\theta}_n - \theta)}{\sqrt{p(1-p)}/F'(\theta)} + o_p(1) > z_{1-\alpha/2}\right) \\
&\to 1 - \Phi(z_{1-\alpha/2}) \\
&= \alpha/2.
\end{aligned}
$$

Similarly, $P(X_{(k_{2n})} < \theta) \to \alpha/2$. Hence the second equality in (7.24) holds. The result for the length of $C(X)$ follows directly from Corollary 7.1. ∎

The confidence interval $[X_{(k_{1n})}, X_{(k_{2n})}]$ given in Corollary 7.2 is called Woodruff's (1952) interval. It has limiting confidence coefficient $1 - \alpha$, a property that is stronger than the $1 - \alpha$ asymptotic correctness.

From Theorem 5.10, if $F'(\theta)$ exists and is positive, then

$$\sqrt{n}(\hat{\theta}_n - \theta) \to_d N\left(0, \frac{p(1-p)}{[F'(\theta)]^2}\right).$$

If the derivative $F'(\theta)$ has a consistent estimator $\hat{d}_n$ obtained using some method such as one of those introduced in §5.1.3, then (7.18) holds with $\hat{V}_n = p(1-p)/\hat{d}_n^2$ and the method introduced in §7.3.1 can be applied to derive the following $1 - \alpha$ asymptotically correct confidence interval:

$$C_1(X) = \left[\hat{\theta}_n - z_{1-\alpha/2}\frac{\sqrt{p(1-p)}}{\hat{d}_n\sqrt{n}}, \ \hat{\theta}_n + z_{1-\alpha/2}\frac{\sqrt{p(1-p)}}{\hat{d}_n\sqrt{n}} \right].$$

The length of $C_1(X)$ is asymptotically almost the same as Woodruff's interval. However, $C_1(X)$ depends on the estimated derivative $\hat{d}_n$ and it is usually difficult to obtain a precise estimator $\hat{d}_n$.

7.3.4 Accuracy of asymptotic confidence sets

In §7.3.1 (Proposition 7.4) we evaluate a $1 - \alpha$ asymptotically correct confidence set $C(X)$ by its volume. We now study another way of assessing $C(X)$ by considering the convergence speed of $P(\theta \in C(X))$.

Definition 7.5. A $1 - \alpha$ asymptotically correct confidence set $C(X)$ for θ is said to be lth-order (asymptotically) accurate if and only if

$$P\big(\theta \in C(X)\big) = 1 - \alpha + O(n^{-l/2}),$$

where l is a positive integer. ∎

We focus on the case where θ is real-valued. For $\theta \in \mathcal{R}$, the confidence set given by (7.19) is the two-sided interval $[\underline{\theta}_{\alpha/2}, \overline{\theta}_{\alpha/2}]$, where $\underline{\theta}_\alpha = \hat{\theta}_n - z_{1-\alpha}\hat{V}_n^{1/2}$, $\overline{\theta}_\alpha = \hat{\theta}_n + z_{1-\alpha}\hat{V}_n^{1/2}$, and $z_a = \Phi^{-1}(a)$. Suppose that the c.d.f. of $\hat{V}_n^{-1/2}(\hat{\theta}_n - \theta)$ admits the Edgeworth expansion (1.106) with $m = 1$. Then

$$P(\theta \geq \underline{\theta}_\alpha) = P\left(\hat{V}_n^{-1/2}(\hat{\theta}_n - \theta) \leq z_{1-\alpha}\right)$$
$$= \Phi(z_{1-\alpha}) + n^{-1/2}p_1(z_{1-\alpha})\Phi'(z_{1-\alpha}) + o(n^{-1/2})$$
$$= 1 - \alpha + O(n^{-1/2}),$$

i.e., the lower confidence bound $\underline{\theta}_\alpha$ or the one-sided confidence interval $[\underline{\theta}_\alpha, \infty)$ is first-order accurate. Similarly,

$$P(\theta \leq \overline{\theta}_\alpha) = 1 - P\left(\hat{V}_n^{-1/2}(\hat{\theta}_n - \theta) < -z_{1-\alpha}\right)$$
$$= 1 - \Phi(-z_{1-\alpha}) - n^{-1/2}p_1(-z_{1-\alpha})\Phi'(-z_{1-\alpha}) + o(n^{-1/2})$$
$$= 1 - \alpha + O(n^{-1/2}),$$

i.e., the upper confidence bound $\overline{\theta}_\alpha$ is first-order accurate. Combining these results and using the fact that $\Phi'(x)$ and $p_1(x)$ are even functions (Theorem

1.16), we obtain that

$$P(\underline{\theta}_{\alpha/2} \leq \theta \leq \overline{\theta}_{\alpha/2}) = 1 - \alpha + o(n^{-1/2}),$$

which indicates that the coverage probability of $[\underline{\theta}_{\alpha/2}, \overline{\theta}_{\alpha/2}]$ converges to $1 - \alpha$ at a rate faster than $n^{-1/2}$. In fact, if we assume that the c.d.f. of $\hat{V}_n^{-1/2}(\hat{\theta}_n - \theta)$ admits the Edgeworth expansion (1.106) with $m = 2$, then

$$P(\underline{\theta}_{\alpha/2} \leq \theta \leq \overline{\theta}_{\alpha/2}) = 1 - \alpha + 2n^{-1}p_2(z_{1-\alpha/2})\Phi'(z_{1-\alpha/2}) + o(n^{-1}), \quad (7.25)$$

i.e., the equal-tail two-sided confidence interval $[\underline{\theta}_{\alpha/2}, \overline{\theta}_{\alpha/2}]$ is second-order accurate.

Can we obtain a confidence bound that is more accurate than $\underline{\theta}_\alpha$ (or $\overline{\theta}_\alpha$) or a two-sided confidence interval that is more accurate than $[\underline{\theta}_{\alpha/2}, \overline{\theta}_{\alpha/2}]$? The answer is affirmative if a higher order Edgeworth expansion is available. Assume that the conditions in Theorem 1.16 are satisfied for an integer $m \geq 2$. Using the arguments in deriving the polynomial q_j in (1.108), we can show (exercise) that

$$P\left(\hat{V}_n^{-1/2}(\hat{\theta}_n - \theta) \leq z_{1-\alpha} + \sum_{j=1}^{m-1} \frac{q_j(z_{1-\alpha})}{n^{j/2}}\right) = 1 - \alpha + O(n^{-m/2}). \quad (7.26)$$

If the coefficients of polynomials $q_1, ..., q_{m-1}$ are known, then

$$\hat{\theta}_n - \hat{V}_n^{-1/2}\left[z_{1-\alpha} + \sum_{j=1}^{m-1} \frac{q_j(z_{1-\alpha})}{n^{j/2}}\right]$$

is an mth-order accurate lower confidence bound for θ. In general, however, some coefficients of q_j's are unknown. Let $\hat{q}_j$ be the same as q_j with all unknown coefficients in the polynomial q_j replaced by their estimators, $j = 1, ..., m - 1$. Assume that $\hat{q}_1(z_{1-\alpha}) - q_1(z_{1-\alpha}) = O_p(n^{-1/2})$, i.e., $\hat{q}_1(z_{1-\alpha})$ is $\sqrt{n}$-consistent. Then, the lower confidence bound

$$\underline{\theta}_\alpha^{(2)} = \hat{\theta}_n - \hat{V}_n^{1/2}[z_{1-\alpha} + n^{-1/2}\hat{q}_1(z_{1-\alpha})]$$

is second-order accurate (Hall, 1992). A second-order accurate upper confidence bound $\overline{\theta}_{\alpha/2}^{(2)}$ can be similarly defined. However, the two-sided confidence interval $[\underline{\theta}_{\alpha/2}^{(2)}, \overline{\theta}_{\alpha/2}^{(2)}]$ is only second-order accurate, i.e., in terms of the convergence speed, it does not improve the confidence interval $[\underline{\theta}_{\alpha/2}, \overline{\theta}_{\alpha/2}]$.

Higher order accurate confidence bounds and two-sided confidence intervals can be obtained using Edgeworth and Cornish-Fisher expansions. See Hall (1992).

Example 7.25. Let $X_1, ..., X_n$ be i.i.d. random d-vectors. Consider the problem of setting a lower confidence bound for $\theta = g(\mu)$, where $\mu = EX_1$ and g is five times continuously differentiable from $\mathcal{R}^d$ to $\mathcal{R}$ with $\nabla g(\mu) \neq 0$. Let $\bar{X}$ be the sample mean and $\hat{\theta}_n = g(\bar{X})$. Then, $\hat{V}_n$ in (7.19) can be chosen to be $n^{-2}(n-1)[\nabla g(\bar{X})]^\tau S^2 \nabla g(\bar{X})$, where S^2 is the sample covariance matrix. Let X_{ij} be the jth component of X_i and Y_i be a vector containing components of the form X_{ij} and $X_{ij}X_{ij'}$, $j = 1, ..., d$, $j' \geq j$. It can be shown (exercise) that $\hat{V}_n^{-1/2}(\hat{\theta}_n - \theta)$ can be written as $\sqrt{n}h(\bar{Y})/\sigma_h$, where $\bar{Y}$ is the sample mean of Y_i's, h is a five times continuously differentiable function, $h(EY_1) = 0$, and $\sigma_h^2 = [\nabla h(EY_1)]^\tau \text{Var}(Y_1)\nabla h(EY_1)$. Assume that $E\|Y_1\|^4 < \infty$ and that Cramér's continuity condition (1.105) is satisfied. By Theorem 1.16, the distribution of $\hat{V}_n^{-1/2}(\hat{\theta}_n - \theta)$ admits the Edgeworth expansion (1.106) with $m = 2$ and $p_1(x)$ given by (1.107). Since $q_1(x) = -p_1(x)$ (§1.5.6), we obtain the following second-order accurate lower confidence bound for θ:

$$\underline{\theta}_\alpha^{(2)} = \hat{\theta}_n - \hat{V}_n^{1/2}\{z_{1-\alpha} + n^{-1/2}[\hat{c}_1\hat{\sigma}_h^{-1} - 6^{-1}\hat{c}_2\hat{\sigma}_h^{-3}(z_{1-\alpha}^2 - 1)]\},$$

where $\hat{\sigma}_h^2 = [\nabla h(\bar{Y})]^\tau S_Y^2 \nabla h(\bar{Y})$, S_Y^2 is the sample covariance matrix based on Y_i's, and $\hat{c}_j$ is the estimator of c_j in (1.107) obtained by replacing the moments of Y_1 with the corresponding sample moments.

In particular, if $d = 1$ and $g(x) = x$, then it follows from Example 1.34 that

$$\underline{\theta}_\alpha^{(2)} = \bar{X} - n^{-1/2}\hat{\sigma}[z_{1-\alpha} - 6^{-1}n^{-1/2}\hat{\kappa}_3(2z_{1-\alpha}^2 + 1)],$$

where $\hat{\sigma}^2 = n^{-1}\sum_{i=1}^n (X_i - \bar{X})^2$ and $\hat{\kappa}_3 = n^{-1}\sum_{i=1}^n (X_1 - \bar{X})^3/\hat{\sigma}^3$ is $\sqrt{n}$-consistent for $\kappa_3 = E(X_1 - \mu)^3/\sigma^3$ with $\sigma^2 = \text{Var}(X_1)$. A second-order accurate lower confidence bound for σ^2 can be similarly derived (Exercise 89). ∎

7.4 Bootstrap Confidence Sets

In this section, we study how to use the bootstrap method introduced in §5.5.3 to construct asymptotically correct confidence sets. There are two main advantages of using the bootstrap method. First, as we can see from previous sections, constructing confidence sets having a given confidence coefficient or being asymptotically correct requires some theoretical derivations. The bootstrap method replaces these derivations by some routine computations. Second, confidence intervals (especially one-sided confidence intervals) constructed using the bootstrap method may be asymptotically second-order accurate (§7.3.4).

We use the notation in §5.5.3. Let $X = (X_1, ..., X_n)$ be a sample from P. We focus on the case where X_i's are i.i.d. so that P is specified by a c.d.f. F

on $\mathcal{R}^d$, although some results discussed here can be extended to non-i.i.d. cases. Also, we assume that F is estimated by the empirical c.d.f. F_n defined in (5.1) (which means that no assumption is imposed on F and the problem is nonparametric) so that $\hat{P}$ in §5.5.3 is the population corresponding to F_n. A bootstrap sample $X^* = (X_1^*, ..., X_n^*)$ is obtained from $\hat{P}$, i.e., X_i^*'s are i.i.d. from F_n. Some other bootstrap sampling procedures are described in Exercises 92 and 95-97. Let θ be a parameter of interest, $\hat{\theta}_n$ be an estimator of θ, and $\hat{\theta}_n^*$ be the bootstrap analogue of $\hat{\theta}_n$, i.e., $\hat{\theta}_n^*$ is the same as $\hat{\theta}_n$ except that X is replaced by the bootstrap sample X^*.

7.4.1 Construction of bootstrap confidence intervals

We now introduce several different ways of constructing bootstrap confidence intervals for a real-valued θ. Some ideas can be extended to the construction of bootstrap confidence sets for multivariate θ. We mainly consider lower confidence bounds. Upper confidence bounds can be similarly obtained and equal-tail two-sided confidence intervals can be obtained using confidence bounds.

The bootstrap percentile

Define
$$K_B(x) = P_*(\hat{\theta}_n^* \leq x), \tag{7.27}$$
where P_* denotes the distribution of X^* conditional on X. For a given $\alpha \in (0, \frac{1}{2})$, the *bootstrap percentile* method (Efron, 1981) gives the following lower confidence bound for θ:
$$\underline{\theta}_{BP} = K_B^{-1}(\alpha). \tag{7.28}$$
The name percentile comes from the fact that $K_B^{-1}(\alpha)$ is a percentile of the bootstrap distribution K_B in (7.27).

For most cases, the computation of $\underline{\theta}_{BP}$ requires numerical approximations such as the Monte Carlo approximation described in §5.5.3. A description of the Monte Carlo approximation to bootstrap confidence sets can be found in §7.4.3 (when bootstrap prepivoting is discussed).

We now provide a justification of the bootstrap percentile method that allows us to see what assumptions are required for a good performance of a bootstrap percentile confidence set. Suppose that there exists an increasing transformation $\phi_n(x)$ such that
$$P(\hat{\phi}_n - \phi_n(\theta) \leq x) = \Psi(x) \tag{7.29}$$
holds for all possible F (including $F = F_n$), where $\hat{\phi}_n = \phi_n(\hat{\theta}_n)$ and Ψ is a c.d.f. that is continuous, strictly increasing, and symmetric about 0.

When $\Psi = \Phi$, the standard normal distribution, the function ϕ_n is called the normalizing and variance stabilizing transformation. If ϕ_n and Ψ in (7.29) can be derived, then the following lower confidence bound for θ has confidence coefficient $1 - \alpha$:

$$\underline{\theta}_E = \phi_n^{-1}(\hat{\phi}_n + z_\alpha),$$

where $z_\alpha = \Psi^{-1}(\alpha)$.

We now show that $\underline{\theta}_{BP} = \underline{\theta}_E$ and, therefore, we can still use this lower confidence bound without deriving ϕ_n and Ψ. Let $w_n = \phi_n(\underline{\theta}_{BP}) - \hat{\phi}_n$. From the fact that assumption (7.29) holds when F is replaced by F_n,

$$\Psi(w_n) = P_*(\hat{\phi}_n^* - \hat{\phi}_n \le w_n) = P_*(\hat{\theta}_n^* \le \underline{\theta}_{BP}) = \alpha,$$

where $\hat{\phi}_n^* = \phi_n(\hat{\theta}_n^*)$ and the last equality follows from the definition of $\underline{\theta}_{BP}$ and the assumption on Ψ. Hence $w_n = z_\alpha = \Psi^{-1}(\alpha)$ and

$$\underline{\theta}_{BP} = \phi_n^{-1}(\hat{\phi}_n + z_\alpha) = \underline{\theta}_E.$$

Thus, the bootstrap percentile lower confidence bound $\underline{\theta}_{BP}$ has confidence coefficient $1 - \alpha$ for all n if assumption (7.29) holds exactly for all n. If assumption (7.29) holds approximately for large n, then $\underline{\theta}_{BP}$ is $1 - \alpha$ asymptotically correct (see Theorem 7.9 in §7.4.2) and its performance depends on how good the approximation is.

The bootstrap bias-corrected percentile

Efron (1981) considered the following assumption that is more general than assumption (7.29):

$$P(\hat{\phi}_n - \phi_n(\theta) + z_0 \le x) = \Psi(x), \qquad (7.30)$$

where ϕ_n and Ψ are the same as those in (7.29) and z_0 is a constant that may depend on F and n. When $z_0 = 0$, (7.30) reduces to (7.29). Since $\Psi(0) = \frac{1}{2}$, z_0 is a kind of "bias" of $\hat{\phi}_n$. If ϕ_n, z_0, and Ψ in (7.30) can be derived, then a lower confidence bound for θ with confidence coefficient $1 - \alpha$ is

$$\underline{\theta}_E = \phi_n^{-1}(\hat{\phi}_n + z_\alpha + z_0).$$

Applying assumption (7.30) to $F = F_n$, we obtain that

$$K_B(\hat{\theta}_n) = P_*(\hat{\phi}_n^* - \hat{\phi}_n + z_0 \le z_0) = \Psi(z_0),$$

where K_B is given in (7.27). This implies

$$z_0 = \Psi^{-1}(K_B(\hat{\theta}_n)). \qquad (7.31)$$

Also from (7.30),

$$1 - \alpha = \Psi(-z_\alpha)$$
$$= P_*\big(\hat\phi_n^* - \hat\phi_n + z_0 \le -z_\alpha\big)$$
$$= P_*\big(\hat\theta_n^* \le \phi_n^{-1}(\hat\phi_n - z_\alpha - z_0)\big),$$

which implies

$$\phi_n^{-1}(\hat\phi_n - z_\alpha - z_0) = K_B^{-1}(1 - \alpha).$$

Since this equation holds for any α, it implies that for $0 < x < 1$,

$$K_B^{-1}(x) = \phi_n^{-1}\big(\hat\phi_n + \Psi^{-1}(x) - z_0\big). \qquad (7.32)$$

By the definition of $\underline{\theta}_E$ and (7.32),

$$\underline{\theta}_E = K_B^{-1}\big(\Psi(z_\alpha + 2z_0)\big).$$

Assuming that Ψ is known (e.g., $\Psi = \Phi$) and using (7.31), Efron (1981) obtained the bootstrap bias-corrected (BC) percentile lower confidence bound for θ:

$$\underline{\theta}_{BC} = K_B^{-1}\big(\Psi\big(z_\alpha + 2\Psi^{-1}(K_B(\hat\theta_n))\big)\big), \qquad (7.33)$$

which is a percentile of the bootstrap distribution K_B. Note that $\underline{\theta}_{BC}$ reduces to $\underline{\theta}_{BP}$ if $K_B(\hat\theta_n) = \frac{1}{2}$, i.e., $\hat\theta_n$ is the median of the bootstrap distribution K_B. Hence, the bootstrap BC percentile method is a bias-corrected version of the bootstrap percentile method and the bias-correction is represented by $2\Psi^{-1}(K_B(\hat\theta_n))$. If (7.30) holds exactly, then $\underline{\theta}_{BC}$ has confidence coefficient $1 - \alpha$ for all n. If (7.30) holds approximately, then $\underline{\theta}_{BC}$ is $1 - \alpha$ asymptotically correct.

The bootstrap BC percentile method improves the bootstrap percentile method by taking a bias into account. This is supported by the theoretical result in §7.4.2. However, there are still many cases where assumption (7.30) cannot be fulfilled nicely and the bootstrap BC percentile method does not work well. Efron (1987) proposed a bootstrap accelerated bias-corrected (BC_a) percentile method (see Exercise 93) that improves the bootstrap BC percentile method. However, applications of the bootstrap BC_a percentile method involve some derivations that may be very complicated. See Efron (1987) and Efron and Tibshirani (1993) for details.

The hybrid bootstrap

Suppose that $\hat\theta_n$ is asymptotically normal, i.e., (7.18) holds with $V_n = \sigma_F^2/n$. Let H_n be the c.d.f. of $\sqrt{n}(\hat\theta_n - \theta)$ and

$$\hat H_B(x) = P_*\big(\sqrt{n}(\hat\theta_n^* - \hat\theta_n) \le x\big)$$

be its bootstrap estimator defined in (5.121). From the results in Theorem 5.20, for any $t \in (0, 1)$, $\hat{H}_B^{-1}(t) - H_n^{-1}(t) \to_p 0$. Treating the quantile of $\hat{H}_B$ as the quantile of H_n, we obtain the following *hybrid bootstrap* lower confidence bound for θ:

$$\underline{\theta}_{HB} = \hat{\theta}_n - n^{-1/2} \hat{H}_B^{-1}(1 - \alpha). \qquad (7.34)$$

The bootstrap-t

Suppose that (7.18) holds with $V_n = \sigma_F^2/n$ and $\hat{\sigma}_F^2$ is a consistent estimator of σ_F^2. The bootstrap-t method is based on $t(X, \theta) = \sqrt{n}(\hat{\theta}_n - \theta)/\hat{\sigma}_F$, which is often called a studentized "statistic". If the distribution G_n of $t(X, \theta)$ is known (i.e., $t(X, \theta)$ is pivotal), then a confidence interval for θ with confidence coefficient $1 - \alpha$ can be obtained (§7.1.1). If G_n is unknown, it can be estimated by the bootstrap estimator

$$\hat{G}_B(x) = P_*\big(t(X^*, \hat{\theta}_n) \leq x\big),$$

where $t(X^*, \hat{\theta}_n) = \sqrt{n}(\hat{\theta}_n^* - \hat{\theta}_n)/\hat{\sigma}_F^*$ and $\hat{\sigma}_F^*$ is the bootstrap analogue of $\hat{\sigma}_F$. Treating the quantile of $\hat{G}_B$ as the quantile of G_n, we obtain the following *bootstrap-t* lower confidence bound for θ:

$$\underline{\theta}_{BT} = \hat{\theta}_n - n^{-1/2} \hat{\sigma}_F \hat{G}_B^{-1}(1 - \alpha). \qquad (7.35)$$

Although it is shown in §7.4.2 that $\underline{\theta}_{BT}$ in (7.35) is more accurate than $\underline{\theta}_{BP}$ in (7.28), $\underline{\theta}_{BC}$ in (7.33), and $\underline{\theta}_{HB}$ in (7.34), the use of the bootstrap-t method requires a consistent variance estimator $\hat{\sigma}_F^2$.

7.4.2 Asymptotic correctness and accuracy

From the construction of the hybrid bootstrap and bootstrap-t confidence bounds, $\underline{\theta}_{HB}$ is $1 - \alpha$ asymptotically correct if $\varrho_\infty(\hat{H}_B, H_n) \to_p 0$, and $\underline{\theta}_{BT}$ is $1 - \alpha$ asymptotically correct if $\varrho_\infty(\hat{G}_B, G_n) \to_p 0$. On the other hand, the asymptotic correctness of the bootstrap percentile (with or without bias-correction or acceleration) confidence bounds requires slightly more.

Theorem 7.9. Suppose that $\varrho_\infty(\hat{H}_B, H_n) \to_p 0$ and

$$\lim_{n \to \infty} \rho_\infty(H_n, H) = 0, \qquad (7.36)$$

where H is a c.d.f. on $\mathcal{R}$ that is continuous, strictly increasing, and symmetric about 0. Then $\underline{\theta}_{BP}$ in (7.28) and $\underline{\theta}_{BC}$ in (7.33) are $1 - \alpha$ asymptotically correct.

Proof. The result for $\underline{\theta}_{BP}$ follows from

$$
\begin{aligned}
P\big(\underline{\theta}_{BP} \leq \theta\big) &= P\big(\alpha \leq K_B(\theta)\big) \\
&= P\big(\alpha \leq H_B\big(\sqrt{n}(\theta - \hat{\theta}_n)\big)\big) \\
&= P\big(\sqrt{n}(\hat{\theta}_n - \theta) \leq -\hat{H}_B^{-1}(\alpha)\big) \\
&= P\big(\sqrt{n}(\hat{\theta}_n - \theta) \leq -H^{-1}(\alpha)\big) + o(1) \\
&= H\big(-H^{-1}(\alpha)\big) + o(1) \\
&= 1 - \alpha + o(1).
\end{aligned}
$$

The result for $\underline{\theta}_{BC}$ follows from the previous result and

$$
z_0 = \Psi^{-1}\big(K_B(\hat{\theta}_n)\big) = \Psi^{-1}\big(\hat{H}_B(0)\big) \to_p \Psi^{-1}\big(H(0)\big) = 0. \quad \blacksquare
$$

Theorem 7.9 can be obviously extended to the case of upper confidence bounds or two-sided confidence intervals. The result also holds for the bootstrap BC_a percentile confidence intervals.

Note that H in (7.36) is not the same as Ψ in assumption (7.30). Usually $H(x) = \Phi(x/\sigma_F)$ for some $\sigma_F > 0$, whereas $\Psi = \Phi$. Also, condition (7.36) is much weaker than assumption (7.30), since the latter requires variance stabilizing.

It is not surprising that all bootstrap methods introduced in §7.3.1 produce asymptotically correct confidence sets. To compare various bootstrap confidence intervals and other asymptotic confidence intervals, we now consider their asymptotic accuracy (Definition 7.5).

Consider the case of $\theta = g(\mu)$, $\mu = EX_1$, and $\hat{\theta}_n = g(\bar{X})$, where $\bar{X}$ is the sample mean and g is five times continuously differentiable from $\mathcal{R}^d$ to $\mathcal{R}$ with $\nabla g(\mu) \neq 0$. The asymptotic variance of $\sqrt{n}(\hat{\theta}_n - \theta)$ can be estimated by $\hat{\sigma}_F^2 = \frac{n-1}{n}[\nabla g(\bar{X})]^\tau S^2 \nabla g(\bar{X})$, where S^2 is the sample covariance matrix. Let G_n be the distribution of $\sqrt{n}(\hat{\theta}_n - \theta)/\hat{\sigma}_F$. If G_n is known, then a lower confidence bound for θ with confidence coefficient $1 - \alpha$ is

$$
\underline{\theta}_E = \hat{\theta}_n - n^{-1/2}\hat{\sigma}_F G_n^{-1}(1 - \alpha), \tag{7.37}
$$

which is not useful if G_n is unknown.

Assume that $E\|X_1\|^8 < \infty$ and condition (1.105) is satisfied. Then G_n admits the Edgeworth expansion (1.106) with $m = 2$ and, by Theorem 1.17, $G_n^{-1}(t)$ admits the Cornish-Fisher expansion

$$
G_n^{-1}(t) = z_t + \frac{q_1(z_t, F)}{\sqrt{n}} + \frac{q_2(z_t, F)}{n} + o\left(\frac{1}{n}\right), \tag{7.38}
$$

where $q_j(\cdot, F)$ is the same as $q_j(\cdot)$ in Theorem 1.17 but the notation $q_j(\cdot, F)$ is used to emphasize that the coefficients of the polynomial q_j depend on F, the c.d.f. of X_1. Let $\hat{G}_B$ be the bootstrap estimator of G_n defined in §7.4.1. Under some conditions (Hall, 1992), $\hat{G}_B^{-1}$ admits expansion (7.38) with F replaced by the empirical c.d.f. F_n for almost all sequences $X_1, X_2, \ldots$. Hence the bootstrap-t lower confidence bound in (7.35) can be written as

$$\underline{\theta}_{BT} = \hat{\theta}_n - \frac{\hat{\sigma}_F}{\sqrt{n}} \left[z_{1-\alpha} + \sum_{j=1}^{2} \frac{q_j(z_{1-\alpha}, F_n)}{n^{j/2}} + o\left(\frac{1}{n}\right) \right] \quad \text{a.s.} \tag{7.39}$$

Under some moment conditions, $q_j(x, F_n) - q_j(x, F) = O_p(n^{-1/2})$ for each x, $j = 1, 2$. Then, comparing (7.37), (7.38), and (7.39), we obtain that

$$\underline{\theta}_{BT} - \underline{\theta}_E = O_p(n^{-3/2}). \tag{7.40}$$

Furthermore,

$$P(\underline{\theta}_{BT} \leq \theta) = P\left(\frac{\hat{\theta}_n - \theta}{\hat{\sigma}_F/\sqrt{n}} \leq \hat{G}_B^{-1}(1-\alpha) \right)$$

$$= P\left(\frac{\hat{\theta}_n - \theta}{\hat{\sigma}_F/\sqrt{n}} \leq z_{1-\alpha} + \sum_{j=1}^{2} \frac{q_j(z_{1-\alpha}, F_n)}{n^{j/2}} \right) + o\left(\frac{1}{n}\right)$$

$$= 1 - \alpha + \frac{\psi(z_{1-\alpha})\Phi'(z_{1-\alpha})}{n} + o\left(\frac{1}{n}\right), \tag{7.41}$$

where $\psi(x)$ is a polynomial whose coefficients are functions of moments of F and the last equality can be justified by a somewhat complicated argument (Hall, 1992).

Result (7.41) implies that $\underline{\theta}_{BT}$ is second-order accurate according to Definition 7.5. The same can be concluded for the bootstrap-t upper confidence bound and the equal-tail two-sided bootstrap-t confidence interval for θ.

Next, we consider the hybrid bootstrap lower confidence bound $\underline{\theta}_{HB}$ given by (7.34). Let $\tilde{H}_B$ be the bootstrap estimator of $\tilde{H}_n$, the distribution of $\sqrt{n}(\hat{\theta}_n - \theta)/\sigma_F$. Then $\hat{H}_B^{-1}(1-\alpha) = \hat{\sigma}_F \tilde{H}_B^{-1}(1-\alpha)$ and

$$\underline{\theta}_{HB} = \hat{\theta}_n - n^{-1/2}\hat{\sigma}_F \tilde{H}_B^{-1}(1-\alpha),$$

which can be viewed as a bootstrap approximation to

$$\underline{\theta}_H = \hat{\theta}_n - n^{-1/2}\hat{\sigma}_F \tilde{H}_n^{-1}(1-\alpha).$$

Note that $\underline{\theta}_H$ does not have confidence coefficient $1 - \alpha$, since it is obtained by muddling up $G_n^{-1}(1-\alpha)$ and $\tilde{H}_n^{-1}(1-\alpha)$. Similar to G_n^{-1}, $\tilde{H}_n^{-1}$ admits the Cornish-Fisher expansion

$$\tilde{H}_n^{-1}(t) = z_t + \frac{\tilde{q}_1(z_t, F)}{\sqrt{n}} + \frac{\tilde{q}_2(z_t, F)}{n} + o\left(\frac{1}{n}\right) \tag{7.42}$$

and $\tilde{H}_B^{-1}$ admits the same expansion (7.42) with F replaced by F_n for almost all $X_1, X_2, \ldots$. Then

$$\underline{\theta}_{HB} = \hat{\theta}_n - \frac{\hat{\sigma}_F}{\sqrt{n}}\left[z_{1-\alpha} + \sum_{j=1}^{2} \frac{\tilde{q}_j(z_{1-\alpha}, F_n)}{n^{j/2}} + o\left(\frac{1}{n}\right)\right] \quad \text{a.s.} \tag{7.43}$$

and, by (7.37),

$$\underline{\theta}_{HB} - \underline{\theta}_E = O_p(n^{-1}), \tag{7.44}$$

since $q_1(x, F)$ and $\tilde{q}_1(x, F)$ are usually different. Results (7.40) and (7.44) imply that $\underline{\theta}_{HB}$ is not as close to $\underline{\theta}_E$ as $\underline{\theta}_{BT}$. Similarly to (7.41), we can show that (Hall, 1992)

$$P(\underline{\theta}_{HB} \leq \theta) = P\left(\frac{\hat{\theta}_n - \theta}{\hat{\sigma}_F/\sqrt{n}} \leq \tilde{H}_B^{-1}(1-\alpha)\right)$$

$$= P\left(\frac{\hat{\theta}_n - \theta}{\hat{\sigma}_F/\sqrt{n}} \leq z_{1-\alpha} + \sum_{j=1}^{2} \frac{\tilde{q}_j(z_{1-\alpha}, F_n)}{n^{j/2}}\right) + o\left(\frac{1}{n}\right)$$

$$= P\left(\frac{\hat{\theta}_n - \theta}{\hat{\sigma}_F/\sqrt{n}} \leq z_{1-\alpha} + \sum_{j=1}^{2} \frac{\tilde{q}_j(z_{1-\alpha}, F)}{n^{j/2}}\right) + O\left(\frac{1}{n}\right)$$

$$= 1 - \alpha + \frac{\tilde{\psi}(z_{1-\alpha})\Phi'(z_{1-\alpha})}{\sqrt{n}} + O\left(\frac{1}{n}\right), \tag{7.45}$$

where $\tilde{\psi}(x)$ is an even polynomial. This implies that when $\tilde{\psi} \neq 0$, $\underline{\theta}_{HB}$ is only first-order accurate according to Definition 7.5.

The same conclusion can be drawn for the hybrid bootstrap upper confidence bounds. However, the equal-tail two-sided hybrid bootstrap confidence interval

$$[\underline{\theta}_{HB}, \overline{\theta}_{HB}] = [\hat{\theta}_n - n^{-1/2}\hat{H}_B^{-1}(1-\alpha), \ \hat{\theta}_n - n^{-1/2}\hat{H}_B^{-1}(\alpha)]$$

is second-order accurate (and $1 - 2\alpha$ asymptotically correct), as is the equal-tail two-sided bootstrap-t confidence interval, since

$$P(\underline{\theta}_{HB} \leq \theta \leq \overline{\theta}_{HB}) = P(\theta \leq \overline{\theta}_{HB}) - P(\theta < \underline{\theta}_{HB})$$
$$= 1 - \alpha + n^{-1/2}\tilde{\psi}(z_{1-\alpha})\Phi'(z_{1-\alpha})$$
$$\quad - \alpha - n^{-1/2}\tilde{\psi}(z_\alpha)\Phi'(z_\alpha) + O(n^{-1})$$
$$= 1 - 2\alpha + O(n^{-1})$$

by the fact that $\tilde{\psi}$ and Φ' are even functions and $z_{1-\alpha} = -z_\alpha$.

For the bootstrap percentile lower confidence bound in (7.28),

$$\underline{\theta}_{BP} = K_B^{-1}(\alpha) = \hat{\theta}_n + n^{-1/2}\hat{H}_B^{-1}(\alpha).$$

Comparing $\underline{\theta}_{BP}$ with $\underline{\theta}_{BT}$ and $\underline{\theta}_{HB}$, we find that the bootstrap percentile method muddles up not only $\tilde{H}_B^{-1}(1-\alpha)$ and $\hat{G}_B^{-1}(1-\alpha)$, but also $\hat{H}_B^{-1}(\alpha)$ and $-\hat{H}_B^{-1}(1-\alpha)$. If $\hat{H}_B$ is asymptotically symmetric about 0, then the bootstrap percentile method is equivalent to the hybrid bootstrap method and, therefore, one-sided bootstrap percentile confidence intervals are only first-order accurate and the equal-tail two-sided bootstrap percentile confidence interval is second-order accurate.

Since $\hat{\theta}_n$ is asymptotically normal, we can use $\Psi = \Phi$ for the bootstrap BC percentile method. Let $\tilde{\alpha}_n = \Phi(z_\alpha + 2z_0)$. Then the bootstrap BC percentile lower confidence bound given by (7.33) is just the $\tilde{\alpha}_n$th quantile of K_B in (7.27). Using the Edgeworth expansion, we obtain that

$$K_B(\hat{\theta}_n) = \tilde{H}_B(0) = \Phi(0) + \frac{\tilde{q}(0, F_n)\Phi'(0)}{\sqrt{n}} + O_p\left(\frac{1}{n}\right)$$

with some function $\tilde{q}$ and, therefore,

$$\tilde{\alpha}_n = \alpha + \frac{2\tilde{q}(0, F_n)\Phi'(z_\alpha)}{\sqrt{n}} + O_p\left(\frac{1}{n}\right).$$

This result and the Cornish-Fisher expansion for $\tilde{H}_B^{-1}$ imply

$$\tilde{H}_B^{-1}(\tilde{\alpha}_n) = z_\alpha + \frac{2\tilde{q}(0, F_n) + \tilde{q}_1(z_\alpha, F_n)}{\sqrt{n}} + O_p\left(\frac{1}{n}\right).$$

Then from (7.33) and $K_B^{-1}(\tilde{\alpha}_n) = \hat{\theta}_n + n^{-1/2}\hat{\sigma}_F\tilde{H}_B^{-1}(\tilde{\alpha}_n)$,

$$\underline{\theta}_{BC} = \hat{\theta}_n + \frac{\hat{\sigma}_F}{\sqrt{n}}\left[z_\alpha + \frac{2\tilde{q}(0, F_n) + \tilde{q}_1(z_\alpha, F_n)}{\sqrt{n}} + O_p\left(\frac{1}{n}\right)\right]. \qquad (7.46)$$

Comparing (7.37) with (7.46), we conclude that

$$\underline{\theta}_{BC} - \underline{\theta}_E = O_p(n^{-1}).$$

It also follows from (7.46) that

$$P(\underline{\theta}_{BC} \leq \theta) = 1 - \alpha + \frac{\tilde{\psi}(z_{1-\alpha})\Phi'(z_{1-\alpha})}{\sqrt{n}} + O\left(\frac{1}{n}\right) \qquad (7.47)$$

with an even polynomial $\tilde{\psi}(x)$. Hence $\underline{\theta}_{BC}$ is first-order accurate in general. In fact,

$$\underline{\theta}_{BC} - \underline{\theta}_{BP} = 2\tilde{q}(0, F_n)\hat{\sigma}_F n^{-1} + O_p(n^{-3/2})$$

and, therefore, the bootstrap BC percentile and the bootstrap percentile confidence intervals have the same order of accuracy. The bootstrap BC percentile method, however, is a partial improvement over the bootstrap percentile method in the sense that the absolute value of $\bar{\psi}(z_{1-\alpha})$ in (7.47) is smaller than that of $\tilde{\psi}(z_{1-\alpha})$ in (7.45) (see Example 7.26).

While the bootstrap BC percentile method does not improve the bootstrap percentile method in terms of accuracy order, Hall (1988) showed that the bootstrap BC_a percentile method in Efron (1987) produces second-order accurate one-sided and two-sided confidence intervals and that (7.40) holds with $\underline{\theta}_{BT}$ replaced by the bootstrap BC_a percentile lower confidence bound.

We have considered the order of asymptotic accuracy for all bootstrap confidence intervals introduced in §7.4.1. In summary, all two-sided confidence intervals are second-order accurate; the one-sided bootstrap-t and bootstrap BC_a percentile confidence intervals are second-order accurate, whereas the one-sided bootstrap percentile, bootstrap BC percentile, and hybrid bootstrap confidence intervals are first-order accurate; however, the latter three are simpler than the former two.

Note that the results in §7.3.4 show that asymptotic confidence intervals obtained using the method in §7.3.1 have the same order of accuracy as the hybrid bootstrap confidence intervals.

Example 7.26. Suppose that $d = 1$ and $g(x) = x$. It follows from the results in §1.5.6 that expansions (7.38) and (7.42) hold with $q_1(x, F) = -\gamma(2x^2 + 1)/6$, $q_2(x, F) = x[(x^2 + 3)/4 - \kappa(x^2 - 3)/12 + 5\gamma^2(4x^2 - 1)/72]$, $\tilde{q}_1(x, F) = \gamma(x^2 - 1)/6$, and $\tilde{q}_2(x, F) = x[\kappa(x^2 - 3)/24 - \gamma^2(2x^2 - 5)/36]$, where $\gamma = \kappa_3 = E(X_1 - \mu)^3/\sigma^3$ (skewness), $\kappa = E(X_1 - \mu)^4/\sigma^4 - 3$ (kurtosis), and $\sigma^2 = \text{Var}(X_1)$.

The function ψ in (7.41) is equal to $x(1+2x^2)(\kappa-3\gamma^2/2)/6$; the function $\tilde{\psi}$ in (7.45) is equal to $\gamma x^2/2$; and the function $\bar{\psi}(x)$ in (7.47) is equal to $\gamma(x^2 + 2)/6$ (see Liu and Singh (1987)). If $\gamma \neq 0$, then $\underline{\theta}_{HB}$, $\underline{\theta}_{BC}$, and the asymptotic lower confidence bound $\underline{\theta}_N = \hat{\theta}_n - n^{-1/2}\hat{\sigma}_F z_{1-\alpha}$ are first-order accurate. In this example, we can still compare their relative performances in terms of the convergence speed of the coverage probability. Let

$$e(\underline{\theta}) = P(\underline{\theta} \leq \theta) - (1 - \alpha)$$

be the error in coverage probability for the lower confidence bound $\underline{\theta}$. It can be shown (exercise) that

$$|e(\underline{\theta}_{HB})| = |e(\underline{\theta}_N)| + C_n(z_\alpha, F) + o(n^{-1/2}) \tag{7.48}$$

and

$$|e(\underline{\theta}_N)| = |e(\underline{\theta}_{BC})| + C_n(z_\alpha, F) + o(n^{-1/2}), \tag{7.49}$$

where $C_n(x, F) = |\gamma|(x^2 - 1)\Phi'(x)/(6\sqrt{n})$. Assume $\gamma \neq 0$. When $z_\alpha^2 > 1$, which is usually the case in practice, $C_n(z_\alpha, F) > 0$ and, therefore, $\underline{\theta}_{BC}$ is better than $\underline{\theta}_N$, which is better than $\underline{\theta}_{HB}$. The use of $\underline{\theta}_N$ requires a variance estimator $\hat{\sigma}_F^2$, which is not required by the bootstrap BC percentile and hybrid bootstrap methods. When a variance estimator is available, we can use the bootstrap-t lower confidence bound, which is second-order accurate even when $\gamma \neq 0$. ∎

7.4.3 High-order accurate bootstrap confidence sets

The discussion in §7.3.4 shows how to derive second-order accurate confidence bounds. Hall (1992) showed how to obtain higher order accurate confidence sets using higher order Edgeworth and Cornish-Fisher expansions. However, the theoretical derivation of these high order accurate confidence sets may be very complicated (see Example 7.25). The bootstrap method can be used to obtain second-order or higher order accurate confidence sets without requiring any theoretical derivation but requiring some extensive computations.

The bootstrap prepivoting and bootstrap inverting

The hybrid bootstrap and the bootstrap-t are based on the bootstrap distribution estimators for $\sqrt{n}(\hat{\theta}_n - \theta)$ and $\sqrt{n}(\hat{\theta}_n - \theta)/\hat{\sigma}_F$, respectively. Beran (1987) argued that the reason why the bootstrap-t is better than the hybrid bootstrap is that $\sqrt{n}(\hat{\theta}_n - \theta)/\hat{\sigma}_F$ is more pivotal than $\sqrt{n}(\hat{\theta}_n - \theta)$ in the sense that the distribution of $\sqrt{n}(\hat{\theta}_n - \theta)/\hat{\sigma}_F$ is less dependent on the unknown F. The bootstrap-t method, however, requires a variance estimator $\hat{\sigma}_F^2$. Beran (1987) suggested the following method called *bootstrap prepivoting*. Let $\Re_n^{(0)}$ be a random function (such as $\sqrt{n}(\hat{\theta}_n - \theta)$ or $\sqrt{n}(\hat{\theta}_n - \theta)/\hat{\sigma}_F$) used to construct a confidence set for $\theta \in \mathcal{R}^k$, $H_n^{(0)}$ be the distribution of $\Re_n^{(0)}$, and let $\hat{H}_B^{(0)}$ be the bootstrap estimator of $H_n^{(0)}$. Define

$$\Re_n^{(1)} = \hat{H}_B^{(0)}(\Re_n^{(0)}). \tag{7.50}$$

If $H_n^{(0)}$ is continuous and if we replace $\hat{H}_B^{(0)}$ in (7.50) by $H_n^{(0)}$, then $\Re_n^{(1)}$ has the uniform distribution $U(0, 1)$. Hence, it is expected that $\Re_n^{(1)}$ is more pivotal than $\Re_n^{(0)}$. Let $\hat{H}_B^{(1)}$ be the bootstrap estimator of $H_n^{(1)}$, the distribution of $\Re_n^{(1)}$. Then $\Re_n^{(2)} = \hat{H}_B^{(1)}(\Re_n^{(1)})$ is more pivotal than $\Re_n^{(1)}$. In general, let $H_n^{(j)}$ be the distribution of $\Re_n^{(j)}$ and $\hat{H}_B^{(j)}$ be the bootstrap estimator of $H_n^{(j)}$, $j = 0, 1, 2, ...$ Then we can use the following confidence sets for θ:

$$C_{PREB}^{(j)}(X) = \{\theta : \Re_n^{(j)} \leq (\hat{H}_B^{(j)})^{-1}(1 - \alpha)\}, \quad j = 0, 1, 2, ... \tag{7.51}$$

Note that for each j, $C^{(j)}_{PREB}(X)$ is a hybrid bootstrap confidence set based on $\Re^{(j)}_n$. Since $\Re^{(j+1)}_n$ is more pivotal than $\Re^{(j)}_n$, we obtain a sequence of confidence sets with increasing accuracies. Beran (1987) showed that if the distribution of $\sqrt{n}(\hat{\theta}_n - \theta)$ has a two-term Edgeworth expansion, then the one-sided confidence interval $C^{(1)}_{PREB}(X)$ based on $\Re^{(0)}_n = \sqrt{n}(\hat{\theta}_n - \theta)$ is second-order accurate, and the two-sided confidence interval $C^{(1)}_{PREB}(X)$ based on $\Re^{(0)}_n = \sqrt{n}|\hat{\theta}_n - \theta|$ is third-order accurate. Hence, bootstrap prepivoting with one iteration improves the hybrid bootstrap method. It is expected that the one-sided confidence interval $C^{(2)}_{PREB}(X)$ based on $\Re^{(0)}_n = \sqrt{n}(\hat{\theta}_n - \theta)$ is third-order accurate, i.e., it is better than the one-sided bootstrap-t or bootstrap BC_a percentile confidence interval. More detailed discussion can be found in Beran (1987).

It seems that, using this iterative method, we can start with a $\Re^{(0)}_n$ and obtain a bootstrap confidence set that is as accurate as we want it to be. However, more computations are required for higher stage bootstrapping and, therefore, the practical implementation of this method is very hard, or even impossible, with current computational ability. We explain this with the computation of $C^{(1)}_{PREB}(X)$ based on $\Re^{(0)}_n = \Re(X, F)$. Suppose that we use the Monte Carlo approximation. Let $\{X^*_{1b}, ..., X^*_{nb}\}$ be i.i.d. samples from F_n, $b = 1, ..., B_1$. Then $\hat{H}^{(0)}_B$ is approximated by $\hat{H}^{(0,B_1)}_B$, the empirical distribution of $\{\Re^{(0)*}_{nb} : b = 1, ..., B_1\}$, where $\Re^{(0)*}_{nb} = \Re(X^*_{1b}, ..., X^*_{nb}, F_n)$. For each b, let F^*_{nb} be the empirical distribution of $X^*_{1b}, ..., X^*_{nb}$, $\{X^{**}_{1bj}, ..., X^{**}_{nbj}\}$ be i.i.d. samples from F^*_{nb}, $j = 1, ..., B_2$, H^*_b be the empirical c.d.f. of $\{\Re_n(X^{**}_{1bj}, ..., X^{**}_{nbj}, F^*_{nb}), j = 1, ..., B_2\}$, and $z^*_b = H^*_b(\Re^{(0)*}_{nb})$. Then $\hat{H}^{(1)}_B$ can be approximated by $\hat{H}^{(1,B_1B_2)}_B$, the empirical distribution of $\{z^*_b, b = 1, ..., B_1\}$, and the confidence set $C^{(1)}_{PREB}(X)$ can be approximated by

$$\left\{ \theta : \Re(X, F) \leq (\hat{H}^{(0,B_1)}_B)^{-1}((\hat{H}^{(1,B_1B_2)}_B)^{-1}(1 - \alpha)) \right\}.$$

The second-stage bootstrap sampling is nested in the first-stage bootstrap sampling. Thus the total number of bootstrap data sets we need is $B_1 B_2$, which is why this method is also called the double bootstrap. If each stage requires 1,000 bootstrap replicates, then the total number of bootstrap replicates is 1,000,000! Similarly, to compute $C^{(j)}_{PREB}(X)$ we need $(1,000)^{j+1}$ bootstrap replicates, $j = 2, 3, ...$, which limits the application of the bootstrap prepivoting method.

A very similar method, *bootstrap inverting*, is given in Hall (1992). Instead of using (7.51), we define

$$C^{(j)}_{INVB}(X) = \{\theta : \Re^{(j)}_n \leq (\hat{H}^{(j)}_B)^{-1}(1 - \alpha)\}, \quad j = 0, 1, 2, ...,$$

where
$$\Re_n^{(j)} = \Re_n^{(j-1)} - (\hat{H}_B^{(j-1)})^{-1}(1-\alpha), \quad j = 1, 2, ...,$$

and $\hat{H}_B^{(j)}$ is the bootstrap estimator of the distribution of $\Re_n^{(j)}$. For each $j \geq 1$, $C_{INVB}^{(j)}(X)$ and $C_{PREB}^{(j)}(X)$ in (7.51) have the same order of accuracy and require the same amount of computation. They are special cases of a general iterative bootstrap introduced by Hall and Martin (1988). Hall (1992) showed that confidence sets having the same order of accuracy as $C_{PREB}^{(j)}(X)$ can also be obtained using Edgeworth and Cornish-Fisher expansions. Thus, the bootstrap method replaces the analytic derivation of Edgeworth and Cornish-Fisher expansions by extensive computations.

Bootstrap calibrating

Suppose that we want a confidence set $C(X)$ with confidence coefficient $1 - \alpha$, which is called the *nominal level*. The basic idea of *bootstrap calibrating* is to improve $C(X)$ by adjusting its nominal level. Let π_n be the actual coverage probability of $C(X)$. The value of π_n can be estimated by a bootstrap estimator $\hat{\pi}_n$. If we find that $\hat{\pi}_n$ is far from $1 - \alpha$, then we construct a confidence set $C_1(X)$ with nominal level $1 - \tilde{\alpha}$ so that the coverage probability of $C_1(X)$ is closer to $1 - \alpha$ than π_n. Bootstrap calibrating can be used iteratively as follows. Estimate the true coverage probability of $C_1(X)$; if the difference between $1 - \alpha$ and the estimated coverage probability of $C_1(X)$ is still large, we can adjust the nominal level again and construct a new calibrated confidence set $C_2(X)$.

The key for bootstrap calibrating is how to determine the new nominal level $1 - \tilde{\alpha}$ in each step. We now discuss the method suggested by Loh (1987, 1991) in the case where the initial confidence sets are obtained by using the method in §7.3.1. Consider first the asymptotic lower confidence bound $\underline{\theta}_N = \hat{\theta}_n - n^{-1/2}\hat{\sigma}_F z_{1-\alpha}$ considered in Example 7.26. The coverage probability $\pi_n = P(\underline{\theta}_N \leq \theta)$ can be estimated by the bootstrap estimator (approximated by Monte Carlo if necessary)

$$\hat{\pi}_n = \hat{G}_B(z_{1-\alpha}) = P_*\big(\sqrt{n}(\hat{\theta}_n^* - \hat{\theta}_n)/\hat{\sigma}_F^* \leq z_{1-\alpha}\big).$$

When the bootstrap distribution admits the Edgeworth expansion (1.106) with $m = 3$, we have

$$\hat{\pi}_n = 1 - \alpha + \left[\frac{q_1(z_{1-\alpha}, F_n)}{\sqrt{n}} + \frac{q_2(z_{1-\alpha}, F_n)}{n}\right]\Phi'(z_{1-\alpha}) + O_p\left(\frac{1}{n^{3/2}}\right).$$

Let h be any increasing, unbounded, and twice differentiable function on the interval $(0, 1)$ and

$$\tilde{\alpha} = 1 - h^{-1}\big(h(1-\alpha) - \delta\big),$$

where

$$\delta = h(\hat{\pi}_n) - h(1 - \alpha)$$

$$= \left[\frac{q_1(z_{1-\alpha}, F_n)}{\sqrt{n}} + \frac{q_2(z_{1-\alpha}, F_n)}{n} \right] \Phi'(z_{1-\alpha}) h'(1 - \alpha)$$

$$+ \frac{[q_1(z_{1-\alpha}, F_n) \Phi'(z_{1-\alpha})]^2}{2n} h''(1 - \alpha) + O_p\left(\frac{1}{n^{3/2}} \right). \quad (7.52)$$

The bootstrap calibration lower confidence bound is

$$\underline{\theta}_{CLB} = \hat{\theta}_n - n^{-1/2} \hat{\sigma}_F z_{1-\tilde{\alpha}}.$$

By (7.52),

$$1 - \tilde{\alpha} = 1 - \alpha + \frac{q_1(z_{1-\alpha}, F_n) \Phi'(z_{1-\alpha})}{\sqrt{n}} + O_p\left(\frac{1}{n} \right) \quad (7.53)$$

and

$$z_{1-\tilde{\alpha}} = z_{1-\alpha} + \frac{q_1(z_{1-\alpha}, F_n)}{\sqrt{n}} + O_p\left(\frac{1}{n} \right) \quad (7.54)$$

(exercise). Thus,

$$\underline{\theta}_{CLB} = \hat{\theta}_n - \frac{\hat{\sigma}_F}{\sqrt{n}} \left[z_{1-\alpha} + \frac{q_1(z_{1-\alpha}, F_n)}{\sqrt{n}} + O_p\left(\frac{1}{n} \right) \right]. \quad (7.55)$$

Comparing (7.55) with (7.39), we find that

$$\underline{\theta}_{CLB} - \underline{\theta}_{BT} = O_p(n^{-3/2}).$$

Thus, $\underline{\theta}_{CLB}$ is second-order accurate.

We can take $[\underline{\theta}_{CLB}, \overline{\theta}_{CLB}]$ as a two-sided confidence interval; it is still second-order accurate. By calibrating directly the equal-tail two-sided confidence interval

$$[\underline{\theta}_N, \overline{\theta}_N] = [\hat{\theta}_n - n^{-1/2} \hat{\sigma}_F z_{1-\alpha}, \ \hat{\theta}_n + n^{-1/2} \hat{\sigma}_F z_{1-\alpha}], \quad (7.56)$$

we can obtain a higher order accurate confidence interval. Let $\hat{\pi}_n$ be the bootstrap estimator of the coverage probability $P(\underline{\theta}_N \leq \theta \leq \overline{\theta}_N)$, $\delta = h(\hat{\pi}_n) - h(1 - 2\alpha)$, and $\tilde{\alpha} = [1 - h^{-1}(h(1 - 2\alpha) - \delta)]/2$. Then the two-sided bootstrap calibration confidence interval is the interval given by (7.56) with α replaced by $\tilde{\alpha}$. Loh (1991) showed that this confidence interval is fourth-order accurate. The length of this interval exceeds the length of the interval in (7.56) by an amount of order $O_p(n^{-3/2})$.

7.5 Simultaneous Confidence Intervals

So far we have studied confidence sets for a real-valued θ or a vector-valued θ with a finite dimension k. In some applications, we need a confidence set for real-valued θ_t with $t \in \mathcal{T}$, where $\mathcal{T}$ is an index set that may contain infinitely many elements, for example, $\mathcal{T} = [0,1]$ or $\mathcal{T} = \mathcal{R}$.

Definition 7.6. Let X be a sample from $P \in \mathcal{P}$, let θ_t, $t \in \mathcal{T}$, be real-valued parameters related to P, and let $C_t(X)$, $t \in \mathcal{T}$, be a class of (one-sided or two-sided) confidence intervals.
(i) Intervals $C_t(X)$, $t \in \mathcal{T}$, are level $1 - \alpha$ *simultaneous confidence intervals* for θ_t, $t \in \mathcal{T}$, if and only if

$$\inf_{P \in \mathcal{P}} P\big(\theta_t \in C_t(X) \text{ for all } t \in \mathcal{T}\big) \geq 1 - \alpha. \qquad (7.57)$$

The left-hand side of (7.57) is the confidence coefficient of $C_t(X)$, $t \in \mathcal{T}$.
(ii) Intervals $C_t(X)$, $t \in \mathcal{T}$, are simultaneous confidence intervals for θ_t, $t \in \mathcal{T}$, with asymptotic significance level $1 - \alpha$ if and only if

$$\lim_{n \to \infty} P\big(\theta_t \in C_t(X) \text{ for all } t \in \mathcal{T}\big) \geq 1 - \alpha. \qquad (7.58)$$

Intervals $C_t(X)$, $t \in \mathcal{T}$, are $1 - \alpha$ asymptotically correct if and only if the equality in (7.58) holds. ∎

If the index set $\mathcal{T}$ contains $k < \infty$ elements, then $\theta = (\theta_t, t \in \mathcal{T})$ is a k-vector and the methods studied in the previous sections can be applied to construct a level $1 - \alpha$ confidence set $C(X)$ for θ. If $C(X)$ can be expressed as $\prod_{t \in \mathcal{T}} C_t(X)$ for some intervals $C_t(X)$, then $C_t(X)$, $t \in \mathcal{T}$, are level $1 - \alpha$ simultaneous confidence intervals. This simple method, however, does not always work. In this section, we introduce some other commonly used methods for constructing simultaneous confidence intervals.

7.5.1 Bonferroni's method

Bonferroni's method, which works when $\mathcal{T}$ contains $k < \infty$ elements, is based on the following simple inequality for k events $A_1, ..., A_k$:

$$P\left(\bigcup_{i=1}^{k} A_i\right) \leq \sum_{i=1}^{k} P(A_i) \qquad (7.59)$$

(see Proposition 1.1). For each $t \in \mathcal{T}$, let $C_t(X)$ be a level $1 - \alpha_t$ confidence interval for θ_t. If α_t's are chosen so that $\sum_{t \in \mathcal{T}} \alpha_t = \alpha$ (e.g., $\alpha_t = \alpha/k$ for all t), then Bonferroni's simultaneous confidence intervals are $C_t(X)$,

$t \in \mathcal{T}$. It can be shown (exercise) that Bonferroni's intervals are of level $1 - \alpha$, but they are not of confidence coefficient $1 - \alpha$ even if $C_t(X)$ has confidence coefficient $1 - \alpha_t$ for any fixed t. Note that Bonferroni's method does not require that $C_t(X)$, $t \in \mathcal{T}$, be independent.

Example 7.27 (Multiple comparison in one-way ANOVA models). Consider the one-way ANOVA model in Example 6.18. If the hypothesis H_0 in (6.53) is rejected, one typically would like to compare μ_i's. One way to compare μ_i's is to consider simultaneous confidence intervals for $\mu_i - \mu_j$, $1 \leq i < j \leq m$. Since X_{ij}'s are independently normal, the sample means $\bar{X}_{i\cdot}$ are independently normal $N(\mu_i, \sigma^2/n_i)$, $i = 1, ..., m$, respectively, and they are independent of $SSR = \sum_{i=1}^{m} \sum_{j=1}^{n_i} (X_{ij} - \bar{X}_{i\cdot})^2$. Consequently, $(\bar{X}_{i\cdot} - \bar{X}_{j\cdot})/\sqrt{v_{ij}}$ has the t-distribution t_{n-m}, $1 \leq i < j \leq m$, where $v_{ij} = (n_i^{-1} + n_j^{-1})SSR/(n - m)$. For each (i, j), a confidence interval for $\mu_i - \mu_j$ with confidence coefficient $1 - \alpha$ is

$$C_{ij,\alpha}(X) = [\, \bar{X}_{i\cdot} - \bar{X}_{j\cdot} - t_{n-m,\alpha/2}\sqrt{v_{ij}}, \ \bar{X}_{i\cdot} - \bar{X}_{j\cdot} + t_{n-m,\alpha/2}\sqrt{v_{ij}}\,], \quad (7.60)$$

where $t_{n-m,\alpha}$ is the $(1 - \alpha)$th quantile of the t-distribution t_{n-m}. One can show that $C_{ij,\alpha}(X)$ is actually UMAU (exercise). Bonferroni's level $1 - \alpha$ simultaneous confidence intervals for $\mu_i - \mu_j$, $1 \leq i < j \leq m$, are $C_{ij,\alpha_*}(X)$, $1 \leq i < j \leq m$, where $\alpha_* = 2\alpha/[m(m - 1)]$. When m is large, these confidence intervals are very conservative in the sense that the confidence coefficient of these intervals may be much larger than the nominal level $1 - \alpha$ and these intervals may be too wide to be useful.

If the normality assumption is removed, then $C_{ij,\alpha}(X)$ is $1 - \alpha$ asymptotically correct as $\min\{n_1, ..., n_m\} \to \infty$ and $\max\{n_1, ..., n_m\}/\min\{n_1, ..., n_m\} \to c < \infty$. Therefore, $C_{ij,\alpha_*}(X)$, $1 \leq i < j \leq m$, are simultaneous confidence intervals with asymptotic significance level $1 - \alpha$.

One can establish similar results for the two-way balanced ANOVA models in Example 6.19 (exercise). ∎

7.5.2 Scheffé's method in linear models

Since multiple comparison in ANOVA models (or, more generally, linear models) is one of the most important applications of simultaneous confidence intervals, we now introduce Scheffé's method for problems in linear models. Consider the normal linear model

$$X = N_n(Z\beta, \sigma^2 I_n), \quad (7.61)$$

where β is a p-vector of unknown parameters, $\sigma^2 > 0$ is unknown, and Z is an $n \times p$ known matrix of rank $r \leq p$. Let L be an $s \times p$ matrix of rank $s \leq r$. Suppose that $\mathcal{R}(L) \subset \mathcal{R}(Z)$ and we would like to construct simultaneous confidence intervals for $t^\tau L\beta$, where $t \in \mathcal{T} = \mathcal{R}^s - \{0\}$.

Let $\hat{\beta}$ be the LSE of β. Using the argument in Example 7.15, for each $t \in \mathcal{T}$, we can obtain the following confidence interval for $t^\tau L\beta$ with confidence coefficient $1 - \alpha$:

$$\left[t^\tau L\hat{\beta} - t_{n-r,\alpha/2}\hat{\sigma}\sqrt{t^\tau Dt}, \; t^\tau L\hat{\beta} + t_{n-r,\alpha/2}\hat{\sigma}\sqrt{t^\tau Dt} \right],$$

where $\hat{\sigma}^2 = \|X - Z\hat{\beta}\|^2/(n-r)$, $D = L(Z^\tau Z)^- L^\tau$, and $t_{n-r,\alpha}$ is the $(1-\alpha)$th quantile of the t-distribution t_{n-r}. However, these intervals are not level $1 - \alpha$ simultaneous confidence intervals for $t^\tau L\beta$, $t \in \mathcal{T}$.

Scheffé's (1959) method of constructing simultaneous confidence intervals for $t^\tau L\beta$ is based on the following equality (exercise):

$$x^\tau A^{-1}x = \max_{y \in \mathcal{R}^k, y \neq 0} \frac{(y^\tau x)^2}{y^\tau Ay}, \tag{7.62}$$

where $x \in \mathcal{R}^k$ and A is a $k \times k$ positive definite matrix.

Theorem 7.10. Assume normal linear model (7.61). Let L be an $s \times p$ matrix of rank $s \leq r$. Assume that $\mathcal{R}(L) \subset \mathcal{R}(Z)$ and $D = L(Z^\tau Z)^- L^\tau$ is of full rank. Then

$$C_t(X) = \left[t^\tau L\hat{\beta} - \hat{\sigma}\sqrt{sc_\alpha t^\tau Dt}, \; t^\tau L\hat{\beta} + \hat{\sigma}\sqrt{sc_\alpha t^\tau Dt} \right], \quad t \in \mathcal{T},$$

are simultaneous confidence intervals for $t^\tau L\beta$, $t \in \mathcal{T}$, with confidence coefficient $1 - \alpha$, where $\hat{\sigma}^2 = \|X - Z\hat{\beta}\|^2/(n-r)$, $\mathcal{T} = \mathcal{R}^s - \{0\}$, and c_α is the $(1 - \alpha)$th quantile of the F-distribution $F_{s,n-r}$.
Proof. Note that $t^\tau L\beta \in C_t(X)$ for all $t \in \mathcal{T}$ is equivalent to

$$\frac{(L\hat{\beta} - L\beta)^\tau D^{-1}(L\hat{\beta} - L\beta)}{s\hat{\sigma}^2} = \max_{t \in \mathcal{T}} \frac{(t^\tau L\hat{\beta} - t^\tau L\beta)^2}{s\hat{\sigma}^2 t^\tau Dt} \leq c_\alpha. \tag{7.63}$$

Then the result follows from the fact that the quantity on the left-hand side of (7.63) has the F-distribution $F_{s,n-r}$. ∎

If the normality assumption is removed but conditions in Theorem 3.12 are assumed, then Scheffé's intervals in Theorem 7.10 are $1 - \alpha$ asymptotically correct (exercise).

The choice of the matrix L depends on the purpose of the analysis. One particular choice is $L = Z$, in which case $t^\tau L\beta$ is the mean of $t^\tau X$. When Z is of full rank, we can choose $L = I_p$, in which case $\{t^\tau L\beta : t \in \mathcal{T}\}$ is the class of all linear functions of β. Another L commonly used when Z is of full rank is the following $(p-1) \times p$ matrix:

$$L = \begin{pmatrix} 1 & 0 & 0 & \cdots & 0 & -1 \\ 0 & 1 & 0 & \cdots & 0 & -1 \\ \cdots & \cdots & \cdots & \cdots & \cdots & \cdots \\ 0 & 0 & 0 & \cdots & 1 & -1 \end{pmatrix}. \tag{7.64}$$

It can be shown (exercise) that when L is given by (7.64),

$$\{t^\tau L\beta : t \in \mathcal{R}^{p-1} - \{0\}\} = \{c^\tau \beta : c \in \mathcal{R}^p - \{0\}, c^\tau J = 0\}, \tag{7.65}$$

where J is the p-vector of ones. Functions $c^\tau \beta$ satisfying $c^\tau J = 0$ are called *contrasts*. Therefore, setting simultaneous confidence intervals for $t^\tau L\beta$, $t \in \mathcal{T}$, with L given by (7.64) is the same as setting simultaneous confidence intervals for all nonzero contrasts.

Although Scheffé's intervals have confidence coefficient $1 - \alpha$, they are too conservative if we are only interested in $t^\tau L\beta$ for t in a subset of $\mathcal{T}$. In a one-way ANOVA model (Example 7.27), for instance, multiple comparison can be carried out using Scheffé's intervals with $\beta = (\mu_1, ..., \mu_m)$, L given by (7.64), and $t \in \mathcal{T}_0$ that contains exactly $m(m-1)/2$ vectors (Exercise 110). The resulting Scheffé's intervals are (Exercise 110)

$$[\bar{X}_{i\cdot} - \bar{X}_{j\cdot} - \sqrt{sc_\alpha v_{ij}}, \ \bar{X}_{i\cdot} - \bar{X}_{j\cdot} + \sqrt{sc_\alpha v_{ij}}], \quad t \in \mathcal{T}_0, \tag{7.66}$$

where $\bar{X}_{i\cdot}$ and v_{ij} are given in (7.60). Since $\mathcal{T}_0$ contains a much smaller number of elements than $\mathcal{T}$, the simultaneous confidence intervals in (7.66) are very conservative. In fact, they are often more conservative than Bonferroni's intervals derived in Example 7.27 (see Example 7.29). In the following example, however, Scheffé's intervals have confidence coefficient $1 - \alpha$, although we consider $t \in \mathcal{T}_0 \subset \mathcal{T}$.

Example 7.28 (Simple linear regression). Consider the special case of model (7.61) where

$$X_i = N(\beta_0 + \beta_1 z_i, \sigma^2), \quad i = 1, ..., n,$$

and $z_i \in \mathcal{R}$ satisfying $S_z = \sum_{i=1}^n (z_i - \bar{z})^2 > 0$, $\bar{z} = n^{-1}\sum_{i=1}^n z_i$. In this case, we are usually interested in simultaneous confidence intervals for the regression function $\beta_0 + \beta_1 z$, $z \in \mathcal{R}$. Note that the result in Theorem 7.10 (with $L = I_2$) can be applied to obtain simultaneous confidence intervals for $\beta_0 y + \beta_1 z$, $t \in \mathcal{T} = \mathcal{R}^2 - \{0\}$, where $t = (y, z)$. If we let $y \equiv 1$, Scheffé's intervals in Theorem 7.10 are

$$[\hat{\beta}_0 + \hat{\beta}_1 z - \hat{\sigma}\sqrt{2c_\alpha D(z)}, \ \hat{\beta}_0 + \hat{\beta}_1 z + \hat{\sigma}\sqrt{2c_\alpha D(z)}], \quad z \in \mathcal{R} \tag{7.67}$$

(exercise), where $D(z) = n^{-1} + (z - \bar{z})^2/S_z$. Unless

$$\max_{z \in \mathcal{R}} \frac{(\hat{\beta}_0 + \hat{\beta}_1 z - \beta_0 - \beta_1 z)^2}{D(z)} = \max_{t=(y,z) \in \mathcal{T}} \frac{(\hat{\beta}_0 y + \hat{\beta}_1 z - \beta_0 y - \beta_1 z)^2}{t^\tau (Z^\tau Z)^{-1} t} \quad (7.68)$$

holds with probability 1, where Z is the $n \times 2$ matrix whose ith row is the vector $(1, z_i)$, the confidence coefficient of the intervals in (7.67) is larger than $1 - \alpha$. We now show that (7.68) actually holds with probability 1 so that the intervals in (7.67) have confidence coefficient $1 - \alpha$. First,

$$P(n(\hat{\beta}_0 - \beta_0) + n(\hat{\beta}_1 - \beta_1)\bar{z} \neq 0) = 1.$$

Second, it can be shown (exercise) that the maximum on the right-hand side of (7.68) is achieved at

$$t = \begin{pmatrix} y \\ z \end{pmatrix} = \frac{Z^\tau Z}{n(\hat{\beta}_0 - \beta_0) + n(\hat{\beta}_1 - \beta_1)\bar{z}} \begin{pmatrix} \hat{\beta}_0 - \beta_0 \\ \hat{\beta}_1 - \beta_1 \end{pmatrix}. \quad (7.69)$$

Finally, (7.68) holds since y in (7.69) is equal to 1 (exercise). ∎

7.5.3 Tukey's method in one-way ANOVA models

Consider the one-way ANOVA model in Example 6.18 (and Example 7.27). Note that both Bonferroni's and Scheffé's simultaneous confidence intervals for $\mu_i - \mu_j$, $1 \leq i < j \leq m$, are not of confidence coefficient $1 - \alpha$ and often too conservative. Tukey's method introduced next produces simultaneous confidence intervals for all nonzero contrasts (including the differences $\mu_i - \mu_j$, $1 \leq i < j \leq m$) with confidence coefficient $1 - \alpha$.

Let $\hat{\sigma}^2 = SSR/(n - m)$, where SSR is given in Example 7.27. The *studentized range* is defined to be

$$R_{st} = \max_{1 \leq i < j \leq m} \frac{|(\bar{X}_{i\cdot} - \mu_i) - (\bar{X}_{j\cdot} - \mu_j)|}{\hat{\sigma}}. \quad (7.70)$$

Note that the distribution of R_{st} does not depend on any unknown parameter (exercise).

Theorem 7.11. Assume the one-way ANOVA model in Example 6.18. Let q_α be the $(1 - \alpha)$th quantile of R_{st} in (7.70). Then Tukey's intervals

$$[c^\tau \hat{\beta} - q_\alpha \hat{\sigma} c_+, \ c^\tau \hat{\beta} + q_\alpha \hat{\sigma} c_+], \qquad c \in \mathcal{R}^m - \{0\}, c^\tau J = 0,$$

are simultaneous confidence intervals for $c^\tau \beta$, $c \in \mathcal{R}^m - \{0\}$, $c^\tau J = 0$, with confidence coefficient $1 - \alpha$, where c_+ is the sum of all positive components of c, $\beta = (\mu_1, ..., \mu_m)$, $\hat{\beta} = (\bar{X}_{1\cdot}, ..., \bar{X}_{m\cdot})$, and J is the m-vector of ones.

Proof. Let $Y_i = (\bar{X}_{i\cdot} - \mu_i)/\hat{\sigma}$ and $Y = (Y_1, ..., Y_m)$. Then the result follows if we can show that

$$\max_{1 \leq i < j \leq m} |Y_i - Y_j| \leq q_\alpha \qquad (7.71)$$

is equivalent to

$$|c^T Y| \leq q_\alpha c_+ \qquad \text{for all } c \in \mathcal{R}^m \text{ satisfying } c^T J = 0, c \neq 0. \qquad (7.72)$$

Let $c(i,j) = (c_1, ..., c_m)$ with $c_i = 1$, $c_j = -1$, and $c_l = 0$ for $l \neq i$ or $l \neq j$. Then $c(i,j)_+ = 1$ and $|[c(i,j)]^T Y| = |Y_i - Y_j|$ and, therefore, (7.72) implies (7.71). Let $c = (c_1, ..., c_m)$ be a vector satisfying the conditions in (7.72). Define $-c_-$ to be the sum of negative components of c. Then

$$|c^T Y| = \frac{1}{c_+} \left| c_+ \sum_{j:c_j < 0} c_j Y_j + c_- \sum_{i:c_i > 0} c_i Y_i \right|$$

$$= \frac{1}{c_+} \left| \sum_{i:c_i > 0} \sum_{j:c_j < 0} c_i c_j Y_j - \sum_{j:c_j < 0} \sum_{i:c_i > 0} c_i c_j Y_i \right|$$

$$= \frac{1}{c_+} \left| \sum_{i:c_i > 0} \sum_{j:c_j < 0} c_i c_j (Y_j - Y_i) \right|$$

$$\leq \frac{1}{c_+} \sum_{i:c_i > 0} \sum_{j:c_j < 0} |c_i c_j| |Y_j - Y_i|$$

$$\leq \max_{1 \leq i < j \leq m} |Y_j - Y_i| \left(\frac{1}{c_+} \sum_{i:c_i > 0} \sum_{j:c_j < 0} |c_i||c_j| \right)$$

$$= \max_{1 \leq i < j \leq m} |Y_j - Y_i| c_+,$$

where the first and the last equalities follow from the fact that $c_- = c_+ \neq 0$. Hence (7.71) implies (7.72). ∎

Tukey's method works well when n_i's are all equal to n_0, in which case values of $\sqrt{n_0} q_\alpha$ can be found using tables or statistical software. When n_i's are unequal, some modifications are suggested; see Tukey (1977) and Milliken and Johnson (1992).

Example 7.29. We compare the t-type confidence intervals in (7.60), Bonferroni's, Scheffé's, and Tukey's simultaneous confidence intervals for $\mu_i - \mu_j$, $1 \leq i < j \leq 3$, based on the following data X_{ij} given in Mendenhall and Sincich (1995):

$j = 1$	2	3	4	5	6	7	8	9	10
$i = 1$ 148	76	393	520	236	134	55	166	415	153
2 513	264	433	94	535	327	214	135	280	304
3 335	643	216	536	128	723	258	380	594	465

In this example, $m = 3$, $n_i \equiv n_0 = 10$, $\bar{X}_1. = 229.6$, $\bar{X}_2. = 309.8$, $\bar{X}_3. = 427.8$, and $\hat{\sigma} = 168.95$. Let $\alpha = 0.05$. For the t-type intervals in (7.60), $t_{27,0.975} = 2.05$. For Bonferroni's method, $\alpha_* = \alpha/3 = 0.017$ and $t_{27,0.983} = 2.55$. For Scheffé's method, $c_{0.05} = 3.35$ and $\sqrt{2c_{0.05}} = 2.59$. From Table 13 in Mendenhall and Sincich (1995, Appendix II), $\sqrt{n_0} q_{0.05} = 3.49$. The resulting confidence intervals are given as follows.

| Method | Parameter | | | Length |
	$\mu_1 - \mu_2$	$\mu_1 - \mu_3$	$\mu_2 - \mu_3$	
t-type	$[-235.2, 74.6]$	$[-353.1, -43.3]$	$[-272.8, 37.0]$	309.8
Bonferroni	$[-273.0, 112.4]$	$[-390.9, -5.5]$	$[-310.6, 74.8]$	385.4
Scheffé	$[-276.0, 115.4]$	$[-393.9, -2.5]$	$[-313.6, 77.8]$	391.4
Tukey	$[-267.3, 106.7]$	$[-385.2, -11.2]$	$[-304.9, 69.1]$	374.0

Apparently, t-type intervals have the shortest length, but they are not simultaneous confidence intervals. Tukey's intervals in this example have the shortest length among simultaneous confidence intervals. Scheffé's intervals have the longest length. ∎

7.5.4 Confidence bands for c.d.f.'s

Let $X_1, ..., X_n$ be i.i.d. from a continuous c.d.f. F on $\mathcal{R}$. Consider the problem of setting simultaneous confidence intervals for $F(t)$, $t \in \mathcal{R}$. A class of simultaneous confidence intervals indexed by $t \in \mathcal{R}$ is called a confidence band. For example, the class of intervals in (7.67) is a confidence band with confidence coefficient $1 - \alpha$.

First, consider the case where F is in a parametric family, i.e., $F = F_\theta$, $\theta \in \Theta \subset \mathcal{R}^k$. If θ is real-valued and $F_\theta(t)$ is nonincreasing in θ for every t (e.g., when the parametric family has monotone likelihood ratio; see Lemma 6.3) and if $[\underline{\theta}, \overline{\theta}]$ is a confidence interval for θ with confidence coefficient (or significance level) $1 - \alpha$, then

$$[F_{\overline{\theta}}(t), F_{\underline{\theta}}(t)], \qquad t \in \mathcal{R},$$

are simultaneous confidence intervals for $F(t)$, $t \in \mathcal{R}$, with confidence coefficient (or significance level) $1 - \alpha$. One-sided simultaneous confidence intervals can be similarly obtained.

When $F = F_\theta$ with a multivariate θ, there is no simple and general way of constructing a confidence band for $F(t)$, $t \in \mathcal{R}$. We consider an example.

Example 7.30. Let $X_1, ..., X_n$ be i.i.d. from $N(\mu, \sigma^2)$. Note that $F(t) = \Phi\left(\frac{t-\mu}{\sigma}\right)$. If μ is unknown and σ^2 is known, then, from the results in Example 7.14, a confidence band for $F(t)$, $t \in \mathcal{R}$, with confidence coefficient

$1 - \alpha$ is

$$\left[\Phi\left(\frac{t-\bar{X}}{\sigma} - \frac{z_{1-\alpha/2}}{\sqrt{n}}\right),\ \Phi\left(\frac{t-\bar{X}}{\sigma} + \frac{z_{1-\alpha/2}}{\sqrt{n}}\right)\right],\quad t \in \mathcal{R}.$$

A confidence band can be similarly obtained if σ^2 is unknown and μ is known.

Suppose now that both μ and σ^2 are unknown. In Example 7.18, we discussed how to obtain a lower confidence bound $\underline{\theta}$ for $\theta = \mu/\sigma$. An upper confidence bound $\bar{\theta}$ for θ can be similarly obtained. Suppose that both $\underline{\theta}$ and $\bar{\theta}$ have confidence coefficient $1 - \alpha/4$. Using inequality (7.59), we can obtain the following level $1 - \alpha$ confidence band for $F(t)$, $t \in \mathcal{R}$:

$$\left[\Phi\left(\frac{a_{n,\alpha}t}{S} - \bar{\theta}\right),\ \Phi\left(\frac{b_{n,\alpha}t}{S} - \underline{\theta}\right)\right],\quad t \in \mathcal{R},$$

where $a_{n,\alpha} = [\chi^2_{n-1,1-\alpha/4}/(n-1)]^{1/2}$, $b_{n,\alpha} = [\chi^2_{n-1,\alpha/4}/(n-1)]^{1/2}$, and $\chi^2_{n-1,\alpha}$ is the $(1-\alpha)$th quantile of the chi-square distribution χ^2_{n-1}. ∎

Consider now the case where F is in a nonparametric family. Let $D_n(F) = \sup_{t \in \mathcal{R}} |F_n(t) - F(t)|$, which is related to the Kolmogorov-Smirnov test statistics introduced in §6.5.2, where F_n is the empirical c.d.f. given by (5.1). From Theorem 6.10(i), there exists a c_α such that

$$P\big(D_n(F) \le c_\alpha\big) = 1 - \alpha. \tag{7.73}$$

Then a confidence band for $F(t)$, $t \in \mathcal{R}$, with confidence coefficient $1 - \alpha$ is given by

$$[\, F_n(t) - c_\alpha,\ F_n(t) + c_\alpha \,]\quad t \in \mathcal{R}. \tag{7.74}$$

When n is large, we may approximate c_α using the asymptotic result in Theorem 6.10(ii), i.e., we can replace (7.73) by

$$\sum_{j=1}^{\infty}(-1)^{j-1}e^{-2j^2 c_\alpha^2} = \frac{\alpha}{2}. \tag{7.75}$$

The resulting intervals in (7.74) have limiting confidence coefficient $1 - \alpha$.

Using $D_n^+(F) = \sup_{t \in \mathcal{R}}[F_n(t) - F(t)]$ and the results in Theorem 6.10, we can also obtain one-sided simultaneous confidence intervals for $F(t)$, $t \in \mathcal{R}$, with confidence coefficient $1 - \alpha$ or limiting confidence coefficient $1 - \alpha$.

When n is small, it is possible that some intervals in (7.74) are not within the interval $[0, 1]$. This is undesirable since $F(t) \in [0, 1]$ for all t. One way to solve this problem is replacing $F_n(t) - c_\alpha$ and $F_n(t) + c_\alpha$ by, respectively, $\max\{F_n(t) - c_\alpha, 0\}$ and $\min\{F_n(t) + c_\alpha, 1\}$. But the resulting intervals have a confidence coefficient larger than $1-\alpha$. The limiting confidence coefficient of these intervals is still $1 - \alpha$ (exercise).

7.6 Exercises

1. Let $X_{i1}, ..., X_{in_i}$, $i = 1, 2$, be two independent samples i.i.d. from $N(\mu_i, \sigma_i^2)$, $i = 1, 2$, respectively, where all parameters are unknown. Let $\bar{X}_i$ and S_i^2 be the sample mean and sample variance of the ith sample, $i = 1, 2$.
   (a) Let $\theta = \mu_1 - \mu_2$. Assume that $\sigma_1 = \sigma_2$. Show that

$$t(X, \theta) = \frac{(\bar{X}_1 - \bar{X}_2 - \theta)/\sqrt{n_1^{-1} + n_2^{-1}}}{\sqrt{[(n_1 - 1)S_1^2 + (n_2 - 1)S_2^2]/(n_1 + n_2 - 2)}}$$

   is a pivotal quantity and construct a confidence interval for θ with confidence coefficient $1 - \alpha$, using $t(X, \theta)$.
   (b) Let $\theta = \sigma_2^2/\sigma_1^2$. Show that $\Re(X, \theta) = S_2^2/(\theta S_1^2)$ is a pivotal quantity and construct a confidence interval for θ with confidence coefficient $1 - \alpha$, using $\Re(X, \theta)$.

2. Let X_i, $i = 1, 2$, be independent with the p.d.f.'s $\lambda_i e^{-\lambda_i x} I_{(0,\infty)}(x)$, $i = 1, 2$, respectively.
   (a) Let $\theta = \lambda_1/\lambda_2$. Show that $\theta X_1/X_2$ is a pivotal quantity and construct a confidence interval for θ with confidence coefficient $1 - \alpha$, using this pivotal quantity.
   (b) Let $\theta = (\lambda_1, \lambda_2)$. Show that $\lambda_1 X_1 + \lambda_2 X_2$ is a pivotal quantity and construct a confidence set for θ with confidence coefficient $1 - \alpha$, using this pivotal quantity.

3. In Example 7.1,
   (a) obtain a pivotal quantity when $\theta = (\mu, \sigma)$ and discuss how to use it to construct a confidence set for θ with confidence coefficient $1 - \alpha$;
   (b) obtain the confidence set in part (a) when f is the p.d.f. of the exponential distribution $E(0, 1)$.

4. In Example 7.3, show that the equation $n[\bar{Y}(\theta)]^2 = t_{n-1,\alpha/2}^2 S^2(\theta)$ defines a parabola in θ and discuss when $C(X)$ is a finite interval, the complement of a finite interval, or the whole real line.

5. Let X be a sample from P in a parametric family indexed by θ. Suppose that $T(X)$ is a real-valued statistic with p.d.f. $f_\theta(t)$ and that $\Re(t, \theta)$ is a monotone function of t for each θ. Show that if

$$f_\theta(t) = g(\Re(t, \theta)) \left| \frac{\partial}{\partial t} \Re(t, \theta) \right|$$

   for some function g, then $\Re(T(X), \theta)$ is a pivotal quantity.

6. Let $X_1, ..., X_n$ be i.i.d. from $N(\theta, \theta)$ with an unknown $\theta > 0$. Find a pivotal quantity and use it to construct a confidence interval for θ.

7. Prove (7.3).

8. Let $X_1, ..., X_n$ be i.i.d. from the exponential distribution $E(0, \theta)$ with an unknown $\theta > 0$.
   (a) Using the pivotal quantity $\bar{X}/\theta$, construct a confidence interval for θ with confidence coefficient $1 - \alpha$.
   (b) Apply Theorem 7.1 with $T = \bar{X}$ to construct a confidence interval for θ with confidence coefficient $1 - \alpha$.

9. Let $X_1, ..., X_n$ be i.i.d. random variables with the Lebesgue p.d.f. $\frac{a}{\theta} \left(\frac{x}{\theta}\right)^{a-1} I_{(0,\theta)}(x)$, where $a \geq 1$ is known and $\theta > 0$ is unknown.
   (a) Apply Theorem 7.1 with $T = X_{(n)}$ to construct a confidence interval for θ with confidence coefficient $1 - \alpha$. Compare the result with that in Example 7.2 when $a = 1$.
   (b) Show that the confidence interval in (a) can also be obtained using a pivotal quantity.

10. Let $X_1, ..., X_n$ be i.i.d. from the exponential distribution $E(a, 1)$ with an unknown a.
    (a) Construct a confidence interval for a with confidence coefficient $1 - \alpha$ by using Theorem 7.1 with $T = X_{(1)}$.
    (b) Show that the confidence interval in (a) can also be obtained using a pivotal quantity.

11. Let X be a single observation from the uniform distribution $U(\theta - \frac{1}{2}, \theta + \frac{1}{2})$, where $\theta \in \mathcal{R}$.
    (a) Show that $X - \theta$ is a pivotal quantity and that a confidence interval of the form $[X + c, X + d]$ with some constants $-\frac{1}{2} < c < d < \frac{1}{2}$ has confidence coefficient $1 - \alpha$ if and only if its length is $1 - \alpha$.
    (b) Show that the c.d.f. $F_\theta(x)$ of X is nonincreasing in θ for any x and apply Theorem 7.1 to construct a confidence interval for θ with confidence coefficient $1 - \alpha$.

12. Let $X_1, ..., X_n$ be i.i.d. from the Pareto distribution $Pa(a, \theta)$, $\theta > 0$, $a > 0$.
    (a) When θ is known, derive a confidence interval for a with confidence coefficient $1 - \alpha$ by applying Theorem 7.1 with $T = X_{(1)}$, the smallest order statistic.
    (b) When both a and θ are unknown and $n \geq 2$, derive a confidence interval for θ with confidence coefficient $1 - \alpha$ by applying Theorem 7.1 with $T = \prod_{i=1}^{n} (X_i/X_{(1)})$.
    (c) Show that the confidence intervals in (a) and (b) can be obtained using pivotal quantities.
    (d) When both a and θ are unknown, construct a confidence set for (a, θ) with confidence coefficient $1 - \alpha$ by using a pivotal quantity.

13. Let $X_1, ..., X_n$ be i.i.d. from the Weibull distribution $W(a, \theta)$, where $a > 0$ and $\theta > 0$ are unknown. Show that $\Re(X, a, \theta) = \prod_{i=1}^{n}(X_i^a/\theta)$ is pivotal. Construct a confidence set for (a, θ) with confidence coefficient $1 - \alpha$ by using $\Re(X, a, \theta)$.

14. Consider Exercise 17 in §6.6. Suppose that F and G are continuous. Construct a confidence interval for θ with confidence coefficient $1 - \alpha$, based on the observation X.

15. Prove (7.4).

16. Let $X_1, ..., X_n$ be i.i.d. binary random variables with $P(X_i = 1) = p$. Using Theorem 7.1 with $T = \sum_{i=1}^{n} X_i$, show that a level $1 - \alpha$ confidence interval for p is

$$\left[\frac{1}{1 + \frac{n-T+1}{T}F_{2(n-T+1),2T,\alpha_2}}, \frac{\frac{T+1}{n-T}F_{2(T+1),2(n-T),\alpha_1}}{1 + \frac{T+1}{n-T}F_{2(T+1),2(n-T),\alpha_1}} \right],$$

where $F_{a,b,\alpha}$ is the $(1 - \alpha)$th quantile of the F-distribution $F_{a,b}$, and $F_{a,0,\alpha}$ is defined to be ∞. (Hint: show that $P(T \geq t) = P(Y \leq p)$, where Y has the beta distribution $B(t, n - t + 1)$.)

17. Let X be a sample of size 1 from the negative binomial distribution $NB(p, r)$ with a known r and an unknown $p \in (0, 1)$. Using Theorem 7.1 with $T = X$, show that a level $1 - \alpha$ confidence interval for p is

$$\left[\frac{1}{1 + \frac{T+1}{r+1}F_{2(T+1),2(r+1),\alpha_2}}, \frac{\frac{r+1}{T}F_{2(r+1),2T,\alpha_1}}{1 + \frac{r+1}{T}F_{2(r+1),2T,\alpha_1}} \right],$$

where $F_{a,b,\alpha}$ is the same as that in the previous exercise.

18. Let T be a statistic having the noncentral chi-square distribution $\chi_r^2(\theta)$ (see §1.3.1), where $\theta > 0$ is unknown and r is a known positive integer. Show that the c.d.f. $F_{T,\theta}(t)$ of T is nonincreasing in θ for each fixed t and use this result to construct a confidence interval for θ with confidence coefficient $1 - \alpha$.

19. Repeat the previous exercise when $\chi_r^2(\theta)$ is replaced by the noncentral F-distribution $F_{r_1,r_2}(\theta)$ (see §1.3.1) with unknown $\theta > 0$ and known positive integers r_1 and r_2.

20. Consider the one-way ANOVA model in Example 6.18. Let $\bar{\mu} = n^{-1}\sum_{i=1}^{m} n_i\mu_i$ and $\theta = \sigma^{-2}\sum_{i=1}^{m} n_i(\mu_i - \bar{\mu})^2$. Construct an upper confidence bound for θ that has confidence coefficient $1 - \alpha$ and is a function of $T = (n - m)(m - 1)^{-1}SST/SSR$.

21. Prove Proposition 7.2 and provide a sufficient condition under which the test $T(X) = 1 - I_{A(\theta_0)}(X)$ has size α.

22. In Example 7.7,
    (a) show that $c(\theta)$ and $c_i(\theta)$'s are nondecreasing in θ;
    (b) show that $(\underline{p}, 1]$ with $\underline{p}$ given by (7.5) is a level $1 - \alpha$ confidence interval for p;
    (c) compare the interval in (b) with the interval obtained using the result in Exercise 16 with $\alpha_1 = 0$.

23. Show that the confidence intervals in Example 7.14 and Exercise 1 can also be obtained by inverting the acceptance regions of the tests for one-sample and two-sample problems in §6.2.3.

24. Let X_i, $i = 1, 2$, be independently distributed as the binomial distributions $Bi(p_i, n_i)$, $i = 1, 2$, respectively, where n_i's are known and p_i's are unknown. Show how to invert the acceptance regions of the UMPU tests in Example 6.11 to obtain a level $1-\alpha$ confidence interval for the odds ratio $\frac{p_2(1-p_1)}{p_1(1-p_2)}$.

25. Let $X_1, ..., X_n$ be i.i.d. from $N(\mu, \sigma^2)$.
    (a) Suppose that $\sigma^2 = \gamma\mu^2$ with unknown $\gamma > 0$ and $\mu \in \mathcal{R}$. Obtain a confidence set for γ with confidence coefficient $1 - \alpha$ by inverting the acceptance regions of LR tests for $H_0 : \gamma = \gamma_0$ versus $H_1 : \gamma \neq \gamma_0$.
    (b) Repeat (a) when $\sigma^2 = \gamma\mu$ with unknown $\gamma > 0$ and $\mu > 0$.

26. Consider the problem in Example 6.17. Discuss how to construct a confidence interval for θ with confidence coefficient $1 - \alpha$ by
    (a) inverting the acceptance regions of the tests derived in Example 6.17;
    (b) applying Theorem 7.1.

27. Let $X_1, ..., X_n$ be i.i.d. from the uniform distribution $U(\theta - \frac{1}{2}, \theta + \frac{1}{2})$, where $\theta \in \mathcal{R}$. Construct a confidence interval for θ with confidence coefficient $1 - \alpha$.

28. Let $X_1, ..., X_n$ be i.i.d. binary random variables with $P(X_i = 1) = p$. Using the p.d.f. of the beta distribution $B(a, b)$ as the prior p.d.f., construct a level $1 - \alpha$ HPD credible set for p.

29. Let $X_1, ..., X_n$ be i.i.d. from $N(\mu, \sigma^2)$ with an unknown $\theta = (\mu, \sigma^2)$. Consider the prior Lebesgue p.d.f. $\pi(\theta) = \pi_1(\mu|\sigma^2)\pi_2(\sigma^2)$, where $\pi_1(\mu|\sigma^2)$ is the p.d.f. of $N(\mu_0, \sigma_0^2\sigma^2)$,

$$\pi_2(\sigma^2) = \frac{1}{\Gamma(a)b^a} \left(\frac{1}{\sigma^2}\right)^{a+1} e^{-1/(b\sigma^2)} I_{(0,\infty)}(\sigma^2),$$

and μ_0, σ_0^2, a, and b are known.

(a) Find the posterior of μ and construct a level $1 - \alpha$ HPD credible set for μ.

(b) Show that the credible set in (a) converges to the confidence interval obtained in Example 7.14 as σ_0^2, a, and b converge to some limits.

30. Let $X_1, ..., X_n$ be i.i.d. with a Lebesgue p.d.f. $\frac{1}{\sigma} f\left(\frac{x-\mu}{\sigma}\right)$, where f is a known p.d.f. and μ and $\sigma > 0$ are unknown. Let X_0 be a future observation that is independent of X_i's and has the same distribution as X_i. Find a pivotal quantity $\Re(X, X_0)$ and construct a level $1 - \alpha$ prediction set for X_0.

31. Let $X_1, ..., X_n$ be i.i.d. from a continuous c.d.f. F on $\mathcal{R}$ and X_0 be a future observation that is independent of X_i's and has the c.d.f. F. Suppose that F is strictly increasing in a neighborhood of $F^{-1}(\alpha/2)$ and a neighborhood of $F^{-1}(1 - \alpha/2)$. Let F_n be the empirical c.d.f. defined by (5.1). Show that the prediction interval $C(X) = [F_n^{-1}(\alpha/2), F_n^{-1}(1 - \alpha/2)]$ for X_0 satisfies $P(X_0 \in C(X)) \to 1 - \alpha$, where P is the joint distribution of $(X_0, X_1, ..., X_n)$.

32. Let $X_1, ..., X_n$ be i.i.d. with a Lebesgue p.d.f. $f(x - \mu)$, where f is known and μ is unknown.

(a) If f is the p.d.f. of the standard normal distribution, show that the confidence interval $[\bar{X} - c_1, \bar{X} + c_1]$ is better than $[X_1 - c_2, X_1 + c_2]$ in terms of their lengths, where c_i's are chosen so that these confidence intervals have confidence coefficient $1 - \alpha$.

(b) If f is the p.d.f. of the Cauchy distribution $C(0, 1)$, show that the two confidence intervals in (a) have the same length.

33. Let $X_1, ..., X_n$ ($n > 1$) be i.i.d. from the exponential distribution $E(\theta, \theta)$, where $\theta > 0$ is unknown.

(a) Show that both $\bar{X}/\theta$ and $X_{(1)}/\theta$ are pivotal quantities, where $\bar{X}$ is the sample mean and $X_{(1)}$ is the smallest order statistic.

(b) Obtain confidence intervals (with confidence coefficient $1 - \alpha$) for θ based on the two pivotal quantities in (a).

(c) Discuss which confidence interval in (b) is better in terms of the length.

34. Prove Theorem 7.3(ii).

35. Show that the expected length of the interval in (7.13) is shorter than the expected length of the interval in (7.12).

36. Consider Example 7.14.

(a) Suppose that $\theta = \sigma^2$ and μ is known. Let a_* and b_* be constants

satisfying $a_*^2 g(a_*) = b_*^2 g(b_*) > 0$ and $\int_{a_*}^{b_*} g(x)dx = 1 - \alpha$, where g is the p.d.f. of the chi-square distribution χ_n^2. Show that the interval $[b_*^{-1}T, a_*^{-1}T]$ has the shortest length within the class of intervals of the form $[b^{-1}T, a^{-1}T]$, $\int_a^b g(x)dx = 1 - \alpha$, where $T = \sum_{i=1}^n (X_i - \mu)^2$.
(b) Show that the expected length of the interval in (a) is shorter than the expected length of the interval in (7.14).
(c) Find the shortest-length interval for $\theta = \sigma$ within the class of confidence intervals of the form $[b^{-1/2}\sqrt{n-1}S, a^{-1/2}\sqrt{n-1}S]$, where $0 < a < b < \infty$, $\int_a^b f(x)dx = 1 - \alpha$, and f is the p.d.f. of the chi-square distribution χ_{n-1}^2.

37. Assume the conditions in Theorem 7.3(i). Assume further that f is symmetric. Show that a_* and b_* in Theorem 7.3(i) must satisfy $a_* = -b_*$.

38. Let f be a Lebesgue p.d.f. that is nonzero in $[x_-, x_+]$ and is 0 outside $[x_-, x_+]$, $-\infty \le x_- < x_+ \le \infty$.
(a) Suppose that f is strictly decreasing. Show that, among all intervals $[a, b]$ satisfying $\int_a^b f(x)dx = 1 - \alpha$, the shortest interval is obtained by choosing $a = x_-$ and b so that $\int_{x_-}^b f(x)dx = 1 - \alpha$.
(b) Obtain a result similar to that in (a) when f is strictly increasing. Use the result to show that the interval $[X_{(n)}, \alpha^{-1/n}X_{(n)}]$ in Example 7.13 has the shortest length among all intervals $[b^{-1}X_{(n)}, a^{-1}X_{(n)}]$.

39. Let $X_1, ..., X_n$ be i.i.d. from the exponential distribution $E(a, 1)$ with an unknown a. Find a confidence interval for a having the shortest length within the class of confidence intervals $[X_{(1)} + c, X_{(1)} + d]$ with confidence coefficient $1 - \alpha$.

40. Consider the HPD credible set $C(x)$ in (7.7) for a real-valued θ. Suppose that $p_x(\theta)$ is a unimodal Lebesgue p.d.f. and is not monotone. Show that $C(x)$ is an interval having the shortest length within the class of intervals $[a, b]$ satisfying $\int_a^b p_x(\theta)d\theta = 1 - \alpha$.

41. Let X be a single observation from the gamma distribution $\Gamma(\alpha, \gamma)$ with a known α and an unknown γ. Find the shortest-length confidence interval within the class of confidence intervals $[b^{-1}X, a^{-1}X]$ with a given confidence coefficient.

42. Let $X_1, ..., X_n$ be i.i.d. with the Lebesgue p.d.f. $\theta x^{\theta-1}I_{(0,1)}(x)$, where $\theta > 0$ is unknown.
(a) Construct a confidence interval for θ with confidence coefficient $1 - \alpha$, using a sufficient statistic.
(b) Discuss whether the confidence interval obtained in (a) has the

shortest length within a class of confidence intervals.
(c) Discuss whether the confidence interval obtained in (a) is UMAU.

43. Let X be a single observation from the logistic distribution $LG(\mu, 1)$ with an unknown $\mu \in \mathcal{R}$. Find a Θ'-UMA upper confidence bound for μ with confidence coefficient $1 - \alpha$, where $\Theta' = (\mu, \infty)$.

44. Let $X_1, ..., X_n$ be i.i.d. from the exponential distribution $E(0, \theta)$ with an unknown $\theta > 0$. Find a Θ'-UMA lower confidence bound for θ with confidence coefficient $1 - \alpha$, where $\Theta' = (0, \theta)$.

45. Let X be a single observation from $N(\theta - 1, 1)$ if $\theta < 0$, $N(0, 1)$ if $\theta = 0$, and $N(\theta + 1, 1)$ if $\theta > 0$.
(a) Show that the distribution of X is in a family with monotone likelihood ratio.
(b) Construct a Θ'-UMA lower confidence bound for θ with confidence coefficient $1 - \alpha$, where $\Theta' = (-\infty, \theta)$.

46. Show that the confidence set in Example 7.9 is unbiased.

47. In Example 7.13, show that the confidence interval $[X_{(n)}, \alpha^{-1/n} X_{(n)}]$ is UMA and has the shortest expected length among all confidence intervals for θ with confidence coefficient $1 - \alpha$.

48. Let $X_1, ..., X_n$ be i.i.d. from the exponential distribution $E(a, \theta)$ with unknown a and θ. Find a UMAU confidence interval for a with confidence coefficient $1 - \alpha$.

49. Let Y and U be independent random variables having the binomial distribution $Bi(p, n)$ and the uniform distribution $U(0, 1)$, respectively.
(a) Show that $W = Y + U$ has the Lebesgue p.d.f. $f_p(w)$ given by (7.17).
(b) Show that the family $\{f_p : p \in (0, 1)\}$ has monotone likelihood ratio in W.

50. Extend the results in the previous exercise to the case where the distribution of Y is the power series distribution defined in Exercise 13 of §2.6.

51. Let $X_1, ..., X_n$ be i.i.d. from the Poisson distribution $P(\theta)$ with an unknown $\theta > 0$. Find a randomized UMA upper confidence bound for θ with confidence coefficient $1 - \alpha$.

52. Let X be a nonnegative integer-valued random variable from a population $P \in \mathcal{P}$. Suppose that $\mathcal{P}$ is parametric and indexed by a real-valued θ and has monotone likelihood ratio in X. Let U be a

random variable from the uniform distribution $U(0,1)$ that is independent of X. Show that a UMA lower confidence bound for θ with confidence coefficient $1 - \alpha$ is the solution of the equation

$$U F_\theta(X) + (1 - U) F_\theta(X - 1) = 1 - \alpha$$

(assuming that a solution exists), where $F_\theta(x)$ is the c.d.f. of X.

53. Let X be a single observation from the hypergeometric distribution $HG(r, n, \theta - n)$ (Table 1.1) with known r, n, and an unknown $\theta = n + 1, n + 2, \ldots$. Derive a randomized UMA upper confidence bound for θ with confidence coefficient $1 - \alpha$.

54. Let $X_1, \ldots, X_n$ be i.i.d. from $N(\mu, \sigma^2)$ with unknown μ and σ^2.
    (a) Show that $\bar\theta = \bar{X} + t_{n-1,\alpha} S/\sqrt{n}$ is a UMAU upper confidence bound for μ with confidence coefficient $1 - \alpha$, where $t_{n-1,\alpha}$ is the $(1 - \alpha)$th quantile of the t-distribution t_{n-1}.
    (b) Show that the confidence bound in (a) can be derived by inverting acceptance regions of LR tests.

55. Prove Theorem 7.7 and Proposition 7.3.

56. Let $X_1, \ldots, X_n$ be i.i.d. with p.d.f. $f(x - \theta)$, where f is a known Lebesgue p.d.f. Show that the confidence interval $[\bar{X} - c_1, \bar{X} + c_2]$ has constant coverage probability, where c_1 and c_2 are constants.

57. Prove the claim in Example 7.18.

58. In Example 7.19, show that
    (a) the testing problem is invariant under $\mathcal{G}_{\mu_0}$, but not $\mathcal{G}$;
    (b) the nonrandomized test with acceptance region $A(\mu_0)$ is UMP among unbiased and invariant tests of size α, under $\mathcal{G}_{\mu_0}$;
    (c) $\mathcal{G}$ is the smallest group containing $\cup_{\mu_0 \in \mathcal{R}} \mathcal{G}_{\mu_0}$.

59. In Example 7.17, show that intervals (7.13) and (7.14) are UMA among unbiased and invariant confidence intervals with confidence coefficient $1 - \alpha$, under $\mathcal{G}_1$ and $\mathcal{G}$, respectively.

60. Let X_i, $i = 1, 2$, be independent with the exponential distributions $E(0, \theta_i)$, $i = 1, 2$, respectively.
    (a) Show that $[\alpha Y/(2-\alpha), (2-\alpha)Y/\alpha]$ is a UMAU confidence interval for θ_2/θ_1 with confidence coefficient $1 - \alpha$, where $Y = X_2/X_1$.
    (b) Show that the confidence interval in (a) is also UMAI.

61. Let $X_1, \ldots, X_n$ be i.i.d. from a bivariate normal distribution with unknown mean and covariance matrix and let $R(X)$ be the sample correlation coefficient. Define $\underline{\rho} = C^{-1}(R(X))$, where $C(\rho)$ is determined by

$$P\big(R(X) \le C(\rho)\big) = 1 - \alpha$$

and ρ is the unknown correlation coefficient. Show that ρ is a Θ'-UMAI lower confidence bound for ρ with confidence coefficient $1 - \alpha$, where $\Theta' = (-1, \rho)$.

62. Let $X_{i1}, ..., X_{in_i}$, $i = 1, 2$, be two independent samples i.i.d. from $N(\mu_i, \sigma^2)$, $i = 1, 2$, respectively, where μ_i's are unknown. Find a UMAI confidence interval for $\mu_2 - \mu_1$ with confidence coefficient $1 - \alpha$ when (a) σ^2 is known; (b) σ^2 is unknown.

63. Consider Exercise 1. Let $\theta = \mu_1 - \mu_2$.
    (a) Show that $\Re(X, \theta) = (\bar{X}_1 - \bar{X}_2 - \theta)/\sqrt{n_1^{-1} S_1^2 + n_2^{-1} S_2^2}$ is asymptotically pivotal, assuming that $n_1/n_2 \to c \in (0, \infty)$. Construct a $1 - \alpha$ asymptotically correct confidence interval for θ using $\Re(X, \theta)$.
    (b) Show that $t(X, \theta)$ defined in Exercise 1(a) is asymptotically pivotal if either $n_1/n_2 \to \frac{1}{2}$ or $\sigma_1 = \sigma_2$ holds.

64. In Example 7.23, show that $C_3(X) = [p_-, p_+]$ with the given $p_{\pm}$. Compare the lengths of the confidence intervals $C_2(X)$ and $C_3(X)$.

65. Show that the confidence intervals in Example 7.14 can be derived by inverting acceptance regions of LR tests.

66. Let $X_1, ..., X_n$ be i.i.d. from the exponential distribution $E(0, \theta)$ with an unknown $\theta > 0$.
    (a) Show that $\Re(X, \theta) = \sqrt{n}(\bar{X} - \theta)/\theta$ is asymptotically pivotal. Construct a $1 - \alpha$ asymptotically correct confidence interval for θ, using $\Re(X, \theta)$.
    (b) Show that $\Re_1(X, \theta) = \sqrt{n}(\bar{X} - \theta)/\bar{X}$ is asymptotically pivotal. Construct a $1 - \alpha$ asymptotically correct confidence interval for θ, using $\Re_1(X, \theta)$.
    (c) Obtain $1 - \alpha$ asymptotically correct confidence intervals for θ by inverting acceptance regions of LR tests, Wald's tests, and Rao's score tests.

67. Let $X_1, ..., X_n$ be i.i.d. from the Poisson distribution $P(\theta)$ with an unknown $\theta > 0$.
    (a) Show that $\Re(X, \theta) = (\bar{X} - \theta)/\sqrt{\theta/n}$ is asymptotically pivotal. Construct a $1 - \alpha$ asymptotically correct confidence interval for θ, using $\Re(X, \theta)$.
    (b) Show that $\Re_1(X, \theta) = (\bar{X} - \theta)/\sqrt{\bar{X}/n}$ is asymptotically pivotal. Construct a $1 - \alpha$ asymptotically correct confidence interval for θ, using $\Re_1(X, \theta)$.
    (c) Obtain $1 - \alpha$ asymptotically correct confidence intervals for θ by inverting acceptance regions of LR tests, Wald's tests, and Rao's score tests.

68. Suppose that $X_1, ..., X_n$ are i.i.d. from the negative binomial distribution $NB(p,r)$ with a known r and an unknown p. Obtain $1 - \alpha$ asymptotically correct confidence intervals for p by inverting acceptance regions of LR tests, Wald's tests, and Rao's score tests.

69. Suppose that $X_1, ..., X_n$ are i.i.d. from the log-distribution $L(p)$ with an unknown p. Obtain $1 - \alpha$ asymptotically correct confidence intervals for p by inverting acceptance regions of LR tests, Wald's tests, and Rao's score tests.

70. In Example 7.24, obtain $1 - \alpha$ asymptotically correct confidence sets for μ by inverting acceptance regions of LR tests, Wald's tests, and Rao's score tests. Are these sets always intervals?

71. Let $X_1, ..., X_n$ be i.i.d. from the gamma distribution $\Gamma(\theta, \gamma)$ with unknown θ and γ. Obtain $1 - \alpha$ asymptotically correct confidence sets for θ by inverting acceptance regions of LR tests, Wald's tests, and Rao's score tests. Discuss whether these confidence sets are intervals or not.

72. Consider the problem in Example 3.21. Construct an asymptotically pivotal quantity and a $1 - \alpha$ asymptotically correct confidence set for μ_y/μ_x.

73. Consider the problem in Example 3.23. Construct an asymptotically pivotal quantity and a $1 - \alpha$ asymptotically correct confidence set for $R(t)$ with a fixed t.

74. Let U_n be a U-statistic based on i.i.d. $X_1, ..., X_n$ and the kernel $h(x_1, ..., x_m)$, and let $\theta = E(U_n)$. Construct an asymptotically pivotal quantity based on U_n and a $1 - \alpha$ asymptotically correct confidence set for θ.

75. Let $X_1, ..., X_n$ be i.i.d. from a c.d.f. F on $\mathcal{R}$ that is continuous and symmetric about θ. Let $\hat{\theta} = W/n - \frac{1}{2}$ and $T(F) = \theta + \frac{1}{2}$, where W and T are given by (6.83) and (5.53), respectively. Construct a confidence interval for θ that has limiting confidence coefficient $1 - \alpha$.

76. Consider the problem in Example 5.15. Construct an asymptotically pivotal quantity and a $1 - \alpha$ asymptotically correct confidence set for θ. Compare this confidence set with those in Example 7.24.

77. Consider the linear model $X = Z\beta + \varepsilon$, where ε has independent components with mean 0 and Z is of full rank. Assume the conditions in Theorem 3.12.
    (a) Suppose that $\text{Var}(\varepsilon) = \sigma^2 D$, where D is a known diagonal matrix and σ^2 is unknown. Find an asymptotically pivotal quantity and

construct a $1 - \alpha$ asymptotically correct confidence set for β.

(b) Suppose that $\text{Var}(\varepsilon)$ is an unknown diagonal matrix. Find an asymptotically pivotal quantity and construct a $1 - \alpha$ asymptotically correct confidence set for β.

78. In part (a) of the previous exercise, obtain a $1 - \alpha$ asymptotically correct confidence set for β/σ.

79. Consider a GEE estimator $\hat{\theta}$ of θ described in §5.4.1. Discuss how to construct an asymptotically pivotal quantity and a $1 - \alpha$ asymptotically correct confidence set for θ. (Hint: see §5.5.2.)

80. Let $X_1, ..., X_n$ be i.i.d. from the exponential distribution $E(a, \theta)$ with unknown a and θ. Find a $1 - \alpha$ asymptotically correct confidence set for (a, θ) by inverting acceptance regions of LR tests.

81. Let $X_{i1}, ..., X_{in_i}$, $i = 1, 2$, be two independent samples i.i.d. from $N(\mu_i, \sigma_i^2)$, $i = 1, 2$, respectively, where all parameters are unknown.

(a) Find $1 - \alpha$ asymptotically correct confidence sets for (μ_1, μ_2) by inverting acceptance regions of LR tests, Wald's tests, and Rao's score tests.

(b) Repeat (a) for the parameter $(\mu_1, \mu_2, \sigma_1^2, \sigma_2^2)$.

(c) Repeat (a) under the assumption that $\sigma_1^2 = \sigma_2^2 = \sigma^2$.

(d) Repeat (c) for the parameter (μ_1, μ_2, σ^2).

82. Let $X_{i1}, ..., X_{in_i}$, $i = 1, 2$, be two independent samples i.i.d. from the exponential distributions $E(0, \theta_i)$, $i = 1, 2$, respectively, where θ_i's are unknown. Find $1 - \alpha$ asymptotically correct confidence sets for (θ_1, θ_2) by inverting acceptance regions of LR tests, Wald's tests, and Rao's score tests.

83. Consider the problem in Example 7.9. Find $1 - \alpha$ asymptotically correct confidence sets for θ by inverting acceptance regions of LR tests, Wald's tests, and Rao's score tests. Which one is the same as that derived in Example 7.9?

84. Let $X_1, ..., X_n$ be i.i.d. from a continuous c.d.f. F on $\mathcal{R}$ and let $\theta = F^{-1}(p)$, $p \in (0, 1)$.

(a) Show that $P(X_{(k_1)} \leq \theta \leq X_{(k_2)}) = P(U_{(k_1)} \leq p \leq U_{(k_2)})$, where $X_{(k)}$ is the kth order statistic and $U_{(k)}$ is the kth order statistic based on a sample $U_1, ..., U_n$ i.i.d. from the uniform distribution $U(0, 1)$.

(b) Show that

$$P(U_{(k_1)} \leq p \leq U_{(k_2)}) = B_p(k_1, n - k_1 + 1) - B_p(k_2, n - k_2 + 1),$$

where

$$B_p(i, j) = \frac{\Gamma(i + j)}{\Gamma(i)\Gamma(j)} \int_0^p t^{i-1}(1 - t)^{j-1} dt.$$

(c) Discuss how to obtain a confidence interval for θ with confidence coefficient $1 - \alpha$.

85. Prove Corollary 7.1.

86. Assume the conditions in Corollary 7.1.
(a) Show that $\sqrt{n}(X_{(k_n)} - \theta)F'(\theta) \to_d N(c, p(1 - p))$.
(b) Prove result (7.24) using the result in part (a).
(c) Construct a consistent estimator of the asymptotic variance of the sample median (see Example 6.27), using Woodruff's interval.

87. Prove (7.25) and (7.26).

88. In Example 7.25, prove that $\hat{V}_n^{-1/2}(\hat{\theta}_n - \theta)$ can be written as $\sqrt{n}h(\bar{Y})/\sigma_h$ and find the explicit form of the function h.

89. Let $X_1, ..., X_n$ be i.i.d. from an unknown c.d.f. F with $E|X_1|^8 < \infty$. Suppose that condition (1.105) is satisfied. Derive a second-order accurate lower confidence bound for $\sigma^2 = \mathrm{Var}(X_1)$.

90. Using the Edgeworth expansion given in Example 7.26, construct a third-order accurate lower confidence bound for μ.

91. Show that $\underline{\theta}_{HB}$ in (7.34) is equal to $2\hat{\theta}_n - K_B^{-1}(1 - \alpha)$, where K_B is defined in (7.27).

92. (Parametric bootstrapping in location-scale families). Let $X_1, ..., X_n$ be i.i.d. random variables with p.d.f. $\frac{1}{\sigma}f\left(\frac{x-\mu}{\sigma}\right)$, where f is a known Lebesgue p.d.f. and μ and $\sigma > 0$ are unknown. Let $X_1^*, ..., X_n^*$ be i.i.d. bootstrap data from the p.d.f. $\frac{1}{s}f\left(\frac{x-\bar{x}}{s}\right)$, where $\bar{x}$ and s^2 are the observed sample mean and sample variance, respectively.
(a) Suppose that we construct the bootstrap-t lower confidence bound (7.35) for μ using the parametric bootstrap data. Show that $\underline{\theta}_{BT}$ has confidence coefficient $1 - \alpha$.
(b) Suppose that we construct the hybrid bootstrap lower confidence bound (7.34) for μ using the parametric bootstrap data. Show that $\underline{\theta}_{HB}$ does not necessarily have confidence coefficient $1 - \alpha$.
(c) Suppose that f has mean 0 and variance 1. Show that $\underline{\theta}_{HB}$ in (b) is $1 - \alpha$ asymptotically correct.

93. (The bootstrap BC_a percentile method). Suppose that we change assumption (7.30) to

$$P\left(\frac{\hat{\phi}_n - \phi_n(\theta)}{1 + a\phi_n(\theta)} + z_0 \le x\right) = \Phi(x),$$

where a is an extra parameter called the acceleration constant and Φ is the c.d.f. of the standard normal distribution.

(a) If $\hat{\phi}_n$, z_0, and a are known, show that the following lower confidence bound for θ has confidence coefficient $1 - \alpha$:

$$\underline{\theta}_E = \phi_n^{-1}\big(\hat{\phi}_n + (z_\alpha + z_0)(1 + a\hat{\phi}_n)/[1 - a(z_\alpha + z_0)]\big).$$

(b) Show that $K_B^{-1}(x) = \phi_n^{-1}\big(\hat{\phi}_n + [\Phi^{-1}(x) - z_0](1 + a\hat{\phi}_n)\big)$, where K_B is defined in (7.27).

(c) Let $\underline{\theta}_{BC}(a) = K_B^{-1}\big(\Phi(z_0 + (z_\alpha + z_0)/[1 - a(z_\alpha + z_0)])\big)$. Show that $\underline{\theta}_{BC}(a) = \underline{\theta}_E$ in part (a). (The bootstrap BC_a percentile lower confidence bound for θ is $\underline{\theta}_{BC}(\hat{a})$, where $\hat{a}$ is an estimator of a.)

94. (Automatic bootstrap percentile). Let $\mathcal{P} = \{P_\theta : \theta \in \mathcal{R}\}$ be a parametric family. Define $K_\theta(x) = P_\theta(\hat{\theta}_n \le x)$, where $\hat{\theta}_n$ is an estimator of θ. Let θ_0 be a given value of θ and $\theta_1 = K_{\theta_0}^{-1}(1 - \alpha)$. The *automatic bootstrap percentile* lower confidence bound for θ is defined to be

$$\underline{\theta}_{ABP} = K_{\hat{\theta}_n}^{-1}(K_{\theta_1}(\theta_0)).$$

Assume the assumption in the previous exercise. Show that $\underline{\theta}_{ABP}$ has confidence coefficient $1 - \alpha$.

95. (Bootstrapping residuals). Consider linear model (3.25): $X = Z\beta + \varepsilon$, where Z is of full rank and ε is a vector of i.i.d. random variables having mean 0 and variance σ^2. Let $r_i = X_i - Z_i^\tau\hat{\beta}$ be the ith residual, where $\hat{\beta}$ is the LSE of β. Assume that the average of r_i's is always 0. Let $\varepsilon_1^*, ..., \varepsilon_n^*$ be i.i.d. bootstrap data from the empirical c.d.f. putting mass n^{-1} on each r_i. Define $X_i^* = Z_i^\tau\hat{\beta} + \varepsilon_i^*$, $i = 1, ..., n$.

(a) Find an expression for $\hat{\beta}^*$, the bootstrap analogue of $\hat{\beta}$. Calculate $E(\hat{\beta}^*|X)$ and $\mathrm{Var}(\hat{\beta}^*|X)$.

(b) Using $l^\tau(\hat{\beta} - \beta)$ and the idea in §7.4.1, construct a hybrid bootstrap lower confidence bound for $l^\tau\beta$, where $l \in \mathcal{R}^p$.

(c) Discuss when the lower confidence bound in (b) is $1 - \alpha$ asymptotically correct.

(d) Describe how to construct a bootstrap-t lower confidence bound for $l^\tau\beta$.

(e) Describe how to construct a hybrid bootstrap confidence set for β, using the idea in §7.4.1.

96. (Bootstrapping pairs). Consider linear model (3.25): $X = Z\beta + \varepsilon$, where Z is of full rank and ε is a vector of independent random variables having mean 0 and finite variances. Let $(X_1^*, Z_1^*), ..., (X_n^*, Z_n^*)$ be i.i.d. bootstrap data from the empirical c.d.f. putting mass n^{-1} on each (X_i, Z_i). Define $\hat{\beta}^* = (Z^\tau Z)^{-1}\sum_{i=1}^n Z_i^* X_i^*$. Repeat (a)-(e) of the previous exercise.

97. (External bootstrapping or wild bootstrapping). Assume the model in the previous exercise. Let $\varepsilon_1^*, ..., \varepsilon_n^*$ be i.i.d. random variables with mean 0 and variance 1. Define the bootstrap data as $X_i^* = Z_i^\tau \hat\beta + |t_i|\varepsilon_i^*$, $i = 1, ..., n$, where $\hat\beta$ is the LSE of β, $t_i = (X_i - Z_i^\tau \hat\beta)/\sqrt{1 - h_i}$, and $h_i = Z_i^\tau (Z^\tau Z)^{-1} Z_i$. Repeat (a)-(e) of Exercise 95.

98. Prove (7.48) and (7.49).

99. Describe how to approximate $C_{PREB}^{(3)}(X)$ in (7.51), using the Monte Carlo method.

100. Prove (7.53) and (7.54).

101. Show that Bonferroni's simultaneous confidence intervals are of level $1 - \alpha$.

102. Let $C_{t,\alpha}(X)$ be a confidence interval for θ_t with confidence coefficient $1 - \alpha$, $t = 1, ..., k$. Suppose that $C_{1,\alpha}(X), ..., C_{k,\alpha}(X)$ are independent for any α. Show how to construct simultaneous confidence intervals for θ_t, $t = 1, ..., k$, with confidence coefficient $1 - \alpha$.

103. Show that $C_{ij,\alpha}(X)$ in (7.60) is UMAU for $\mu_i - \mu_j$.

104. Consider the two-way balanced ANOVA model in Example 6.19. Using Bonferroni's method, obtain level $1 - \alpha$ simultaneous confidence intervals for
    (a) α_i, $i = 1, ..., a - 1$;
    (b) μ_{ij}, $i = 1, ..., a$, $j = 1, ..., b$.

105. Prove (7.62). (Hint: use the Cauchy-Schwarz inequality.)

106. Let $x \in \mathcal{R}^k$, $y \in \mathcal{R}^k$, and A be a $k \times k$ positive definite matrix.
    (a) Suppose that $y^\tau A^{-1} x = 0$. Show that

    $$x^\tau A^{-1} x = \max_{c \in \mathcal{R}^k, c \neq 0, c^\tau y = 0} \frac{(c^\tau x)^2}{c^\tau A c}.$$

    (b) Assume model (7.61) with a full rank Z. Using the result in (a), construct simultaneous confidence intervals (with confidence coefficient $1 - \alpha$) for $c^\tau \beta$, $c \in \mathcal{R}^p$, $c \neq 0$, $c^\tau y = 0$, where $y \in \mathcal{R}^p$ satisfies $Z^\tau Z y = 0$.

107. Assume the conditions in Theorem 3.12. Show that Scheffé's intervals in Theorem 7.10 are $1 - \alpha$ asymptotically correct.

108. Assume the conditions in Theorem 3.12 and Theorem 7.10. Derive $1 - \alpha$ asymptotically correct simultaneous confidence intervals for $t^\tau L\beta/\sigma$.

109. Prove (7.65).

110. Find explicitly the $m(m-1)/2$ vectors in the set T_0 in (7.66) so that $\{t^\tau L\beta : t \in T_0\}$ is exactly the same as $\mu_i - \mu_j$, $1 \le i < j \le m$. Show that the intervals in (7.66) are Scheffé's simultaneous confidence intervals.

111. In Example 7.28, show that
   (a) Scheffé's intervals in Theorem 7.10 with $t = (1, z)$ and $L = I_2$ are of the form (7.67);
   (b) the maximum on the right-hand side of (7.68) is achieved at t given by (7.69);
   (c) y in (7.69) is equal to 1 and (7.68) holds.

112. Consider the two-way balanced ANOVA model in Example 6.19. Using Scheffé's method, obtain level $1 - \alpha$ simultaneous confidence intervals for α_i's, β_j's, and γ_{ij}'s.

113. Let $X_{ij} = N(\mu + \alpha_i + \beta_j, \sigma^2)$, $i = 1, ..., a$, $j = 1, ..., b$, be independent, where $\sum_{i=1}^a \alpha_i = 0$ and $\sum_{j=1}^b \beta_j = 0$. Construct level $1 - \alpha$ simultaneous confidence intervals for all linear combinations of α_i's and β_j's, using
   (a) Bonferroni's method;
   (b) Scheffé's method.

114. Assume model (7.61) with $\beta = (\beta_0, \beta_1, \beta_2)$ and $Z_i = (1, t_i, t_i^2)$, where $t_i \in \mathcal{R}$, $\sum_{i=1}^n t_i = 0$, $\sum_{i=1}^n t_i^2 = 1$, and $\sum_{i=1}^n t_i^3 = 0$.
   (a) Construct a confidence ellipsoid for (β_1, β_2) with confidence coefficient $1 - \alpha$;
   (b) Construct simultaneous confidence intervals for all linear combinations of β_1 and β_2, with confidence coefficient $1 - \alpha$.

115. Show that the distribution of R_{st} in (7.70) does not depend on any unknown parameter.

116. For $\alpha = 0.05$, obtain numerically the t-type confidence intervals in (7.60), Bonferroni's, Scheffé's, and Tukey's simultaneous confidence intervals for $\mu_i - \mu_j$, $1 \le i < j \le 4$, based on the following data X_{ij} from a one-way ANOVA model ($q_{0.05} = 4.45$):

	$j = 1$	2	3	4	5	6
$i = 1$	0.08	0.10	0.09	0.07	0.09	0.06
2	0.15	0.09	0.11	0.10	0.08	0.13
3	0.13	0.10	0.15	0.09	0.09	0.17
4	0.05	0.11	0.07	0.09	0.11	0.08

117. (Dunnett's simultaneous confidence intervals). Let X_{0j} $(j = 1, ..., n_0)$ and X_{ij} $(i = 1, ..., m, j = 1, ..., n_0)$ represent independent measurements on a standard and m competing new treatments. Suppose that $X_{ij} = N(\mu_i, \sigma^2)$ with unknown μ_i and $\sigma^2 > 0$, $i = 0, 1, ..., m$. Let $\bar{X}_{i\cdot}$ be the sample mean based on X_{ij}, $j = 1, ..., n_0$, and $\hat{\sigma}^2 = [(m+1)(n_0 - 1)]^{-1} \sum_{i=0}^{m} \sum_{j=1}^{n_0} (X_{ij} - \bar{X}_{i\cdot})^2$.
(a) Show that the distribution of

$$R_{st} = \max_{i=1,...,m} |(\bar{X}_{i\cdot} - \mu_i) - (\bar{X}_{0\cdot} - \mu_0)|/\hat{\sigma}$$

does not depend on any unknown parameter.
(b) Show that

$$\left[\sum_{i=0}^{m} c_i \bar{X}_{i\cdot} - q_\alpha \hat{\sigma} \sum_{i=1}^{m} |c_i|, \ \sum_{i=0}^{m} c_i \bar{X}_{i\cdot} + q_\alpha \hat{\sigma} \sum_{i=1}^{m} |c_i| \right]$$

for all $c_0, c_1, ..., c_m$ satisfying $\sum_{i=0}^{m} c_i = 0$ are simultaneous confidence intervals for $\sum_{i=0}^{m} c_i \mu_i$ with confidence coefficient $1 - \alpha$, where q_α is the $(1 - \alpha)$th quantile of R_{st}.

118. Let $X_1, ..., X_n$ be i.i.d. from the uniform distribution $U(0, \theta)$, where $\theta > 0$ is unknown. Construct a confidence band for the c.d.f. of X_1 with confidence coefficient $1 - \alpha$.

119. Let $X_1, ..., X_n$ be i.i.d. with the p.d.f. $\frac{1}{\sigma} f\left(\frac{t-\mu}{\sigma}\right)$, where f is a known Lebesgue p.d.f. (a location-scale family). Let F be the c.d.f. of X_1.
(a) Suppose that $\mu \in \mathcal{R}$ is unknown and σ is known. Construct simultaneous confidence intervals for $F(t)$, $t \in \mathcal{R}$, with confidence coefficient $1 - \alpha$.
(b) Suppose that μ is known and $\sigma > 0$ is unknown. Construct simultaneous confidence intervals for $F(t)$, $t \in \mathcal{R}$, with confidence coefficient $1 - \alpha$.
(c) Suppose that $\mu \in \mathcal{R}$ and $\sigma > 0$ are unknown. Construct level $1 - \alpha$ simultaneous confidence intervals for $F(t)$, $t \in \mathcal{R}$.

120. Let $X_1, ..., X_n$ be i.i.d. from F on $\mathcal{R}$ and F_n be the empirical c.d.f. Show that the intervals

$$\left[\max\{F_n(t) - c_\alpha, 0\}, \ \min\{F_n(t) + c_\alpha, 1\} \right], \qquad t \in \mathcal{R},$$

form a confidence band for $F(t)$, $t \in \mathcal{R}$, with limiting confidence coefficient $1 - \alpha$, where c_α is given by (7.75).

References

We provide some references for further readings on the topics covered in this book.

For general probability theory, Billingsley (1986) and Chung (1974) are suggested, although there are many standard textbooks. An asymptotic theory for statistics can be found in Serfling (1980), Shorack and Wellner (1986), Sen and Singer (1993), Barndorff-Nielsen and Cox (1994), and van der Vaart (1998).

More discussions of fundamentals of statistical decision theory and inference can be found in many textbooks on mathematical statistics, such as Cramér (1946), Wald (1950), Savage (1954), Ferguson (1967), Rao (1973), Rohatgi (1976), Bickel and Doksum (1977), Lehmann (1986), Casella and Berger (1990), and Barndorff-Nielsen and Cox (1994). Discussions and proofs for results related to sufficiency and completeness can be found in Rao (1945), Blackwell (1947), Hodges and Lehmann (1950), Lehmann and Scheffé (1950), and Basu (1955). More results for exponential families are given in Barndorff-Nielsen (1978).

The theory of UMVUE in §3.1.1 and §3.1.2 is mainly based on Chapter 2 of Lehmann (1983). More results on information inequalities can be found in Cramér (1946), Rao (1973), Lehmann (1983), and Pitman (1979). The theory of U-statistics and the method of projection can be found in Hoeffding (1948), Randles and Wolfe (1979), and Serfling (1980). The related theory for V-statistics is given in von Mises (1947), Serfling (1980), and Sen (1981). Three excellent textbooks for the theory of LSE are Scheffé (1959), Searle (1971), and Rao (1973). Additional materials for sample surveys can be found in Basu (1958), Godambe (1958), Cochran (1977), Särndal, Swensson, and Wretman (1992), and Ghosh and Meeden (1997).

Excellent textbooks for the Bayesian theory include Lindley (1965), Box and Tiao (1973), Berger (1985), and Schervish (1995). For Bayesian computation and Markov chain Monte Carlo, more discussions can be found in references cited in §4.1.4. More general results on invariance in estimation and testing problems are provided by Ferguson (1967) and Lehmann (1983,

1986). The theory of shrinkage estimation was established by Stein (1956) and James and Stein (1961); Lehmann (1983) and Berger (1985) provide excellent discussions on this topic. The method of likelihood has more than 200 years of history (Edwards, 1974). An excellent textbook on the MLE in generalized linear models is McCullagh and Nelder (1989). Asymptotic properties for MLE can be found in Cramér (1946), Serfling (1980), and Sen and Singer (1993). Asymptotic results for the MLE in generalized linear models are provided by Fahrmeir and Kaufmann (1985).

An excellent book containing results for empirical c.d.f.'s and their properties is Shorack and Wellner (1986). References for empirical likelihoods are provided in §5.1.2 and §6.5.3. More results in density estimation can be found, for example, in Rosenblatt (1971) and Silverman (1986). Discussions of partial likelihoods and proportional hazards models are given in Cox (1972) and Fleming and Harrington (1991). More discussions on statistical functionals can be found in von Mises (1947), Serfling (1980), Fernholz (1983), Sen and Singer (1993), and Shao and Tu (1995). Two textbooks for robust statistics are Huber (1981) and Hampel et al. (1986). A general discussion of L-estimators and sample quantiles can be found in Serfling (1980) and Sen (1981). L-estimators in linear models are covered by Bickel (1973), Puri and Sen (1985), Welsh (1987), and He and Shao (1996). Some references on generalized estimation equations and quasi-likelihoods are Godambe and Heyde (1987), Godambe and Thompson (1989), McCullagh and Nelder (1989), and Diggle, Liang, and Zeger (1994). Two textbooks containing materials on variance estimation are Efron and Tibshirani (1993) and Shao and Tu (1995).

The theory of UMP, UMPU, and UMPI tests in Chapter 6 is mainly based on Lehmann (1986) and Chapter 5 of Ferguson (1967). Berger (1985) contains a discussion on Bayesian tests. Results on large sample tests and chi-square tests can be found in Serfling (1980) and Sen and Singer (1993). Two textbooks on nonparametric tests are Lehmann (1975) and Randles and Wolfe (1979).

Further materials on confidence sets can be found in Ferguson (1967), Bickel and Doksum (1977), Lehmann (1986), and Casella and Berger (1990). More results on asymptotic confidence sets based on likelihoods can be found in Serfling (1980). The results on high order accurate confidence sets (§7.4.3) are based on Hall (1992). The theory of bootstrap confidence sets is covered by Hall (1992), Efron and Tibshirani (1993), and Shao and Tu (1995). Further discussions on simultaneous confidence intervals can be found in Scheffé (1959), Lehmann (1986), and Tukey (1977).

The following references are those cited in this book. Many additional references can be found in Lehmann (1983, 1986).

Arvesen, J. N. (1969). Jackknifing U-statistics. *Ann. Math. Statist.*, **40**, 2076-2100.

Bahadur, R. R. (1957). On unbiased estimates of uniformly minimum variance. *Sankhyā*, **18**, 211-224.

Bahadur, R. R. (1964). On Fisher's bound for asymptotic variances. *Ann. Math. Statist.*, **35**, 1545-1552.

Bahadur, R. R. (1966). A note on quantiles in large samples. *Ann. Math. Statist.*, **37**, 577-580.

Barndorff-Nielsen, O. E. (1978). *Information and Exponential Families in Statistical Theory*. Wiley, New York.

Barndorff-Nielsen, O. E. and Cox, D. R. (1994). *Inference and Asymptotics*. Chapman & Hall, London.

Basag, J., Green, P., Higdon, D., and Mengersen, K. (1995). Bayesian computation and stochastic systems. *Statist. Sci.*, **10**, 3-66.

Basu, D. (1955). On statistics independent of a complete sufficient statistic. *Sankhyā*, **15**, 377-380.

Basu, D. (1958). On sampling with and without replacement. *Sankhyā*, **20**, 287-294.

Beran, R. (1987). Prepivoting to reduce level error of confidence sets. *Biometrika*, **74**, 151-173.

Berger, J. O. (1976). Inadmissibility results for generalized Bayes estimators of coordinates of a location vector. *Ann. Statist.*, **4**, 302-333.

Berger, J. O. (1980). Improving on inadmissible estimators in continuous exponential families with applications to simultaneous estimation of gamma scale parameters. *Ann. Statist.*, **8**, 545-571.

Berger, J. O. (1985). *Statistical Decision Theory and Bayesian Analysis*, second edition. Springer-Verlag, New York.

Bickel, P. J. (1973). On some analogues to linear combinations of order statistics in the linear model. *Ann. Statist.*, **1**, 597-616.

Bickel, P. J. and Doksum, K. A. (1977). *Mathematical Statistics*. Holden Day, San Francisco.

Bickel, P. J. and Yahav, J. A. (1969). Some contributions to the asymptotic theory of Bayes solutions. *Z. Wahrsch. Verw. Geb.*, **11**, 257-276.

Billingsley, P. (1986). *Probability and Measure*, second edition. Wiley, New York.

Blackwell, D. (1947). Conditional expectation and unbiased sequential estimation. *Ann. Math. Statist.*, **18**, 105-110.

Box, G. E. P. and Tiao, G. C. (1973). *Bayesian Inference in Statistical Analysis*. Addison-Wesley, Reading, MA.

Brown, L. D. (1966). On the admissibility of invariant estimators of one or more location parameters. *Ann. Math. Statist.*, **37**, 1087-1136.

Brown, L. D. and Fox, M. (1974). Admissibility in statistical problems involving a location or scale parameter. *Ann. Statist.*, **2**, 248-266.

Carroll, R. J. (1982). Adapting for heteroscedasticity in linear models. *Ann. Statist.*, **10**, 1224-1233.

Carroll, R. J. and Cline, D. B. H. (1988). An asymptotic theory for weighted least-squares with weights estimated by replication. *Biometrika*, **75**, 35-43.

Casella, G. and Berger, R. L. (1990). *Statistical Inference*. Wadsworth, Belmont, CA.

Chan, K. S. (1993). Asymptotic behavior of the Gibbs sampler. *J. Amer. Statist. Assoc.*, **88**, 320-325.

Chen, J. and Qin, J. (1993). Empirical likelihood estimation for finite populations and the effective usage of auxiliary information. *Biometrika*, **80**, 107-116.

Chen, J. and Shao, J. (1993). Iterative weighted least squares estimators. *Ann. Statist.*, **21**, 1071-1092.

Chung, K. L. (1974). *A Course in Probability Theory*, second edition. Academic Press, New York.

Clarke, B. R. (1986). Nonsmooth analysis and Fréchet differentiability of M-functionals. *Prob. Theory and Related Fields*, **73**, 197-209.

Cochran, W. G. (1977). *Sampling Techniques*, third edition. Wiley, New York.

Cox, D. R. (1972). Regression models and life tables, *J. R. Statist. Soc.*, B, **34**, 187–220.

Cramér, H. (1946). *Mathematical Methods of Statistics*. Princeton University Press, Princeton, NJ.

Diggle, P. J., Liang, K.-Y., and Zeger, S. L. (1994). *Analysis of Longitudinal Data.* Clarendon Press, Oxford.

Draper, N. R. and Smith, H. (1981). *Applied Regression Analysis,* second edition. Wiley, New York.

Durbin, J. (1973). *Distribution Theory for Tests Based on the Sample Distribution Function.* SIAM, Philadelphia, PA.

Dvoretzky, A., Kiefer, J., and Wolfowitz, J. (1956). Asymptotic minimax character of the sample distribution function and of the classical multinomial estimator. *Ann. Math. Statist.,* **27**, 642-669.

Edwards, A. W. F. (1974). The history of likelihood. *Internat. Statist. Rev.,* **42**, 4-15.

Efron, B. (1979). Bootstrap methods: Another look at the jackknife. *Ann. Statist.,* **7**, 1-26.

Efron, B. (1981). Nonparametric standard errors and confidence intervals (with discussions). *Canadian J. Statist.,* **9**, 139-172.

Efron, B. (1987). Better bootstrap confidence intervals (with discussions). *J. Amer. Statist. Assoc.,* **82**, 171-200.

Efron, B. and Morris, C. (1973). Stein's estimation rule and its competitors — An empirical Bayes approach. *J. Amer. Statist. Assoc.,* **68**, 117-130.

Efron, B. and Tibshirani, R. J. (1993). *An Introduction to the Bootstrap.* Chapman & Hall, New York.

Esseen, C. and von Bahr, B. (1965). Inequalities for the rth absolute moment of a sum of random variables, $1 \leq r \leq 2$. *Ann. Math. Statist.,* **36**, 299-303.

Fahrmeir, L. and Kaufmann, H. (1985). Consistency and asymptotic normality of the maximum likelihood estimator in generalized linear models. *Ann. Statist.,* **13**, 342-368.

Farrell, R. H. (1964). Estimators of a location parameter in the absolutely continuous case. *Ann. Math. Statist.,* **35**, 949-998.

Farrell, R. H. (1968a). Towards a theory of generalized Bayes tests. *Ann. Math. Statist.,* **38**, 1-22.

Farrell, R. H. (1968b). On a necessary and sufficient condition for admissibility of estimators when strictly convex loss is used. *Ann. Math. Statist.,* **38**, 23-28.

Ferguson, T. S. (1967). *Mathematical Statistics.* Academic Press, New York.

Fernholz, L. T. (1983). *Von Mises Calculus for Statistical Functionals.* Lecture Notes in Statistics, **19**, Springer-Verlag, New York.

Fleming, T. R. and Harrington, D. P. (1991). *Counting Processes and Survival Analysis.* Wiley, New York.

Fuller, W. A. (1996). *Introduction to Statistical Time Series*, second edition. Wiley, New York.

Gelfand, A. E. and Smith, A. F. M. (1990). Sampling-based approaches to calculating marginal densities. *J. Amer. Statist. Assoc.*, **85**, 398-409.

Geweke, J. (1989). Bayesian inference in econometric models using Monte Carlo integration. *Econometrica*, **57**, 1317-1339.

Geyer, C. J. (1994). On the convergence of Monte Carlo maximum likelihood calculations. *J. R. Statist. Soc.*, B, **56**, 261-274.

Ghosh, M. and Meeden, G. (1997). *Bayesian Methods in Finite Population Sampling.* Chapman & Hall, London.

Godambe, V. P. (1958). A unified theory of sampling from finite populations. *J. R. Statist. Soc.*, B, **17**, 269-278.

Godambe, V. P. and Heyde, C. C. (1987). Quasi-likelihood and optimal estimation. *Internat. Statist. Rev.*, **55**, 231-244.

Godambe, V. P. and Thompson, M. E. (1989). An extension of quasi-likelihood estimation (with discussion). *J. Statist. Plan. Inference*, **22**, 137-172.

Hall, P. (1988). Theoretical comparisons of bootstrap confidence intervals (with discussions). *Ann. Statist.*, **16**, 927-953.

Hall, P. (1992). *The Bootstrap and Edgeworth Expansion.* Springer-Verlag, New York.

Hall, P. and Martin, M. A. (1988). On bootstrap resampling and iteration. *Biometrika*, **75**, 661-671.

Hampel, F. R. (1974). The influence curve and its role in robust estimation. *J. Amer. Statist. Assoc.*, **62**, 1179-1186.

Hampel, F. R., Ronchetti, E. M., Rousseeuw, P. J., and Stahel, W. A. (1986). *Robust Statistics: The Approach Based on Influence Functions.* Wiley, New York.

He, X. and Shao, Q.-M. (1996). A general Bahadur representation of M-estimators and its application to linear regression with nonstochastic designs. *Ann. Statist.*, **24**, 2608-2630.

Hodges, J. L., Jr. and Lehmann, E. L. (1950). Some problems in minimax point estimation. *Ann. Math. Statist.*, **21**, 182-197.

Hoeffding, W. (1948). A class of statistics with asymptotic normal distribution. *Ann. Math. Statist.*, **19**, 293-325.

Hogg, R. V. and Tanis, E. A. (1993). *Probability and Statistical Inference*, fourth edition. Macmillan, New York.

Huber, P. J. (1964). Robust estimation of a location parameter. *Ann. Math. Statist.*, **35**, 73-101.

Huber, P. J. (1981). *Robust Statistics*. Wiley, New York.

Ibragimov, I. A. and Has'minskii, R. Z. (1981). *Statistical Estimation: Asymptotic Theory*. Springer-Verlag, New York.

James, W. and Stein, C. (1961). Estimation with quadratic loss. *Proc. Fourth Berkeley Symp. Math. Statist. Prob.*, **1**, 311-319. University of California Press, CA.

Jeffreys, H. (1939, 1948, 1961). *The Theory of Probability*. Oxford University Press, Oxford.

Jones, M. C. (1991). Kernel density estimation for length biased data. *Biometrika*, **78**, 511-519.

Kalbfleisch, J. D. and Prentice, R. T. (1980). *The Statistical Analysis of Failure Time Data*. Wiley, New York.

Kaplan, E. L. and Meier, P. (1958). Nonparametric estimation from incomplete observations. *J. Amer. Statist. Assoc.*, **53**, 457-481.

Kiefer, J. and Wolfowitz, J. (1956). Consistency of the maximum likelihood estimator in the presence of infinitely many nuisance parameters. *Ann. Math. Statist.*, **27**, 887-906.

Kolmogorov, A. N. (1933). Sulla determinazione empirica di una legge di distribuzione. *Giorn. Inst. Ital. Attuari*, **4**, 83-91.

Le Cam, L. (1953). On some asymptotic properties of maximum likelihood estimates and related Bayes' estimates. *Univ. of Calif. Publ. in Statist.*, **1**, 277-330.

Lehmann, E. L. (1975). *Nonparametrics: Statistical Methods Based on Ranks*. Holden Day, San Francisco.

Lehmann, E. L. (1983). *Theory of Point Estimation*. Springer-Verlag, New York.

Lehmann, E. L. (1986). *Testing Statistical Hypotheses*, second edition. Springer-Verlag, New York.

Lehmann, E. L. and Scheffé, H. (1950). Completeness, similar regions and unbiased estimation. *Sankhyā*, **10**, 305-340.

Liang, K.-Y. and Zeger, S. L. (1986). Longitudinal data analysis using generalized linear models. *Biometrika*, **73**, 13-22.

Lindley, D. V. (1965). *Introduction to Probability and Statistics from a Bayesian Point of View*. Cambridge University Press, London.

Liu, R. Y. and Singh, K. (1987). On a partial correction by the bootstrap. *Ann. Statist.*, **15**, 1713-1718.

Loève, M. (1977). *Probability Theory I*, fourth edition. Springer-Verlag, New York.

Loh, W.-Y. (1987). Calibrating confidence coefficients. *J. Amer. Statist. Assoc.*, **82**, 155-162.

Loh, W.-Y. (1991). Bootstrap calibration for confidence interval construction and selection. *Statist. Sinica*, **1**, 479-495.

McCullagh, P. and Nelder, J. A. (1989). *Generalized Linear Models*, second edition. Chapman & Hall, London.

Mendenhall, W. and Sincich, T. (1995). *Statistics for Engineering and the Sciences*, fourth edition. Prentice-Hall, Englewood Cliffs, NJ.

Metropolis, N., Rosenbluth, A. W., Rosenbluth, M. N., Teller, A. H., and Teller, E. (1953). Equations of state calculations by fast computing machines. *J. Chemical Physics*, **21**, 1087-1091.

Milliken, G. A. and Johnson, D. E. (1992). *Analysis of Messy Data, Vol. 1: Designed Experiments*. Chapman & Hall, London.

Moore, D. S. and Spruill, M. C. (1975). Unified large-sample theory of general chi-squared statistics for test of fit. *Ann. Statist.*, **3**, 599-616.

Müller, H.-G. and Stadrmüller, U. (1987). Estimation of heteroscedasticity in regression analysis. *Ann. Statist.*, **15**, 610-625.

Natanson, I. P. (1961). *Theory of Functions of a Real Variable*, Vol. 1, revised edition. Ungar, New York.

Nummelin, E. (1984). *General Irreducible Markov Chains and Non-Negative Operators*. Cambridge University Press, New York.

Owen, A. B. (1988). Empirical likelihood ratio confidence intervals for a single functional. *Biometrika*, **75**, 237-249.

Owen, A. B. (1990). Empirical likelihood confidence regions. *Ann. Statist.*, **18**, 90-120.

Owen, A. B. (2001). *Empirical Likelihood*. Chapman & Hall/CRC, Boca Raton.

Parthasarathy, K. P. (1967). *Probability Measures on Metric Spaces*. Academic Press, New York.

Petrov, V. V. (1975). *Sums of Independent Random Variables*. Springer-Verlag, Berlin-Heidelberg.

Pitman, E. J. G. (1979). *Some Basic Theory for Statistical Inference*. Chapman & Hall, London.

Puri, M. L. and Sen, P. K. (1985). *Nonparametric Methods in General Linear Models*. Wiley, New York.

Qin, J. (1993). Empirical likelihood in biased sample problems. *Ann. Statist.*, **21**, 1182-1196.

Qin, J. and Lawless, J. (1994). Empirical likelihood and general estimating equations. *Ann. Statist.*, **22**, 300-325.

Qin, J., Leung, D., and Shao, J. (2002). Estimation with survey data under nonignorable nonresponse or informative sampling. *J. Amer. Statist. Assoc.*, **97**, 193-200.

Quenouille, M. (1949). Approximation tests of correlation in time series. *J. R. Statist. Soc.*, B, **11**, 18-84.

Randles, R. H. and Wolfe, D. A. (1979). *Introduction to the Theory of Nonparametric Statistics*. Wiley, New York.

Rao, C. R. (1945). Information and the accuracy attainable in the estimation of statistical parameters. *Bull. Calc. Math. Soc.*, **37**, 81-91.

Rao, C. R. (1947). Large sample tests of statistical hypotheses concerning several parameters with applications to problems of estimation. *Proc. Comb. Phil. Soc.*, **44**, 50-57.

Rao, C. R. (1973). *Linear Statistical Inference and Its Applications*, second edition. Wiley, New York.

Rohatgi, V. K. (1976). *An Introduction to Probability Theory and Mathematical Statistics.* Wiley, New York.

Rosenblatt, M. (1971). Curve estimates. *Ann. Math. Statist.*, **42**, 1815-1842.

Royden, H. L. (1968). *Real Analysis*, second edition. Macmillan, New York.

Särndal, C. E., Swensson, B., and Wretman, J. (1992). *Model Assisted Survey Sampling.* Springer-Verlag, New York.

Savage, S. L. (1954). *The Foundations of Statistics.* Wiley, New York.

Scheffé, H. (1959). *Analysis of Variance.* Wiley, New York.

Schervish, M. J. (1995). *Theory of Statistics.* Springer-Verlag, New York.

Searle, S. R. (1971). *Linear Models.* Wiley, New York.

Sen, P. K. (1981). *Sequential Nonparametrics: Invariance Principles and Statistical Inference.* Wiley, New York.

Sen, P. K. and Singer, J. M. (1993). *Large Sample Methods in Statistics.* Chapman & Hall, London.

Serfling, R. J. (1980). *Approximation Theorems of Mathematical Statistics.* Wiley, New York.

Shao, J. (1989). Monte Carlo approximations in Bayesian decision theory. *J. Amer. Statist. Assoc.*, **84**, 727-732.

Shao, J. (1993). Differentiability of statistical functionals and consistency of the jackknife. *Ann. Statist.*, **21**, 61-75.

Shao, J. and Tu, D. (1995). *The Jackknife and Bootstrap.* Springer-Verlag, New York.

Shorack, G. R. and Wellner, J. A. (1986). *Empirical Processes with Applications to Statistics.* Wiley, New York.

Silverman, B. W. (1986). *Density Estimation for Statistics and Data Analysis.* Chapman & Hall, London.

Smirnov, N. V. (1944). An approximation to the distribution laws of random quantiles determined by empirical data. *Uspehi Mat. Nauk,* **10**, 179-206.

Smyth, G. K. (1989). Generalized linear models with varying dispersion. *J. R. Statist. Soc.*, B, **51**, 47-60.

Stein, C. (1956). Inadmissibility of the usual estimator for the mean of a multivariate distribution. *Proc. Third Berkeley Symp. Math. Statist. Prob.*, **1**, 197-206. University of California Press, Berkeley, CA.

Stein, C. (1959). The admissibility of Pitman's estimator for a single location parameter. *Ann. Math. Statist.*, **30**, 970-979.

Stone, C. J. (1974). Asymptotic properties of estimators of a location parameter. *Ann. Statist.*, **2**, 1127-1137.

Stone, C. J. (1977). Consistent nonparametric regression (with discussion). *Ann. Statist.*, **5**, 595-645.

Strawderman, W. E. (1971). Proper Bayes minimax estimators of the multivariate normal mean. *Ann. Statist.*, **42**, 385-388.

Tanner, M. A. (1996). *Tools for Statistical Inference*, third edition. Springer-Verlag, New York.

Tate, R. F. and Klett, G. W. (1959). Optimal confidence intervals for the variance of a normal distribution. *J. Amer. Statist. Assoc.*, **54**, 674-682.

Tierney, L. (1994). Markov chains for exploring posterior distributions (with discussions). *Ann. Statist.*, **22**, 1701-1762.

Tsiatis, A. A. (1981). A large sample study of Cox's regression model. *Ann. Statist.*, **9**, 93-108.

Tsui, K.-W. (1981). Simultaneous estimation of several Poisson parameters under squared error loss. *Ann. Inst. Statist. Math.*, **10**, 299-326.

Tukey, J. (1958). Bias and confidence in not quite large samples. *Ann. Math. Statist.*, **29**, 614.

Tukey, J. (1977). *Exploratory Data Analysis*. Addison-Wesley, Reading, MA.

van der Vaart, A. W. (1998). *Asymptotic Statistics*. Cambridge University Press, Cambridge.

Vardi, Y. (1985). Empirical distributions in selection bias models. *Ann. Statist.*, **13**, 178-203.

von Mises, R. (1947). On the asymptotic distribution of differentiable statistical functionals. *Ann. Math. Statist.*, **18**, 309-348.

Wahba, G. (1990). *Spline Models for Observational Data.* SIAM, Philadelphia, PA.

Wald, A. (1943). Tests of statistical hypotheses concerning several parameters when the number of observations is large. *Trans. Amer. Math. Soc.*, **54**, 426-482.

Wald, A. (1950). *Statistical Decision Functions.* Wiley, New York.

Weerahandi, S. (1995). *Exact Statistical Methods for Data Analysis.* Springer-Verlag, New York.

Welsh, A. H. (1987). The trimmed mean in the linear model. *Ann. Statist.*, **15**, 20-36.

Woodruff, R. S. (1952). Confidence intervals for medians and other position measures. *J. Amer. Statist. Assoc.*, **47**, 635-646.

Wu, C. F. J. (1986). Jackknife, bootstrap and other resampling methods in regression analysis (with discussions). *Ann. Statist.*, **14**, 1261-1350.

List of Notation

$\mathcal{R}$: the real line.

$\mathcal{R}^k$: the k-dimensional Euclidean space.

$c = (c_1, ..., c_k)$: a vector (element) in $\mathcal{R}^k$, which is considered as a $k \times 1$ matrix (column vector) when matrix algebra is involved.

c^τ: the transpose of a vector c, which is considered as a $1 \times k$ matrix (row vector) when matrix algebra is involved.

$\|c\|$: the Euclidean norm of a vector $c \in \mathcal{R}^k$, $\|c\|^2 = c^\tau c$.

$\mathcal{B}$: the Borel σ-field on $\mathcal{R}$.

$\mathcal{B}^k$: the Borel σ-field on $\mathcal{R}^k$.

(a, b) and $[a, b]$: the open and closed intervals from a to b.

$\{a, b\}$: the set consisting of the elements a and b.

I_k: the $k \times k$ identity matrix.

A^τ: the transpose of a matrix A.

$\text{Det}(A)$: the determinant of a matrix A.

$\text{tr}(A)$: the trace of a matrix A.

$\|A\|$: the norm of a matrix A defined as $\|A\|^2 = \text{tr}(A^\tau A)$.

A^{-1}: the inverse of a matrix A.

A^-: the generalized inverse of a matrix A.

$A^{1/2}$: the square root of a nonnegative definite matrix A defined by $A^{1/2} A^{1/2} = A$.

$A^{-1/2}$: the inverse of $A^{1/2}$.

A^c: the complement of the set A.

$P(A)$: the probability of the set A.

I_A: the indicator function of the set A.

δ_x: the point mass at x or the c.d.f. degenerated at x.

$\{a_n\}$: a sequence of vectors or random vectors $a_1, a_2, ...$

$a_n \to a$: $\{a_n\}$ converges to a as n increases to ∞.

$\to_{a.s.}$: convergence almost surely.

$\to_p$: convergence in probability.

$\rightarrow_d$: convergence in distribution.

g', g'', and $g^{(k)}$: the first-, second-, and kth-order derivatives of a function g on $\mathcal{R}$.

$g(x+)$ or $g(x-)$: the right or left limit of the function g at x.

$\partial g/\partial x$ or ∇g: the partial derivative of the function g on $\mathcal{R}^k$.

$\partial^2 g/\partial x \partial x^\tau$ or $\nabla^2 g$: the second-order partial derivative of the function g on $\mathcal{R}^k$.

$F^{-1}(p)$: the pth quantile of a c.d.f. F, $F^{-1}(t)=\inf\{x : F(x) \geq t\}$.

$E(X)$ or EX: the expectation of a random variable (vector or matrix) X.

$\mathrm{Var}(X)$: the variance (covariance matrix) of a random variable (vector) X.

$\mathrm{Cov}(X,Y)$: the covariance between random variables X and Y.

$\mathcal{P}$: a family containing the population P that generates data

$b_T(P)$: the bias of an estimator T under population P.

$\tilde{b}_T(P)$: an asymptotic bias of an estimator T under population P.

$\mathrm{mse}_T(P)$: the mse of an estimator T under population P.

$R_T(P)$: the risk of an estimator T under population P.

$\mathrm{amse}_T(P)$: an asymptotic mse of an estimator T under population P.

$e_{T'_n,T_n}(P)$: the asymptotic relative efficiency of T'_n w.r.t. T_n.

$\alpha_T(P)$: probability of type I error for a test T.

$\beta_T(P)$: power function for a test T.

$X_{(i)}$: the ith order statistic of X_1, ..., X_n.

$\bar{X}$: the sample mean of $X_1, ..., X_n$, $\bar{X} = \frac{\sum_{i=1}^n X_i}{n}$.

S^2 : the sample variance (covariance matrix) of $X_1, ..., X_n$, $S^2 = \frac{\sum_{i=1}^n (X_i-\bar{X})(X_i-\bar{X})^\tau}{n-1}$.

F_n: the empirical c.d.f. based on $X_1, ..., X_n$.

$N(\mu,\sigma^2)$: the one-dimensional normal distribution or random variable with mean μ and variance σ^2.

$N_k(\mu,\Sigma)$: the k-dimensional normal distribution or random vector with mean vector μ and covariance matrix Σ.

$\Phi(x)$: the standard normal c.d.f.

z_α: the αth quantile of the standard normal distribution.

χ_r^2: a random variable having the chi-square distribution χ_r^2.

$\chi_{r,\alpha}^2$: the $(1-\alpha)$th quantile of the chi-square distribution χ_r^2.

$t_{r,\alpha}$: the $(1-\alpha)$th quantile of the t-distribution t_r.

$F_{a,b,\alpha}$: the $(1-\alpha)$th quantile of the F-distribution $F_{a,b}$.

List of Abbreviations

a.e.: almost everywhere.

amse: asymptotic mean squared error.

ANOVA: analysis of variance.

a.s.: almost surely.

BC: bias-corrected.

BC_a: accelerated bias-corrected.

BLUE: best linear unbiased estimator.

c.d.f.: cumulative distribution function.

ch.f.: characteristic function.

CLT: central limit theorem.

GEE: generalized estimation equation.

GLM: generalized linear model.

HPD: highest posterior density.

i.i.d.: independent and identically distributed.

LR: likelihood ratio.

LSE: least squares estimator.

MCMC: Markov chain Monte Carlo.

MELE: maximum empirical likelihood estimator.

m.g.f.: moment generating function.

MLE: maximum likelihood estimator.

MQLE: maximum quasi-likelihood estimator.

MRIE: minimum risk invariant estimator.

mse: mean squared error.

p.d.f.: probability density function.

RLE: root of likelihood equation.

SLLN: strong law of large numbers.

UMA: uniformly most accurate.

UMAI: uniformly most accurate invariant.

UMAU: uniformly most accurate unbiased.

UMP: uniformly most powerful.

UMPI: uniformly most powerful invariant.

UMPU: uniformly most powerful unbiased.

UMVUE: uniformly minimum variance unbiased estimator.

WLLN: weak law of large numbers.

w.r.t.: with respect to.

Index of Definitions, Main Results, and Examples

559

Examples

Author Index

Subject Index

Printed by Printforce, the Netherlands